Radiocarbon After Four Decades

RADIOCARBON

R.E. Taylor A. Long R.S. Kra
Editors

Radiocarbon After Four Decades

An Interdisciplinary Perspective

With 148 Illustrations

Springer-Verlag
New York Berlin Heidelberg London Paris
Tokyo Hong Kong Barcelona Budapest

R.E. Taylor
Department of Anthropology
Institute of Geophysics and Planetary Physics
University of California, Riverside
Riverside, CA 92521-0418 USA

Austin Long
Department of Geosciences
The University of Arizona
Tuscon, AZ 85721 USA

Renee S. Kra
Department of Geosciences
The University of Arizona
Tuscon, AZ 85721 USA

Library of Congress Cataloging-in-Publication Data
Radiocarbon after four decades : an interdisciplinary perspective /
[editors], R.E. Taylor, Austin Long, Renee S. Kra.
p. cm.
Includes bibliographical references and index.
ISBN 0-387-97714-7. — ISBN 3-540-97714-7
1. Radiocarbon dating. I. Taylor, R. E. (Royal Ervin), 1938–
II. Long, Austin. III. Kra, Renee S.
QC798.D3R3 1992
546′.6815884—dc20 91-44448

Printed on acid-free paper.

Production managed by Ellen Seham; manufacturing supervised by Robert Paella.
Camera-ready copy provided by the editors.
Printed and bound by Edwards Brothers, Inc., Ann Arbor, MI.
Printed in the United States of America.

9 8 7 6 5 4 3 2 1

ISBN 0-387-97714-7 Springer-Verlag New York Berlin Heidelberg
ISBN 3-540-97714-7 Springer-Verlag Berlin Heidelberg New York

FOREWORD

Four decades have passed since the beginning of the use of radiocarbon as a dating, biological and environmental tracer isotope. More than thirty years ago, in December 1960, the Nobel Prize for Chemistry was bestowed on Willard Frank Libby (1909–1980) for the development of the radiocarbon (^{14}C) method. One of the scientists who proposed Libby for the Nobel laureate characterized the significance of the ^{14}C method in these words: "Seldom has a single discovery in chemistry had such an impact on the thinking of so many fields of human endeavor. Seldom has a single discovery generated such wide public interest" (Nobel Foundation 1964).

On 4–8 June 1990, a conference at the University of California Lake Arrowhead Conference Center marked the major contributions that ^{14}C has made to research in archaeology, biochemistry, environmental studies, geochemistry, geology, geophysics, hydrology and oceanography over the last forty years. Originally, plans for this conference had been discussed 15 years ago, following Libby's retirement in 1976 from the University of California, Los Angeles (UCLA). The proposed seminar was envisioned as a vehicle around which a *Festschrift* volume would have been developed. Libby's death in 1980, at the age of 71, terminated these plans. The concept of the Lake Arrowhead Conference arose from discussions among R E Taylor (University of California, Riverside), James Arnold (University of California, San Diego) and Ernest Anderson (retired from the University of California/Los Alamos National Laboratory).

During 1948 to 1949 at the University of Chicago, James R Arnold – as a postdoctoral fellow – and Ernest C Anderson – as Libby's first graduate student at Chicago – carried out the critical experiments that established the essential validity of the ^{14}C method. The involvement of R E Taylor resulted from the opportunity Willard Libby gave to a graduate student in archaeology/anthropology at UCLA to serve as a research assistant in Libby's Isotope Laboratory during the 1970s.

In 1987, an International Advisory Committee was formed to provide counsel and direction in the organization of the conference. The members of this committee were Ernest C Anderson, James R Arnold, Paul E Damon, Gordon J Fergusson, Frederick Johnson, Kunihiko Kigoshi, Willem G Mook, Hans Oeschger, Ingrid U Olsson, Henry A Polach, Meyer Rubin, Hans E Suess, Minze Stuiver, Henrik Tauber and John C Vogel. A Local Organizing Committee, consisting of Rainer Berger (UCLA), Jonathon E Ericson (University of California, Irvine) and R E Taylor, as Chair, was responsible for conference arrangements.

The moderators for the six conference sessions were Paul E Damon (The University of Arizona) and Minze Stuiver (University of Washington) for the Natural Carbon Cycle; Henry A Polach (Australian National University) for Instrumentation and Sample Preparation; Willem G Mook (University of Groningen) for Hydrology; Fred Wendorf (Southern Methodist University) for Old World Archaeology and Paleoanthropology; Rainer Berger and R E Taylor for New World Archaeology; Meyer Rubin (United States Geological Survey) for Earth Sciences; and Lloyd A Currie (National Institute of Standards and Technology) for Environmental Sciences. Historical and New Approaches Theme Lectures, held in the evenings, inspired the participants with anecdotal reminiscences and great expectations for future trends.

The moderators all did excellent jobs of organizing their sessions, inviting scientists who are experts in their respective fields. "New faces" blended with the "old guard" of the international radiocarbon community, as feelings of pride, admiration and accomplishment permeated the beautiful summer setting of Lake Arrowhead.

This volume, a culmination of the discussions and presentations at Lake Arrowhead, is a unique tribute to the achievements of radiocarbon dating over the last forty years. Few 20th century scientific discoveries have had such a profound interdisciplinary impact. The chapters represent overviews of the history, development and future trends in the field, documenting the far-reaching influence that radiocarbon has had on a wide spectrum of scientific disciplines – including archaeology, astrophysics, biology, chemistry, dendrochronology, environmental sciences, geosciences, hydrology, oceanography, palynology and physics. For the first time, specialists have gathered to assess comprehensively the broad nature and consequences of the introduction of radiocarbon into the carbon cycle.

R E Taylor, Austin Long and Renee S Kra

REFERENCE

Nobel Foundation 1964 *Nobel Lectures, Chemistry 1942–1962.* Amsterdam, Elsevier: 587–588.

Acknowledgments

The conference organizers are indebted, for generous financial support of the conference, to the Center for Accelerator Mass Spectrometry (CAMS), University of California/Lawrence Livermore National Laboratory (UC/LLNL), as well as to the Willard F Libby Foundation, the Systemwide University of California Administration, Systemwide Institute of Geophysics and Planetary Physics (IGPP), UCLA Branch of the IGPP, University of California, Riverside (UCR) campus administration, University of California, Irvine (UCI) campus administration, UCLA campus administration, the UCLA Department of Chemistry and the Society for Archaeological Sciences. We particularly wish to thank Jay C Davis (CAMS) and Orson L Anderson (Systemwide IGPP) for their support and encouragement, and acknowledge the helpful suggestions of Minze Stuiver and other members of the International Advisory Committee for the design of the conference logo.

We wish to express our sincere appreciation to Donna Kirner for her efficient management of the Lake Arrowhead Conference, as well as the very helpful assistance of Jeffrey Lehman, Louis Payen, Joye Sage and Anna Weaver.

The editors most gratefully acknowledge Harry E Gove and Donald O Henry, who did not attend the conference, but nonetheless, contributed significant chapters to this volume. Assistant Editor of *RADIOCARBON*, Frances D Moskovitz, played a key role in the physical production of this book, and we sincerely thank her for it. For their invaluable help in shaping the final content of this volume, we wholeheartedly acknowledge the work of the reviewers of the original manuscripts:

Brian Atwater
Bryant Bannister
Mike F Barbetti
Edouard Bard
Edward Boyle
Wallace S Broecker
Michael D Coe
Tyler B Coplen
Lloyd A Currie
Paul E Damon
George A Dawson
Owen K Davis
James M Devine
Robert Earl Dickinson
Douglas J Donahue
David Elmore
George W Farwell
Gordon J Fergusson
Joan Feynman
George Frison
Mebus A Geyh
Herbert Haas
P Edgar Hare
Calvin J Heusser
Frank Hole
Gordon Jacobi
Jack R Jokipii
A J T Jull
Mark J Kenoyer
George A Klouda
J L Lanzerotti
John H Law
David C Lowe
Richard H Meadow
Clement W Meighan
Willem G Mook
Andrew Moore
John E Noakes
Tsung-Hung Peng
Fred M Phillips
Jim Phillips
L Neil Plummer
Henry A Polach
George C Reid
Don S Rice
Fred N Robertson
Meyer Rubin
Garth Sampson
Jim G Shaffer
Michael B Schiffer
Kerry Sieh
Charles P Sonett
Thomas W Stafford, Jr
Jerry J Stipp
Michael Thurman
Pierre M Vermeersch
John S Vogel
Martin Wahlen
Al Yang

Kungliga Svenska Vetenskapsakademien har vid sin sammankomst den 3 november 1960 i enlighet med föreskrifterna i det av

ALFRED NOBEL

den 27 november 1895 upprättade testamentet beslutat att överlämna det pris som detta år bortgives för den viktigaste kemiska upptäckt eller förbättring till

WILLARD F. LIBBY

för hans upptäckt av kol-14 som tidmätare inom arkeologi, geologi, geofysik m. fl. vetenskaper.

Stockholm den 10 december 1960.

Bertil Lindblad
Akademiens preses

Erik Rudberg
Akademiens sekreterare

KUNGL. VETENSK. AKAD. SIGILL

Libby's Nobel Award

Photo courtesy of Leona Marshall Libby and the Libby Foundation;

Willard Frank Libby (1905–1980)
Nobel Laureate in Chemistry for 1960
for the development of the radiocarbon method of dating
(Photo by Edward F Greer, December 1970; Courtesy of the Libby Foundation)

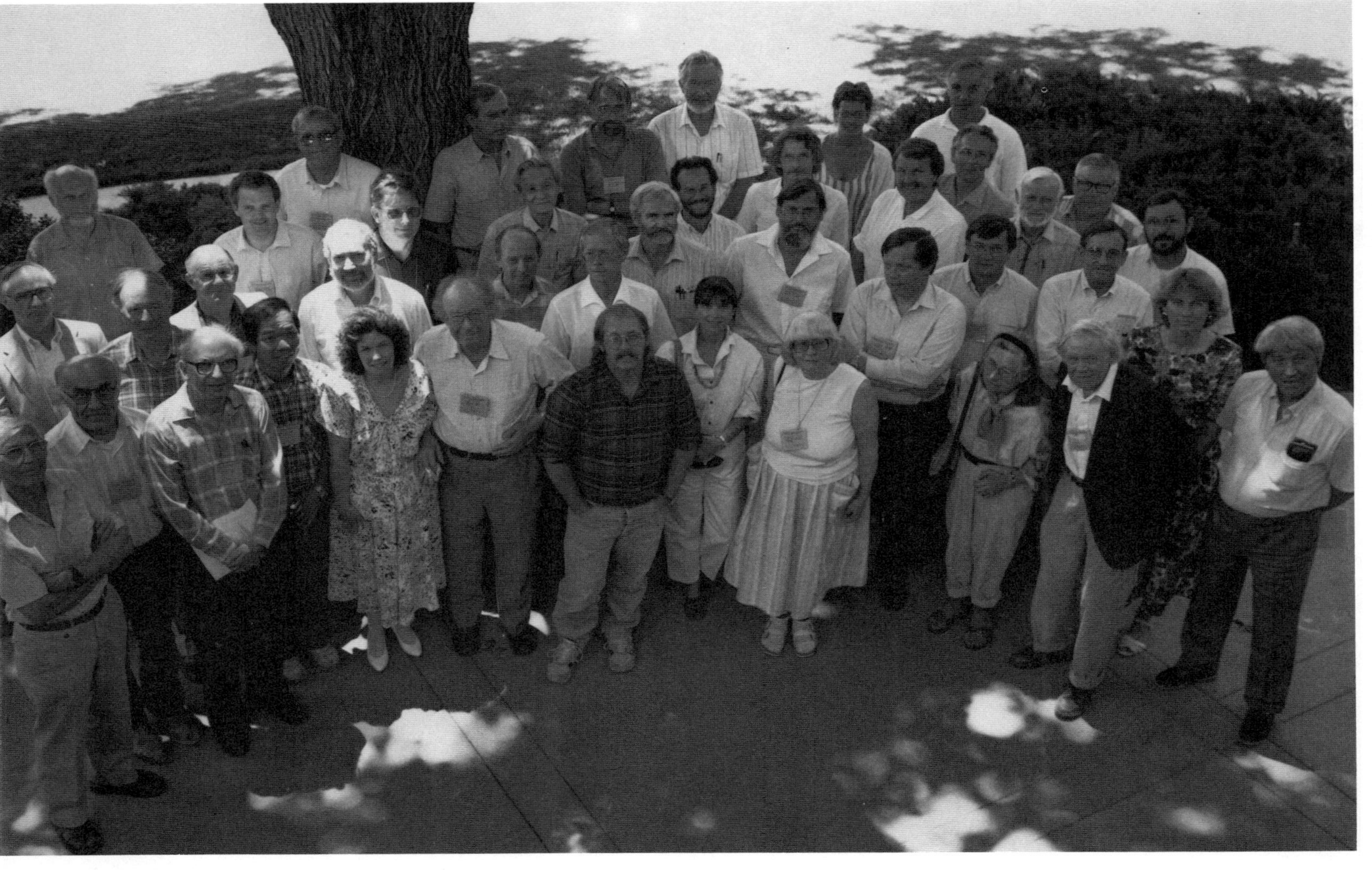

Participants of the conference, "Four Decades of Radiocarbon Studies: An Interdisciplinary Perspective," 4–8 June 1990, Lake Arrowhead, California

CONTENTS

CONTRIBUTORS

James R Arnold, Department of Chemistry, B-017, University of California, San Diego, La Jolla, California 92093 USA

Maurice Arnold, Laboratoire Mixte CNRS-CEA, Centre des Faibles Radioactivités, Parc du CNRS, F-91198 Gif sur Yvette Cédex, France

Edouard Bard, Laboratoire Mixte CNRS-CEA, Centre des Faibles Radioactivités, Parc du CNRS, F-91198 Gif sur Yvette Cédex, France

Bernd Becker, Institut für Botanik, Universität Hohenheim, Garbenstrasse 30, Postfach 700562, D-7000 Stuttgart 70 Germany

Rainer Berger, Institute of Geophysics and Planetary Physics, University of California, Los Angeles, California 90024 USA

Roelf P Beukens, IsoTrace Laboratory, University of Toronto, 60 St George Street, Toronto, Ontario M5S 1A7 Canada

Georges Bonani, Institut für Mittelenergiephysik, ETH-Hönggerberg, CH-8093 Zürich, Switzerland

Rainer Bösinger, Institut für Umweltphysick, Universität Heidelberg, Im Neuenheimer Feld 366, D-6900 Heidelberg, Germany

Wallace S Broecker, Lamont-Doherty Geological Observatory, Columbia University, Palisades, New York 10964 USA

Christopher D Charles, Lamont-Doherty Geological Observatory and Department of Geological Sciences, Columbia University, Palisades, New York 10964 USA

Lloyd A Currie, Center for Analytical Chemistry, National Institute of Standards and Technology, Gaithersburg, Maryland 20899 USA

Paul E Damon, Department of Geosciences, The University of Arizona, Tucson, Arizona 85721 USA

Jay C Davis, Center for Accelerator Mass Spectrometry, Lawrence Livermore National Laboratory, PO Box 808, Livermore, California 94550 USA

Stanley N Davis, Department of Hydrology and Water Resources, The University of Arizona, Tucson, Arizona 85721 USA

Jean-Claude Duplessy, Laboratoire Mixte CNRS-CEA, Centre des Faibles Radioactivités, Parc du CNRS, F-91198 Gif sur Yvette Cédex, France

Josette Duprat, Laboratoire de Géologie et Océanographie, Université de Bordeaux-1, Avenue des Facultés, 33405 Talence, France

Richard G Fairbanks, Lamont-Doherty Geological Observatory and Department of Geological Sciences, Columbia University, Palisades, New York 10964 USA

Scott L Fedick, Department of Anthropology, University of California, Riverside, California 92521-0418 USA

Jean-Charles Fontes, Laboratoire d'Hydrologie et de Géochimie Isotopique, Université Paris-Sud, F-91405 Orsay Cédex, France

Roger J Francey, CSIRO, Division for Atmospheric Research, PO Box 77, Aspendale, Victoria, Australia

M J Fulker, British Nuclear Fuels, Ltd, Sellafield, Seascale, Cumbria CA20 1PG England

Mebus A Geyh, Niedersächsisches Landesamt für Bodenforschung, ^{14}C und Tritium Laboratorium, Stilleweg 2, Postfach 51 01 53, D-3000 Hannover 53 Germany

Harry E Gove, Nuclear Structure Research Laboratory, University of Rochester, Rochester, New York 14627 USA

C Vance Haynes, Jr, Departments of Anthropology and Geosciences, The University of Arizona, Tucson, Arizona 85721 USA

Robert E M Hedges, Radiocarbon Accelerator Unit, Research Laboratory for Archaeology and the History of Art, Oxford University, 6 Keble Road, Oxford OX1 3QJ England

Donald O Henry, Department of Anthropology, The University of Tulsa, 600 South College Ave, Tulsa, Oklahoma 74104-3189 USA

John L Jirikowic, Department of Geosciences, The University of Arizona, Tucson, Arizona 85721 USA

Martyn Jope and Margaret Jope, 1 Chalfont Road, Oxford, England

Robert M Kalin, Department of Geosciences, The University of Arizona, Tucson, Arizona 85721 USA

G E Kocharov, AF Ioffe Physical-Technical Institute, Academy of Sciences, Polytechnicheskaya 26, St Petersburg 194021 Russia

Bernd Kromer, Institut für Umweltphysik, Universität Heidelberg, Im Neuenheimer Feld 366, D-6900 Heidelberg, Germany

Laurent Labeyrie, Laboratoire Mixte CNRS-CEA, Centre des Faibles Radioactivités, Parc du CNRS, F-91198 Gif sur Yvette Cédex, France

K R LaJoie, Branch of Engineering, Seismology and Geology, U S Geological Survey, 345 Middlefield Road, Menlo Park, California 94025 USA

Devendra Lal, Geological Research Division A-020, Scripps Institution of Oceanography, University of California, San Diego, La Jolla, California 92093-0220 USA

Ingeborg Levin, Institut für Umweltphysick, Universität Heidelberg, Im Neuenheimer Feld 366, D-6900 Heidelberg, Germany

Austin Long, Department of Geosciences, The University of Arizona, Tucson, Arizona 85721 USA

Willem G Mook, Netherlands Institute for Sea Research (NIOZ), PO Box 59, 1790 Ab Den Burg Texel, The Netherlands

Jean Moyes, Laboratoire de Géologie et Océanographie, Université de Bordeaux-1, Avenue des Facultés, 33405 Talence, France

Karl Otto Münnich, Institut für Umweltphysik, Universität Heidelberg, Im Neuenheimer Feld 366, D-6900 Heidelberg, Germany

Ellyn M Murphy, Battelle Pacific Northwest Laboratories, PO Box 999, Richland, Washington 99352 USA

Robert L Otlet, RadioCarbon Dating, Downs Croft, The Holloway, Harwell, Oxfordshire OX11 0LS England

Fred J Pearson, Ground-Water Geochemistry, 1304 Walnut Hill Lane, Suite 210, Irving, Texas 75038 USA

Gordon W Pearson, Palaeoecology Centre, The Queen's University of Belfast, Belfast BT7 1NN Northern Ireland

Tsung-Hung Peng, Environmental Science Division, Oak Ridge National Laboratory, Bldg 1000, MS-6335, Oak Ridge, Tennessee 37831-6335 USA

Dorothy M Peteet, Goddard Institute for Space Studies, National Aeronautics and Space Administration, 2880 Broadway, New York, New York 10025 USA

George Plafker, Branch of Alaskan Geology, MS-904, US Geological Survey, 345 Middlefield Road, Menlo Park, California 94025 USA

Henry A Polach, Research Consultations, Inc, PO Box 43, Garran, ACT 2605 Australia

Peter Robertshaw, Department of Anthropology, California State University, San Bernardino, California 02407 USA

Meyer Rubin, Radiocarbon Laboratory, U S Geological Survey, National Center 971, Reston, Virginia 22092 USA

Robert L Schuch, Sandia National Laboratories, Albuquerque, New Mexico 87185 USA

Charles P Sonett, Department of Planetary Sciences and Lunar and Planetary Laboratory, The University of Arizona, Tucson, Arizona 85721 USA

Robert S Sternberg, Department of Geology, Franklin and Marshall College, PO Box 3003, Lancaster, Pennsylvania 17604-3003 USA

Minze Stuiver, Quaternary Isotope Laboratory AJ-20, Department of Geological Sciences, University of Washington, Seattle, Washington 98195 USA

Hans E Suess, Scripps Institution of Oceanography, University of California, San Diego, La Jolla, California 92037 USA

Martin Suter, Paul Scherrer Institut, c/o ETH-Hönggerberg, CH-8093 Zürich, Switzerland

Karl A Taube, Department of Anthropology, University of California, Riverside, California 92521-0418 USA

R E Taylor, Radiocarbon Laboratory, Department of Anthropology and Institute of Geophysics and Planetary Physics, University of California, Riverside, California 92521-0418 USA

N B A Trivett, Atmospheric Environment Service, 4905 Dufferin Street, Downsview, Ontario M3H ST4 Canada

A J Walker, RadioCarbon Dating, Downs Croft, The Holloway, Harwell, Oxfordshire OX11 0LS England

Fred Wendorf, Department of Anthropology, Southern Methodist University, Dallas, Texas 75275 USA

Willy Wölfli, Institut für Mittelenergiephysik, ETH-Hönggerberg, CH-8093 Zürich, Switzerland

James D Wright, Lamont-Doherty Geological Observatory and Department of Geological Sciences, Columbia University, Palisades, New York 10964 USA

EDITORS

R E Taylor is an archaeologist/archaeometrist in the Department of Anthropology and Institute of Geophysics and Planetary Physics, University of California, Riverside. His principal interest involves the application of ^{14}C and other Quaternary dating methods to provide temporal placement for archaeological materials. Over the last decade, his research has focused on the ^{14}C dating of bone as specifically applied to the dating of New World human skeletal materials in the context of controversies concerning the character and timing of the peopling of the western hemisphere. He was involved in early applications of AMS technology in ^{14}C measurements of human bone. He is the author of *Radiocarbon Dating: An Archaeological Perspective* (Academic Press 1987) and co-editor (with C W Meighan) of *Chronologies in New World Archaeology* (Academic Press 1978). His current appointment at UCR is Professor of Anthropology and Director, Radiocarbon Laboratory. He came to UCR in 1969 following the completion of graduate studies at UCLA, where he served as a Research Assistant in the Isotope Laboratory of the late Willard F Libby.

Austin Long became interested in geochemistry in 1956, as an undergraduate summer intern at Lamont Geological Observatory. While working on his PhD under Paul Damon at The University of Arizona, he helped convert the Libby-style solid carbon dating laboratory to a ^{14}C gas proportional laboratory. In 1963, Long joined the staff of the Smithsonian Institution, and with Joel Sigalove, turned out their first methane gas-proportional dates. In 1968, Long returned to Arizona to become director of the Radiocarbon Laboratory, and added a stable isotope facility as well. Long has supervised a number of MS and PhD students, primarily in isotope geohydrology. In 1987, coincident to moving it to a new underground counting chamber, the laboratory was converted to exclusively liquid scintillation counting. Long became Editor of *RADIOCARBON* late in 1988, and in January 1989, the editorial offices moved to Tucson as part of the Geosciences Department. As Editor, he has been active in involving eastern Europe in the radiocarbon community, and he is stressing quality assurance in ^{14}C dating.

Renee S Kra has been Managing Editor of *RADIOCARBON* for 25 years. She started her long-term affair with *RADIOCARBON* in the Geology and Geophysics Department at Yale University, under the aegis of Edward S Deevey, Richard Foster Flint and Irving Rouse, the founders of the journal. With Minze Stuiver, who became Senior Editor in 1976, she co-edited the Proceedings of the 10th, 11th and 12th International Radiocarbon Conferences. With Austin Long, Editor since 1988, she has edited the 13th, and is currently working on the Proceedings of the 14th conference, held in Tucson in May 1991. In 1989, she moved with the journal to The University of Arizona, and became Assistant Research Scientist in the Geosciences Department. Kra has a Bachelor's degree in English and a Master's degree in Anthropology. She is Executive Director of the International Radiocarbon Data Base (IRDB), and is currently co-editing, with Ofer Bar-Yosef, a volume on *Late Quaternary Chronology and Paleoclimates of the Eastern Mediterranean.*

HISTORICAL PERSPECTIVES

PREFACE

R E TAYLOR

In one of his early recollections – or "ruminations" as he once characterized them – on the origins of the ^{14}C method, Libby observed that it was "difficult to know exactly when the idea was born." He suggested that its beginnings lay in his realization that cosmic-ray ^{14}C production led to ". . . a continuous labeling of the biosphere . . . which is terminated at death" (Libby 1967: 5–6). In an interview conducted much later, he recalled that the origin involved his reading of a publication (Korff & Danforth 1939) which reported finding neutrons in the atmosphere: "As soon as I read Korff's paper . . . that's carbon dating" (Libby 1979: 33, 40).

As James Arnold notes in the opening chapter, Libby typically was working on dozens of ideas at any one time. We can then perhaps understand why he did not know himself precisely where and when the idea of the ^{14}C method actually occurred to him. Fortunately, we have the recall of Libby's major collaborator during the critical period in the early development of the method. Arnold's account is an important primary historical source, in some cases, providing new data and perspectives heretofore undocumented, in other cases, supplementing information he previously had shared with colleagues (Marlowe 1980; Taylor 1985, 1987: 147–170). An example of the new information is the role of Robert Schuch in the development of the Chicago solid carbon counting instrumentation in the face of the Libby rule: "if it ain't broke, don't fix it." Those who entered the ^{14}C field many years later can only imagine what it must have felt like when, as Arnold relates, he was, for a few hours, the "only person in the world who knew that ^{14}C dating worked." One does indeed live ones scientific life for moments like this. It is fortunate (some would call it "Libby's Luck") that the second "known-age" sample – wood from a modern fake "Ptolemaic" mummy case – was not measured first.

Hans Suess was one of the individuals that came to Chicago to learn the mysteries of measuring natural ^{14}C using the recalcitrant solid carbon method. He left to establish the first US Geological Survey (USGS) ^{14}C laboratory – and later the University of California, San Diego (UCSD) Mt Soledad facility at La

Jolla – using acetylene as a counting gas. At the USGS, Suess provided the initial documentation of the effects of fossil-fuel combustion on recent ^{14}C levels (Suess effect). At UCSD, he became the prime mover in extending the scope of the "deVries effects." He used the bristlecone pine tree-ring record to establish the reality of multiple short-term (at first, "wriggles," then "wiggles") as well as the long-term, major trend variations in natural ^{14}C activity in the atmosphere. His work was instrumental in establishing the distinction between calendar and radiocarbon time and the need to "calibrate" ^{14}C dates.

In an article published in 1981 that undertook a retrospective view of the "early days" of accelerator mass spectrometry (AMS), another Nobel Prize laureate, the late Luis Alvarez, commented that he generally declined invitations to conferences discussing the history of his field. The reason was because he would rather "remember my friends who did pioneering work in nuclear physics as the vigorous young men they were, four and a half decades ago, instead of the white-haired old men I would meet at such conferences and see in my shaving mirror" (Alvarez 1981: 1). In remembering, we celebrate the great scientific legacy of these vigorous young men who established the radiocarbon field.

REFERENCES

Alvarez, LW 1981 The early days of accelerator mass spectrometry. In *Kutschera, W, ed,* Proceedings of the Symposium on Accelerator Mass Spectrometry. *ANL/PHY 81-1. Springfield, National Technical Information Service: 1–15.*

Korff, SA and Danforth, WE 1939 Neutron measurements with boron-trifluoride counters. Physical Review *55: 980.*

Libby, WF 1967 History of radiocarbon dating. In Radiocarbon Dating and Methods of Low Level Counting. *Vienna, IAEA: 3–25.*

_____*1979 Interview with WF Libby, April 12, 1979, on file at the Center for the History of Physics, American Institute of Physics.*

Marlowe, G 1980 WF Libby and the archaeologists, 1946–1948. In *Stuiver, M and Kra, RS, eds, Proceedings of the 10th International ^{14}C Conference.* Radiocarbon *22(3): 1005–1014.*

Taylor, RE 1985 The beginnings of radiocarbon dating in American Antiquity: *An historical perspective.* American Antiquity *50: 309–325.*

_____*1987* Radiocarbon Dating: An Archaeological Perspective. *Orlando, Florida, Academic Press.*

1

THE EARLY YEARS WITH LIBBY AT CHICAGO: A RETROSPECTIVE

JAMES R ARNOLD
With comments by Robert L Schuch

Like other scientific advances, radiocarbon dating did not happen in a moment. The fortieth anniversary could have been celebrated as early as 1986 – since, as we shall see, the idea was certainly in Willard Libby's mind in 1946 – or as late as 1991, since the first date list of unknowns appeared in *Science* in 1951 (Arnold & Libby 1951). In fact, what is surprising in retrospect is not that it took so long, but that it was accomplished so quickly. In this age of team research, centers and consortia, it is remarkable how much of the accomplishment was due to one man.

It may help to gain perspective to summarize the state of knowledge needed for ^{14}C dating in early 1946. These are what seem to be the key points: 1) The nuclide had been discovered at Berkeley in 1937 by Martin Kamen and Libby's student, Samuel Ruben (Ruben & Kamen 1941; *cf* Kamen 1963, 1985: 122–146). A few weak counting sources had been prepared by Kamen at the Berkeley cyclotron; 2) the half-life was listed in the tables as "between 1,000 and 25,000 years"; 3) it was made by the reaction $^{14}N(n,p)^{14}C$, and Serge Korff had noted that cosmic-ray-produced neutrons in the atmosphere should mostly disappear by that reaction (Korff & Danforth 1939); 4) no counting methods existed which could measure ^{14}C at natural levels, although Libby's earlier invention of the screen-wall counter was a step in that direction; 5) while there had been progress in quantitative understanding of the cosmic rays – for example, they were known to be mostly high energy protons – estimates of neutron production, in particular, were still crude. Thus, there was great uncertainty as to the production rate. Libby's 1946 estimate was, in fact, low by roughly a factor of ten; 6) Nothing was known about the behavior and distribution of natural ^{14}C – or even whether it existed. Finally, it should be noted that there were no regular established programs for support of basic science by the United States government, and certainly not for a project like this, which, if it had not been done earlier, would be hard to fund in today's tightly compartmented scientific support structure.

But the year 1946 was a year of excitement and hope, especially for Willard Libby. Let us begin with a bit of biography. Libby described himself to us as a farmer's son, sent to Berkeley to learn agricultural science, who was drawn into chemistry by the great teachers there. After receiving his BS degree, he was encouraged to go on, which he did, receiving his PhD in the depression year, 1933 (Libby 1933). He told me once that he had two job offers at that point, a half-time instructor's position at a junior college, and an instructorship at Berkeley. Of course he stayed. In the thirties, he published a number of interesting papers, most significantly, for our present purpose, several on long-lived natural radioactivities. It was this work that caused him to develop the screen-wall counter to permit windowless counting of weak beta activities, with excellent geometry and sample-background comparison. When, in 1939, he received a Guggenheim Fellowship to go to Princeton, he was still an assistant professor. He had acquired a reputation for originality and daring, but was certainly not yet a "big name."

However, in 1945, at the end of World War II, when the University of Chicago assembled a brilliant cast of nuclear scientists in what was then the Institute for Nuclear Studies and now the Fermi Institute – including Fermi, Urey, Teller and others – Libby was clearly one of the stars. How did this come about? The key is in wartime research on the Manhattan Project. Harold Urey was the inevitable choice to lead the effort to separate ^{235}U from bulk uranium. Libby was one of the first two people he hired, in what became a huge, industrial-scale effort. He made many contributions, but the one that put him on the map was the invention of the diffusion barrier ultimately used for the huge Oak Ridge plant. He also earned Urey's friendship and gratitude by taking over an important share of the heavy administrative load of the project. He enjoyed being boss and did it well; Urey hated it.

I first arrived in Chicago in early 1946, bearing a brand new PhD obtained using Manhattan Project research, and spent the first few months of my stay working with Libby (who had also just arrived). His first publication on cosmic-ray effects was published in the *Physical Review* in June of that year (Libby 1946). It mentions ^{14}C as one product of cosmic-ray bombardment of the atmosphere and states that "an attempt (at detection) is intended in these laboratories." My first memory of hearing about the dating idea, however, goes back to a party (one of many) at the Libby home later in the year.

In a way, my upbringing had prepared me for ^{14}C dating. My father, an attorney by profession, was a serious amateur archaeologist. He was, for example, the American secretary of the England-based Egypt Exploration Society. This helped me to see the point concerning dating very quickly. Here begins my bit part in the early history. I went home for Christmas in 1946 and passed on my excitement successfully to my father. By the time I returned to Chicago in January 1947, Libby had a package from my father's friend, Ambrose Lansing, Curator of Egyptian Archaeology at the Metropolitan Museum in New York. It contained ten samples from various periods of Egyptian history, all well dated, historically. Lansing offered them as a set on which to check the method – a little early.

I was quite embarrassed when Libby showed me the package and letter. I apologized for the over-enthusiasm, offered to make a full explanation of the state of things to Lansing, and to return the package. Characteristically, Bill said not a word in reply. He simply took back the package and letter, and put them on a shelf just above his desk. I realized suddenly that he was serious! For me, that was the moment the dating project was born.

The next year I spent at Harvard, so my account for this active period is based on hearsay. Libby acquired a remarkable graduate student named Ernie Anderson, who had also spent years on the Manhattan Project at Los Alamos, but had not been lucky enough to get a PhD out of it. At first, while still a paid assistant, he helped Libby prove the existence of natural ^{14}C. His assigned PhD thesis problem was to verify, if possible, the idea of the "contemporary assay," *ie*, that living things all over the world show the same specific activity of ^{14}C. With the aid of the City of Baltimore, which sold methane produced in its sewage plant (hence "living" carbon) and of Aristid von Grosse, who was in charge of a plant for isotopic enrichment of ^{13}C in methane, Libby and Anderson accomplished the first task ahead of schedule – in the latter half of 1947 (Anderson *et al* 1947). They also began to develop the technique for direct measurement of ^{14}C at natural levels, without isotopic enrichment. Finally, with strong backing from Harold Urey, Libby received funding from the Viking Fund (later Wenner-Gren Foundation) to support the dating effort. One happy result for me was my return in early 1948 to Chicago to join the effort.

This is perhaps the most suitable point to attempt a very difficult task: to describe Bill Libby as he was then, at the height of his powers. He was about 40 years old. A tall man, pale of complexion and usually wearing a light-colored suit, he made a strong first impression and radiated authority. His manner was direct and informal. He was astonishingly original and creative; he had dozens of ideas at work at any given time. With all that, he was well organized and neat (in his office at least). With young coworkers, students and postdocs, he was a mixture of warm father figure and holy terror. He was a quick and generally very accurate judge of such people, and he cared intensely about developing their talents. This meant that the better you did, the tougher and more unreasonable he became. He adopted the Vince Lombardi football coach model. I was often furious at him, but his training has served me well for over forty years. To the outside world, we were his children and could do no wrong. However different he and Hans Suess were in personality, they are the two most intuitive scientists I have known, both able to arrive at the right answer quickly by processes others could not follow. Others, meeting him for the first time, tended to be either attracted or repelled. He made both friends and enemies easily.

In early 1948, the first true counting system for natural ^{14}C had been invented and was taking shape. There were two major components. At the center was the screen-wall counter, with a coating of sample carbon on an active area of about 400 cm^2, and a comparable blank (anthracite coal) on a matching area. The original Libby design required a skillful, delicate touch to assemble. The Schuch redesign of the following year was a liberation. "Living carbon" produced a net count of 4 counts per minute (cpm) in this first device, with a background (inside a lead shield) of about 150 cpm. The second key component was the electronic anticoincidence shield, an idea borrowed by Libby from the cosmic-ray physicists, and put together in first crude form by Anderson. Libby was very difficult about letting him improve it. When the system was working – perhaps 10% of the time in early 1948 – the background was reduced to about 40 cpm. At this level, natural ^{14}C was detectable, a major step forward.

With the key inventions demonstrated, and two of us to carry out Libby's daily instructions, the pace began to pick up. The improvements from that point were incremental. Bit by bit, with setbacks, the reliability of the system improved. One step at a time, the background was reduced and even the counter efficiency slowly increased. Some time in mid-1948, we achieved – briefly at first – a duty cycle above 50%. Then, the Lansing box was opened again, and one sample, wood from the step pyramid of Zoser at Sakkara, was removed and processed. Anderson moved forward with increasing momentum toward the completion of his dissertation (Anderson 1949, 1953), and I had the exciting (read frightening) responsibility of preparing and counting the first ancient sample. On a hot Saturday afternoon that summer, I had enough counts from our still temperamental system to calculate a preliminary result. I will never forget the impact of finding that it was about halfway between Ernie's contemporary wood values and the blank. For a couple of heady hours, I was the only person in the world who knew that ^{14}C dating worked. One lives for such moments.

How did we know it should be about half? In parallel with our work in 217 Jones Laboratory, Libby had enlisted Antonia Engelkemeir at the Argonne

National Laboratory and Mark Inghram in the Physics Department to collaborate with him in determining the half-life by measuring the activity of gas samples prepared from a source of mass-spectrometrically measured $^{14}C/^{12}C$ in counters of varying diameter and length. The results of two successive determinations remained the standard for many years; the so-called Libby or official "dating" half-life of ^{14}C, 5568 years, is still theirs (Engelkemeir *et al* 1949).

It was, I believe, some months later that we acquired our supporting committee of archaeologists and single geologist (Richard Foster Flint), headed by Frederick Johnson. They gave us invaluable support in the period up through the first date list, both in selecting samples of known age for the initial calibration, and in screening samples for importance and proper documentation when we moved on to unknowns. The need for such assistance had been vividly underlined for us when our second ancient sample, a piece of purported Ptolemaic mummy case, exhibited essentially the same activity as Anderson's contemporary average. It was, of course, a fake (Libby 1967: 17; Arnold in Marlowe 1980: 1012–1013). The committee also had its problems at first in educating the profession in the need to avoid paraffin impregnation and other preservation techniques in samples for ^{14}C measurement, and in the meaning of the reported errors. But in the end, these things worked out quite well.

The first presentation of our data on the comparison of historical and ^{14}C dates on a set of documented samples was made by Libby at a seminar in New York sponsored by the Viking Fund. The *New York Times* story on it the next day was headlined "Scientist Stumbles on New Radioactive Dating Method." Quite a stumble, lasting about three years up to that point! After that, I was delegated to report these data at several meetings – I remember particularly the Society for American Archaeology session.

One important contribution of Ernie Anderson to ^{14}C dating has not received wide recognition: it was Ernie who brought Bob Schuch into the effort as senior lab technician. Libby was a stern believer in the principle now known as "if it ain't broke, don't fix it." As a result, lab procedures and equipment were typically frozen at the stage where they first worked. This was particularly painful in the case of the screen-wall counter. Bob had not only the practical skill to design and build simple, reliable versions of these items, but the diplomatic talent to persuade Bill to let him do it.

At the mature stage of the project, with cylinders Matthew, Mark, Luke and John in operation (giving the "Gospel Truth"), and before Bob Schuch's move to Los Alamos ("The Lord giveth and the Lord taketh away" was Ernie's lofty comment), we had the counter background down to about 4 cpm, with the contemporary carbon count rate at about 6 cpm (Anderson, Arnold & Libby

1951) This was equivalent to a 1-sigma error on the order of 200 years for a fairly young sample. At this level, the deviations from the strict exponential decay curve were suggestive but not dramatic.

As the method gained acceptance, Libby began to receive requests for help from others who wanted to set up ^{14}C labs. He was delighted at the prospect. The picture of holding the academic fate of a generation of assistant professors of archaeology in his hands had no charm for him. "I don't want to be the pope of archaeological dating," was one frequent remark. Virtually the last job he gave me was to train the live prospects in all aspects of the technique. I remember particularly Beth Ralph, Larry Kulp and Hans Suess. Hans, being himself, did not pay much attention to the details, since he already had ideas for doing it better.

Because of stories I had heard from my father, I persuaded Bill to avoid samples with strong religious significance in the early phases, at least until the method was well established. He stuck with the bargain, but the very first sample measured after I left the project was a piece of cloth used to wrap one of the Dead Sea Scrolls. He would have enjoyed the Shroud of Turin exercise.

Not many scientific achievements get, or deserve, fortieth anniversary conferences, or their own journals. Radiocarbon dating was developed, I believe, at the earliest time permitted by the general state of background knowledge, and in something not far from the best imaginable way, because of the skill and daring of one man. It is a pleasure to be able to pay tribute to him, before an audience of people who know the value of the achievement.

ADDITIONAL COMMENTS: ROBERT L SCHUCH

When Ernie Anderson first took me into Room 217 of the Jones Laboratory, I was surprised to see this tall person with thinning hair whom everyone described as being a young man. Later, the fact was emphasized that he had been elected to the National Academy of Sciences and I thought to myself, "After all he *is* forty-one." My position was rather unusual as I was a brash technician with little knowledge of physical chemistry. And yet, Bill Libby and I got along very well. (It took me at least twenty years before I could call him Bill and I am still not comfortable with it.)

Dr Libby had certain rules and regulations. The first was for me to know the location of every item in the laboratory. He had me spend the first two days going through every drawer and cabinet. Jim Arnold would help me identify items so I would not need to respond "Is that one of those shiny cup things?" when Libby asked for a Dewar. I am still annoyed at any store clerk that does not know the stock. The second rule was "Don't touch any of the counting apparatus while in operation." The third rule was, "Don't change any design, technique or configuration without first convincing me it is necessary." I had trouble with the third regulation because it has always been my nature to try to improve things – to find a better, more efficient and more reliable way.

Our first encounter was over the de Khotinsky wax used to seal the end caps of the screen-wall counters. I considered this to be some sort of ancient alchemy which caused out-gassing in the counter and burned fingers. After a series of o-ring-fitted end caps were tried, he accepted the idea and I did not need to mess with de Khotinsky any more. I did not have to but Libby continued to do so. Another item was the installation of 33 grid wires on the screen-wall counters – a tedious and difficult task, which only Libby and a few others could accomplish. This job was made more manageable by installing miniature (no. 1-72) size screws to tighten the grid wires. Libby accepted this technique somewhat reluctantly probably because it made the job too easy.

One idea that received his immediate approval was to seal unprocessed samples using state-of-the-art materials and techniques – one of which was wrapping the samples in clear polyethylene and sealing the ends with a soldering iron. I also questioned why the lead bricks to access the counting chamber were being loaded and unloaded by hand? "To keep things simple" Libby would answer. When I visited him some time later, Libby had a counter-weighted door on his new apparatus, which he described to me almost apologetically. As time went by, I was given more responsibility in the preparation of samples and training graduate students and visitors from other laboratories in the art of assembling screen-wall counters. The ultimate arrived one weekend when no graduate student was available and Libby allowed me to take on one of the 3-hour sample changes. Dr Libby always introduced me as his Chief Technician, and it was during this time at Chicago that I met Maria and Joseph Mayer, Harold Urey, W D Harkins and others who would stop in the lab to talk. I was never asked to leave as subjects such as The Viking Fund, space science, thermal diffusion barriers and worldwide radioactive fallout were discussed.

Libby and Jim Arnold assembled an excellent group of graduate students and associates including Bill Johnston, Tom Sugihara, Richard Wolfgang, Ed Martell, Marv Kalkstein, Andy Suttle, Maury Fox, Sherry Roland and Howard Hornig. Occasionally, Libby would get a little up-tight when the graduate students would draft me into extra-curricular activities such as repairing boats or fixing cars. On one occasion, we had a breakdown while towing a boat. He gave me a lecture and then offered the use of his car.

In 1950, I went on to Los Alamos with Ernie Anderson and we possibly made the final improvement on the early Chicago screen-wall counter – a double screen-wall arrangement capable of measuring sample and background simultaneously. Jim Arnold predicted our early problems with this system when he commented "Double double toil and trouble." Jim Arnold at Chicago and Newt Hayes at Los Alamos continued to pursue ^{14}C measurements using liquid scintillation systems, and I spent a lot of time cutting holes in refrigerators to install the instrumentation. Libby seemed somewhat reluctant to accept this technique, especially when Hayes complained that the natural radiocarbon background was affecting his measurements of lemon grass oils.

One seemingly minor incident sticks in my mind. While cleaning up one day, I found a dime and placed it on his desk. Later, Libby asked me about the dime and I told him I found it on the floor. He shook my hand, looked me straight in the eye and said, "Bob, you're an honest man." I really did not understand what that was all about but, with a comment such as that coming from a person of his stature, who needed anything more? Young graduate students would ask me how I got along so well with Libby. I would respond: "Maybe it's because I gave him back his dime."

References

Anderson, EC (ms) 1949 Natural radiocarbon. PhD dissertation, University of Chicago.

_____1953 The production and distribution of natural radiocarbon. *Annual Review of Nuclear Science* 2: 63–89.

Anderson, EC, Arnold, JR and Libby, WF 1951 Measurement of low level radiocarbon. *Review of Scientific Instruments* 22: 225–230.

Anderson, EC, Libby, WF, Weinhouse, S, Reid, AF, Kirschenbaum, A and Grosse, AV 1947 Natural radiocarbon from cosmic radiation. *Physical Review* 72: 931–936.

Arnold, JR and Libby, WF 1951 Radiocarbon dates. *Science* 113: 111–120.

Engelkeimeir, AG, Hamill, WH, Inghram, MG and Libby WF 1949 The half-life of radiocarbon (C^{14}). *Physical Review* 75: 1825–1833.

Kamen, MD 1963 Early history of carbon-14. *Science* 140: 584-590.

_____1985 *Radiant Science, Dark Politics, A Memoir of the Nuclear Age*. Berkeley: University of California Press.

Korff, SA and Danforth, WE 1939 Neutron measurements with boron-trifluoride counters. *Physical Review* 55: 980.

Libby, WF (ms) 1933 Radioactivity of ordinary elements, especially samarium and neodymium: method of detection. PhD dissertation, University of California, Berkeley.

_____1946 Atmospheric helium three and radiocarbon from cosmic radiation. *Physical Review* 69: 671–672.

_____1967 History of radiocarbon dating. In *Radioactive Dating and Methods of Low Level Counting*. Vienna, IAEA: 745–751.

Marlowe, G 1980 WF Libby and the archaeologists, 1946–1948. *In* Stuiver, M and Kra, RS, eds, Proceedings of the 10th International ^{14}C Conference. *Radiocarbon* 22(3): 1005–1014.

Rubin, S and Kamen, MD 1941 Long-lived radioactive carbon: ^{14}C. *Physical Review* 59: 349–354.

2

THE EARLY RADIOCARBON YEARS: PERSONAL REFLECTIONS

HANS E SUESS

LIBBY AT CHICAGO

In 1949, Harrison Brown arranged an invitation for me to spend a limited time at the Institute for Nuclear Studies at the University of Chicago. I considered myself most fortunate to be invited to this institute, not only because of the large number of Nobel laureates working there, but also because Willard Libby had just developed a method for age determinations on wood and other substances which are formed by photosynthesis from atmospheric CO_2 and water. Libby's method made it possible to obtain well-defined values for the time elapsed since the respective organic substance had formed. It seemed obvious to me that, by concentrating on this method, rather spectacular results in a variety of fields could be obtained. In 1950, soon after my arrival in Chicago from Europe, I had seen Willard Libby (called Bill by his colleagues) at the Institute, walking along the corridor. One day I approached him bravely and asked him if I could see him in his office and talk about his ^{14}C dating method. Libby nodded, pulled out a notebook, "Next week, Tuesday, three in the afternoon," he said, and without waiting for an answer, walked away.

It was most likely due to Libby's influence that shortly thereafter I was offered a position with the United States Geological Survey in Washington, DC. However, before I could obtain such a federal position, I had to undergo time-consuming political "clearances." In the meantime, I was approached by Professor Edward Deevey of Yale University. Deevey had managed to obtain complete funding for a Libby-type (*ie*, solid carbon) ^{14}C laboratory, including the necessary space. However, he was unable to find a qualified person to operate the laboratory for the money he had available. When I finally was hired by the USGS in Washington, I had my salary, but no laboratory. So I asked for permission to use the equipment at the "Geochronometric Laboratory" at Yale for the time being. Earl Ingerson, then Branch Chief, granted my request and I spent several fruitful months getting Deevey's instruments operational and discussing unsolved problems of ^{14}C dating with Professor Evelyn Hutchinson at Yale.

AT THE USGS

Shortly after Christmas 1952, Ruth and I, with our two children, returned to Washington, DC. I was temporarily assigned office space at the Naval Gun Factory in Anacostia. There, I completed a paper that I had begun to write in Chicago with Professor Harold Urey (Suess & Urey 1956). Also, without any improvising and preliminary testing, I designed and ordered the electronic instrumentation as well as a double-mantle stainless-steel counter, anti-coincidence counters and other items.

By the time space for my laboratory became available in the basement of the Department of the Interior building, I had decided to use acetylene as counting gas. In the late 1930s at the University of Hamburg, I had used acetylene to investigate radiochemical products obtained by neutron activation of HBr gas (Suess 1939). These experiments indicated that acetylene was not only promising as a counting gas, but also that its properties made it relatively insensitive to electronegative impurities – contrary to CO_2, which, as a counting gas, is extremely sensitive to the presence of such impurities.

There is a myth, apparently spread by physicists, that frozen acetylene tends to explode spontaneously. I have frozen and thawed acetylene hundreds of times and never managed to get it to detonate, except by mixing it with oxygen (or ozone) and igniting it. Perhaps, if one is very lucky, one might observe a powerful detonation after freezing it down with wet cupric oxide and/or dirty mercury. At gas pressures of more than two atmospheres of acetylene, it might well explode. However, I never tried.

When, as planned, the space for the laboratory became available in early 1953, the branch chief assigned two more persons to the radiocarbon laboratory – Corrine Alexander and a geologist, Dr Meyer Rubin. Routine ^{14}C measurements were begun in the summer of 1953. The first results were presented at the NSF conference in October 1953 in Williams Bay, Wisconsin (Suess 1953).

The late Professor Richard Foster Flint of Yale University had asked his students to collect wood samples from glacial moraines for several years. Flint distinguished four major substages of the last glaciation, which he called the "Wisconsin," and its substages, Cochrane, Mankato, Cary and Tazwell. The names were analogous to four mountain ranges that represent the four relatively recent glacial moraines in North America. Indeed, the samples gave four distinct groups of ages. This then led to a fierce controversy between believers and non-believers in ^{14}C dating. The sample ages corresponded to their age sequence; their absolute age values, however, appeared younger than expected (eg, Antevs 1957).

Radiocarbon measurements made during 1953 and 1954 at the USGS ^{14}C laboratory in Washington, DC were the basis of the following suggestions:

1. The maximum extent of the North American ice sheet occurred around 18,000 radiocarbon years ago;
2. The anthropogenic drop in ^{14}C activity in air occurred during the industrial revolution, when ^{14}C-free CO_2 from fossil-fuel combustion was added to the atmosphere;
3. *Homo neanderthalensis* survived in selected areas until about 30,000 years ago.

A total of approximately 200 ^{14}C age determinations were carried out during the two years during which I was head of the USGS Radiocarbon Dating Laboratory (Suess 1954; Rubin & Suess 1955). A large fraction of them, such as most results pertaining to samples from Alaska, were primarily of local interest. Thereafter, Dr Rubin became head of the USGS Laboratory, which later moved to Reston, Virginia, a suburb of Washington, DC.

AT LA JOLLA

In 1954, Dr Roger Revelle, then Director of the Scripps Institution of Oceanography, University of California, San Diego (La Jolla) persuaded me to leave the Geological Survey for a position at his institution. I happily accepted. I had always hoped for an appointment that would allow me to pursue my personal research interests and to teach students. It was not difficult for Revelle to arouse my interest in basic oceanographic problems that could be solved by ^{14}C and tritium measurements. Revelle promised to get me money and space for another ^{14}C dating laboratory and a tenured position as a Professor at the University of California. Regrettably, for political reasons, Revelle left La Jolla soon after my laboratory was completed.

I suggested to oceanographers from other institutions that they should concentrate on the Atlantic and Mediterranean Oceans, whereas we would mainly investigate the Pacific and Indian Oceans. I was joined by Professor Norris Rakestraw, a marine chemist, and by Dr George Bien, who, in his younger years, had been doing oceanography in China. They wanted to supervise sampling of deep water on expeditions. The ^{14}C content of several hundred samples of surface and deep ocean water was then measured in the new radiocarbon laboratory at Scripps (Hubbs, Bien & Suess 1960). The necessary ^{13}C values were determined, in part, in the laboratory of Professor Harmon Craig, and, in part, in Professor Sam Epstein's lab at the California Institute of Technology in Pasadena.

Then, to my surprise, I heard through the grapevine, that a colleague at Scripps had obtained more than $20,000,000 from the National Science Foundation for ship time and ^{14}C determinations. Since Roger Revelle had left, and I was under no obligation to continue work on oceanographic problems, this terminated my involvement in ^{14}C in the oceans. I became a member of the Department of Chemistry at UCSD, although the ^{14}C laboratory remained administrated by the Scripps Institution of Oceanography. This arrangement later led to administrative difficulties.

TREE RINGS AND THE RADIOCARBON TIME SPECTRUM

We immediately began measurements of dendrochronologically dated wood (*Sequoia gigantea*) obtained from T L Smiley of the Tree-Ring Laboratory at The University of Arizona in Tucson. The wood consisted of 25 samples that had grown between 1200 BC and AD 1950. The measurements showed, for the first time, the systematic rise with age in the ^{14}C content of samples in BC times, and the need to calibrate the Libby ages by comparing them with the ^{14}C in samples of known ages (Suess 1961).

For several years, Dr C W Ferguson of the Tree-Ring Laboratory had been working on an annual tree-ring sequence using bristlecone pine (originally *Pinus aristata*, now known as *Pinus longaeva*). We had occasionally dated a few samples before their precise tree-ring age had been established. So we knew of Ferguson's progress and were prepared to obtain samples. Indeed, one day we received from Ferguson a letter addressed to three laboratories (University of Pennsylvania, University of Arizona and the Scripps laboratory) offering tree-ring-dated wood samples for ^{14}C measurements, containing no more than ten rings and weighing at least 20 grams. Using two ultra-low-background counters of the Houtermans-Oeschger type, we immediately undertook to measure all of the samples that Ferguson could supply.

By the time of the Twelfth Nobel Symposium on "Radiocarbon Variations and Absolute Chronology," organized by Dr Ingrid Olsson and held in Uppsala, Sweden in August 1969, Dr Ferguson had established a continuous series of tree rings of bristlecone pine wood going back more than 7000 years (Ferguson 1970). By that time, we already knew that the so-called "deVries wiggle" around AD 1700 was not the only irregularity in the ^{14}C content of atmospheric CO_2 (Suess 1970). An important question that could not be answered at the time was whether fluctuations (or wiggles) of this type were random variations or followed some rigorous law.

The son of the late physicist Fritz Houtermans, Jan Cornelius Houtermans, had been studying mathematics at Munich, Bavaria and Madison, Wisconsin. I tried to get him interested in a statistical analysis of our measured ^{14}C values as a function of time. While working with me in La Jolla, he contemplated submitting a doctoral dissertation dealing with this subject, but this turned out to be impossible at UCSD for financial reasons. However, using the La Jolla ^{14}C data, he submitted a thesis entitled "Geophysical Interpretations of Bristlecone Pine Radiocarbon Measurements" to Professor Hans Oeschger at the University of Bern. This thesis showed that the radiocarbon variations clearly corresponded to a line spectrum. Among other lines, a prominent 200-year line was present, one that had been observed before (N Kruse, personal communication, 1970). When I showed the results to Libby, he exclaimed, "If this is true, then the radiocarbon values should be a most interesting geophysical parameter."

Indeed, it seems remarkable that the ^{14}C spectrum, obtained by Jan Houtermans in 1971 from only 350 ^{14}C values measured in La Jolla, already showed all the essential features that could be seen in data published much later by several investigators who managed to recognize a spectrum from other less complete series. It seems that a spectrum of ^{14}C variations with time is the best way to recognize to what degree a ^{14}C time series is accurate and to what extent it is in error (Suess & Linick 1990).

Certainly, from a geophysical or astrophysical viewpoint, the ^{14}C spectrum is a most interesting geophysical global parameter. It may take many years before it is fully understood (Sonett & Finney 1990). Not less interesting to me will be an understanding of the psychological causes that led the great majority of investigators to deny, for many years, the existence of regular deviations of the ^{14}C values, so-called "wiggles" from a smooth line, and the fact that a clearly defined spectrum existed. How was it possible that many, in fact the great majority of workers, for nearly 20 years – from before 1970 until close to 1990 – did not recognize the existence of a time spectrum? In general, wiggles were considered non-existent or rare phenomena. There must have been a psychological reason – perhaps the "beauty of the smooth line" or the work of a powerful lobby? In any case, the importance of psychological effects cannot be denied. For this reason, it seems safer if measurement and evaluation of results are not done by one and the same person.

REFERENCES

Antevs, E 1957 Geological tests of the varve and radiocarbon chronologies. *Journal of Geology* 65: 124–138.

Ferguson, CW 1970 Dendrochronology of the bristlecone pine, *Pinus aristata*. Establishment of a 7484-year chronology in the White Mountains of eastern-central California, USA. *In* Olsson, IU, ed, *Radiocarbon Variations and Absolute Chronology*. Proceedings of the 12th Nobel Symposium. Stockholm, Almqvist & Wiksell: 237–260.

Hubbs, CL, Bien, GS and Suess, HE, 1960 La Jolla natural radiocarbon measurements I. *American Journal of Science Radiocarbon Supplement* 2: 197–223.

Rubin, M and Suess, HE, 1955 U S Geological Survey radiocarbon dates II. *Science* 121: 481–488.

Sonett, CP and Finney, SA 1990 The spectrum of radiocarbon. *Philosophical Transactions of the Royal Society* [*London*] A330: 413–426.

Suess, HE 1939 Das Verholten von Bromwasserstoff bei Bromkern-Prozessen. *Zeitschrift fuer Physicalische Chemie* 45: 297.

_____1953 Natural radiocarbon and the rate of exchange of carbon dioxide between the atmosphere and the sea. Proceedings of the Conference on Nuclear Processes in Geologic Settings. *NAS-NSF Publication*: 52.

_____1954 U S Geological Survey radiocarbon dates I. *Science* 120: 467–473.

_____1961 Secular changes in the concentration of atmospheric radiocarbon. Proceedings of the Conference on Problems Related to Interplanetary Matter. *NAS-NSF Publication* 845(33): 90–95.

_____1970 Bristlecone-pine calibration of the radiocarbon time-scale 5200 BC to the present. *In* Olsson, IU, ed, *Radiocarbon Variations and Absolute Chronology*. Proceedings of the 12th Nobel Symposium. Stockholm, Almqvist & Wiksell: 303–312.

Suess, HE and Linick, TW 1990 The ^{14}C record in bristlecone pine wood of the past 8000 years based on the dendrochronology of the late CW Ferguson. *Philosophical Transactions of the Royal Society [London]* A330: 403–412.

Suess, HE and Urey, HC 1956 Abundance of the Elements. *Review of Modern Physics* 28: 53.

THE NATURAL CARBON CYCLE

PREFACE

PAUL E DAMON

During the four decades since the first measurement, in 1949, of natural ^{14}C in a shielded, screen-wall counter with anticoincidence, radiocarbon research has gone through five stages of unequal duration. The first stage began with that first measurement of natural ^{14}C and ended, after three years of intense work by Libby's research group, with the publication of the first date list in 1951 and the first edition of Libby's book, Radiocarbon Dating, *in 1952. The second stage, from 1952 to 1955, saw radiocarbon dating come of age with the burgeoning of seven additional radiocarbon dating laboratories. The third stage culminated in 1962 with the determination of a more precise half-life by three laboratories, and the recognition that secular variation of radiocarbon required calibration of the radiocarbon time scale to convert radiocarbon ages to calendar years. During the fourth stage, first-order calibration resolved important dating problems, and researchers strived to understand the causes of secular variation, and to improve the calibrated time scale. This stage terminated with the publication of the special Calibration Issue of* RADIOCARBON *in 1986, with abundant high-precision data and user-friendly computer calibration programs covering the most recent nine millennia.*

The fifth and ongoing stage involves attempts to push the calibration back into the last glacial period, an increasing understanding of the causes of secular variation of radiocarbon and the advent of tandem accelerator mass spectrometry (TAMS). TAMS made it possible to analyze for ^{14}C in milligram- and submilligram-size carbon samples, allowed for greater selectivity, and extended the range of radiocarbon dating to samples previously undatable, such as the blood stain on an arrowhead or a single grain of wheat.

During all of these stages, radiocarbon dating continued with growing sophistication, and the use of radiocarbon as a natural tracer of global environmental processes became increasingly important. Currently, the most significant archive of radiocarbon is dendrochronologically dated tree rings. This natural archive of radiocarbon not only allows for precise calibration of the ^{14}C time scale, but also provides an invaluable source of information concerning the *causes of secular ^{14}C variations in nature. Thus, it seems fitting to begin our section with the history of the use of dendrochronology in calibrating the radiocarbon time scale by Bernd Becker and the state of the calibration process by Minze Stuiver and Gordon Pearson.*

Secular variation of ^{14}C is not random. There is a long-term trend that is generally attributed to changes in the Earth's dipole moment. This subject is covered by Robert Sternberg. When the trend curve is removed, the resulting spectrum contains cyclical variations on a century to millennial scale. Charles Sonett, an expert on time-series analysis, addresses this aspect of the secular ^{14}C variation, providing evidence from the ^{14}C spectrum and other sources implicating solar activity as the major causal factor. He discusses the evidence for not only a hydromagnetic but also a bolometric aspect to solar activity that modulates both ^{14}C and climate. Paul Damon and John Jirikowic make use of the evidence discussed by Sonett to suggest that the Earth is entering a new warm epoch that will exacerbate the Greenhouse Effect through the 21st century.

Jean-Claude Duplessy et al and Tsung-Hung Peng and Wallace Broecker illustrate the use of ^{14}C as a tracer of global processes by applying it to the oceans. Devendra Lal brings the subject down to solid earth with a chapter on in-situ production in rocks, whereas Grant Kocharov extends the subject to the stars by using cosmogenic isotopes to provide evidence for supernova explosions.

The second-order discrepancies in radiocarbon ages have become as important as the first-order agreement was thought to be. Recognition of their existence has led to a more precise and accurate calibrated radiocarbon time scale. Information concerning secular variation of radiocarbon and other cosmogenic isotopes stored in their natural archives has become one of the most important proxy indicators of global environmental change. The father of radiocarbon dating came to realize this potential, and stated, in 1967, that "a phase is now being entered where radiocarbon dates are used to test the validity of the geophysical parameters involved" (Libby 1967: 17). We are well within this phase, as illustrated by the chapters in this section of Radiocarbon After Four Decades: An Interdisciplinary Perspective.

REFERENCE

Libby, WF 1967 History of radiocarbon dating. In Radiocarbon Dating and Methods of Low-Level Counting. *Vienna, IAEA: 3–26.*

CALIBRATION OF THE RADIOCARBON TIME SCALE, 2500–5000 BC

MINZE STUIVER and GORDON W PEARSON

INTRODUCTION

The need for age calibration of conventional radiocarbon ages is widely recognized in the Earth sciences. This need arises from assumptions made when calculating a conventional radiocarbon date. Implied in the calculation is the postulated constancy of atmospheric ^{14}C levels during the past, a condition that is only very marginally fulfilled. Therefore, the proper interpretation of a conventional radiocarbon age (Stuiver & Polach 1977) not only depends on the precision and accuracy of the measurement, but also on the extent to which the assumption of constancy of past atmospheric ^{14}C levels is met.

The preferred check on conventional radiocarbon dates is made by measuring the ^{14}C activity of wood samples that have known (dendrochronologically determined) ages. The calibration curves used for the correction of the radiocarbon ages are generated by plotting the wood radiocarbon ages *versus* the calibrated (cal) ages. Many calibration measurements have been made during the thirty-plus years since de Vries (1958, 1959) demonstrated the variability of atmospheric ^{14}C over the past centuries. Compilations of these studies can be found, for instance, in Damon, Lerman and Long (1978), Klein *et al* (1982) and Stuiver and Kra (1986). A high-precision bi-decadal calibration curve back to 2500 BC is available (Stuiver & Pearson 1986; Pearson & Stuiver 1986), and the material discussed here extends this work to 5000 BC.

In some instances, corrections have to be applied to the published data sets of 1986. Tree-ring ^{14}C determinations back to 2500 BC made at the Quaternary Isotope Laboratory of the University of Washington (eg, Stuiver & Becker 1986) need corrections because previously undetected minor amounts of radon released by ovens during sample preparation have to be taken into account. The correction tends to increase, by 10 to 20 years, the radiocarbon ages of most of the previously published post-2500 BC tree-ring samples of the Seattle laboratory. Minor corrections (for counting efficiency change at Belfast and $\delta^{13}C$

normalization at Pretoria) need to be applied to the Belfast (eg, Pearson *et al* 1986) and Pretoria (Vogel, Fuls & Visser 1986; Vogel, personal communication) data sets. The details on these corrections will be published elsewhere. In regard to the data used below for the construction of the 2500–5000 BC calibration curve, and for the interlaboratory comparisons, all necessary corrections were applied.

Our discussion focuses on the incremental calibration knowledge gained since 1986 for the pre-2500 BC era. For such a discussion, the precision and accuracy of the ^{14}C measurements have to be known. The precision is derived from the internal consistency of repeat measurements, and information on accuracy is obtained by comparing the results (offset) of different laboratories on identical samples. Many laboratories, as an approximation, derive an age error from Poisson counting statistics only. Substantial under-reporting of the age error occurs (International Study Group 1982; Scott *et al* 1989). It is obvious that for the time-scale calibration efforts, the adherence to desired performance criteria is extremely important.

STANDARD ERRORS IN MEASUREMENT

An evaluation of the degree of measuring precision and accuracy of the laboratories involved in time-scale calibration can be made relatively easily, because wood of the same age is often measured by more than one laboratory. Although the sources of variance are additive, the ratio, K (actual standard error/quoted standard error), is a convenient expression of the degree to which the quoted error is representative of the overall error in a radiocarbon date. K values have a tendency to increase with sample age (lower sample activities). The following evaluation is for a restricted sample age (0–8000 ^{14}C yrs) and fairly static K.

The quoted standard error, σ, in the results reported by the Quaternary Isotope Laboratory are based on counting-rate statistics. A modified form of Poisson counting error is used for the calculations in that, in addition to the Poisson standard deviation in the sample count, the largest of 1) the Poisson standard deviation in the average of multiple standard runs, and 2) the standard deviation derived from the observed scatter in these multiple runs, is entered in the calculations of the quoted standard error. At the Belfast laboratory, several sources of variance are taken into account in the quoted error (Pearson *et al* 1986).

Based on the Seattle measurements for radiocarbon ages of pairs of contemporaneous wood samples originating in different parts of the world, $K_{Seattle}$ is estimated at 1.5 (n = 30). This K value is an upper limit for laboratory reproducibility, because part of the differences may also be due to differences in tree ^{14}C activity. Duplicate Belfast samples of wood of the same age yielded $K_{Belfast}$ = 1.23 (n = 55); duplicate Seattle counts of the same sample gas yielded $K_{Seattle}$ = 1.5 (n = 169). Another evaluation of $K_{Seattle}$ can be obtained from the ten-year data set, published back to 2500 BC (Stuiver & Becker 1986). Two data points are part of each bi-decadal average. Whereas the standard deviation in the radiocarbon age differences of 215 decadal pairs (AD 1800–2500 BC) is 26.2 years, the average standard deviation (derived from the quoted standard deviations in the 1986 calibration volume) in the age differences of these pairs is 18.2 years. If all sample pairs would have had identical radiocarbon ages, $K_{Seattle}$ would be 26.2/18.2 = 1.4. This again is an upper limit for this age range, as not all samples in a sample pair have identical ^{14}C activity.

Interlaboratory comparisons are needed to identify any offsets, and these lead to independent K information as well. The Seattle data set was compared to those of other laboratories (after appropriate corrections were applied to the Seattle, Belfast and Pretoria series) and yielded the information in Table 3.1. The Belfast-Seattle comparison is for bi-decadal wood samples that differ in dendrochronological origin (respectively, Irish oak *versus* German oak / US sequoia wood). The midpoint cal ages of the 20-year wood block pairs differ by 1.5 years maximally (Stuiver & Pearson 1986). The Pretoria (Vogel, Fuls & Visser 1986), La Jolla (Linick, Suess & Becker 1985) and Groningen (de Jong, Becker & Mook 1986) samples each cover one or more cal years, and here the radiocarbon age average of all available samples in each 20 cal-year interval was taken for the comparison with the bi-decadal Seattle data. The German oak chronology (Becker & Kromer 1986) is the basis for most of the Seattle, Groningen, Heidelberg (Kromer *et al* 1986), La Jolla and Pretoria measurements listed in Table 3.1. Seattle and Tucson (Linick *et al* 1986) bristlecone pine wood measurements resulted in the Seattle-Tucson comparison.

When calculating interlaboratory K values, the average standard deviation in the age differences, σ (derived from the quoted errors in the radiocarbon ages), is compared to the observed standard deviation, σ_{tot}, in the radiocarbon age differences, ($K = \sigma_{tot}/\sigma$). As part of the observed age, differences may be due to use of different wood samples (eg, Irish *versus* German wood, or wood from different trees when comparing results from the same chronology). The derived K value is an upper limit for laboratory reproducibility.

The largest offset in Table 3.1 (between Heidelberg and Seattle) is barely above the 2σ level. Applying an error multiplier greater than 1.0 to the Seattle and Heidelberg results brings the offset within the critical 2σ range. The Seattle-Belfast offset, which is the most relevant to the construction of the calibration

TABLE 3.1. A comparison of radiocarbon ages of samples with approximately identical cal age (see text). The radiocarbon ages were determined at Seattle and one of the following laboratories: Belfast (Pearson *et al* 1986, personal communication); Groningen (de Jong, Becker & Mook 1986); Heidelberg (Kromer *et al* 1986, adjusted to a 7177 BC zero point for the unified Donau 6-Main 4/11 German chronology); La Jolla (Linick, Suess & Becker 1985, adjusted to same zero point as Heidelberg); Pretoria (Vogel, Fuls & Visser 1986, personal communication); and Tucson (Linick *et al* 1986). Positive numbers indicate older Seattle radiocarbon ages. The standard deviations in the offsets are based on quoted sigmas.

Laboratory	Belfast/ Seattle	Groningen/ Seattle	Heidelberg/ Seattle	La Jolla/ Seattle	Pretoria/ Seattle	Tucson/ Seattle
No. of comparisons	302	36	6	97	72	13
Time interval	AD 1000– 5000 BC	3210– 3910 BC	~7000 BC	2500– 5000 BC	1930– 3350 BC	5685– 5805 BC
Average standard deviation σ in age differences (yrs) derived from quoted errors	18.7	13.6	31.7	35.8	16.1	31.1
Observed standard deviation σ_{tot} in age differences (yrs)	30.6	24.3	28.3	46.9	19.3	39.9
Offset (mean difference, yrs)	2 ± 1	−4 ± 2	28 ± 13	−4 ± 3	4 ± 2	−8 ± 8
$K_{\text{Seattle-Lab X}}$ ($K = \sigma_{tot}/\sigma$)	1.6	1.8	0.9	1.3	1.2	1.3

curves below, averages only a few years for sample materials in the AD 1000–5000 BC interval.

The average error multiplier, ($K_{\text{Seattle-Lab x}}$), for the interlaboratory differences is 1.35. Thus, if individual lab error multipliers had been identical, the lab error multiplier of each laboratory would have an upper limit of 1.35. It is not our intention to estimate larger or smaller than average K values for each laboratory. Although one case (see below), where the offset is much too large to be statistically justifiable, was excluded from Table 3.1, routine doubling of the quoted errors of the individual laboratories appears to be too pessimistic an estimate of laboratory reproducibility for samples in the Table 3.1 age range.

For the 2500–5000 BC interval, the Seattle-Belfast offset is −5.5 ± 1.6 years (n = 125), with $K_{\text{Seattle-Belfast}}$ = 1.7. As noted, K values will have a tendency to increase with sample age. Figure 3.1 compares the observed distribution in the age differences of the bi-decadal data *versus* the Gaussian distribution compatible with the average quoted error.

The Seattle and Belfast bi-decadal radiocarbon ages are compared to those of Groningen (de Jong, Mook & Becker 1989) and Pretoria (Vogel, Fuls & Visser

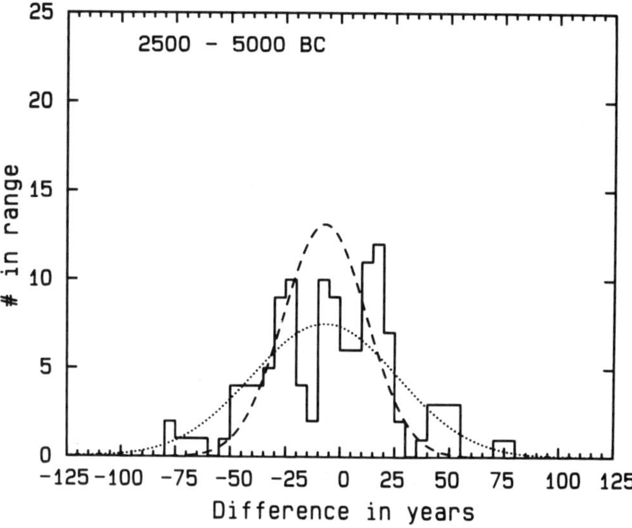

Fig 3.1. ^{14}C age differences of 125 pairs (with one member each measured in Seattle and Belfast) of contemporaneously formed (bi-decadal) wood samples covering the 2500–5000 BC interval. The Belfast and Seattle laboratory errors predict a narrower Gaussian distribution (---) for the age differences (average σ = 19.0 yrs) than actually observed (···, σ = 33.2 yrs), resulting in an error multiplier $K_{\text{Seattle-Belfast}}$ = 1.7. This K value is an upper limit for the (joint) laboratory reproducibility of the measurements (see text).

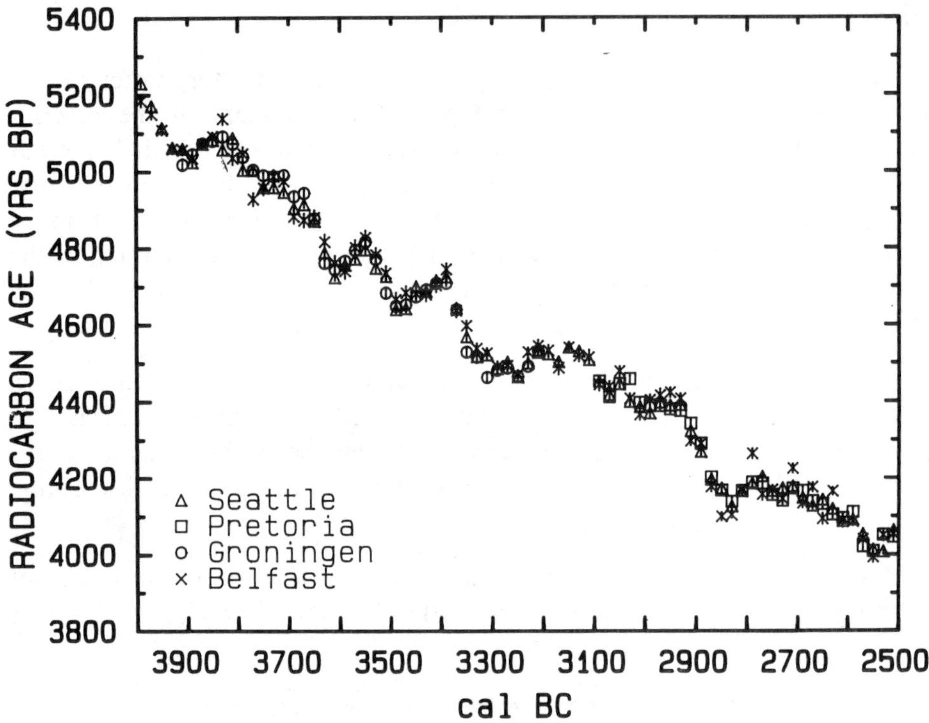

Fig 3.2. Comparison of ^{14}C age determinations at four laboratories. The Belfast (Pearson *et al* 1986) and Seattle results each consist of 125 bi-decadal samples. For Groningen (de Jong, Becker & Mook 1986, 3210–3910 BC) and Pretoria (Vogel, Fuls & Visser 1986, 1930–3350 BC), the samples covering shorter cal age intervals were averaged when falling in corresponding bi-decades. Vertical error bars are based on the individual laboratory errors and do not include an error multiplier. Irish oak samples were measured in Belfast; the other laboratories used materials from the German oak chronology.

1986) in Figure 3.2. The conformity of the records (Figs 3.2, 3.3) would have been impossible if the cal ages of the Irish and German dendrochronologies had not been 100% reliable back to 5000 BC.

Not included in the above comparisons (Table 3.1, Fig 3.3) is a 4075–4925 BC data set of the German chronology (Kromer *et al* 1986) with adjusted zero point of 7177 BC for the start of the unified German oak chronology. Here, $K_{Seattle\text{-}Heidelberg}$ has an acceptable value of 1.7 (n = 43), but the offset of −35 ± 4 years (Heidelberg dates being older) on wood of the same chronology (albeit from different trees) cannot be explained. Additional measurements by the Heidelberg, Belfast and Seattle laboratories are needed to solve this problem.

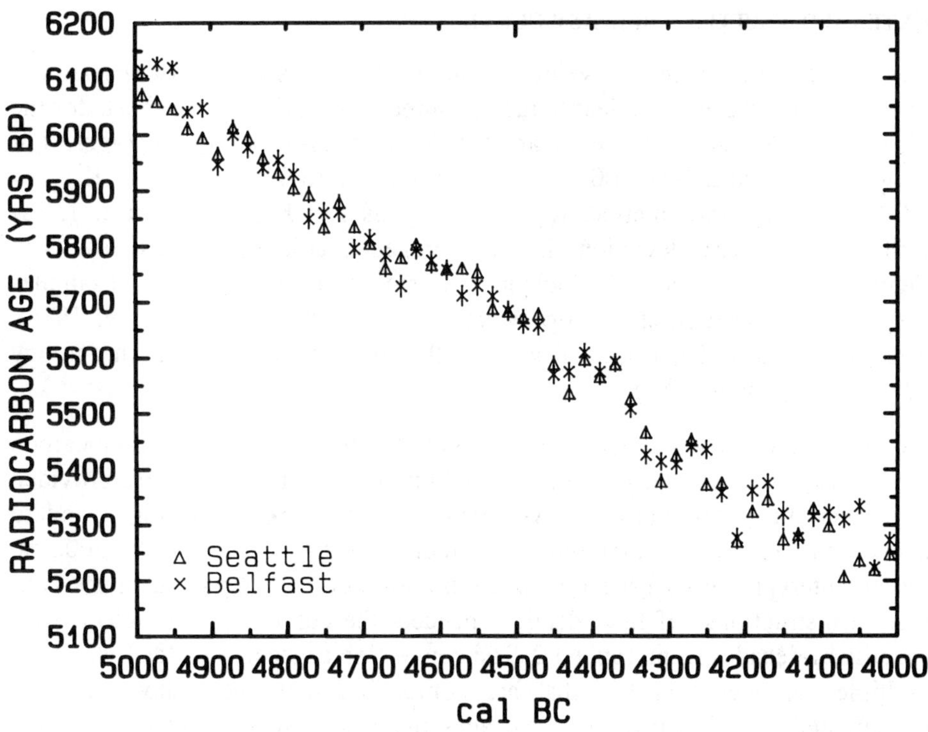

Fig 3.3. Comparison of the ^{14}C ages of bi-decadal samples measured in Seattle and Belfast. Vertical error bars are based on the individual laboratory errors and do not include an error multiplier. The materials investigated were Irish oak (Belfast) and German oak (Seattle).

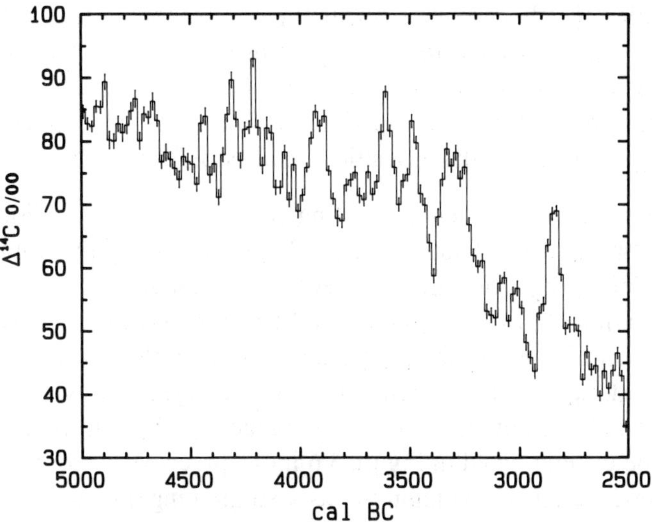

Fig 3.4. Bar diagram of $\Delta^{14}C$ values obtained by averaging the Seattle and Belfast ^{14}C ages of bi-decadal wood samples from the German (Seattle) and Irish (Belfast) oak chronologies. The standard deviation in each bi-decadal average (|) includes a 1.7 error multiplier for each laboratory.

EXTENDED SEATTLE - BELFAST CURVE

The bi-decadal averaged conventional radiocarbon ages of Belfast (Pearson *et al* 1986, as corrected) and Seattle (as measured since 1986) are the basis for the Figure 3.4 $\Delta^{14}C$ values, Table 3.2 and the Figure 3.5A–E radiocarbon ages. For samples from the 2500–5000 BC cal age range, we took a "generous" $K_{Seattle}$ = 1.7, because, for this interval, $K_{Seattle\text{-}Belfast}$ equals 1.7; $K_{Belfast}$ was set at 1.7 as well. The standard deviations in the averaged Seattle and Belfast results in Figures 3.4–1.5 and Table 3.2 include a 1.7 error multiplier. Bi-decadal samples derived from the Irish chronology were used by Belfast (Pearson *et al* 1986), whereas Seattle utilized decadal wood cellulose from the German chronology (Becker & Kromer 1986).

The calibration curves (Figs 3.5A–E) depict the transformation of radiocarbon ages to calibrated AD/BC (or BP) ages. Calibration instructions were provided in Stuiver and Pearson (1986). As a first approximation, the average standard deviation of 14.8 years in the bi-decadal radiocarbon ages of the 2500–5000 cal yrs BC interval can be used for the calibration curve σ employed in the calibration instructions. If finer detail is needed, the curve σ can be selected for specific bi-decades from Table 3.2. The conversion procedure yields 1) single or multiple cal ages (yrs BC) that are compatible with the radiocarbon age investigated, and 2) range(s) of cal ages that take into account the standard deviation in the radiocarbon age (and calibration curve). (Extended tables and expanded figures are being prepared for a forthcoming issue of *Radiocarbon.)*

In view of the scholarly theme of this book, we have refrained from adding numerical tables of radiocarbon ages with their corresponding cal ages and range of cal ages. The cal ages and cal age ranges can be obtained utilizing a computerized calibration program of the Quaternary Isotope Laboratory (available upon request, IBM-compatible version 3.0). This updated program version incorporates the Belfast/Seattle averages of Table 3.2.

The cal ranges obtained when converting ideal (zero error) radiocarbon ages coinciding with the 2500–5000 BC interval are graphically presented in Figure 3.6. Ranges of cal ages are produced even for point source radiocarbon dates because the variability and uncertainty in the calibration curve have to be taken into account. In Figure 3.6, the ranges are plotted on the vertical scale relative to the youngest cal date of the range (eg, the youngest cal age is set at zero). For a specific radiocarbon age, the cal age ranges resulting from the conversion process can be differentiated in Figure 3.6 into a precise cal age range (the areas of the curve where age conversion yields a single range) or a more calamitous multiple range (the areas where the wiggly nature of the calibration curve are encountered).

TABLE 3.2. Averages of Seattle and Belfast radiocarbon age (and $\Delta^{14}C$) determination of bi-decadal wood samples from the German (Seattle) and Irish (Belfast) oak chronologies. The BC ages listed in the table are the bi-decadal mid-points. The standard deviations σ include a 1.7 error multiplier for both laboratories (see text).

Year BC (cal yr)	^{14}C age ± σ (^{14}C yr)	$\Delta^{14}C$ ± σ (‰)	Year BP (cal yr) 0 BP = AD 1950
2510 BC	4058 ± 12	34.9 ± 1.4	4459 BP
2530 BC	4015 ± 11	42.9 ± 1.4	4479 BP
2550 BC	4007 ± 12	46.5 ± 1.5	4499 BP
2570 BC	4048 ± 11	43.7 ± 1.4	4519 BP
2590 BC	4088 ± 11	40.9 ± 1.4	4539 BP
2610 BC	4087 ± 15	43.7 ± 1.9	4559 BP
2630 BC	4137 ± 11	39.7 ± 1.3	4579 BP
2650 BC	4119 ± 16	44.5 ± 2.0	4599 BP
2670 BC	4143 ± 12	43.9 ± 1.5	4619 BP
2690 BC	4141 ± 13	46.7 ± 1.7	4639 BP
2710 BC	4194 ± 12	42.3 ± 1.5	4659 BP
2730 BC	4155 ± 12	50.0 ± 1.5	4679 BP
2750 BC	4167 ± 15	51.0 ± 1.9	4699 BP
2770 BC	4186 ± 16	51.0 ± 2.1	4719 BP
2790 BC	4210 ± 13	50.4 ± 1.6	4739 BP
2810 BC	4165 ± 14	58.9 ± 1.7	4759 BP
2830 BC	4108 ± 12	68.9 ± 1.4	4779 BP
2850 BC	4131 ± 14	68.5 ± 1.7	4799 BP
2870 BC	4188 ± 13	63.5 ± 1.7	4819 BP
2890 BC	4278 ± 15	54.2 ± 1.9	4839 BP
2910 BC	4309 ± 15	52.8 ± 1.8	4859 BP
2930 BC	4398 ± 16	43.6 ± 2.0	4879 BP
2950 BC	4401 ± 13	45.8 ± 1.7	4899 BP
2970 BC	4402 ± 17	48.2 ± 2.2	4919 BP
2990 BC	4380 ± 14	53.6 ± 1.7	4939 BP
3010 BC	4376 ± 16	56.7 ± 2.0	4959 BP
3030 BC	4402 ± 15	55.8 ± 1.9	4979 BP
3050 BC	4454 ± 12	51.5 ± 1.5	4999 BP
3070 BC	4421 ± 12	58.4 ± 1.5	5019 BP
3090 BC	4447 ± 17	57.5 ± 2.1	5039 BP
3110 BC	4509 ± 16	52.0 ± 2.0	5059 BP
3130 BC	4525 ± 16	52.5 ± 2.0	5079 BP
3150 BC	4540 ± 16	53.1 ± 2.0	5099 BP
3170 BC	4498 ± 15	61.1 ± 1.9	5119 BP
3190 BC	4524 ± 13	60.2 ± 1.6	5139 BP
3210 BC	4530 ± 14	61.9 ± 1.7	5159 BP
3230 BC	4513 ± 15	66.8 ± 1.8	5179 BP
3250 BC	4465 ± 15	75.9 ± 1.9	5199 BP
3270 BC	4498 ± 17	74.0 ± 2.1	5219 BP
3290 BC	4486 ± 16	78.2 ± 2.0	5239 BP

TABLE 3.2 (continued)

Year BC (cal yr)	^{14}C age ± σ (^{14}C yr)	$\Delta^{14}C$ ± σ (‰)	Year BP (cal yr) 0 BP = AD 1950
3310 BC	4521 ± 14	76.1 ± 1.8	5259 BP
3330 BC	4521 ± 12	78.8 ± 1.5	5279 BP
3350 BC	4577 ± 14	73.9 ± 1.7	5299 BP
3370 BC	4640 ± 18	68.0 ± 2.2	5319 BP
3390 BC	4730 ± 15	58.7 ± 1.9	5339 BP
3410 BC	4710 ± 18	63.9 ± 2.3	5359 BP
3430 BC	4685 ± 15	69.9 ± 1.8	5379 BP
3450 BC	4691 ± 19	71.7 ± 2.3	5399 BP
3470 BC	4650 ± 14	79.7 ± 1.7	5419 BP
3490 BC	4644 ± 14	83.1 ± 1.8	5439 BP
3510 BC	4726 ± 14	74.7 ± 1.8	5459 BP
3530 BC	4753 ± 14	73.7 ± 1.8	5479 BP
3550 BC	4801 ± 14	69.9 ± 1.8	5499 BP
3570 BC	4776 ± 14	75.9 ± 1.8	5519 BP
3590 BC	4752 ± 13	81.6 ± 1.6	5539 BP
3610 BC	4726 ± 12	87.7 ± 1.5	5559 BP
3630 BC	4793 ± 16	81.4 ± 2.0	5579 BP
3650 BC	4871 ± 14	73.5 ± 1.8	5599 BP
3670 BC	4905 ± 13	71.5 ± 1.6	5619 BP
3690 BC	4898 ± 14	75.1 ± 1.7	5639 BP
3710 BC	4950 ± 14	70.7 ± 1.7	5659 BP
3730 BC	4965 ± 16	71.4 ± 2.0	5679 BP
3750 BC	4957 ± 13	75.1 ± 1.7	5699 BP
3770 BC	4985 ± 14	73.8 ± 1.7	5719 BP
3790 BC	5011 ± 13	73.0 ± 1.6	5739 BP
3810 BC	5073 ± 15	67.4 ± 1.8	5759 BP
3830 BC	5089 ± 16	67.8 ± 2.0	5779 BP
3850 BC	5085 ± 12	70.9 ± 1.5	5799 BP
3870 BC	5071 ± 10	75.3 ± 1.3	5819 BP
3890 BC	5027 ± 13	83.9 ± 1.6	5839 BP
3910 BC	5058 ± 12	82.4 ± 1.5	5859 BP
3930 BC	5060 ± 14	84.6 ± 1.7	5879 BP
3950 BC	5111 ± 13	80.5 ± 1.6	5899 BP
3970 BC	5165 ± 10	75.8 ± 1.2	5919 BP
3990 BC	5217 ± 13	71.4 ± 1.6	5939 BP
4010 BC	5255 ± 15	68.9 ± 1.8	5959 BP
4030 BC	5219 ± 13	76.3 ± 1.6	5979 BP
4050 BC	5281 ± 15	70.7 ± 1.9	5999 BP
4070 BC	5244 ± 14	78.3 ± 1.8	6019 BP
4090 BC	5305 ± 14	72.7 ± 1.7	6039 BP
4110 BC	5325 ± 15	72.6 ± 1.9	6059 BP
4130 BC	5279 ± 16	81.3 ± 2.0	6079 BP
4150 BC	5293 ± 24	82.0 ± 3.0	6099 BP
4170 BC	5357 ± 17	76.1 ± 2.2	6119 BP
4190 BC	5332 ± 16	82.1 ± 2.0	6139 BP

TABLE 3.2 (continued)

Year BC (cal yr)	^{14}C age ± σ (^{14}C yr)	Δ^{14}C ± σ (‰)	Year BP (cal yr) 0 BP = AD 1950
4210 BC	5271 ± 18	92.9 ± 2.2	6159 BP
4230 BC	5370 ± 14	82.2 ± 1.7	6179 BP
4250 BC	5392 ± 16	81.8 ± 1.9	6199 BP
4270 BC	5448 ± 16	76.9 ± 2.1	6219 BP
4290 BC	5419 ± 16	83.4 ± 1.9	6239 BP
4310 BC	5393 ± 17	89.6 ± 2.1	6259 BP
4330 BC	5453 ± 15	84.2 ± 1.9	6279 BP
4350 BC	5519 ± 15	77.8 ± 1.9	6299 BP
4370 BC	5589 ± 16	71.1 ± 2.0	6319 BP
4390 BC	5568 ± 18	76.4 ± 2.2	6339 BP
4410 BC	5601 ± 18	74.7 ± 2.2	6359 BP
4430 BC	5552 ± 18	83.9 ± 2.3	6379 BP
4450 BC	5579 ± 18	82.8 ± 2.2	6399 BP
4470 BC	5671 ± 15	73.1 ± 1.9	6419 BP
4490 BC	5666 ± 18	76.4 ± 2.2	6439 BP
4510 BC	5683 ± 18	76.7 ± 2.2	6459 BP
4530 BC	5696 ± 18	77.5 ± 2.3	6479 BP
4550 BC	5742 ± 19	73.9 ± 2.3	6499 BP
4570 BC	5748 ± 14	75.7 ± 1.8	6519 BP
4590 BC	5757 ± 16	77.1 ± 2.0	6539 BP
4610 BC	5769 ± 18	78.2 ± 2.2	6559 BP
4630 BC	5799 ± 14	76.7 ± 1.8	6579 BP
4650 BC	5771 ± 12	83.2 ± 1.5	6599 BP
4670 BC	5768 ± 18	86.2 ± 2.3	6619 BP
4690 BC	5806 ± 13	83.7 ± 1.6	6639 BP
4710 BC	5821 ± 15	84.2 ± 1.9	6659 BP
4730 BC	5872 ± 18	80.0 ± 2.3	6679 BP
4750 BC	5842 ± 18	86.6 ± 2.3	6699 BP
4770 BC	5876 ± 18	84.7 ± 2.3	6719 BP
4790 BC	5912 ± 19	82.5 ± 2.3	6739 BP
4810 BC	5940 ± 19	81.3 ± 2.3	6759 BP
4830 BC	5949 ± 16	82.7 ± 2.1	6779 BP
4850 BC	5989 ± 16	80.0 ± 2.0	6799 BP
4870 BC	6007 ± 19	80.1 ± 2.3	6819 BP
4890 BC	5959 ± 17	89.3 ± 2.1	6839 BP
4910 BC	6008 ± 13	85.3 ± 1.7	6859 BP
4930 BC	6026 ± 13	85.4 ± 1.6	6879 BP
4950 BC	6069 ± 12	82.3 ± 1.4	6899 BP
4970 BC	6085 ± 13	82.7 ± 1.6	6919 BP
4990 BC	6091 ± 14	84.6 ± 1.8	6939 BP

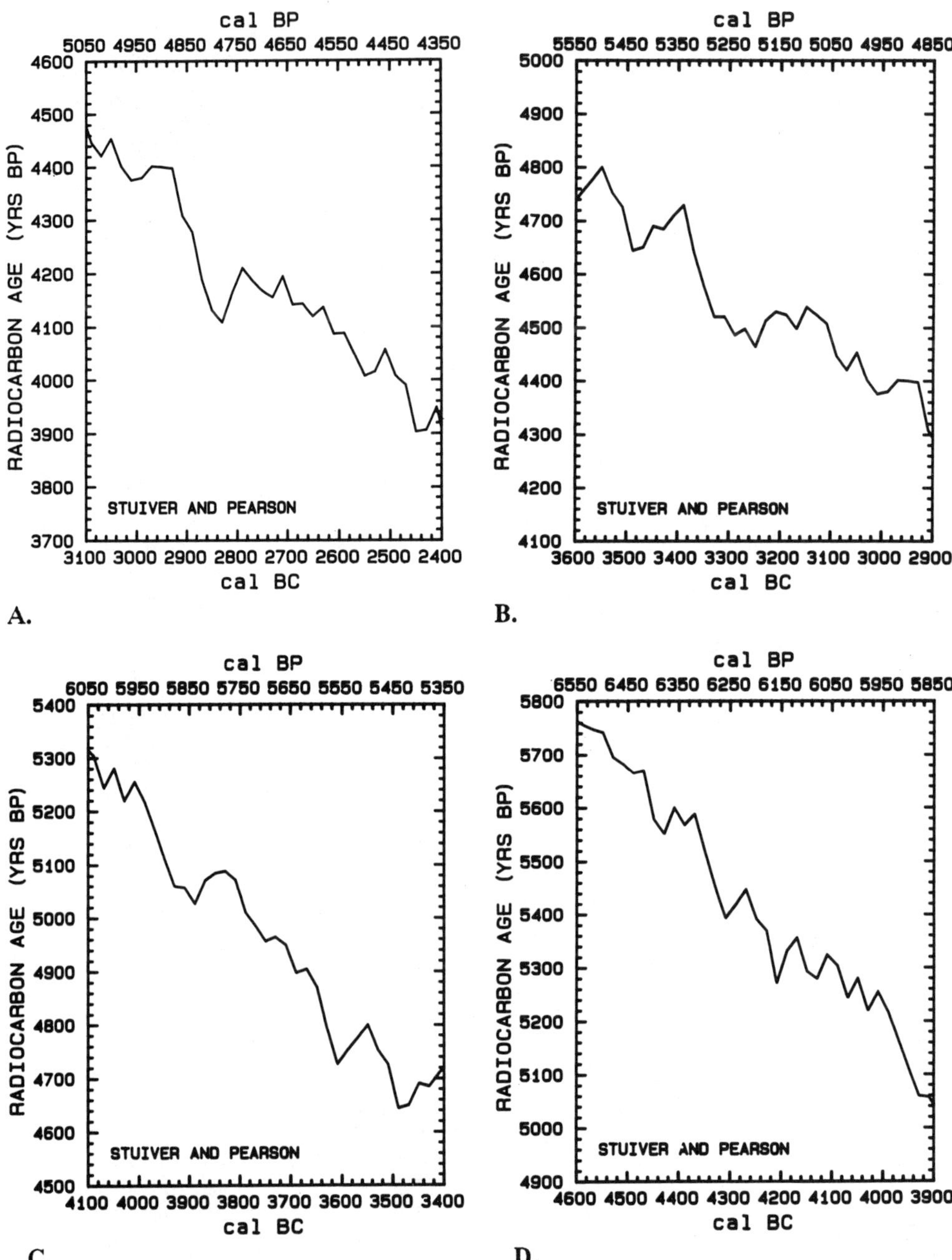

Fig 3.5 A–E. Radiocarbon ages of bi-decadal wood samples. The ages were obtained by averaging the Seattle and Belfast determinations on wood from the German (Seattle) and Irish (Belfast) oak chronologies. The average standard deviation for the bi-decadal data points in the figures is 15 ^{14}C yrs. This standard deviation accounts for the full range of laboratory variability (K = 1.7 was applied to the Belfast as well as the Seattle results; see text).

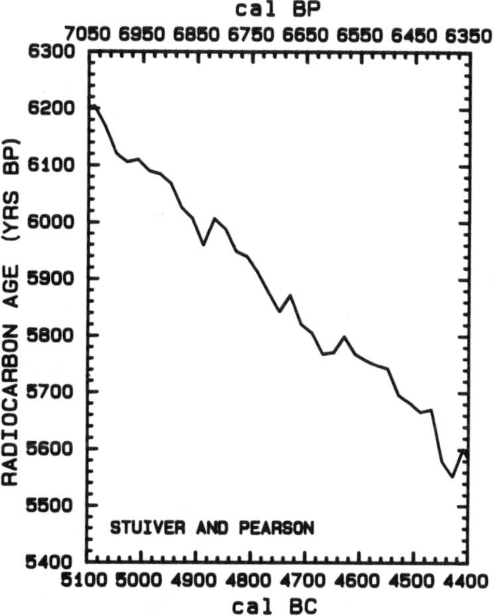

Fig 3.5E.

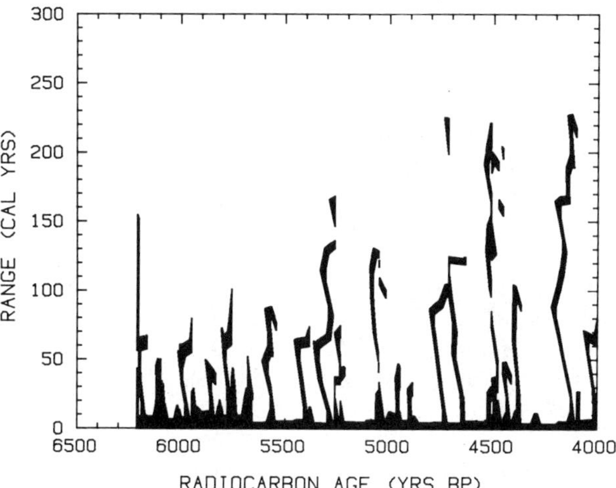

Fig 3.6. The ranges, in cal years, obtained from the calibration of radiocarbon ages in the 4000–6300 BP interval (0 BP = AD 1950). The calibration process yields one or more discrete cal ages, with a range around these discrete ages that represents the standard deviation in the calibration curve. The youngest cal age of the range(s) obtained for each radiocarbon age was set at zero. This distribution of ranges is for a hypothetical case where the standard deviation in the radiocarbon age is zero.

For the calibration of radiocarbon dates of which the cal age is older than 5000 BC, the reader is referred to, for example, Kromer *et al* (1986), Linick, Suess and Becker (1985) and Linick *et al* (1986). Belfast measurements covering the 5000–7000 BC interval will be released shortly, and integration with additional Seattle data should extend the joint Belfast-Seattle bi-decadal calibration curve to 7000 BC by 1993.

Advances in $^{234}U/^{230}Th$ dating (Edwards, Chen & Wasserburg 1987) of corals yield additional information on the offset between radiocarbon ages and U/Th ages (Bard *et al* 1990). For a discussion of the adjustment of the radiocarbon time scale back to 28,000 ^{14}C years BP, the reader is referred to Stuiver *et al* (1991).

Acknowledgments

This research was funded by the National Science Foundation grant BNS-9004492. We thank Dr Bernd Becker from the University of Hohenheim, Stuttgart for providing German oak (unified Donau 6/ Main 4/11 series) samples, and P J Reimer and P Wilkinson for crucial technical and analytical support.

References

Bard, E, Hamelin, B, Fairbanks, RG and Zindler, A 1990 Calibration of the ^{14}C timescale over the past 30,000 years using mass spectrometric U-Th ages from Barbados corals. *Nature* 345: 405–410.

Becker, B and Kromer, B 1986 Extension of the Holocene dendrochronology by the preboreal Pine series, 8800 to 10,100 BP. *In* Stuiver, M and Kra, RS, eds, Proceedings of the 12th International ^{14}C Conference. *Radiocarbon* 28(2B): 961–967.

Damon, PE, Lerman, JC and Long, A 1978 Temporal fluctuations of atmospheric ^{14}C, causal factors and implications. *Annual Review of Earth and Planetary Sciences* 6: 457–494.

Edwards, RL, Chen, JH and Wasserburg, GJ 1987 ^{238}U-^{234}U-^{230}Th-^{232}Th systematics and the precise measurement of time over the past 500,000 years. *Earth and Planetary Science Letters* 81: 175–192.

International Study Group 1982 An inter-laboratory comparison of radiocarbon measurements in tree rings. *Nature* 298: 619–623.

Jong, AFM de, Becker, B and Mook, WG 1986 High-precision calibration of the radiocarbon time scale. *In* Stuiver, M and Kra, RS, eds, Proceedings of the 12th International ^{14}C Conference. *Radiocarbon* 28(2B): 939–942.

Jong, AFM de, Mook, WG and Becker, B 1989 Corrected calibration of the radiocarbon time scale, 3904–3203 cal BC. *Radiocarbon* 31(2): 201–205.

Klein, J, Lerman, JC, Damon, PE and Ralph, EK 1982 Calibration of radiocarbon dates: Tables based on the consensus of the Workshop on Calibrating the Radiocarbon Time Scale. *Radiocarbon* 24(2): 103–150.

Kromer, B, Rhein, M, Bruns, M, Schoch-Fischer, H, Münnich, KO, Stuiver, M and Becker, B 1986 Radiocarbon calibration data for the 6th to the 8th millennia BC. *In* Stuiver, M and Kra, RS, eds, Proceedings of the 12th International ^{14}C Conference. *Radiocarbon* 28(2B): 954–960.

Linick, TW, Long, A, Damon, PE and

Ferguson, CW 1986 High-precision radiocarbon dating of bristlecone pine from 6554 to 5350 BC. *In* Stuiver, M and Kra, RS, eds, Proceedings of the 12th International ^{14}C Conference. *Radiocarbon* 28(2B): 943–953.

Linick, TW, Suess, HE and Becker, B 1985 La Jolla measurements of radiocarbon in south German oak tree-ring chronologies. *Radiocarbon* 27(1): 20–32.

Pearson, GW, Pilcher, JR, Baillie, MGL, Corbett, DM and Qua, F 1986 High-precision ^{14}C measurement of Irish oaks to show the natural ^{14}C variations from AD 1840–5210 BC. *In* Stuiver, M and Kra, RS, eds, Proceedings of the 12th International ^{14}C Conference. *Radiocarbon* 28(2B): 911–934.

Pearson, GW and Stuiver, M 1986 High-precision calibration of the radiocarbon time scale, 500–2500 BC. *In* Stuiver, M and Kra, RS, eds, Proceedings of the 12th International ^{14}C Conference. *Radiocarbon* 28(2B): 839–862.

Scott, EM, Aitchison, TC, Harkness, DD, Baxter, MS and Cook, GT 1989 An interim progress report on stages 1 and 2 of the international collaborative program. *In* Long, A and Kra, RS, eds, Proceedings of the 13th International ^{14}C Conference. *Radiocarbon* 31(3): 414–421.

Stuiver, M and Becker, B 1986 High-precision decadal calibration of the radiocarbon time scale, AD 1950–2500 BC. *In* Stuiver, M and Kra, RS, eds, Proceedings of the 12th International ^{14}C Conference. *Radiocarbon* 28(2B): 863–910.

Stuiver, M, Braziunas, TF, Becker, B and Kromer, B 1991 Climatic, solar, oceanic, and geomagnetic influences on Late-Glacial and Holocene atmospheric $^{14}C/^{12}C$ change. *Quaternary Research* 35: 1–24.

Stuiver, M and Kra, RS, eds 1986 Calibration issue. *In* Stuiver, M and Kra, RS, eds, Proceedings of the 12th International ^{14}C Conference. *Radiocarbon* 28(2B): 805–1030.

Stuiver, M and Pearson, GW 1986 High-precision calibration of the radiocarbon time scale, AD 1950–500 BC. *In* Stuiver, M and Kra, RS, eds, Proceedings of the 12th International ^{14}C Conference. *Radiocarbon* 28(2B): 805–838.

Stuiver, M and Polach, H 1977 Discussion: Reporting of ^{14}C data. *Radiocarbon* 19(3): 355–363.

Vogel, JC, Fuls, A and Visser, E 1986 Radiocarbon fluctuations during the third millennium BC. *In* Stuiver, M and Kra, RS, eds, Proceedings of the 12th International ^{14}C Conference. *Radiocarbon* 28(2B): 935–938.

Vries, Hessel de 1958 Variation in the concentration of radiocarbon with time and location on earth. *Koninklijke Nederlandse Akademie van Wetenschappen* Series B 61: 94–102.

_____1959 Measurement and use of natural radiocarbon. *In* Abelson, PH, ed, *Researches in Geochemistry*. New York, John Wiley & Sons: 169–189.

4

THE HISTORY OF DENDROCHRONOLOGY AND RADIOCARBON CALIBRATION

BERND BECKER

Dedicated to the memory of C W Ferguson

INTRODUCTION

The field of dendrochronology has developed during the first decades of this century. Tree-ring dating was the only method of precise age determination until the 1950s, when the radiocarbon dating method was developed. It is a fascinating chapter in the history of research to observe the immediate interdisciplinary cooperation between tree-ring dating and radiocarbon dating that started in the 1950s and continues up to this day with increasing importance. This chapter presents a brief historical overview of the development of superlong Holocene tree-ring calendars and the calibration of the radiocarbon time scale derived from tree-ring measurements, with a synopsis of the actual stage of research.

SUPERLONG HOLOCENE TREE-RING RECORDS – A REVIEW OF THE 80-YEAR HISTORY OF DENDROCHRONOLOGY

The fact that the tree-ring structure of wood can be used for dating purposes was first recognized by A E Douglass (1867–1962) (Fig 4.1). This outstanding pioneer of dendrochronology established a method of tree-ring dating and reconstruction of past climate by tree-ring variations. He started his career at the Lowell Astronomical Observatory in Flagstaff, Arizona. As a young astronomer, he studied the relationship between sunspot cycles, climate and tree-ring patterns. Douglass detected a surprising similarity between tree-ring width curves over large geographical distances, and discovered that conifer trees form, coincidently, very small annual growth layers during drought years known from the historical record of the southwest United States. From this observation, Douglass derived a basic assumption of dendrochronology. He concluded that tree-ring patterns of wood samples of unknown age must be precisely dated by cross-matching to a tree-ring calendar, which comprises the signature years in tree-ring sequences back in time. However, it took about a quarter of the century, from his first successful attempts in 1906, when he dated cutting years of stumps of modern trees, until his sensational absolute dating of prehistoric Pueblo ruins.

In 1929, Douglass finally found the missing link between a floating 585-year tree-ring sequence of timbers from Indian settlements and his absolutely dated tree-ring calendar, which he had extended from modern conifer trees back to the year, AD 1260. This linkage dated, at once, 40 prehistoric sites, including the famous cliff dwellings of Mesa Verde. This successful discovery introduced dendrochronology definitively as a method of absolute dating in archaeology (Douglass 1935).

A further milestone in the development of the long dendrochronological records was the discovery of the ancient bristlecone pines (*Pinus aristata* and *P longaeva*). In 1953, Edmund Schulman (Fig 4.1), a coworker of Douglass at the Tree-Ring Laboratory of the University of Arizona, Tucson, found conifer stands at high-elevation sites in the White Mountains of California where trees older than 4000 years exist. Schulman used the patterns of the ancient trees for the construction of an unbroken 4600-year sequence (Schulman 1958).

After Schulman's untimely death in 1958, the work on the bristlecone pine chronology in Tucson was successfully continued by the late C W Ferguson (Fig 4.2). He extended Schulman's chronology of living bristlecone pines by sub-

Fig 4.1. A E Douglass (right) and Edmund Schulman (left); photo taken by Bruno Huber

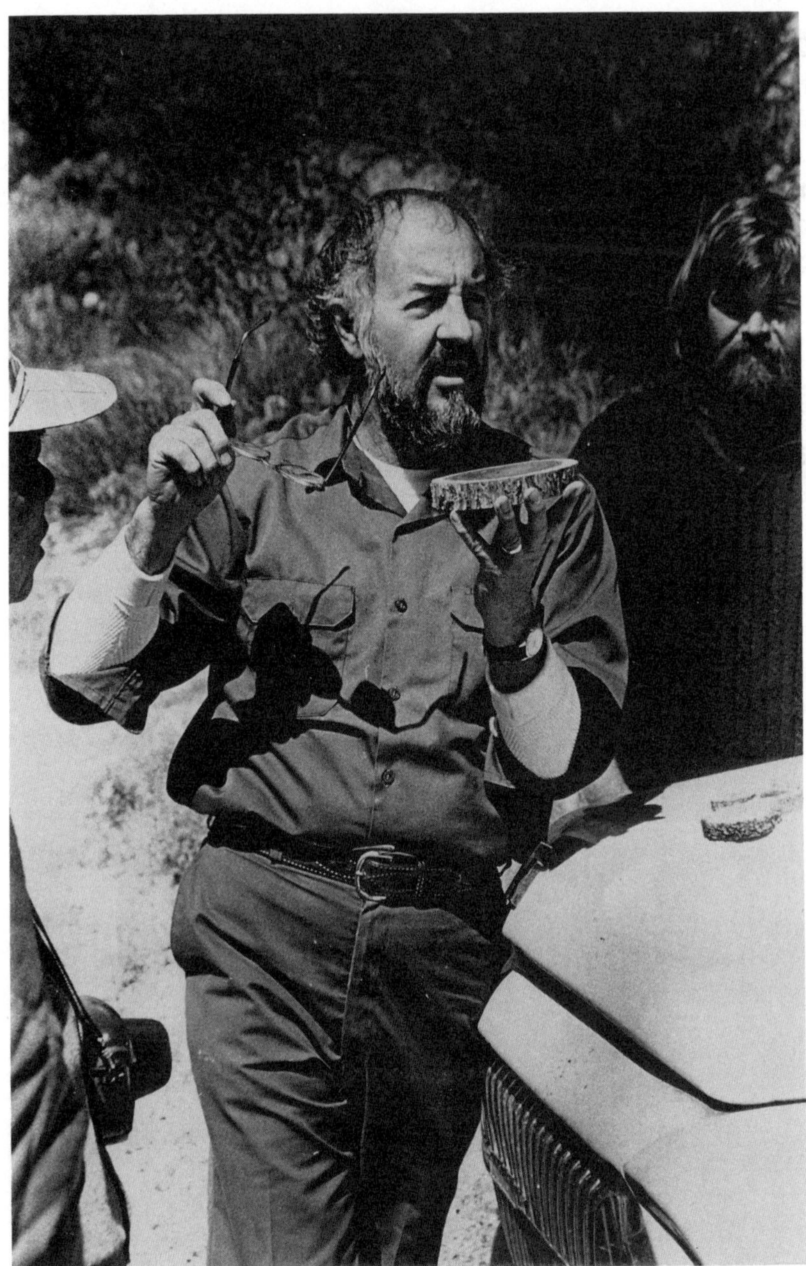

Fig 4.2. C W Ferguson (Photo courtesy of the Laboratory of Tree-Ring Research, The University of Arizona)

fossil trunk remnants, which were preserved over millennia in high-elevation forest borderlines of the White Mountains. In 1969, Ferguson presented a continuous 7104-year tree-ring record (Ferguson 1969). For two decades this sequence was the longest worldwide dendrochronology. Ferguson finally enlarged the series up to 6700 BC (Ferguson & Graybill 1983).

The 7104-year bristlecone pine dendrochronology was one of the most important steps in the absolute chronology of the Holocene. As presented at the 12th Nobel Symposium in Uppsala, in 1969, knowledge of natural long-term variations of atmospheric ^{14}C was derived basically from analyses of bristlecone pine wood samples (Olsson 1970). Since that time, the tree-ring dating method became increasingly important, not only through its application for dating in archaeology and medieval architectural history, but also in geophysical research.

Development of Dendrochronology in Germany

Stimulated by the success of A E Douglass, the German botanist, Bruno Huber (Fig 4.3), started tree-ring studies at the University of Tharandt in the 1940s. After the end of World War II, Huber continued this research at the University of München. A first step in the long German dendrochronology was the construction of the tree-ring record of the famous "Spessart oaks." Huber and his coworkers extended this chronology by linking oak tree-ring patterns from historical frame buildings back to AD 832 (Huber & Giertz-Siebenlist 1969). Early dendrodates derived from this more than 1000-year oak and from an 1100-year fir (*Abies alba*) chronology, derived by Huber's coworkers (Becker & Giertz-Siebenlist 1970), demonstrated the potential of this new method for dating medieval architectural history in central Europe.

Very similar to the general acceptance of the tree-ring dating method in America as a result of A E Douglass' datings of the Pueblo ruins, Bruno Huber had spectacular success in deriving dendrochronological age determinations of prehistoric Neolithic sites in Europe. During the early 1960s, Huber began a cooperative project with Swiss and German archaeologists, who, at that time, were excavating Neolithic Swiss lake dwelling sites. One has to consider that when Huber started his tree-ring analyses on dwelling samples, there was no realistic chance of absolute dendrodating of these prehistoric sites, since the absolute chronology still ended at AD 832. However, Huber found the cross-match between the floating oak chronologies of the prehistoric sites, Thayngen-Weier and Burgäschisee, and from some samples from Niederwil. Today, we know from our absolute dendrodates of Swiss lake dwelling sites that Neolithic settlements lasted, in that region, at least over two millennia, from 4400 to 2400 BC (Becker *et al* 1985; Becker, Krause & Kromer 1989). Luckily, the three sites that Huber analyzed were all inhabited during the same century.

Fig 4.3. Bruno Huber (Reprinted with permission from Verlag der Zeitschrift für Naturforschung, Tubingen 1966. See Ferguson, Huber & Suess 1966.)

With the cross-dating of these floating oak series, Huber could demonstrate, in 1963, the synchronous existence of different Neolithic cultures, eg, the Michelsberger culture of Thayngen-Weier, the Cortaillod culture of Burgäschisee Süd/Südwest, and the Pfyn culture of Niederwil (Huber & Merz 1963).

Further, together with C W Ferguson and H E Suess, Bruno Huber, already in 1966, derived an absolute age determination of these Neolithic sites by dendrochronologically calibrated radiocarbon dates. Ferguson, Huber and Suess had matched the ^{14}C variations of the floating Thayngen-Weier oak sequence to those of the bristlecone pine calibration curve of 4100 to 3600 BC (see Fig 4.4). The dates differ from the actual dendrodates by not more than 15 years! (Ferguson, Huber & Suess 1966). This result was so revolutionary to archaeologists, that some leading German and Swiss archaeologists objected to the ^{14}C age correction (Milojcic 1957, 1958, 1961). However, final confirmation of this sequence came 15 years later, with our cross-match of the Thayngen-Burgäschisee-Niederwil sequences and the German oak calendar.

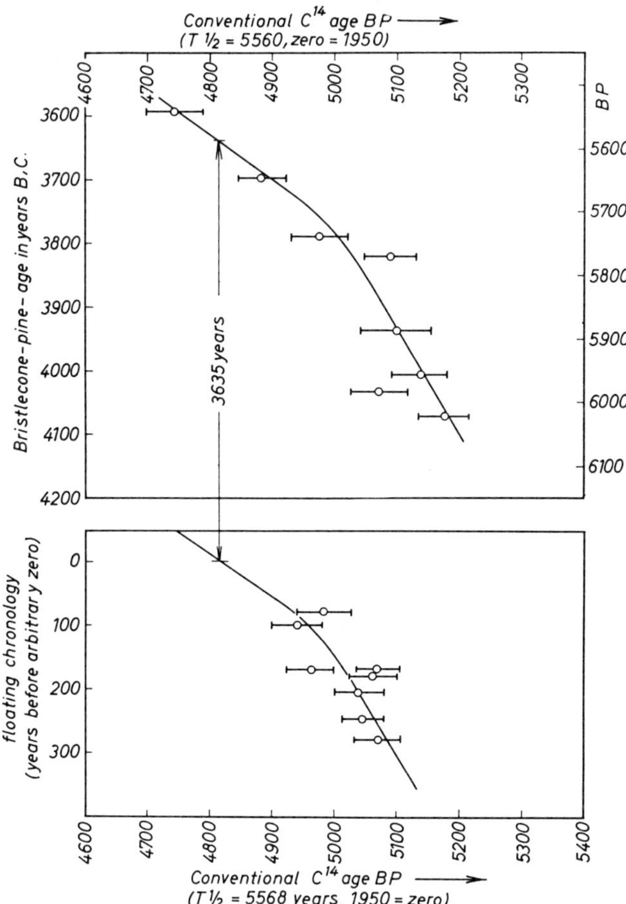

Fig 4.4. Wiggle matching for the Thayngen-Weier ^{14}C measurements with the bristlecone pine calibration curve, as published by Ferguson, Huber and Suess (1966). Reprinted by permission of Verlag der Zeitschrift der Naturforschung.

Development of Long Tree-Ring Records in Western and Central Europe

As far as archaeologic dating is concerned, European tree-ring research has focused on oak (*Quercus robur, Q petraea*). In the western and central European lowland regions, oak was the most frequently used timber for building construction since Neolithic times. Central and western Europe, then, had the most to gain from the development of the oak dendrochronology.

During the early stages of tree-ring dating, research concentrated on archaeological samples. Unfortunately, prehistoric settlement sites tend to cluster in time during certain periods, whereas other periods remain undocumented by wood settlement remains. For example, tremendous numbers of oak timbers are available in central Europe from Neolithic and Bronze Age dwelling sites, or from Roman sites. During the Hallstatt period, on the other hand, especially between 600 and 400 BC, and during the period of population migrations, especially between 250 to 500 BC, in Germany, timber structures have been excavated only rarely. This may explain why the Holocene dendrochronology of Europe has been derived not from archaeologic samples, but from subfossil oak timbers preserved in natural deposits.

During the early 1970s, several research groups began tree-ring research on subfossil oak trunk deposits. In Belfast, M G L Baillie and J R Pilcher started sampling oak cross-sections from Irish peat bogs (Pilcher *et al* 1977). In Germany, Bernd Becker (Stuttgart-Hohenheim), Axel Delorme (Göttingen) and Burkhart Schmidt (Köln) began to collect oak trunks dredged out of alluvial gravel deposits along the major river valleys (Becker, Delorme & Schmidt 1977) (Fig 4.5).

The construction of the European Holocene oak tree-ring calendar became a project of decades and is still unfinished. The reason is that oak cross-sections from bogs and gravel deposits consist of not more than 100 to 400 tree rings, which is a very short time span, compared with up to 4000 tree rings of bristlecone pine patterns. Therefore, the European oak dendrochronology had to link together a puzzle of hundreds of stepwise overlapping, very short tree-ring

Fig 4.5. Subfossil oak trunks, the major source of the Holocene tree-ring chronology

curves. This may also explain why some of the longstanding gaps of the European Holocene chronology finally have been bridged only after international cooperation. In 1984, the laboratories of Belfast, Köln and Hohenheim carried out a stepwise cross-match of oak chronologies from Northern Ireland, Great Britain and northern and southern Germany, which closed a final gap of the Irish chronology, as well as one of the gaps of the German oak chronology. From this cross-match, the Irish chronology has been extended to an unbroken sequence of 7272 tree rings, whereas the German oak chronology at that point, had reached the year, 4000 BC (Pilcher *et al* 1984).

Further progress in Holocene dendrochronology has evolved from the joining of two floating river oak series to the German absolute oak sequence. The cross-match of our 3242-year mid-Holocene series has been achieved by successful cross-dating with a north German oak series of the period, 3500 to 4500 BC. This extended the German oak calendar to 7200 BC (Becker & Schmidt, in press). Recently, in Hohenheim, two even older chronologies of the Main River valley have been joined to the absolute oak chronology. The Main 9 sequence, which was already the subject of radiocarbon calibration measurements of the La Jolla, Heidelberg, Seattle and Belfast laboratories, now extends the Holocene oak calendar over 9929 years back to 7938 BC. This German oak master chronology actually represents the largest unbroken Holocene tree-ring record.

I should add here that the Göttingen tree-ring laboratory has developed independently a long German oak tree-ring chronology. By tying together northern German bog oaks and western German river oak series, Hubertus Leuschner established a continuous 8246-year chronology back to 6255 BC (Leuschner & Delorme 1988). This sequence cross-dates with the Hohenheim oak chronology. Thus, both Holocene oak series are replicated externally (M Baillie, personal communication).

Table 4.1 shows the present stage of superlong dendrochronologies.

TABLE 4.1. Superlong Holocene Tree-Ring Chronologies

Species	Region	Beginning date (BC)	Length (yrs)	Reference
Oak*	W Europe	5282	7272	Pilcher *et al* (1984)
Oak	C Europe	6255	8246	Leuschner & Delorme (1988)
Pine**	SW USA	6700	8691	Ferguson & Graybill (1986)
Oak	C Europe	7938	9929	Becker (this chapter)
Pine†	C Europe	9420‡	1604	Becker (this chapter)

**Quercus robur, Quercus petraea*; ***Pinus aristata, Pinus longaeva*; †*Pinus sylvestris*; ‡Minimum age according to radiocarbon calibration, see text.

Further extension of the European oak dendrochronology will not exceed 200 to 400 years. According to the earliest ^{14}C-dated subfossil oak trunks of the Hohenheim collection, oak forests migrated back to central Europe not before 9300 BP (Becker & Kromer 1991).

Using subfossil pines, which spread over central European valley plains since 12,200 BP, we have established a very precisely dated tree-ring sequence, which leads back to the end of the Late Glacial period (see the last section, below).

CORRECTION AND CALIBRATION OF THE RADIOCARBON TIME SCALE BY HOLOCENE TREE-RING CHRONOLOGIES

In 1986, at the international colloquium, "Chronologies relatives et chronologie absolu dans le Proche Orient de 16,000 à 4000 BP," of the Centre National de la Recherche Scientifique (CNRS), in Lyon, France, P E Damon presented a paper entitled "The history of the calibration of radiocarbon dates by dendrochronology." The paper (Damon 1987) gives a comprehensive summary of the history of radiocarbon research and dendrochronology.

I will point out here only certain aspects of the ^{14}C age calibration, mainly as far as it is related to the long Holocene chronology and to the prehistoric archaeological time scale.

The Early Stages of Calibration Until the Twelfth Uppsala Nobel Symposium

The first hints of natural variations of the atmospheric ^{14}C level were reported by K O Münnich in 1957 from German oak samples dated by Bruno Huber (Münnich 1957). At the same time, Hessel de Vries began measuring dendrodated wood samples. He found ^{14}C variations of 2% from the average level of the period, AD 1525 to 1935, and observed two minima of atmospheric ^{14}C concentration at AD 1600 and 1800 (de Vries 1959). These variations were confirmed and extended by a further study of the Cambridge, Copenhagen and Heidelberg laboratories, which, in 1960, presented measurements on a 1300-year sequoia tree-ring sequence (Willis, Tauber & Münnich 1960).

As mentioned before, between 1954 and 1958, the California bristlecone pine and sequoia tree-ring calendars became available for long-term ^{14}C calibration studies. H E Suess stated in 1961 that the de Vries-effect fluctuations must be superimposed by a more general long-term trend.

Systematic measurements of the ^{14}C concentrations of long tree-ring series were carried out by the laboratories at Groningen (J C Lerman, W G Mook and J C Vogel), Pennsylvania (E K Ralph and H N Nader), La Jolla (H E Suess) and

Tucson (P E Damon, A Long and D C Grey). Based on measurements on bristlecone pine and sequoia tree-ring samples of the period, 5300 to 2000 BC, Suess introduced the concept of "wiggles," that is, the occurrence of medium-term ^{14}C variations. A major result of early calibration measurements was the proof of a long-term decrease of the ^{14}C content of tree-ring series since 5000 BC, and the conclusion that this variation obviously required corrections of conventional ^{14}C dates to older ages, up to 700 to 1000 years. The early calibration research was perfectly summarized in Uppsala in 1969, at the 12th Nobel Symposium on Radiocarbon Variations and Absolute Chronology (Olsson 1970).

In western Europe, archaeologists immediately turned to the newly available calibration curve for the correction of the Neolithic and Bronze Age time scales. One of the consequences of the correction of the radiocarbon time scale to older ages is that European prehistoric cultures generally do not postdate Middle East cultures, which previously had been assumed, on the basis of conventional ^{14}C dates (Renfrew 1973, 1977).

Confirmation of the Bristlecone Pine Calibration Curve

Research on natural atmospheric radiocarbon variations continued and intensified after the Uppsala symposium when the European oak chronologies became available. In La Jolla, H E Suess and T W Linick started a calibration project on German oaks. At that time, still floating, the prehistoric oak chronology had the advantage of a more precise time solution of the sampling (compared with the bidecadal block samples of bristlecone pine, oak samples can be prepared in single years up to pentadal blocks). The La Jolla calibration measurements of German oak finally confirmed the La Jolla bristlecone pine calibration curve over its entire length (Linick, Suess & Becker 1985).

However, during the 1970s, the Suess wiggles in the bristlecone pine curve with "cosmic swung" (Suess 1965) were still being questioned. Unfortunately, a first series of high-precision ^{14}C measurements on the Irish oak chronology covered just a period of low atmospheric variation, resulting in a more smoothed calibration curve (3200–2100 BC, Pearson *et al* 1977), which seemed to contradict the observation of large Holocene wiggles.

The discussion on the validity of the wiggles was finally terminated by the Groningen laboratory, which presented high-precision ^{14}C measurements on German oak samples for the period, 3800 to 3200 BC (de Jong, Mook and Becker 1979). The measurements had been carried out with remarkably improved precision of the sampling, as well as of the ^{14}C measurements. The wood samples were prepared on single-ring to three-year blocks, and the measurements had a precision on a 2‰ level. The series covered a period when the Suess bristlecone pine ^{14}C curve showed major oscillations. The de Jong series replicated perfectly the three wiggles of that 600-year period, and confirmed, as well, the La Jolla measurements on German oak of the period, 3800 to 3200 BC. A comparison between the Groningen high-precision calibration and the La Jolla measurements on bristlecone pine and German oak, published by Suess in 1980 (Fig 4.6), demonstrates very impressively the accuracy of the earlier La Jolla calibration measurements.

High-Precision Calibration of the Past 11,000 Years

During the 1980s, the Irish and German oak tree-ring calendars advanced further. Based on these series, high-precision ^{14}C measurement projects started in Belfast, Groningen, Heidelberg, Pretoria and Seattle (which also used American Douglas fir). The Tucson laboratory contributed high-precision measurements on the enlarged bristlecone pine chronology.

At the 12th Radiocarbon Conference in Trondheim in 1985, the high-precision ^{14}C calibration data were presented by the intercalibration group. The progress of the calibration as reported in Trondheim (Stuiver & Kra 1986), is comparable to that of the Uppsala Symposium.

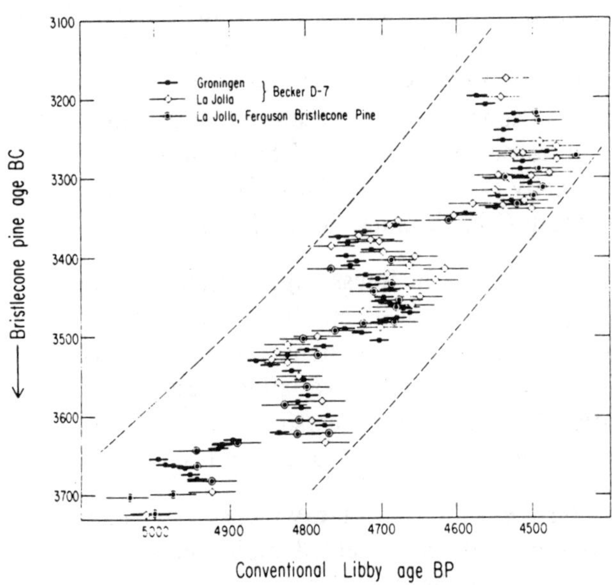

Fig 4.6. ^{14}C wiggles, 3800 to 3200 BC. Compared are La Jolla measurements on bristlecone pine and German oak with the high-precision Groningen measurement on German oak (after Suess 1980). Reprinted by permission of *Endeavour*. © 1980 Pergamon Press PLC.

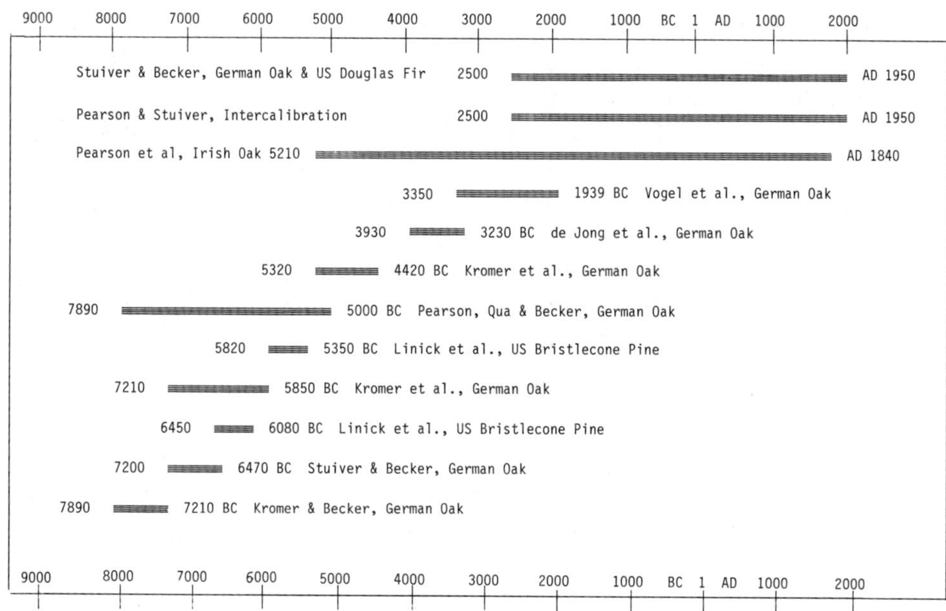

Fig 4.7. Actual stage of high-precision ^{14}C calibration measurements derived from absolutely dated tree-ring series. Data were compiled by the intercalibration group at a meeting in Groningen, organized by W G Mook and Johannes van der Plicht in 1989.

The actual stage of interlaboratory-calibrated high-precision radiocarbon measurements on absolutely dated tree-ring samples is presented in Figure 4.7. The diagram is a result of a meeting of the calibration group in Groningen in 1989 organized by W G Mook and Johannes van der Plicht.

To summarize, the following aspects of the high-precision ^{14}C calibration are:

1. The correction of conventional ^{14}C dates can be applied on the basis of interlaboratory-calibrated high-precision ^{14}C curves from the present to 2500 BC (Pearson & Stuiver 1986; Stuiver & Pearson 1986). Up to 5210 BC, a bidecadal high-precision calibration curve of Irish oak can be used (Pearson *et al* 1986), and from 5100 to 7980 BC, a calibration curve on German oak from the Belfast laboratory was established, but not yet published (Pearson, Becker & Qua, in press). Major parts of the mid-Holocene period were measured by the Heidelberg laboratory, and can also be used for calibration (Kromer *et al* 1986). The method of calibrating conventional ^{14}C ages is now simplified by computer programs available from Seattle (Stuiver & Reimer 1986) and Groningen (van der Plicht & Mook 1989).

2. High-precision ^{14}C measurements of dendrodated wood have improved the knowledge of the structure of medium-term variations and also of the long-term trend of ^{14}C variations of the atmosphere. From the earlier bristlecone pine ^{14}C curve, a sinusoidal long-term trend was suggested, and a possible decrease from the high mid-Holocene level towards the Late Glacial was assumed by researchers. The recent Heidelberg, Belfast and Seattle calibration data on German oak back to 7938 BC, in combination with the recent Early Holocene pine measurements, clearly demonstrate the existence of a long-term decrease of the atmospheric ^{14}C level, at least since the end of the Late Glacial. This trend is superimposed by century time-scale variations (Kromer *et al* 1986; Stuiver *et al* 1991).

   Most of the variance in the Holocene atmospheric ^{14}C record can be attributed to these two effects: the geomagnetic influence on the flux of cosmic rays entering the atmosphere is recorded in the millennia time-scale variation, whereas the solar influence can be seen from the century time-scale oscillations (Stuiver *et al* 1991).

3. High-precision ^{14}C calibration curves derived from Holocene oak and pine tree-ring records provide a detailed picture on the structure of the wiggles. Stuiver and Braziunas (1989) found cyclicities in the atmospheric ^{14}C record derived from maximum-entropy spectral analysis of high-precision ^{14}C calibration data over the past 9600 years. The Holocene century-type ^{14}C oscillations (consisting of approximately equal rates of $\Delta^{14}C$ increase and decrease) are described as Maunder oscillation type (with a period of about 180 years) and Spörer oscillation type (nearly 220 years long). They are evidently produced by solar activity (Stuiver & Braziunas 1989).

EXTENSION OF DENDROCHRONOLOGY AND RADIOCARBON CALIBRATION TO LATE GLACIAL TIMES

The European dendrochronological record actually reaches to the end of the Late Glacial. This was not achieved by further enlargement of the absolute oak chronology beyond its 10,000-year range. According to the dates of the earliest findings of trunks, the oak forest migrated back to central Europe not before 9300 ^{14}C years BP. The extension of the long chronology was achieved through subfossil pine trunks from alluvial river terraces in southern Germany and northern Italy (Becker & Kromer 1991). Subfossil pine trees date from the Bølling and Allerød interstadials, (between 12,400 and 10,800 ^{14}C years BP), from the end of the Younger Dryas and the entire Preboreal. We have established a 1604-year floating pine (*Pinus sylvestris*) chronology, which extends the absolute European tree-ring record over 11,370 dendroyears BP. The age determination of this pine sequence results from Heidelberg high-precision

^{14}C measurements near the overlaps of the floating pine and the absolute oak series. From stable isotopic analyses (^{2}H, ^{13}C measurements, GSF München) of the pine wood samples, we found clear indications that the chronology exceeds even the Younger Dryas/Preboreal boundary, and comprises the last part of the Younger Dryas cold phase back to 11,370 dendroyears BP (Becker, Kromer & Trimborn 1991). This most recent progress in the long dendrochronology provides information on the radiocarbon calibration of the Early Holocene and Late Glacial:

1. From high-precision measurements of the Heidelberg laboratory, a continuation of the decrease in atmospheric radiocarbon concentration since the Late Glacial can be observed. At 11,000 dendroyears BP, the atmospheric ^{14}C level must have been 12 to 14% higher than today.

2. Remarkable century-type ^{14}C oscillations occurred during the Early Holocene. This phenomenon causes intervals of apparently constant ^{14}C ages in the calibration curve for periods of up to 450 dendroyears. These ^{14}C age plateaus occur at 10,000, 9600, 8750 and 8200 ^{14}C years BP. The calibration curve indicates severe problems in ^{14}C dating, especially at the Holocene/Pleistocene boundary. ^{14}C dates of 10,000 BP *versus* 9600 BP can be related to a real time interval of minimally 90 to maximally 860 dendroyears, depending on the true age position in the two ^{14}C age plateaus!

3. The long German tree-ring record is used as a link for ^{14}C age calibrations as derived from other time series, such as lake varves or corals. A cross-match of the ^{14}C patterns of the German pine and the Lake of the Clouds sequences (Stuiver 1971) with uranium/thorium-dated ^{14}C measurements of coral samples drilled from reefs off the shore of Barbados (Bard *et al* 1990) may extend the ^{14}C calibration curve up to 20,000 years (Stuiver *et al* 1991).

REFERENCES

Bard, E, Hamelin, B, Fairbanks, RG and Zindler, A 1990 Calibration of the ^{14}C time scale over the past 30,000 years using mass spectrometric U-Th ages from Barbados corals. *Nature* 345: 405–410.

Becker, B, Billamboz, A, Egger, H, Gassmann, P, Orcel, A, Orcel, C and Ruoff, U 1985 Dendrochronologie in der Ur- und Frühgeschichte. *Antiqua* 11: 1–68.

Becker, B, Delorme, A and Schmidt, B 1977 Koordination der Jahrringforschung beim Aufbau einer postglazialen Eichenchronologie. *In* Frenzel, B ed, Dendrochronologie und postglaziale Klimaschwankungen. *Erdwissenschaftliche Forschungen* XIII: 143–146.

Becker, B and Giertz-Siebenlist, V 1970 Eine über 1100-jährige mitteleuropäische Tannenchronologie. *Flora* 159: 310–346.

Becker, B, Krause, R and Kromer, B 1989 Zur absoluten Chronologie der Frühen Bronzezeit. *Germania* 67: 421–442.

Becker, B and Kromer, B 1991 Dendrochronology and radiocarbon calibration of the Early Holocene. *In* Barton, N, Roe, D and Roberts, A, eds, The late glacial of northwest Europe. *BAR International Series*, in press.

Becker, B, Kromer, B and Trimborn, P 1991 A stable isotope tree-ring timescale of the Late Glacial/Holocene boundary. *Nature*, in press.

Becker, B and Schmidt, B, in press, Extension of the European oak chronology to the past 9924 years. *In* Mook, WG and Waterbolk, HT, eds, 2nd International Symposium, Archaeology and ^{14}C. *PACT* 29.

Damon, PE 1987 The history of the calibration of radiocarbon dates by dendrochronology. *In* Aurenche, O, Evin, J and Hours, F, eds, Chronologies du Proche Orient. *BAR International Series* 379(i): 61–104.

Douglass, AE 1935 Dating Pueblo Bonito and other ruins of the Southwest. National Geographical Society, *Contributed Technical Papers, Pueblo Bonito Series* 1: 1–74.

Ferguson, CW 1969 A 7104-year annual tree-ring chronology for bristlecone pine, *Pinus aristata*, from the White Mountains, California. *Tree Ring Bulletin* 29(3–4): 1–29.

Ferguson, CW and Graybill, DA 1983 Dendrochronology of Bristlecone Pine: A progress report. *In* Stuiver, M and Kra, RS, eds, Proceedings of the 11th International ^{14}C Conference. *Radiocarbon* 25(2): 287–288.

Ferguson, CW, Huber, B and Suess, HE 1966 Determination of the age of Swiss lake dwellings as an example of dendrochronologically-calibrated radiocarbon dating. *Zeitschrift für Naturforschung* 21 (34): 1173–1177.

Huber, B and Giertz-Siebenlist, V 1969 Unsere tausendjährige Eichen-Jahrringchronologie durchschnittlich 57(10-159) fach belegt: Sitzungsbericht der Österr. Akademie der Wissenschaften. *Mathmatesch-Naturwissenschaftliche Klasse* 178 (1–4): 37–45.

Huber, B and Merz, W 1963 Jahrringchronologische Synchronisierung der jungsteinzeitlichen Siedlungen Thayngen-Weier und Burgäschisee-Süd und Südwest. *Germania* 41: 1–9.

Jong, AFM de, Mook, WG and Becker, B 1979 Confirmation of the Suess wiggles 3200–3800 BC. *Nature* 280: 48–49.

Kromer, B, Rhein, M, Bruns, M, Schoch-Fischer, H, Münnich, KO, Stuiver, M and Becker, B 1986 Radiocarbon calibration data for the 6th to the 8th millennia BC. *In* Stuiver, M and Kra, RS, eds, Proceedings of the 12th International ^{14}C Conference. *Radiocarbon* 28(2B): 954–960.

Leuschner, HH and Delorme, A 1988 Treering work in Göttingen. Absolute oak chronologies back to 6255 BC. *PACT* 22 (II.5): 123–132.

Linick, TW, Suess, HE and Becker, B 1985 La Jolla measurements of radiocarbon on south German oak tree-ring chronologies. *Radiocarbon* 27(1): 20–30.

Milojcic, V 1957 Zur Anwendbarkeit der 14-C Datierungen. *Germania* 35: 102–125.

_____ 1958 Zur Anwendbarkeit der 14-C Datierungen. *Germania* 36: 409–420.

_____ 1961 Zur Anwendbarkeit der 14-C Datierungen. *Germania* 39: 434–455.

Münnich, KO 1957 Heidelberg natural radiocarbon measurements I. *Science* 126: 194–199.

Olsson, IU, ed 1970 *Radiocarbon Variations and Absolute Chronology*. Proceedings of the 12th Nobel Symposium. Stockholm, Almqvist & Wiksell: 652 p.

Pearson, GW, Becker, B and Qua, F, in press, High-precision ^{14}C measurement of German oaks to show the natural ^{14}C variations from 7980 to 5000 BC. *Radiocarbon*.

Pearson, GW, Pilcher, JR, Baillie, MGL, Corbett, DM and Qua, F 1986 High-precision ^{14}C measurement of Irish oak to show the natural ^{14}C variations from AD 1840 to 5210 BC. *In* Stuiver, M and Kra RS, eds, Proceedings of the 12th International ^{14}C Conference. *Radiocarbon* 28 (2B): 911–934.

Pearson, GW, Pilcher, JR, Baillie, MGL and Hillam, J 1977 Absolute radiocarbon dating using a low altitude European tree-ring calibration. *Nature* 270: 25–28.

Pearson, GW and Stuiver, M 1986 High-precision calibration of the radiocarbon time scale, AD 1959–2500 BC. *In* Stuiver, M and Kra, RS, eds, Proceedings of the 12th International ^{14}C Conference. *Radiocarbon* 28(2B): 839–862.

Pilcher, JR, Baillie, MGL, Schmidt, B and Becker B 1984 A 7,272 year tree-ring chronology for western Europe. *Nature* 312: 150–152.

Pilcher, JR, Hillam, J, Baillie, MGL and Pearson, GW 1977 A long sub-fossil oak tree-ring chronology from the north of Ireland. *New Phytologist* 79: 713–729.

Plicht, J, van der and Mook, WG 1989 Calibration of radiocarbon ages by computer. *In* Long, A and Kra, RS, eds, Proceedings of the 13th International ^{14}C Conference. *Radiocarbon* 31(3): 805–816.

Renfrew, C 1973 *Before Civilization*. New York, Alfred Knopf: 292 p.

_____1977 Ancient Europe is older then we thought. *National Geographic Magazine* 152: 615–662.

Schulman, E 1958 Bristlecone pine, oldest living thing. *National Geographic Magazine* 123(3): 355–392.

Stuiver, M and Braziunas, TF 1989 Atmospheric ^{14}C and century-scale solar oscillations. *Nature* 338(6214): 405–408.

Stuiver, M, Braziunas, TF, Becker, B and Kromer, B 1991 Climatic, solar, oceanic and geomagnetic influences on the Late Glacial and Holocene atmospheric $^{14}C/12C$ change. *Quaternary Research* 35:1–24.

Stuiver, M and Kra, RS, eds 1986 Calibration Issue. Proceedings of the 12th International ^{14}C Conference. *Radiocarbon* 28(2B): 805–1030.

Stuiver, M and Pearson, GW 1986 High-precision calibration of the radiocarbon time scale, AD 1950 to 500 BC. *In* Stuiver, M and Kra, RS, eds, Proceedings of the 12th International ^{14}C Conference. *Radiocarbon* 28(2B): 805–838.

Stuiver, M and Reimer, PJ 1986 A computer program for radiocarbon age calculation. *In* Stuiver, M and Kra, RS, eds, Proceedings of the 12th International ^{14}C Conference. *Radiocarbon* 28(2B): 1022–1030.

Suess, HE 1961 Secular variations in the concentration of atmospheric radiocarbon. *In* Proceedings of an informal conference, *Problems Related to Interplanetary Matter*. Highland Park: 90–95.

_____1965 Secular variations of the cosmic-ray-produced carbon-14 in the atmosphere and their interpretations. *Journal of Geophysical Research* 70: 5937–5952.

_____1980 Radiocarbon geophysics. *Endeavour, New Series* 4(3): 113–117.

Vries, H de 1959 Measurement and use of natural radiocarbon. *In* Abelson, PH, ed, *Research in Geochemistry*. New York, John Wiley & Sons: 169–189.

Willis, EH, Tauber, H and Münnich, KO 1960 Variations in the atmospheric radiocarbon concentration over the past 1300 years. *American Journal of Science Radiocarbon Supplement* 2: 1–4.

5

THE PRESENT STATUS OF UNDERSTANDING OF THE LONG-PERIOD SPECTRUM OF RADIOCARBON

CHARLES P SONETT

INTRODUCTION

Libby's (1952) discovery of the ubiquitous distribution of radiocarbon in the biosphere provided the basis for a new means of dating of archaeological and paleontological finds. His principle was based upon the then reasonable supposition that the atmospheric reservoir from which radiocarbon was drawn was uniform and unchanging. But with Suess' discovery of the dilution of the atmospheric reservoir from burning of fossil fuel, evidence for variability began to solidify. For a synopsis, see, eg, Walker (1977). In 1958, some six years after Libby's discovery, de Vries (1958) and Barker (1958) disclosed variations on the order of 1% during the past 400 years in the atmospheric reservoir. Shortly thereafter, Willis *et al* (1960) showed that over the past 1200 years the 'fundamental assumptions' of radiocarbon dating were empirically correct to ~ 1.5%, but they surmised more. Their results suggested that a residual variability was related to solar physics through modulation of cosmic-ray (CR) activity. Willis *et al* (1960) also recognized this, as well as the possibility of 150–200-year and 1000-year periods in the atmospheric inventory of radiocarbon. If the CR flux were constant, the atmospheric ^{14}C inventory would be in secular equilibrium. Then radiocarbon would be an absolute archaeological clock, but it would be of lesser interest geophysically, though certifying Libby's (1952) thesis of a constant atmospheric inventory. Today, a major challenge of radiocarbon research is understanding the spectrum of the radiocarbon variability. The apparently periodic spectral features are suggestive of the presence of natural oscillators in the terrestrial environment, in the Sun or both.

The ultimate source of terrestrial radiocarbon is traceable to the CR flux upon the top of the atmosphere from which an atmospheric neutron sea is generated by spallations. Radioactive ^{14}C is produced terrestrially primarily by the specific nuclear reaction

$$^{14}N(n,p) \longrightarrow {}^{14}C \qquad (5.1)$$

where ^{14}N is atmospheric. ^{14}C decays by

$$^{14}C \longrightarrow {}^{14}N + \nu^- + \beta^- \tag{5.2}$$

where ν^- is the antineutrino and β^- the electron. Neutrons participate in the N(n,p) reaction yielding ^{14}C (Lingenfelter & Ramaty 1970; O'Brien *et al*, in press). The half-life of radiocarbon, $\tau_{1/2}$ = 5730 years (Lederer, Hollander & Perlman 1967).

THE HISTORICAL RECORD

Variations in the atmospheric inventory of radiocarbon are usually designated by $\Delta^{14}C$, a function of time, and sometimes for brevity, as 'radiocarbon' although that term is misleading. Determination of $\Delta^{14}C$ requires an absolute time scale given by tree-ring chronologies. The deviation of the specific activity in wood *vs* the activity expected on the basis of absolute chronology defines the atmospheric $\Delta^{14}C$, which is usually expressed in per mil values that represent the fluctuation of the $^{14}C/^{12}C$ ratio in the atmosphere.

Stuiver (1961, 1965) showed early that a relationship appeared to be present between the atmospheric inventory and sunspot activity (see also Damon, Long & Grey 1966.) From a formal standpoint, a major step in understanding the variation was computation of the spectrum of time variations by Houtermans (1971). His work disclosed both ~ 200-year and ~ 2000-year periods in the radiocarbon data from Suess' Mt Soledad laboratory at the University of California at San Diego (La Jolla), and represents the first codification of the variability problem in terms of a spectrum. Although acceptance of the idea of an inconstant reservoir was growing, the belief that it was periodic was apparently still fragmented at this time.

In an attempt to establish whether the variations were real, de Jong and Mook (1980) undertook a precise examination of the record over the specific time interval from 3900–3200 BC. Their work also demonstrated a significant variability; the Fourier transform shows a strong feature with a period of about 150 years and a weak signature at ~ 200 years (Sonett 1984). Since, for the full record, the 150-year period appears much weaker than that at 200 years, this somewhat troublesome result suggests that perhaps the spectrum of radiocarbon is not time-independent. A possible minimum in the amplitude of the shorter period, due to the modulation mentioned below during the time when de Jong and Mook took their data, may explain why their 150-year period appears more prominently that the more usual 209-year value, but this conjecture has not been tested.

Neftel, Oeschger and Suess (1981), following the analysis of Houtermans (1971), also found the major features at ~ 200 and ~ 2000 years reported earlier by Houtermans. The spectrum of radiocarbon from the La Jolla record is also given from a calculation of Kruse (reported by Suess 1980). Sonett (1984) later reported on the spectral decomposition of the radiocarbon record, noting marginal evidence for a connection between the long period (lately computed to be ~ 2300 years) and the 200-year (by now more accurately 209-year) period, further supporting the view that the 209-year period is forced by the Sun. Comparison of the time records of the La Jolla data and the new Belfast record (Pearson *et al* 1986) lent further credibility to a real period at 209 years. Today, acceptance is general that the atmospheric inventory is inconstant and is reflected in radiocarbon determined from wood.

THE ATMOSPHERIC RADIOCARBON RECORD

The radiocarbon sequence used here (Table 5.1) is a composite of segments, some overlapping, from records of Pearson *et al* (1986), Linick *et al* (1986) and Linick and Suess (personal communication). To make up a continuous sequence, all records are interleaved in ascending time (Fig 5.1A). The composite sequence extends from 7202 BC to AD 1900. The non-stationary detrended radiocarbon time series consists of sections with differing time spacing and the data sequence is non-stationary.

The major trend in the radiocarbon data is from about 115‰ to less than 20‰ over the record, and is commonly attributed to secular changes in the intensity of the geomagnetic dipole moment (Buchta 1969; Barton, Merrill & Barbetti 1979; McElhinney & Senenyake 1982; Creer 1988). Using a 5th order polynomial (Fig 5.1B), the trend is removed, and the smaller and older dip in the record is seen in the more recent 9000-year record. But the subject of the secular variability lies in the special province of geomagnetism, outside the scope of this review.

TABLE 5.1. Sources of the Composite Radiocarbon Sequence

Laboratory	Time period
La Jolla*	6505 BC–AD 1900
Belfast**	5210 BC–AD 1840
La Jolla/Stuttgart†	7199 BC–AD 245

*Linick and Suess (personal communication)
**Pearson *et al* (1986)
‡Linick *et al* (1986)

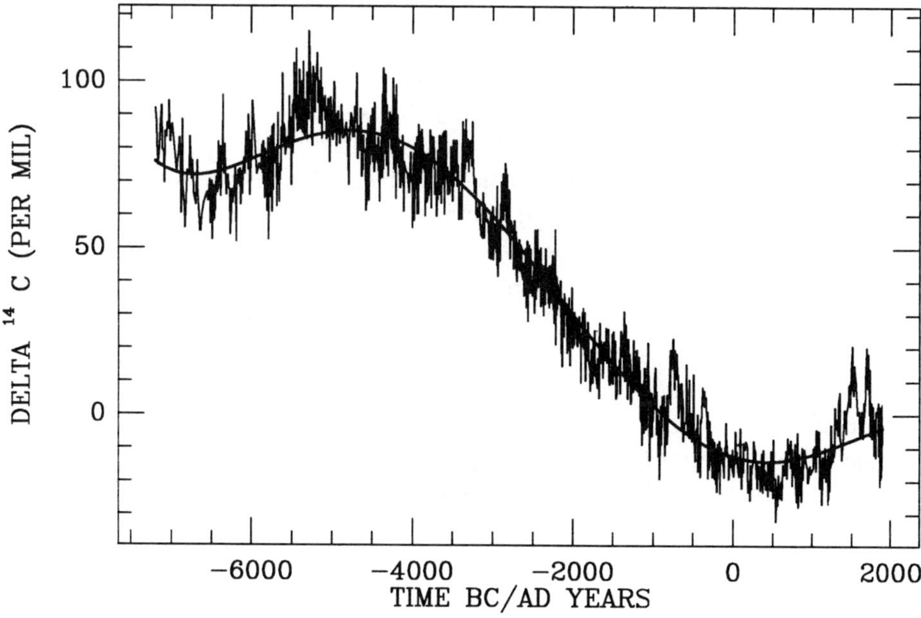

Fig 5.1A. Composite $\Delta^{14}C$ record (Table 5.1) with model 5th order detrend polynomial; dark line is the detrend function.

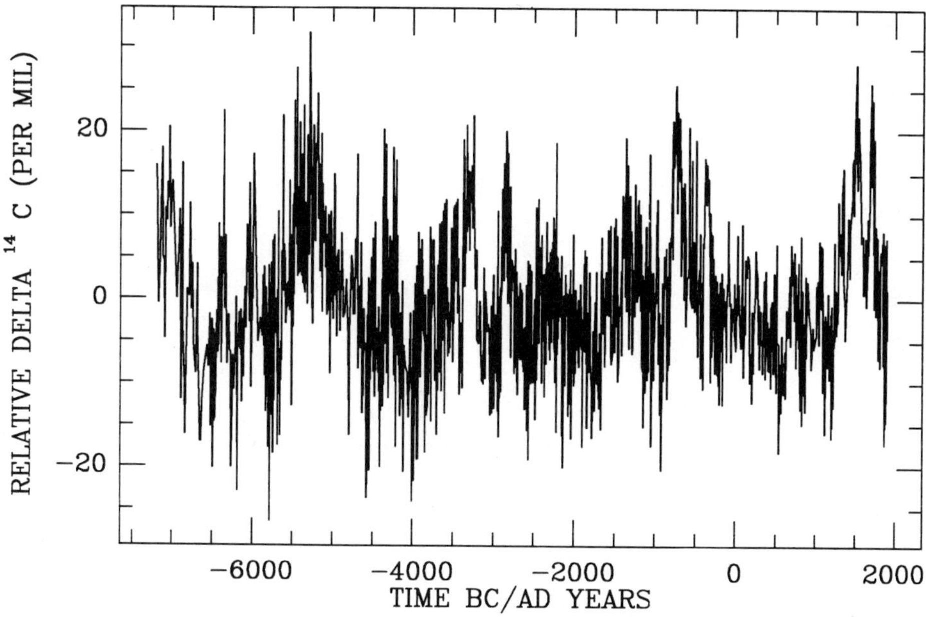

Fig 5.1B. Detrended composite of Figure 5.1A

The periodogram (Fig 5.2) discloses a complex spectrum in the multihundred-year and millennial range of periods. (Periods shorter than 100 years, *ie*, Gleissberg (1944, 1966), appear in the full record, eg, Damon and Sonett (in press), but space restricts this chapter to the longer periods.) The calculated periodogram, by itself, is an inexact representation of the spectrum for several reasons: 1) the spectrum may be non-stationary; 2) side-lobe interference from the convolution of the data and window spectra; 3) calculation of Fourier components is at discrete times. These may correspond to positions in Fourier space, which lie on a line shoulder leading to underestimation of spectral peak value and period. This can be especially severe for narrow, sharply tuned (high Q) spectral features (Sonett 1984; Sonett & Finney 1990). Because it gives 'windowless' spectral estimates, maximum entropy (MEM) has enjoyed a certain vogue, but the problem of instabilities, optimum order (which is not guaranteed by the Aikike criterion) and the unsolved problem of assessing statistical quality restrict its value; MEM calculations should be backed by the Fourier transform.

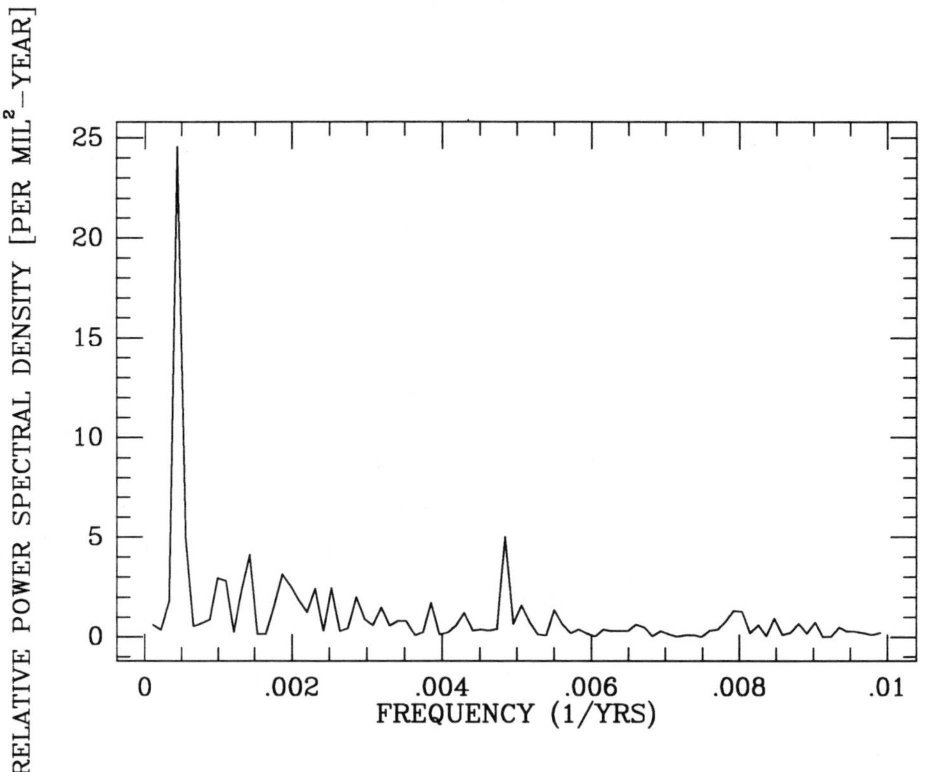

Fig 5.2. ^{14}C periodogram of the Figure 5.1 composite radiocarbon sequence. Sequence is 5th order detrended per Figure 5.1 and interpolated using a cubic spline.

A recent approach to estimating the spectrum that is conceptually more secure is the Bayesian algorithm developed by Jaynes (1983) and Bretthorst (1988), based upon earlier research of Jeffreys (1939, 1973). It is complex and cannot be adequately described here. Very briefly, it is an application of Bayes' theorem; as used here with constant prior probability, it is essentially a maximum likelihood estimator, and is used to compute a probability distribution for a model consisting here of nine sinusoids. Jaynes and Bretthorst show the conditions under which the probability distribution is basically a power spectrum. The probability of the model under consideration relative to white noise is also a complex topic; it is related to the relative amplitude of the likelihood function over the frequency space spanned by the calculation. Better computation of model radiocarbon periods and amplitudes than those shown in Table 5.2 may be possibly made but require some previous knowledge of the properties of the data.[1] Further, the apparent non-stationarity of the $\Delta^{14}C$ record suggests that a purely multiperiodic model may not adequately describe the spectral properties of radiocarbon.

Table 5.2 shows calculations of the probability of a model including fifth-order detrender and nine sinusoids. These estimates represent about as far as one can go in frequency estimation of variations of the atmospheric radiocarbon inventory, without taking into account the time dependence of the spectrum and a non-constant (Jeffreys 1939, 1973) prior. Major features include the ubiquitous 209- and 2300-year periods and several physically unassigned lines in the multihundred-year range of period. Line identifications in Table 5.2 are thought to contain some side-lobe contamination and harmonics of the 2400-year period.

TABLE 5.2. ^{14}C Bayesian Model Line Periods and Amplitudes

Frequency (yrs^{-1})	Period (yrs)*	Amplitude
$0.0019510 \pm 0.2765 \times 10^{-4}$	513, 505, 520	2.188 ± 0.4449
$0.0003255 \pm 0.8593 \times 10^{-4}$	3072, 2431, 4174	2.644 ± 0.4338
$0.0048039 \pm 0.2220 \times 10^{-4}$	209, 207, 209	2.463 ± 0.2593
$0.0028472 \pm 0.1685 \times 10^{-4}$	351, 349, 353	2.782 ± 0.1823
$0.0007477 \pm 0.4318 \times 10^{-4}$	1337, 1264, 1419	1.879 ± 0.1763
$0.0044540 \pm 0.1713 \times 10^{-4}$	225, 224, 225	1.527 ± 0.4212
$0.0022730 \pm 0.2177 \times 10^{-4}$	440, 436, 444	2.357 ± 0.7360
$0.0010371 \pm 0.2196 \times 10^{-4}$	964, 944, 985	2.533 ± 0.3593
$0.0004327 \pm 0.2106 \times 10^{-4}$	2311, 2204, 2429	5.308 ± 0.3779

*Two-sigma error limits; center value is mean.

[1]This is the often discussed 'prior' of Bayesian statistics corresponding to the idea that the likelihood is subject to earlier statistical knowledge about the data.

THE RADIOCARBON RECORD: VARIATIONS IN TIME

There is good reason to believe that the power spectrum may yield only an incomplete understanding of potentially vital information carried in the radiocarbon data. To expose this problem, we return to the time series representation of the data, bandpass filtered so that periods of less than approximately 50 years and greater than approximately one millennium are removed. (The data are interpolated to ten-year intervals.) Bandpass filtering is carried out by using two lowpass filters, one for removing periods less than about 50 years, and the other, periods less than about 1000 years. The two outputs are differenced to yield the bandpassed response. End effects (from the long-period filter) reduce the effective series length by 1000 years; hence, the reduced length to about 8000 years of the filtered data shown in Figure 5.3 ending at about AD 1000. The smallness of the ordinate ($\Delta^{14}C$) of Figure 5.3 is an indication that most of the variance in the data is removed by the filtering with only a residuum persisting.

The variability in spacing of maxima in the filtered sequence is large and shows the period to be variable. The question of the meaning of these maxima is currently under study and is not clear. As the maxima are similar to the four Maunder-like minima (radiocarbon maxima) of the last millennium studied by Stuiver and Quay (1981), we make the working hypothesis that at least some are

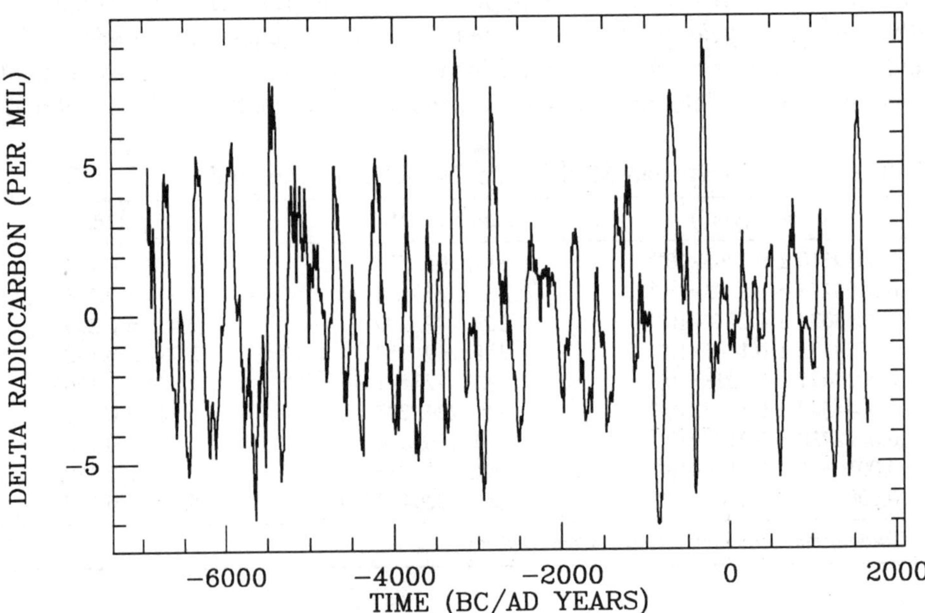

Fig 5.3. $\Delta^{14}C$ narrow banded to emphasize the dominant period characterized by $\lambda_m = 209$ years. The 2300-year variability can be seen as an amplitude modulation of the sequence peaks; vestigial suggestion of ~ 1000 years is also present.

Maunder-like, in qualitative agreement with the results of Stuiver and Braziunas (1989). If these are Maunder minima, two significant issues arise. First, the Sun would appear to be hydromagnetically variable over long periods, but with irregular 'period', and secondly, based upon the admittedly sparse but suggestive evidence that the present millennia Maunder minima are accompanied by climatic change (inferring variable solar irradiance), the climatic record for the entire Holocene is at least partially exposed by these data.

In the very complex time sequence of Figure 5.3, the number of radiocarbon maxima is uncertain, and noise may well be a factor in both period and amplitude. On the other hand, we do not yet know the spectrum of the noise contribution to $\Delta^{14}C$ nor the noise content of the data shown in Figure 5.3, recalling also that it has been heavily filtered, with the effect that noise tends to masquerade as a periodic signal. Inspection of the maxima shows that 3–5 of them stand out. Indeed, their average separation is ~ 2000 years, which suggests the long period found in the spectrum of radiocarbon. These maxima also suggest the existence of amplitude modulation, *ie*, the long-period modulation of the ~ 209-year period inferred by Sonett (1984). With the evidence for amplitude modulation, the argument that the data contain a real set of maxima is strengthened; at modulation minima, a signal of 200-year period would tend periodically to be diminished, and at least some of the small maxima would be real and merely modulated to small values.

Evidence from the histogram of interpeak spacings suggests that it is Gaussian, including a long-period tail. Stuiver and Braziunas (1988, 1989) have computed sinusoidal models for maxima divided into two groups, labeled Spörer (period = 220 years) and Maunder (period = 180 years) (Maunder 1922). But the histogram shows no such bifurcation or evidence for sinusoidal variability. Using all 34 maxima and the filtered 8000-year record, the average spacing is $\bar{\lambda}_s$ = 235 years and the standard deviation $\sigma^{1/2}$ 130 years. Finally, the record of spacings in Figure 5.3 is inconsistent with the persistent 209-year period found in the radiocarbon spectrum. This disparity is a most important unresolved issue surrounding the data of Figure 5.3. Until further analysis can be carried out, the source of the variability (with the exception of the 2000-year period) should be regarded as speculative and possibly even noise.

Stuiver has argued that the historic Maunder minimum cannot have been due solely to cessation of solar activity, but that an additional factor was required. Following this line of reasoning, the maxima of Figure 5.3 are so large compared to those of the past two millennia, that even if the historic minimum did represent complete switch-off of solar activity, the earlier maxima would require a mechanism in addition to solar activity to explain the very large radiocarbon maxima. Kocharov (1987) reports that during the historic Maunder period, rather than a cessation of solar activity, the 11-year cycle temporarily transformed into a 22-year period.

The periodicity of approximately 2300 years is also found in numerous records of $\Delta^{18}O$ in glacial ice cores and foraminifera from ocean cores as well as in the Middle European oak dendroclimatic record. The most prominent period in the Camp Century $\Delta^{18}O$ core is reported to be a line at 2550 years. If the 2300-year source were extraterrestrial, it should interact non-linearly with the geomagnetic secular trend; thus far, no such effect has been detected. Also, if the ~ 2300-year period were extraterrestrial, a non-linear mechanism would be required to account for its modulation of the 209-year period. Characteristic deep-water residence times for the global oceans range from 1000 years, for the Atlantic, to 2000 years for the full Atlantic-Pacific circulation (Broecker & Peng 1970; Broecker & Li 1970). In themselves, these time periods do not explain the global ocean-atmosphere resonance required to drive a 2300-year total CO_2 cycle, but the periods are consistent with such a resonance. Yiou *et al* (in press) report periods, primarily Milankovich-derived, from deuterium analysis of the Vostok ice core. Sonett (1984), Sonett and Finney (1990), and Damon and Sonett (in press) find a 2300-year period in radiocarbon. (Damon and Sonett also report vestigial evidence of a 4000-year period using the Bretthorst algorithm.) It may never be possible to isolate the 2300-year forcing with confidence. If due to orbit-insolation, it may still drive global CO_2 and thus *dilution* of radiocarbon. If so, a search for an ocean-atmosphere resonance independent of orbit-insolation may be meaningless.

THE SOURCE OF THE INCONSTANT 209-YEAR PERIOD

Taking the 209-year 'period' to be of solar origin, can it be attributed to a specific physical cause on the Sun? Although a case might be made for a noise-excited global hydromagnetic eigenmode system coupled to the solar dynamo, this conjecture seems flawed because of the high variance, 61% of the mean period, requiring a very noisy solar oscillator. The only kind of self-excited natural system in which the 'periodicity' appears to be so perversely irregular is a chaotic oscillator (Weiss & Cattaneo 1984). Until better information becomes available, or this aspect of the radiocarbon record is shown to be stochastic, it is as good a guess of what is going on as we have. If correct, it does mean that predictions of future solar activity and intensity on time scales meaningful to everyday life are a rather fruitless way to spend one's time.

In summary, if the solar activity cycle is chaotic (and this remains to be shown), the route to understanding through the power spectrum is not especially informative. A return to the simpler view of the appropriately filtered time series may provide a more rational view of the underlying physics driving the radiocarbon variability.

A major unresolved question is whether a signal so characteristically 'random' can still cast out as robust a period as the 209-year line expressed only vaguely at best in the time series. Based upon identification of the 209-year period with the Maunder minimum, the earlier evidence of a bolometric component is present in the solar variability. It is expressed as a variation in air temperature suggested by Bristlecone pine growth (Sonett & Suess 1984). Although this association appears in the associated MEM spectra, it is not identified in the simpler correlation plot of radiocarbon *vs* tree-ring growth. Other manifestations of solar activity can be detected in the Gleissberg cycle, in aurorae and similar geophysical records, and in the vestigial but suggestive variation in solar radius with the Gleissberg cycle, but these data are marginal. The delta radiocarbon spectrum is by no means fully understood.

REFERENCES

Barker, H 1958 Radiocarbon dating: its scope and limitations. *Antiquity* 32: 253–263.

Barton, CE, Merrill, RT and Barbetti, M 1979 Intensity of the earth's magnetic field over the last 10,000 years. *Physics of the Earth and Planetary Interiors* 20: 96.

Bretthorst, LG 1988 Bayesian spectral analysis and parameter estimation. *In* Berger, J, Fienberg, S, Gani, J, Krickeberg, K and Singer, B, eds, *Lecture Notes in Statistics* 48. Springer-Verlag.

Broecker, WS and Li, YH 1970 Interchange of water between the major oceans. *Journal of Geophysical Research* 75: 3545–3552.

Broecker, WS and Peng, TH 1970 *Tracers in the Sea*. Palisades, New York, Eldigio Press: 690 p.

Buchta, V 1969 Changes of the Earth's magnetic moment and radiocarbon dating. *Nature* 224: 681.

Creer, KM 1988 Geomagnetic field and radiocarbon activity through Holocene time. *In* Stephenson, FR and Wolfendale, AW, eds, NATO advanced research workshop on secular, solar, and geomagnetic variations through the last 10,000 years. *NATO ARW Series*. Dordrecht, The Netherlands, Kluwer.

Damon, PE, Long, A and Grey, DC 1966 Fluctuations of C^{14} during the last six millennia. *Journal of Geophysical Research* 71: 1055–1063.

Damon, PE and Sonett, CP, in press, Solar and terrestrial components of the atmospheric ^{14}C variation spectrum. *In* Sonett, CP, Giampapa, MS and Matthews, MS, eds, *The Sun in Time*. Tucson, University of Arizona Press.

Gleissberg, W 1944 A table of secular variations of the solar cycle. *Terrestrial Magnetism and Atmospheric Electricity* 49: 243–244.

_____1966 Ascent and descent in the eight year cycle of solar activity. *Journal of the British Astronomical Association* 76: 265–268.

Houtermans, JC (ms) 1971 Geophysical interpretation of bristlecone pine radiocarbon measurements using a method of Fourier analysis of unequally spaced data. PhD thesis, University of Bern, Switzerland.

Jaynes, ET 1983 *Papers on Probability, Statistics, and Statistical Physics*. Reprint collection. D Reidel.

Jeffreys, H 1939 *Theory of Probability*.

Oxford, Oxford University Press.

Jeffries, H 1973 *Scientific Inference*, 3rd edition. Oxford, Oxford University Press.

Jong, AFM de and Mook, WG 1980 Medium-term atmospheric ^{14}C variations. *In* Stuiver, M and Kra, RS, eds, Proceedings of the 10th International ^{14}C Conference. *Radiocarbon* 22(2): 267–272.

Kocharov, GE 1987 Nuclear processes in the solar atmosphere and the particle-acceleration problem. *Astrophysics and Space Science* 6: 155–262.

Lederer, CM, Hollander, JM and Perlman, I 1967 *Table of Isotopes*, 6th edition. New York, John Wiley & Sons.

Libby, W 1952 *Radiocarbon Dating*. Chicago, University of Chicago Press.

Lingenfelter, RE and Ramaty, R 1970 Astrophysical and geophysical variations in C 14 production. *In* Olsson, IU, ed, *Radiocarbon Variations and Absolute Chronology*. Proceedings of the 12th Nobel Symposium. New York, John Wiley & Sons: 513–537.

Linick, TW, Long, A, Damon, PE and Ferguson, CW 1986 High-precision radiocarbon dating of Bristlecone pine from 6554 to 5350 BC. *In* Stuiver, M and Kra, RS, eds, Proceedings of the 12th International ^{14}C Conference. *Radiocarbon* 28(2B): 943–953.

Maunder, EW 1922 The prolonged sunspot minimum, 1645–1715. *British Astronomical Association Journal* 32: 140–145.

McElhinny, MW and Senanyake, WE 1982 Variations in the geomagnetic dipole: The past 50,000 years. *Geomagnetism and Geoelectricity* 34: 39.

Neftel, A, Oeschger, H and Suess, HE 1981 Secular non-random variations of cosmogenic carbon-14 in the terrestrial atmosphere. *Earth and Planetary Science Letters* 56: 127.

O'Brien, K, Shea, MA, Smart, DF and Zerda Lerner, A de la, in press, The production in the Earth's atmosphere of cosmogenic isotopes and their inventories. *In* Sonett, CP, Giampapa, MS and Matthews, MS, eds, *The Sun in Time*. Tucson, University of Arizona Press.

Pearson, GW, Pilcher, JR, Baillie, MGL, Corbett, DM and Qua, F 1986 High-precision ^{14}C measurement of Irish oaks to show the natural ^{14}C variations from AD 1840 to 5210 BC. *In* Stuiver, M and Kra, RS, eds, Proceedings of the 12th International ^{14}C Conference. *Radiocarbon* 28(2B): 911–934.

Sonett, CP 1984 Very long solar periods and the radiocarbon record. *Reviews of Geophysics and Space Physics* 22: 239–254.

Sonett, CP and Finney, SA 1990 The spectrum of radiocarbon. *Philosophical Transactions of the Royal Society*. A330: 413–426

Sonett, CP and Suess, HE 1984 Correlation of bristlecone pine ring widths with atmospheric carbon-14 variations: A climate-Sun relation. *Nature* 307: 141–143.

Stuiver, M 1961 Variations in radiocarbon concentration and sunspot activity. *Journal of Geophysical Research* 66: 273–276.

_____1965 Carbon-14 content of 18th and 19th century wood, variations correlated with sunspot activity. *Science* 149: 533–535.

Stuiver, M and Braziunas, TF 1988 The solar component of the atmospheric ^{14}C record. *In* Stephenson, FR and Wolfendale, AW, eds, *Secular, Solar and Geomagnetic Variations in the Last 10,000 Years*. Dordrecht, The Netherlands, Kluwer: 245–266.

_____1989 Atmospheric ^{14}C and century-scale solar oscillations. *Nature* 338: 405–408.

Stuiver, M and Quay, PD 1981 Atmospheric ^{14}C changes resulting from fossil fuel CO_2 and cosmic ray variability. *Earth and Planetary Science Letters* 53: 349–362.

Suess, HE 1980 The radiocarbon record in tree rings of the last 8000 years. *In* Stuiver, M and Kra, RS, eds, Proceedings of the 10th International ^{14}C Conference. *Radiocarbon* 22(2): 1–4.

Vries, H, de 1958 Variations in concentration of radiocarbon with time and location on Earth. *Koninklijke Nederlandse Akademie van Wetenschappen Series B* 61: 94.

Walker, JCG 1977 *Evolution of the Atmosphere*. New York, MacMillan.

Weiss, NO and Cattaneo, F 1984 Periodic and aperiodic dynamo waves. *Geophysical and Astrophysical Fluid Dynamics* 30: 305–341.

Willis, EH, Tauber, H and Münnich, KO 1960 Variations in the atmospheric radiocarbon concentration over the past 1300 years. *American Journal of Science Radiocarbon Supplement* 2: 1–4.

Yiou, P, Genthon, C, Jouozel, J, Ghil, M, Le Treut, H, Barnola, JM, Lorius, C and Korotkevitch, YN, in press, High-frequency paleovariability in climate and in CO_2 levels from Vostok ice-core records. *In* Keir, R, ed, *Interaction of the Global Carbon and Climatic Systems*. Electric Power Research Institute.

6

GLACIAL-TO-INTERGLACIAL CHANGES IN OCEAN CIRCULATION

JEAN-CLAUDE DUPLESSY, MAURICE ARNOLD, EDOUARD BARD LAURENT LABEYRIE, JOSETTE DUPRAT and JEAN MOYES

INTRODUCTION

Only a few years after its discovery by Anderson and Libby (1947), radiocarbon has revolutionized our understanding of the climatic history of the upper Quaternary. Flint and Rubin (1955) demonstrated that the moraines deposited over the northern United States by the last great ice sheets contained organic material with measurable ^{14}C activity. Numerous measurements performed on moraines and glacial sediments from northern America and northern Europe (see a review in Denton and Hughes (1981)) showed that continental ice sheets reached their maximum extension only 18,000 years ago, and retreated slowly to disappear about 6500 years ago.

Climatic changes are best studied in marine sediments because the oxygen isotope record provides a common stratigraphic framework to all deep-sea cores (Emiliani 1955; Shackleton & Opdyke 1973), and because empirical transfer functions provide estimates of past climatic parameters (Imbrie & Kipp 1971). Moreover, the development of new geochemical tracers, such as the Cd content (Boyle & Keigwin 1987) and the $^{13}C/^{12}C$ ratio of benthic foraminifera (Duplessy *et al* 1988), offer new approaches to reconstruct the past deep-water circulation and its variations during the Quaternary.

However, until recent years, the finesse of marine paleoclimatic reconstructions has not benefited from the progress in radiocarbon geochemistry. This was due to the fact that, although the standard error of ^{14}C dating is less than one century for samples younger than 20,000 years, the amount of carbon stored in foraminiferal shells and deposited during one century in deep-sea sediments was much smaller than that which was required for one single analysis. As a consequence, only the main chronological trends of the climatic changes of the last 30,000 years were studied. Broecker, Ewing and Heezen (1960) demonstrated that the climatic warming at the end of the glaciation was abrupt and occurred about 11,000 years ago, but it was not possible to estimate the duration of the climatic transition. Using a small proportional counter, Duplessy *et al* (1981) measured a high sedimentation rate core collected in the northeastern Atlantic Ocean and showed that the deglaciation was not a smooth transition. These authors evidenced two major phases of ice melting, which were well correlated with the retreat and advance phases of the North Atlantic polar front (Ruddiman & McIntyre 1973, 1981). However, the accuracy of their measurements was poor (the error at the 1σ level was > 500 years for ages older than 10,000 years). In most other oceanic basins, the dating of the deglaciation record was impossible, because the size of the samples to be analyzed was too large, compared to the potential resolution of the climatic record of deep-sea sediments.

The development of a new technique, counting of $^{14}C/^{12}C$ ratios by accelerator mass spectrometry (AMS) in only one milligram of carbon, permits the assignment of precise ages to monospecific foraminiferal samples (Andrée *et al* 1984) with an upper limit of 41,000 years (Arnold *et al* 1987). Two main applications have been developed: 1) the comparison of the timing of major temperature changes and meltwater input in the various oceanic basins; 2) the reconstitution of past changes of the CO_2 ventilation rate in the deep ocean, which may be deduced from the ^{14}C concentration difference between benthic (bottom dwelling) and planktonic (near surface dwelling) foraminifera. In this chapter, we shall review the main results of these studies, which illustrate that the glacial oceanic circulation was very different from the modern circulation, and that the shift from one mode to another may have occurred very abruptly, within a few centuries.

TIMING OF THE REORGANIZATION OF THE OCEAN-ATMOSPHERE SYSTEM

At the end of the last glaciation, surficial ocean and atmosphere experienced a major reorganization. This was marked by an abrupt sea-surface warming in high latitude areas and other circulation changes at lower latitudes, such as variations in the deep and intermediate water upwelling rate.

Several cores from the North Atlantic Ocean have been analyzed. Cores CH 73-139 C off Ireland and SU 81-18 off Portugal have sedimentation rates of 10 and 20 cm/kyr, respectively, and both cores record the details of the deglaciation. Core Su 81-18 exhibits an abrupt temperature increase by more than 10°C in less than 400 years at about 12,500 BP, emphasizing the extreme rapidity of the retreat of the polar front during the last deglaciation (Fig 6.1). The resolution is somewhat lower in Core CH 73-139 C than in Core SU 81-18, and the sea-surface warming is dated at about 13,000 BP. Taking into account the important

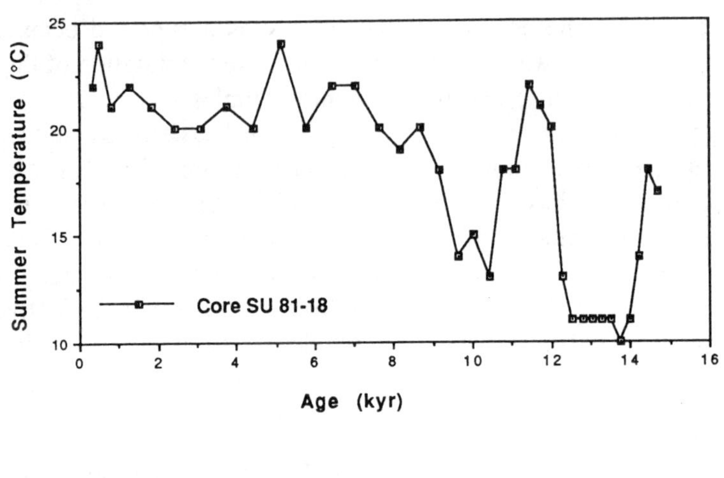

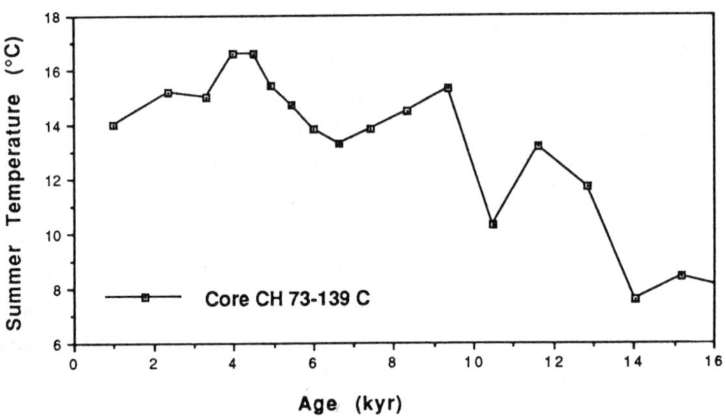

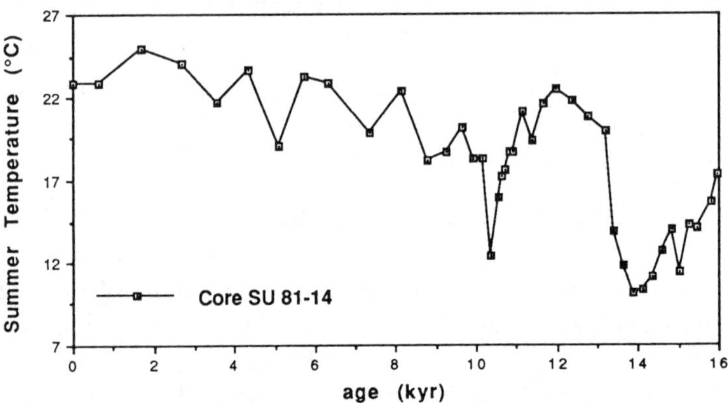

Fig 6.1. Sea-surface temperature records of the North Atlantic Ocean during the last deglaciation. Core SU 81-18: 37°46′N, 10°11′W, 3135 m; Core CH 73-139 C: 54°38′N, 16°21′W, 2209 m; Core SU 81-14: 36°46′N, 9°51′W, 2795 m.

faunal changes correlated with the warming of the polar waters and the bioturbation that continually mixes the recently deposited sediment, it is impossible to date accurately the mid-point of the temperature change in Core CH 73-139 C. The data demonstrate only the presence of a well-established warm-water mass off Ireland at 11,500 BP (Bard *et al* 1987a, b).

However, in Core V 23-81, which was also collected off Ireland and has a sedimentation rate of 15 cm/kyr, the abundance decrease of the cold polar species, *Neogloboquadrina pachyderma* (left coiling), is accurately dated at about 13,600 BP (Broecker *et al* 1988a, c, d). This dates the surface-water warming about 1000 years before that recorded in Core SU 81-18, and we do not know whether this age difference is real or is an artifact. Further, Core SU 81-14, (Bard *et al* 1989) collected close to SU 81-18, also dates the warming 1000 years before that recorded in SU 81-18 (Fig 6.1). However, as this core exhibits large reworking, and is contaminated by older sediments at least during part of the glacial-to-interglacial transition, we are not confident of its chronology.

In several northwestern Atlantic cores, Keigwin and Jones (1989) depicted sharp $\delta^{18}O$ decrease in planktonic foraminifera about 13,500 BP, but it is not clear whether these surface-water changes reflect a major reorganization of the ocean-atmosphere system or simply the flux of meltwater brought into the Atlantic ocean by the Mississippi River. As $\%CaCO_3$ in the sediment increases only later in the sedimentary record, the western Atlantic record is probably dominated by variations of the meltwater discharge.

In the North Pacific, the surface-water warming along the Japanese continental margin has been dated at about 13,000 BP (Labeyrie *et al* 1990). In the South China Sea, no temperature estimates are presently available. However, Broecker *et al* (1988b) discovered an abrupt rise of the shell number of all foraminiferal species at the close of glacial times. Broecker *et al* estimated the age of this change to be about 13,500 BP and interpreted it as proof of a major oceanic reorganization. It should be stressed that this age estimate is derived from the ^{14}C dating of one foraminiferal species, *Pulleniatina obliquiloculata*, whereas the age of the same sediment levels is 1000 years younger if the ages of another species, *G sacculifer*, are taken into account. No explanations have been offered for this difference. Recently, Broecker *et al* (1988e) considered the ages given by *G sacculifer* to be most reliable in this core. This suggests that the oceanic changes recorded in Core SU 81-18 from the North Atlantic and V 35-5 from the South China Sea were synchronous. However, the analysis of the nearby Core SONNE 50-37 KL added more confusion, because the ^{14}C ages of *G sacculifer* and *P obliquiloculata* agree, but faunal counts indicate that the productivity increase occurred about 14,000 years ago (Broecker *et al* 1990a).

Thus, it would be important to determine whether this event has a local or a large-scale significance and its exact timing.

In the Southern Ocean, most cores, such as MD 84-527 (Fig 6.2), exhibit a noticeable sedimentation rate decrease during isotope stage 2, which probably reflects the presence of sea ice at the core site during the last glacial maximum. Faunal analyses of Cores MD 73-025, MD 84-527 and MD 84-551 show that open conditions reappeared at about 12,500–13,000 BP (Fig 6.3). This result suggests that the high-latitude warming was synchronous in both hemispheres (Labracherie *et al* 1989).

In conclusion, the ocean-atmosphere system experienced dramatic changes (sea-surface temperature warming, sea-ice retreat, productivity variations) during the period 13,500–12,500 BP, but the available data do not yet have the accuracy required to compare the timing of the climatic evolution of the various oceanic basins with a precision better than 5 to 10 centuries.

VENTILATION AGE OF DEEP AND INTERMEDIATE WATERS

Since the recent discovery that the global circulation of deep and intermediate waters was strongly dependent on the earth's climate (Boyle & Keigwin 1987; Duplessy & Shackleton 1985; Duplessy *et al* 1988; Kallel *et al* 1988a; Curry *et al* 1988), one of the most attractive projects of AMS radiocarbon determination has been that of measuring changes in the radiocarbon age of the deep ocean.

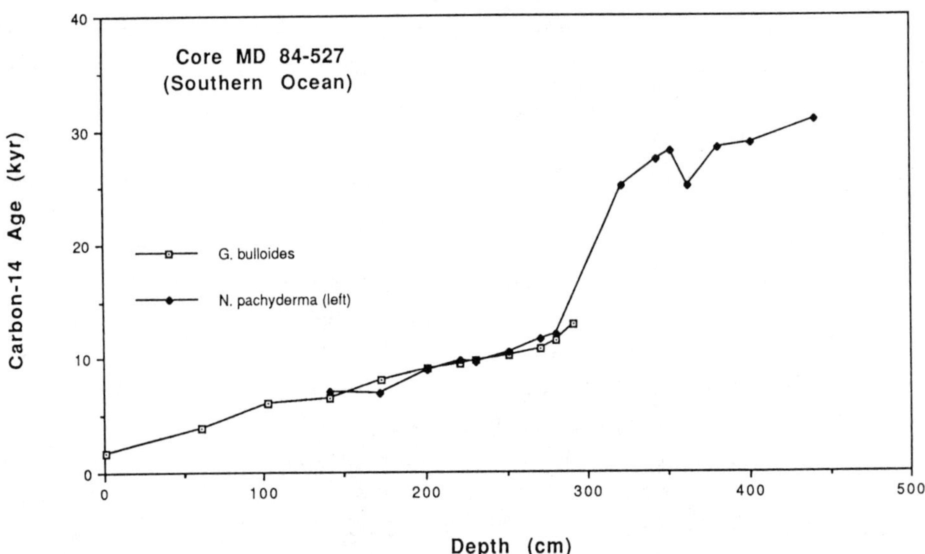

Fig 6.2. Radiocarbon age *vs* depth of two planktonic foraminiferal species in Core MD 84-527.

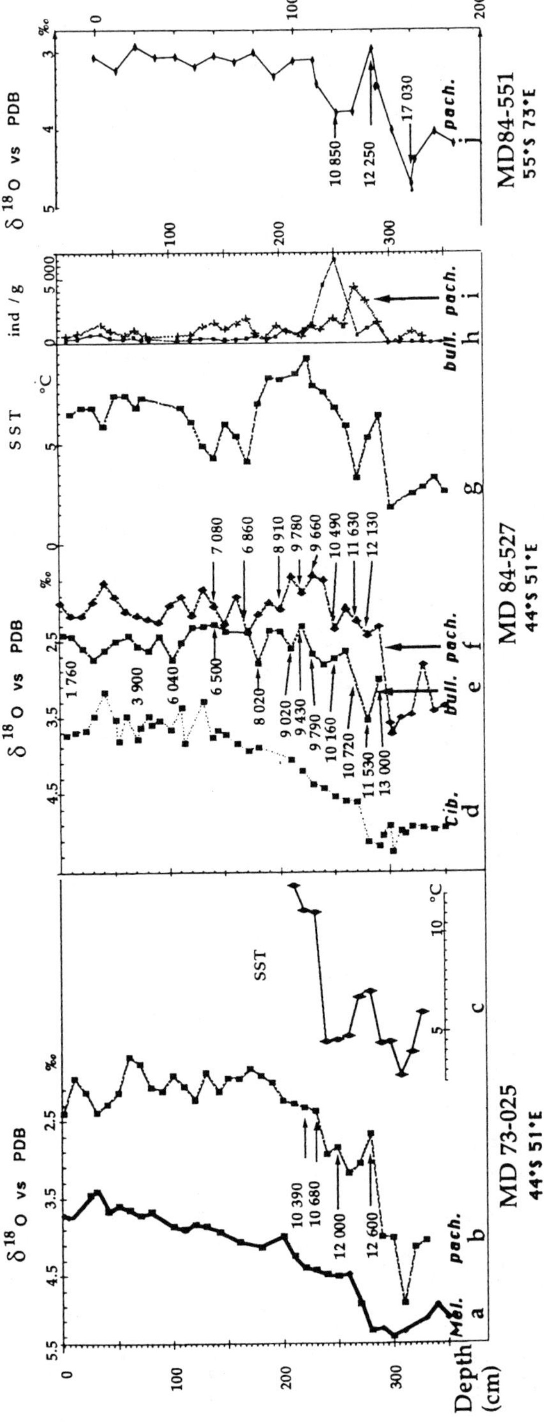

Fig 6.3. Oxygen isotope records, sea-surface temperature estimates and radiocarbon ages of three Southern Ocean cores. *Mel* = *Melonis pompilioides; pach* = *Neogloboquadrina pachyderma* (left coiling); *Cib* = *Cibicides wuellerstorfii; bull* = *Globigerina bulloides.*

^{14}C measurements on shells of foraminifera hand-picked from deep-sea sediments offer the possibility that this reconstruction can be established for the last 30,000 years (Broecker *et al* 1984). The idea is that planktonic and benthic foraminifera coexisting at the same sediment level were deposited at the same time. Hence, the radiocarbon age difference between benthic and planktonic foraminifera is a direct measurement of the radiocarbon age of the deep water at the depth from which the core was collected, relative to the radiocarbon age of surface water. At low latitude, it is generally assumed that the radiocarbon surface water age was not significantly different from the modern one (Bard 1988).

Broecker *et al*'s (1984) initial attempts to achieve this were thwarted by the difficulties imposed by bioturbation, which permanently mixes the upper ten centimeters of sediment or more. Shackleton *et al* (1988) and Duplessy *et al* (1989) have minimized this difficulty by making measurements in cores with a sufficiently high accumulation rate (≥ 10 cm/kyr), so that bioturbation is less of a problem. Broecker *et al* (1988e, 1990b) analyzed cores with lower sedimentation rates and used, as a criterion for the validity of each determination of benthic-planktonic age difference, that agreement exists between the ages obtained on two separate species of planktonic foraminifera. At present, only seven cores (4 from the Atlantic, and 3 from the Pacific) fulfill one of these criteria and were used for these reconstructions.

In the case of the Atlantic, all the data were generated by Broecker's group (1990a, b). They show that the age of glacial deep water averaged 650 years during peak glacial time as compared to 350 years today, whereas the age of the well-ventilated intermediate waters is only about 200 years. This result is in agreement with the reconstruction of the glacial deep-water circulation (Boyle & Keigwin 1987; Duplessy *et al* 1988; Boyle 1988; Oppo & Fairbanks 1987). Broecker (1989) pointed out that the older age for the glacial deep water in the Atlantic may be explained either by a slowdown of the flux of North Atlantic Deep Water (NADW) or by an inversion of the Atlantic circulation, but this circulation pattern is inconsistent with the ^{14}C and Cd data.

In the Pacific Ocean, the situation is more confusing: planktonic-benthic age difference measured in a core from the South China Sea and representing Pacific deep waters at a depth of about 2100 m indicates a ventilation age not significantly different from the modern one (Broecker *et al* 1990b). By contrast, detailed analysis of Core TR 163-31 raised from a water depth of about 3000 m, shows that the age difference between the planktonic foraminifera, *N dutertrei*, and the benthic foraminifera, *Uvigerina*, has significantly changed during the last 30,000 years (Fig 6.4A). During the last glaciation, it was close

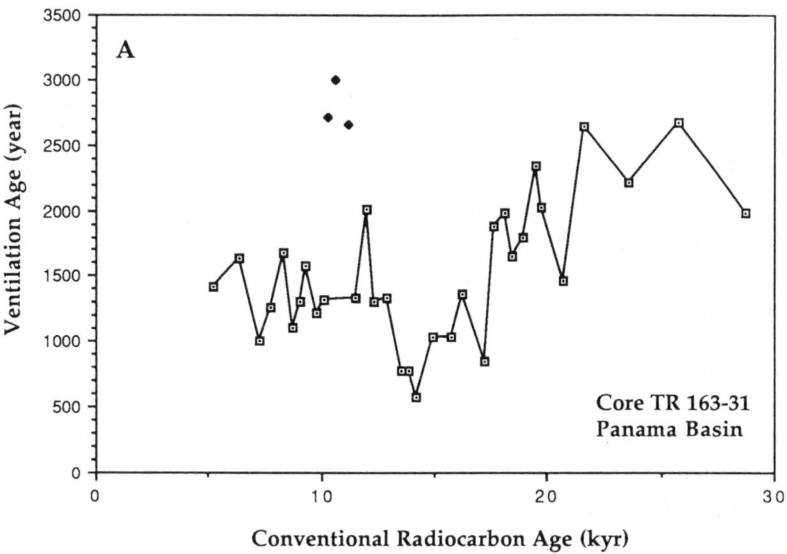

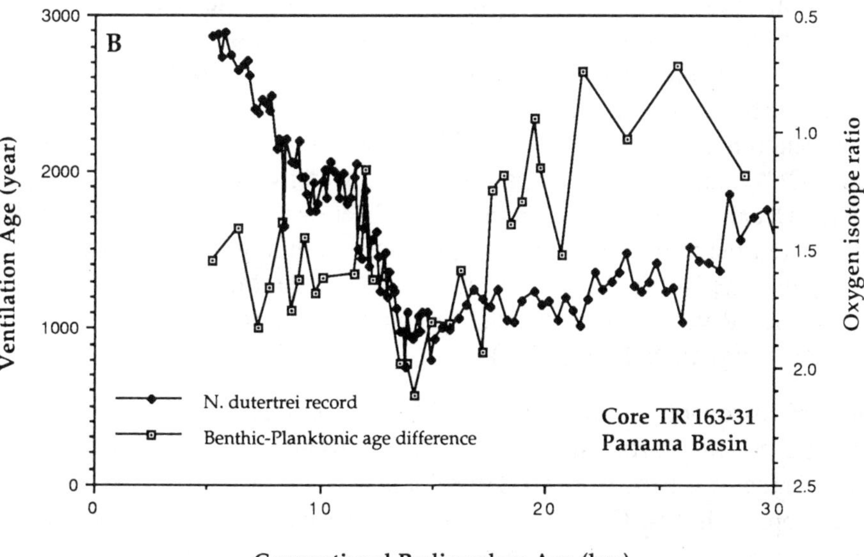

Fig 6.4. A. Variations of the ventilation age of the Pacific deep water in the Panama basin (benthic-planktonic ^{14}C age difference) *vs* radiocarbon age. B. Comparison of the variations of the ventilation age of the Pacific deep water with the oxygen isotope record.

to 2100 years (with a rather large scatter in the data), indicating that the ventilation age of the deep Pacific was about 500 years greater than it is today. The ventilation age decreased by the end of isotope stage 2 and exhibited a noticeable minimum at the beginning of the deglaciation (Fig 6.4B). Finally, it reached the modern value during the end of the deglaciation. This core also exhibits some large planktonic-benthic age differences in the middle of the deglaciation (Fig 6.4A). However, the apparent ventilation age increase would be so high that we have no physical process to explain it. As this large planktonic-benthic age difference is associated with anomalously young ^{14}C ages of planktonic foraminifera, causing a major (> 1000 years) age reversal in the stratigraphic sequence, we believe that it mainly reflects a sedimentation artifact.

The ventilation age of intermediate waters recorded in Core CH 84-14 off Japan exhibits variations similar to those of Core TR 163-31 during the deglaciation. It was noticeably smaller than under modern conditions during most of the glacial-to-interglacial transition. Unfortunately, this core did not reach the last glacial maximum (Kallel *et al* 1988b; Duplessy *et al* 1989).

Propagation of the Meltwater Signal during the Deglaciation

At the end of the last ice age, about 15,000 years ago, the ice sheets which covered the high latitudes of northern America and Eurasia began to melt. During the following 8000 years, about 50 million cubic kilometers of glacial meltwater were introduced into the ocean, resulting in a sea-level rise of about 120 m (Fairbanks 1989). As ice caps are highly impoverished in heavy isotopes of oxygen, this melting event resulted in a significant decrease of the global sea water $^{18}O/^{16}O$ ratio. The major part of the meltwater was released into the North Atlantic Ocean, because the Laurentide ice sheet, which was the biggest of those that covered the northern hemisphere, released both large quantities of icebergs directly into the North Atlantic and meltwater via the Mississippi River and the St Lawrence River drainage systems (Kennett & Shackleton 1975; Broecker *et al* 1988a; Keigwin & Jones 1989).

The comparison of the chronology of the planktonic and benthic records of Cores SU 81-18 (northeastern Atlantic), TR 163-31 (Panama Basin), V 35-5 (South China Sea, Broecker *et al* 1988e) and MD 79-257 (mouth of the Zambezi River, tropical Indian Ocean) shows that the variations of the oxygen isotopic composition of surface and deep waters of the major oceanic basins exhibited only a lagtime shorter than 1000 years against those of the North Atlantic Ocean, which directly received most of the water resulting from the melting of the Laurentide and European ice sheets (Fig 6.5). The fast transfer of the isotope signal was mainly due to an enhanced ocean thermohaline circulation at

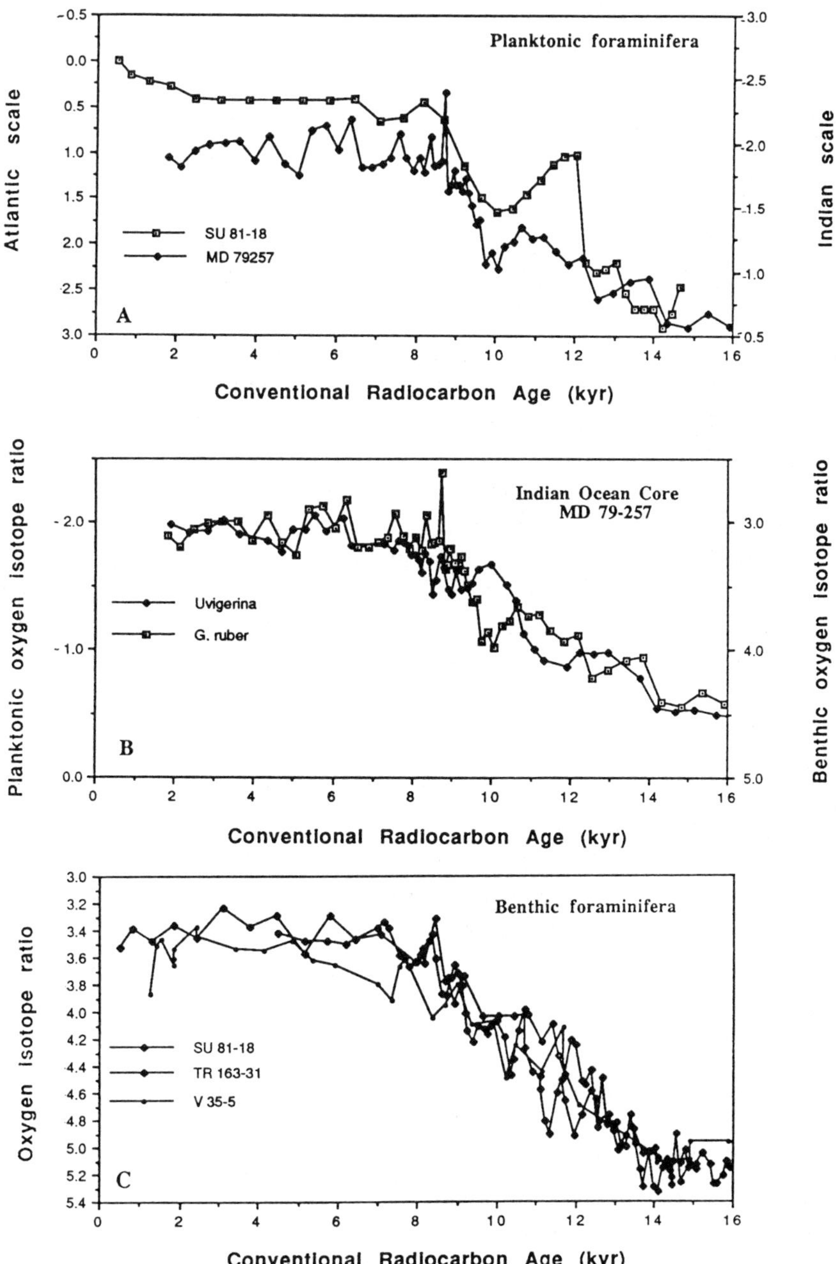

Fig 6.5. A. Oxygen isotope records of the planktonic foraminifera in Cores SU 81-18 and MD 79-257. B. Oxygen isotope records of the planktonic and benthic foraminifera in Indian Ocean Core MD 79-257 (20°24'S, 36°20'E, 1262 m). C. Oxygen isotope records of the benthic foraminifera in Cores SU 81-18 (37°46'N, 10°11'W, 3135 m), TR 163-31 (3°37'S, 83°58'W, 3210 m) and V35-5 (7°12'N, 112°05'E, 1953 m).

the beginning of the glacial-to-interglacial transition. Our observations suggest that significant changes in the atmospheric and surface-water circulation may generate instabilities affecting the global ocean, and that the last deglaciation provides a major example of such changes.

WHERE DO WE STAND?

^{14}C measurements of individual foraminiferal species performed by AMS have provided reliable ages on samples deposited during a short time interval, which is close to the standard error of ^{14}C dating. This new technique has revealed high amplitude changes of both sea-surface temperature and ocean circulation, which occurred abruptly, within a few centuries. Intensive dating of deep-sea cores from the various basins have shown that the sedimentary record is very often blurred by numerous phenomena, such as bioturbation, sediment reworking and redeposition of older sediments. As a consequence, the factor limiting the determination of the ages of the major changes that occurred during the deglaciation is usually the quality of the sedimentary record and not the accuracy of radiocarbon dating. More cores have to be analyzed in order to reconstruct the climatic history of the last 30,000 years and the ventilation of the different water masses, within the theoretical accuracy of AMS.

ACKNOWLEDGMENTS

Thanks are due D de Zertucha, A Castera, J Dudouit, E Kaltnecker and P Maurice for technical assistance with the Tandetron AMS facility. Useful discussions with N J Shackleton are gratefully acknowledged. This work has been supported by CNRS (Dept TOAE-INSU, Programme National d'Etude de la Dynamique du Climat), CEA and EEC Grant EV 4C-0072-F. This is CFR contribution No. 1130.

REFERENCES

Anderson, ED and Libby, WF 1947 Natural radiocarbon from cosmic radiation. *Physical Review* 72: 931–936.

Andrée, M, Beer, J, Oeschger, H, Bonani, G, Hofmann, HJ, Morenzoni, E, Nessi, M, Suter, M and Wölfli, W 1984 Target preparation for milligram sized ^{14}C samples and data evaluation for AMS measurements. *Nuclear Instruments and Methods* 233: 274–279.

Arnold, M, Bard, E, Maurice, P and Duplessy, JC 1987 ^{14}C dating with the Gif-sur-Yvette Tandetron accelerator: Status report. *Nuclear Instruments and Methods* B29: 120–123.

Bard, E 1988 Correction of accelerator mass spectrometry ^{14}C ages measured in planktonic foraminifera: Paleoceanographic applications. *Paleoceanography* 3: 635–645.

Bard, E, Arnold, M, Duprat, J, Moyes, J and Duplessy, JC 1987a Reconstruction of the last deglaciation: Deconvolved records of $\delta^{18}O$ profiles, micropaleontological variations and accelerator mass spectrometric ^{14}C dating. *Climate Dynamics* 1: 101–112.

Bard, E, Arnold, M, Maurice, P, Duprat, J,

Moyes, J and Duplessy, JC 1987b Retreat velocity of the North Atlantic polar front during the last deglaciation determined by ^{14}C accelerator mass spectrometry. *Nature* 328: 791–794.

Bard, E, Fairbanks, RG, Arnold, M, Maurice, P, Duprat, J, Moyes, J and Duplessy, JC 1989 Sea-level estimates during the last deglaciation based on $\delta^{18}O$ and accelerator mass spectrometry ^{14}C ages measured in *Globigerina bulloides. Quaternary Research* 31: 381-391.

Boyle, EA 1988 Cadmium: Chemical tracer of deepwater paleoceanography. *Paleoceanography* 3: 471–489.

Boyle, EA and Keigwin, LD 1987 North Atlantic thermohaline circulation during the past 20,000 years linked to high latitude surface temperature. *Nature* 330: 35–40.

Broecker, WS 1989 Some thoughts about the radiocarbon budget for the glacial Atlantic. *Paleoceanography* 4: 213–220.

Broecker, WS, Andrée, M, Bonani, G, Wölfli, W, Oeschger, H and Klas, M 1988a Can the Greenland climatic jumps be identified in records from ocean and land? *Quaternary Research* 30: 1–16.

Broecker, WS, Andrée, M, Bonani, G, Wölfli, W, Oeschger, H, Klas, M, Mix, A and Curry, W 1988b Preliminary estimates for the radiocarbon age of deep water in the glacial ocean. *Paleoceanography* 3: 659–669.

Broecker, WS, Andrée, M, Klas, M, Bonani, G, Wölfli, W and Oeschger, H 1988c New evidence from the South China Sea for an abrupt termination of the last glacial period. *Nature* 333: 156–158.

Broecker, WS, Andrée, M, Wölfli, W, Oeschger, H, Bonani, G, Kennett, J and Peteet, D 1988d The chronology of the last deglaciation: Implications to the cause of the Younger Dryas event. *Paleoceanography* 3: 1–19.

Broecker, WS, Ewing, M and Heezen, BC 1960 Evidence for an abrupt change in climate close to 11,000 years ago. *American Journal of Science* 258: 429–440.

Broecker, WS, Klas, M, Clark, E, Trumbore, S, Bonani, G and Wölfli, W 1990a Accelerator mass spectrometry radiocarbon measurements on foraminifera shells from deep-sea cores. *Radiocarbon* 32(2): 119–133.

Broecker, WS, Klas, M, Ragano-Beavan, N, Mathieu, G, Mix, A, Andrée, M, Oeschger, H, Wölfli, W, Suter, M, Bonani, G, Hofmann, HJ, Nessi, M and Morenzoni, E 1988e Accelerator mass spectrometry radiocarbon measurements on marine carbonate samples from deep sea cores and sediment traps. *Radiocarbon* 30(3): 261–295.

Broecker, WS, Mix, A, Andrée, M and Oeschger, H 1984 Radiocarbon measurements on coexisting benthic and planktonic foraminifera shells: Potential for reconstructing ocean ventilation times over the past 20,000 years. *Nuclear Instruments and Methods* B5: 331–339.

Broecker, WS, Peng, TH, Trumbore, S, Bonani, G and Wölfli, W 1990b The distribution of radiocarbon in the glacial ocean. *Global Biogeochemical Cycles*, in press.

Curry, WB, Duplessy, JC, Labeyrie, LD and Shackleton, NJ 1988 Changes in the distribution of $\delta^{13}C$ of deepwater $\sum CO_2$ between the last glaciation and the Holocene. *Paleoceanography* 3: 317-341.

Denton, GH and Hughes, TJ 1981 *The Last Great Ice Sheets.* New York, John Wiley & Sons: 484 p.

Duplessy, JC, Arnold, M, Bard, E, Juillet-Leclerc, A, Kallel, N and Labeyrie, LD 1989 AMS ^{14}C study of transient events and of the ventilation rate of the Pacific Intermediate water during the last deglaciation. *In* Long, A and Kra, RS, eds, Proceedings of the 13th International ^{14}C Conference. *Radiocarbon* 31(3): 493–502.

Duplessy, JC, Delibrias, G, Turon, JL, Pujol, C and Duprat, J 1981 Deglacial warming of the Northeastern Atlantic Ocean: Correlation with the paleoclimatic evolution of the European continent. *Palaeogeography, Palaeoclimatology, Palaeoecology* 35:

121–144.

Duplessy, JC and Shackleton, NJ 1985 Response of deep water circulation to Earth's climatic change 135,000–107,000 years ago. *Nature* 316: 500–507.

Duplessy, JC, Shackleton, NJ, Fairbanks, RG, Labeyrie, L, Oppo, D and Kallel, N 1988 Deep water source variations during the last climatic cycle and their impact on the global deep water circulation. *Paleoceanography* 3: 343–360.

Emiliani, C 1955 Pleistocene temperatures. *Journal of Geology* 63: 538–578.

Fairbanks, RG 1989 A 17,000 year glacio-eustatic sea level record: Influence of glacial melting rates on the Younger Dryas event and deep ocean circulation. *Nature* 342: 637–642.

Flint, RF and Rubin, M 1955 Radiocarbon dates of pre Mankato events in eastern and central North America. *Science* 121: 649–658.

Imbrie, J and Kipp, NG 1971 A new micro-paleontological method for quantitative paleoclimatology: Application to a Late Pleistocene Caribbean core. *In* Turekian, KK, ed, *The Late Cenozoic Glacial Ages*. New Haven, Connecticut, Yale University Press: 71–181.

Kallel, N, Labeyrie, LD, Arnold, M, Okada, H, Dudley, WC and Duplessy, JC 1988a Evidence of cooling during the Younger Dryas in the western North Pacific. *Oceanologica Acta* 11: 369–375.

Kallel, N, Labeyrie, LD, Juillet-Leclerc, A, and Duplessy, JC 1988b A deep hydrological front between intermediate and deep water masses in the glacial Indian Ocean. *Nature* 333: 651–655.

Keigwin, LD and Jones, GA 1989 Glacial-Holocene stratigraphy, chronology and paleoceanographic observations on some North Atlantic sediment drifts. *Deep Sea Research*: 36: 845–867.

Kennett, JP and Shackleton, NJ 1975 Laurentide ice sheet meltwater recorded in Gulf of Mexico deep sea cores. *Science* 188: 147–150.

Labeyrie, LD, Kallel, N, Arnold, M, Juillet-Leclerc, A, Maitre, F, Duplessy, JC and Shackleton, NJ 1990 Variabilité des oeaux intermédiaires et profondes dans l'océan Pacifique Nord Ouest pendant la dernière déglaciation. *Oceanologica Acta* 10: 329–339.

Labracherie, M, Labeyrie, LD, Duprat, J, Bard, E, Arnold, M, Pichon, JJ and Duplessy, JC 1989 The last deglaciation in the Southern Ocean. *Paleoceanography* 4: 629–638.

Oppo, D and Fairbanks, RG 1987 Variability in the deep and intermediate water circulation during the past 25,000 years: Northern hemisphere modulation of the Southern Ocean. *Earth and Planetary Science Letters* 86: 1–15.

Ruddiman, WF and McIntyre, A 1973 Time-transgressive deglacial retreat of polar waters from the North Atlantic. *Quaternary Research* 3: 117–130.

_____ 1981 The North Atlantic Ocean during the last deglaciation. *Palaeogeography, Palaeoclimatology, Palaeoecology* 35: 145–214.

Shackleton, NJ, Duplessy, JC, Arnold, M, Maurice, P, Hall, M and Cartlidge, J 1988 Radiocarbon age of last glacial Pacific deep water. *Nature* 335: 708–711.

Shackleton, NJ and Opdyke, ND 1973 Oxygen isotope and paleomagnetic stratigraphy of equatorial pacific core V 28-238: Oxygen isotope temperature and ice volumes on a 10^5 and 10^6 year. *Quaternary Research* 3: 39–55.

7

RECONSTRUCTION OF RADIOCARBON DISTRIBUTION IN THE GLACIAL OCEAN

TSUNG-HUNG PENG and WALLACE S BROECKER

INTRODUCTION

Since its discovery about four decades ago, radiocarbon has been an important isotope tracer for studying the ventilation rates of deep seawater and the rain rate of calcite toward the deep-sea floor (Broecker & Peng 1982). After their production in the atmosphere, ^{14}C atoms invade the surface water of the ocean. The ^{14}C in surface water is then slowly transported to the deeper ocean as surface water is mixed with deep water. The difference in $^{14}C/^{12}C$ ratios between surface water and deep water depends on how fast the deep water is replaced by the surface water through ocean circulation and mixing processes. The distribution of ^{14}C in the water column of the present-day ocean can be measured by counting the ^{14}C atoms in CO_2 gas extracted from the dissolved inorganic carbon in seawater. The ^{14}C atoms in seawater are also preserved in calcite shells of marine microorganisms. For example, planktonic foraminifera that live in near-surface waters produce calcite shells that contain ^{14}C derived from water that is close to equilibrium with the atmosphere. These shells eventually sink to the sea floor and provide valuable materials for dating deep-sea sediments. The rain rate of calcite can thus be estimated from results of such ^{14}C datings.

The distribution of ^{14}C in the present-day ocean has provided vital information about the current rates of deep-sea ventilation. Because ventilation rates were assumed constant during the Holocene period, variations in atmospheric ^{14}C content over the last 8000 years, as recorded in tree rings, were attributed solely to changes in ^{14}C production rates caused mainly by changes in the magnetic field of the earth (Damon, Lerman & Long 1978; Andrée *et al* 1986). It is also possible that the mode of operation of the ocean system was perturbed by the great warming that accompanied deglaciation 11,000 years ago (Broecker, Peteet & Rind 1985). Peng (1989) reported possible changes in deep-sea ventilation rates in the last 7000 years, based on ^{14}C variations in the atmosphere as well as in the ocean. The ocean-atmosphere system apparently changes continuously in response to climate changes. At present, CO_2 in the atmosphere continues to increase each year and has already reached the highest level of the last 160,000 years (Barnola *et al* 1987; Keeling *et al* 1989) due to anthropogenic releases of CO_2 into the atmosphere over the last two centuries. The potential greenhouse warming caused by high levels of CO_2 and other increasing radiative gases related to human activity could change the current climate, and hence, the sea's mode of operation during the next 100 years or so.

During the peak glacial period (20,000–14,000 yrs ago), the earth was colder, more covered by ice, dustier and poorer in greenhouse gases. For example, the partial pressure of CO_2 in the glacial atmosphere was about 200 μatm, whereas the Holocene pre-anthropogenic value was 280 μatm (Neftel *et al* 1988). About 14,000 years ago, a climatic change occurred that created conditions similar to those of today. Broecker and Denton (1989) postulated that this change was brought about by a reorganization of the entire ocean-atmosphere system. If so, we must learn how the ocean-atmosphere system operated during glacial time and before a major climatic change. We need this baseline information for comparison with the Holocene ocean if we want to learn, understand and, most of all, predict how the current ocean-atmosphere system would respond to potential climatic change as CO_2 and other greenhouse gases increase in the atmosphere. To study the glacial ocean, we need to examine the deep-sea sediments in which a wealth of information is stored.

Measurements of cadmium-to-calcium ratios in shells of benthic foraminifera (Boyle & Keigwin 1987; Boyle 1988) from deep-sea sediments reveal that, during glacial time, the pattern of circulation was much different from today's. The Cd/Ca ratio is known to vary with the concentration of nutrients. These results indicate that the nutrient constituent maxima, which, in today's ocean, lie at intermediate depths, were shifted toward the bottom. The strong contrast observed today between the nutrient contents of deep water in the Atlantic and the Pacific Oceans was smaller during glacial time.

We can use the information stored in the deep-sea sediments, such as the ^{14}C age in calcite shells of foraminifera, to reconstruct the distribution of ^{14}C in the glacial ocean over the last 20,000 years or so (Broecker *et al* 1984). The $^{14}C/^{12}C$ ratio in surface water is preserved in planktonic shells and that ratio in bottom water of various depths is preserved in benthic shells. Let us assume that the ratio of $^{14}C/^{12}C$ for benthic foraminifera to planktonic foraminifera coexisting in deep-sea sediments did not change with time during the glacial period. We can use ^{14}C measurements of coexisting benthic and planktonic foraminifera shells for the $^{14}C/^{12}C$ difference between surface water and water at the depth from which the core was taken. This record provides a means for estimating the rate of deep seawater ventilation in the glacial ocean.

The accelerator mass spectrometry (AMS) ion counting technique for ^{14}C measurements makes possible the dating of a single foraminifera species. Since 1984, a number of deep-sea cores have been chosen for AMS ^{14}C dating of foraminifera shells (Andrée *et al* 1986; Shackleton *et al* 1988; Broecker *et al* 1988a,b 1990). In this chapter, we summarize the relevant published AMS ^{14}C measurements of deep-sea sediments and reconstruct the ^{14}C distribution in the glacial ocean. We also discuss implications of circulation in the glacial ocean.

DISTRIBUTION OF ^{14}C IN THE PRESENT-DAY OCEAN

To appreciate the importance of ^{14}C distribution in the ocean, it is first necessary to look at the data collected in the present-day ocean. Broecker and Peng (1982) present a good picture of the pre-anthropogenic $^{14}C/^{12}C$ distribution in the ocean. Bien, Rakestraw and Suess (1960) and Broecker *et al* (1960) made some of these measurements on surface waters collected prior to major bomb tests. Stuiver and Östlund (1980, 1983) and Östlund and Stuiver (1980) made other measurements on deep waters free of tritium (hence, free of bomb ^{14}C) during the GEOSECS program in 1973 to 1978. Between 40°N and 40°S, the surface waters of the ocean have nearly uniform $\Delta^{14}C$ values, averaging close to −50‰. All of the deep-water values ($\Delta^{14}C$ shown in Figure 7.1 for waters at a depth of 3 km) are lower (*ie*, more negative) than the values for warm surface water. The values decrease systematically from the northern Atlantic to the Antarctic and from the Antarctic to the northern Indian and Pacific Oceans. The trend is closely related to the path followed by deep waters formed in the northern Atlantic (*ie*, North Atlantic Deep Water (NADW)).

For convenience, we converted to radiocarbon ages the differences between the $^{14}C/^{12}C$ ratios for surface and bottom waters at any given point in the ocean (*ie*, the time required for the $^{14}C/^{12}C$ ratio of surface water to decay to that of deep water). Although these ages lack the physical meaning attached, they do provide a measure of the isolation time for the carbon in bottom water; the older its ^{14}C age, the greater is the isolation time. Shown in Figure 7.2 are ^{14}C ages of the water at a depth of 3 km. These ages generally follow the $\Delta^{14}C$ pattern for deep water. The major exception is the Antarctic Ocean, where the age differences are much smaller than expected from the deep-water $\Delta^{14}C$ values because the surface waters of the Antarctic have low $\Delta^{14}C$ values. In the tropical Pacific, the average age difference is about 1600 years, whereas that in the tropical Atlantic is about 350 years.

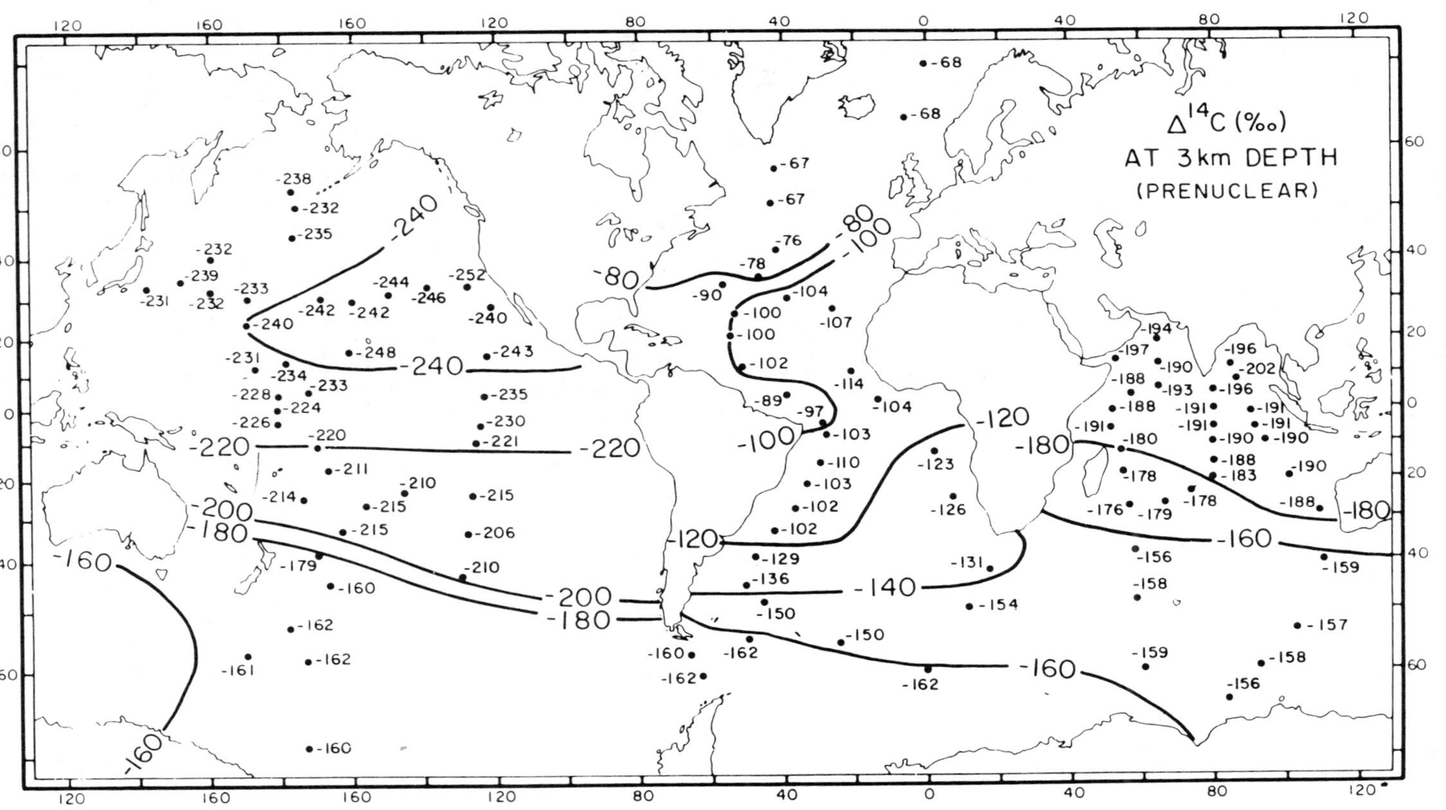

Fig 7.1. The distribution of $\Delta^{14}C$ values at 3 km depth in the ocean based on results obtained as part of the 1972–1978 GEOSECS program

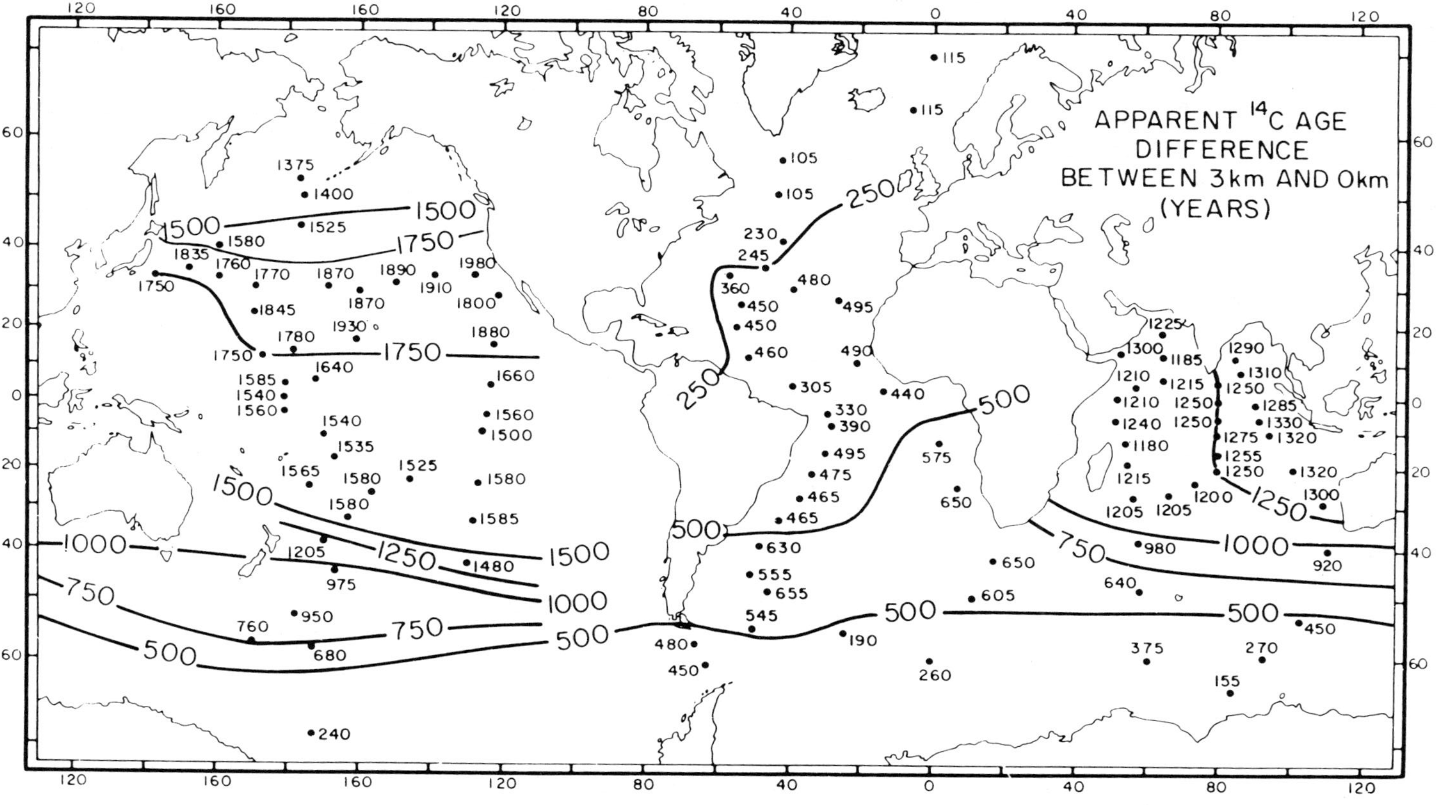

Fig 7.2. The distribution of the age difference between dissolved carbon from waters at 3 km depth and dissolved carbon from surface waters prior to nuclear testing

RECONSTRUCTION OF AGE DIFFERENCES FROM DEEP-SEA SEDIMENTS AND POSSIBLE SOURCES OF AGE BIAS

To study the distribution of ^{14}C in the glacial ocean, we need to reconstruct the past ^{14}C age differences preserved in deep-sea sediments. In an ideal situation where perturbing influences are absent, the ^{14}C age difference between planktonic and benthic forams coexisting at some depth in a deep-sea core should provide a measure of the apparent ^{14}C age for deep water at the time the layer of sediment containing forams was deposited. If so, we could conceivably reconstruct the apparent age distribution of bottom water for various times in the past by studying cores from different geographic locations and depths.

However, as Broecker *et al* (1984) describe in detail, this ideal situation is disrupted by several processes occurring on the sea floor that tend to alter the planktonic-benthic age difference. The most serious disturbing factors include bioturbation, abundance and dissolution effects. Radiocarbon age measurements of many deep-sea cores indicate that the upper 8 or so centimeters of sediments are almost completely mixed (a layer this thick under typical oceanic conditions takes several thousand years to accumulate). Peng and Broecker (1984) describe the impact of bioturbation on the age difference. The most obvious problem created by bioturbation is that the discernible record of sediment is smoothed into units of several thousand years. Using typical mid-ocean sediments (accumulation rate of 1–2 cm every thousand years), we will not be able to learn of short-term changes in deep-sea ventilation rates.

A combination of bioturbation and abundance effects introduces serious bias in the age difference. For example, if the planktonic abundance remains constant and the benthic abundance decreases by a factor of 4 from the glacial to the postglacial period, the observed ^{14}C age difference between the benthic and the planktonic forams will be as much as 2000 years larger than the ^{14}C age difference between bottom water and surface water. On the other hand, if the benthic abundance is higher in the Holocene than in glacial time, the measured ^{14}C age difference becomes smaller than the ^{14}C age difference between bottom water and surface water. This age bias could range from near zero to double the age difference between bottom water and surface water. To avoid such age biases, we must search for areas on the sea floor where the ratio of bioturbation depth to sedimentation rate is much smaller than 4×10^3 years (a value typical for open-ocean sediments).

Because organisms living within the sediment release CO_2 through respiration, the pore waters in the upper portion of the sediment may be undersaturated with calcite. Thus, despite the supersaturation of the overlying water, the calcite found in these sediments may partially dissolve. Coupled with bioturbation, dissolution can create a bias in the age difference between planktonic and benthic shells. The source of the bias lies in the difference between the degree of solution for the planktonic and benthic species chosen for dating. If the effects were identical, no bias would be created. Studies of calcite in marine sediments reveal that shells of foraminifera disintegrate into fragments as they dissolve, and some species do so more readily than others. As a result, the whole shells of breakage-prone species will yield a lower ^{14}C age than those of a breakage-resistant species. Also, whole shells of a given species should have younger ages than shell fragments of that species if, indeed, the fragments were created by dissolution. When benthic shells appear to be more resistant to dissolution than planktonic shells, the dissolution will tend to increase the age difference between these shells (and hence, increase the apparent age of deep seawater). To reduce such age biases, we must select regions of the sea floor where dissolution effects are minimal and sedimentation rates are high, and we must pick planktonic species with shells that are resistant to breakage.

Other factors, such as the growth habitats of the species of interest and possible contamination effects, could also cause significant bias. Examples follow:

1. The material present may have been reworked.
2. The $^{14}C/^{12}C$ ratio in the calcite may have been modified by secondary growth of calcite in sediment pore waters and by exchange between calcite and carbonate ions in pore waters.
3. Not all planktonic species live exclusively in the surface mixed layer of the oceans; hence, deeper dwelling species may contain less ^{14}C than surface-dwelling species.
4. Areas of upwelling may have a lower $^{14}C/^{12}C$ ratio.
5. Benthic forams may live within the sediment and use inorganic carbon from pore waters to form their shells, although the basic assumption is that they live at the sediment-water interface and use inorganic carbon from bottom water to form their shells.

Despite these complicating factors, many deep-sea cores have been selected carefully for sampling to determine ^{14}C ages of hand-picked planktonic and benthic foraminifera shells by using AMS.

RESULTS OF RECONSTRUCTION

Andrée *et al* (1986) first reported the results of AMS ^{14}C dating from cores with high sedimentation rates. These cores, V35-5 and V35-6, with sedimentation rates of 15 and 10 cm/10^3 years, respectively, were raised from a water depth of 2 km in the southern part of the South China Sea. This study, aimed at Holocene time only, provided evidence that the age difference for waters at 2 km depth in the western Pacific remained constant throughout the Holocene to within the measurement uncertainty (*ie*, ~200 yrs).

Broecker *et al* (1988a) reported the first estimates for the radiocarbon age of deep water in the glacial ocean. Based on ^{14}C measurements on a core from Ceara Rise, Broecker *et al* (1988a) suggested that deep water in the western tropical Atlantic was roughly twice as old during glacial time as it is today. Results from the East Pacific Rise and the South China Sea (Broecker *et al* 1988a) and from the eastern tropical Pacific (Shackleton *et al* 1988) suggested that deep water in the tropical Pacific was about 500 years older during glacial time than it is today. However, a question has been raised regarding the Pacific results. The core studied by Shackleton *et al* (1988) came from an area where upwelling currently influences the $^{14}C/^{12}C$ ratio in the photic zone. The planktonic shells in this case do not necessarily represent conditions typical of the surface waters. The East Pacific Rise core (Broecker *et al* 1988a) came from an area that had a low sediment accumulation rate (2 cm/10^3 yrs). The area also showed changes in foraminifera abundance from the glacial period to the Holocene, so significant age bias was probably introduced by bioturbation-abundance effects. The core from the South China Sea (Broecker *et al* 1988a) showed a systematic and unexplained discordance between the ages of two planktonic species, *Globigerinoides sacculifera* and *Pulleniatina obliquiloculata*.

Broecker *et al* (1990) recently reported new radiocarbon results with improvements to overcome the uncertainties associated with previous results from the tropical Pacific. One important criterion used for verifying the validity of each measurement of benthic-planktonic age difference is that an agreement must exist (at the 2 σ level) between the ages obtained on two separate species of planktonic foraminifera. They presented results that meet this criterion from five deep-sea cores: one from the South China Sea (representing 2.1 km-deep water in the western tropical Pacific), one from the Caribbean (representing 1.8 km-deep water from the western tropical Atlantic), and three from the Ceara Rise (representing 2.8 to 4.0 km-deep water in the western tropical Atlantic).

Table 7.1 summarizes benthic-planktonic age differences derived from these cores. For the tropical Pacific, only a slightly discernible change is apparent with time in the $^{14}C/^{12}C$ difference between surface and deep waters. Since this is the only core analyzed for the Pacific Ocean, more measurements must confirm this small difference. For the reconstruction of ^{14}C distribution in the

TABLE 7.1. Summary of Mean Benthic - Planktonic Age Differences for Glacial Oceans

Location	Core	Water depth (km)	Glacial sed rate (cm/kyr)	Number B,P pairs	Mean (B-P) age (yrs)	Today's (B-P) age (yrs)
Atlantic (12°N, 79°W)	V28-122	1.8*	11.9	3	195 ± 140	300
Atlantic (4°N, 43°W)	K110-82	2.8	5.3	5	780 ± 220	350
Atlantic (5°N, 43°W)	K110-66	3.5	4.7	4	600 ± 105	350
Atlantic (5°N, 43°W)	K110-50	4.0	3.2	5	705 ± 150	400
Pacific (19°N, 116°E)	S50-37	2.1	17.9	4	1670 ± 105	1600

*Sill depth of the Caribbean

tropical Pacific, the mean benthic-planktonic age difference is taken to be 1670 years during the peak glacial time, as opposed to 1600 years today. For the tropical Atlantic, a sufficient number of measurements in the depth range of 2.8–4.0 km indicate that the age of NADW was twice as great during glacial time as it is today. As Figure 7.3 shows, the age of Atlantic deep water averaged 675 years during the peak glacial time as opposed to 350 years today.

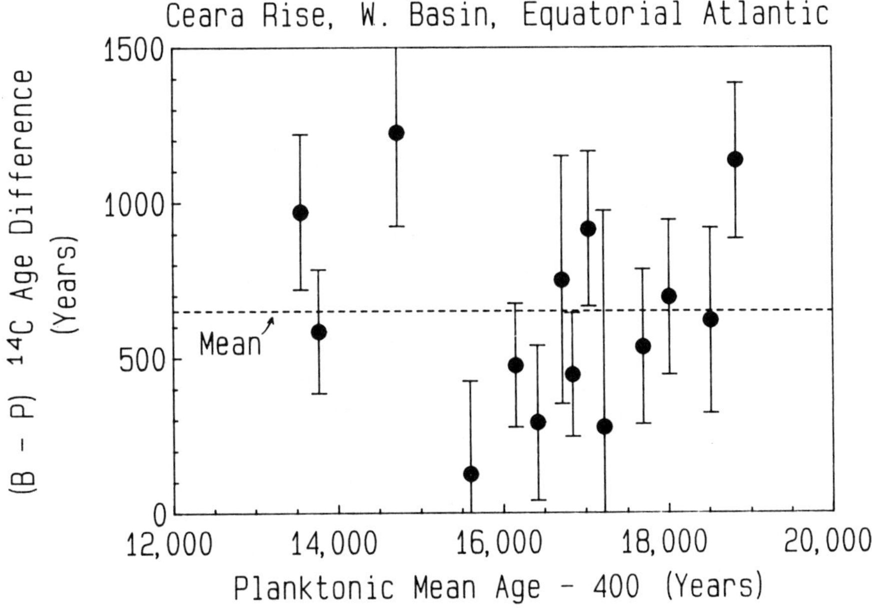

Fig 7.3. Benthic-planktonic age differences *vs* horizon age (mean planktonic age minus 400 yrs) for the three cores from the Ceara Rise in the western basin of the equatorial Atlantic. The mean age difference is 675 years.

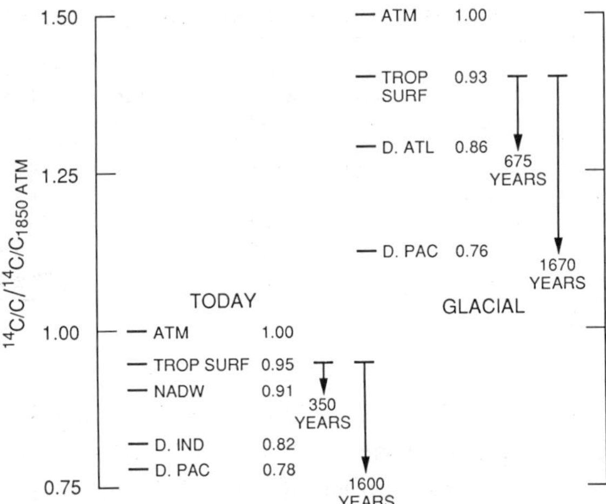

Fig 7.4. Reconstructed radiocarbon content of various parts of the ocean-atmosphere system for glacial time as compared with today's distribution. The estimate of the absolute $^{14}C/^{12}C$ value for the glacial system is based on the ^{230}Th – ^{14}C age comparisons on Barbados coral by Bard *et al* (1990). The numbers in the diagram are the ratios of the $^{14}C/^{12}C$ in the water of interest to those in the atmosphere.

From these measurements, we can attempt to reconstruct the distribution of ^{14}C in the glacial ocean. First, we need to determine the absolute ^{14}C level in the earth system during the last glacial time. Bard *et al* (1990) reported that during glacial time the ^{14}C ages of a coral recovered from Barbados were about 3200 years younger than the radiothorium ages of the same coral. The authors believed this difference represented a ^{14}C inventory 1.5 times higher on earth during glacial time (presumably reflecting a lower magnetic field on earth and, consequently, a higher ^{14}C production rate). If we accept this result, the atmospheric $^{14}C/^{12}C$ ratio for the peak glacial period should be placed at 1.5 times the AD 1850 value (Fig 7.4).

To reconstruct the ^{14}C content of various parts of the ocean-atmosphere system for glacial time, we assume a steady state in which the $^{14}C/^{12}C$ difference between the atmosphere and the surface ocean must be of the proper magnitude so that the influx of ^{14}C into the ocean from air-sea CO_2 exchange is equal to the decay flux within the sea. Such ^{14}C influx depends not only on the air-sea $^{14}C/^{12}C$ difference but also on the air-sea CO_2 partial pressure difference, the CO_2 partial pressure in the atmosphere and the gas exchange rate, which in turn depends on the wind velocity over the ocean. Ice-core studies (Neftel *et al* 1988) show the partial pressure of CO_2 (pCO_2) was ~200 μatm during glacial time as opposed to 280 μatm in AD 1850. Although the mean wind velocity over the ocean may have been different during glacial time, there is no way to know even the sign of this difference. We can only assume that the wind speed was similar to today's. Hence, we also assume the air-sea CO_2 exchange rate is similar to today's. The glacial-to-Holocene differences in sea surface temperature need not be considered because the temperature effects on the CO_2 concentration and the diffusivity of CO_2 in seawater largely compensate one another in the CO_2 exchange.

From the assumptions above, we can say:

$$[1-\frac{(^{14}C/C)_{surf}}{(^{14}C/C)_{atm}}]_{glacial} = \frac{280}{200}[1-\frac{(^{14}C/C)_{surf}}{(^{14}C/C)_{atm}}]_{1850} \quad .$$

The $^{14}C/^{12}C$ ratio for surface water of the tropical ocean in AD 1850 was about 95% of the atmospheric value (Broecker & Peng 1982). If we assume the geographic pattern of the $^{14}C/^{12}C$ ratio in the glacial ocean is similar to today's, we calculate, with the above equation, that the $^{14}C/^{12}C$ ratio in the glacial tropical surface ocean is 93% that in the glacial atmosphere. This value is independent of the absolute ^{14}C inventory. The age difference of 675 years during glacial time between tropical surface water and deep Atlantic water gives a $^{14}C/^{12}C$ ratio of 0.86 for the deep Atlantic. In the same way, the age difference of 1670 years during glacial time in the tropical Pacific gives a ratio of 0.76 for deep Pacific waters. We compare this reconstruction with the distribution of ^{14}C in the present-day ocean in Figure 7.4. For the present-day ocean, the $^{14}C/^{12}C$ ratio of NADW is 91% that in the atmosphere, whereas deep Pacific water has a value of 78%.

MODELING THE CIRCULATION IN THE GLACIAL OCEAN

To understand the implications and relationships between the reconstructed ^{14}C age of deep water and the mode of ocean circulation during the glacial period, we use the PANDORA box model, a global geochemical model that attempts to simulate the real ocean (Broecker & Peng 1985, 1987). As shown in Figure 7.5, PANDORA divides the global ocean into ten boxes, with warm surface, intermediate thermocline, and cold deep boxes for the Atlantic and the combined Pacific-Indian Oceans. The Antarctic Ocean is placed between them with surface and deep boxes. Figure 7.5 also shows the surface area of those boxes in contact with the atmosphere and the volume of each box. The biological cycle is represented by the phosphorus cycle using a mean residence time of P in the

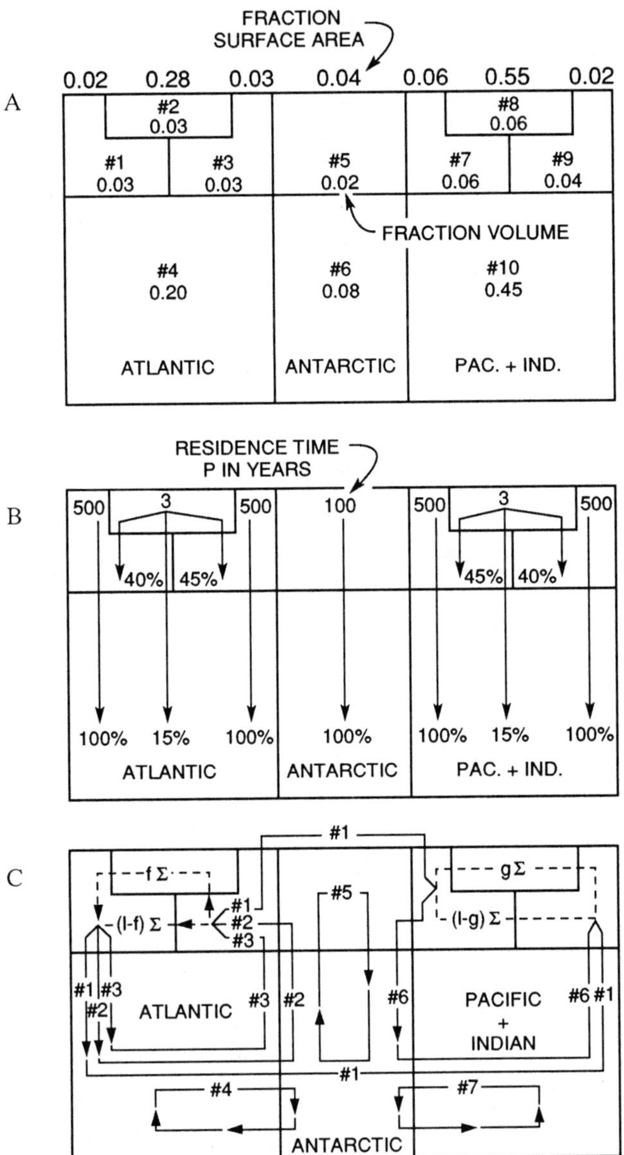

Fig 7.5. A revised version of PANDORA adopted for exploration of circulation patterns capable of reproducing the nutrient and ^{14}C distributions reconstructed for the glacial ocean. In A, we give the volume fractions of all boxes and the surface area fractions of seven boxes. In B are the residence times (in yrs) for phosphate in each of the reservoirs receiving sunlight. Also shown are the fate of the phosphorus atoms removed from surface water in particulate form. In C are the seven circulation loops. The eighth loop is not shown, but since this loop connects box 3 with box 7 and has a high flux of water (circumpolar current), these two boxes remain nearly identical.

surface boxes and a fraction of its remineralization in the lower boxes. Figure 7.5 shows values for these parameters. The circulation is depicted by various flow loops. In addition to the seven circulation loops (Fig 7.5), an eighth one connects the southern thermocline reservoir of the Atlantic (box 3) with the southern thermocline reservoir of the Pacific-Indian (box 7). A rapid mixing between these two reservoirs *via* circumpolar flow is expected, and therefore a sufficiently high flow *via* the eighth loop is given to maintain nearly the same composition in these two reservoirs.

A special feature of this loop structure is the branching in the upper ocean. A fraction of flow, f, for the Atlantic and g for the Pacific-Indian, moves through the warm surface layer, and the remainder, 1 - f, for the Atlantic and 1 - g for the Pacific-Indian, moves along the thermocline. This scheme is designed to reproduce effectively the observed properties of the surface of today's ocean. With such a limited number of boxes and flow loops, the real ocean cannot be truly represented. To simulate the present-day ocean closely, best values of f, g and fluxes of each circulation loop are chosen and given in Table 7.2. Resulting model distributions of pCO_2, PO_4, CO_3^{2-}, $\Delta^{14}C$ and the age of deep water are also given in the table.

To simulate the reconstructed distribution of ^{14}C and nutrients in the glacial ocean, the model flow of NADW is reduced by slowing down loops 1, 2 and 3. Table 7.3 shows magnitudes of water fluxes for all loops during these simulations with different degrees of NADW reduction. To maintain a nearly uniform $^{14}C/^{12}C$ difference between the surface and the deep Pacific-Indian boxes, decreases in the strength of loop 1 are matched by corresponding increases in the strength of loop 6. By reducing the NADW flow, we can increase the surface-to-deep ^{14}C difference as expected.

The response of the nutrient distribution to changes in NADW is unexpected. As Figure 7.6 shows, the phosphate content of the Atlantic deep water does not change significantly with decreasing NADW. The reason is that the nutrient content of water entering the Atlantic must balance that of water leaving the Atlantic. Because the residence time of phosphorus in the ocean (~100,000 yrs) is about 100 times longer than the ocean mixing time (~1000 yrs), no significant gain or loss of this nutrient can occur during a few passes of water through the Atlantic. Since all the water leaving the Atlantic in the model does so *via* the export of deep water, the phosphate content of this exported water must be equal to that in the water entering the Atlantic. As the phosphate input is nearly independent of the flux of water through the Atlantic, very little change in the phosphate content of Atlantic deep water occurs when the rate of NADW production decreases. These results reveal that ^{14}C and nutrient distributions in

TABLE 7.2. Standard Configuration Used in the PANDORA Model for the Holocene

Parameter assignment	Reservoir No.	Reservoir Identity	pCO_2 (μatm)	PO_4 (μmol/kg)	CO_3^{2-} (μmol/kg)	^{14}C (‰)	Age of deep water (yrs)
f = 0.65	-	Atmosphere	311	-	-	−14	-
g = 0.65	1	North Atlantic thermocline	282	0.86	119	−77	-
L1 = 6.0 sverdrups	2	Atlantic surface	308	0.06	193	−57	-
L2 = 13.5 sverdrups	3	South Atlantic thermocline	338	1.40	108	−104	-
L3 = 10.5 sverdrups	4	Atlantic deep	327	1.20	110	−111	500
L4 = 4.0 sverdrups	5	Antarctic surface	303	1.39	108	−117	-
L5 = 3.0 sverdrups	6	Antarctic deep	410	1.80	95	−146	-
L6 = 15.0 sverdrups	7	South Pacific-Indian thermocline	325	1.34	111	−104	-
L7 = 20.0 sverdrups	8	Pacific-Indian surface	309	0.03	198	−54	-
L8 = 100 sverdrups	9	North Pacific thermocline	369	1.92	125	−169	-
	10	Pacific-Indian deep	576	2.58	82	−185	1240

TABLE 7.3. Transports (in Sverdrups) Used in Calculations for Changes in NADW Flux

Water route	Holocene standard	NADW ×0.8	NADW ×0.6	NADW ×0.4	NADW ×0.2
Loop 1	6.0	4.8	3.6	2.4	1.2
Loop 2	13.5	10.8	8.1	5.4	2.7
Loop 3	10.5	8.4	6.3	4.2	2.1
Sum of loops 1-3	30.5	24.0	18.0	12.0	6.0
Loop 4	4.0	4.0	4.0	4.0	4.0
Loop 5	3.0	3.0	3.0	3.0	3.0
Loop 6	15.0	16.2	17.4	18.6	19.8
Loop 7	20.0	20.0	20.0	20.0	20.0
Loop 8	100.0	100.0	100.0	100.0	100.0

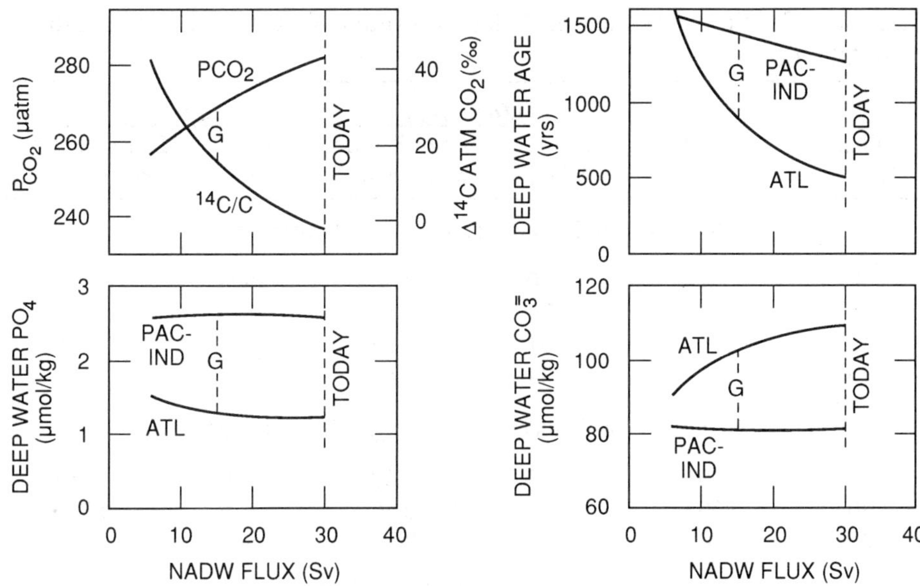

Fig 7.6. Response of PANDORA to changes in the flux of the combined NADW loops (*ie*, loops 1, 2 and 3 in Fig 7.5) from 30 sverdrups for the Holocene standard to values as low as 6 sverdrups. --G-- marks the flux at which the age of deep water in the Atlantic is increased by a factor of two.

the sea are not as closely related as one might expect. Therefore, the information gained from ^{14}C measurements on foraminifera is not necessarily redundant with that obtained from cadmium, barium or $^{13}C/^{12}C$ measurements on foraminifera.

The input of Antarctic Bottom Water (AABW) to the deep Atlantic also plays an important role in influencing the surface-to-deep ^{14}C difference in the Atlantic. This input should also affect the nutrient content of deep water in the Atlantic. In this model, AABW is simulated by loop 4. Table 7.4 gives results of the simulation using various AABW water fluxes combined with varying strengths of NADW. These results indicate that even a very large increase in AABW flux (eg, from 4 to 20 sverdrups) has created only a very small change on both the age and the nutrient content of deep water in the Atlantic.

Failure of this model to simulate the nutrient distribution in the Atlantic Ocean during glacial time suggests that other mechanisms ignored in this model must be taken into consideration. Hence, a new circulation pattern in the PANDORA model needs to be developed. One important element recently reported by Broecker and Peng (ms) is the surface circulation pattern. They suggest that the nutrient difference between deep Atlantic and deep Pacific waters is set by global conveyor that describes linkages between deep-water flow and surface-

TABLE 7.4. Sensitivity to Flux of Antarctic Bottom Water (*ie*, Loop 4) into the Deep Atlantic

Flux Loop 4	pCO_2 (μatm)	PO_4(μmol/kg) Deep Atlantic	PO_4(μmol/kg) Deep Pacific and Indian	Age (yrs) Deep Atlantic	Age (yrs) Deep Pacific and Indian
NADW = Holocene Standard (*ie*, 30 sverdrups)					
4	311	1.20	2.58	500	1240
8	311	1.26	2.55	540	1230
12	311	1.31	2.53	560	1225
16	311	1.34	2.52	580	1225
NADW = 0.6 Holocene Standard (*ie*, 18 sverdrups)					
4	302	1.23	2.60	770	1350
10	302	1.35	2.56	820	1340
20	301	1.45	2.52	880	1330
NADW = 0.4 Holocene Standard (*ie*, 12 sverdrups)					
4	296	1.31	2.60	1030	1420
10	295	1.44	2.55	1080	1420
20	295	1.53	2.51	1120	1410

water flow of global oceans. This global conveyor emphasizes that export of deep water from the deep Atlantic must be balanced by a return flow of upper waters. The magnitude of interocean differences in nutrient contents of deep waters depends on the fraction of the return water that comes *via* Aguelhas retroflection (poor in nutrients) as opposed to the fraction that enters as Antarctic surface, intermediate and bottom flow (rich in nutrients). Broecker and Peng's (ms) modeling results suggest that the important factor is the pattern of return flow and that the rate of exchange of water between the oceans is of secondary importance.

CONCLUSIONS

Distribution of ^{14}C in the ocean is a crucial property for characterizing the rate of deep-sea ventilation. The recent development of an AMS ion counting technique has made possible the extension of ^{14}C distribution from the present-day ocean into the glacial ocean by dating the planktonic and benthic foraminifera in the deep-sea sediments. Although many pitfalls from this approach have been discovered, a sampling strategy that minimizes the ^{14}C age bias caused by effects of bioturbation-abundance, bioturbation-dissolution and other potential factors has been designed. The strategy for such studies includes, for example, selecting regions of the sea floor where dissolution effects are minimal and sedimentation rates are high, and selecting planktonic species with shells that are resistant to breakage. Recent results of ^{14}C measurements taken with these precautions show that change in the mean benthic-planktonic age difference in the tropical Pacific was small, 1670 years during the peak glacial period as compared to 1600 years today. The age of Atlantic deep water was twice as great during glacial time (675 yrs) as it is today (350 yrs). Reconstruction of ^{14}C distribution in the glacial ocean indicates that the $^{14}C/^{12}C$ ratio in the tropical surface ocean was 93% that in the atmosphere. This value was 86% for the deep Atlantic and 76% for the deep Pacific. Although reconstruction of these ^{14}C ratios can be simulated by reducing the strength of glacial NADW in the PANDORA model, the reconstructed nutrient distribution cannot be explained. Taken together with nutrient distribution in the glacial ocean, model results suggest that the dependence on the rate of exchange of water between the oceans is second order, the important factor being the pattern of return flow in the global conveyor.

ACKNOWLEDGMENTS

Research sponsored by the Carbon Dioxide Research Program, Atmospheric and Climate Research Division, Office of Health and Environmental Research, US Department of Energy, under contract DE-AC05-84OR21400 with Martin Marietta Energy Systems, Inc Publication No. 3511, Environmental Sciences Division, ORNL.

REFERENCES

Andrée, M, Oeschger, H, Broecker, WS, Beavan, N, Klas, M, Mix, A, Bonani, G, Hofmann, HJ, Suter, M, Wölfli, W and Peng, T-H 1986 Limits on the ventilation rate for the deep ocean over the last 12,000 years. *Climate Dynamics* 1: 53–62.

Bard, E, Hamelin, B, Fairbanks, R and Zindler, A 1990 Calibration of the ^{14}C timescale over the past 30,000 years using mass spectrometric U-Th ages from Barbados corals. *Nature* 345: 405–409.

Barnola, JM, Raynaud, D, Korotkevich, YS and Lorius, C 1987 Vostok ice core provides 160,000-year record of atmospheric CO_2. *Nature* 329: 408–414.

Bien, GS, Rakestraw, NW and Suess, HE 1960 Radiocarbon concentration in Pacific Ocean water. *Tellus* 12: 436–443.

Boyle, EA 1988 Cadmium chemical tracer of deep water paleoceanography. *Paleoceanography* 3: 471–489.

Boyle, EA and Keigwin, L 1987 North Atlantic thermohaline circulation during the past 20,000 years linked to high-latitude surface temperature. *Nature* 330: 35–40.

Broecker, WS, Andrée, M, Bonani, G, Wölfli, W, Oeschger, H, Klas, M, Mix, A and Curry, W 1988a Preliminary estimates for the radiocarbon age of deep water in the glacial ocean. *Paleoceanography* 3: 659–669.

Broecker, WS and Denton, GH 1989 The role of ocean-atmosphere reorganizations in glacial cycles. *Geochimica et Cosmochimca Acta* 53: 2465–2501.

Broecker, WS, Gerard, R, Ewing, M and Heezen, BC 1960 Natural radiocarbon in the Atlantic Ocean. *Journal of Geophysical Research* 65: 2903–2931.

Broecker, WS, Klas, M, Beavan, N, Mathieu, G, Mix, A, Andrée, M, Oeschger, H, Wölfli, W, Suter, M, Bonani, G, Hofmann, FJ, Nessi, M and Morenzoni, E 1988b Accelerator mass spectrometry radiocarbon measurements on marine carbonate samples from deep sea cores and sediment traps. *Radiocarbon* 30(3): 261–295.

Broecker, WS, Mix, A, Andrée, M and Oeschger, H 1984 Radiocarbon measurements on coexisting benthic and planktonic foraminifera shells: Potential for reconstructing ocean ventilation times over the past 20,000 years. *Nuclear Instruments and Methods* B5: 331–339.

Broecker, WS and Peng, T-H 1982 *Tracers in the Sea*. Palisades, New York, Eldigio Press: 690 p.

_____1985 Carbon cycle: 1985: Glacial to interglacial changes in the operation of the global carbon cycle. *In* Stuiver, M and Kra, RS, eds, Proceedings of the 12th International ^{14}C Conference. *Radiocarbon* 28(2A): 309–327.

_____1987 The role of $CaCO_3$ compensation in the glacial to interglacial atmospheric CO_2 change. *Global Biogeochemical Cycle* 1: 15–391.

_____(ms) Factors controlling the distribution of phosphate in the deep ocean. Unpublished manuscript.

Broecker, WS, Peng, T-H, Trumbore, S, Bonani, G and Wölfli, W 1990 The distribution of radiocarbon in the glacial ocean. *Global Biogeochemical Cycle* 4: 103–117.

Broecker, WS, Peteet, D and Rind, D 1985 Does the ocean-atmosphere system have more than one stable mode of operation? *Nature* 315: 21–25.

Damon, PE, Lerman, JC and Long, A 1978 Temporal fluctuations of atmospheric ^{14}C: causal factors and implications. *Annual Review of Earth and Planetary Sciences* 6: 457–494.

Keeling, CD, Bacastow, RB, Carter, AF, Piper, SC, Whorf, TP, Heimann, M, Mook, WG and Roeloffzen, H 1989 A three-dimensional model of atmospheric CO_2 transport based on observed winds: 1. Analysis of observational data. AGU, *Geophysical Monograph* 55: 165–231.

Neftel, A, Oeschger, H, Staffelbach, T and Stauffer, B 1988 CO_2 record in the Byrd ice core 50,000–5,000 years BP. *Nature* 331: 609–611.

Östlund, HG and Stuiver, M 1980 GEOSECS Pacific radiocarbon. *Radiocarbon* 22(1): 25–53.

Peng, T-H 1989 Changes in ocean ventilation rates over the last 7000 years based on ^{14}C variations in the atmosphere and oceans. *In* Long, A and Kra, RS, eds, Proceedings of the 13th International ^{14}C Conference. *Radiocarbon* 31(3): 481–492.

Peng, T-H and Broecker, WS 1984 The impacts of bioturbation on the age difference between benthic and planktonic foraminifera in deep sea sediments. *Nuclear Instruments and Methods* B5: 346–352.

Shackleton, NJ, Duplessy, J-C, Arnold, M, Maurice, P, Hall, MA and Cartlidge, J 1988 Radiocarbon age of last glacial Pacific deep water. *Nature* 335: 708–711.

Stuiver, M and Östlund, HG 1980 GEOSECS Atlantic radiocarbon. *Radiocarbon* 22(1): 1–24.

_____1983 GEOSECS Indian Ocean and Mediterranean radiocarbon. *Radiocarbon* 25(1): 1–29.

8

RADIOCARBON FLUCTUATIONS AND THE GEOMAGNETIC FIELD

ROBERT S STERNBERG

INTRODUCTION

Knowledge of the long-term secular variations of the global geomagnetic field strength and atmospheric radiocarbon activity have proceeded in tandem over the past 40 years. The simultaneous pursuit of these fields of research was an important factor in the recognition of long-term radiocarbon fluctuations because a plausible physical mechanism was provided. This chapter will review both the paleomagnetic and radioisotope research that bears on the problem of geomagnetic modulation of radiocarbon fluctuations.

THE GEOMAGNETIC FIELD

The structure of the internal part of the geomagnetic field (see eg, Merrill & McElhinny 1983: 15–49) can be separated into the dipole moment (DM), which describes about 90% of the variation over the earth's surface, and the nondipole moment, which incorporates the rest. The field at any one point on the surface includes dipole and nondipole components, so the dipole field can only be accurately determined by properly averaging results from well-distributed points on the surface.

An external contribution to the Earth's magnetic field due to the solar wind accounts for small amplitude fluctuations in the field with periods from milliseconds to a few years. Measures of geomagnetic activity on this time scale, such as the aa indices, are correlated with other measures of heliomagnetic activity, such as sunspot numbers.

The source of the internal part of the field is a dynamo action in the liquid outer core, so that relatively rapid changes with periods of years to a few millennia can occur. These changes are known as secular variations (SV). Detailed global results on geomagnetic structure and SV are available for the past few decades, but data become sparse going back into the 19th century and earlier. A few longer historic records from single sites, such as the 400-year record from

London (Malin & Bullard 1981), provide important SV information. Historic data on absolute strength of the field are only available since 1832, when Gauss (1832) determined how field strength could be measured.

Analysis of historic secular variation reveals changes in direction and intensity of both the dipole and nondipole fields (Lund & Olson 1987; Courtillot & Le Mouël 1988). During the past few hundred years, the dipole field has, on the average, drifted westward at about 0.08° yr^{-1} and southward at about 0.01° yr^{-1}. The strength of the DM has decreased by 8% in a nearly linear fashion since the time of Gauss. The nondipole field also drifts predominantly westward over time, at a rate of about 0.18° yr^{-1}, or 2000 years for a complete revolution. Besides this drifting part of the nondipole field, stationary nondipole anomalies also fluctuate in amplitude.

Paleomagnetism, Paleointensity and Events

To examine the full spectrum of geomagnetic SV, a proxy indicator of prehistoric SV is required. This is important in understanding the geomagnetic dynamo (McFadden 1984; McFadden, Merrill & McElhinny 1985; Lund & Olson 1987), and in probing other properties of the mantle and core (Gubbins 1989). Paleomagnetism and radioisotopes produced by cosmic rays provide two different types of proxy records.

Titanomagnetite and titanohematite minerals in rocks, sediments and archaeologic features and artifacts are capable of carrying a stable remanent magnetization, thereby acting as tape recorders of the magnetic field. When these minerals are heated above their Curie temperatures, 580°C for magnetite and 680°C for hematite, all previous remanence is erased, and a total thermoremanent magnetization (TRM) is acquired. The direction of the TRM is parallel to the ambient field at the time of firing; this forms the basis for studies of SV of direction (Creer, Tucholka & Barton 1983; Sternberg 1989b) and the most commonly used types of archaeomagnetic dating (Wolfman 1984; Eighmy & Sternberg 1990).

Paleointensity Method

Determination of the paleointensity (PI) of the magnetic field would be straightforward under ideal circumstances. Theory and experiment indicate that TRM is linearly proportional to the strength of the ambient field (Stacey & Banerjee 1974: 103–120). The constant of proportionality, which we may call the TRM susceptibility, is a complex function of the ferromagnetic mineralogy, grain size and shape. However, if the ferromagnetic grains of the sample were stable, the PI could be simply determined by measuring the original remanence of the sample, then reheating it in a known laboratory field and measuring the newly acquired remanence. Unfortunately, the grains commonly undergo some alteration either *in situ* or during the laboratory heating itself, so that the TRM susceptibility will have changed between the time of the original magnetization of the sample and its final heating in the laboratory. Another potential problem is the acquisition by the sample of secondary components of magnetization subsequent to the acquisition of the primary TRM.

Thellier and Thellier (1959) (also Thellier 1977) devised an ingenious experiment to deal with these problems, whereby the sample is repeatedly heated to sequentially higher temperatures, and the TRM susceptibility is sought in each discrete temperature range. Only that temperature interval which is free of secondary components of magnetization and laboratory-induced alteration is used. When the remaining original thermoremanence is plotted against the thermoremanence acquired in the lab, the slope for the selected temperature interval multiplied by the intensity of the lab field gives the PI (Nagata, Arai & Momose 1963). Sample preselection criteria and reliability checks within the experiment (Fucugauchi 1979; Prévot *et al* 1985; Aitken *et al* 1988b; Salis, Bonhommet & Levi 1989) can be used to enhance the success rate of the Thellier-Thellier experiment. In recent years, accuracy has also been improved by adding correction factors for cooling-rate dependence of TRM (Fox & Aitken 1980) and anisotropy (Aitken *et al* 1981). Nonetheless, the traditional PI experiment is time consuming, and sample alteration is common enough so that the success rate is sometimes disappointing. The accuracy of conventional PI results has been questioned (Walton 1988a, b); this pitfall can be remedied by more careful studies (Aitken *et al* 1988a, b).

Variants of the Thellier-Thellier PI method have been used with small samples and automated magnetometers containing built-in furnaces to reduce the time involved for the experiment (Walton 1977; Aitken *et al* 1986). This approach is advantageous because samples with various properties can be rapidly assessed for their suitability for PI determination. The required instrumentation is not, however, commercially available. Other PI methods have been used (Thomas 1983), most notably the Shaw (1974) method, where the original remanence and lab TRM are demagnetized using an alternating magnetic field, and demagnetization spectra of laboratory-induced anhysteretic remanence are used to assess alteration. Results of the two methods compare favorably (Kono 1978; Senanayake, McElhinny & McFadden 1982; Dunlop, Reid & Hyodo 1987; Aitken *et al* 1988b). Using standard equipment, the Shaw method is less time consuming than the Thellier-Thellier method. Possible disadvantages are the use of alternating-field demagnetization spectra for a thermoremanence, and the necessity of heating the sample to a high temperature before any results can be obtained.

Paleointensity Records

Regional archaeointensity results (Creer, Tucholka & Barton 1983) can be quite coherent when careful methods by a single laboratory are used (Aitken *et al* 1989). Data compilations are notoriously noisy, most likely due to a combination of dating errors and bad PI results. Some of these "bad" results might be screened out by more stringent selection criteria and reliability checks, but no single test or simple combination of tests is likely to reject all spurious results. Walton (1988b), for example, has used Mesoamerican archaeointensity data compiled by Sternberg (1983) as an example of a noisy data set. I believe that meaningful patterns can still be extracted from such data by smoothing techniques (Sternberg 1989a).

Regional PI curves include both the dipole and nondipole field. If the intensities are corrected by magnetic latitude or inclination for the dipole effect (Smith 1967; Barton, Merrill & Barbetti 1979), averages of contemporaneous PIs from different geographic areas should approximate the DM. Several such compilations have been made for the Holocene (Bucha & Neustupný 1967; Smith 1967; Cox 1968; Bucha 1970; Burlatskaya 1970, 1978; Barton, Merrill & Barbetti 1979; Champion 1980; McElhinny & Senanayake 1982; Burlatskaya 1985). These compilations suggest a single quasi-sinusoidal fluctuation during the last 10,000 years, having a maximum of about 40% above the present-day DM in the first couple of centuries BC and AD, a minimum of 80% of the present-day DM from 4000–5000 BC, and the present-day DM at 10,000 BC (Fig 8.1). The PI data are of uneven quality, and are biased towards Europe and western Asia, and to the past 3000 years (Barton, Merrill & Barbetti 1979). Nonetheless, the general quasi-sinusoidal pattern has remained rather robust to the different global analyses over the last 20 years. Shorter DM periodicities, on the order of centuries, have been suggested (Champion 1980; Burlatskaya *et al* 1970; Burlatskaya 1978, 1985) but are not convincing. On the other hand, Creer (1988) and Aitken *et al* (1989) have suggested that drifting components and oscillating standing components of the nondipole field are largely responsible for much of the observed archaeointensity patterns.

Paleointensity Prior to 12,000 BP

Cox (1968) proposed a statistical model for geomagnetic field reversals, whereby the DM would continuously oscillate in a sinusoidal fashion with a period of about 10,000 years, and random fluctuations of the nondipole field would occasionally generate a field greater than and opposite to the dipole field, trig-

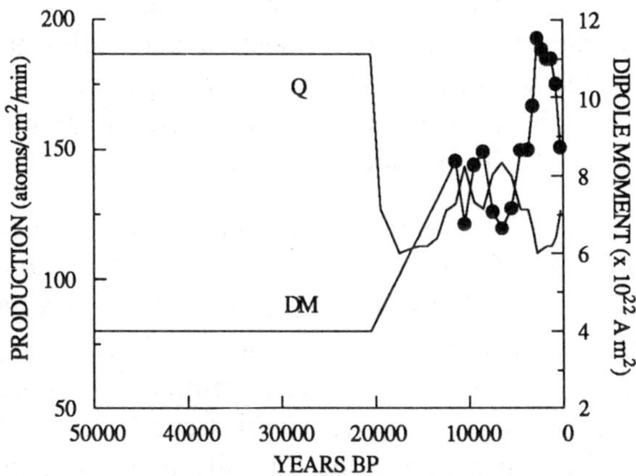

Fig 8.1. Dipole-moment model based on the 500- and 1000-year averages of McElhinny and Senanayake (1982), and a simple curve approximating the earlier results. Also shown is corresponding ^{14}C production using an inverse square-root relationship between DM and ^{14}C production, and calibrated with a production of 132 atoms/cm^2/min for a DM of 8×10^{22} A m^2.

gering a reversal. This model made the idea of a continuously oscillating sinusoidal dipole moment more appealing, but the data no longer support it.

The PI data from 12,000–50,000 years are sparse because of the increasing scarcity of suitable archaeological material and the difficulty in dating volcanic rocks of this age (Barbetti & Flude 1979; McElhinny & Senanayake 1982; Salis 1987). The PIs for this period are generally lower than for the last 12,000 years, with DM values generally hovering about 4×10^{22} A m^2, or about half of the average value for the last 12,000 years. Figure 8.1 shows one field model consistent with the current data. Anomalously high PIs were reported by Barbetti and McElhinny (1976) from aboriginal fireplaces by Lake Mungo, Australia, dating to 29,000 years, but have yet to be confirmed elsewhere. All the PI diagrams shown associated with these high PIs are concave upwards, which suggests the possibility of thermal alteration. Neither are the Lake Mungo results incompatible with a lightning-induced isothermal remanence in the baked clay (Barbetti & McElhinny 1976; Roperch, Bonhommet & Levi 1988). Both of these situations would imply that these PIs, calculated under the assumption of a stable thermoremanence, would be spurious. It is also possible that the high Lake Mungo PIs are real, but represent only a regional manifestation of the nondipole field (Coe 1977).

Sediment Paleointensities

Sediments also can acquire a depositional magnetization that provides a record of secular variation. These records complement thermoremanent records; they have the advantage of being quasi-continuous and potentially going back further in time, but the recording is generally not as accurate or precise as thermoremanent records. Similarity of archaeomagnetic and sedimentary records adds to their credibility because of the different recording processes involved. Such congruence has now been seen in directional records from several regions (Barton & Barbetti 1982; Thompson 1982; Verosub & Mehringer 1984).

Paleointensities derived from sediments are more problematic. Two methods have been used, each of which is capable of yielding only relative PIs. Replicating the process of magnetization is not as easily done as with thermoremanence, but artificial redeposition has been used (Tucker 1981; Thouveny 1987) to infer relative PIs from sediments. If the remanence in sediments is proportional to geomagnetic field strength (Kent 1973), relative PIs can also be derived by normalizing the intensity of remanence by a magnetic property that accounts for the variable concentration of magnetic minerals. Criteria have been developed to assess whether normalization by anhysteretic remanence (ARM) is valid (Levi & Banerjee 1976; King, Banerjee & Marvin 1983). These criteria have been used in several studies to obtain reasonable agreement between relative sedimentary PIs from marine and lacustrine sediments with archaeomagnetic or volcanic records from nearby areas (King, Banerjee & Marvin 1983; Constable 1985; Constable & Tauxe 1987; Thouveny 1987; Tauxe & Valet 1989).

Events

A final paleomagnetic topic relevant to radiocarbon geophysics is that of geomagnetic reversals and excursions. The polarity of the dipole field occasionally reverses, with an average reversal frequency of one per 0.14 Ma during the past 10 Ma (McFadden & Merrill 1984). The reversal process lasts several thousand years, and includes a reduction in the strength of the DM to perhaps 10% of normal (Prévot *et al* 1985). The last unequivocal reversal was the transition from the reversed Matuyama chron to the present normal Brunhes chron at 720,000 BP.

The existence of geomagnetic events, relatively short periods of polarity inversion, during the late Matuyama and Brunhes chrons is still problematic (Champion, Lanphere & Kuntz 1988). The one event within the reach of radiocarbon time that is most strongly supported by the extant data is the Laschamp event, identified in volcanic rocks of the Chaîne des Puys, France (Bonhommet & Babkine 1967). Roperch, Bonhommet and Levi (1988) found that PIs in the flows with anomalous directions were less than 15% of the present field. An important test of the validity of Laschamp as a global event will be whether it can be found in rocks of similar age from other areas of the world. Levi *et al* (1990) find that the K/Ar dates of Laschamp, 46.6 ± 2.4 ka (2 sigmas) and for the Skalamaelifell event identified in volcanic rocks of Iceland (Kristjánsson & Gudmundsson 1980), 42.9 ± 7.8 ka, were not statistically different. Low PIs, about 10% of present, have also been associated with the Skalamaelifell (Marshall, Chauvin & Bonhommet 1988; Levi *et al* 1990).

The case for the reality of Laschamp as either a regional or global event is not unambiguous. Heller and Petersen (1982) have suggested that the anomalous directions in the Laschamp flows can result from the phenomenon of self-reversal, although the rock magnetism is not inconsistent with a true reversed magnetization (Roperch, Bonhommet & Levi 1988). The Laschamp event has not been observed ubiquitously in rocks of this age, which could suggest it is a regional excursion due to secular variation (Verosub & Banerjee 1977) rather than a globally observed reversal (Harrison & Ramirez 1975). This fails to explain, however, why the Laschamp event was not seen in the nearby sedimentary record of Lac du Bouchet (Thouveny, Creer & Blunk 1990). Alternative explanations would include a real but short event/excursion that is not resolvable in the Lac du Bouchet record, self-reversal in Laschamp, or recording problems in Lac du Bouchet.

Radiocarbon Fluctuations and the Geomagnetic Field

Radiocarbon Fluctuations

It is now well accepted that the atmospheric concentration of ^{14}C varies systematically over time (Olsson 1970), necessitating the calibration of ^{14}C dates (Stuiver & Kra 1986). The tree-ring record for the past 10,000 years shows a variation of about 120‰, with a dominant long-term periodicity of 10,000–12,000 years and shorter-term periods of about 2000 and 200 years (Houtermans 1971; Suess 1980; Neftel, Oeschger & Suess 1981; Sonett 1984; Damon, Cheng & Linick 1989). Somewhat more problematic is the variation before the beginning of the tree-ring record (Stuiver 1970; Stuiver 1978; Barbetti 1980). Perhaps the best record is that based on ^{14}C and mass spectrometric U/Th dates on corals from Barbados (Bard *et al* 1990). Discrepancies of as much as 3500 years in this record are consistent with a monotonic decrease of atmospheric ^{14}C activity from about 400‰ at 20,000 BP to 100‰ at 10,000 BP.

In general, possible causes of fluctuations in atmospheric ^{14}C are variations in the production of ^{14}C in the atmosphere, or changes in the carbon/radiocarbon geochemical system (Olsson 1970; Damon, Lerman & Long 1978; Stuiver *et al* 1991). Radiocarbon is produced when neutrons produced by cosmic rays interact with ^{14}N in the atmosphere; thus, it is affected by the galactic primary cosmic-ray flux, modulation of this flux by the heliomagnetic field and by the geomagnetic field, or by production due to solar cosmic rays. Systemic changes include changes in geochemical reservoir sizes or exchange rates between them, or in the amount of ^{14}C in the reservoirs or system as a whole.

Modeling

To quantify the results of any of these changes, geochemical box models have been used (Arnold & Anderson 1957; Craig 1957; Revelle & Suess 1957). Electrical circuits can be used as an analogue of the coupled linear differential equations and first-order exchange processes commonly used to model this system (de Vries 1958; Houtermans, Suess & Oeschger 1973). Computer software packages designed to analyze electrical circuits have since been used for modeling ^{14}C problems (Damon, Sternberg & Radnell 1983). The geochemical system has been treated as a linear system with first-order exchange between well-mixed reservoirs in box models of varying complexity (Houtermans 1966; Ramaty 1967; Houtermans, Suess & Oeschger 1973; Lazear, Damon & Sternberg 1980), such as the three-box model shown in Figure 8.2, and with a

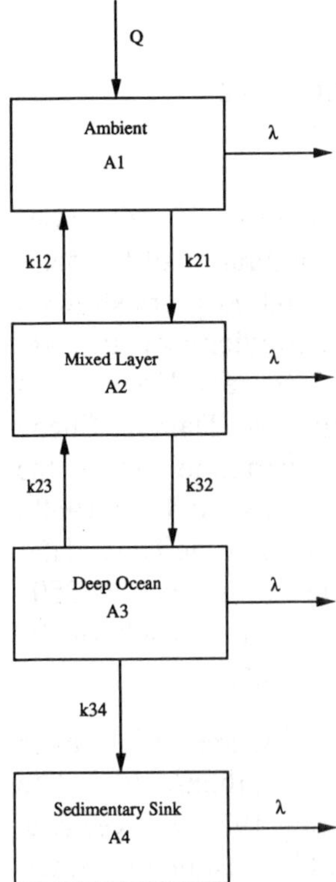

Fig 8.2. A three-box exchange model with sedimentary sink. The ambient reservoir would consist of the atmosphere and rapidly exchanging biosphere. Q represents ^{14}C production, λ is ^{14}C decay, A is activity, and k_{ij} is the exchange constant reservoir i to j.

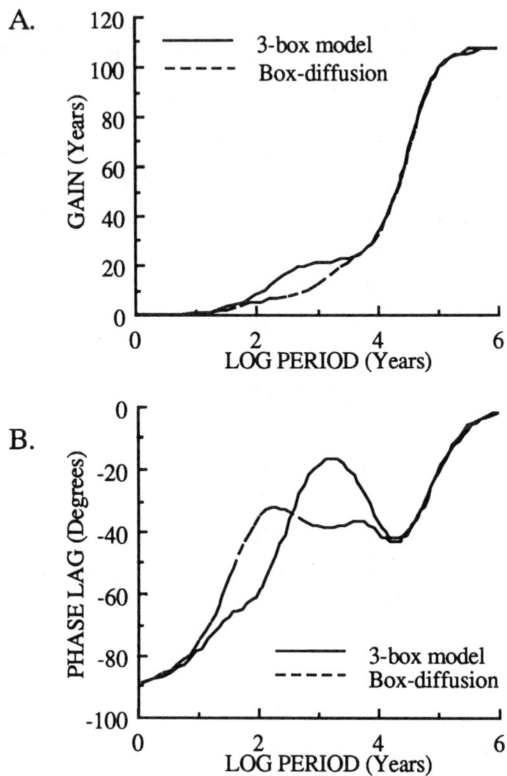

Fig 8.3. The system response of ambient reservoir for the three-box and box-diffusion models (after Damon, Sternberg & Radnell 1983). A. Gain *vs* log of the period. B. Phase lag *vs* log of the period.

box-diffusion model to model carbon transport via diffusion to the deep oceans (Oeschger *et al* 1975). The system acts as a low-pass filter, such that changes in ^{14}C production are attenuated to a greater extent for higher frequency variations (Fig 8.3). The complexity of an appropriate model thus depends on the problem being examined and the frequency content of the relevant fluctuations. A 2- or 3-box model is generally adequate for the lower frequencies of interest in geomagnetic modulation (Fig 8.3).

Production

An important component of the modeling process is the production function used to introduce ^{14}C into the system. The shorter-term ^{14}C fluctuations are usually attributed to heliomagnetic modulation of the primary galactic-ray flux (see Damon, Cheng & Linick 1989; Stuiver *et al* 1991). The longer-term fluctuations are attributed to geomagnetic modulation – the charged cosmic rays that create ^{14}C are shielded to a greater or lesser degree as the dipole moment increases or decreases. The ^{14}C is oxidized, and global mixing of the resulting $^{14}CO_2$ is fast relative to the residence time in the atmosphere. Thus, long-term ^{14}C fluctuations act as global integrators of the magnetic field.

For DM variations within a factor of about 4 relative to the present field, the ^{14}C production varies inversely with the square root of DM variations (Elsasser, Ney & Winckler 1956; Wada & Inoue 1966; Ramaty 1967; Lingenfelter & Ramaty 1970; O'Brien 1979; Blinov 1988; Lal 1988b). This relationship is calibrated using the current value of the DM, 8.065 x 10^{22} A m^2 in 1950 (McDonald & Gunst 1968), and the current value for the ^{14}C production. This value is more problematic; recent evaluations vary by 35% (Damon 1988; Damon & Sternberg 1989). Because of the inverse square-root relationship, production variations are asymmetric relative to DM changes (Sternberg & Damon 1979). Figure 8.1 shows a model for dipole moment variation over the last 50,000 years based on the compilation of McElhinny & Senanayake (1982), and also shows my calculation of the corresponding production function, assuming that Lingenfelter and Ramaty's (1970) value of 2.2 atoms/cm^2/sec averaged over three recent solar cycles corresponds to the 1950 DM.

Inventory

The ^{14}C inventory, herein identified as the total amount of ^{14}C in all geochemical reservoirs, depends only on the ^{14}C production and not on the details of the reservoir partitioning. It is simply equal to the time integral of the production history, weighted by the radioactive decay (Elsasser, Ney & Winckler 1956; Houtermans 1966; Libby 1967; Ramaty 1967; Lingenfelter & Ramaty 1970; Sternberg & Damon 1979). For steady-state production, the inventory will be equal to the production rate; the fact that this does not seem to be true for the present epoch is evidence for production rate variation in the past. Nonetheless, with his usual insight, Libby (1967) commented that the close agreement between present production rates and the inventory suggests that ^{14}C fluctuations have not been extreme in the past few thousand years. This agreed with the conclusion drawn from calibration studies. Ramaty (1967) and Lingenfelter and Ramaty (1970) used the inventory somewhat differently by modeling atmospheric ^{14}C fluctuations in terms of a parameter R_0 representing the ratio of present-day production to inventory. The fluctuations can be matched if R_0 varies by about ± 25%, suggesting a corresponding variation in geomagnetically modulated ^{14}C production. Sternberg and Damon (1979) used the inventory as a constraint on sinusoidal dipole moment models. This forced the conclusion that if such a model were appropriate, the peak field should occur several hundred years BC, earlier than was then being suggested by the PI data. Although the idea of a sinusoidal dipole moment has largely been dropped, it is noteworthy that subsequent PI compilations (Champion 1980; McElhinny &

Senanayake 1982) have shown a broad DM peak occurring during the first few centuries BC.

Figure 8.4 shows the same dipole moment function as in Figure 8.1, with my calculation for the present-day inventory, integrating back from the present to an asymptotic value of 126 dpm/cm^2. Of course, a limitation in using the inventory as a constraint on ^{14}C production is the uncertainty in any assay of the present inventory. An example of such an assay is given in Damon, Lerman & Long (1978). One of the largest uncertainties is the estimate of the amount of ^{14}C in the sedimentary sink (Lingenfelter & Ramaty 1970; Damon 1988; Damon & Sternberg 1989). As estimates of the amount of ^{14}C in sediments have increased, so have inventory values, from 119 dpm/cm^2 (Damon, Lerman & Long 1978) to 135 dpm/cm^2 (Damon 1988), to 143 ± 8 dpm/cm^2 (Damon & Sternberg 1989). The assay and modeled values are not generally inconsistent, although the latest assays are high relative to the value above. The modeled inventory value can be increased by about 5 dpm/cm^2 if the dipole moment drops more suddenly to a lower value at 12,000 years. A dipole moment low accompanying proposed excursions would not have a great effect on the present inventory. For a "Laschamp" event with a DM equal to 0.8 x 10^{22} A m^2 (10% of present) lasting from 40,000–50,000 years, the inventory increases by only 1 dpm/cm^2 (Sternberg & Damon 1991).

Long-Term Fluctuations

Several studies have investigated the effect of geomagnetically modulated production on atmospheric ^{14}C activity fluctuations. Most of these have considered

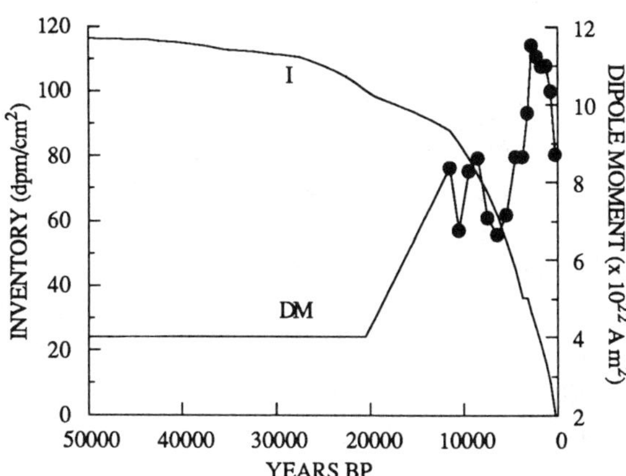

Fig 8.4. Integration back in time of the total ^{14}C inventory using the production function of Fig 8.1

the problem in terms of forward modeling, where a DM model has been used to calculate the corresponding ^{14}C fluctuations. Sinusoidal inputs to the models permit analytical solutions to be obtained for the corresponding set of linear differential equations (Ramaty 1967; Bucha 1970; Damon 1970; Lal & Venkatavaradan 1970; Ralph 1972; Yang & Fairhall 1972; Sternberg & Damon 1979), but numerical solutions to irregular forcing functions have also been used (Kigoshi & Hasegawa 1966; Grey 1969; Siegenthaler & Beer 1988; Tauxe & Valet 1989; Bard *et al* 1990; Lal 1989; Stuiver *et al* 1991). Several of these studies incorporated more detailed sensitivity analyses to account for uncertainties in the geochemical system and/or the DM forcing function (Kigoshi & Hasegawa 1966; Sternberg & Damon 1979).

Using DMs that continued to oscillate sinusoidally for many cycles into the past at least circumvented the problem of initial conditions. If the calculation is started as long ago as 50,000 years, using for example the pre-12,000 BP data set of McElhinny and Senanayake (1982), then the initial conditions are also unimportant. However, if the DM calculation is begun at 12,000 BP, the initial conditions chosen for atmospheric activity will affect the resulting activity record. Of course, these two approaches give the same results if the appropriate initial conditions are chosen (Stuiver *et al* 1991). The high initial conditions required to match the data (Siegenthaler & Beer 1988; Stuiver *et al* 1991) are consistent with the low DM pattern prior to 12,000 years (McElhinny & Senanayake 1982).

Selection of a particular model and production function will have a corresponding effect on the modeled results. Similar results can be generated if a change in the production function is counteracted by a change in the model. Sedimentation, which has not been explicitly included in some models, can be significant in this respect. For example, if the ^{14}C removed by sedimentation is not accounted for in an inventory assay, a correspondingly lower production rate must be used, whereas higher production is needed to explicitly supply this ^{14}C (Libby 1967; Lingenfelter & Ramaty 1970; Damon 1988; Damon & Sternberg 1989). Addition of a sedimentary sink also lowers the DC gain of the system, so that a higher production rate is required to generate correct activities (Damon, Sternberg & Radnell 1983). The higher DC gain without the sink will also mean smaller relative fluctuations at higher frequencies (Houtermans, Suess & Oeschger 1973), which can explain why models without the sink do not reveal the 11-year heliomagnetic cycle (Damon, Sternberg & Radnell 1983).

Most investigators see the agreement between the ^{14}C data and the modeled fluctuations as close enough to confirm the predominant role of the geomagnetic field in controlling long-term ^{14}C changes, at least during the last 10,000 years.

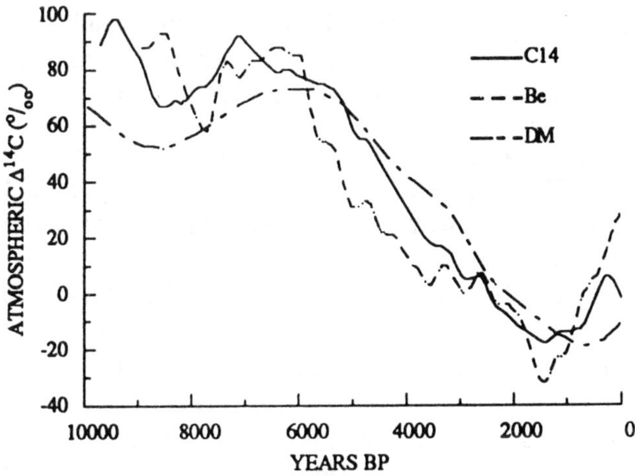

Fig 8.5. Long-term atmospheric ^{14}C fluctuations, equivalent fluctuations derived from ^{10}Be data, and ^{14}C fluctuations modeled from the dipole moment data of McElhinny and Senanayake (1982) with a box-diffusion model (after Stuiver *et al* 1991)

The good agreement between the long-term ^{14}C fluctuation trend and the modeled results of Stuiver *et al* (1991) are shown in Figure 8.5. The effects of climatic changes during the Holocene on the long-term record are generally thought to be secondary (Damon 1970; Stuiver *et al* 1991), and heliomagnetic modulation of production is generally considered to be significant only for shorter periods of less than 200 years. Lal (1985) has shown that the long-term ^{14}C fluctuations could be caused by prescribed climatic effects on reservoir sizes or exchange rates. He suggests that the quality and geographic distribution of the archaeointensity data are insufficient to conclude confidently that DM variations are the cause of the long-term ^{14}C fluctuations. This is in accord with the paleomagnetic conclusions of Creer (1988) and Aitken *et al* (1989). However, modeling by Siegenthaler, Heimann & Oeschger (1980) and comparative ^{14}C dating of benthic and planktonic foraminifera (Andrée *et al* 1986) suggest that climatic effects on the ^{14}C record have not been great during the Holocene. A climatic effect during the Pleistocene, or during the transition from Pleistocene to Holocene, appears more plausible (Siegenthaler, Heimann & Oeschger 1980; Keir 1983). Of course, although certain causes of variation may predominate during certain time periods or in certain frequency bands, there will always be multiple mechanisms at work simultaneously, and the real task is to deconvolve them (Lal 1985, 1989), that is, to separate their various effects.

The geomagnetic modulation of the ^{14}C record has been difficult to analyze for times prior to 12,000 years because of the paucity of available PI and calibrated ^{14}C data. Using the older PI data in the McElhinny and Senanayake (1982) data set, Bard *et al* (1990) have now shown that the resulting modeled ^{14}C fluctuations back to 30,000 years agree within errors to their ^{14}C results calibrated against U/Th dates on the same coral samples.

Another interesting question is whether a recent reversal would be evident in the ^{14}C record. Grey (1971) examined this problem, and estimated that the resulting increase in atmospheric activity relative to a sinusoidal dipole moment would be approximately 100‰, clearly enough to be seen if calibrated data existed for the period of the reversal. Damon and Sternberg (1978) calculated that if the DM were reduced to about 25% of its recent value for a duration of 2000 years, an activity spike of 200‰ would be generated; a duration of 5000 years would cause a spike of about 500‰. If such a spike had occurred at 45,000 BP, the date suggested for the Laschamp event, it would have essentially decayed away by the beginning of the tree-ring record. The predicted trend would be consistent with the decreasing activity trend observed in the coral record of Bard *et al* (1990). Sternberg and Damon (1991) carried out additional modeling showing the effects of a low DM before 12,000 years and geomagnetic excursions during the Pleistocene.

Radiocarbon Dipole Moment

The ^{14}C production function could be reconstructed in the frequency domain by operating on the Fourier transform of the output activities with the system transfer function (Houtermans 1966; Houtermans, Suess & Oeschger 1973). Stuiver and Quay (1980) calculated the production function and sunspot numbers in the time domain by writing the box model differential equations as finite difference equations, and working the output back through the model. The longer-term geomagnetic effect was removed by detrending the ^{14}C data. Barton, Merrill and Barbetti (1979) directly calculated production and DM values from ^{14}C activities, calling their result the radiocarbon dipole moment (RCDM). Sternberg and Damon (1981) followed the procedure of Stuiver and Quay (1980) to calculate the RCDM, expressing their model with finite difference calculations, then using the ^{14}C activities to calculate production and dipole moment values. Because the system acts as a low-pass filter, higher frequency heliomagnetic fluctuations and noise are greatly amplified by this inversion procedure, so Sternberg and Damon (1983) applied a low-pass filter to these results. Although the longest term (10,000 years) and shortest term (<200 years) ^{14}C fluctuations are generally attributed to geomagnetic and heliomagnetic modulation, respectively, the source of intermediate-term fluctuations (~2000 years) is still unclear. Design of the low-pass filter was somewhat arbitrary, but it was assumed that periods shorter than about 200 years should be filtered out of the RCDM. The RCDM values agreed fairly well with the paleomagnetic compilations back to about 7000 BP, before which the RCDM

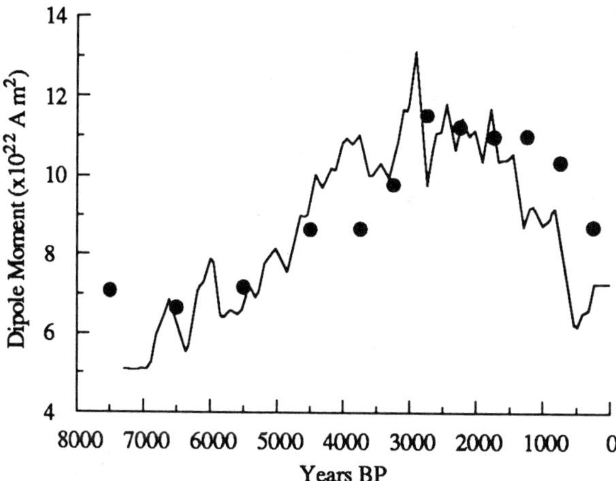

Fig 8.6. The radiocarbon dipole moment curve calculated from ^{14}C data (after Sternberg & Damon 1983). Paleomagnetic averages (●) are from McElhinny and Senanayake (1982).

values become lower than the paleomagnetic results (Fig 8.6). This discrepancy is in the same sense as that shown by Siegenthaler and Beer (1988) and Stuiver *et al* (1991). Although procedures were not detailed, Veksler (1988) also computed RCDMs.

^{10}Be

Both the paleomagnetic and ^{14}C data independently suggest long-term fluctuations of the DM. A third source of proxy data comes from another radioisotope, ^{10}Be (Raisbeck & Yiou 1984, 1988; Oeschger, Beer & Andrée 1987; Oeschger & Beer 1990). ^{10}Be is also created in the atmosphere by cosmic-ray activity, but it is precipitated as an aerosol after an atmospheric residence time of 1 or 2 years, which is shorter than the approximate ten-year residence time of gaseous $^{14}CO_2$. Thus, attenuation of production changes are much less for ^{10}Be. On the other hand, the shorter residence time of ^{10}Be probably means that it is not well mixed in the atmosphere, so the concentration is more related to the local vertically integrated production rate than to the global rate. Accumulation of ^{10}Be in sediments is complicated by remobilization and deposition patterns. The records in polar ice may be less complicated by these factors, but precipitation will still be greatly affected by climate. *In situ* production of ^{10}Be in tree rings offers another possible proxy record of DM variations (Lal, Arnold & Nishiizumi 1985; Lal 1988a).

Beer *et al* (1984, 1988) and Beer, Siegenthaler and Blinov (1988) reported that the ^{10}Be concentration record in the Camp Century, Greenland ice core does not clearly reflect the suggested long-term DM and ^{14}C changes, although it does reflect short-term heliomagnetic modulation. However, I think the case can be made that the correlation between long-term DM, ^{14}C and ^{10}Be records is reasonable. Figure 8.5 shows a ^{14}C fluctuation record derived by Stuiver *et al* (1991) by converting ^{10}Be data to an equivalent ^{14}C production record and feeding it into a box-diffusion model (Beer *et al* 1988); this record clearly correlates with the tree-ring-derived ^{14}C record. The initial conditions necessary to produce a good correlation (Beer *et al* 1988; Siegenthaler & Beer 1988) are consistent with the lower DM prior to 12,000 years. The analysis is complicated by the fact that geomagnetic modulation of production is greatest at the equator, and least at the poles where the ice-core data exist. Brown *et al* (1987) analyzed the ^{10}Be record for 1000–10,000 BP in mid-latitude lake sediments where geomagnetic modulation would be greater, but the picture was greatly clouded by uncertainties over ^{10}Be transport. Depending on the transport model adopted, the inferred production rate could be either consistent or inconsistent with geomagnetic modulation.

The longer half-life of ^{10}Be, 1.5 Ma, also allows it to see further into the past than ^{14}C. Raisbeck *et al* (1985) detected a ^{10}Be spike at the time of the Brunhes-Matuyama magnetic reversal at 0.7 Ma, presumably due to a decrease in the DM. Raisbeck *et al* (1985) have also noted two ^{10}Be peaks at about 35,000 and 60,000 years in the Dome C, Antarctic ice core. They again suggest the possible cause of geomagnetic reversals, several of which have been suggested for these general dates. Sonett, Morfill and Jokipii (1987) alternatively suggest that these spikes could have been caused by cosmic-ray bursts due to supernova explosions.

Finally, Kocharov *et al* (1989) suggest that the ^{10}Be record can be inverted to obtain an equivalent DM function, similar to the procedure for inferring the RCDM. This incorporates the assumption that long-term ^{10}Be fluctuations are largely due to geomagnetic modulation.

RESEARCH FOR THE FUTURE

Several areas of research are suggested by the conclusions that one can (or cannot) draw concerning the issues raised in this paper. I would recommend the following:

Paleomagnetism

1. Compilation of a publicly available and easily useful archaeomagnetic data base (Sternberg 1988)
2. A critical analysis of the extant archaeointensity data

3. New, more reliable archaeointensities, including better geographic coverage and more results prior to 5000 BP
4. Additional reliable paleointensity records derived from sediments
5. Further work on the existence, geographic extent and paleointensities of Quaternary reversals or events.

^{14}C and ^{10}Be

1. Extension of the tree-ring calibration record
2. Compilation of other, reliable calibration records (eg, varves), especially extending into the Pleistocene
3. Calibration records from non-ambient reservoirs (eg, corals)
4. Additional ^{10}Be records, including low-latitude glacial and sedimentary records, with correlative paleomagnetic studies, and further study of the ^{10}Be geochemical system
5. Acquisition of necessary data, and modeling to deconvolve the various contributions to radioisotope fluctuations.

ACKNOWLEDGMENTS

This work was done during a sabbatical leave supported by Franklin and Marshall College, and NSF grant EAR-8721466. I would like to thank the organizers and sponsors of the meeting that resulted in the publication of this volume. Support services were kindly provided by the University of Konstanz, Federal Republic of Germany and the Swiss Technological Institute, Zurich, Switzerland. I appreciate helpful discussions with Paul Damon, Friedrich Heller and Jurg Beer. Minze Stuiver shared a very useful preprint. Comments of Edouard Bard, one anonymous reviewer and the editors are appreciated.

References

Aitken, MJ, Alcock, PA, Bussell, GD and Shaw, CJ 1981 Archaeomagnetic determination of the past geomagnetic intensity using ancient ceramics: Allowance for anisotropy. *Archaeometry* 23(1): 53–64.

Aitken, MJ, Allsop, AL, Bussell, GD and Winter, M 1986 Paleointensity determination using the Thellier technique: Reliability criteria. *Journal of Geomagnetism and Geoelectricity* 38: 1353–1363.

_____1988a Comment on the "Lack of reproducibility in experimentally determined intensities of the earth's magnetic field" by D Walton. *Reviews of Geophysics* 26(1): 23–25.

_____1988b Determination of the intensity of the earth's magnetic field during archaeological times: Reliability of the Thellier technique. *Reviews of Geophysics* 26(1): 3–12.

_____1989 Geomagnetic intensity variation during the last 4000 years. *Physics of the Earth and Planetary Interiors* 56: 49–58.

Andrée, M, Oeschger, H, Broeker, W, Beavan, N, Klas, M, Mix, A, Bonani, G, Hofmann, HJ, Suter, M, Woelfli, W and Peng, T-H 1986 Limits on the ventilation rate for the deep ocean over the last 12,000 years. *Climate Dynamics* 1: 53–62.

Arnold, JR and Anderson, EC 1957 The distribution of carbon-14 in nature. *Tellus* 9(1): 28–32.

Barbetti, M 1980 Geomagnetic strength over the last 50,000 years and changes in atmospheric ^{14}C concentration: Emerging trends. *In* Stuiver, M and Kra, RS, eds, Proceedings of the 10th International ^{14}C Conference. *Radiocarbon* 22(2): 192–199.

Barbetti, M and Flude, K 1979 Geomagnetic variation during the late Pleistocene period and changes in the radiocarbon time scale. *Nature* 279: 202–205.

Barbetti, MF and McElhinny, MW 1976 The Lake Mungo geomagnetic excursion. *Philosophical Transactions of the Royal Society of London* A181: 515–542.

Bard, E, Hamelin, B, Fairbanks, RG and Zindler, A 1990 Calibration of the ^{14}C timescale over the past 30,000 years using mass spectrometric U-Th ages from Barbados corals. *Nature* 345: 405–410.

Barton, CE and Barbetti, M 1982 Geomagnetic secular variation from recent lake sediments, ancient fireplaces and historical measurements in southeastern Australia. *Earth and Planetary Science Letters* 59: 375–387.

Barton, CE, Merrill, RT and Barbetti, M 1979 Intensity of the earth's magnetic field over the last 10,000 years. *Physics of the Earth and Planetary Interiors* 20: 96–110.

Beer, J, Andrée, M, Oeschger, H, Siegenthaler, U, Bonani, G, Hofmann, H, Morenzoni, E, Nessi, M, Suter, M, Woelfli, W, Finkel, R and Langway, C, Jr 1984 The Camp Century ^{10}Be record: Implications for long-term variations of the geomagnetic dipole moment. *Nuclear Instruments and Methods* B5: 380–384.

Beer, J, Siegenthaler, U and Blinov, A 1988 Temporal ^{10}Be variations in ice: Information on solar activity and geomagnetic field intensity. *In* Stephenson, FR and Wolfendale, AW, eds, *Secular, Solar and Geomagnetic Variations in the Last 10,000 Years*. Dordrecht, The Netherlands, Kluwer: 297–313.

Beer, J, Siegenthaler, U, Bonani, G, Finkel, RC, Oeschger, H, Suter, M and Woelfli, W 1988 Information on past solar activity and geomagnetism from ^{10}Be in the Camp Century ice core. *Nature* 331: 675–679.

Blinov, A 1988 The dependence of cosmogenic isotope production rate on solar activity and geomagnetic field variations. *In* Stephenson, FR and Wolfendale, AW, eds, *Secular, Solar and Geomagnetic Variations in the Last 10,000 Years*. Dordrecht, The Netherlands, Kluwer: 329– 340.

Bonhommet, N and Babkine, J 1967 Sur la présence d'aimantations inversées dans la Chaîne de Puys. *Comptes Rendus de l'Academie des Sciences* B264: 92–94.

Brown, TA, Nelson, DE, Southon, JR and

Vogel, JS 1987 ^{10}Be production rate variations as recorded in a mid-latitude lake sediment. *Nuclear Instruments and Methods* B29: 232–237.

Bucha, V 1970 Influence of the earth's magnetic field on radiocarbon dating. *In* Olsson, IU, ed, *Radiocarbon Variations and Absolute Chronology*. Proceedings of the 12th Nobel Symposium. Stockholm, Almqvist & Wiksell: 501–510.

Bucha, V and Neustupný, E 1967 Changes of the earth's magnetic field and radiocarbon dating. *Nature* 215: 261–263.

Burlatskaya, SP 1970 Change in geomagnetic field intensity in the last 8500 years according to global archeomagnetic data. *Geomagnetism and Aeronomy* (English translation) 10: 544–548.

_____1978 Characteristics of the spectrum of secular geomagnetic field variations in the past 8500 years. *Geomagnetism and Aeronomy* (English translation) 18: 621–623.

_____1985 Variations of the virtual dipole moment of the geomagnetic field. *Izvestiya, Earth Physics* (English translation) 21(2): 153–156.

Burlatskaya, SP, Nachasova, IE, Uechaeva, TB, Rusakov, OM, Zagniv, GF, Tarhov, EN and Tchelidze, ZA 1970 Archaeomagnetic research in the USSR: Recent results and spectral analysis. *Archaeometry* 12: 73–85.

Champion, DE 1980 Holocene geomagnetic secular variation in the western United States: Implications for the global geomagnetic field. *US Geological Survey Open-File Report* 80–824, Denver, Colorado.

Champion, DE, Lanphere, MA and Kuntz, MA 1988 Evidence for a new geomagnetic reversal from lava flows in Idaho: Discussion of short polarity reversals in the Brunhes and Matuyama polarity chrons. *Journal of Geophysical Research* 93(B10): 11,667–11,680.

Coe, RS 1977 Sources and models to account for Lake Mungo paleomagnetic excursion and their implications. *Nature* 269: 49–51.

Constable, CG 1985 Eastern Australian geomagnetic field intensity over the past 14,000 years. *Geophysical Journal of the Royal Astronomical Society* 81: 121–130.

Constable, CG and Tauxe, L 1987 Palaeointensity in the pelagic realm: marine sediment data compared with archaeomagnetic and lake sediment records. *Geophysical Journal of the Royal Astronomical Society* 90: 43–59.

Courtillot, V and Le Mouël, JL 1988 Time variations of the earth's magnetic field: From daily to secular. *Annual Reviews of Earth and Planetary Sciences* 16: 389–476.

Cox, A 1968 Lengths of geomagnetic polarity reversals. *Journal of Geophysical Research* 73: 3,247–3,260.

Craig, H 1957 The natural distribution of radiocarbon and the exchange time of carbon dioxide between atmosphere and sea. *Tellus* 9(1): 1–17.

Creer, KM 1988 Geomagnetic field and radiocarbon activity through Holocene time. *In* Stephenson, FR and Wolfendale, AW, eds, *Secular, Solar and Geomagnetic Variations in the Last 10,000 Years.* Dordrecht, The Netherlands, Kluwer: 381–397.

Creer, KM, Tucholka, P and Barton, CE 1983 *Geomagnetism of Baked Clays and Recent Sediments*. Amsterdam, Elsevier: 323 p.

Damon, PE 1970 Climatic versus magnetic perturbation of the atmospheric C14 reservoir. *In* Olsson, IU, ed, *Radiocarbon Variations and Absolute Chronology*. Proceedings of the 12th Nobel Symposium. Stockholm, Almqvist & Wiksell: 571–593.

_____1988 Production and decay of radiocarbon and its modulation by geomagnetic field-solar activity changes with possible implications for global environment. *In* Stephenson, FR and Wolfendale, AW, eds, *Secular, Solar and Geomagnetic Variations in the Last 10,000 Years*. Dordrecht, The Netherlands, Kluwer: 267– 283.

Damon, PE, Cheng, S and Linick, TW 1989 Fine and hyperfine structure in the spectrum of secular variations of atmospheric ^{14}C. *In* Long, A and Kra, RS, eds, Proceedings of the 13th International ^{14}C Conference. *Radiocarbon* 31(3): 704–718.

Damon, PE, Lerman, JC and Long, A 1978 Temporal fluctuations of atmospheric ^{14}C: Causal factors and implications. *Annual Reviews of Earth and Planetary Sciences* 6: 457–494.

Damon, PE and Sternberg, RS 1978 The use of radiocarbon as a boundary condition for geomagnetic field models (abstract). *EOS, Transactions of the American Geophysical Union* 59: 1057.

_____1989 Global production and decay of radiocarbon. *In* Long, A and Kra, RS, eds, Proceedings of the 13th International ^{14}C Conference. *Radiocarbon* 31(3): 697–703.

Damon, PE, Sternberg, RS and Radnell, CJ 1983 Modeling of atmospheric radiocarbon fluctuations for the past three centuries. *In* Stuiver, M and Kra, RS, eds, Proceedings of the 11th International ^{14}C Conference. *Radiocarbon* 25(2): 249–258.

Dunlop, DJ, Reid, B and Hyodo, H 1987 Alteration of the coercivity spectrum and paleointensity determination. *Geophysical Research Letters* 14(11): 1091–1094.

Eighmy, JL and Sternberg, RS 1990 *Archaeomagnetic Dating*. Tucson, University of Arizona Press: 446 p.

Elsasser, W, Ney, EP and Winckler, JR 1956 Cosmic-ray intensity and geomagnetism. *Nature* 178: 1226–1227.

Fox, JMW and Aitken, MJ 1980 Cooling-rate dependence of thermoremanent magnetisation. *Nature* 283(5746): 462–463.

Fucugauchi, JU 1979 Further reliability tests for determination of palaeointensities of the earth's magnetic field. *Geophysical Journal of the Royal Astronomical Society* 61: 243–251.

Gauss, CF 1832 Intensitas vis magneticae terrestris ad mensuram absolutam revocata. *Gauss'sche Werke*, Goettingen, Gesellschaft der Wissenschaften zu Goettingen.

Grey, DC 1969 Geophysical mechanisms for ^{14}C variations. *Journal of Geophysical Research* 74(26): 6333–6340.

Grey, DC 1971 ^{14}C data and the Laschamp reversed event. *Journal of Geomagnetism and Geoelectricity* 23(1): 123–127.

Gubbins, D 1989 Implications of geomagnetism for mantle structure. *Philosophical Transactions of the Royal Society of London* A328: 365–375.

Harrison, CGA and Ramirez, E 1975 Areal coverage of spurious reversals of the earth's magnetic field. *Journal of Geomagnetism and Geoelectricity* 27: 139–151.

Heller, F and Petersen, N 1982 The Laschamp excursion. *Philosophical Transactions of the Royal Society of London* A306(1492): 169–177.

Houtermans, J 1966 On the quantitative relationships between geophysical parameters and the natural ^{14}C inventory. *Zeitschrift für Physik* 193: 1–12.

_____(ms) 1971 Geophysical interpretation of bristlecone pine radiocarbon measurements using a method of Fourier analysis of unequally spaced data. PhD dissertation, University of Bern.

Houtermans, JC, Suess, HE and Oeschger, H 1973 Reservoir models and production rate variations of natural radiocarbon. *Journal of Geophysical Research* 78(12): 1897–1908.

Keir, RS 1983 Reduction of thermohaline circulation during deglaciation: the effect on atmospheric radiocarbon and CO_2. *Earth and Planetary Science Letters* 64: 445–456.

Kent, DV 1973 Post-depositional remanent magnetization in deep-sea sediment. *Nature* 246: 32–34.

Kigoshi, K and Hasegawa, K 1966 Secular variation of atmospheric radiocarbon concentration and its dependence on geomagnetism. *Journal of Geophysical Research* 71: 1065–1071.

King, JW, Banerjee, SK and Marvin, J 1983 A new rock magnetic approach to selecting sediments for geomagnetic paleointensity studies: Application to paleointensity for the last 4000 years. *Journal of Geophysical Research* 88: 5911–5921.

Kocharov, GE, Blinov, AV, Konstantinov, AN and Levchenko, VA 1989 Temporal ^{10}Be and ^{14}C variations: A tool for paleomagnetic research. *Radiocarbon* 31(2): 163–168.

Kono, M 1978 Reliability of palaeointensity methods using alternating field demagnetization and anhysteretic remanence. *Geophysical Journal of the Royal Astronomical Society* 54: 241–261.

Kristjánsson, L and Gudmundsson, A 1980 Geomagnetic excursion in late-glacial basalt outcrops in south-western Iceland. *Geophysical Research Letters* 7: 337–340.

Lal, D 1985 Carbon cycle variations during the past 50,000 years: Atmospheric $^{14}C/^{12}C$ ratio as an isotopic indicator. *In* Sundquist, ET and Broecker, WS, eds, *The Carbon Cycle and Atmospheric CO_2: Natural Variations Archaean to Present*. Washington, DC, American Geophysical Union: 221–233.

_____1988a In situ-produced cosmogenic isotopes in terrestrial rocks. *Annual Review of Earth and Planetary Sciences* 16: 355–388.

_____1988b Theoretically expected variations in the terrestrial cosmic-ray production rates of isotopes. *In* Castagnoli, GC, ed, *Solar-Terrestrial Relationships and the Earth Environment in the Last Millennia*. Amsterdam, Elsevier: 216–233.

_____(ms) 1989 The influence of climatic and oceanic changes on the atmospheric ^{14}C record. Paper presented at the 3rd International CO_2 Conference, Hinterzarten, Germany, October 16–20.

Lal, D, Arnold, JR and Nishiizumi, K 1985 Geophysical records of a tree: New application for studying geomagnetic field and solar activity changes during the past 10^4 years. *Meteoritics* 20(2): 403–414.

Lal, D and Venkatavaradan, VS 1970 Analysis of the causes of C14 variations in the atmosphere. *In* Olsson, IU, ed, *Radiocarbon Variations and Absolute Chronology*. Proceedings of the 12th Nobel Symposium. Stockholm, Almqvist & Wiksell: 549–567.

Lazear, G, Damon, PE and Sternberg, R 1980 The concept of DC gain in modeling secular variations in atmospheric ^{14}C. *In* Stuiver, M and Kra, R, eds, Proceedings of the 11th International ^{14}C Conference. *Radiocarbon* 22(2): 318–327.

Levi, S, Audunsson, H, Duncan, RA, Kristjánnson, L, Gillot, P-Y and Jakobsson, SP 1990 Late Pleistocene geomagnetic excursion in Icelandic lavas: Confirmation of the Laschamp excursion. *Earth and Planetary Science Letters* 96: 443–457.

Levi, S and Banerjee, SK 1976 On the possibility of obtaining relative paleointensities from lake sediments. *Earth and Planetary Science Letters* 29: 219–226.

Libby, WF 1967 Radiocarbon and paleomagnetism. *In* Hindmarsh, WR, Lowes, FJ, Roberts, PH and Runcorn, SK, eds, *Magnetism and the Cosmos*. New York, American Elsevier: 60–65.

Lingenfelter, RE and Ramaty, R 1970 Astrophysical and geophysical variations in C14 production. *In* Olsson, IU, ed, *Radiocarbon Variations and Absolute Chronology*. Proceedings of the 12th Nobel Symposium. Stockholm, Almqvist & Wiksell: 513–535.

Lund, SP and Olson, P 1987 Historic and paleomagnetic secular variation and the earth's core dynamo process. *Reviews of Geophysics* 25(5): 917–928.

Malin, SRC and Bullard, EB 1981 The direction of the earth's magnetic field at London, 1570–1975. *Philosophical Transactions of the Royal Society of London* A299(1450): 357–423.

Marshall, M, Chauvin, A and Bonhommet, N 1988 Preliminary paleointensity measurements and detailed magnetic analyses of basalts from the Skalamaelifell excursion, southwest Iceland. *Journal of Geophysical Research* 93(B10): 11,681–11,698.

McDonald, KL and Gunst, RH 1968 Recent trends in the earth's magnetic field. *Journal of Geophysical Research* 73: 2057–2067.

McElhinny, MW and Senanayake, WE 1982

Variations in the geomagnetic dipole 1: The past 50,000 years. *Journal of Geomagnetism and Geoelectricity* 34: 39–51.

McFadden, PL 1984 A time constant for the geodynamo? *Physics of the Earth and Planetary Interiors* 34: 117–125.

McFadden, PL and Merrill, RT 1984 Lower mantle convection and geomagnetism. *Journal of Geophysical Research* 89: 3354–3362.

McFadden, PL, Merrill, RT and McElhinny, MW 1985 Non-linear processes in the geodynamo: Paleomagnetic evidence. *Geophysical Journal of the Royal Astronomical Society* 83: 111–126.

Merrill, RT and McElhinny, MW 1983 *The Earth's Magnetic Field.* London, Academic Press: 401 p.

Nagata, T, Arai, Y and Momose, K 1963 Secular variation of the geomagnetic total force during the last 5000 years. *Journal of Geophysical Research* 68: 5277–5281.

Neftel, A, Oeschger, H and Suess HE 1981 Secular non-random variations of cosmogenic carbon-14 in the terrestrial atmosphere. *Earth and Planetary Science Letters* 56: 127–147.

O'Brien, K 1979 Secular variation in the production of cosmogenic isotopes in the earth's atmosphere. *Journal of Geophysical Research* 84(A2): 423–431.

Oeschger, H and Beer, J 1990 The past 5000 years of solar modulation of cosmic radiation from ^{10}Be and ^{14}C studies. *Philosophical Transactions of the Royal Astronomical Society* A330: 471–480.

Oeschger, H, Beer, J and Andrée, M 1987 ^{10}Be and ^{14}C in the earth system. *Philosophical Transactions of the Royal Society of London* A323: 45–56.

Oeschger, H, Siegenthaler, U, Schotterer, U and Gugelmann, A 1975 A box diffusion model to study the carbon dioxide exchange in nature. *Tellus* 27(2): 168–192.

Olsson, IU, ed 1970 *Radiocarbon Variations and Absolute Chronology.* Stockholm, Almqvist & Wiksell: 652 p.

Prévot, M, Mankinen, EA, Coe, RS and Grommé, CS 1985 The Steens Mountain (Oregon) geomagnetic polarity transition 2. Field intensity variations and discussion of reversal models. *Journal of Geophysical Research* 90(B12): 10,417–10,448.

Raisbeck, GM and Yiou, F 1984 Production of long-lived cosmogenic nuclei and their applications. *Nuclear Instruments and Methods* B5: 91–99.

_____ 1988 ^{10}Be as a proxy indicator of variations in solar activity and geomagnetic field intensity during the last 10,000 years. *In* Stephenson, FR and Wolfendale, AW, eds, *Secular, Solar and Geomagnetic Variations in the Last 10,000 Years.* Dordrecht, The Netherlands, Kluwer: 287–296.

Raisbeck, GM, Yiou, F, Bourles, D and Kent, DV 1985 Evidence for an increase in cosmogenic ^{10}Be during a geomagnetic reversal. *Nature* 315: 315–317.

Ralph, EK 1972 A cyclic solution for the relationship between magnetic and atmospheric C-14 changes. *In* Rafter, TA and Grant-Taylor, T, eds, *Proceedings of the 8th International ^{14}C Conference.* Wellington, Royal Society of New Zealand: 90–98.

Ramaty, R 1967 The influence of geomagnetic shielding on ^{14}C production and content. *In* Hindmarsh, WR, Lowes, FJ, Roberts, PH and Runcorn, SK, eds, *Magnetism and the Cosmos.* New York, American Elsevier: 66–78.

Revelle, R and Suess, HE 1957 Carbon dioxide exchange between atmosphere and ocean and the question of an increase of atmospheric CO_2 during the past decades. *Tellus* 9(1): 18–27.

Roperch, P, Bonhommet, N and Levi, S 1988 Paleointensity of the earth's magnetic field during the Laschamp excursion and its geomagnetic implications. *Earth and Planetary Science Letters* 88: 209–219.

Salis, J-S 1987 Variation séculaire du champ magnétique terrestre – Direction et paleointensite sur la période 7000 – 70,000 BP dans la Chaîne des Puys. *Mémoires et Documents du Centre Armoricain d'Etude Structurale des Socles,* Universite de Rennes 11: 190 p.

Salis, J-S, Bonhommet, N and Levi, S 1989 Paleointensity of the geomagnetic field from dated lavas of the Chaîne des Puys, France. 1. 7–12 thousand years before present. *Journal of Geophysical Research* 94(B11): 15,771–15,784.

Senanayake, WE, McElhinny, MW and McFadden, PL 1982 Comparison between the Thelliers' and Shaw's palaeointensity methods using basalts less than 5 million years old. *Journal of Geomagnetism and Geoelectricity* 34: 141–161.

Shaw, J 1974 A new method of determining the magnitude of the palaeomagnetic field. *Geophysical Journal of the Royal Astronomical Society* 39: 133–141.

Siegenthaler, U and Beer, J 1988 Model comparison of ^{14}C and ^{10}Be isotope records. *In* Stephenson, FR and Wolfendale, AW, eds, *Secular, Solar and Geomagnetic Variations in the Last 10,000 Years.* Dordrecht, The Netherlands, Kluwer: 315–327.

Siegenthaler, U, Heimann, M and Oeschger, H 1980 ^{14}C variations caused by changes in the global carbon cycle. *In* Stuiver, M and Kra, RS, eds, Proceedings of the 10th International ^{14}C Conference. *Radiocarbon* 22(2): 177–191.

Smith, PJ 1967 The intensity of the ancient geomagnetic field: A review and analysis. *Geophysical Journal of the Royal Astronomical Society* 12: 321–362.

Sonett, CP 1984 Very long solar periods and the radiocarbon record. *Reviews of Geophysics and Space Physics* 22(3): 239–254.

Sonett, CP, Morfill, GE and Jokipii, JR 1987 Interstellar shock waves and ^{10}Be from ice cores. *Nature* 330: 458–460.

Stacey, FD and Banerjee, SK 1974 *The Physical Principles of Rock Magnetism.* Amsterdam, Elsevier: 195 p.

Sternberg, R 1983 Archaeomagnetism in the Southwest of North America. *In* Creer, KM, Tucholka, P and Barton, CE, eds, *Geomagnetism of Baked Clays and Recent Sediments.* Amsterdam, Elsevier: 158–167.

_____(ms) 1988 A computerized archaeomagnetic data base: Plans and prospects. Paper presented at the IAGA Symposium on New Trends in Geomagnetism, Prague, Czechoslovakia, June 27–July 2.

_____1989a Archaeomagnetic paleointensity in the American southwest during the past 2000 years. *Physics of the Earth and Planetary Interiors* 56: 1–17.

_____1989b Secular variation of archaeomagnetic direction in the American southwest, AD 750–1425. *Journal of Geophysical Research* 94(B1): 527–546.

Sternberg, RS and Damon, PE 1979 Sensitivity of radiocarbon fluctuations and inventory to geomagnetic and reservoir parameters. *In* Berger, R and Suess, HE, eds, *Radiocarbon Dating.* Proceedings of the 9th International ^{14}C Conference. Berkeley, University of California Press: 691–717.

_____1981 A finite-difference solution for the radiocarbon dipole moment (abstract). *EOS, Transactions of the American Geophysical Union* 62: 272.

_____1983 Atmospheric radiocarbon: implications for the geomagnetic dipole moment. *In* Stuiver, M and Kra, RS, eds, Proceedings of the 11th International ^{14}C Conference. *Radiocarbon* 25(2): 239–248.

_____(ms) 1991 Implications of dipole moment secular variation from 50–10 ka for the radiocarbon record. Paper presented at the 14th International ^{14}C Conference, Tucson, Arizona, May 20–24.

Stuiver, M 1970 Long-term C14 variations. *In* Olsson, IU, ed, *Radiocarbon Variations and Absolute Chronology.* Proceedings of the 12th Nobel Symposium. Stockholm, Almqvist & Wiksell: 197–213.

_____1978 Radiocarbon timescale tested against magnetic and other dating methods. *Nature* 273: 271–274.

Stuiver, M, Braziunas, TF, Becker, B and Kromer, B 1991 Late-glacial and Holocene atmospheric $^{14}C/^{12}C$ change: Climate, solar, oceanic and geomagnetic influences. *Quaternary Research* 35(1): 1–24.

Stuiver, M and Kra, RS, eds 1986 Proceedings of the 12th International ^{14}C Conference. *Radiocarbon* 28(2B): 805–1030.

Stuiver, M and Quay, PD 1980 Changes in atmospheric carbon-14 attributed to a variable sun. *Science* 207(4426): 11–19.

Suess, HE 1980 The radiocarbon record in tree rings of the last 8000 years. *In* Stuiver, M and Kra, RS, eds, Proceedings of the 10th International ^{14}C Conference. *Radiocarbon* 22(2): 200–209.

Tauxe, L and Valet, J-P 1989 Relative paleointensity of the earth's magnetic field from marine sedimentary rocks: A global perspective. *Physics of the Earth and Planetary Interiors* 56: 59–68.

Thellier, E 1977 Early research on the intensity of the ancient geomagnetic field. *Physics of the Earth and Planetary Interiors* 13: 241–244.

Thellier, E and Thellier, O 1959 Sur l'intensité terrestre du champ magnétique terrestre dans le passé historique et géologique. *Annales Géophysique* 15: 285–376.

Thomas, R 1983 Review of archaeointensity methods. *Geophysical Surveys* 5: 381–393.

Thompson, R 1982 A comparison of geomagnetic secular variation as recorded by historical, archaeomagnetic and palaeomagnetic measurements. *Philosophical Transactions of the Royal Society of London* A306: 103–112.

Thouveny, N 1987 Variations of the relative palaeointensity of the geomagnetic field in western Europe in the interval 25–10 kyr BP as deduced from analyses of lake sediments. *Geophysical Journal of the Royal Astronomical Society* 91: 123–142.

Thouveny, N, Creer, KM and Blunk, I 1990 Extension of the Lac du Bouchet palaeomagnetic record over the last 120,000 years. *Earth and Planetary Science Letters* 97: 140–161.

Tucker, P 1981 Palaeointensities from sediments: Normalization by laboratory redeposition. *Earth and Planetary Science Letters* 56: 398–404.

Veksler, VS 1988 Intensity variations of the earth's magnetic field in the past. *Geomagnetism and Aeronomy* (English translation) 28(6): 902–903.

Verosub, KL and Banerjee, SK 1977 Geomagnetic excursions and their paleomagnetic record. *Reviews of Geophysics* 15: 145–155.

Verosub, KL and Mehringer, PJ, Jr, 1984 Congruent paleomagnetic and archaeomagnetic records from the western United States: AD 750 to 1450. *Science* 224: 387–389.

Vries, H, de 1958 Variation in concentration of radiocarbon with time and location on earth. *Koninklijke Nederlandse Akademie van Wetenschappen* B61: 94–102.

Wada, M and Inoue, A 1966 Relation between the carbon 14 production rate and the geomagnetic moment. *Journal of Geomagnetism and Geoelectricity* 18(4): 485–488.

Walton, D 1977 Archaeomagnetic intensity measurements using a SQUID magnetometer. *Archaeometry* 19(2): 192–200.

_____1988a Comments on "Determination of the intensity of the earth's magnetic field during archaeological times: Reliability of the Thellier Technique." *Reviews of Geophysics* 26(1): 13–14.

_____1988b The lack of reproducibility in experimentally determined intensities of the earth's magnetic field. *Reviews of Geophysics* 26(1): 15–22.

Wolfman, D 1984 Geomagnetic dating methods in archaeology. *In* Schiffer, MB, ed, *Advances in Archaeological Method and Theory*. New York, Academic Press 7: 363–458.

Yang, AIC and Fairhall, AW 1972 Variations of natural radiocarbon during the last 11 millennia and geophysical mechanisms for producing them. *In* Rafter, TA and Grant-Taylor, T, eds, *Proceedings of the 8th International ^{14}C Conference*. Wellington, Royal Society of New Zealand: 60–73.

9

SOLAR FORCING OF GLOBAL CLIMATE CHANGE?

PAUL E DAMON and JOHN L JIRIKOWIC

INTRODUCTION

Precise analysis of the carbon isotopes in dendrochronologically dated tree rings demonstrates that atmospheric ^{14}C has varied ± 5% during the past nine millennia. These data have been summarized in the "Calibration Issue" of *Radiocarbon* (Stuiver & Kra 1986). Discrete Fourier Transform (DFT) power spectral analysis of the atmospheric ^{14}C variations, after removal of the long-term geomagnetic field-induced trend, reveals three fundamental periods. By using other independent spectral methods, Damon & Sonett (in press) corroborated these periods – the 2250-year (Hallstattzeit), the 210-year (Suess), and the 88-year (Gleissberg). Collectively, these are known as the de Vries Effect or "Suess Wiggles" (see Damon 1987, for review). Houtermans (1971) first reported the two longer periods, with the exception of the Gleissberg cycle, from Fourier analysis of the La Jolla ^{14}C data set. Suess (1980) has repeatedly stressed the persistence and solar origin of the 210-year period. Stuiver and Braziunas (1989) observed all of these lines in both the DFT and Maximum Entropy Method (MEM) spectra of the high-precision ^{14}C data, but considered the two short periods to be harmonics of a near 425-year fundamental. However, even if this hypothesis is correct, both DFT and MEM power spectra show the resolved hypothetical fundamental to be weak in comparison to its harmonics. A weak 11-year period (Schwabe cycle) emerges from spectral analysis of high-precision ^{14}C data of single annual tree rings. This cycle is usually obscured, however, by solar-flare-produced ^{14}C, which tends to be out of phase during the Schwabe cycle. The ^{14}C component is produced by galactic cosmic rays modulated by the solar wind (Damon, Cheng & Linick 1989).

The shorter periods (*ie*, 11, 88 and 210 years) can be related to changes in the ^{14}C production rate that are inversely related to solar activity (as measured by the solar index, R_z). Thus, atmospheric ^{14}C, expressed as $\Delta^{14}C$, obtained by measurement of the carbon isotopes in tree rings, increases and decreases with the various solar minima and maxima as shown in Figure 9.1 (see Damon, Lerman & Long 1978; Damon & Sonett, in press, for reviews). The longer

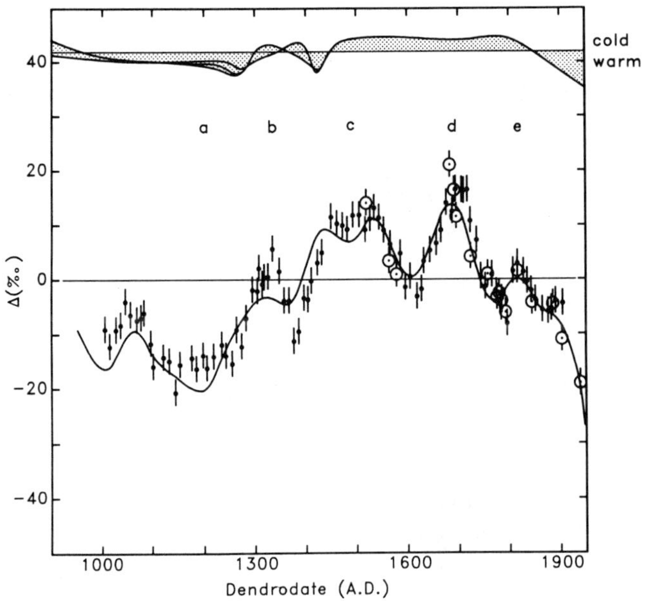

Fig 9.1. $\Delta^{14}C$ variations during the last millennium. ⊙ = high-precision data of de Vries (1958), fractionation corrected by Lerman, Mook and Vogel (1970); • = the high-precision data of Stuiver and Quay (1980); the solid line is derived by Fourier analysis algorithm after polynomial-regression detrending and winsorization based upon an intermediate-precision data set of 1200 analyses from five laboratories (Klein *et al* 1980). The letter, a, marks the Medieval Warm Epoch; b, c, d and e mark the Wolf, Spörer, Maunder and Dalton minima, respectively. The climate curve is based upon global montane glacial advance and retreat date (Rothlisberger 1986). The double curves are the result of uncertainties in calibration of ^{14}C dates.

≈2300-year period is enigmatic in origin but clearly modulates the higher frequencies, as does the longer-term variation ascribed to changes in the geomagnetic field. It is also tempting to relate the longer ≈2300-year period to changes in the geomagnetic field, but this would seem to require a dipole moment variation for which there is currently no evidence. The ≈2300-year period can also be correlated with climate change (Damon, Cheng & Linick 1989; Damon & Sonett, in press). We have referred to the ≈2300-year period as the Hallstattzeit period (Damon & Sonett, in press) because of its apparent correlation with climate epochs documented by Schmidt and Gruhle (1988). The ≈2300-year period appears to act as a gate allowing extraordinarily large Maunder- and Spörer-type $\Delta^{14}C$ peaks to occur every ≈2300 years (Damon 1988). Suess (1968) and Damon (1968), following de Vries (1958, 1959), correlated ^{14}C fluctuations during the last millennium with climate change with low solar activity corresponding to high $\Delta^{14}C$ and colder climate during the Little Ice Age and high solar activity corresponding to low $\Delta^{14}C$ and warmer temperatures during the Medieval Warm Epoch. This hypothetical relationship has been a subject of controversy during subsequent decades.

A DIRECT RELATIONSHIP BETWEEN SOLAR ACTIVITY AND CLIMATE?

Past climates, before the availability of instrumental records, must be reconstructed from historical records and proxy indicators of climate. Unfortunately, historical records are qualitative, subjective and restricted to centers of civilization. Proxy indicators of climate respond to complex regional and local weather parameters and "have been plagued by data quality problems" (Wigley 1988: 216). The timing and correlation of climatic events are not the least of these problems. Stuiver *et al* (1991) have concluded that, "correlation coefficients between climate and Q [^{14}C production rate] records usually are small, and when autocorrelation of both climate and Q are taken into account, proof of a climate-Q relationship readily disappears in a statistical morass."

In addition to the problems encountered with proxy indicators of climate, there may have been, in the past, a psychological barrier to the concept of a changing Sun forcing change in the global climate. Eddy (1976: 1200) states, "The reality of the Maunder Minimum and its implications of basic solar change may be one more defeat in our long losing battle to keep the Sun perfect, or, if not perfect, consistent, and if inconsistent, regular. Why we think the Sun should be any of these when other stars are not is more a question for social than for physical science." Slightly more than two decades of observation at the Mount Wilson Observatory involving 99 stars have established that cyclical variation in stars, including solar-type stars, is a common phenomenon (Soderblom & Baliunas 1988). Recent satellite measurements have demonstrated that total solar irradiance varies during the 11-year Schwabe solar cycle (Willson *et al* 1986; Willson & Hudson 1988). A linear relationship exists between the Wolf Sunspot Index (R_z) and total irradiance (W m^{-2}) during solar cycle 21:

$$S(W{\cdot}m^{-2}) = 1366.82 + 7.71*10^{-3}R_z \quad (9.1)$$

For an extreme solar cycle with maximum $R_z = 200$, the increase in total irradiance would only be about 1.3‰. This is an insufficient change to force a significant climate response without some amplification. The increase, however, is not distributed equally over all wave lengths. For example, a review by Lean (1987) suggests that ultraviolet irradiances increase with solar activity during the 11-year cycle. The increase is a function of wave length varying from less than 1% for the near ultraviolet to from 80 to 155% at 121.6 nm (Lyman alpha). Despite the small change in total solar irradiance, atmospheric elements at all levels, from the surface to the top of the atmosphere, show a surprisingly strong correlation with solar activity (R_z or 10.7 cm solar flux) when the data are divided according to the phase of the quasi-biennial cycle (see for example, Labitzke & van Loon 1990). The quasi-biennial cycle (QBO) is the tendency for the wind in the equatorial stratosphere to change from westward to eastward about every 27 months.

As a consequence of the new evidence for solar variability and evidence for a solar-weather correlation involving the QBO, the possible role of a changing Sun affecting climate must be seriously considered. Evidence that montane glacier advances and retreats are synchronous between the northern and southern hemispheres (Rothlisberger 1986) confirms the global extent of the cooling and warming during climate episodes such as the Little Ice Age and Medieval Warm Epoch (Fig 9.1). The correlation between relatively high $\Delta^{14}C$ during the Little Ice Age and low $\Delta^{14}C$ during the Medieval Warm Epoch suggests that solar activity not only modulates ^{14}C production but also forces global climate change, as first suggested over two decades ago (Suess 1968; Damon 1968; de Vries 1958, 1959). Recently, a statistically significant correlation has been found between two earlier episodes of glacial advance and retreat and $\Delta^{14}C$ (Wigley 1988; Wigley & Kelly 1990). Sonett and Suess (1984) have also found a significant correlation between $\Delta^{14}C$ and tree-ring width in bristlecone pine from the White Mountains of California. Wide tree-ring widths correlated with low $\Delta^{14}C$ during the 5290-year record. Temperature may be the most important limiting factor of tree-ring growth for these high-altitude trees, so again, warm temperatures may be associated with low $\Delta^{14}C$ and *vice versa.*

The Camp Century, Greenland ice core (Dansgaard *et al* 1971) oxygen isotope record correlates fairly well with the $\Delta^{14}C$ record. The more negative $\delta^{18}O$ values during the Little Ice Age occur close to the Dalton, Maunder and Spörer minima (Fig 9.2). Thus, for this ice core, colder Greenland polar temperatures may be associated with high $\Delta^{14}C$ production rates and low solar activity (R_z) suggesting a direct variation with solar irradiation (S, $W{\cdot}m^{-2}$). Periodicities observed in power spectra are also very similar (Table 9.1). Cooling during the Maunder Minimum seems to be due more to cool summers rather than extremely cold winters (see Figure 2 of Johnsen *et al* 1972). It is not surprising that lower latitude ice cores (eg, Dye3 @ 65°12′N *vs* Camp Century @ 77°10′N) do not show a significant correlation of $\delta^{18}O$ with the solar minima, although North Atlantic Sea Ice shows similar variation to the Camp Century $\delta^{18}O$ (Broecker, personal correspondence). Seasonal variations of $\delta^{18}O$ are as extreme as the shift of average $\delta^{18}O$ from the full Holocene interglacial to the full Wisconsin glacial epochs. Seasonal rates of accumulation may also vary. Thus,

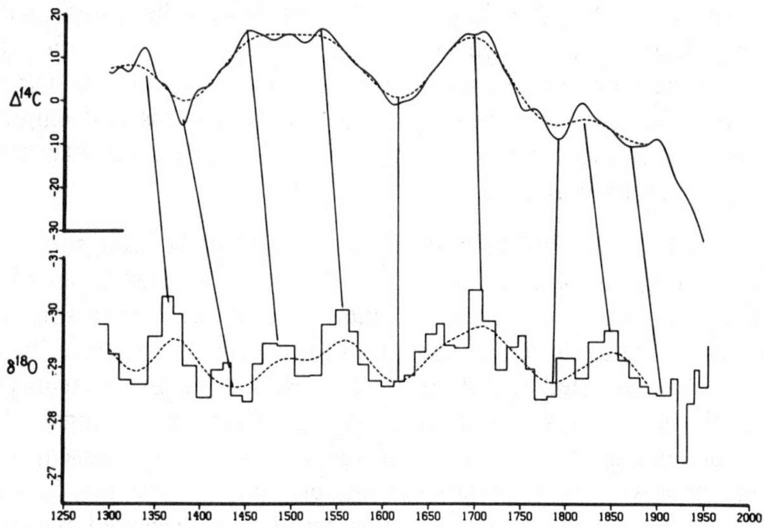

Fig 9.2. $\delta^{18}O$ *vs* $\Delta^{14}C$ in the Camp Century, Greenland ice core (Dansgaard *et al* 1971). The $\Delta^{14}C$ date is based on the Calibration Issue of ***Radiocarbon*** (Stuiver & Kra 1986). The ice core chronology is normalized to the Gleissberg cycle (Table 9.1), and the $\Delta^{14}C$ curve is corrected for phase lag. The dashed curves are smoothed by use of a cubic spline. Correlation lines are between peaks and valleys in the two spline-fitted curves.

TABLE 9.1. Significant Periodicities in $\Delta^{14}C$ (Tree Rings) and $\delta^{18}O$ (Camp Century Ice)

Method	Significant periods	Reference
$\delta^{18}O$ (MEM)	78, 181, 2000	Dansgaard *et al* (1971)
*$\delta^{18}O$ (normalized)	88, 205, 2260	Damon & Sonett (1991)
$\Delta^{14}C$ (MEM, AR120)	88, 208, 2240	Damon & Sonett (1991)
$\Delta^{14}C$ (DFT)	88, 207, 2272	Damon & Sonett (1991)

*$\delta^{18}O$ normalized to Gleissberg period in sunspot record (R_z)

$\delta^{18}O$ values may reflect seasonal differences in precipitation and temperature as well as interannual variation. The southerly Greenland sites are also well within the westerly storm track resulting in precipitation with different isotopic compositions.

WORKING HYPOTHESIS

Summarizing, four periods have been observed in both the $\Delta^{14}C$ and climate records: the ≈2300-year (Hallstattzeit), 210-year (Suess), 88 (Gleissberg) and

11-year (Schwabe). A fifth period, at 150 years, has not been observed in the climate record, and is much weaker in the $\Delta^{14}C$ spectrum than the 210-year period during the most recent millennium. The 210-, 88- and 11-year periods appear to be related to solar activity. Though its cause is still enigmatic, a reasonable although inconclusive argument has been made for a solar origin for the ≈2300-year period (Hood & Jirikowic 1990).

The dependence of $S(W{\cdot}m^{-2})$ on R_z observed during the 21st solar cycle is insufficient to explain the Little Ice Age. Clearly, there must be an additional factor. Eddy (1977a, b) suggested that the effect of a change in solar activity might be much greater for longer episodes of low or high solar activity such as the Little Ice Age and Medieval Warm Epoch. He states that, "It seems possible to me that the slowly varying envelope could reflect slow changes of a few percent in total energy output of the Sun" (Eddy 1977b: 92). Consequently, we will assume as a working hypothesis to interpret these events that $S(W{\cdot}m^{-2})$ is a function of period, T, as well as solar activity as measured by the Wolf Sunspot Index, R_z

$$S(W{\cdot}m^{-2})\ \alpha\ f(T, R_z) \quad . \tag{9.2}$$

THE SENSITIVITY OF CLIMATE TO SOLAR IRRADIANCE

According to Reid and Gage (1988), the relationship between the global equilibrium surface temperature of the Earth, ϕ_{eq}, and S is

$$\phi_{eq} = \frac{(S/4)\ (1-\alpha)\ -\ A}{B} \tag{9.3}$$

where α is the planetary albedo, B is the infrared cooling parameter taken to be 2.2 $W{\cdot}m^{-2}{\cdot}K^{-1}$, and A is fixed to obtain ϕ_{eq} = 14.8°C (Hoffert, Calligari & Hsieh 1980). Estimating the steady-state R_z to be approximately 40, from equation 9.1, S = 1367 $W{\cdot}m^{-2}$ and solving for A = 206.6 $W{\cdot}m^{-2}$. Since α = 0.3, equation 9.3 becomes

$$\phi_{eq} = \frac{S/5.7\ -\ 206.6}{2.2} \tag{9.4}$$

and

$$\Delta\phi_{eq} = \frac{\Delta S/5.7}{2.2} \quad . \tag{9.5}$$

Thus,

$$\frac{\Delta S}{\Delta \phi_{eq}} = 12.6 \mathrm{W \cdot m^{-2} K^{-1}} \qquad (9.6)$$

and, dividing by 5.7, we obtain, tautologically, the infrared cooling factor, 2.2 $\mathrm{W \cdot m^{-2} \cdot K^{-1}}$, which would be the forcing in $\mathrm{W \cdot m^{-2}}$ to obtain an equilibrium change in global surface temperature change of 1°C.

Hence, for an extreme 11-year solar cycle increasing from $R_z = 0$ to 200, from equation 9.1, the change in $S(\mathrm{W \cdot m^{-2}})$ would be 1.54 $\mathrm{W \cdot m^{-2}}$. Such a change, if sustained, would result in an equilibrium global surface temperature change of 0.12°C. However, an 11-year cycle repeating from $R_z = 0$ to 200 would be dampened as a result of ocean-atmosphere interaction by a factor of about 0.24 (Wigley 1988), and so would yield a temperature increase of only 0.03°C.

SOLAR FORCING OF TWENTIETH CENTURY GLOBAL WARMING?

As a *Gedanken* (thought) experiment, we will extend the dominant century-scale solar periods into the 20th and 21st centuries, and assign modest temperature changes to each cycle that are comparable with the climatic record during the Christian Era (see Table 11.1, Bradley 1985, for review). The two extremes during this interval were the Medieval Warm Epoch and the Little Ice Age. The total change in global temperature from extreme warm to extreme cold between these epochs was certainly less than 2°C and probably about 1.5 ± 0.5°C. The near 2300-year Hallstattzeit cycle associated with cold epochs (Schmidt & Gruhle 1988) was at a maximum during the Little Ice Age and characteristically should be followed by climate amelioration. Thus, another Little Ice Age would be anticipated only in a future too distant to be of concern in the present context. We need be concerned with only the amelioration. Following Eddy's suggestion, we assigned for our *Gedanken* experiment, a modest 0.30°C to the Suess 210-year cycle and half that, 0.15°C, to the Gleissberg 88-year cycle, *ie*, considerably (≈50%) less than the estimated extremes for the Medieval Warm Epoch and Little Ice Age fluctuation. Following Eddy, we fixed the phase of the 210-year cycle at the mid-point of the Maunder Minimum, AD 1690 (Eddy 1976). The phase of the 88-year Gleissberg was fixed at the Dalton Minimum, AD 1810. Thus, the next minimum would be anticipated at AD 1898. In fact, a solar activity minimum does occur between AD 1875 and 1915. This minimum is not as extreme as the Maunder Minimum, which is compatible with a solar origin for the Hallstattzeit cycle (Hood & Jirikowic 1990).

Having set the conditions, the cycles were then projected into the 20th and 21st centuries. We included a lag time between solar forcing and climatic response equal to 1.6 years for the 11-year cycle, 7.5 years for the 88-year cycle, and 13.2 years for the 210-year cycle (Wigley 1988). The required change in the solar energy must be divided by a damping factor (Table 9.2) because equilibrium will not be attained during the duration of the cycling periods (Wigley 1988). The parameters for solar cycle 21 provide a frame of reference. The 88- and 210-year cycles require 5.6 and 9.3 times the changes in the energy observed during 11-year solar cycle 21 to yield respective changes of 0.150°C and 0.300°C. This can also be compared to the current estimate of Greenhouse forcing of 2.04 $W{\cdot}m^{-2}$ (Hansen & Lacis 1990a), which, at equilibrium, would be sufficient to raise global temperatures by about 0.9°C, assuming the 2.2 $W{\cdot}m^{-2}{\cdot}K^{-1}$ infrared cooling factor used in this chapter. However, equilibrium would not have been attained and our analysis suggests that Greenhouse Warming is lagging forcing by more than 25 years (eg, see Figure 1 of Wigley & Raper 1987) neglecting other anthropogenic effects such as increasing aerosol perturbation and natural feedback mechanisms such as increased cloud cover.

Our calculations are based upon an assumed infrared cooling factor or climate sensitivity of 2.2 $W{\cdot}m^{-2}{\cdot}K^{-1}$. According to Hansen and Lacis (1990b), the possible range of this parameter is from 0.7 to 2.7 $W{\cdot}m^{-2}{\cdot}K^{-1}$. The lower values, such as 1.33 $W{\cdot}m^{-2}{\cdot}K^{-1}$ used by Hansen and Lacis (1990b), imply very high climate sensitivity, and would require a long lag time between forcing and climate response to be compatible with the observed temperature increase (more than 45 years for 1.33 $W{\cdot}m^{-2}{\cdot}K^{-1}$).

Keeping in mind the above uncertainties, the result predicted by our *Gedanken* experiment is shown in Figure 9.3. The model predicts a rapid rise in temperature from the beginning of the 20th century to the mid-1950s of slightly over 0.5°C, followed by a slight decrease through the 1970s, after which the temperature rapidly rises again to a maximum increase relative to the beginning of the century of a little over 0.8°C in the third decade of the 21st century. Subsequent cooling during the latter part of the 21st century would not exceed 0.7°C or a little less than the previous warming. Figure 9.4 shows that the

TABLE 9.2. Variation of Total Solar Irradiance (ΔS) Required to Obtain 0.6° Global Warming From Solar Cycles (210-yr, 88-yr and 11-yr)

Cycle yrs	Damping factor*	Lag-time yrs*	ΔΘ °C	ΔS Wm^{-2}**	% of S_o	Forcing Wm^{-2}†
11	0.24	1.6	±0.012	±0.55‡	±0.04	±0.096
88	0.61	7.5	±0.150	±3.10	±0.23	±0.54
208	0.74	13.2	±0.300	±5.11	±0.37	±0.90

*Wigley (1988); **Change in solar irradiance; †ΔS ÷ 5.7; ‡Solar cycle 21 (R_z = 12 to 155)

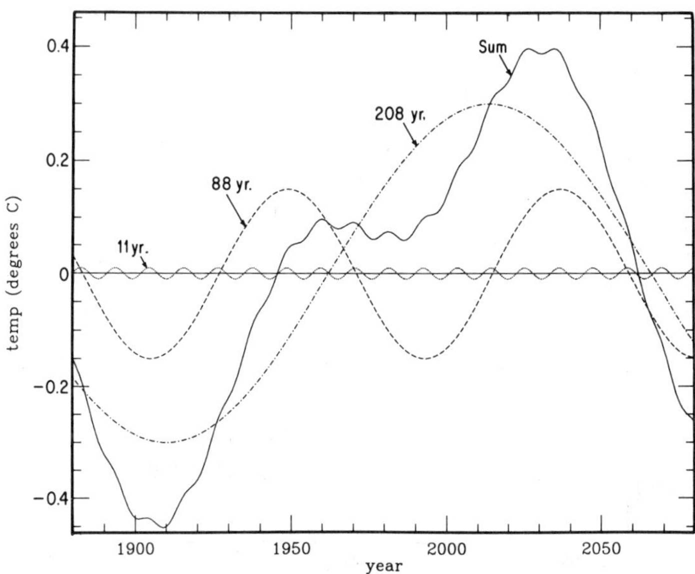

Fig 9.3. Model curve by assigning forcings of ±0.1, ±0.5 and ±0.9 W·m^{-2} to the 11-, 88- and 210-year solar cycles, respectively. The forcings are compatible with past climate change.

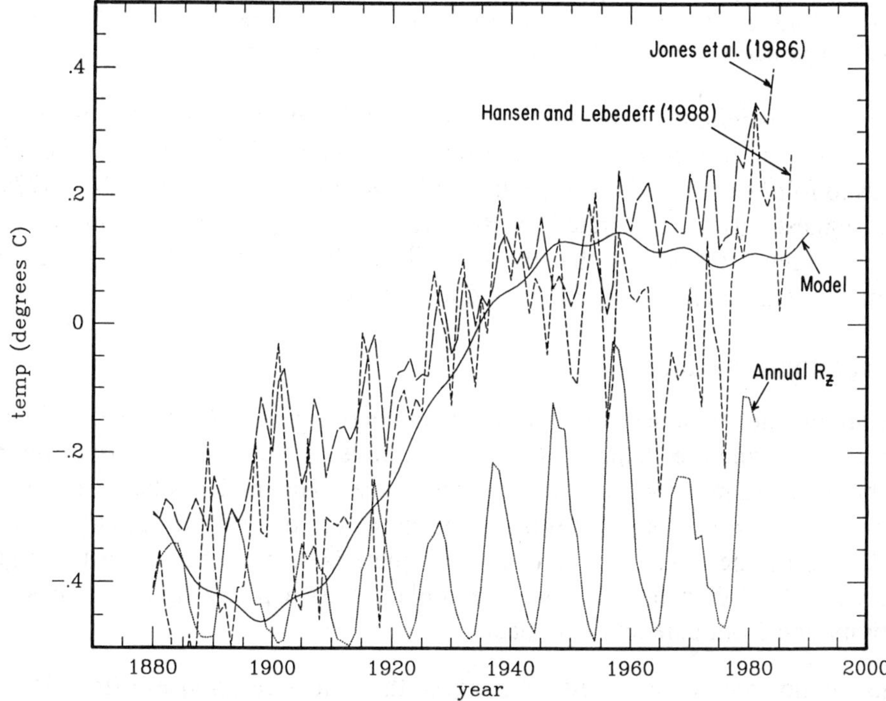

Fig 9.4. Model curve compared to estimates of 20th-century global temperature change. Note that the envelope of the 11-year running average of the solar cycle also tends to follow the global temperature change as pointed out by Reid and Gage (1988).

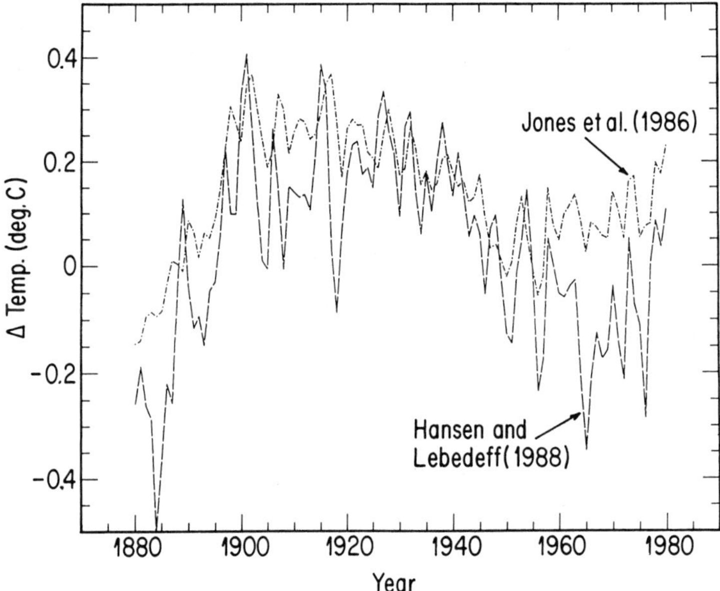

Fig 9.5. Estimates of the difference between global temperature change and the solar cycle model curve. Note that if the model were assumed to have validity, a long lag time (≥ 90 years) would be implied between Greenhouse forcing and climate response, because most of the observed global temperature change would be the natural result of solar forcing.

shape and magnitude of the model curve compare favorably with the estimated global temperature increase. The estimated global temperature increase also tends to follow the envelope of solar activity, R_z, as suggested by Eddy (1977a), and suggestively demonstrated by Reid and Gage (1988; see our Figure 9.4). Finally, Figure 9.5 shows that the discrepancy between the model curve and estimated global temperature curves does not exceed 0.2°C.

Concluding Remarks

We are inclined to believe that the agreement between our *Gedanken* experiment model curve and the shape as well as magnitude of the estimated 20th century global temperature rise is not mere coincidence. The model is based upon the assumption that solar energy follows the long periods of solar activity as well as R_z during the 11-year cycle. The parameterization derives from the high-precision $\Delta^{14}C$ data for the last nine millennia as a proxy indicator of solar activity and from paleoclimatic data.

This in no way is intended to suggest that the Greenhouse Effect is not important. The effect can be observed in the laboratory and determines the equilibrium temperatures of Earth, Mars and Venus. Only the observable thermal effect of the current Greenhouse nonequilibrium forcing is in dispute. Perhaps the greatest uncertainties involve the infrared cooling factor (climate sensitivity) and the lag time. Thus, the effect of solar energy may be masking the Greenhouse Effect in temperature record and may continue to increment it far into the 21st century. It may not be possible to distinguish the Greenhouse Effect from solar warming until the first half of the next century. One thing seems certain, based on the $\Delta^{14}C$ proxy record of solar activity, we cannot rely on an early 21st century new Little Ice Age to offset the Greenhouse Effect, as suggested by Seitz *et al* (1989). If our evaluation of the past is a reliable key to the future, an event as severe as the Little Ice Age should not be expected until the 34th century.

ACKNOWLEDGMENTS

The authors benefitted from discussions with Professors Charles P Sonett and Alexander T Wilson of The University of Arizona. This effort was stimulated by a suggestion to provide a solar scenario for 20th century global warming by James E Hansen of the Goddard Institute for Space Studies. We thank the following for helpful reviews of this paper: Wallace Broecker, Andrew Cohen, Owen Davis, James Hansen, Malcolm Hughes, Robert Sternberg, George Reid and Charles Sonett. We would also thank Aran Armstrong for his help with library work and word processing. This work was supported by NSF Grants EAR-88-22292 and ATM-8919535 and the State of Arizona.

References

Bradley, RS 1985 *Quaternary Paleoclimatology Methods of Paleoclimatic Reconstruction*. Boston, Allen & Unwin: 472 p.

Damon, PE 1968 Radiocarbon and Climate. *Meteorological Monographs* 8: 151–154.

_____1987 The history of the calibration of radiocarbon dates by dendrochronology. *In* Aurenche, O, Evin, J and Hours, F, eds, Chronologies du Proche Orient. *BAR International Series* 379: 61–104.

_____1988 Production and decay of radiocarbon and its modulation by geomagnetic field-solar activity changes with possible implications for global environment. *In* Stephenson, FR and Wolfdale, AW, eds, *Secular, Solar, and Geomagnetic Variations in the Last 10,000 Years*. Dordrecht, The Netherlands, Kluwer: 267–285.

Damon, PE, Cheng, S and Linick TW 1989 Fine and hyperfine structure in the spectrum of secular variations of the atmospheric ^{14}C. *In* Long, A and Kra, RS, eds, Proceedings of the 13th International ^{14}C Conference. *Radiocarbon* 31(3): 704–718.

Damon, PE, Lerman, JC and Long, A 1978 Temporal fluctuations of atmospheric ^{14}C: Causal factors and implications. *Annual Review of Earth and Planetary Science* 6: 457–494.

Damon, PE and Sonett, CP, in press, Solar and terrestrial components of the atmospheric ^{14}C variation spectrum. *In* Sonett, CP, Giampapa, MS and Matthews, MS, eds, *The Sun in Time*. Tucson, Arizona, University of Arizona Press.

Dansgaard, W, Johnsen, SJ, Clausen, HB and Langway, CC 1971 Climatic record revealed by the Camp Century ice core. *In* Turekian, K, ed, *The Late Glacial Ages*. New Haven, Connecticut, Yale University Press: 37–46.

Eddy, JA 1976 The Maunder Minimum. *Science* 192: 1189–1202.

_____1977a Climate and the changing Sun. *Climate Change* 1: 173–190.

_____1977b The case of the missing sunspots. *Scientific American* 236: 80–92.

Hansen, JE and Lacis, AA (ms) 1990a Sun and dust versus the Greenhouse. Submitted to *Nature*.

_____1990b Comparison of solar and other influences on long-term climate. *In* Schatten, K and Arking, A, eds, *Climate Impact of Solar Variability*. Proceedings of Goddard Conference NASA CP Document, in press.

Hansen, JE and Lebedeff, S 1988 Global surface air surface temperatures: Update through 1987. *Geophysical Research Letters* 15(4): 323–326.

Hoffert, MI, Callegari, AJ and Hsieh, CT 1980 The role of deep sea heat storage in the secular response to climate forcing. *Journal of Geophysical Research* 88: 6667.

Hood, LL and Jirikowic, JJ 1990 Recurring variations of a probable solar origin on the atmospheric Δ^{14}C record. *Geophysical Research Letters* 17(1): 85–88.

Houtermans, JC (ms) 1971 Geophysical interpretation of bristlecone pine radiocarbon measurements using a method of Fourier analysis of unequally spaced data. PhD Thesis, University of Bern, Switzerland.

Johnsen, SJ, Dansgaard, W, Clausen, HB and Langway, CC, Jr 1972 Oxygen isotope profiles through the Antarctic and Greenland ice sheets. *Nature* 235(5339): 429–434.

Jones, PD, Wigley, TML and Wright, PB 1986 Global temperature variation between 1862 and 1984. *Nature* 322: 430–434.

Klein, J, Lerman, JC, Damon, PE and Linick, T 1980 Radiocarbon concentration in the atmosphere: 8000-year record of variations in tree rings. First results of a USA workshop. *In* Stuiver, M and Kra, RS, eds, Proceedings of the 10th International ^{14}C Conference. *Radiocarbon* 22(3): 950–961.

Labitzke, K and van Loon, H 1990 Associations between the 11-year solar cycles, the quasi-biennial oscillation and the atmosphere: A summary of recent work. *Philo-*

sophical Transactions of the Royal Society (London) A330: 577–589.

Lean, J 1989 Contribution of ultraviolet irradiance variations to changes in the Sun's total irradiance. *Science* 244: 197–200.

Lerman, JC, Mook, WG and Vogel, JC 1970 ^{14}C in tree rings from different localities. *In* Olsson, IU, ed, *Radiocarbon Variations and Absolute Chronology*. Proceedings of the 12th Nobel Symposium. Stockholm, Almqvist & Wiksell: 275–301.

Reid, GC and Gage, KS 1988 The climatic impact of secular variations in solar irradiance. *In* Stephenson, FR and Wolfdale, AW, eds, *Secular, Solar, and Geomagnetic Variations in the Last 10,000 Years.* Dordrecht, The Netherlands, Kluwer: 225–243.

Rothlisberger, F 1986 *10,000 Jahre Gletschargeschicte der Erde.* Aarau, Verlag Sauerlander: 416 p.

Schmidt, B and Gruhle, W 1988 Radiokohlenstoffgehalt und dendrochronologie. *Naturwissenshaftliche Rundschau* 5: 177–182.

Seitz, F, Bendetsen, K, Jastrow, R and Nierenberg, WA 1989 *Scientific Perspective on the Greenhouse problem.* Washington, DC, George C Marshall Institute: 34 p.

Soderblom, DR and Baliunas, SL 1988 The Sun and the stars: What stars indicate about solar variability. *In* Stephenson, FR and Wolfdale, AW, eds, *Secular, Solar, and Geomagnetic Variations in the Last 10,000 years*. Dordrecht, The Netherlands, Kluwer: 25–48.

Sonett, CP and Suess, HE 1984 Correlation of bristlecone pine ring width with atmospheric ^{14}C variations: A climate-Sun relation. *Nature* 308: 141–143.

Stuiver, M and Braziunas, TF 1989 Atmospheric ^{14}C and century-scale solar oscillations. *Nature* 338: 405–408.

Stuiver, M, Braziunas, TF, Becker, B and Kramer, B 1991 Climatic, solar oceanic and geomagnetic influences on Late-Glacial and Holocene atmospheric ^{14}C/^{12}C change. *Quaternary Research* 35: 1–24.

Stuiver, M and Kra, RS, eds 1986 Calibration issue. Proceedings of the 12th International ^{14}C Conference. *Radiocarbon* 28(2B): 805– 1030.

Stuiver, M and Quay, PD 1980 Changes in atmospheric carbon-14 attributed to a variable Sun. *Science* 207: 11–19.

Suess, HE 1968 Climate changes, solar activity, and cosmic-ray production rate of natural radiocarbon. *Meteorological Monographs*: 146–150.

_____1980 The radiocarbon record in tree rings of the last 8000 years. *In* Stuiver, M and Kra, RS, eds, Proceedings of the 10th International ^{14}C Conference. *Radiocarbon*: 22(2): 200–209.

Vries, H, de 1958 Variations in concentration of radiocarbon with time and location on Earth. *Koninklijke Nederlandse Akademie Wetenschappen Series B* 61: 94–102.

_____1959 Measurement and use of radiocarbon. *In* Abelson, PH, ed, *Researches in Geochemistry*. New York, John Wiley & Sons: 169–189.

Wigley, TML 1988 The climate of the past 10,000 years and the role of the Sun. *In* Stephenson, FR and Wolfdale, AW, eds, *Secular, Solar, and Geomagnetic Variations in the Last 10,000 Years.* Dordrecht, The Netherlands, Kluwer: 209–224.

Wigley, TML and Kelly, PM 1990 Holocene climate change, ^{14}C wiggles and variations in solar irradiance. *Philosophical Transactions of the Royal Society (London)* A330: 547–560.

Wigley, TML and Raper, SCB 1987 Thermal expansion of sea water associated with global warming. *Nature* 330(12): 127–131.

Willson, RC and Hudson, HS 1988 Solar luminosity variations in the solar cycle. *Nature* 332: 810–812.

Willson, RC, Hudson, HS, Frohlich, C and Brusa, RW 1986 Long-term downward trend in total solar irradiance. *Science* 234: 1114–1117.

10

RADIOCARBON AND ASTROPHYSICAL-GEOPHYSICAL PHENOMENA

G E KOCHAROV

INTRODUCTION

On 29 May 1965, the well-known journal, *Nature*, published the article, "Possible Antimatter Content of the Tunguska Meteor of 1908" by C Cowan, C R Atluri and W F Libby. The Director of the Physico-Technical Institute, Professor Boris P Konstantinov, showed me this paper on 10 July 1965. Naturally, he was very excited by this paper. In the early 1960s, Professor Konstantinov proposed and realized some special experiments to check the possibility of antimeteorites entering the solar system. For 20 days, we regularly discussed the problem of retrospective monitoring of different natural phenomena by measuring ^{14}C content in tree rings. As a result, we proposed some unique possibilities for investigating astrophysical, geophysical and ecological problems (Konstantinov & Kocharov 1965).

It is a great honor for me to devote this summary chapter to the memory of W F Libby and B P Konstantinov.

COSMOGENIC ISOTOPES ON THE EARTH

Cosmogenic isotopes are produced in the galactic (GCR) and solar (SCR) cosmic-ray interactions with the material of the Moon and planets, meteorites, cosmic dust and the Earth's atmosphere and surface. The production rate of cosmogenic isotopes depends on the properties of cosmic rays, *ie,* their generation time profile, total flux, chemical composition and energy spectrum. The cosmic-ray characteristics in the vicinity of the Earth are governed both by the properties of cosmic-ray sources and by the specific features of their propagation in interstellar space and the heliosphere. By studying cosmogenic radioisotope concentration as a function of time in natural archives such as tree rings, polar ice cores, oceanic and sea-bottom sediments, we can reconstruct temporal variations of the cosmic rays in the past. We can also determine the properties of primary cosmic rays generated in such sources as supernova explosion and solar flares, as well as solar activity and its impact on various processes on the Earth, the characteristics of monotonic variations of the geomagnetic field and its reversals over time scales inaccessible by any other method.

The most detailed data presently available on variations of the cosmic-ray flux in the past have been obtained by studying the ^{14}C and ^{10}Be isotopes produced in the Earth's atmosphere. ^{14}C and ^{10}Be are the radioisotopes with the highest atmospheric production rate (^{14}C: 2.2 atoms cm^{-2} sec^{-1}; ^{10}Be: 0.02 atoms cm^{-2} sec^{-1}) of all relatively long-lived ($T \geq 100$ yrs) isotopes. The most significant advantage of these isotopes over any other cosmogenic radionuclide is the existence of natural archives to record the atmospheric concentration of an isotope during any year in the past, and faithfully store the information for times exceeding their half-lives, namely, 5730 years for ^{14}C and 1.5 Myr for ^{10}Be.

The major source of radiocarbon is the capture of thermal neutrons released in cosmic-ray interactions in the atmosphere by nitrogen, $^{14}N(n,p)^{14}C$. ^{10}Be is produced in spallation reactions of nitrogen and oxygen by cosmic rays. These isotopes are rapidly oxidized to form CO_2 and BeO, and enter subsequently into various geochemical and geophysical processes. CO_2 distributes itself throughout the global carbon exchange system, whereas BeO becomes attached to atmospheric aerosols, and is precipitated on the Earth's surface with them. Thus, the samples used to study these isotopes are essentially different, namely, ^{14}C measurements are primarily carried out on tree rings, whereas the ^{10}Be content is determined in polar ice and oceanic sediments.

The high production rate of ^{14}C and its comparatively short half-life permit us to use well-developed radiometric techniques. Application of such techniques to measure ^{10}Be content in samples would involve formidable difficulties due to its low production rate and long half-life. Thus, a more sophisticated and expensive method, accelerator mass spectrometry (AMS), is used. On the other hand, ^{14}C studies require the consideration of complex transport processes damping out fast processes and shifting the phase. Such behavior of the carbon system makes it difficult to study the high-frequency component of ^{14}C production-rate variations. However, the ^{14}C record facilitates the isolation of variations with scale times longer than 11 years, which is the strongest period in the cosmic-ray flux. On the other hand, the production rate of ^{10}Be is linearly related to its concentration in a sample. The long half-life of ^{10}Be enables us to study variations over even longer time scales.

The best precision routinely obtained in single-sample natural radiocarbon measurements is 0.2%. Accuracy of ^{10}Be measurements in natural samples is typically ~5%. ^{10}Be content in ice samples is being studied in Switzerland (eg, see Beer, Siegenthaler & Blinov 1988) and France (eg, see Raisbeck & Yiou 1988).

In this chapter, we make use of our year-by-year measurements of ^{14}C abundance in tree rings for the last 400 years. For the time interval, 20,000 to 40,000 BP, we used the data on ^{14}C abundance in stalagmites (Vogel 1983). For ^{10}Be, we used the experimental data obtained by the French and Swiss researchers mentioned above. We base the transfer of cosmogenic isotope abundance in a dated sample to production rates of ^{14}C and ^{10}Be and, thus, to the cosmic-ray source, solar activity and geomagnetic field, on the results of investigations of the Space Research Department of the Leningrad State Technical University and the Astrophysical Department of the Physico-Technical Institute of the Academy of Sciences of the USSR (see Kocharov *et al* 1989, 1990 and references therein).

MANIFESTATION OF THE SOLAR SCHWABE (11-YEAR) CYCLE IN THE TEMPORAL VARIATION OF ATMOSPHERIC RADIOCARBON

Many experimental efforts were undertaken during the last 20 years to observe the 11-year solar cycle in ^{14}C variations in dated samples. The Schwabe cycle is the most prominent in solar activity, and is important for studying the history of solar activity and solar-terrestrial relationships. It is difficult to observe the 11-year cycle in ^{14}C variation because, as noted above, the carbon system acts as a low-pass filter, and strongly dampens short-term variations. This results in the radiocarbon series long-term cycles being stronger and masking Schwabe-cycle manifestations of the radiocarbon abundance in the Earth's atmosphere. Thus, we need high-precision, annual measurements of radiocarbon abundance in tree rings for a long time interval. The longer the interval, the higher the accuracy of subtracting long-term variations. Theoretical estimates of the amplitude of the 11-year cycle of radiocarbon variations in dated samples are about 1% or even much less. Thus, measurement accuracy has to be much better than 1%. Another important factor is strong solar flares. Flares, such as on 23 February 1956 or 29 September 1989, can increase radiocarbon abundance in tree rings ~1%. Thus, they can essentially change the time profile of radiocarbon abundance in dated samples, and considerably complicate the 11-year-cycle manifestation (Lingenfelter & Ramaty 1970; Damon, Cheng & Linick 1989). It is important to note that the solar-flare effect may depend on tree location (namely latitude effect, eg, Fan *et al* 1983). To obtain the characteristics of short-term variations of radiocarbon abundance, we should have at least 200–300 years of very accurate annual measurements. It is desirable to obtain independent data from different laboratories for different tree locations.

Presently, more than 300 years of annual data on ^{14}C abundance in tree rings are available from our laboratories. These data were obtained during the last 20 years. We have two series of data. The first is a combination of experimental data from several Soviet laboratories (Kocharov *et al* 1985); the other contains the results of only one laboratory (Zhorzholiani *et al* 1988a; Kocharov *et al* 1990). Our first ^{14}C record was analyzed earlier (Kocharov *et al* 1985; Galli *et al* 1987a, b). Here, I will consider the latter.

Figure 10.1 presents our results in a three-point smoothing procedure. The first part, containing the Maunder minimum interval, will be discussed below. Now I consider only the interval, AD 1775–1940. Figure 10.1 shows short-term (~10 yr) variations and, at the same time, a long-term variation. To obtain the character of the latter, I used a 33-point smoothing procedure that excludes short-term fluctuations (Fig 10.2). For comparison, the result of a 33-point smoothing procedure for the Wolf number is also shown in Figure 10.2.

The decrease of radiocarbon abundance in the late part of the time sequence is due to the well-known Suess effect. Other peculiarities before the 20th century may be of solar and/or climatic origin. Thus, the trend curve (Fig 10.2) was removed from the original series to distinguish short-term variations (Fig 10.3).

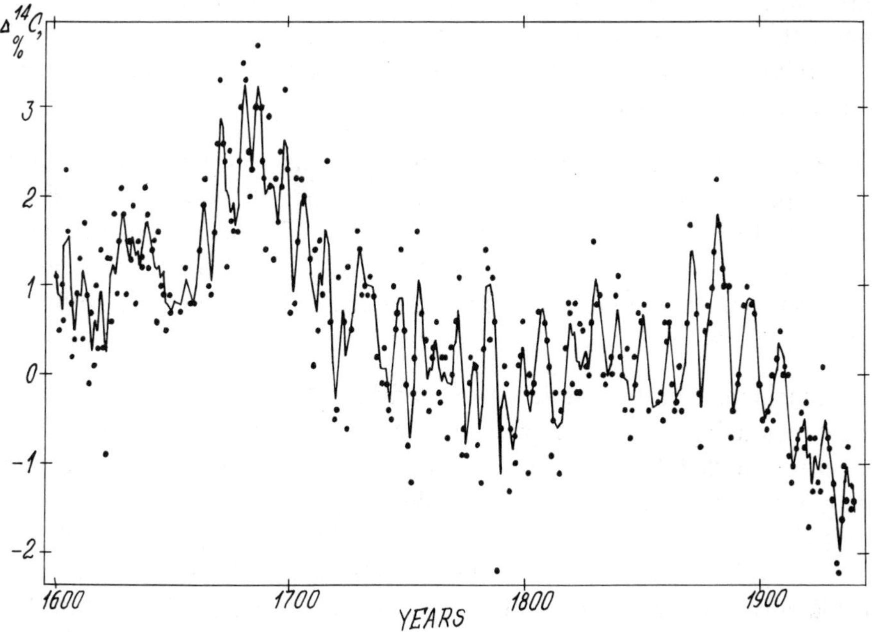

Fig 10.1. Radiocarbon abundance in tree rings for the time interval, AD 1600–1940 years. Counting precision is 0.20–0.30%.

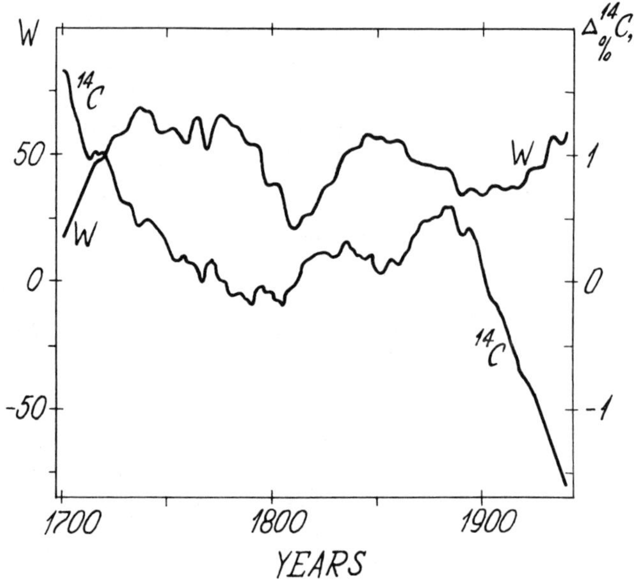

Fig 10.2. Long-term variations of atmospheric radiocarbon abundance and the Wolf number (1700–1940 yrs)

Hyperfine structure consists of 10- to 11-year components, which naturally connect this to the Schwabe solar cycle. The mean amplitude of the cycle is 0.57 ± 0.06%, which is practically the same as the 0.48% obtained for our first ^{14}C record (Galli *et al* 1987a).

The next step was to find different harmonic signals in the radiocarbon data. Fourier spectral analysis shows the existence of the following periods: 6.5, 7.8, 10.0, 11.0, 12.5, 14.0, 15.5, 20.0 and 24.0 years. The largest are 10.0 years – 0.29%, 11.0 – 0.36%, 12.5 – 0.23%, 15.5 – 0.20%. Thus, the 11-year harmonic is the principal one, and other harmonics increase the total amplitude up to the value, 0.57%, given above.

The coincidence of two series of experimental data, obtained by different laboratories using different experimental equipment and tree-ring samples, led us to conclude that the Schwabe solar cycle in radiocarbon abundance is present during more than 20 solar cycles. We need special consideration of the amplitude of the ^{14}C Schwabe cycle, because, due to the phase shift of the 11-year ^{14}C cycle, flare and modulation effects in the ^{14}C production rate are considerably overlapped (Damon, Cheng & Linick 1989). Another important point is the reduction of effective cutoff rigidities due to geomagnetic storms following major solar flares (Lingenfelter & Ramaty 1970). In the next step of analysis, we hope to distinguish pure Schwabe cycle, using the method of sep-

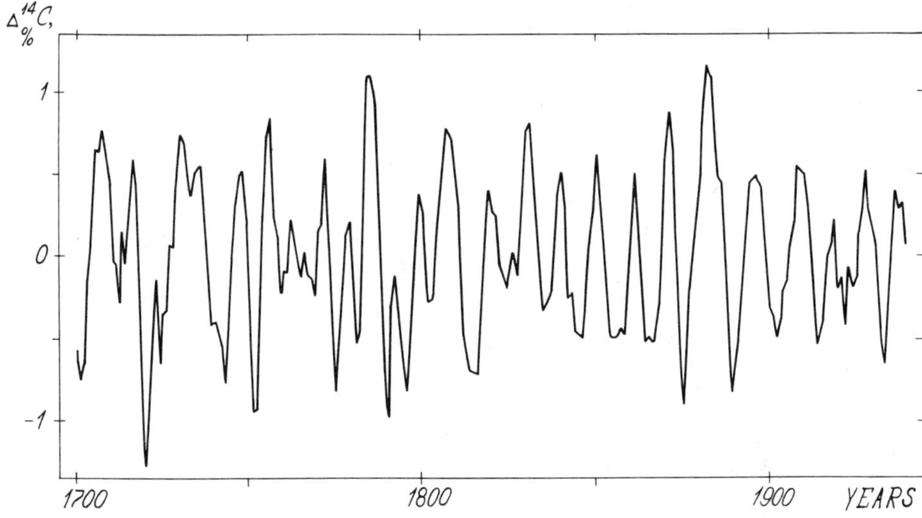

Fig 10.3. Short-term variations of ^{14}C abundance in tree rings (three-point smoothing)

aration of the galactic and solar components effects developed by Damon, Cheng and Linick (1989). An identification of solar-flare-produced radiocarbon in tree rings provides a unique opportunity to obtain the upper limit of total energy contained in flare-accelerated particles, which is the key problem in solar astrophysics. It is obvious that the higher the total energy of accelerated particles, the lower the probability of a corresponding flare – thus, the importance of observing the longest time interval in the cosmogenic isotope record.

Variation of Radiocarbon Content in Tree Rings During the Maunder Minimum (AD 1640–1715) of Solar Activity

By analyzing the available data on radiocarbon concentration in tree rings, Kocharov *et al* (1983) and Stuiver and Braziunas (1988) showed that, during the last 10,000 years, solar activity regularly passed through deep minima. During such periods (100–200 yrs), galactic cosmic-ray flux grows significantly (1.5–2.0 times higher than now), and tree-ring growth is depressed considerably (Kocharov 1986, 1987). The nature of deep solar minima and of solar influence on cosmic rays during such minima are not yet known.

Now I will discuss the problem of heliomagnetic modulation of the cosmic-ray flux during the Maunder minimum using high-precision measurements of atmospheric ^{14}C concentration.

As noted above, two cycles of measurements were made in the USSR. In the first case, the measurements were made for even years (Vasiliev & Kocharov 1983; Kocharov 1986) by the Physical-Technical Institute (FTI). In the second case, we obtained radiocarbon abundance for each tree ring (Zhorzholiani *et al* 1988a, b; Kocharov, Peristykh & Chesnokov 1989), using the facilities of Tbilisi State University (TU). Pine-tree samples collected at 54°N, 56°E were used in FTI, and fir-tree samples from 50°N, 24°E and 48°N, 25°E at TU. The time resolution in FTI measurements is also the highest (1 yr), but ^{14}C abundance is obtained for only even years. It was concluded (Vasiliev & Kocharov 1983; Kocharov 1986) that, during the Maunder minimum, a periodicity close to 20 years is more characteristic. Galli *et al* (1987b) used a cyclogram method to study both the ^{14}C data of Vasiliev and Kocharov (1983) and the ^{10}Be data of Beer *et al* (1983b). It was shown that the 20-year cycle exists in both series.

No ^{13}C values were obtained at FTI, and thus, the samples were not corrected for isotopic fractionation. Dmitriev, Peristykh and Kharchenko (1987) measured the ^{13}C abundance in 40 of 55 samples used by Vasiliev and Kocharov (1983), and studied, in detail, the periodicity problem. The authors showed that the 11-year cycle is the principal one before the Maunder minimum, but there are also 16-year periodicities. The authors also found 16-year, 22-year and 11-year periods during the Maunder minimum, with the amplitude of the 22-year (Hale) cycle being the highest.

Taking into account the significance of the problem of cosmic-ray modulation during the pre-Maunder, Maunder and post-Maunder periods in new cycle investigations, ^{14}C and ^{13}C abundance were measured in each annual tree ring. The data obtained are presented in Figure 10.1 (primary experimental results) and Figure 10.4 (results of moving-average processing by 3 and 11 points). As shown, ^{14}C abundance in tree rings varies considerably during the Maunder minimum. We obtained short-period variations by removing the trend curve from the original series. For periodicity determination, Buys-Ballot, Wittaker and Schuster periodogram methods were used. For the interval, 1645–1715, the following periods are obtained: 17-year; 26–27-year; 22-year; 7-year; 10-year; 12-year. The amplitudes of 7-year, 10-year and 11-year periods are low. For the pre-Maunder period, the highest is the 7-year period, then 18-year and 11-year.

Summary of Maunder Minimum ^{14}C Variations

1. During the Maunder minimum, radiocarbon abundance in annual tree rings changed in time. This means that the Sun modulates intensity of cosmic rays, even during times of extra-low sunspot numbers. The inverse relationship between the ^{14}C production rate and sunspots (solar activity) during the 11-year solar cycle also holds during the deep solar minima.

2. For the Maunder minimum, both series of our experiments, FTI and TU, show that the periods in the time interval, 17–26 years, are more pro-

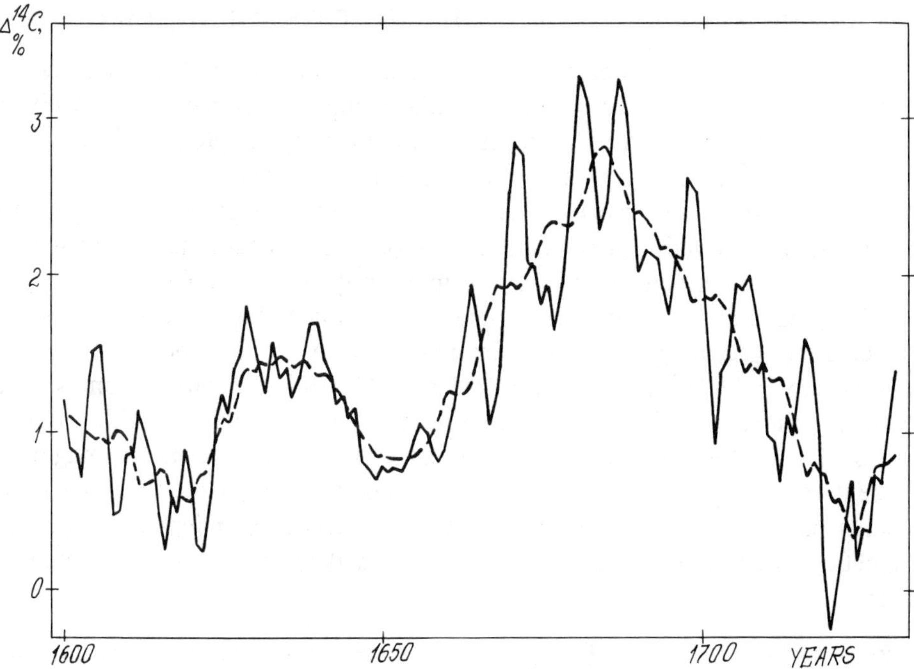

Fig 10.4. Radiocarbon abundance variation in the Earth's atmosphere during the Maunder minimum; — 3-point moving average smoothing; – – – 11-point moving average smoothing

nounced. The obtained periodicity is undoubtedly related to the Hale cycle.

3. For the pre-Maunder period, there is a difference in the character of periodograms for the two series of the ^{14}C record. For FTI data, the main amplitude is the 11-year periodicity, and for TU data, the highest amplitude is the 7-year periodicity. The reason for this discrepancy is unknown. New measurements are needed, using samples from different kinds of trees, from different places (latitudes, longitudes, altitudes, far from and close to the sea, etc).

4. When the Sun enters and emerges from the deep minimum, a change of heliomagnetic modulation of the cosmic-ray flux occurs, namely, the periods higher than 11 years become more important.

These data are essential for the theory of modulation, and it is important to study the possibilities of changing modulation character during the transition of the Sun to and from the deep minimum. From an experimental point of view, it is important to measure radiocarbon abundance in annual tree rings for the Spörer minimum (AD 1420–1570).

COSMOGENIC ISOTOPES AND THE GEOMAGNETIC FIELD IN THE PAST

A new method of paleomagnetic investigation based on the analysis of cosmogenic isotope abundance in terrestrial samples was recently conceived (Kocharov *et al* 1989). The idea of this method is very simple and clear. Near the geomagnetic pole, the magnetic field has no influence on the cosmogenic isotope production rate because the geomagnetic cutoff energy becomes less than the effective threshold energy of the corresponding nuclear reactions in which the given isotope is generated. For the modern geomagnetic field, this region for ^{10}Be production is located on geomagnetic latitudes higher than 68°. With an increase of geomagnetic dipole moment (GDM) by 50%, the boundary moves from 68° to 70°. If we decrease GDM by 50%, the boundary becomes 64°. Thus, if we measure ^{10}Be abundance in ice in two places with latitudes higher and lower than the boundary value, we can calculate the global geomagnetic moment by a comparative analysis of obtained data. Of course, we can also use another pair: ^{10}Be from a high latitude (abundance is independent of the magnetic field influence) and ^{14}C abundance in dated samples that contain a geomagnetic record.

A very important feature of this method is the possibility of measuring the geomagnetic dipole moment. Traditional methods allow us to measure local geomagnetic records, which may be significantly influenced by non-dipole fields.

Based on the new method, we reconstructed temporal variations of the geomagnetic field in the past, including the epoch of the Brunhes-Matuyama magnetic reversal. We used experimental data on ^{10}Be concentrations in deep-sea sediments obtained by Raisbeck *et al* (1985). Figure 10.5 presents the results. There is an obvious minimum around 720,000 BP, which corresponds, according to geological data, to the Brunhes-Matuyama reversal. The minimum of the magnetic field occurs 3000–4000 years after 720,000 BP, and quasi-periodic variations occur in the 10,000-year period with an amplitude of 10–20%.

Figure 10.5 illustrates rich possibilities of a developed method for paleomagnetic investigations: long time scale ($\geq 10^6$ yrs), high accuracy and the unique opportunity to measure the geomagnetic dipole moment.

FIRST EXPERIMENTAL CONFIRMATION OF THE HYPOTHESIS ON COSMIC-RAY GENERATION DURING THE SUPERNOVA EXPLOSION

The main principles and theoretical models that enable us to identify the supernova (SN) characteristics by means of the temporal evolution of radiocarbon abundance in the Earth's atmosphere has been developed by Konstantinov and Kocharov (1965, 1967), Kocharov (1970), Lingenfelter and Ramaty (1970),

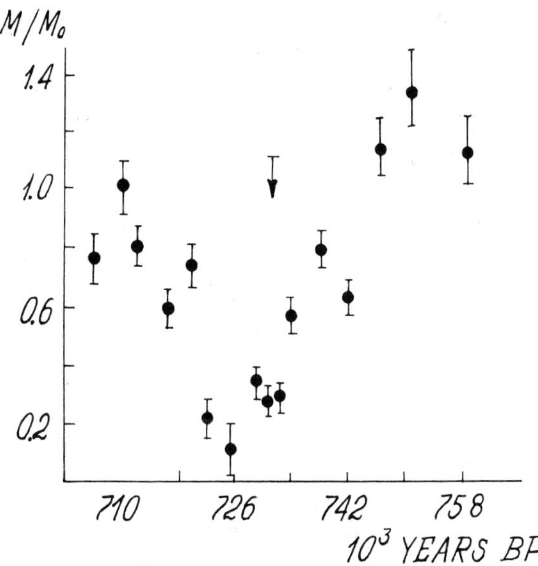

Fig 10.5. The time profile of the Brunhes-Matuyama magnetic reversal. The arrow marks the reversal position.

and Kocharov and Rumyantsev (1971). Several components can be distinguished in a supernova-produced variation in the generation rate of cosmogenic isotopes: rapid (due to γ rays) and slow (due to CR). Simultaneous analysis of the different components allows us to determine the explosion time (t_e), distance (r) to the SN, CR total energy (E_{CR}) and dynamics of cosmic-ray generation during the explosion.

The analysis (Kocharov 1970) of experimental data on ^{14}C concentration for blocks of tree rings over an 8000-year time scale obtained in the late 1960s showed that, along with time variation of geomagnetic origin, another component might be explained in terms of cosmic-ray effect of SNe ($r \leq 100$ pc, $E_{CR} \simeq 5 \cdot 10^{50}$ erg, t_e – a few tens of thousands years ago).

The high-precision annual ^{14}C measurements in tree rings (Kocharov *et al* 1974) allowed us to estimate the upper limit of total energy contained in the gamma-ray component at 10^{49} erg for SN Tycho Brage, Kepler and Cassiopeia A.

Based on ^{10}Be abundance data in the basal sediments of the Pacific Ocean, Lingenfelter (1976) pointed out a three-fold ^{10}Be generation-rate increase in the Earth's atmosphere a few million years BP. He connected such an increase to cosmic rays generated in nearby SN_e ($r \simeq 60$ pc, $E_{CR} \simeq 10^{50}$ erg).

Kocharov (1982) and Konstantinov and Kocharov (1984) analyzed the results of ^{10}Be abundance measurements by Raisbeck *et al* (1981) and Beer *et al* (1983a) in the ice cores from Antarctica and Greenland, and found a considerable increase of CR intensity over the period 10,000–40,000 years ago. The researchers interpreted the time profile and amplitude of the effect as a result of a nearby SN_e ($r \approx 30$ pc, $E_{CR} \approx 10^{50}$ erg). Measurements for the time interval under consideration by Raisbeck *et al* (1987) and Vogel (1983), involving ^{10}Be in ice cores and ^{14}C in stalagmites, respectively, considerably increased the volume of experimental data, which enables us to conclude that cosmic rays from SN_e (a few tens of thousands years BP) were really detected.

Time profiles of cosmic-ray-produced ^{10}Be in deep ice cores from Dome C and Vostok stations, Antarctica, and from Dye-3 station, Greenland, are practically the same. According to Vogel's (1983) data, there is a synchronous increase of ^{14}C abundance in stalagmites. Thus, there is very good agreement between experimental data obtained by two independent scientific groups, who used ice cores from the North and South Poles. It is of principal importance that, for two cosmogenic isotopes (^{14}C and ^{10}Be), with different geochemical and geophysical behavior, the time profiles of the abundances are practically identical. Thus, the observed effect is undoubtedly global, and during 30,000–40,000 BP, the production rate of cosmogenic isotopes in the Earth's atmosphere was considerably higher than now. The question is why was this so?

Sonett, Morfil and Jokipii (1987) gave a qualitative explanation for the increased abundance of ^{10}Be in ice cores obtained by Raisbeck *et al* (1987) as a consequence of supernova shock waves. All the experimental data on the ^{10}Be and ^{14}C production rate from 30,000–40,000 BP were considered quantitatively by Konstantinov and Levchenko (1988), Kocharov, Konstantinov and Levchenko (1990), and Konstantinov, Kocharov and Levchenko (1990). Figure 10.6 presents the results on the time variation of ^{10}Be and ^{14}C production rates. Possible explanations are the following: the absence of solar and geomagnetic modulation of cosmic rays during the period, 30,000–40,000 BP; the flux of solar cosmic rays was 100 times greater than now; the intensity of cosmic rays in the interstellar medium increased during that time interval.

The production-rate enhancement of ^{14}C and ^{10}Be, in principle, may be caused by the geomagnetic field reduction (eg, geomagnetic reversal). But ^{10}Be abundance in ice at Vostok station is not sensitive to magnetic-field variations, because the station is close to the geomagnetic pole, assuming the magnetic pole was, at the time, in the same place as it is now. But, if we do not make this assumption, and the magnetic pole was farther from Vostok than it is now, ^{10}Be

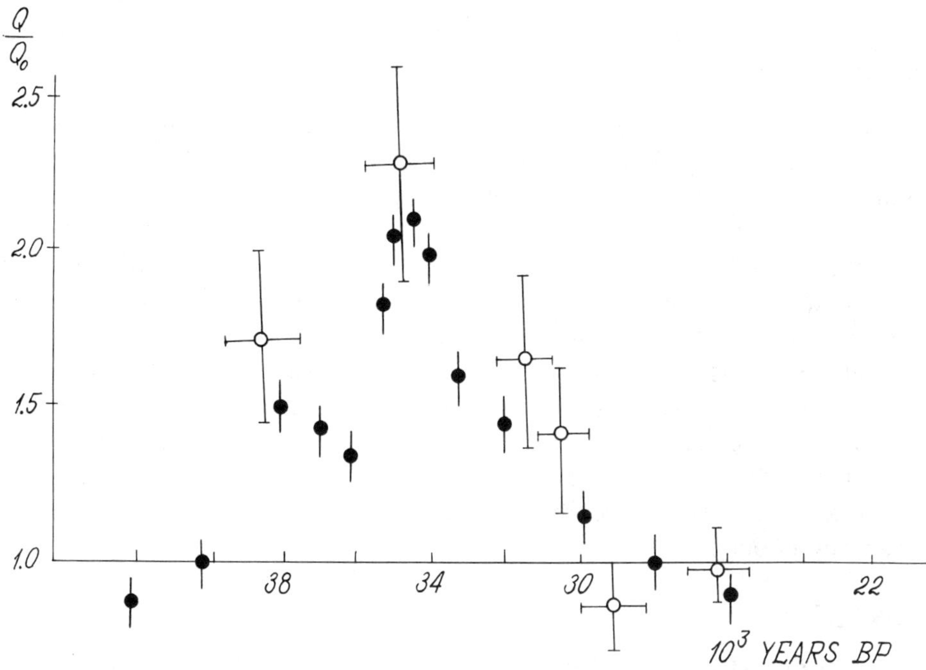

Fig 10.6. ^{10}Be and ^{14}C production rate, 30,000–40,000 BP; ● - ^{10}Be; ○ - ^{14}C

production rate, at that time, had to be less and not higher than now, unless the geomagnetic field was simultaneously significantly weaker.

Now let us consider the solar modulation effect. For the modern cosmic-ray energy spectrum, an effective energy region for ^{10}Be production is 1 to 5 GeV. We cannot explain the observed effect, even if we assume that, during the whole time interval of 30,000–40,000 BP, solar activity was practically zero. An alternative possibility is related to the assumption that, at that time, the interstellar CR spectrum was considerably enriched in the low (< 5 GeV) energy region, in comparison to the present. The considered effect, in principle, can be governed by solar cosmic rays, if we assume that, at that time, the Sun was very active, and the mean flux of solar cosmic rays was 100 times higher than the modern value. We cannot exclude such a possibility. But even in this case, to explain a proportional increase of ^{14}C production rate, we should assume that, in the considered time interval, the solar cosmic-ray energy spectrum was harder than now, and was very close to the galactic cosmic-ray spectrum. It is clear that it is a less extreme assumption to suppose that during the period, 30,000–40,000 BP, the galactic cosmic-ray flux was higher than now. Thus, the most likely reason is increased intensity of cosmic rays in the interstellar medium. To explain the obtained time profile of the ^{10}Be and ^{14}C generation rate, a burst-like cosmic-ray source should be located at the distance, 55 pc, and the total energy of cosmic rays has to be $5 \cdot 10^{49}$ erg.

Figure 10.6 shows increased values of ^{10}Be and ^{14}C generation rate over the time interval, 35,000–40,000 years ago, as well as the larger increase between 30,000–35,000 BP. Such an increase may also be linked to a supernova explosion. If one assumes the generation of accelerated particles started 40,000 years ago, then the increased cosmogenic isotope generation rate during the first 5000 years may be a result of high-energy gamma-ray or neutron emission generated by cosmic rays in supernova material. On the other hand, we cannot rule out a two-stage cosmic-ray-flux increase as a result of supernova dynamics and shock-wave propagation. It seems of great importance to carry out a more detailed measurement of cosmogenic isotope abundance over the time interval, 35,000–40,000 BP. Note the importance of an identification of at least one more SN_e as a cosmic-ray source. In this case, it would enable us to derive the SN_e frequency in our galaxy and estimate the SN_e cosmic-ray total energy.

Conclusion

During the last 40 years, the efforts of the international radiocarbon community made this new branch of scientific investigation important. Radiocarbon dating is a unique source of information on fundamental and applied problems in astrophysics, geophysics and ecology. Now radiocarbon investigations consist of two promising directions of scientific research, *ie,* dating and measuring ^{14}C abundance in independently dated samples.

An analysis of high-precision measurements of the cosmogenic isotope concentration in accurately dated samples has yielded a number of new and important results. Among them:

1. Variations in the ^{14}C content during the Maunder minimum of solar activity were revealed for the first time, and cosmic-ray modulation was shown to exist even during the near total absence of sunspots. This latter discovery is principally important for the solar-activity problem and for the theory of solar modulation of cosmic rays.

2. The states of exceedingly low solar activity were regular rather than exceptional events in the recent history of the Sun. During the last 10,000 years, solar activity passed through at least 15 deep minima.

3. High-precision annual ^{14}C measurements in tree rings for the past 300 years enabled the discovery of the Schwabe-cycle manifestation in time variations of radiocarbon abundance in the Earth's atmosphere.

4. A synchronous increase of the ^{10}Be and ^{14}C production rate in the Earth's atmosphere 30,000–40,000 years ago has been evaluated. The time profile and amplitude increase show that a proper source of increased intensity of cosmic rays was located not far from the solar system ~ 50 pc. This is the first experimental confirmation of the longstanding fundamental hypothesis on cosmic-ray generation during a supernova explosion.

5. A new powerful method of geomagnetic dipole moment reconstruction in the past was presented in this paper. We used this method to extend the period of investigation from 10,000 years, accessible through archaeomagnetic and paleomagnetic methods, up to 1,000,000 years. Temporal variation of the geomagnetic field during the Brunhes-Matuyama reversal (720,000) was obtained for the first time.

The new method offers a value of the geomagnetic dipole moment instead of the local value of the geomagnetic field, which can be strongly modified by the non-dipole component. This method allows us not only to measure temporal variations of the geomagnetic field in the past, but also to establish the nature of short-scale variations and, in principle, to solve the geomagnetic poles migration problem.

Acknowledgments

The results considered in this chapter were obtained through active collaboration with many of my colleagues and friends in the USSR. I would like to thank all of them.

I thank Professors Paul E Damon and R E Taylor for the invitation to the conference, "Four Decades of Radiocarbon Studies: An Interdisciplinary Perspective," at which this review of the field was presented.

References

Beer, J, Andrée, M, Oeschger, H, Stauffer, B, Balzer, R, Bonani, G, Stoller, C, Suter, M, Wölfli, W and Finkel, RC 1983a Temporal ^{10}Be variations in ice. *In* Stuiver, M and Kra, RS, eds, Proceedings of the 11th International ^{14}C Conference. *Radiocarbon* 25(2): 269–278.

Beer, J, Oeschger, H, Andrée, M, Bonani, G, Hofman, HJ, Nessi, M, Suter, M, Wölfli, W, Finkel, R, Langway, CC and Siegenthaler, U 1983b Temporal ^{10}Be variations. *In* Ramana Murthy, P, ed, Proceedings of the 18th International Cosmic Ray Conference. Bangalore. *IUPAP* 9: 317–320.

Beer, J, Siegenthaler, U and Blinov, A 1988 Temporal ^{10}Be variations in ice: Information on solar activity and geomagnetic variations in the last 10,000 years. In *Secular, Solar and Geomagnetic Variations in the Last 10,000 Years.* Dordrecht, The Netherlands, Kluwer: 297–317.

Cowan, C, Atluri, CR and Libby, WF 1965 Anti-matter content of the Tunguska meteor of 1908. *Nature* 206: 861–865.

Damon, PE, Cheng, S and Linick, TW 1989 Fine and hyperfine structure in the spectrum of secular variations of atmospheric ^{14}C. *In* Long, A and Kra, RS, eds, Proceedings of the 13th International ^{14}C Conference. *Radiocarbon* 31(3): 704–718.

Dmitriev, PB, Peristykh, AN and Kharchenko, AA 1987 Variation of radiocarbon abundance in the Earth's atmosphere during 1600–1730 years. *Physical-Technical Institute Preprint* 1104: 18 p (in Russian).

Fan, CY, Tei-Mei, C, Si-Xun, Y and Kai-Mei, D 1983 Radiocarbon activity variation in dated tree rings grown in MacKenzie Delta. *In* Stuiver, M and Kra, RS, eds, Proceedings of the 11th International ^{14}C Conference. *Radiocarbon* 25(2): 205–212.

Galli, M, Castagnoli, CC, Attolini, MR, Cechini, S, Nanni, T, Kocharov, GE, Mikheeva, IB, Bitvinskas, TT, Konstantinov, A and Metskhvarishvili, R Ya 1987a A 400-year ^{14}C record: 11-year and longer cycles. *In* Chudakov, E, ed, Proceedings of the 20th International Cosmic Ray Conference, Moscow. *IUPAP* 4: 280–283.

Galli, M, Castagnoli, CC, Attolini, MR, Cechini, S, Nanni, T, Kocharov, GE, Vasiliev, VA and Konstantinov, AN 1987b The 20 year cycle of solar activity in ^{14}C and ^{10}Be (before and during Maunder minimum). *In* Chudakov, A, ed, Proceedings of the 20th International Cosmic Ray Conference, Moscow. *IUPAP* 4: 284–287.

Kocharov, GE 1970 Variations of radiocarbon abundance in the Earth's atmosphere. *Atomnaya Energia* 28: 444–446 (in Russian).

____1982 Burst of cosmic radiation and cosmogenic isotopes. *In* Kocharov, GE, ed *Integrated Investigation of the Sun.* Leningrad, Ioffe Physical-Technical Institute: 203–207.

____1986 Cosmic ray archaeology, solar activity and supernova explosions. *In* Proceedings of the International School and Workshop on Plasma Astrophysics, Sukhumi, USSR. *European Space Agency* (ESA) SP-251: 259–270.

____1987 Tree rings: A unique source of information on processes on the Earth and in space. *Annales Academica Scientianum Fennicae* A3-145: 137–147.

Kocharov, GE, Bitvinskas, TT, Vasiliev, VA, Dergachev, VA, Konstantinov, AN, Metskhvarishvili, R Ya, Ostryakov, VM and Stupneva, AV 1985 Cosmogenic isotopes and astrophysical phenomena. *In* Kocharov, GE, ed, *Astrophysical Phenomena and Radiocarbon.* Leningrad, Ioffe Physical-Technical Institute: 9–142 (in Russian).

Kocharov, GE, Blinov, AV, Konstantinov, AN and Levchenko, VA 1989 Temporal ^{10}Be and ^{14}C variations: A tool for paleomagnetic research. *Radiocarbon* 31(2): 163–168.

Kocharov, GE, Dergachev, VA, Sementsov, AA, Romanova, EN, Rumyantsev, SA and Malanova, NS 1974 Concentration of radiocarbon in tree rings 1564–1583, 1593–1615, 1688–1712. *In* Kocharov, GE, ed, *Proceedings of the 5th Conference on Astrophysical Phenomena and Radiocarbon.* Tbilisi: 47–60.

Kocharov, GE, Konstantinov, AN and Levchenko, VA 1990 Cosmogenic ^{10}Be: Cosmic rays over the past 150,000 years. *In* Proceedings of the 12th International Cosmic Ray Conference, Adelaida. *IUPAP* 7: 120–123.

Kocharov, GE, Peristykh, AN and Chesnokov, VI 1989 Radiocarbon content in tree rings in the period of Maunder minimum. *Physical-Technical Institute Preprint* N1374: 17 p.

Kocharov, GE and Rumyantsev, SA 1971 γ-radiation of supernova and atmospheric radiocarbon. *In* Kocharov, GE, ed, *Radiocarbon.* Vilnius, Latvia, Raide Publications: 35–38 (in Russian).

Kocharov, GE, Vasiliev, VA, Dergachev, VA and Ostryakov, VM 1983 An 8000-year sequence of galactic cosmic-ray fluctuations. *Pis'ma Astronomichesky Zhurnal* 9: 206–210 (in Russian).

Kocharov, GE, Zhorzholiani, IV, Lomtatidze, ZV, Peristykh, AN, Tsereteli, SL and Chesnokov, VI 1990 On the solar activity characteristics during the past 400 yr. *Pis'ma Astronomichesky Zhurnal* 16: 723–728 (in Russian).

Konstantinov, AN and Kocharov, GE 1984 A 30,000 year record of the cosmic-ray intensity. *Pis'ma Astronomichesky Zhurnal* 10: 94–97 (in Russian).

Konstantinov, AN, Kocharov, GE and Levchenko, VA 1990 On the supernova explosion 35 ky ago. *Pis'ma Astronomichesky Zhurnal* 16: 799–803 (in Russian).

Konstantinov, AN and Levchenko, VA 1988 Cosmic ray variations over the last 150 thousand years. *In* Kocharov, GE, ed, *Cosmic Rays and Isotope Ecology.* Leningrad, Ioffe Physical-Technical Institute: 48–64 (in Russian).

Konstantinov, BP and Kocharov, GE 1965 Astrophysical phenomena and radiocarbon: *Soviet Academy Reports* 165: 63–64 (in Russian).

_____1967 Astrophysical phenomena and radiocarbon. *Physical-Technical Institute Preprint* N64: 43 p.

Lingenfelter, RE 1976 Cosmic ray produced neutrons and nuclides in the Earth's atmosphere. In *Spallation Nuclear Reactions and Their Applications.* Dordrecht, The Netherlands, D Reidel: 193–205.

Lingenfelter, RE and Ramaty, R 1970 Astrophysical and geophysical variations in ^{14}C production. *In* Olson, IU, ed, *Radiocarbon Variations and Absolute Chronology.* Proceedings of the 12th Nobel Symposium. New York, John Wiley & Sons: 513–537.

Raisbeck, GM and Yiou, F 1988 ^{10}Be as a proxy indicator of variations in solar activity and geomagnetic field intensity during the last 10,000 years. In *Secular, Solar and Geomagnetic Variations in the Last 10,000 Years.* Dordrecht, The Netherlands, Kluwer: 287–296.

Raisbeck, GM, Yiou, F, Bourles, D and Kent, D 1985 Evidence for an increase in cosmogenic ^{10}Be during a geomagnetic reversal. *Nature* 315: 315–317.

Raisbeck, GM, Yiou, F, Bourles, D, Lorius, C, Joursel, J and Barkov, NI 1987 Evidence for two intervals of enhanced ^{10}Be deposition in Antarctic ice during the last glacial period. *Nature* 326: 273–277.

Raisbeck, GM, Yiou, F, Fruneau, M, Loiseaux, JM and Ravel, JC 1981 Cosmogenic $^{10}Be/^{9}Be$ as a probe of atmospheric transport processes. *Geophysical Research Letters* 8: 1015–1018.

Sonett, CP, Morfil, GE and Jokipii, JR 1987 Interstellar waves and ^{10}Be from ice core. *Nature* 330: 458–460.

Stuiver, M and Braziunas, T 1988 The solar component of the atmospheric ^{14}C record in secular solar and geomagnetic variations in the last 10,000 years. In *Secular, Solar and Geomagnetic Variations in the Last 10,000 Years.* Dordrecht, The Netherlands, Kluwer: 245–260.

Vasiliev, VA and Kocharov, GE 1983 On the solar activity dynamics during the Maunder minimum. *In* Kocharov, GE, ed, *International Seminar on Cosmophysics.* Leningrad, Ioffe Physical-Technical Institute: 75–100 (in Russian).

Vogel, JC 1983 ^{14}C variations during the upper Pleistocene. *In* Stuiver, M and Kra, RS, eds, Proceedings of the 11th International ^{14}C Conference. *Radiocarbon* 25(2): 213–218.

Zhorzholiani, IV, Kereselidze, PG, Kocharov, GE, Lomtatidze, ZN, Marchilashvili, NM, Metskhvarishvili, R Ya, Tagauri, ZA, Tsereteli, SL and Chesnokov, VI 1988a Measurement of radiocarbon content in tree rings for the period 1600–1940 yy with corrections for isotope fractionation. *In* Kocharov, GE, ed, *Experimental Methods of Investigations of Astrophysical and Geophysical Phenomena.* Leningrad, Ioffe Physical-Technical Institute: 93–113 (in Russian).

Zhorzholiani, IV, Kocharov, GE, Lomtatidze, ZV, Metskhvarishvili, R Ya and Chesnokov, VI 1988b Processing and comparison of radiocarbon data for 1600–1800 years obtained in the Physical-Technical Institute and Tbilisi University using the moving average technique. *In* Kocharov, GE, ed, *Experimental Methods of Investigations of Astrophysical and Geophysical Phenomena.* Leningrad, Ioffe Physical-Technical Institute: 114–124 (in Russian).

11

COSMOGENIC *IN SITU* RADIOCARBON ON THE EARTH

DEVENDRA LAL

INTRODUCTION

Radiocarbon is continuously produced on the Earth by nuclear interactions of cosmic rays in the Earth's atmosphere, and because of its dynamic circulation in the carbon reservoirs, it finds applications in diverse branches of science (Libby 1946, 1967). The scope of applications of *natural radiocarbon* has widened considerably in the last decade with the advent of accelerator mass spectrometry (AMS). This technique allows a high-precision measurement of 10^6–10^7 atoms ^{14}C, thereby allowing its measurement in samples of ~ 0.1 mg carbon from the dynamic reservoirs. This advance makes it possible, for example, to date bones by measuring ^{14}C in proteins (Hedges & Law 1989) and in individual species of planktonic and benthic calcareous shells deposited in the sediments (Andrée *et al* 1986).

The AMS advance also now makes it possible to measure ^{14}C and other long-lived radionuclides expected to be produced *in situ* by cosmic rays in terrestrial materials (Davis & Schaeffer 1955; Lal & Peters 1967; Lal & Arnold 1985; Phillips *et al* 1986; Lal, Nishiizumi & Arnold 1987) and by radiogenic neutrons and α-particles (Wetherill 1953; Davis & Schaeffer 1955; Zito *et al* 1980). In this chapter, I discuss the terrestrial production rates of ^{14}C due to these nuclear reactions, but I consider specifically the applications made to date and those possible using the *in situ* cosmogenic ^{14}C in terrestrial solids. The *in situ* cosmogenic ^{14}C has been detected unambiguously in terrestrial rocks (Jull *et al* 1990) and in polar ice (Lal *et al* 1990), and successfully applied to determining the ablation and accumulation rates of polar ice (Lal *et al* 1990; Lal & Jull 1990) according to expectations (Lal, Nishiizumi & Arnold 1987).

PRODUCTION OF RADIOCARBON ON THE EARTH

Principal Nuclear Reactions

The principal mode of production of natural radiocarbon in the Earth's atmosphere is the thermal neutron capture by atmospheric nitrogen (Korff 1940).

More than 99% of the ^{14}C produced in the atmosphere arises from this reaction (Lingenfelter 1963; Lal & Peters 1967). I estimate that the rest (~ 0.7%) is produced primarily from the spallation of atmospheric oxygen. The average concentration of nitrogen in terrestrial solids is < 50 ppm, and therefore, the *in situ* ^{14}C production in solids from cosmic-ray spallation of oxygen (typical concentration ~ 50%) is at least one order of magnitude greater than that due to capture of cosmic-ray thermal neutrons in nitrogen, and can be ignored generally.

The radiogenic *in situ* production of ^{14}C on the Earth occurs (*cf* Zito *et al* 1980) primarily by the exothermic reactions with neutrons:

$$^{17}O + n = {}^{14}C + {}^{4}He \tag{11.1}$$

$$^{14}N + n = {}^{14}C + {}^{1}H \tag{11.2}$$

$$^{13}C + n = {}^{14}C + \gamma \tag{11.3}$$

and with alpha particles:

$$^{11}B + {}^{4}He = {}^{14}C + {}^{1}H \ . \tag{11.4}$$

In the upper few meters of rocks exposed at sea level, the *in situ* production rate of ^{14}C due to cosmic-ray neutrons exceeds the radiogenic production of ^{14}C by several orders of magnitude because of the very low abundance of target elements in the rocks (those in reactions 11.1–11.4), and also due to the lower radiogenic neutron flux (Lal 1987a) compared to the cosmogenic neutron flux. The (^{4}He, p) reaction in ^{11}B does not constitute an important source of ^{14}C because of the low abundance of ^{11}B and the low energies of the radiogenic ^{4}He particles. Zito *et al* (1980) estimate correctly that reactions (11.1) and (11.2) produce about 66% and 34% of non-spallogenic *in situ* ^{14}C and reaction (11.3) about 10^{-3}%. These are, of course, typical estimates; in certain mineral compositions, the ^{13}C and ^{11}B reactions may be more important.

Cosmogenic and Radiogenic Production Rates of ^{14}C in Terrestrial Materials

Because of geomagnetic screening of the cosmic-ray flux at the top of atmosphere, the secondary cosmic-ray flux in the atmosphere is latitude-dependent. It is altitude dependent because the cosmic-ray energy is dissipated in nuclear reactions with the atmospheric nuclei. The rate of nuclear disintegrations in the atmosphere varies with absorption mean free paths of

(150–200) g cm^{-2}, depending on the latitude and altitude (Lal & Peters 1967). The energy spectrum and flux of neutrons in the atmosphere are well known (Lal & Peters 1967; Lingenfelter 1963). In Figure 11.1, we reproduce the estimated rates of nuclear disintegrations in the atmosphere (Lal & Peters 1967) for 0–10 km altitude and 0°–90° geomagnetic latitudes, due to nucleons of energy ≥ 40 MeV. In a rock exposed at any given location in the atmosphere, the total rate of nuclear disintegrations (per g rock) at the rock surface would be higher than that in the atmosphere by the scaling factor $A^{-1/3}$, where A is the mean atomic mass number of the target; the nuclear interaction cross-section varies as $A^{2/3}$, whereas the number of target atoms g^{-1} as A^{-1}.

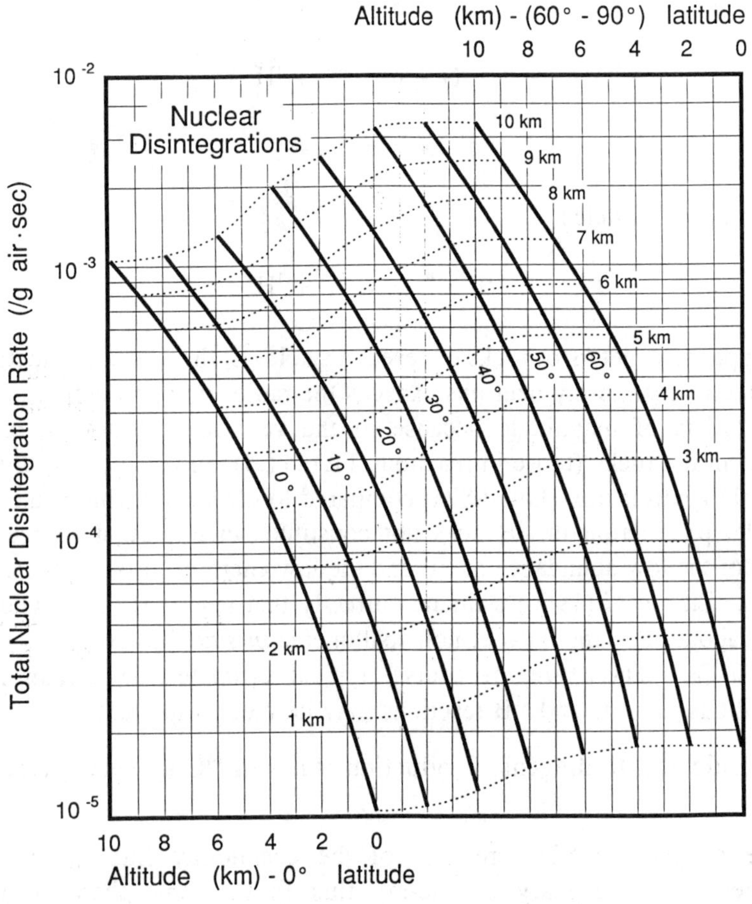

Fig 11.1. The total rate of nuclear disintegrations in the atmosphere (with energy release > 40 MeV) is plotted as a function of altitude and geomagnetic latitude for 0–10 km altitude and 0°–90° latitudes, in steps of 10°. Curves for different latitudes have been successively displaced by 1 km along the abscissa (based on Lal & Peters 1967).

Within a rock, the rate of nuclear disintegrations at latitudes > 50° decreases exponentially with a mean free path of ~ 150 g cm^{-2} (equivalent to ~ 0.5 m rock). The production rate of neutrons in the nuclear disintegrations is obtained by multiplying the rate by the average yield of neutrons per disintegration. In air, the average yield is 2.9, whereas in a typical rock it is 3.2 (Lal 1987a). Most of the neutrons produced in the nuclear disintegrations slow down to thermal energies (Lingenfelter 1963). The rate of production of ^{14}C from thermal neutron reactions (11.1)–(11.3) can therefore be calculated using Figure 11.1.

As mentioned above, the principal target element in rocks and ice, from which ^{14}C is produced in nuclear spallations produced by energetic neutrons, is oxygen. The yield of ^{14}C per nuclear disintegration in oxygen is estimated to be 3% (Lal, Nishiizumi & Arnold 1987). The average ^{14}C yield per disintegration in a rock is estimated to be smaller by about (30–40)%. In quartz, the production rate of ^{14}C has been determined (Jull *et al* 1990) to be 99 ± 12 ^{14}C atoms per gram of quartz per year at geomagnetic latitude ~ 60° and altitude 2000 m. This corresponds to a yield of 3.5% in nuclear disintegrations in quartz, indicating that the value of 3% for ^{14}C yield in oxygen may be somewhat of an underestimate.

The *in situ* production rate of ^{14}C in a rock at any latitude and altitude can be obtained by multiplying the ^{14}C yield per nuclear disintegration in the rock at that location with the nuclear disintegration rate. We have already discussed how these rates can be obtained from the corresponding rates given for the atmosphere (Lal & Peters 1967). The latter are given in the form of a third-degree polynomial in altitude for 0–10 km altitudes by Lal (1990). Until a better production rate estimate is available, the value of 2% may be adopted for the yield of ^{14}C in nuclear disintegrations in normal rock types.

Taking 3% for the yield of ^{14}C in nuclear disintegrations in oxygen, and including production of ^{14}C due to thermal neutron capture in nitrogen present in trapped air in ice (about 10% by volume), Lal *et al* (1990) estimate *in situ* ^{14}C production rates of 15.3, 38.7, 85.3, 167 and 300 atoms ^{14}C g^{-1} ice yr^{-1} at geomagnetic latitude ≥ 60° and altitude 0, 1, 2, 3 and 4 km, respectively. (Note that thermal neutron capture in nitrogen is important only for the case of ablating ice; in accumulating ice, air is trapped within ice crystals at depths significantly below that where *in situ* production occurs (see Fig 11.4).)

As mentioned before, at mid- and high latitudes, the cosmic radiation is quickly absorbed in underground materials, with an exponential mean free path of ~ 150 g cm^{-2}. At depths exceeding 10^3 g cm^{-2} (~ 3 m rock), the flux of radiogenic neutrons becomes comparable to the cosmogenic neutrons. At greater depths, low-energy (~ MeV) thermal neutron reactions are primarily radiogenic. However, since most of the cosmic-ray energy is absorbed rapidly in the atmo-

TABLE 11.1. Approximate Radiocarbon Production Rates on the Earth in a Column of 1 Cm^2 Cross-Section

	^{14}C atoms cm^{-2} sec^{-1}
1. Global average cosmic-ray production in the Earth's atmosphere	~ 2*
2. Integrated *in situ* cosmogenic production rate in a rock exposed at:	
a) 5 km	3×10^{-3}
b) 2 km	5×10^{-4}
3. Integrated *in situ* cosmogenic production rate in the oceans (Lal *et al* 1988)	10^{-4}
4. Integrated radiogenic production rate in the crust**	10^{-3}

*Lingenfelter (1963)

**Based on the assumption of a mean radiogenic neutron flux of 4 neutrons cm^{-2} day^{-1}

sphere, the atmospheric cosmogenic production far exceeds the total ^{14}C production in the crust, as seen from Table 11.1, where I have listed the approximate *columnar production rates of ^{14}C on the Earth due to cosmogenic and radiogenic reactions*. (The cosmogenic reactions include muon reactions.) Even if the latter was significant, the *in situ* cosmogenic production in terrestrial surface materials would be much more important geophysically. This is due primarily to two facts: 1) the cosmogenic production is depth dependent; and 2) it occurs in the upper layers of the Earth, which are dynamic, subject to mixing/move-ment/erosion.

IN SITU ^{14}C ON THE EARTH; OBSERVATIONS AND COMPARISON WITH THEORY

A number of cosmogenic *in situ* radioactive and stable nuclides have recently been studied with a view to using them as clocks for studying geophysical processes (*cf* Lal 1988a). The basis of this application is the continuous production of nuclides in terrestrial materials in a geometry-sensitive manner. Because of the production rate dependence on depth, one can construct a suitable model for the evolution of a horizon, for example, sedimentation on or erosion of a horizon. One can derive the relevant model parameters if the concentrations of cosmogenic nuclides of different half-lives could be studied in a rock horizon.

This central idea has been used successfully in meteoritics for studying the evolutionary history of meteorites and surficial erosion rates (*cf* Lal 1972). The nuclide buildup in the sample, a cumulative record, is given by the differential equation:

$$\frac{dC}{dt} = -C\lambda + q(p,x(t)) \tag{11.5}$$

where λ is the disintegration constant of the radionuclide and q is the *in situ* production rate; x(t) is the time-dependent depth from the exposed surface and p is the atmospheric pressure at the rock surface. The nuclide concentration, C may or may not be in a steady state; it would not generally be expected to be so. The solution of Eq (11.5) may be a complex function of p, x and q; the complexity is dictated by how these parameters vary with time.

Steady-state solutions of Eq (11.1) have been considered by Lal and Arnold (1985) and Lal, Nishiizumi and Arnold (1987) for erosion/ablation and accumulation. The *in situ* method has been successfully applied for studies of erosion of rocks, based on production of ^{10}Be and ^{26}Al in quartz, and of ^{3}He in olivines (*cf* Lal 1988a, 1990).

I discuss below the expected concentration of ^{14}C in terrestrial solids, and briefly review the experimental data to date.

In Situ ^{14}C in Polar Ice

The potential applications of *in situ* ^{14}C in polar ice for studies of ice ablation and accumulation rates were outlined by Lal, Nishiizumi and Arnold (1987). The first suggestion of *in situ*-produced ^{14}C in ice was made by Fireman and Norris (1982), who studied ^{14}C in the carbon dioxide extracts from both accumulation and ablation ice samples. Conclusive proof for presence of *in situ* ^{14}C in ice was provided by Lal *et al* (1990) in ablation ice samples from the Allan Hills region. These authors extracted both CO and CO_2 from the samples and found ^{14}C in the two fractions in a fixed proportion, 0.6:0.4, respectively. The total measured concentrations of ^{14}C are shown in Figure 11.2 for samples from two locations, as a function of depth; the fractional activities in the CO phase for the same samples are plotted in Figure 11.3.

The ablation ice in the Allan Hills main ice field is expected to be quite "old," > 100,000 years based on the U-series dating method, which is consistent with the finding of old terrestrial-age meteorites in the Allan Hills main ice field (for references, see Lal *et al* 1990). The expected ^{14}C concentration in *zero age* ice due to trapped atmospheric CO_2 in the ice samples is about 950 atoms g^{-1} ice, taking the typical CO_2 concentration to be 0.03 cm^3 CO_2 kg^{-1} ice (Andrée *et al* 1986). Assuming the age of the ice to be 50,000 years (a lower limit), we expect the ^{14}C concentration due to the atmospheric trapped CO_2 to be only about 2 atoms ^{14}C g^{-1} ice. Similarly, we expect the residual *in situ* ^{14}C produced in the ice during its accumulation to be < 2 atoms ^{14}C g^{-1} ice. Thus, the signals

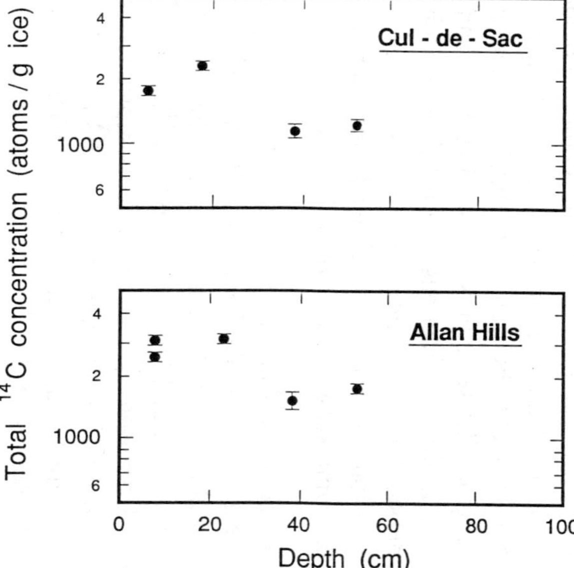

Fig 11.2. The measured total ^{14}C concentration in ice (atom/g ice) from CO_2 and CO gas phases combined are plotted as function of depth for Allan Hills main ice field station #10 and Cul-de-Sac. Data are from Lal *et al* (1990).

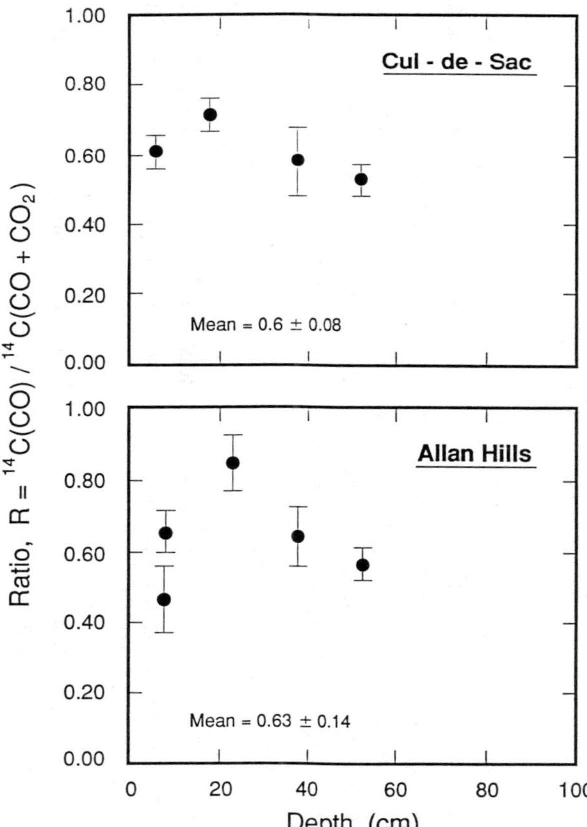

Fig 11.3. The measured ratios of concentrations of ^{14}C in the CO gas phase, relative to the total in the CO_2 and CO gas phases in ice in the samples plotted in Fig 11.2. Data are from Lal *et al* (1990).

observed (Figs 11.2 and 11.3) in the CO_2 and CO fractions are due nearly entirely to the *in situ*-produced ^{14}C in the ice during its outcropping stage. The depth profiles of total ^{14}C concentrations (Fig 11.2) are consistent with the expected 1/e slope of ~ 170 cm for *in situ* ^{14}C, taking the mean density of ice to be 0.9 g cm^{-3}. Further, the mean fraction of *in situ*-produced ^{14}C in the CO phase is ~ 0.6. This is quite consistent with laboratory experiments conducted with artificially produced ^{11}C. The ratio $^{11}CO/^{11}CO_2$ is found to depend on the physical nature of the target material irradiated to form ^{11}C. For solid phases (solid CO_2 and $NaHCO_3$ targets), the yields of ^{11}CO and $^{11}CO_2$ were found to be the same (for references and discussion, see Lal *et al* 1990).

The above results provide unambiguous proof for the *in situ* production of ^{14}C in Allan Hills ice samples. *The expected concentration of in situ* ^{14}C formed in ablating ice, during its outcropping, based on the steady-state ablation model (Lal, Nishiizumi & Arnold 1987), is given by

$$C(x) = \frac{P_0 e^{-\rho x/\Lambda}}{\lambda + \rho a/\Lambda} \tag{11.6}$$

where P_0 is the *in situ* production rate of ^{14}C at the exposed ice surface, ρ (g cm^{-3}) is the density of ice, λ is the disintegration constant of ^{14}C, Λ is the absorption mean free path (g cm^{-2}) of cosmic rays in ice (= 150 g cm^{-2}), and a is the rate of ablation of ice (cm yr^{-1}). For a given measured value of C(x), the equation is solved numerically for the ablation rate, a. Using the estimated *in situ* ^{14}C production rates, the measured concentrations of ^{14}C yield ablation rates of 5.8 ± 0.7 and 7.6 ± 0.8 cm yr^{-1} for Allan Hills Station 10 and Cul-de-Sac, respectively.

In ice accumulating at a constant rate, s (cm yr^{-1}) the total ^{14}C concentration of atmospheric trapped and in situ-produced $C_T(x)$ at depth x (cm) is given by Lal, Nishiizumi and Arnold (1987)

$$C_T(x) = C_0 e^{-\lambda(T-t_0)} + \frac{P_0}{\rho\frac{s}{\Lambda}-\lambda}\left(e^{-\lambda\frac{x}{s}} - e^{-\rho\frac{x}{\Lambda}}\right) \tag{11.7}$$

where T is the age of the ice since its accumulation as firn, and t_0 is the time interval required for the formation of ice; C_0 is the concentration of ^{14}C trapped with air during the firn to ice transition, λ is the disintegration constant of ^{14}C, P_0 the *in situ* production rate of ^{14}C in ice at the site at the surface and Λ is as defined above the absorption mean free path for nuclear active particles of the cosmic radiation in ice, ~ 150 g cm^{-2} (Lal, Nishiizumi & Arnold 1987). In

Equation (11.7), the first term on the RHS is for the atmospheric trapped ^{14}C. This equation is valid if air in the firn is in diffusive equilibrium with the surface air; otherwise an appropriate value smaller than t_o must be used. Stauffer (1981) discussed the accumulation of atmospheric gases by firn and ice. Greenland Ice Sheet Project (GISP) research shows that atmospheric gases are in exchange with the atmosphere until closure of gas pore volumes. Closure occurs at a critical density of firn, about 0.82 g cm^{-3}, equivalent to a depth of 60–80 m of firn. This is confirmed by the ^{39}Ar data of Loosli (1979). The second term in Equation (11.7) gives the concentration of the *in situ* ^{14}C, most of which is produced when the weight of the overlying ice is < 3–4 Λ, *ie*, during the firn stage. I assume that all *in situ* ^{14}C is produced inside ice crystals and remains there, *ie*, it is not in exchange with atmospheric CO_2. A schematic diagram outlining the stages at which *in situ* ^{14}C is produced and *in situ* and atmospheric ^{14}C are trapped, is shown in Figure 11.4.

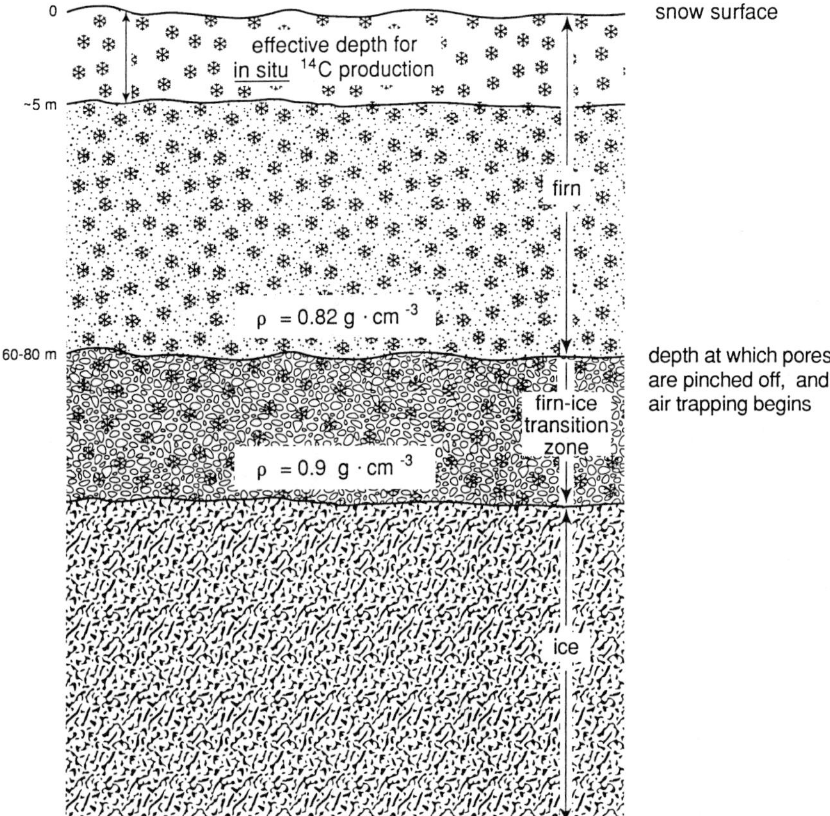

Fig 11.4. A schematic diagram of an ice accumulation zone showing the characteristic depth zones where 1) *in situ* production occurs during accumulation, 2) pores in firn close, and 3) ice forms, locking in the atmospheric and *in situ* ^{14}C.

In Equation (11.7), implicit is the assumption that $x = s$T, and we can rewrite

$$C_T(x) = e^{-\lambda T}\left[C_0 e^{\lambda t_0} + \frac{P_0}{D(s)}\left(1-e^{-TD(s)}\right)\right] \tag{11.8}$$

$$D\,(s) = (\rho s/\Lambda) - \lambda \ . \tag{11.9}$$

The ratio of *in situ* ^{14}C to trapped atmospheric ^{14}C activity in an accumulating ice sample is then given by R,

$$R = \frac{C_x(in\ situ)}{C_x(\text{atmospheric})} = \frac{P_0 e^{-\lambda t_0}}{C_0 D(s)}\left(1-e^{-TD(s)}\right) \tag{11.10}$$

nearly independent of the age of the ice, as would be expected. For a range of ice accumulation rates, say (1–100) cm yr^{-1}, the value of D(s) lies between $(6\times10^{-3} - 6\times10^{-1})yr^{-1}$; hence, for samples of age exceeding 1000 years, the term within the brackets in Equation (10) is close to unity. Also since the value of t_0 ranges between 100 and 300 years, the term $e^{-\lambda t_0}$ is close to unity within a few percent. The relation (11.10) then simplifies to:

$$R \sim \frac{P_0}{C_0 D(s)} \tag{11.11}$$

$$\sim \frac{P_0 \Lambda}{C_0 \rho s} = 0.176\frac{P_0}{s} \ . \tag{11.11'}$$

For more exact calculations, we must of course use Equation (11.10), but the approximate identity is useful to see the relative contributions of *in situ* and trapped atmospheric ^{14}C in accumulation samples. Typical value of trapped CO_2 concentration in ice is 0.03 $cm^3\ kg^{-1}$ ice (Andrée *et al* 1986). Taking the $^{14}C/^{12}C$ ratio in "modern" pre-bomb carbon to be 1.17×10^{-12}, the value of C_0 for modern carbon is found to be 9.43×10^2 atoms $^{14}C\ g^{-1}$ ice. The corresponding calculated values of the ratio for four altitudes of accumulation of ice at geomagnetic latitudes, ≥ 60° is given in Figure 11.5. The *in situ* ^{14}C production rates in Figure 11.1 at 0, 1, 2 and 3 km altitudes are taken as 15.3, 38.7, 85.3 and 167 atoms $^{14}C\ g^{-1}$ ice yr^{-1} (Lal *et al* 1990); the density of ice was fixed at 0.9 g cm^{-2}.

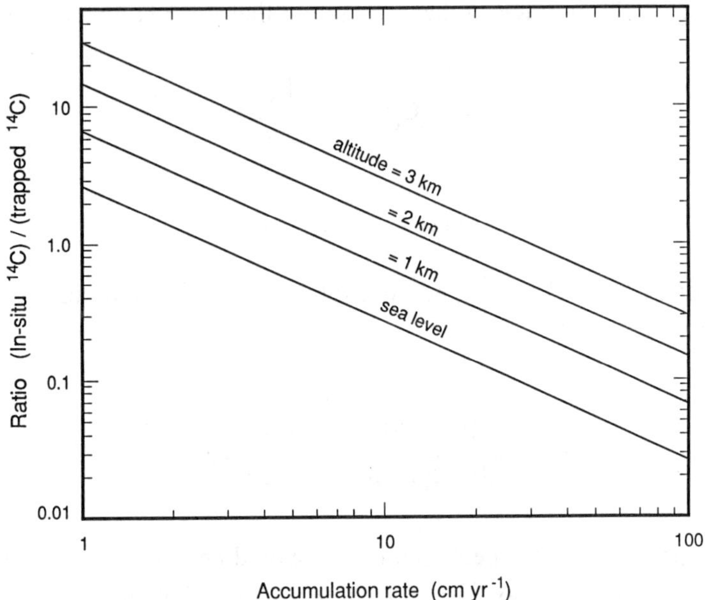

Fig 11.5. Estimated ratios of trapped *in situ* and atmospheric ^{14}C in accumulation ice at depths exceeding the firn-ice transformation depth, as a function of accumulation altitude, for four altitudes. Calculations are due to Lal and Jull (1990).

In Situ ^{14}C in Terrestrial Rocks

The measurements of *in situ* rocks is limited so far to the work of Jull *et al* (1990). They measured total ^{14}C concentrations in a suite of rocks where the effect of erosion is minimal. In the future, measurements of the ratio $^{14}CO/^{14}CO_2$ would prove very useful to identify unambiguously the *in situ* ^{14}C concentration. The equation governing the expected ^{14}C concentrations in a rock steadily eroding at a rate, ε (cm yr^{-1}) is given by Equation (11.6), if we replace a by ε. In the rocks studied by Jull *et al* (1990), presumably $\mu\varepsilon \ll \lambda$. In case erosion is appreciable, *ie*, $\mu\varepsilon \gg \lambda$, Equation (11.6) reduces to

$$C(x) = \frac{P_o \Lambda}{\rho \varepsilon} \cdot e^{-\rho x/\Lambda} \quad . \qquad (11.12)$$

The optimum erosion rates which can be studied using *in situ* ^{14}C are in the range $10^{-3} - 3 \times 10^{-2}$ cm yr^{-1}, obtained by setting $\mu\varepsilon \sim \lambda$.

POTENTIAL STUDIES OF IN SITU ^{14}C IN TERRESTRIAL MATERIALS

I have discussed the applications of cosmogenic *in situ* ^{14}C in ice for studying accumulation and ablation rates of ice and erosion rates of rocks. A variety of geophysical problems can be studied using ^{14}C produced *in situ* in rocks/soils/sands, along the lines discussed earlier for *in situ* nuclides in general (Lal 1988a, 1990). I would like to point out here a special advantage for measuring *in situ* ^{14}C produced in accumulation ice to study changes in the cosmic-ray flux on the Earth, over other nuclides produced in the Earth's atmosphere or *in situ* in rocks.

As can be seen from Figure 11.4, the *in situ* ^{14}C is produced in the top 300–400 g cm^{-2} before it gets buried by accumulating snow. This corresponds to *in situ* ^{14}C accumulation over periods of 6 to 40 years for the range of accumulation rates 10–50 cm yr^{-1} commonly encountered in polar regions. No further *in situ* production of ^{14}C then occurs, and both the atmospheric trapped ^{14}C and the *in situ* ^{14}C in this layer decay with time, as this layer sinks due to snow accumulation on the top. This results in a "freezing-in" of the information on the cosmic-ray flux at the accumulation during the 6–40-year period mentioned above. This information is a *quasi-differential* type in the sense it represents integrated cosmic-ray flux during the period Δt in which the layer of interest sinks below a depth of ~ 400 g cm^{-2}. Changes in the cosmic-ray flux at different epochs in the past (< 40,000 yr) can therefore be studied from analyses of ice samples deposited during corresponding time periods, provided that the accumulation rate is known.

Similar studies can be carried out in documented terrestrial rock samples which were shielded from cosmic radiation in the past up to some time and thereafter exposed to cosmic radiation thereafter in a fixed geometry. This condition can hopefully be realized in a variety of circumstances. Lal (1986) suggested studies of stones from Sphinxes and pyramids. Rocks with glacial polish in Sierra Nevada mountain ranges, exposed to cosmic radiation since the last glacial retreat, were studied by Nishiizumi *et al* (1989) and Phillips *et al* (1986). Note, however, that the information on cosmic-ray flux obtained from such studies is an integral type, since one measures the resulting nuclide concentration in the sample after exposure for a finite period. To see changes in the cosmic-ray flux, one has to study nuclides of different half-lives in the sample. Even then, one cannot hope to deduce accurate information on any short-term changes (decadal to century) in the cosmic-ray flux from this integral low-pass record. A quasi-differential cosmic-ray flux record similar to that in the ice can be found in tree rings (Lal, Arnold & Nishiizumi 1985), provided the *in situ*-produced ^{10}Be in tree rings can be studied without any interference from atmospheric ^{10}Be transported to the ring, eg, via groundwater.

The information on changes in cosmic-ray flux in the past, based on records of atmospheric $^{14}C/^{12}C$ ratios in tree rings, is also an integral type. It is appreciably modulated by changes in the carbon cycle induced by climatic change (Lal

1985). As such, both the observed short- and long-term changes in $^{14}C/^{12}C$ ratios are in part due to short-term changes in solar activity (Stuiver 1961) and long-term changes in geomagnetic field (Elsasser, Ney & Winckler 1956). These issues have been discussed extensively in recent years but no consensus has yet been reached on how best to understand and interpret the ^{14}C tree-ring record (*cf* Stuiver & Brazuinas 1989; Damon 1989; Damon & Sonnett, in press; Lal 1985).

Another cosmogenic record is that of ^{10}Be (atmospheric) precipitated along with snow, which apparently points towards an enhanced ^{10}Be production in the atmosphere during certain periods in the past 140,000 years of the record (Beer *et al* 1988; Raisbeck *et al* 1987; Sonett, Morfill & Jokipii 1987; Raisbeck *et al* 1990). The concentrations of ^{10}Be in ice depend on meteorological factors, as well as changes in its production rate. The polar ice record of ^{10}Be is, in fact, heavily modulated by climate changes and the consequent changes in meteorological factors (Lal 1987b and references therein). However, it appears that it may be possible to correct for meteorological variations by comparing the ^{10}Be and δ^2H or $\delta^{18}O$ concentrations in the samples (Raisbeck *et al* 1990).

In summary, at present, ambiguous evidences point to both short- and long-term changes in the cosmic-ray flux in the past. It would be interesting to determine the changes in the flux using an independent method. I suggest that studies of *in situ* ^{14}C in accumulation ice would adequately meet this need. For example, if analyses of ice samples showing excess ^{10}Be concentrations also show excess *in situ* ^{14}C concentration, this would constitute an unambiguous case for an increase in cosmic-ray flux, either due to a long period of low solar activity or a nearby supernova explosion (Kocharov 1992; Sonnett, Morfill & Jokipii 1987), both of which would result in an increased cosmic-ray flux. In that case, one would be able to rule out a lower geomagnetic field as the cause of the increased ^{10}Be production during the period, because any changes in the geomagnetic field would not affect the cosmic-ray flux at high latitudes in either Antarctic or Arctic ice samples. On the other hand, studies of "old" samples accumulated in low latitudes would provide information on changes in the geomagnetic field (*cf* Lal, Arnold & Nishiizumi 1985).

Of particular interest would also be the study of epochs when the tree-ring data show evidence for large rapid changes in $^{14}C/^{12}C$ ratios, within periods of the order of 10–100 years (*cf* Stuiver & Braziunas 1989). If these changes are primarily due to changes in the cosmic-ray flux, due to decreased modulation of cosmic-ray flux during periods of "quiet" Sun, one would also observe a substantial increase in *in situ* ^{14}C concentrations in accumulation ice. An absence of the transient (or the presence of a weak transient) in the ice record would argue for a strong climatic control on the tree-ring $^{14}C/^{12}C$ record. For a discussion of the expected changes in the atmospheric production rate of ^{14}C, see Castagnoli and Lal (1980) and Lal (1988b).

In the above, I have discussed the studies to date on *in situ* ^{14}C in ice and rock samples. Much more work remains to be done in this direction. For a proper application, the *in situ* production rates of ^{14}C have to be determined accurately. Both studies of ^{14}C concentrations in samples exposed in known configurations and artificial irradiation experiments to determine the proportion of ^{14}CO and $^{14}CO_2$ formed in ice and in rock minerals, should be performed. The *in situ* production of ^{14}CO and its application to discriminate against the ubiquitous $^{14}CO_2$ on the Earth can be taken advantage of in the case of rock minerals also.

I have pointed out here an interesting future application of *in situ* ^{14}C accumulation ice, namely that of determining changes in the cosmic-ray flux due to changes either in solar activity or in the Earth's geomagnetic field.

ACKNOWLEDGMENTS

It is a pleasure to record my fruitful collaboration with Drs J R Arnold, D J Donahue, A J T Jull and K Nishiizumi. I am very thankful to Paul Damon and David Elmore for very helpful comments on the paper. The work was funded in part by NSF grants NSF EAR89-04677 and NSF DPP84-09526, and in part by a donation from Professor Roger Revelle.

REFERENCES

Andrée, M, Beer, J, Loetscher, HP, Moor, E, Oeschger, H, Bonani, G, Hoffman, HJ, Morenzoni, E, Nessi, M, Suter, M and Wölfli, W 1986 Dating polar ice by ^{14}C accelerator mass spectrometry. *In* Stuiver, M and Kra, RS, eds, Proceedings of the 12th International ^{14}C Conference. *Radiocarbon* 28(2A): 417–423.

Beer, J, Siegenthaler, U, Bonani, G, Finkel, RC, Oeschger, H, Suter, M and Wölfli, W 1988 Information on past solar activity and geomagnetism from ^{10}Be in the Camp Century ice core. *Nature* 331: 675–679.

Castagnoli, G and Lal, D 1980 Solar modulation effects in terrestrial production of carbon-14. *In* Stuiver, M and Kra, RS, eds, Proceedings of the 10th International ^{14}C Conference. *Radiocarbon* 22(2): 133–158.

Damon, PE 1989 Radiocarbon, solar activity and climate. *In* Avery, SK and Tinsley, BA, eds, *Proceedings of Mechanisms for Tropospheric Effects of Solar Variability and the Quasi-Biennial Oscillation.* Boulder, Colorado, NCAR: 198–208.

Damon, PE and Sonett, CP, in press, Solar and terrestrial components of the atmospheric ^{14}C variation spectrum. *In* Sonett, CP, Giampapa, MS and Matthews, MS, eds, *The Sun in Time.* Tucson, University of Arizona Press.

Davis, R, Jr and Schaeffer, OA 1955 Chlorine-36 in nature. *Annals of the New York Academy of Science* 62: 105–122.

Elsasser, W, Ney, EP and Winckler, JR 1956 Cosmic-ray intensity and geomagnetism. *Nature* 178: 1226–1227.

Fireman, EL and Norris, TL 1982 Ages and composition of gas trapped in Allan Hills and Byrd Core. *Earth and Planetary Science Letters* 60: 339–350.

Hedges, REM and Law, IA 1989 The Radiocarbon dating of bones. *Applied Geochemistry* 4: 249–253.

Jull, AJT, Donahue, DJ, Linick, TW and Wilson, GC 1990 Spallogenic ^{14}C in high-altitude rocks and in Antarctic meteorites. *In* Long, A and Kra, RS, eds, Proceedings of the 13th International ^{14}C Conference. *Radiocarbon* 31(3): 719–724.

Kocharov, GE, 1992 Radiocarbon and astrophysical-geophysical phenomena. *In* Taylor, RE, Long, A and Kra, RS, eds, *Radiocarbon after Four Decades: An Interdisciplinary Perspective.* New York, Springer-Verlag, this volume.

Korff, SA 1940 On the contribution to the ionization at sea level produced by the neutrons in the cosmic radiation. *Journal of Geophysical Research* 45: 133–134.

Lal, D 1972 Hard rock cosmic ray archaeology. *Space Science Review* 14: 3–102.

_____1986 Cosmic ray interactions in the ground: temporal variations in cosmic ray intensities and geophysical studies. *In* Reedy, RC and Englert, P, eds, Proceedings of the Workshop on Cosmogenic Nuclides. *LPI Technical Report* 86–06: 43–45.

_____1985 Carbon cycle variations during the past 50,000 years: Atmospheric $^{14}C/^{12}C$ ratio as an isotopic indicator. *In* Sundquist, ET and Broecker, WS, eds, The carbon cycle and atmospheric CO_2: Natural variations, Archean to Present. American Geophysical Union, Washington DC, *Geophysical Monographs* 32: 321–333.

_____1987a Production of ^{3}He in terrestrial rocks. *Chemical Geology (Isotope Geoscience Section)* 66: 89–98.

_____1987b ^{10}Be in polar ice: Data reflect changes in cosmic ray flux or polar meteorology? *Geophysical Research Letters* 14: 785–788.

_____1988a In situ-produced cosmogenic isotopes in terrestrial rocks. *Annual Review Earth and Planetary Sciences* 16: 355–388.

_____1988b Theoretically expected variations in the terrestrial cosmic ray production rates of isotopes. *In* Castagnoli, G, ed, *Solar-Terrestrial Relationships and the Earth Environment in the Last Millennia.* Bologna, Italy, Soc Italiana di Fisica Bologna: 216–233.

Lal, D 1991 Cosmic ray tagging of erosion surfaces: In situ production rates and erosion models. *Earth and Planetary Science Letters*, in press.

Lal, D and Arnold, JR 1985 Tracing quartz through the environment. *Proceedings of the Indian Academy of Sciences (Earth and Planetary Science)* 94: 1–5.

Lal, D, Arnold JR and Nishiizumi, K 1985 Geophysical records of a tree: new application for studying geomagnetic field and solar activity changes during the past 10^4 years. *Meteoritics* 20: 403–414.

Lal, D, Chung, Y, Platt, T and Lee, T 1988 Twin cosmogenic radiotracer studies of phosphorus recycling and chemical fluxes in the upper ocean. *Limnology and Oceanography* 33: 1559–1567.

Lal, D and Jull, AJT 1990 On determining ice accumulation rates in the past 40,000 years using *in situ* cosmogenic ^{14}C. *Geophysical Research Letters* 17: 1303–1306.

Lal, D, Jull, AJT, Donahue, DJ, Burtner, D and Nishiizumi, K 1990 Polar ice ablation rates measured using *in situ* cosmogenic ^{14}C. *Nature* 346: 350–352.

Lal, D, Nishiizumi, K and Arnold, JR 1987 *In situ* cosmogenic ^3H, ^{14}C and ^{10}Be for determining the net accumulation and ablation rates of ice sheets. *Journal of Geophysical Research* 92(B6): 4947–4952.

Lal, D and Peters, B 1967 Cosmic ray produced radioactivity on the Earth. *In* Flügge, S, ed, *Handbook of Physics*. Berlin, Springer-Verlag 46(2): 551–612.

Libby, WF 1946 Atmospheric helium-three and radiocarbon from cosmic radiation. *Physical Review* 69: 671–672.

_____ 1967 *Radiocarbon Dating*, 2nd edition. Chicago, University of Chicago Press: 175 p.

Lingenfelter, RE 1963 Production of carbon-14 by cosmic-ray neutrons. *Reviews of Geophysics* 1: 35–55.

Loosli, HH (ms) 1979 Ein Altersbestimmungsmethode mit Ar-39. Habilitations schrift, Universitat Bern. Thesis.

Nishiizumi, K, Winterer, EL, Kohl, CP, Lal, D, Arnold, JR, Klein, J and Middleton, R 1989 Cosmic ray production rates of ^{10}Be and ^{26}Al in quartz from glacially polished rocks. *Journal of Geophysical Research* 94: 17,907–17,916.

Phillips, FM, Leavy, BD, Jannik, NO, Elmore, D and Kubik, PW 1986 The accumulation of cosmogenic chlorine-36 in rocks: a method for surface exposure dating. *Science* 231: 41–43.

Raisbeck, GM, Yiou, F, Bowles, D, Lorius, C, Jonzel, J and Barkov, NI 1987 Evidence for two intervals of enhanced ^{10}Be deposition in Antarctic ice during the last glacial period. *Nature* 326: 273–277.

Raisbeck, GM, Yiou, F, Jouzel, J and Petit, JR 1990 ^{10}Be and δ^2H in polar ice cores as a probe of the solar variability's influence on climate. *Philosophical Transactions of the Royal Society of London* A 330: 463–470.

Sonnett, CP, Morfill, GE and Jokipii, JR 1987 Interstellar shock waves and ^{10}Be from ice cores. *Nature* 330: 458–460.

Stauffer, B 1981 Mechanismen des lufteinschlusses in naturlichem eis. *Zeitschrift fuer Gletscherkunde und Glazialgeologie* 17: 17–56.

Stuiver, M 1961 Variations in radiocarbon concentration and sunspot activity. *Journal of Geophysical Research* 66: 273–276.

Stuiver, M and Braziunas, TF 1989 Atmospheric ^{14}C and century-scale solar oscillations. *Nature* 338: 405–408.

Wetherill, GW 1953 Spontaneous fission yields from Uranium and Thorium. *Physical Review* 92: 907–912.

Zito, R, Donahue, DJ, Davis, SN, Bentley, HW and Fritz, P 1980 Possible subsurface production of carbon 14. *Geophysical Research Letters* 7: 235–238.

INSTRUMENTATION AND SAMPLE PREPARATION

PREFACE

AUSTIN LONG

From the earliest days of natural radiocarbon measurement through the accelerator mass spectrometry (AMS) revolution, accurate and precise radiocarbon dating has developed along with and depended upon technical innovations in physics, chemistry and electronics. The technology has not yet reached a state of routine or fully automated analysis. Even in laboratories employing a level of automation at some stages of the procedure, highly skilled personnel keep track of sample behavior and equipment performance. The art and science of the craft is often passed from professor to student in a "master/apprentice" style, and every three years, the practitioners of radiocarbon dating gather to exchange experiences at an International Radiocarbon Conference. Procedures and instrumentation have continually evolved. Some of the originators have passed the baton to younger scientists. But research and development remain a strong common denominator among radiocarbon labs. This section is a distillate of the craft of radiocarbon dating.

Appropriately, the first chapter deals with the first procedures a sample encounters: isolation and purification of a chemical phase or phases that represent the event or archaeological culture or geologic stratum to be dated. If a sample is contaminated with carbon that is younger or older than the sample's true age, the measured date will be wrong. Removal of contaminants, and proof of their removal is clearly critical to the integrity of the date. Moreover, owing to the 1000-times difference in sample size, sample pretreatment procedures may employ different techniques for conventional β-counting than for AMS. Robert Hedges' contribution reviews strategies and techniques for sample pretreatment both for conventional and for AMS radiocarbon dating.

The most demanding accomplishment in the measurement of radiocarbon is the achievement of high precision (precision ≤ 2‰, or about 16 years). It demands enormous attention to all aspects of the procedure, quantitative understanding of all parameters that affect the counting, and a large amount of counting time.

Only a handful of laboratories have succeeded in routinely producing high-precision ^{14}C dates. Why this is such a small club is apparent from Bernd Kromer's paper.

For the first decade of ^{14}C dating, gas counting dominated the detection techniques. Liquid scintillation became viable only after the development of high-yield synthesis of benzene from CO_2, with all six benzene carbons derived from the sample. Now, despite its more complex chemical preparation, more than half of the active labs use the liquid-scintillation technique, and most of the β-counting dates produced in the world are done by liquid scintillation. Henry Polach presents a detailed summary of the rise of liquid scintillation that includes a condensed textbook on operating a laboratory.

Whereas the development of gas-proportional counting and liquid scintillation were significant mileposts in radiocarbon dating, the practical application of AMS to routine radiocarbon dating has been a true revolution. Harry Gove presents a historical synopsis of early experiences, parallel work in different labs, and why AMS rather than cyclotrons.

At present, β-counting methodologies have evidently reached a high level of development. Both liquid scintillation and gas-proportional counting can achieve high precision. Their present rate of improvement is real but incremental. AMS, on the other hand, is still rapidly advancing in higher signal strength, lower chemical blanks and machine background, automation. All AMS facilities are rapidly improving their systems and procedures. Roelf Beukens discusses the parameters that affect precision and accuracy of AMS ^{14}C dating, and illustrates how the IsoTrace Laboratory is tackling the challenge of improving both.

SAMPLE TREATMENT STRATEGIES IN RADIOCARBON DATING

R E M HEDGES

INTRODUCTION

The chemical conversion of a sample in its pristine state to the form in which the abundance of radiocarbon can be directly measured is the main link between the "radiocarbon date" and the basic assumptions upon which this date rests. The importance of this link was recognized from the beginning (Libby 1954), but as radiocarbon dating became more routine, sample pretreatment and subsequent conversion to a medium suitable for ^{14}C detection has, more and more, been taken for granted. The changes in radiocarbon dating brought about by the accelerator mass spectrometry (AMS) measuring technique (Gove 1978; Kutschera 1981; Wölfli, Polach & Anderson 1984; Gove, Litherland & Elmore 1987; Yiou & Raisbeck 1990), and the consequent reduction in sample size by a factor of at least a thousand, has stimulated a re-examination of many pretreatment methods, and thereby enabled the link between assumption and date to be more closely controlled. This account describes how that has happened.

Assumptions and Errors in Radiocarbon Dating

A radiocarbon date includes an associated error estimated by the measuring laboratory. Implicit in the estimate is the restriction that errors can be (at least approximately) quantified, and have a Gaussian distribution. These restrictions apply reasonably well to laboratory measurements of the ^{14}C abundance, but take no account of the degree to which the particular sample being measured violates the basic assumptions of radiocarbon dating. It is the job of sample treatment chemistry to minimize this violation, to make apparent those situations where severe violation is unavoidable, and to attempt to estimate the likely effect on the validity of the date. A corollary of this is the need to be able to validate a particular result, or at least a particular method in a given context.

The basic assumptions in radiocarbon dating include the following:

1. The planetary distribution of ^{14}C in the biosphere is uniform in time and space;

2. The sample, as measured, contains carbon that came only from a living organism (or similar system that ceases to be in chemical exchange with the biosphere), and that the living organism took its carbon only from the biosphere.

We know that Assumption 1 is only approximately true, but corrections, which apply to major classes of samples as a whole, can be fairly easily applied. These include the tree-ring correction curve for atmospheric CO_2, which is applicable over the whole planet for the last 10,000 years (and largely compensates for variations in biospheric ^{14}C in time), and various 'reservoir' corrections, such as the different ages of marine shells, which largely account for the spatial variation of ^{14}C found in the earth's oceans. Assumption 2 is much more problematic, because it may be, and often is, specific to the particular sample and/or context.

The sample- or context-specific factors that can alter the $^{14}C/^{12}C$ ratio are the following:

1. Isotopic fractionation

2. Contamination (*ie*, admixture from the sample's environment) by carbon with a different $^{14}C/^{12}C$ ratio (*ie*, contamination by ^{14}C, or by ^{12}C)

3. The 'event' being dated is not represented by the $^{14}C/^{12}C$ ratio because of the way the sample was formed.

Each of these factors is discussed in turn below.

Differences Between Radiocarbon Measurements by 'Conventional' and 'AMS' Methods

What advantages beta-counting methods may have (eg, in cost and in precision) are rather lacking when it comes to sample chemistry. The need to prepare typically one gram of carbon restricts the sorts of samples available, and also the kind of chemistry (and therefore degree of rigor) that can be carried out. One advantage, however, appears to be that handling larger samples induces less laboratory contamination (see Contamination, below, for a discussion of this), enabling, in rare circumstances, older dates than those obtained by AMS to be measured (Grootes *et al* 1975).

The range of materials measured by AMS has now become very wide, requiring a corresponding range of chemical methods. Many of these methods can employ relatively 'advanced' techniques of purification (such as semi-preparative high performance liquid chromatography (HPLC)), which can easily handle milligram-sized samples. Two dangers that AMS dating faces are that many samples can be physically small, and thus more mobile in a deposit, so that stratigraphic interpretation becomes more difficult. (On the other hand, the greater choice available in what is dated places more emphasis on the significance of the object itself, eg, an archaeological artifact rather than a piece of charcoal.) Second, the potential for dating microscopic traces of carbon permits exotic but poorly understood materials to be measured, even if the requisite sample pretreatment is not yet able thoroughly to characterize them. The AMS technique does frequently have some very important advantages when it comes to validating a date, however, and these should be emphasized. First, different chemical fractions can often be extracted from the same sample, and their dates can be compared for consistency. Second, many samples are large enough to permit repeat dates to be made if the first measurement is suspect in any way. Both these approaches are invaluable in increasing confidence in the reliability of a date.

PROCESSES THAT MAY INVALIDATE A DATE

Fractionation

It is usual to consider isotopic fractionation in two stages; that occurring in the natural environment of the sample, and that occurring in the laboratory. This is because 'natural' fractionation is an interesting study in its own right (van der Merwe 1982). The fractionation involved is between atmospheric CO_2 and the living organism. Most of this occurs during metabolism of the CO_2 itself, since the difference in mass of one atom has a relatively larger effect in a small molecule like CO_2 than in, say, an amino acid. This review is not the place to summarize the extensive body of knowledge on carbon isotope fractionation in the biosphere, but the following points should be noted:

1. Fractionation in the stable isotopes, $^{13}C/^{12}C$, can be measured very precisely (to a standard deviation of 0.1‰).
2. Fractionation of $^{14}C/^{12}C$ can be estimated from that in $^{13}C/^{12}C$ by applying statistical thermodynamic arguments to show that, to first order, the effect is linear with mass difference (*ie*, the fractionation is twice as much) (Craig 1959).
3. The variation in $^{14}C/^{12}C$ ratio from natural isotopic fractionation can be at least 3%, and so some degree of correction is necessary if the accuracy of the $^{14}C/^{12}C$ measurement is to be upheld.

Fractionation can also take place during the diagenesis of organic material while buried. This is usually small, however, because of the large size of most of the biomolecules involved, but its effect should be borne in mind (Bada, Schoeninger & Schimmelmann 1989).

Laboratory fractionation is inevitably incurred in the chemical processing of the submitted sample to whatever material is most easily measured (*ie*, CO_2, benzene or graphite). It can be corrected in the same way as natural fractionation (although noting that to measure $^{13}C/^{12}C$ ratios, any sub-sample taken must be converted to CO_2). Most laboratories make a $^{13}C/^{12}C$ measurement since, in many cases, the correction required is significant in terms of the accuracy sought in the date. The magnitude of the fractionation depends on the chemical process, and particularly on the yield, since any process with 100% yield cannot fractionate. Therefore, one aim, in devising chemical procedures for sample conversion, is to achieve yields of over 90% in a well-mixed material. Note that fractionation in the laboratory is most serious in the later stages of sample conversion, where the molecules being studied are smaller. Since the $^{14}C/^{12}C$ ratio is measured by reference to a standard, which will also have undergone the same chemical conversion, the absolute degree of fractionation is less important than the relative fraction between sample and standard.

AMS measurements are often made on the $^{14}C/^{13}C$ ratio, where the total fractionation is reduced by half. The production of graphite, for AMS ion source targets, can give rise to spatially variable fractionation, but this kind of problem is best addressed through an analysis of the AMS measurement itself (see Beukens 1992). Another complication that AMS can bring is in the application of chromatographic separation methods to milligram samples. Here, fractionation, eg, of an amino acid during the collection of an eluted 'peak', can be quite severe unless great care is taken to obtain close to 100% yields.

Fractionation is, in fact, the most easily corrected of the distortions of the $^{14}C/^{12}C$ ratio, because of the ability to make corrections from $^{13}C/^{12}C$ measurements.

Contamination

First, some general observations. The range of interest of $^{14}C/^{12}C$ goes from 10^{-12} to 10^{-15}, corresponding to a range from modern material to material of about 55,000 years old. It follows that at 55,000 years, modern material will have a thousand times greater ^{14}C concentration, so that *old material is extremely sensitive to contamination by modern material.* Contamination of recent samples by old material is, in many ways, less serious, for the error (as % change in ^{14}C) is now approximately equal to the percentage of contaminant.

However, since much recent material is to be dated at high precision, the presence of 1% of, say, a petroleum-based preservative, would make the date too old by some 80 years – an unacceptably large error. Much of the study of contamination, and of how to remove it – and this is usually the most serious error in radiocarbon dating – consists in trying to find foolproof methods that can give reliable dates within a known range of field situations. However, testing these methods can be misleading, for it is necessary but not sufficient for a given method to achieve the 'right date' (eg, for material of known age), since this does not prove the contamination has been removed; it is quite possible, indeed common, that field contamination is of a similar age to that of the sample (see the end of Examples of Laboratory Processes . . . , below). We might hope that we detect such contamination by performing some kind of chemical or stable isotope analysis. However, this is very difficult, because we are trying to detect the presence of an unknown contaminant at below the 1% level, and this is not usually possible. Progress in methods has come about slowly, through the interaction of much dating experience, modern chemical analysis and rather limited insight into the diagenetic processes whereby contamination is incorporated. Much more needs to be done.

As with fractionation, contamination is best considered in two stages, in the field and in the laboratory. Field contamination can take any or all of several forms:

Mechanical. For example, where rootlets of more recent vegetation intimately penetrate a sample (be it bone or peat). In principle, such contamination can be mechanically separated. Handpicking is not an impossible task for AMS-sized samples. Chemical methods may also be used to remove extraneous material where mechanical sorting is not feasible; an example is that of concentrating pollen grains by treatment principally with HF followed by NaOCl bleach (Brown *et al* 1988).

Chemical. This includes numerous processes, with a corresponding range of complexity. The simplest perhaps is recrystallization, applicable mainly to carbonates. It may be detectable by SEM examination, and by XRD. The straightforward addition of organic material is very common. These are likely to be mobile organics in the deposits, for example, humic and fulvic acids, as well as deliberately added materials such as preservatives, cleaning detergents, etc. A complication is that many organic materials undergo diagenetic processes in which the organic molecules (eg, proteins or polysaccharides) degrade into material that more closely resembles the organic materials of their environment, so that indigenous but degraded material may not easily be differentiated from exogenous incorporated contamination. Further, there may be cross-linking reactions between the indigenous and exogenous material (in a similar manner to which humic acids are chemically built up), so that simple chemical separation methods cannot be applied. (More details on these problems in relation to bone pretreatment are considered in Examples of Laboratory Processes . . . , below).

Microbiological. Decaying organic materials can provide nutrients for bacteria and fungi. For the most part, such organisms recycle carbon atoms, rather than add different ones (although 'modern' contamination by bacteria in sediment cores that reduce atmospheric CO_2 has been documented (Geyh, Krumbein & Kudrass 1970) and no doubt occurs in other analogous situations). Even the recycling of material can be dangerous, however, if there is a mix of material of different ages, and can lead to metabolic products that combine both exogenous and indigenous carbon. The extent of microbiological metabolic activity is apparent at a qualitative rather than quantitative level (the production of branched chain acids in the analysis of lipid components is one indication), and, at present, no chemical methods have been devised specifically to take account of such effects.

It should be evident from the discussion above that the chemical natures of many forms of contamination are not predictable to any useful degree. The deposition of secondary carbonate is to be expected (and is easily removed by treatment with acid). Likewise, the uptake of such mobile high-molecular-weight soil-organic components as humic and fulvic acids is common. (Again, standard pretreatment methods always consider such a possibility by giving soil-derived samples an 'alkali wash'; although this is not necessarily sufficient.) Lipids might be mobile as micelles, or may be immobilized by attachment to clay particles. There is evidence that lipid penetration into potsherds from soil does not take place (Heron, Evershed & Goad, ms), although this may not be the case for all burial environments. Other molecular species have scarcely been sought, since the range of possible choice is so huge, although undoubtedly exogenous amino acids have been found to reside in at least some buried bone. In principle, and indeed in practice to a limited extent, it is possible to take different fractions, including suspected contaminants, and date them separately by AMS. In this way, if a suspected exogenous fraction has the same 'age' as a suspected endogenous fraction, the effect of contamination on the date may be reckoned to be small. Equally, the dating of suspected exogenous fractions can provide evidence for the kind of contaminating environmental chemistry taking place. But, as implied here, the evidence at present remains scrappy and largely anecdotal.

Laboratory contamination remains an important problem for AMS measurements. It manifests as a limitation in the oldest dates that can be measured, since AMS systems themselves, on 'ideal' graphite samples, can measure implied ages in excess of 70,000 years. The problem divides between pretreatment methods and target conversion methods. It appears that the major problem is in the combustion and graphitization techniques of target production, for which most labs report a 'background' (actually a level of equivalent contaminating modern carbon) of 1–3 μg modern carbon, mainly deriving from the combustion. (See Gove 1978; Kutschera 1981; Wölfli, Polach & Anderson 1984; Gove, Litherland & Elmore 1987; Yiou & Raisbeck 1990; Stuiver & Kra 1980, 1983, 1986; Long & Kra 1989; Vogel, Nelson & Southon 1987 for further discussion.) This effectively limits the maximum dates from AMS laboratories to between 40,000 and 50,000 years, and also implies an older maximum date for larger samples. At this age, the effect of pretreatment contamination does not seem to be a limiting feature, but will doubtless become so when target preparation is sufficiently improved. Note that, although a background may be subtracted, the limit in the age is reached because the background cannot be sufficiently accurately estimated. Most AMS labs report a scatter in their contamination background of about 50% (of the actual value). One form of contamination that should be mentioned derives from the use of 'tracer' ^{14}C. This tends to occur in biomedical or geoscience research laboratories rather than radiocarbon dating laboratories, but has proved almost impossible to eliminate. Any samples to be processed in a suspect laboratory should have extensive background controls performed first.

The Relation of Material to the Event

There are almost as many different relationships between the dated 'event' and the sample as there are different types of sample. Most of these relationships are rather subtle, and many call for a chemical approach to simplify the situation. Part of the expertise in radiocarbon dating consists in understanding the particular relationship in order to establish a clear connection between the sample $^{14}C/^{12}C$ ratio and the event dated. Some examples – by no means exhaustive – are given below.

Volcanic Sources. Here there is a local non-biospheric production of CO_2 that can depress the ^{14}C level in plants growing near fumaroles or at thermal springs. (Stable isotope measurements can help evaluate the effect.)

Ocean Reservoirs. The residence time of dissolved inorganic and organic carbon in the oceans is dependent on many complex global parameters (see, eg, Stuiver, Pearson & Braziunas 1986; Duplessy *et al* 1992; Peng & Broecker 1992; Toggweiler (ms) 1990), and significantly affects the 'date' of marine samples.

Groundwater. The exchange of dissolved CO_2 with geological carbonates such as calcite means that groundwater is likely to date 'too old'. (This subject is also taken up in detail by both Fontes (1992) and Geyh (1992).)

Foraminifera Dating. A deep-sea core will contain a mixture of planktonic and benthic foraminifera, each taking their carbon from the ocean surface or the deep ocean. These have different radiocarbon ages. The dating of foraminifera is also notable in requiring examination of other processes such as partial dissolution and bioturbation which can bias the result.

Terrestrial Mollusks. The carbonate of the shell may contain a substantial contribution of carbonate from the local country rock. This depends on the particular species. (The protein of the shell will reflect the radiocarbon age of the vegetation, however, although such a radiocarbon measurement has not been made.) (see Goodfriend & Stipp 1983.)

Turnover Rate in Bone. Bone protein – collagen – may have a residence time in an organism that is comparable to the precision of measurement. (This depends on the particular tissue.)

Mobility in Tree Rings. There is evidence (Long *et al* 1979) that lignins are much more mobile than cellulose in tree rings, so that consistent dates are best achieved on the cellulose fraction.

Reworked Material. If material has been redeposited, its radiocarbon age will not reflect the most recent depositional event.

Complex Materials. Two examples are given here – pottery and sediments. Both contain carbon that probably derived from several different sources, and neither can yet be considered satisfactorily datable unless the sources in a particular sample are reasonably well understood.

1. *Pottery.* At least five different sources of carbon in pottery can be anticipated. These are, first, the 'geological' carbon in the clay, carbonate in added temper or in association with the raw clay, and carbon added as temper in the form of vegetable (eg, rice, straw or adventitious cereal grains) or animal matter (claims have been made for the detection of blood in some pottery). On firing, these sources will decompose and be oxidized to a greater or lesser extent. Firing may also add carbon in the form of absorbed smoke and soot (this may be from wood or, much less likely, coal). Use of the vessel for cooking may add food residues, most characteristically in the form of lipids that are retained in the pottery pores, and also possibly as macroscopic carbonaceous residues adhering to the surfaces. Finally, other compounds containing carbon may come from absorption of mobile organics in the burial environment, from microbiological activity and from diagenesis of the potsherd material. Each of these different sources has been isolated and dated, using mechanical and chemical methods of separation, although sometimes distinctions are difficult to make. By looking at the pattern of all the results, it is usually possible to have

some confidence in publishing a date of firing or pottery use (although dating several fractions is, of course, more expensive). With more experience, it may eventually turn out that a sufficiently reliable date may be obtained simply by measuring characteristic lipids from a potsherd.

2. *Sediments.* Dating sediments is a major activity of radiocarbon measurements, although surprisingly little work has been done to determine individual fractions (Fowler, Gillespie & Hedges 1986; Giger *et al* 1984). Taking lacustrine sediments as an example, recently living organic material is accumulated from terrestrial and aquatic higher plants, and from algae. In hard water lakes, both aquatic higher plants and algae, as well as animals higher up the food chain (Haynes, Damon & Grey 1966: 4–5), are likely to have a carbon contribution from the lake bicarbonate, which itself is likely to contain geological carbonate, and therefore have a reduced ^{14}C level. At present, it does not seem possible to estimate this contribution to any useful accuracy – the differences in stable isotope composition effectively cancel out, for example. The problem is somewhat analogous to that of dating groundwater. In addition to these inputs, reworked and therefore significantly older, material, such as terrigenous inwash, or inert carbonaceous material, such as graphite or coal eroding in, are also common. Finally, the confusing influence of bacterial metabolism in the chemically structured sediment, and of the movement of pore waters, should not be underestimated. However, most radiocarbon dating of sediments has been carried out on 'bulk sediment' in which only the most labile of humics and the carbonates have been removed. It is well known that such an approach can give anomalous dates during erosive episodes, in hard waters, and in low carbon sediments (where the reduced, geological carbon contribution may dominate). By far the best approach is to date terrestrial plant macrofossils, if this is possible (Zbinden *et al* 1989). Otherwise, individual fractions again provide useful information. For example, the chemical survival of many complex lipids enables higher plant cuticle waxes to be isolated and dated in favorable circumstances (Farr *et al* 1990).

The aim of this section is to emphasize that many, perhaps most, situations where radiocarbon dating is applied require both an appreciation of the complex processes by which organic material is deposited as well as a kind of 'deconvolution chemistry' to unravel what time and the environment has tangled up.

LABORATORY PROCEDURES

The laboratory ideally should carry out three processes: 1) the selection of samples from the field (in conjunction with the field scientist or archaeologist); 2) the 'pretreatment' of the sample to extract a compound containing the set of carbon atoms for isotopic measurement; and 3) the conversion of the pretreated sample to a suitable form for measurement.

Sample Selection

The value of laboratory participation in sample selection should be evident from consideration of the processes that may invalidate a date, and especially the relation of the material to the event, as discussed above. Unfortunately, it is still rather uncommon.

Pretreatment in General

The type of pretreatment employed depends on the sample material, on its quantity, on its likely age, and on the likely chemistry of its burial environment. Some generalizations can be made, however. The laboratory has three options in principle (even if these are reduced in practice):

1. To extract from the sample a well-defined chemical compound with carbon atoms that can only have come from the organism;
2. To identify the significant contaminants and specifically remove them;
3. To 'clean' the sample by removing unidentifiable but chemically more unstable material.

Of course, all three approaches can be combined.

The aim is, or should be, to prepare a sample for chemical conversion in such a way that the laboratory has a reasonably clear idea of the chemical nature of the material being converted.

The range of material for radiocarbon dating is impressive, and Table 12.1 lists many of the sample types routinely dated. It is not possible to indicate methods for all of these, for which a search of the literature should be made (eg, especially in *Radiocarbon* and its triennial publications of the Proceedings of International Radiocarbon Conferences: Stuiver & Kra 1980, 1983, 1986; Long & Kra 1989). A detailed account of procedures used at the Oxford AMS laboratory is given in Hedges *et al* 1989.

Two types of material – bone and 'charred' material that includes many of the categories listed in Table 12.1 – have been selected for a more detailed account of pretreatment methods. They are considered in Examples of Laboratory Processes . . . , below, after all the chemical processes in radiocarbon dating have been summarized.

TABLE 12.1. Types of material commonly dated by radiocarbon that require chemical pretreatment

Geological deposits
Carbonates (eg, speleothems, tufas, mollusks, corals, foraminifera)
Ice cores
Meteorites
Sediments (eg, peats, lacustrine deposits, palaeosols, modern soils)
Biological deposits
Wood, charcoal
Seeds, leaves, twigs
Pollen
Insect remains
Fish remains
Bone, antler, horn
Coprolites
Anthropogenic deposits
Burned bone
Pottery
Metal casting cores
Slags, iron
Resins, glues, food remains, blood remains (eg, on tools, pottery, paintings)
Textiles (eg silk, wool, cotton, linen)
Hair, leather, parchment, paper
Wall paintings

Sample Conversion

The principal techniques employed to measure the $^{14}C/^{12}C$ ratio are gas counting (using CO_2) (see Kromer & Münnich 1992), scintillation counting (using benzene) (see Polach 1992), and AMS (using graphite, or in the case of Oxford, CO_2) (see Beukens 1992), each requiring a specific compound to be made from the pretreated material. To simplify procedures, virtually all laboratories first oxidize the pretreated material to CO_2, usually with O_2, or (for most AMS samples) with CuO. For gas counting, the CO_2 has to be rigorously purified, especially to remove O_2. The CO_2 at this stage is generally sub-sampled to measure the stable carbon isotope ratio. Reduction is carried out using lithium to make acetylene, for subsequent catalytic polymerization to benzene, or, in the case of the IsoTrace Laboratory of Toronto, decomposition to graphite in a radio-frequency discharge. A popular route for the production of graphite by AMS laboratories is to reduce CO_2 by either hydrogen or zinc (to CO), followed by the deposition of graphite by catalyzed disproportionation of the CO onto an iron substrate. To a certain extent, each laboratory has developed its own version of both oxidation and reduction processes, taking into account yield, sample size, background contamination, ease of routine and/or non-skilled operation, etc. Documentation, not necessarily up-to-date, is to be found in the Radiocarbon and Accelerator Mass Spectrometry Conference Proceedings (to which reference has already been made). Experimental details are best obtained from the laboratories themselves.

The main features of sample conversion relevant to the control of errors in a radiocarbon date have already been discussed (*ie*, fractionation and laboratory contamination). Because this process is much more under the laboratory's control, it is a less contentious and 'difficult' operation than pretreatment, but although routine, it nevertheless requires the greatest care, since it is the direct input to the radiocarbon measurement.

Examples of Laboratory Processes in Sample Pretreatment: Bone and 'Charred' Material

Bone is a complex material that has a definite composition in the living organism. This definition is lost during subsequent diagenesis and interaction with the organic constituents of its burial environment. About 20% of modern bone by weight is protein, of which 90% is the insoluble protein collagen. The approach to dating bone (Hedges & Law 1989) depends upon the degree of degradation of its protein content, on its age and on the contaminating influences of its environment. The last aspect is rarely available for study, although the application of consolidating chemicals, which is quite common on badly preserved bone, can be monitored, for example, by infrared spectrometry.

Well-preserved (more than 30% collagen remaining), recent, scarcely contaminated bone can be reliably pretreated by decalcification in dilute HCl, and washing the extracted collagen with acid and alkali. More reliable is the subsequent conversion of insoluble collagen to soluble gelatin, which is separated by filtration. This, the standard 'Longin' (1971) method is widely used for non-critical situations. As conditions become more critical, it is useful to be able to characterize the state of degradation of the bone, for example, by measuring the insoluble protein content, by measuring the infrared spectrum of the extracted collagen or gelatin (which can detect a few percent of 'humic' and chemically related contamination (DeNiro & Weiner 1988a)), and by measuring the glycine/aspartate or hydroxyproline/aspartate ratio (Weiner & Bar-Yosef 1990) (which is much higher for collagen than other proteins). We have found that the best next step is to purify the extracted gelatin by ion exchange (Law & Hedges 1990), retaining most of the humic-type material on the column.

Further types of chromatography can be carried out subsequently on the gelatin fraction. With infrared spectrometry, the 'clean-up' of badly contaminated collagen to ion-exchanged gelatin can be monitored and correlated with increasing credibility in the date (Law *et al* 1991). This approach appears to give 'reliable' results (*ie*, in accord with archaeological expectation) for collagen preservation levels down to about 1–3% of the modern level.

Other methods of purification are available, from ultrafiltration to select higher molecular weight fractions (Brown *et al* 1989) (thereby retaining less degraded gelatin), to specific enzymic proteolysis (DeNiro & Weiner 1988b) to peptides. This latter method may have an advantage in cleaving cross-linked humic-type contaminating compounds from the amino acid chain. Total hydrolysis to amino acids may have a similar advantage, but it also has its dangers since, if it takes place in the presence of free contaminating humics and/or polysaccharides, the subsequent formation of aminosugars may not be removed easily by any subsequent chromatographic separation.

To summarize at this point:

1. A variety of increasingly sophisticated methods is available to deal with increasingly degraded/contaminated bone. But note that the approaches outlined above fail when the collagen content of a bone falls below about 1% of its original value. Usually, the amino-acid composition at this stage no longer corresponds to that for collagen.

2. The degree of degradation, and to a rough extent of contamination, can be characterized by measurements on bone extracts.

3. While the characterization from (2) might suggest a method from (1), there is still no way to *guarantee* that such a method will give a reliable result in principle. Further, there is no way to determine in a given sample, if the actual result obtained is to be relied upon. Obviously, the better preserved the bone, the higher the reliability.

4. Most laboratories have to balance the cost and effort of a more reliable and more elaborate technique against the possible improvement in quality of the date. Certainly, it is not feasible to use the most sophisticated methods for every date.

Some further experimental developments are worth reporting. The abundance in collagen of the amino acid hydroxyproline, and its rarity in other proteins, has been remarked for many years. Where hydroxyproline can be isolated from bone, it would appear to offer a guarantee that its constituents are indigenous to the bone. However, Long *et al* (1989), have reported the presence of hydroxyproline in sediments, and it is also known to be an important constituent in structural proteins in the cell walls of plants and fungi, so that the guarantee must be regarded as less than absolute. Several hydroxyproline extract dates have been reported, but the extraction methods used are time-consuming and difficult. Also, the method is applicable only when at least some reasonably well-preserved collagen survives, and then the hydroxyproline content is only about 10% of such collagen. It is not known whether the occurrence of hydroxyproline in the environment of the bone is likely to present a serious form of contamination. Preparative HPLC has been used to purify the major amino acids in collagen (Stafford *et al* 1990; van Klinken & Mook 1990), and although the technique is difficult and experimental, it has already shown that, in some bones, some of the amino acids present and extracted by standard methods are quite clearly not indigenous. On the other hand, consistency in dates from individual amino acids greatly strengthens the reliability of a date. But such consistency becomes increasingly unlikely for bones with 'no' collagen. Such bones are common from hot climates (Weiner & Bar-Yosef 1990). There is evidence (Masters 1987) that non-collagenous acidic proteins are better preserved in bone than collagen under these conditions, perhaps through stabilization by being strongly adsorbed by the hydroxyapatite matrix (although much of this undergoes recrystallization).

Taylor (1992) reports on attempts to isolate and date osteocalcin. This small protein, like hydroxyproline, occurs more or less uniquely in bone and appears to survive longer than collagen. The task of isolating the protein in purified, non-crosslinked and preferably undegraded form from bone that has lost all its collagen is, however, a major one and its achievement would make a most important advance in the dating of badly preserved bone.

'Charred' Material. The main component in charred material – elemental carbon – is chemically simple and inert (hence, its survival). It cannot therefore be extracted by a specific method; rather all impurities must be removed. In practice, charred material covers a wide spectrum of composition, from the more-or-less pure carbon of macroscopic masses of charcoal, through finely disseminated carbon, to the much more humic-like composition of partially burned and subsequently degraded plant remains.

Environmental contaminants may be strongly adsorbed onto the very high surface area of charcoal. The usual method of cleaning is to treat with acid (carbonates are also likely in plant ash), with alkali to remove humic adsorbents, and, more optionally, with oxidizing agents (eg, sodium hypochlorite) to break down additional contaminants. It is hard to gauge the success of these methods – massive charcoals often have older dates than their associated cultural or depositional event – and the number of very old charcoal dates is fairly small (largely because of scarcity). However, attempts to achieve very old dates on coals suggest that environmental contaminants can be very tenacious, at least for the more hydrocarbon-rich composition found in many coals. It may be useful to measure the H/C ratios for critical charcoals after pretreatment in order to assess the propensity for chemically linked contamination.

Massive charcoals are straightforward in comparison with more finely disseminated or degraded material. Many sediments and deposits contain such material, often with questionable stratigraphic control and/or place of origin. This makes validation of the dates more difficult. While the chemical methods outlined above are still applicable, they may well be too aggressive and hasten the dissolution of the entire sample. A recent paper (Gillespie 1990) addresses this problem, and recommends the use of nitric acid and sodium chlorate. The presence of clays in sediments is an additional difficulty, because the clay may strongly bind humic acids, which are then released on dissolution of the clay by HF. Many sediments of recent origin may also contain cellulose, and the same methods that purify carbon also tend to purify cellulose. This may be welcome if both components must have the same age, but often the elemental carbon component represents reworked eroded material and will have a greater age. While cellulose can by removed after extensive acid hydrolysis, it is not possible to remove carbon and leave purified cellulose to be dated. Of course, these considerations have only arisen as the small sample size of the AMS technique has enabled the dating of sediment fractions of this type.

Where organic material is only partially burned, a continuum between the carbon and the humic content may remain. For example, attempts to clean up burned seeds by washing with alkali can easily lead to their disappearance, and it is therefore wise to make a preliminary test. It is often possible to make a rather arbitrary kind of fractionation, with a 'humic' extract and a carbonaceous residue. Where samples are too small for this, it is usually only possible, at best, to rely on acid and very brief alkaline extraction. An illustrative and interesting application of these methods is to be found in the approaches of two AMS laboratories (Housley *et al* 1990; Nelson, Vogel & Southon 1990) in dating charred seeds from Akrotiri (which dates the eruption of Thera during the Greek Bronze Age). The radiocarbon dates at issue require particularly accurate determination. The question of the radiocarbon age of any contaminating material and the extent to which this can be removed is crucial, and depends ultimately on controversial differences in laboratory pretreatment. In the case of charred bone, the carbonaceous residue can account for 1–5% of the total weight, and in general, appears to give reliable results, even on charred bone older than 30,000 years. However, it is very difficult to demonstrate independently that the carbonaceous residue is not contaminated. Charred bone appears to contain usefully high levels of lipids, and work is proceeding to test whether extracted lipids can be used to secure a second fraction for dating. Dating charred bone is particularly important for the many sites where bone collagen does not survive. Cremated bone, on the other hand, has been too strongly oxidized to contain sufficient residual carbon.

The ability to obtain dates on both a 'humic' and 'carbonaceous' fraction for many charred materials has provided information on the reliability of 'humic' dates, and the extent to which environmental contamination may invalidate a date (Batten *et al* 1986). In this study, it was found that the 'humic' date generally agreed with the 'carbon' date, unless the sample had become seriously stratigraphically misplaced – for example, a Neolithic charred seed found in a Palaeolithic level. In such cases, the 'humic' component date tended to represent the age of the stratigraphic level rather than the sample itself. This is perhaps not surprising; it suggests, first, that at least some of the 'humic' component in a charred sample is from the environment, and second, that mobile humic components in the local depositional environment generally have the same radiocarbon age as the depositional event. This implies that much humic acid contamination will be of similar age to that of the sample being dated, and reinforces the warning that a pretreatment method that gives the 'correct' date has not necessarily removed environmental contamination.

CONCLUSION

The enormous variety of situations to which radiocarbon dating can be usefully applied makes a detailed account of the chemistry of sample treatment impossible in a short chapter. However, the aims of such treatment are common to all radiocarbon dates; that is, to ensure that the carbon atoms actually being measured represent the basic assumptions underlying the principles of radiocarbon dating. The science of sample treatment consists in devising adequate methods to achieve this for a given sample, and at the same time, to estimate the extent to which the method is effective. There is no simple prescription for this. The best means to establish a lasting chronological framework lies in the interplay between increasingly sophisticated chemical procedures, greater understanding of the chemistry of environmental contamination and sample diagenesis, and the validation of methods through feedback from dates obtained under controlled conditions.

REFERENCES

Bada, JL, Schoeninger, MJ and Schimmelmann, A 1989 Isotopic fractionation during peptide bond hydrolysis. *Geochimica et Cosmochimica Acta* 53: 3337–3341.

Batten, RJ, Bronk, CR, Gillespie, R, Gowlett, JAJ, Hedges, REM and Perry, C 1986 A review of the operation of the Oxford Radiocarbon Accelerator Unit. *In* Stuiver, M and Kra, RS, eds, Proceedings of the 12th International ^{14}C Conference. *Radiocarbon* 28(2A): 177–185.

Beukens, RP 1992 Radiocarbon accelerator mass spectrometry: Background, precision and accuracy. *In* Taylor, RE, Long, A and Kra, RS, eds, *Radiocarbon After Four Decades: An Interdisciplinary Perspective.* New York, Springer-Verlag, this volume.

Brown, TA, Nelson, DE, Mathews, RW, Vogel, JS and Southon, JR 1989 Radiocarbon dating of pollen by accelerator mass spectrometry. *Quaternary Research* 32(3): 205–212.

Brown, TA, Nelson, DE, Vogel, JS and Southon, JR 1988 Improved collagen extraction by modified Longin method. *Radiocarbon* 30(2): 171–177.

Craig, H 1959 Carbon 13 in plants and the relationship between carbon 13 and carbon 14 variations in nature. *Journal of Geology* 62: 115–143.

DeNiro, MJ and Weiner, S 1988a A chemical, enzymatic and spectroscopic characterization of "collagen" and other organic fractions from prehistoric bones. *Geochimica et Cosmochimica Acta* 52: 2197–2206.

_____1988b Use of collagenase to purify collagen from prehistoric bones for stable isotopic analysis. *Geochimica et Cosmochimica Acta* 52: 2425–2431.

Duplessy, J-C, Arnold, A, Bard, E, Labeyrie, L, Duprat, J and Mayes, J 1992 Glacial-to-interglacial changes in ocean circulation. *In* Taylor, RE, Long, A and Kra, RS, eds, *Radiocarbon After Four Decades: An Interdisciplinary Perspective.* New York, Springer-Verlag, this volume.

Farr, KM, Jones, DM, O'Sullivan, PE, Eglinton, G, Tarling, DH and Hedges, REM 1990 Paleolimnological studies of laminated sediments from Shropshire-Cheshire meres. *Hydrobiologia*: 279–292.

Fontes, J-Ch 1992 Chemical and isotopic constraints on ^{14}C dating of groundwater. *In* Taylor, RE, Long, A and Kra, RS, eds, *Radiocarbon After Four Decades: An Interdisciplinary Perspective.* New York, Springer-Verlag, this volume.

Fowler, AJ, Gillespie, R and Hedges, REM 1986 Radiocarbon dating of sediments. *In* Stuiver, M and Kra, RS, eds, Proceedings of the 12th International ^{14}C Conference. *Radiocarbon* 28(2A): 441–450.

Geyh, MA 1992 Numerical modeling with groundwater ages. *In* Taylor, RE, Long, A and Kra, RS, eds, *Radiocarbon After Four Decades: An Interdisciplinary Perspective.* New York, Springer-Verlag, this volume.

Geyh, MA, Krumbein, WE and Kudrass, HR 1970 Unreliable radiocarbon dating of long-stored deep-sea sediments due to bacterial activity. *Marine Geology* 17: M45–M50.

Giger, W, Sturm, M, Sturm, H and Schaffner, C 1984 $^{14}C/^{12}C$-ratios in organic matter and hydrocarbons extracted from dated lake sediments. *Nuclear Instruments and Methods* 233(B5): 394–397.

Gillespie, R 1990 On the use of oxidation for AMS sample pretreatment. *In* Yiou, F and Raisbeck, G, eds, Proceedings of the 5th International Conference on AMS. *Nuclear Instruments and Methods* B52(3,4): 345–347.

Goodfriend, GA and Stipp, JJ 1983 Limestone and the problem of radiocarbon dating of land-snail carbonate. *Geology* 11: 575–577.

Gove, H, ed 1978 *Proceedings of the 1st International Conference on AMS.* New York, University of Rochester Press: 1–401.

Gove, HE, Litherland, AE and Elmore, D, eds 1987 Proceedings of the 4th International Conference on AMS. *Nuclear Instruments and Methods* B29: 1–455.

Grootes, PM, Mook, WG, Vogel, JC, de Vries, AE, Haring, A and Kistemaker, J 1975 Enrichment of radiocarbon for dating samples up to 75,000 years. *Zeischrift für Naturforschung* 30: 1–14.

Haynes, CV, Jr, Damon, PE and Grey, DC 1966 Arizona radiocarbon dates VI. *Radiocarbon* 8: 1–21.

Hedges, REM and Law, IA 1989 The radiocarbon dating of bone. *Applied Geochemistry* 4: 249–253.

Hedges, REM, Law, IA, Bronk, CR and Housley, RA 1989 The Oxford accelerator mass spectrometry facility: Technical developments in routine dating. *Archaeometry* 31(2): 99–113.

Heron, C, Evershed, R and Goad, LJ (ms) Effects of diagenesis migration on organic residues associated with buried potsherds. Submitted for publication to *Journal of Archaeological Science.*

Housley, RA, Hedges, REM, Law, IA and Bronk, CR 1990 Radiocarbon dating by AMS of the destruction of Akrotiri. *In* Hardy, DA and Renfrew, AC, eds, *Proceedings of Thera and the Aegean World III.* London: The Thera Foundation: 207–215.

Klinken, GJ, van and Mook, WG 1990 Preparative high-performance liquid chromatographic separation of individual amino acids derived from fossil bone collagen. *Radiocarbon* 32(2): 155-164.

Kromer, B and Münnich, KO 1992 CO_2 gas proportional counting in radiocarbon dating: Review and perspective. *In* Taylor, RE, Long, A and Kra, RS, eds, *Radiocarbon After Four Decades: An Interdisciplinary Perspective.* New York, Springer-Verlag, this volume.

Kutschera, W, ed 1981 *Proceedings of the Symposium on Accelerator Mass Spectrometry.* ANL/PHY 81-1. Springfield, National Technical Information Service: 1–15.

Law, IA and Hedges, REM 1990 A semi-automated bone pretreatment system and the pretreatment of older and contaminated samples. *In* Long, A and Kra, RS, eds, Proceedings of the 13th International ^{14}C Conference. *Radiocarbon* 31(3): 247–253.

Law, IA, Housley, RA, Hammond, N and Hedges, REM 1991 Cuello: Resolving the chronology through direct dating of conserved and low-collagen bone by AMS. *Radiocarbon* 33(3), in press.

Libby, WF 1954 Radiocarbon dating. *Endeavour* 13: 5–16.

Long, A, Arnold, CD, Damon, PC, Ferguson, CW, Lerman, JC and Wilson, AT 1979 Radial translocution of carbon in bristlecone pine. *In* Berger, R and Suess, HE, eds, *Radiocarbon Dating.* Proceedings of the 9th International ^{14}C Conference. Berkeley, University of California Press: 532–537.

Long, A and Kra, RS, eds 1989 Proceedings of the 13th International ^{14}C Conference. *Radiocarbon* 31(3): 229–1082.

Long, A, Wilson, AT, Ernst, RD, Gore, BH and Hare, PE 1989 AMS radiocarbon dating of bones at Arizona. *In* Long, A and Kra, RS, eds, Proceedings of the 13th International ^{14}C Conference. *Radiocarbon* 31(3): 231–238.

Longin, R 1971 New method for collagen extraction for radiocarbon dating. *Nature* 230: 241–247.

Masters, PM 1987 Preferential preservation of noncollagenous protein during bone diagenesis: Implications for chronometric and stable isotopic measurements. *Geochimica et Cosmochimica Acta* 51: 3209–3214.

Merwe, N, van der 1982 Carbon isotopes, photosynthesis and archaeology. *American Scientist* 70: 596–606.

Nelson, DE, Vogel, JS and Southon, JR 1990 Another suite of confusing radiocarbon dates for the destruction of Akrotiri. *In* Hardy, DA and Renfrew, AC, eds, *Thera and the Aegean World III.* London, The Thera Foundation.

Peng, T-H and Broecker, WS 1992 Reconstruction of radiocarbon distribution in the glacial ocean. *In* Taylor, RE, Long, A and Kra, RS, eds, *Radiocarbon After Four*

Decades: An Interdisciplinary Perspective. New York, Springer-Verlag, this volume.

Polach, H 1992 Four decades of progress in ^{14}C dating by liquid scintillation counting and spectrometry. *In* Taylor, RE, Long, A and Kra, RS, eds, *Radiocarbon After Four Decades: An Interdisciplinary Perspective.* New York, Springer-Verlag, this volume.

Stafford TW, Hare, PE, Currie, L, Jull, AJT and Donahue, D 1990 Accuracy of North American human skeleton ages. *Quaternary Research* 34(1): 111–120.

Stuiver, M and Kra, RS, eds 1980 Proceedings of the 10th International ^{14}C Conference. *Radiocarbon* 22(2 & 3): 133–1016.

_____1983 Proceedings of the 11th International ^{14}C Conference. *Radiocarbon* 25(2): 171-795.

_____1986 Proceedings of the 12th International ^{14}C Conference. *Radiocarbon* 28 (2A): 177–804.

Stuiver, M, Pearson, GW and Braziunas, TF 1986 Radiocarbon age calibration of marine samples back to 9000 cal yr BP. *In* Stuiver, M and Kra, RS, eds, Proceedings of the 12th International ^{14}C Conference. *Radiocarbon* 28(2B): 980–1021.

Taylor, RE 1992 Radiocarbon dating of bone: To collagen and beyond. *In* Taylor, RE, Long, A and Kra, RS, eds, *Radiocarbon After Four Decades: An Interdisciplinary Perspective.* New York, Springer-Verlag, this volume.

Toggweiler, JR (ms) 1990 Modeling the radiocarbon distribution in the modern oceans. Paper presented at "Four Decades of Radiocarbon Studies: An Interdisciplinary Perspective." Lake Arrowhead, California, June 5-9.

Vogel, JS, Nelson, DE and Southon, JR 1987 ^{14}C background levels in an accelerator mass spectrometry system. *Radiocarbon* 29(3): 323–333.

Weiner, S and Bar-Yosef, O 1990 States of preservation of bones from prehistoric sites in the Near East: A survey. *Journal of Archaeological Sciences* 17: 187–196.

Wölfli, W, Polach, HA and Anderson, HH, eds 1984 Proceedings of the 3rd International Conference on AMS. *Nuclear Instruments and Methods* 233: (B5).

Yiou, F and Raisbeck, GM, eds 1990 Proceedings of the 5th International Conference on AMS. *Nuclear Instruments and Methods* B52(3,4): 211–630.

Zbinden, H, Andrée, M, Oeschger, H, Ammann, B, Lotter, A, Bonani, G and Wölfli, W 1989 Atmospheric radiocarbon at the end of the last glacial: An estimate based on AMS radiocarbon dates on terrestrial macrofossils from lake sediments. *In* Long, A and Kra, RS eds, Proceedings of the 13th International ^{14}C Conference. *Radiocarbon* 31(3): 795–804.

13

CO_2 GAS PROPORTIONAL COUNTING IN RADIOCARBON DATING - REVIEW AND PERSPECTIVE

BERND KROMER and KARL OTTO MÜNNICH

INTRODUCTION

Gas counting is a mature and powerful technique central to radiocarbon dating. The method was taken from the detection techniques used in nuclear physics and adapted to the special requirements of low-level counting of the carbon gases. A compilation made by W G Mook in 1983 lists 174 gas counters used in ^{14}C dating, the counting gases being CO_2 (115 counters), CH_4 (38), C_2H_2 (20) and C_2H_6 (1). In the present contribution, the current status of CO_2 gas counting is reviewed. The emphasis on CO_2 is justified by several observations: 1) CO_2 is the primary gas to be produced in all methods; 2) routine techniques are able to achieve high purity CO_2 gas, so further conversion to hydrocarbons appears unwarranted; 3) the cryogenic properties of CO_2 facilitates handling and thus minimizes contamination; 4) all gas counting laboratories involved in high-precision work, eg, for calibration, use CO_2 as counting gas. Most of the techniques mentioned in this review were already available around 1975 (see the Proceedings of the Ninth International Radiocarbon Conference (Berger & Suess 1979)); since that time, gas proportional counting has become a routine operating technique in several laboratories. Wherever possible, reference is made to individual laboratories; however, as the more technical aspects of the technique that are central to this chapter are rarely fully documented in the literature, we take most examples from the Heidelberg laboratory, and are fully aware that this description gives heavy weight to a single installation.

COMBUSTION

Organic material is generally converted to CO_2 using either the de Vries-type continuous combustion (de Vries & Barendsen 1953; Münnich 1957) or the combustion 'bomb' method (Burleigh 1972; Dörr, Kromer & Münnich 1989). In the de Vries-type combustion technique, oxides other than CO_2 are removed in several steps (reaction with $KMnO_4$, precipitation of CO_2 to carbonate in a $CaCl_2/NH_4$ solution), followed by the final purification (see below). With this technique the combustion process can be controlled fairly easily, even for volatile material, by adjusting the flow of oxygen and nitrogen and by mixing 'problematic' material, such as oxalic acid, with non-carbonate, organic free sand.

The 'bomb' combustion technique offers the advantage of increased throughput and lower concentration of non-carbon oxides (Dörr, Kromer & Münnich 1989), but it is not well suited for some materials (eg, those of low carbon content).

Gas Purity

Very early in the development of radiocarbon dating, it was realized that the key issue in using CO_2 as counting gas in proportional counters is to achieve and maintain high gas purity. This requirement is imposed by the low mobility of electrons in molecular gases such as CO_2 (Huxley & Crompton 1974; Sauli 1977), making them susceptible to attachment and recombination before they can be collected at the anode.

Requirements

The electrons originating from the decay of radiocarbon and drifting to the anode may become attached to electronegative impurities in the CO_2 gas. Ultimately, even for very pure gas, recombination will limit the performance of a proportional counter filled with CO_2. Both processes have been studied in great detail by several authors (eg, Roether 1961; Brenninkmeijer & Mook 1979; Povinec 1979). The demands on gas purity are severe: the concentration of oxygen must be kept well below a few ppm, and that of SO_2, nitrogen oxides, water vapor and freons will hamper the performance in levels as low as some tens of ppm at pressure levels commonly used in CO_2 counting.

The recombination of electrons with CO_2 sets an upper limit to the pressure of the counting gas at about 12 to 16 atm (Roether 1961; Tans & Mook 1978).

Purification Techniques

Fortunately, proven techniques can routinely purify CO_2 gas to the level required for reliable use in proportional counting. The most widely used method is to circulate CO_2 over hot copper and silver for an extended period of time (typically days). This process will result in sufficiently pure CO_2, but may introduce radon. Therefore, the final sample must be stored for several weeks to allow for the decay of radon.

In the Heidelberg laboratory, we use an alternative method that is equally efficient in trapping impurities and faster than the traditional method. It also removes radon to insignificant levels (Bruns 1976). After passing the initial, 'coarse' separation steps mentioned for the de Vries type combustion, the CO_2

gas is sent through a column filled with activated charcoal at 0°C. In the charcoal, contaminants are adsorbed by a chromatographic process with CO_2 acting as carrier gas. The chromatographic parameters (flow rate of CO_2, pressure drop along the column, terminating pressure) are adjusted to result in optimum purification yet minimal fractionation (below 0.5‰ in $\delta^{13}C$) and little loss of CO_2 (1%) (Schoch & Münnich 1981).

Gas Containers

To maintain high gas purity, CO_2 is usually stored in stainless steel containers at high pressure. In the Heidelberg laboratory, the containers are made of chromatography-grade pipes (OD 12 mm, length 470 mm) and bellows valves.[1] This design has an additional advantage: the total weight of the container (ca 450 g) and CO_2 gas sample (16.8 g) is within the range of one-milligram-resolution balances. Thus, the amount of sample can be determined with high accuracy (10^{-4}) gravimetrically without any long-term drift.

Counting

Counter Design

Basically, two types of proportional counters were developed for CO_2 gas counting: quartz tubes coated with a tin oxide (later gold) film (de Vries, Stuiver & Olsson 1959) and high-purity copper tubes (Münnich 1957; Tans & Mook 1978; Stuiver, Robinson & Yang 1979). Initially, rather sophisticated end designs were used to minimize spectrum distortion due to incomplete charge collection (eg, Groeneveld 1977). More recently, Stuiver, Robinson and Yang (1979) and Schoch and Münnich (1981) showed that flat quartz end plates are sufficient to provide stable operation leading to an economic design for multi-counter operation (Fig 13.1A & 13.1B).

The present status of multiwire counters and their potential for CO_2 gas counting is described by Povinec (1992).

Counting Parameters

The key parameters affecting low-level $^{14}CO_2$ counting are:

- *Gas pressure* – The CO_2 pressure affects background (due to the lower surface area of the counter at higher pressure levels) and the demands on gas purity, as the attachment coefficient has been shown empirically to be dependent almost linearly on pressure (Roether 1961; Tans & Mook 1978).

[1]Nupro BK6-MM

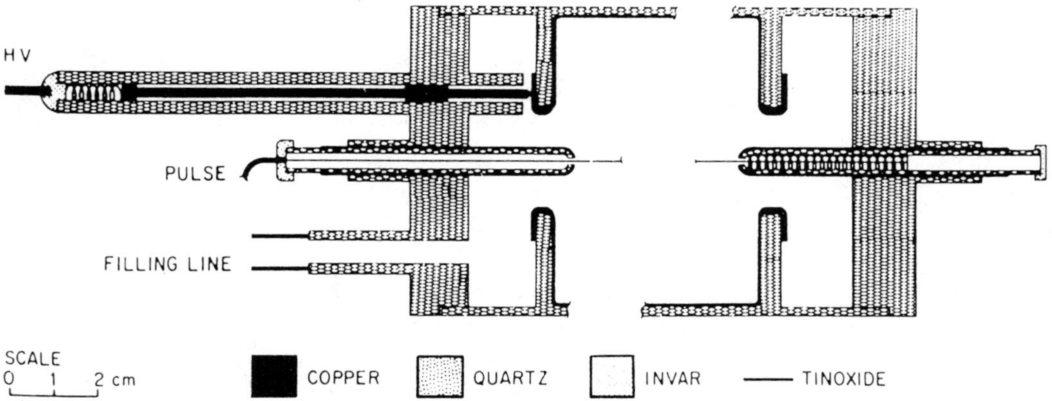

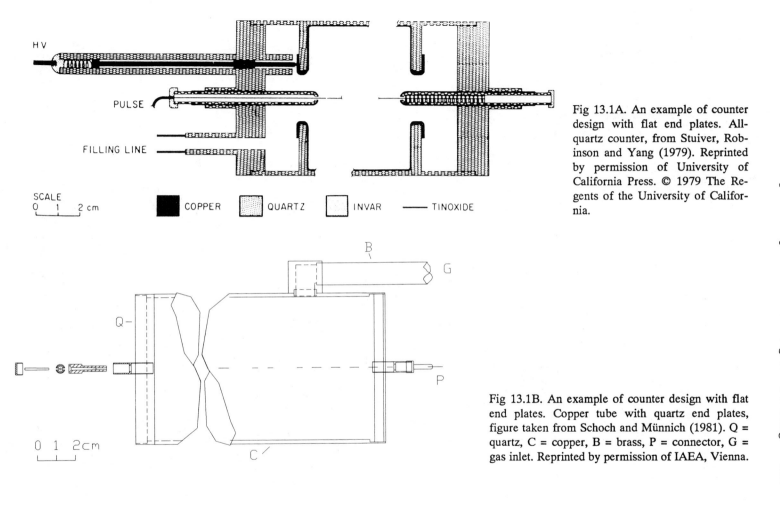

Fig 13.1A. An example of counter design with flat end plates. All-quartz counter, from Stuiver, Robinson and Yang (1979). Reprinted by permission of University of California Press. © 1979 The Regents of the University of California.

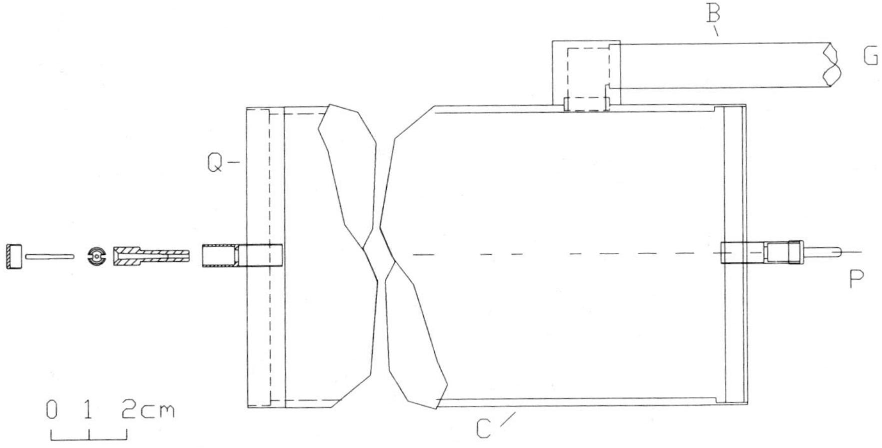

Fig 13.1B. An example of counter design with flat end plates. Copper tube with quartz end plates, figure taken from Schoch and Münnich (1981). Q = quartz, C = copper, B = brass, P = connector, G = gas inlet. Reprinted by permission of IAEA, Vienna.

Stable operation has been demonstrated up to 10 atm (Tans & Mook 1978), but in practice, pressure levels of 1 to 6 atm are used.

- *High voltage vs anode diameter* – Two considerations affect the choice of these parameters, *ie*, 1) keep transit times of electrons drifting towards the anode (Povinec 1979) as short as possible, and 2) keep high voltage low enough to avoid spurious discharges. Thicker wires lead to higher field strength outside the multiplication region and, thus, are preferable, in terms of lower sensitivity to impurities (Brenninkmeijer & Mook 1979). However, they require high values of the high voltage. A practical limit is 10 kV, resulting in anode diameter values of 20 to 50 μm.

- *Gas multiplication* – Generally applicable gas gain formulas for electrons in CO_2 have been developed by several investigators (see Zastawny 1966; Aoyama 1985); these formulas were compared to experimental data by Grootes (1977). Groeneveld (1977) devised a method to determine gas multiplication from the coincidence count rate at different thresholds. Usually, the gas gain is chosen between 5000 and 10,000.

Counter Background

It is important to realize that, for most ^{14}C dating problems, the *stability* of the counting background is more crucial than its magnitude. Magnitude eventually sets limits to the dating range for a given set of counter parameters, whereas it is the stability that affects the dating precision of all samples.

Underground Counters

Moving the counters well below the surface removes most of the variability in the cosmic-ray flux. Several radiocarbon laboratories have underground counting facilities. From the compilation of Mook (1983), it is apparent empirically that a favorable trend in the ratio of the background and muon count rate is 0.8% (0.8 cpm/100 cpm). Most of the reduction in background compared to near-surface installations is achieved by the first few meters of shielding through soil (Stuiver, Robinson & Yang 1979); Mook (1983) calculates from his survey of underground installations a scale height of 25 m water equivalent (ca 10% reduction per meter of soil).

Near-Surface Counters

Even in near-surface counting rooms, background variation caused by variations in cosmic-ray flux can be minimized. In the Heidelberg laboratory, the counters are located in the basement of a five-story building (7 ceilings, in total 92 cm concrete). The background count rate varies with the coincidence count rate as shown in Figure 13.2 for a period of one year. The background count rate in

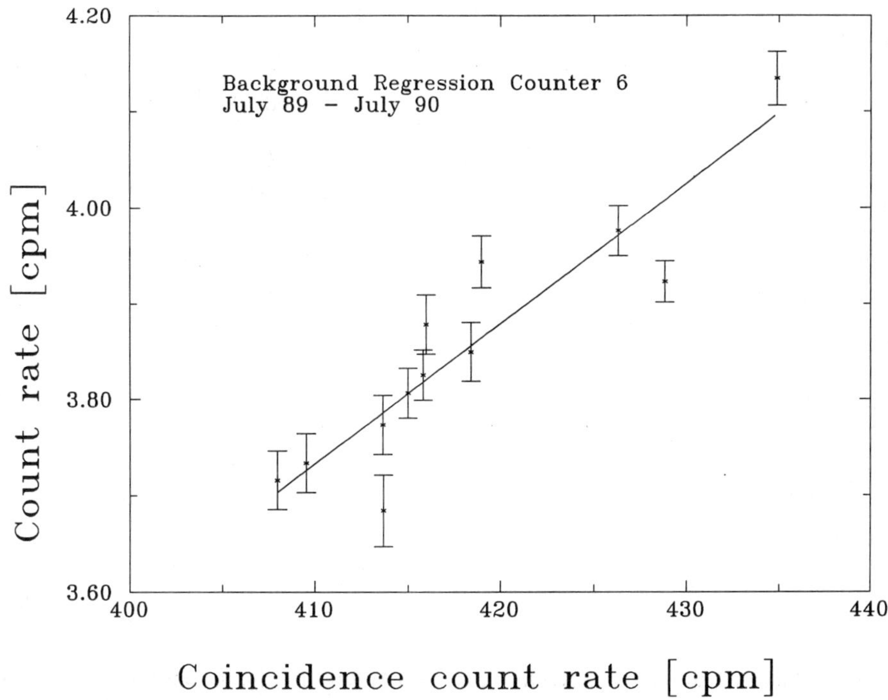

Fig 13.2. Variation of background count rate *vs* coincidence count rate for Counter 6 in the Heidelberg laboratory for the interval July 1989 to July 1990

one counter covering a period of three years is shown in Figure 13.3A. Most of the fluctuation is caused by changes in cosmic-ray flux, as is seen from the residual count rate (Fig 13.3B) obtained by the following procedure: for each one-month interval, we calculate, by linear regression, the correlation of background count rate *vs* coincident count rate, based on the counting of the preceding 10 background measurements. After subtraction of the estimated background based on this sliding correlation, the remaining average variation of the actual background count rate is about 0.07 cpm (Fig 13.3B). This value is only slightly higher than the statistical error of a single background measurement of *ca* 0.05 cpm.

Background Reduction by Electronic Techniques

Recently, Mäntynen *et al* (1987) proposed to apply rise time and pulse-height analysis of the counting signal to CO_2 gas counting in order to reduce background count rate. This technique is well established for low-energy β counting of tritium (Harris & Mathieson 1970, 1971; Oeschger & Wahlen 1975) and in solar neutrino research (see eg, Davis *et al* 1972; Plaga 1989). The reduction in background even for ^{14}C events producing relatively large pulses is very sub-

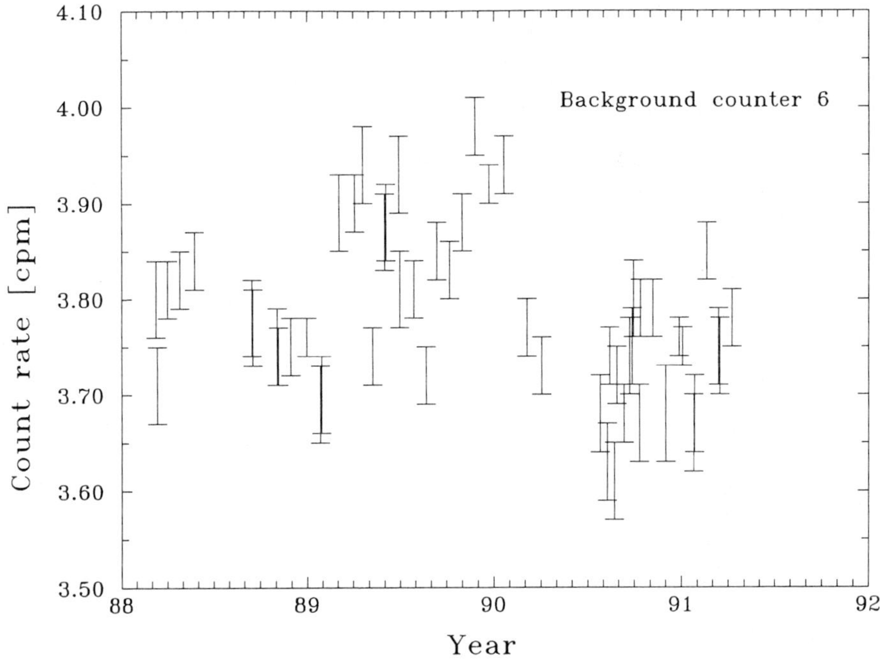

Fig 13.3A. Background correction results for Counter 6: Raw data for the interval 1988 to 1991

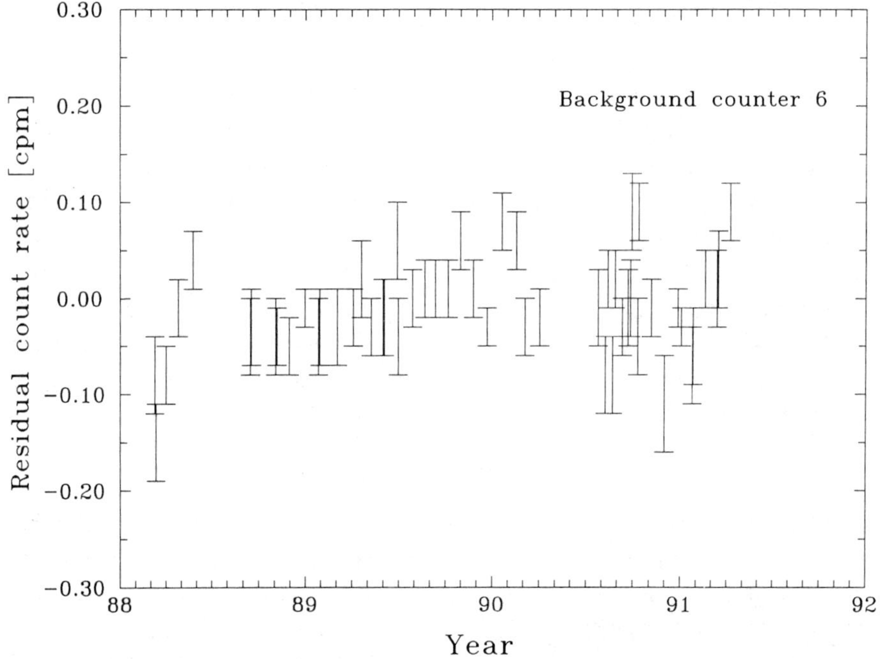

Fig 13.3B. Background correction results for Counter 6: Residual count rate after subtraction of calculated background according to the regression of Figure 13.2 (updated for each month from background determinations of the preceding 10 measurements), and the coincidence count rate of the respective interval. The statistical error of each background measurement is 0.03 to 0.05 cpm.

stantial (to one third of its undiscriminated value). Further studies should determine if the pulse rise discrimination can be made selective enough to keep the overall efficiency to ^{14}C events stable and reproducible to 1 to 2‰.

PASSIVE AND ACTIVE SHIELDING

Low activity lead[2] walls, 10 to 15 cm thick, sometimes supplemented by paraffin-boric-acid absorbers, are used for passive shielding.

Usually, the guard counter system used for the active anticoincidence consists of a ring of Geiger counters or of a concentric arrangement of copper or brass tubes. Mäntynen *et al* (1987) proposed a liquid-scintillation guard as an interesting alternative.

PURITY CHECK AND CORRECTION OF GAS GAIN

Several methods are used to determine the purity of the CO_2 gas prior to or during counting. Direct methods include monitoring the pulse distribution of a mono-energetic source (Roether 1961; Brenninkmeijer & Mook 1979). These methods require a counter fitted with a thin window or the calibration source integrated in or moved into the active region. Alternatively, the muon (coincident) spectrum is used to obtain information about the purity of the CO_2 gas and the absolute magnitude of the gas multiplication (Groeneveld 1977).

In the Heidelberg laboratory, we set a discriminator threshold in the coincident channel, so that the count rate above the threshold is about 50% of the total coincident count rate. Any variation in the gas gain causes a corresponding shift in the ratio of the two channels (purity channel/total channel). As the counts accumulate during the full length of the counting interval, this type of purity information represents the average gas gain with insignificant statistical error.

Figure 13.4 shows an example of the correlation of the 'purity ratio' defined above *vs* the modern count rate (cpm/g CO_2) of one of our counters. If this correlation is applied to the raw measurements of the Heidelberg substandard (see below), their empirical variance becomes comparable to the Poisson error based on the counts only, even on a level well below 2‰ (Table 13.1).

COUNTER ELECTRONICS, DATA HANDLING

Pulse parameters and count rates in ^{14}C gas counting impose only moderate demands on data acquisition and handling compared to the degree of sophisti-

[2]Vieille Montagne, B-4900 Liege-Angleur, Belgium

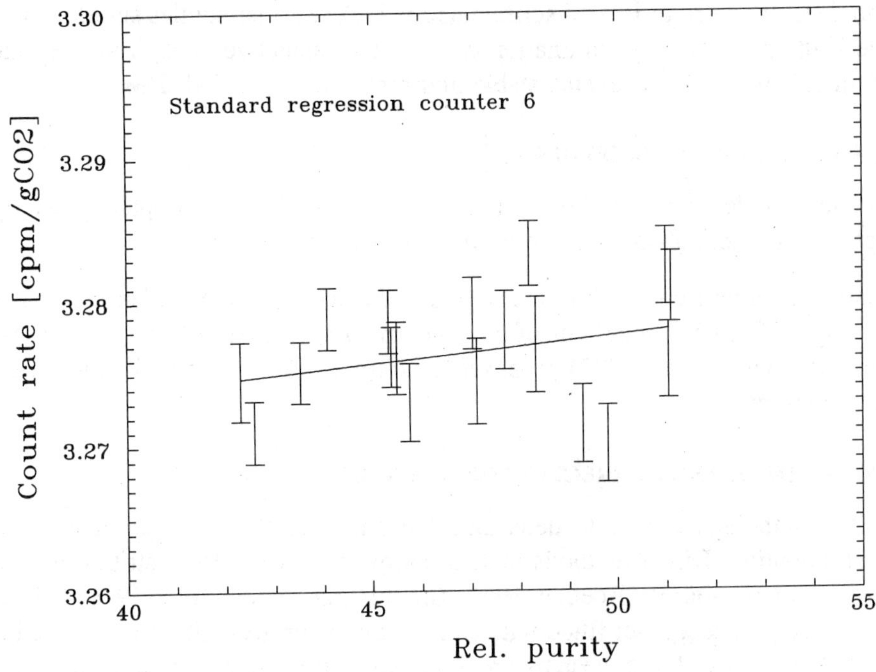

Fig 13.4. Ratio of purity channel (set to record nominally 50% of the coincidence count rate, see text) to the total coincidence count rate *vs* the modern value (cpm/g CO_2) of Counter 6 in the Heidelberg laboratory

TABLE 13.1. Standard deviation of three samples of the Heidelberg substandard, corrected for purity variations according to Figure 13.4, compared to the mean error of a single measurement, based on Poisson statistics of the counts only

Lab no.	n*	σ** (‰)	s† (‰)
Hd-13287	14	1.26	1.63
Hd-13289	20	0.81	1.44
Hd-13365	14	0.81	0.94

*n = number of measurements (2-day interval)
**σ = Poisson error (1/sqrt(N)), N = counts of a 2-day interval
†s = standard deviation of the results of n measurements

cation of electronic instrumentation in nuclear physics. Using state-of-the-art analog components, the signal processing line can be designed to low noise and high DC-stability at low cost (Schlosser, Kromer & Roether 1983; Gomer 1989). Reliable long-term counting can be achieved through distributed processing of the digital signals using dedicated microcontroller boards at each counter connected via serial links to a PC in the laboratory. Using the computational capabilities of a PC, the statistical properties of the counting process can be monitored on-line, for example, by subdivision of the counting interval (Gomer 1989) or by advanced time-series or pulse-parameter analyses (Kaihola, Polach & Kojola 1984; Kaihola *et al* 1984).

PRECISION

The influence of Poisson statistics of decay counting and other sources of error on the confidence interval of a ^{14}C date has been a key issue in ^{14}C dating from the very beginning of the method. Yet, we can recognize two distinct areas where the need for precision well below the commonly accepted 0.5 to 1% level became obvious: 1) in the investigation of the time scales and spatial distribution of the world ocean circulation; 2) in studies of the natural carbon cycle through ^{14}C analyses of tree-ring series. Both fields require an overall precision of a few per mil in the determination of ^{14}C activity. To meet these demands, techniques were developed at several places between 1972 and 1978, mostly using CO_2 gas counting. Analytical laboratory procedures were revised to limit their contribution to overall error to less than 1‰. As the amount of carbon in most cases is limited to less than 10 g of carbon, counting times were extended to intervals of the order of one week or more.

Techniques to Minimize Errors

Techniques to achieve an overall precision of a few per mil in the determination of the $^{14}C/^{12}C$ activity ratio are primarily an extension of existing routine procedures involving more precise equipment or advanced monitoring techniques. Some of these procedures have been cited above (magnitude and stability of gas purity, absolute amount of the sample, on-line controls of the counting process, background and purity corrections).

High CO_2 gas pressure in the counter immediately leads to a better signal-to-background ratio, as the background count rate is only weakly dependent on gas pressure, but requires considerable effort in gas quality control. Pressure levels up to 4 atm are used routinely in several laboratories. Tans and Mook (1978) describe a multicounter system operated at 6 atm; for a modern sample of 2 moles (!) of carbon, they reported an overall error of 1.3‰ after counting the sample for two days.

For high precision, the standard has to be measured frequently in order to keep its contribution to the error low. In this respect, the oxalic acid standard is not optimal; its preparation is somewhat cumbersome, which renders it 'costly' in terms of labor, and the count rate is just comparable to most of the samples to be measured, leading to long counting times of the standard. In Heidelberg, we use a laboratory substandard 'Wilhelm' (see Karlen *et al* 1966), which has a 10.2-fold ^{14}C activity compared to oxalic acid. It is derived from an aqueous solution of sodium carbonate (Merck 6392) prepared in bulk quantities (ca 20 L), to which 100 mL of ^{14}C-enriched sodium carbonate solution is added. The specific activity of the substandard can be controlled very precisely. Commercially available sodium carbonate is almost totally free of ^{14}C (0.5 to 0.7 pMC), and both the amount of 'dead' sodium carbonate (ca 2 kg, after drying in an exsiccator) and that of the active solution can be determined gravimetrically to better than 0.1‰.

SMALL GAS COUNTERS

Through specially designed miniature counters, the range of ^{14}C gas counting has been extended to small samples. Otlet, Huxtable and Sanderson (1986) give an extensive review of the development and characteristics of miniature counters used for ^{14}C dating of 1 to 100 mg samples; thus, only a brief account is given here. Evidently, AMS is the method of choice for small samples, but equally obvious are the advantages of small gas counters in terms of costs of construction and operation (Jelen & Geyh 1986). To put the role of small gas counters in perspective, it is useful to look at counting times required to reach a certain precision for a given sample size (Table 13.2). As counting parameters can be held sufficiently stable over several weeks, sample sizes between 10 and 100 mg are still within the reach of beta counting of moderate precision (1–5%). This range of sample size and precision makes small gas counters attractive for those areas where the precision is limited by external factors or where the precision available by AMS is not needed, as in ^{14}C dating of groundwater or in soil research.

TABLE 13.2. Comparison of counting time (days) *vs* sample size (mg) and Poisson statistics (from sample counts only) for a modern sample and a counter of 75% efficiency

Sample size (mg)	Counting statistics *vs* counting time (days) (5%)	(1%)	(0.5%)	Volume @ 3 atm (ml)
10	2.7	64	–	6
50	0.5	13	51	30
100	0.25	6.5	25	63

The development of small counters has profited much from low-level β counting in rare isotope (^{39}Ar, Loosli, Heimann & Oeschger 1980) and solar neutrino research (for a summary, see Plaga 1989). Through a low-activity NaI guard system, modern to background ratios of 4 to 15 for sample sizes between 10 and 66 mg are obtained (Otlet, Huxtable & Sanderson 1986), even in surface or near-surface installations (Loosli, Forster & Otlet 1986).

Conclusions

Gas counting has acquired its central role in ^{14}C dating for several reasons:

- Gas counters are widely employed in nuclear physics. The drift of electrons in gases has been studied extensively and it is well understood. Gas multiplication results in a high signal-to-noise ratio.
- Low-level gas counters are inherently stable detectors. Their counting efficiency is controlled mainly by geometry; variations in the operational parameters can be monitored accurately, leading to small perturbations only, for which we can compensate, down to a level on the order of 1‰. The counter background is well understood; its main source of variation (cosmic-ray-flux variability) can be reduced to insignificant levels or can be corrected.
- CO_2 gas counting requires the lowest number of steps in sample preparation, and thus minimizes the possibility of contamination and fractionation.
- Gas counting is very flexible in terms of sample size. Counter volumes used in routine dating range from 10 mL to 7.5 L. Several proven techniques routinely process carbonaceous samples. Multicounter operation is facilitated through economic design of the counters, the counting electronics and automated control of the counting process over long, unattended counting intervals.
- With CO_2 gas counting, high-precision ^{14}C dating, at an overall precision of 2‰ or even better, has been demonstrated at several laboratories based on routine laboratory work.

Acknowledgment

We thank Matthias Born for performing the long-term analysis of the Heidelberg counters and for drafting the figures.

REFERENCES

Aoyama, T 1985 Generalized gas gain formula for proportional counters. *Nuclear Instruments and Methods* A234: 125–131.

Berger, R and Suess, HE, eds 1979 *Radiocarbon Dating*. Proceedings of the 9th International ^{14}C Conference. Berkeley/Los Angeles, University of California Press: 787 p.

Brenninkmeijer, CAM and Mook, WG 1979 The effect of electronegative impurities on CO_2 proportional counting: An on-line purity test counter. *In* Berger, R and Suess, HE, eds, *Radiocarbon Dating*. Proceedings of the 9th International ^{14}C Conference. Berkeley/Los Angeles, University of California Press: 185–196.

Bruns, M (ms) 1976 Gaschromatographie mit ^{222}Rn. Thesis, Institut für Umweltphysik, University of Heidelberg, Germany (translated by AERE Harwell, LB/G910/049).

Burleigh, R 1972 Bomb combustion of radiocarbon samples, *In* Rafter, TA and Grant-Taylor, T, eds, *Proceedings of the 8th International ^{14}C Conference*. Wellington, Royal Society of New Zealand: 110–119.

Davis, R, Jr, Evans, JC, Radeka, U and Rogers, LC 1972 Report on the Brookhaven solar neutrino experiment. *Brookhaven National Laboratory Report* 16937.

Dörr, H, Kromer, B and Münnich, KO 1989 Fast sample preparation of organic material. *In* Long, A and Kra, RS, eds, Proceedings of the 13th International ^{14}C Conference. *Radiocarbon* 31(3): 264–268.

Gomer, T (ms) 1989 Elektronik und Datenverarbeitung einer Vielfachzählanlage. Thesis, Institut für Umweltphysik, University of Heidelberg, Germany.

Groeneveld, DJ 1977 Tritium analysis of environmental water. Thesis, University of Groningen, The Netherlands.

Grootes, PM 1977 Thermal diffusion isotopic enrichment and radiocarbon dating beyond 50,000 years BP. PhD thesis, University of Groningen, The Netherlands.

Harris, TJ and Mathieson, E 1970 Pulse shape discrimination in proportional counters – theory of electronic system. *Nuclear Instruments and Methods* 88: 181–192.

_____ 1971 Pulse shape discrimination in proportional counters – experimental results with an optimised electronic system. *Nuclear Instruments and Methods* 96: 397–403.

Huxley, LGH and Crompton, RW 1974 *Diffusion and Drift of Electrons in Gases*. New York, John Wiley & Sons.

Jelen, K and Geyh, MA 1986 A low-cost miniature counter system for radiocarbon dating. *In* Stuiver, M and Kra, RS, eds, Proceedings of the 12th International ^{14}C Conference. *Radiocarbon* 28(2A): 578–585.

Kaihola, L, Polach, H and Kojola, H 1984 Time series analysis of low level gas counting data. *Radiocarbon* 26(2): 159–165.

Kaihola, L, Polach, H, Kojola, H, Tervahauta, J, Mäntynen, P and Soini, E 1984 Low level gas multicounter for ^{14}C dating of small samples. *Nuclear Instruments and Methods* B5: 436–438.

Karlen, I, Olsson, IU, Kallberg, P and Kilicci, S 1966 Absolute determination of the activity of two ^{14}C dating standards. *Arkiv Geofysik* 6: 465–471.

Loosli, HH, Forster, M and Otlet, RL 1986 Background measurement with different shielding and anticoincidence systems. *In* Stuiver, M and Kra, RS, eds, Proceedings of the 12th International ^{14}C Conference. *Radiocarbon* 28(2A): 615–624.

Loosli, HH, Heimann, M and Oeschger, H 1980 Low level gas proportional counting in an underground laboratory. *In* Stuiver, M and Kra, RS, eds, Proceedings of the 10th International ^{14}C Conference. *Radiocarbon* 22(2): 461–469.

Mäntynen, P, Äikää, O, Kankainen, T and Kaihola, L 1987 Application of pulse-shape discrimination to improve the precision of carbon-14 gas proportional counting method. *Applied Radiation and Isotopes* 38(10): 869–873.

Mook, WG 1983 International comparison of proportional counters for ^{14}C activity measurements. *In* Stuiver, M and Kra, RS, eds, Proceedings of the 11th International Radiocarbon Conference. *Radiocarbon* 25(2): 475–478.

Münnich, KO (ms) 1957 Messung natürlichen Radiokohlenstoffs mit einem CO_2-Proportional-Zählrohr. Einige Anwendungen der Methode. PhD thesis, University of Heidelberg, Germany.

Oeschger, H and Wahlen, M 1975 Low level counting techniques. *Annual Review of Nuclear Science* 25: 423–430.

Otlet, RL, Huxtable, G and Sanderson, DCW 1986 The development of practical systems for ^{14}C measurement in small samples using miniature counters. *In* Stuiver, M and Kra, RS, eds, Proceedings of the 12th International ^{14}C Conference. *Radiocarbon* 28(2A): 603–614.

Plaga, R (ms) 1989 Proportionalzähler mit geformter Siliziumkathode und damit zusammenhängende Beiträge zum Gallium-Solar-Neutrino-Experiment. PhD thesis, University of Heidelberg, Germany.

Povinec, P 1979 A study of proportional counter optimization for long-term counting. *Nuclear Instruments and Methods* 163: 363–368.

_____1992 ^{14}C gas counting: Is there still a future? *In* Long, A and Kra, RS, eds, Proceedings of the 14th International ^{14}C Conference. *Radiocarbon*, in press.

Roether, W (ms) 1961 CO_2-Proportionalzählrohre bei höheren Fülldrucken. Thesis, Institut für Umweltphysik, University of Heidelberg, Germany.

Sauli, F 1977 *Principles of Operation of Multiwire Proportional and Drift Chambers.* CERN Academic Training Programme, Geneva, Switzerland.

Schlosser, P, Kromer, B and Roether, W 1983 Electronics for low-level counting using a microcomputer. *Nuclear Instruments and Methods* 216: 155–160.

Schoch, H and Münnich, KO 1981 Routine performance of a new multi-counter system for high-precision dating. In *Methods of Low Level Counting and Spectrometry*. Vienna, IAEA: 361–370.

Stuiver, M, Robinson, SW, and Yang, I 1979 ^{14}C dating up to 60,000 years with high efficiency proportional counters. *In* Berger, R and Suess, HE, eds, *Radiocarbon Dating*. Proceedings of the 9th International ^{14}C Conference. Berkeley/Los Angeles, University of California Press: 202–215.

Tans, PP and Mook, WG 1978 Design, construction and calibration of a high accuracy carbon-14 counting set up. *Radiocarbon* 21 (1): 22–40.

Vries, H, de and Barendsen, GW 1953 Radiocarbon dating by a proportional counter filled with carbon dioxide. *Physica* 19: 987–1003.

Vries, H, de, Stuiver, M and Olsson, I 1959 A proportional counter for low level counting with high efficiency. *Nuclear Instruments and Methods* 5: 111–114.

Zastawny, A 1966 Gas amplification in a proportional counter with carbon dioxide. *Journal of Scientific Instruments* 43: 179–181.

14

FOUR DECADES OF PROGRESS IN ^{14}C DATING BY LIQUID SCINTILLATION COUNTING AND SPECTROMETRY

HENRY A POLACH

"Go forth and build a house of bricks"
(the parent of the third little piggy)

INTRODUCTION

Liquid scintillation (LS) radiometry (the counting of nuclear emanations) has undergone continuous development since attempts were first made in the early 1950s to detect natural ^{14}C by this method. The potential of detection of ionizing radiations from ^{3}H, ^{14}C and ^{32}P, isotopes used extensively as tracers in biomedical and biological research, was the trigger that prompted the commercial development of LS counters, first by Lyle Packard, the founder of the Packard Instrument Company, USA and then by many others in the USA and Europe. Though efficient and technologically advanced for their times, the early counters had very high backgrounds and, hence, offered poor resolution of radioactive isotopes at environmental concentrations.

A number of perceptive and persistent researchers saw the enormous potential that LS counting offered. The steady improvements involved the development of a suitable sample solvent (benzene), scintillation solutes, better photo tubes, innovations in instrument and counting vial design and better data capture, validation and analysis. The achieved background reductions and ease of operations placed radiocarbon dating by LS counting at a par with gas proportional counting by the mid-1970s. However, spectacular advances, which greatly benefited radiocarbon dating, were made only in the last six years when two major manufacturers commenced catering especially to low-level detection of environmental radiation.

This chapter briefly traces the history of developments as seen from a radiocarbon dater's point of view. In the process, it examines the principles of LS counting, the evolution of counters into alpha and beta high-resolution spectrometers, the means and technology associated with major background reductions, the benzene synthesis process, counting vial materials and design, the properties of counting cocktails and data acquisition and validation. The development of low-level alpha and beta spectrometry and attainment of ultra-low backgrounds is very pertinent to radiocarbon dating. Thus, this chapter also describes the achievements of the state-of-the-art technology offered by Canberra-Packard (USA) and Wallac (Finland).

Based on its merit, low-level LS spectrometry has become the method of choice for the detection of radiation at environmental levels and for newly established or refurbished radiocarbon dating laboratories that aim at high precision and resolution ^{14}C determinations.

LIQUID SCINTILLATION COUNTING

Basic Principles

In the early 1940s, Broser and Kallmann (1947) discovered that ionizing radiation causes fluorescence in certain organic compounds called scintillators, which emit very short bursts of photons. These photons can be quantitatively detected by determining the anode current of a photomultiplier tube (PMT) positioned to intercept the photons. The basic principle is that the number of photons, hence, the anode current of the PMT, is proportional to the energy of the ionizing event and a function of the level and energy of radioactivity.

In 1953, Packard Instruments built the first commercial LS counter, using the design of Hiebert and Watts (1953). The detection system was based on the principle that two opposed PMTs see the burst of photons, and coincident electronics enable the sample pulse detection only if both PMTs have seen the photon flux simultaneously, typically within 20 nanoseconds (Reynolds, Harrison & Salvini 1950). This *counting in nanosecond coincidence* had the very desirable effect of significantly reducing counts due to the spontaneous production of spurious pulses (dark current noise) from the PMTs, thus ensuring a favorable signal-to-noise ratio.

Several researchers immediately resolved to try the LS method to detect natural ^{14}C at a time when CO_2 gas proportional counting was in its infancy. Hays, Williams and Rogers (1953), Audric and Long (1954), Arnold (1954) and others (using as carbon sample solvents acetylene, ethanol, methanol or methyl borate) presented their attempts at the Liquid Scintillation Counting conference held in 1957 at Northwestern University. Arnold (1958: 134), in a summary of the then mildly successful results (but now historic efforts and perceptions), concluded:

> The only thing that could be said with certainty about the future of liquid scintillation counting, as applied to radiocarbon dating, was that it awaited an elegant solution of the chemical synthesis problem for the conversion of specimen carbon to suitable organic liquids which preferably should be scintillation solvents.

Further progress, evolution and refinement of conventional multipurpose LS counter and counting technology can be traced by studying the very comprehensive reviews of Rapkin (1970) and Horrocks (1974) and, as applied to radiocarbon dating, the reviews of Tamers (1965), Polach (1969, 1974) and Noakes (1977). These were characterized by reliable and stable counters that, for low-level ^{14}C, optimally worked at 65–75% efficiency with background count rates of around 1/7th to 1/15th of the radiocarbon dating modern reference standard. Such performance in the hands of one expert (Pearson 1979) lead to a precision of ± 0.2% modern carbon (pMC), equivalent to the very best of gas proportional counters (Mook 1983). Nevertheless, the radiocarbon daters' cry that could be heard in the early 1980s was: "*Give us better counters*!"

Elegant Scintillation Solvent and Solutes

For radiocarbon dating, the first breakthrough relating to sample preparation was achieved when Barker (1953) perfected the preparation of acetylene from sample CO_2 in reaction with molten lithium. The first breakthrough for routine LS ^{14}C dating came in 1961 with the B_2H_6 activated silica alumina catalyst, which enabled C_6H_6 synthesis yields of 50% (Noakes *et al* 1963). This was followed in 1962 with the scale-up by Stipp (personal communication) of the Barker (1953) Li_2C_2 reaction. By 1964, Noakes had deciphered the mechanism of C_2H_2 trimerization over the catalyst with transition metals (eg, chromium and vanadium), giving > 90% yields. Florida State University was the first modern LS laboratory that incorporated all then-known developments in a totally enclosed system (Noakes, Kim & Stipp 1965; Stipp, Knauer & Godell 1966). Also at that time, Stipp described experiments with the mechanism of automatic sample, standard and background rotation (Noakes, Kim & Stipp 1965), which, under the name of *Quasi Simultaneous Counting*, was perfected and routinely applied at ANU in 1966 (Polach 1969).

Benzene is the ideal sample molecule for radiocarbon dating. It contains carbon derived totally from the sample, is an excellent solvent for scintillators and has excellent scintillation and light transmission properties (eg, Birks 1974: 27). The preferred primary scintillator solutes then (Tamers 1965), as well as now for many laboratories (Noakes & Valenta 1989), is PPO with POPOP as the secondary wavelengths shifter, both dissolved in toluene and added to the sample benzene. Polach *et al* (1983a) first suggested, for radiocarbon dating, an alternative scintillator, butyl-PBD, added directly in dry form to the sample benzene. It is ideal for LS ^{14}C low-level counting and spectrometry, as it retains its superior properties in the presence of quenchers.

Thus, Jim Arnold's wise words relating to the early experiments in 1954 had to wait 11 years to come partially true in 1965. Since then, as we have already seen in relation to instrumental and scintillator development, there was no turning back. Progress relating to benzene synthesis also remained continuous. The Li reaction processes were better understood (Polach & Stipp 1967; Tamers 1975; Burleigh, Hewson & Matthews 1977) resulting in reproducibly high conversion yields (95 ± 2%) of sample carbon to benzene carbon, high-purity unquenched benzene (Polach, Gower & Fraser 1973) and radon-free benzene (Tamers 1975; Hood et al 1989). Thus, it took another 10 years for the synthesis process to become elegant (if not perfect), and Jim Arnold no longer has to wait to continue his 1954 experiment.

Ideal Vial

The function of a counting vial is to contain the sample solvent and scintillator solute within the space provided for it between (or next to) the PMTs. The International Electrotechnical Commission recommends standard dimensions: height = 60.5 ± 2.5 mm, outside diameter = 28 ± 2 mm and cap configuration. This translates into an effective volume of ~ 22 mL. There are no specifications relating to type and purity of vial materials, wall thickness or uniformity, cap material or type and performance of the seal. Smaller diameter 7 mL mini-vials are now also in use. These can be held in special holders within the standard 22 mL space by all counters.

The evolution of vial designs especially for radiocarbon dating can be traced to Tamers (1965). He cut in half a standard low-^{40}K borosilicate glass 22 mL counting vial to manufacture a vial of ~ 10 mL volume (enabling the dating of 5–9 g of elemental carbon), mounted it on a black pedestal to retain the geometry of the 22 mL vial, and masked with a black shield the excess glass above the counting liquid meniscus and between the cap. Except for the narrow 7 mL mini-vial, this design remains the basis of all subsequent developments. Changes involve the use of teflon (Calf & Polach 1974), quartz (Haas 1979) and delrin (Schotterer & Oeschger 1980). Commercially available teflon/copper vials, from sources in the USA or Finland, owe their design to Noakes (1977) and Kuc and Rozanski (1978). Polach *et al* (1983a) described their superior performance, and using rigorous cleaning procedures (ANU Lab Manual 1990: 35), the teflon vials' long-term performance remained very satisfactory for ^{14}C dating with a precision of 0.7 pMC. However, Hogg *et al* (1991) highlighted some difficulties as well as undesirable long-term instability characteristics of teflon. Hogg thus recommends the use of quartz for high precision (< 0.5 pMC) and resolution ^{14}C dating. The quartz vials, if made from specially selected materials, give performance equal to that of teflon without exhibiting any of the undesirable characteristics. As seen below, ordinary low-^{40}K (widely used in biomedical LS counting, hence, inexpensive) narrow 7 mL mini-vials perform very well in the Canberra-Packard spectrometers, if these are especially selected and fitted out for low-level radioisotope (^{3}H and ^{14}C) counting. Polach *et al* (1988) described a novel, but not widely used, 0.3 mL teflon/black delrin vial and holder assembly. It enables ^{14}C signal resolution of 100 mg of elemental carbon with a precision of 2.5 pMC in 2 days counting time, if used in a modern high-efficiency low-level LS spectrometer.

Liquid Scintillation Counters

Radionuclides most commonly determined by the LS process are ^{3}H, ^{14}C and ^{32}P. However, almost any kind of radioactive decay, be it negatron, positron, electron capture, alpha, gamma, proton, neutron, cosmic muon, etc, interacts with the aromatic solvent molecules. Their excitation energy is transferred to the solute, which acts as an efficient source of photons. These, in the case of all beta-particle emitters (as well as, eg, neutron, muon and gamma, which contribute to the observed background) produce a photon flux, hence, anode current proportional to the energy of the ionizing event. LS counters enable pulse-height analysis by recording the presence of pulses (events) that fall within certain limited energy windows. Two to five such variable-width windows span the energy range from 0–2 MeV. Knowing the continuum of pulse height produced by beta emitters, zero energy to maximum energy, the upper limit of energy windows can be selected, and the isotope(s) identified.

Such fixed-window counting generates total gross counts within specified windows or channels. Because of the excellent energy resolution of the LS counters, they can perform data evaluation and validation parameters, such as selection of counting channels (windows) of interest, channel ratio, external standardization, etc. Nevertheless, the loss of energy information is significant, especially in low-level counting, as we need to define the variable background component of the recorded gross count.

Background

Background is a term applied to the observed count rate when counting a sample containing no radionuclide. Such is the case, for example, when counting analytical reagent benzene, bypassing the possibly contaminating sample-carbon to benzene-carbon synthesis steps. Sources of background in LS counting have been comprehensively studied and reviewed (Horrocks 1974: 198–207; Gupta

& Polach 1985: 56–85; Horrocks 1985; Polach 1987: 4–6). Recognized sources of background contributions and remedial steps are as follows:

- *Sample contamination* by radionuclides to be assayed or traces of other radionuclides will cause an increase in the gross count rate. This can stem from memory effects in the lithium reactor (Radnell & Muller 1980), carbon present in the lithium metal (Geyh 1969: 465; 1990), trace amounts of ^{14}C in the dilution gas used to make up standard volumes of CO_2 prior to synthesis (Hogg, personal communication) and traces of sample radon dissolved in the synthesized benzene due to incorrect benzene-recovery procedures (Hood *et al* 1989). In all cases, both the source and magnitude can be checked by comparing synthesized blanks to reagent grade benzene background counts.
- *Radioactivity of materials* used in the construction of counters, PMTs and vials will contribute to all determinations equally and affect the resolution but not validity.
- *Environmental radiation,* if statistically reproducible, has no effect on validity, only on performance. Cosmic-ray flux is considered to be a variable component, and many shielded environments have been built (or sought) when ultra low-level backgrounds are essential. Examples of these are the shielded ^{14}C laboratories in Canberra, Bern, London, Menlo Park, Lucas Heights and Tucson. All these are well documented and are given here without references. If the environmental radiation is due to contamination from tracer experiments, then the results are not reproducible and often catastrophic (eg, Rubin, Myssen & Polach 1984), and there is no ready or prescribed remedy. Solutions, often drastic, need to be found at a local level.
- *Interfering radiation from the external standard source* (used in all LS counters) can cause both a variable and constant component – variable, if the radioactive source location geometry in relation to the sample is not reproducible, – constant, if the source is inadequately shielded. Removal of the source and verification of background count-rate changes are an elegant way of checking if the problem exists.
- *Interfering fluctuating radiations from accelerators and nuclear reactors* most often are caused by variable omni-directional thermal neutron flux of low intensity. Whereas these present no health hazard, they have favorable capture cross-sections in the scintillation medium, and hence, are totally unacceptable at low-level counting locations. Solutions will range from usage of graded shields to underground shielded low-level counting rooms, to relocation to distant sites.

- *Thermionic and secondary electron emissions* from the photocathode and first stage dynodes are minimized by the LS manufacturer by selection of PMTs, shorter coincidence resolution times, high-voltage reduction and dead times of suitable duration (4 microseconds).
- *Photomultiplier tube cross-talk* can be eliminated by the LS manufacturer by electronic means such as high bias prior to coincidence summing, pulse amplitude comparison prior to coincidence, lesser pulse analysis, etc. If cross-talk has not been minimized, as can be the case for older LS counters, then the user can make improvements by masking the periphery of the PMTs, masking the counting vials to enable a view of the active counting volumes only and by changing the geometry of the optical reflectors.
- *Chemi- and photoluminescence* can be identified and/or suppressed in modern counters, but is best eliminated by selecting a solvent and solute and proper handling prior to counting (eg, do not expose to fluorescent light).
- *Static-electricity-induced* spurious discharges are minimized by temperature/humidity control and grounding. They may be eliminated by positive and negative ionizers and/or a delay of 5–10 minutes before commencing the count after sample vial movement into the counting chamber. The latter can be done automatically in all counters, and some are fitted with static suppression devices.
- *Line noise and transients* are eliminated by 'High Isolation Transformers' placed between each counter and the local power supply. These are relatively expensive but are essential in low-level counting applications.
- *Radio-frequency* (eg, air-transmitted switching noises) can be eliminated when the amplified noise is channelled into an anticoincidence gate that blocks the true signals while interference occurs. Some modern counters have such a device.
- *Electronic noise* in modern counters is very unlikely. Counters benefit from an air-conditioned, humidity-controlled environment out of reach of direct sunlight. The PMTs benefit from a constant low temperature (8–12°C).
- *Photomultiplier tube dark current noise* in the nanosecond coincidence resolution mode is insignificant in modern counters. For low-level counting, manufacturers often specify and use low-noise, quartz-faced tubes and recommend refrigerated counters.

Background Reduction

From the beginning, fixed-window LS counters yielded good isotope detection and resolution efficiencies. Modern counters simply enhance these characteristics. Background has been reduced further by a factor of 10–15, in relation to the average counter of the 1975 vintage. The preferred term, used by manufacturers for comparison of equipment performance, is based on the ratio of the square of efficiency, E, and background, B. E^2/B for a typical radiocarbon dating LS *counter* has been increased, on average, from an original of 400 to 1200, and for specialist low-level *spectrometers* to 9500 and more – a minimum improvement of better than 790% over the average conventional LS counter, and an improvement of more than 2375% over 37 years.

Passive shielding in conventional counters is minimal but adequate at 5 cm thickness of lead for biomedical and biological research using radioactive tracers. The compact design of these counters generally precludes addition of lead by the user. However, whenever lead was added, the background was significantly reduced, and performance was comparable to the average gas proportional counter (Polach *et al* 1983b, Table 1). Commercially available counters including significant amounts of lead (> 10 cm) were first built for ^3H applications using large-volume vials (Noakes, Neary & Spaulding 1973; Iwakura *et al* 1979) or for dissolved $^{14}CO_2$ counting (Eichinger *et al* 1980). Kojola *et al* (1984b: 949) used an asymmetric graded (Pb, Cd and Cu) shield with 20 cm of lead on top but only 10 cm on the sides (to save mass without affecting performance).

Schotterer and Oeschger (1980), Bowman (1986) and Calf and Airey (1987) achieved spectacular background reduction by placing LS counters underground as a form of passive shielding.

Active shielding, primarily to detect cosmic muons, was first applied to radiocarbon dating by Anderson, Arnold and Libby (1951) to reduce the background of their screen counters. (The sample counter ionizations are electronically inhibited in the presence of simultaneous *guard* counts.) Such guard-counter active shielding is standard in all gas proportional systems, and was used in LS counters for the first time by Pietig and Scharpenseel (1964) with spectacular results, which must have been overlooked or forgotten, as it took another 12 years for the experiments to be repeated (Alessio *et al* 1976; Punning & Rajamae 1975; Noakes 1977: 197; Iwakura *et al* 1979; Broda & Radoszewski 1982; Jiang *et al* 1983; Kojola *et al* 1984).

Electronic optimization, other than the elimination of guard-coincident events, aims to reduce spurious noise and increase the long-term stability of LS counters. Examples are high-voltage or spectral stabilization (eg, Wang 1970; Soini 1975a; Berthold 1980), variable-coincidence bias (Polach *et al* 1983b), chemiluminescence monitor (Soini 1977), lesser-pulse analysis (Laney 1971), amplitude-disparity discrimination (Soini 1975b), static-charge elimination (Kananen *et al* 1984) and radio-frequency suppression (Kojola *et al* 1984).

Pulse-burst discrimination (PB) is a form of background reduction by pulse-shape analysis based on the long-known observation (Jerde, Peterson & Stein 1967) that high-energy radiation incident upon a photo tube produces one large pulse with smaller trailing pulses or 'bursts'. Valenta (1987) describes electronic burst-counting circuitry applied to a commercially available LS spectrometer, which uses the number of burst afterpulses to discriminate valid scintillation events from background events. In order to get better background-event recognition (rejection), Noakes and Valenta (1989) experimented with auxiliary scintillators that would increase the number of burst pulses due to environmental ionizing events. They incorporated a slow fluor (long decay constant) in a plastic holder, which can hold the low-^{40}K, narrow 7 mL mini-vial, and modified the LS spectrometer to incorporate a plastic guard/light guide made of the same material placed between the photo tubes. When the holder and its mini-vial are inserted in the plastic guard, the increased number of bursts enables, with good efficiency, sample beta count to environmental background count discrimination. In the case of ^{14}C for radiocarbon dating at a counting efficiency of 64%, the background was reduced from 3.62 cpm (no discrimination) to 0.43 cpm (maximum discrimination using the mini-vial, holder and guard configuration), placing it squarely in the low-level counting bracket.

LIQUID SCINTILLATION SPECTROMETRY

The transition in performance from very good to excellent was achieved only when the results could be acquired, retained, analyzed and reanalyzed in the form of multichannel energy spectra. Towards the end of the 1970s, external multichannel analyzers (MCAs) were being interfaced with LS counters for experimental purposes (eg, Ediss 1980). However, whenever MCAs were incorporated into top-line counters, such as then available from Beckman, Packard, Kontron, Intertechnique, Phillips, LKB-Wallac, etc, the spectral information was stored internally, often with the facility to view them under hardware programs, but without the ability for the user to access the information they contained. Further, only one sample, one spectrum at a time, could be accumulated, and once the calculations were performed, the spectral information was lost, unless it was externally stored. Reviewing the situation, Polach *et al* (1983c: 504) considered that to be a handicap, and suggested a *windowless* approach to LS spectrometry, at all levels of isotope abundance, based on an inbuilt MCA, that can transmit to an external computer for storage and analysis. The user, under software control, can reprogram the counter to optimize parameters of interest, or subsequently reanalyze the spectra using a different approach to their validation or interpretation. This was demonstrated on an experimental setup using a multitude of LS counters, and is of immense value to ^{14}C dating (Polach *et al* 1983c: 487–502).

Pulse-height analysis (PHA) is the basis of LS counting as well as spectrometry. Built-in multichannel analyzers today enable the recording and analysis of the relative pulse heights of individual ionizing events under software control in external computers. All modern LS spectrometers incorporate this facility (Beckman, Canberra-Packard, Wallac). The most sophisticated of these incorporates two multiparameter analog-to-digital converters and two independent 2048 ch MCAs. The splitting of the spectrum into two related halves (eg, coincident and anticoincident guard pulses) enables the simultaneous acquisition of four 1024 channel energy spectra (Polach *et al* 1984). This means that the integrity of each pulse's energy can be preserved and recorded by software selectable gating to MCA memory locations. Applications of such an advanced system encompass complex environmental studies, high-precision or resolution ^{3}H and ^{14}C assays, as well as validation of data or research into the nature of interfering radiations and shielding properties, as all accepted and all rejected energy spectra can be preserved and analyzed.

Pulse-shape analysis (PSA) is based on the observations that the intensity of photons from a scintillation event shows division into prompt and delayed components (eg, Horrocks 1970: 277). The relative amount of light in the prompt and delayed components is dependent on the specific ionization mechanism and, hence, the type of particle causing it. For example, beta particles and Compton scatter electrons from gamma-ray interactions contribute mostly to the prompt component. Muon, neutron and alpha-particle interactions contribute mostly to the delayed component. The experimental application of this to the simultaneous detection of alpha and beta particles in liquid detectors was reviewed, for example, by McKlveen and McDowell (1984). Oikari *et al* (1987) described a software-controlled PSA system applied to a commercially available low-level LS spectrometer enabling full resolution of alpha and beta events while not affecting the high counting efficiency and ultra-low-background characteristics of the system. Kaihola and Oikari (1991) evaluated factors affecting alpha-particle detection performance for radium daughters and ^{241}Am. Using the same instrument, Polach and Kaihola (1988) demonstrated that LS alpha/beta spectrometry enables the detection, at extremely low levels, of the ubiquitous ^{222}Rn, which Nydal (1983) described to be the scourge of radiocarbon daters by radiometry. Experimental, successful background reductions, using low-^{40}K glass vials and PS analysis, were described by Kaihola (1991).

DISCUSSION AND CONCLUSIONS

We have seen that background noise identification and reduction by a guard (cosmic-ray counter or pulse-burst discrimination) enables liquid scintillation spectrometry of ^{14}C beta with unprecedented definition of signal above noise.

Pulse-shape analysis enables simultaneous alpha- and beta-particle energy separation and, hence, identification, and this, with low-level LS spectrometry, extends the detection of isotope species beyond ^{3}H and ^{14}C to ^{36}Cl, $^{89,90}Sr$, ^{85}Kr, Th, Ra and Rn isotopes at environmental abundances.

Low-level liquid scintillation spectrometry has come of age in the last six years with two major manufacturers offering low-level spectrometers, each one with its specific merits, as can be judged from the quoted literature. I understand that Beckman is entering the low-level spectrometry market. This augurs well for the field, and demonstrates the wide interest low-level LS research has stimulated.

Tangible advantages that can be attributed to the described technology for ^{14}C dating and environmental studies are:

- Enhanced validation of ^{14}C dating results through a sequential cycle, quasi-simultaneous counting of samples, modern reference, background and machine standards incorporating tracer isotopes (Polach 1969). Such a cycle function can be performed by all automatic sample-changer-equipped LS counters and spectrometers, irrespective of their age. An improvement to the sequential cycle, enabling long-term programming of an LS spectrometer according to the required precision of each sample, is the ability under software control to load the samples in a user-defined totally random sequence (Kojola *et al* 1984: 949). Thus, sample/standards/machine standard counting times can be optimized. This is especially useful for low-level spectrometry, enhancing validity, precision and resolution of isotope detection and efficiency of production.
- Optimization of LS spectrometers for the highest signal-to-noise ratio, rather than maximum counting efficiency, enhancing stability, reproducibility and resolution of isotopic species determination.
- Permanent storage and subsequent retrieval of all primary spectra for scrutiny (quality control) or reinterpretation (research). This, under user-directed software control, can also provide the ability to manipulate spectra without destroying the primary evidence (long-term quality assurance).

Whereas the ultimate achievable accuracy of a ^{14}C determination depends on the performance of the equipment, many user-related factors affect the results. There is evidence in the literature that some users experience difficulties in handling the modern LS spectrometers and their associated software. They claim that their expectations have not been met in terms of, for example, performance, stability and reproducibility of results, or that the software introduced computational errors. The references are not given here lest these claims are taken to be authoritative statements rather than individual perceptions of true experiences. At a workshop on low-level LS spectrometry, held at the International Conference on New Trends in Liquid Scintillation Counting and Organic Scintillators, Gatlinburg, Tennessee, October 1989, Cook and Polach (1991) concluded that:

- Liquid-scintillation instrumentation is now generally ahead of application technique and user know-how.
- Emphasis needs to be placed not only on user education but also on research leading to improved sample preparation and usage of modern cocktails.
- Exchange of information, technology and know-how transfer from experienced researchers to routine users should be enhanced by manufacturer-supported workshops and the publication of detailed procedural manuals.
- Manufacturers will have to accept much more responsibility in order to overcome the rapidly growing 'black-box' syndrome that is the root of the criticism. This can only be partially alleviated by improved sales backup. Peer and collegiate knowledge transfer is seen as the *modus operandi* leading to full usage of on-board facilities of modern LS spectrometers through an understanding of the principles on which they are founded.
- Quality control clearly lies with the individual producing the results. However, quality assurance can be given only when global validity of the results is assured by open participation in international cross-checks, standardization and free exchange of procedural information. Nowhere is this more important than in low-level environmental radiation assays.

The understanding of principles underlying the modern LS spectrometers is vital, since excellence in technology alone is insufficient to guarantee success.

ACKNOWLEDGMENTS

Gordon Cook, Lauri Kaihola, John Noakes and Jerry Stipp critically reviewed the text and Dilette Polach assisted with the search for and checked the references. I thank them for their valuable contributions.

REFERENCES

Alessio, M, Allegri, L, Bella, F and Improta, S 1976 Study of background characteristics by means of high efficiency liquid scintillation counter. *Nuclear Instruments and Methods* 137: 537–543.

Anderson, EC, Arnold, JR and Libby, WF 1951 Measurement of low level radiocarbon. *Review of Scientific Instruments* 22(4): 225–230.

Arnold, JR 1954 Scintillation counting of natural radiocarbon: 1. The counting method. *Science* 119: 155–157.

_____1958 Archaeology and chemistry. *In* Bell, CG, Jr and Hayes, F, eds, *Liquid Scintillation Counting*. New York, Pergamon Press: 129–134.

Audric, BN and Long, JVP 1954 Use of dissolved acetylene in liquid scintillation counters for the measurement of carbon-14 for low specific activity. *Nature* 173: 992–993.

Barker, H 1953 Radiocarbon dating: Large scale production of acetylene from organic material. *Nature* 172: 631–632.

Berthold, F 1980 A new approach to automatic photomultiplier stabilisation for photon and scintillation counters. *In* Peng, C-T, Horrocks, DL and Alpen, EL, eds, *Liquid Scintillation Counting: Recent Applications and Developments*. New York, Academic Press: 273–280.

Birks, JB 1974 Towards an understanding of scintillation process in organic molecular systems. *In* Stanley, PE and Scoggins, B, eds, *Liquid Scintillation Counting: Recent Developments*. New York, Academic Press: 1–38.

Bowman, S 1986 The potential of the London underground for liquid scintillation counting. *In* Stuiver, M and Kra, RS, eds, Proceedings of the 12th International ^{14}C conferences. *Radiocarbon* 28(2A): 592–596.

Broda, R and Radoszewski, T 1982 Scintillation detector with anticoincidence shield for determination of radioactive concentration of standard solutions. *In* Povinec, P and Usacev, S, eds, *Low Radioactivity Measurements and Applications*. Bratislava, Slovenske Pedagogicke Nakladatelstvo: 329–333.

Broser, I, von and Kallmann, H 1947 Über die Anregung von Leuchtstoffen durch schnelle Korpuskularteilchen. *Zeitschrift für Naturforschung* 2(8): 439–440.

Burleigh, R, Hewson, AD and Matthews, KJ 1977 Synthesis of benzene for low-level ^{14}C measurement: A review. *In* Crook, MA and Johnson, P, eds, *Liquid Scintillation Counting* 5. London, Heyden: 205–209.

Calf, GE and Airey, PE 1987 Liquid scintillation counting of carbon-14 in a heavily shielded site. *In* Ambrose, WR and Mummery, JMJ, eds, *Archaeometry: Further Australian Studies*. Canberra, ANU Press: 351–356.

Calf, GE and Polach, HA 1974 Teflon vials for liquid scintillation counting of carbon-14 samples. *In* Stanley, PE and Scoggins, B, eds, *Liquid Scintillation Counting: Recent Developments*. New York, Academic Press: 224–234.

Cook, GT and Polach, HA 1991 Low-level liquid scintillation counting workshop. *In* Ross, H, Noakes, J and Spaulding, J, eds, *New Trends in Liquid Scintillation Counting and Organic Scintillators*. Chelsea, Michigan, Lewis Publishers: 691–694.

Ediss, C 1980 A multichannel analyser interface for a Beckman 9000 liquid scintillation counter. *In* Peng, C-T, Horrocks, DL and Alpen, EL, eds, *Liquid Scintillation Counting: Recent Applications and Developments*. New York, Academic Press: 281-289.

Eichinger, L, Rauret, W, Salvamoser, J and Wolf, M 1980 Large-volume liquid scintillation counting of carbon-14. *In* Stuiver, M and Kra, RS, eds, Proceedings of the 10th International ^{14}C Conference. *Radiocarbon* 22(2): 417–427.

Geyh, MA 1969 Problems in radiocarbon dating of small samples by means of acetylene, ethane or benzene. *International*

Journal of Applied Radiation and Isotopes 20: 463–466.

_____1990 Radiocarbon dating problems using acetylene as counting gas. *In* Long, A, Kra, RS and Scott, EM, eds, Proceedings of the International Workshop on Intercomparison of ^{14}C Laboratories. *Radiocarbon* 32(3): 321–324.

Gupta, SK and Polach, HA 1985 *Radiocarbon Dating Practices at ANU.* Canberra, Radiocarbon Laboratory, ANU: 176 p.

Haas, H 1979 Specific problems with liquid scintillation counting of small benzene volumes and background count rate estimations. *In* Berger, R and Suess, HE, eds, *Radiocarbon Dating.* Proceedings of the 9th International ^{14}C Conference. Berkeley, University of California Press: 246–255.

Hays, FN, Williams, DL and Rogers, B 1953 Liquid scintillation counting of natural C-14. *Physical Review* 92: 512–513.

Hiebert, RD and Watts, RJ 1953 Fast coincidence circuit for H^3 and C^{14} measurements. *Nucleonics* 11(12): 38–41.

Hogg, A, Polach, H, Robertson, S and Noakes, J 1991 Application of high purity synthetic quartz vials to liquid scintillation low-level ^{14}C counting of benzene. *In* Ross, H, Noakes, J and Spaulding, J, eds, *New Trends in Liquid Scintillation Counting and Organic Scintillators.* Chelsea, Michigan, Lewis Publishers: 123–131.

Hood, D, Hatfield, R, Patrick, C, Stipp, J, Tamers, M, Leidl, R, Lyons, B, Polach, H, Robertson, S and Zhou, W 1989 Radon removal during benzene preparation for radiocarbon dating by liquid scintillation spectrometry. *In* Long, A and Kra, RS, eds, Proceedings of the 13th International ^{14}C Conference. *Radiocarbon*: 31(3): 254–259.

Horrocks, DL 1970 Pulse shape discrimination with organic scintillation solutions. *Applied Spectroscopy* 24: 397–404

_____1974 *Applications of Liquid Scintillation Counting.* New York, Academic Press: 346 p.

_____1985 Studies of background sources in liquid scintillation counting. *International Journal of Applied Radiation and Isotopes* 36(8): 609–617.

Iwakura, T, Kasida, Y, Inoue, Y and Tokunaga, N 1979 A low-background liquid scintillation counter for measurement of low-level tritium. *In* Behaviour of tritium in the environment. *IAEA Proceedings Series.* Vienna, IAEA: 163–171.

Jerde, RL, Peterson, LE and Stein, W 1967 Effects of high energy radiation on noise pulses from photomultiplier tubes. *Review of Scientific Instruments* 38(10): 1387–1395.

Jiang, H, Luu, S, Fu, S, Zhang, W, Zhang, T, Ye, Y, Li, M, Fu, P, Wang, S, Peng, Ch and Jiang, P 1983 Model DYS low-level liquid scintillation counter. *In* McQuarrie, SA, Ediss, C and Wiebe, LI, eds, *Advances in Scintillation Counting.* Alberta, University of Alberta Press: 478–493.

Kaihola, L 1991 Liquid scintillation counting performance using glass vials with the Wallac 1220 Quantulus™. *In* Ross, H, Noakes, J and Spaulding, J, eds, *New Trends in Liquid Scintillation Counting and Organic Scintillators.* Chelsea, Michigan, Lewis Publishers: 495–500.

Kaihola, L and Oikari, T 1991 Some factors affecting alpha particle detection in liquid scintillation spectrometry. *In* Ross, H, Noakes, J and Spaulding, J, eds, *New Trends in Liquid Scintillation Counting and Organic Scintillators.* Chelsea, Michigan, Lewis Publishers: 211–218.

Kananen, K, Ala-Uotila, M, Oikari, T and Soini, E 1984 A study of the effect of humidity and use of an ioniser on static electricity using an LKB-Wallac 1211 RackBeta liquid scintillation counter. *Wallac Report*, Turku, Finland: 1–9.

Kojola, H, Polach, H, Nurmi, J, Oikari, T and Soini, E 1984 High resolution low-level liquid scintillation beta-spectrometer. *International Journal of Applied Radiation and Isotopes* 35(10): 949–952.

Kuc, T and Rozanski, K 1978 A small volume teflon-copper vial for ^{14}C low level liquid scintillation counting. *International Journal of Applied Radiation and Isotopes*

30: 452–454.

Laney, BH 1971 Electronic rejection of optical crosstalk in a twin phototube scintillation counter. *In* Horrocks, DL and Peng, C-T, eds, *Organic Scintillators and Scintillation Counting*. New York, Academic Press: 991–1003.

McKlveen, JW and McDowell, WJ 1984 Liquid scintillation alpha spectrometry techniques. *Nuclear Instruments and Methods* 223: 372–376.

Mook, WG 1983 International comparison of proportional gas counters for ^{14}C activity measurements. *In* Stuiver, M and Kra, RS, eds, Proceedings of the 11th International ^{14}C Conference. *Radiocarbon* 25(2): 475–484.

Noakes, JE 1977 Considerations for achieving low level radioactivity measurements with liquid scintillation counters. *In* Crook, MA and Johnson, P, eds, *Liquid Scintillation Counting* 4. London, Heyden: 189–206.

Noakes, JE, Isbell, AF, Stipp, JJ and Hood, DW 1963 Benzene synthesis by low temperature catalysis for radiocarbon dating. *Geochimica et Cosmochimica Acta* 27: 797–804.

Noakes, JE, Kim, SM and Stipp, JJ 1965 Chemical and counting advances in liquid scintillation counting. *In* Chatters, RM and Olson, EA, eds, *Proceedings of the 6th International Conference on Radiocarbon and Tritium Dating*. USAAEC, CONF - 650652: 68–92.

Noakes, JE, Neary, MP and Spaulding, JD 1973 Tritium measurements with a new liquid scintillation counter. *Nuclear Instruments and Methods* 109: 177–187.

_____1974 A new liquid scintillation counter for measurements of trace amounts of ^{3}H and ^{14}C. *In* Stanley, PE and Scoggins, B, eds, *Liquid Scintillation Counting: Recent Developments*. New York, Academic Press: 53–66.

Noakes, JE and Valenta, RJ 1989 Low background liquid scintillation counting using an active sample holder and pulse discrimination electronics. *In* Long, A and Kra, RS, eds, Proceedings of the 13th International ^{14}C Conference. *Radiocarbon* 31(3): 332–341.

Nydal, R 1983 The radon problem in ^{14}C counting. *In* Stuiver, M and Kra, RS, eds, Proceedings of the 11th International ^{14}C Conference. *Radiocarbon* 25(2): 501–510.

Oikari, T, Kojola, H, Nurmi, J and Kaihola, L 1987 Simultaneous counting of low alpha- and beta-particle with liquid scintillation spectrometry and pulse shape analysis. *International Journal of Applied Radiation and Isotopes* 38(10): 875–878.

Pearson, GW 1979 Precise measurement by liquid scintillation counting. *Radiocarbon* 21(1): 1–21.

Pietig, F, von and Scharpenseel, HW 1964 Alterbestimmung mit Flüssigheits-Scintillations-Spectrometer, über die Wirksamkeit von Abschirmungsabnahmen. *Atompraxis* 7: 1–3.

Polach, HA 1969 Optimisation of liquid scintillation radiocarbon age determinations and reporting of ages. *Atomic Energy in Australia* 12(3): 21–28.

_____1974 Application of liquid scintillation spectrometers to radiocarbon dating. *In* Stanley, PE and Scoggins, B, eds, *Liquid Scintillation Counting: Recent Developments*. New York, Academic Press: 153–171.

_____1987 Evaluation and status of liquid scintillation counting for radiocarbon dating. *Radiocarbon* 29(1): 1–11.

Polach, HA, Gower, J and Fraser, I 1972 Synthesis of high purity benzene for radiocarbon dating by the liquid scintillation method. *In* Rafter, TA and Taylor, T, eds, *Proceedings of the 8th International ^{14}C Conference*. Wellington, Royal Society of New Zealand: 145–157.

Polach, H, Gower, J, Kojola, H and Heinonen, A 1983a An ideal vial and cocktail for low-level scintillation counting. *In* McQuarrie, SA, Ediss, C and Wiebe, LI, eds, *Advances in Scintillation Counting*. Alberta, University of Alberta Press: 508–525.

Polach, H and Kaihola, L 1988 Determination of radon by liquid scintillation alpha/beta particle spectrometry. *Radiocarbon* 30(1): 19–24.

Polach, H, Kaihola, L, Robertson, S and Haas, H 1988 Small sample ^{14}C dating by liquid scintillation spectrometry. *Radiocarbon* 30(2): 153–155.

Polach, H, Kojola, H, Nurmi, J and Soini, E 1984 Multiparameter liquid scintillation spectrometry. *In* Wölfli, W, Polach, H and Anderson, HH, eds, Accelerator mass spectrometry, AMS-84. *Nuclear Instruments and Methods* 233[B5](2): 439–442.

Polach, H, Nurmi, J, Kojola, H and Soini, E 1983b Electronic optimization of scintillation counters for detection of low-level ^{3}H and ^{14}C. *In* McQuarrie, SA, Ediss, C and Wiebe, LI, eds, *Advances in Scintillation Counting*. Alberta, University of Alberta Press: 420–441.

Polach, H, Robertson, S, Butterfield, D, Gower, J and Soini, E 1983c The 'windowless' approach to scintillation counting: low-level ^{14}C as an example. *In* McQuarrie, SA, Ediss, C and Wiebe, LI, eds, *Advances in Scintillation Counting*. Alberta, University of Alberta Press: 494–507.

Polach, HA and Stipp, JJ 1967 Improved synthesis techniques for methane and benzene radiocarbon dating. *International Journal of Applied Radiation and Isotopes* 18: 359–364.

Punning, JM and Rajamäe, R 1975 Some possibilities for decreasing the background of liquid scintillation beta-ray counter. *In* Povinec, P and Usacev, S, eds, *Low Radioactivity Measurements and Applications*. Bratislava, Slovenske Pedagogicke Nakladatelstvo: 169–171.

Radnell, CJ and Muller, AB 1980 Memory effects in the production of benzene for radiocarbon dating. *In* Stuiver, M and Kra, RS, eds, Proceedings of the 10th International ^{14}C conference. *Radiocarbon* 22(2): 479–486.

Rapkin, E 1970 Development of the modern liquid scintillation counter. *In* Bransome, ED, Jr, ed, *The Current Status of Liquid Scintillation Counting*. New York, Grune and Stratton: 45–68.

Reynolds, GT, Harrison, FB and Salvini, G 1950 Liquid scintillation counters. *Physics Review* 78: 488–493.

Rubin, M, Mysen, B and Polach, H 1984 Graphite sample preparation for AMS in a high pressure and high temperature press. *In* Wölfli, W, Polach, H and Anderson, HH, eds, Accelerator mass spectrometry, AMS-84. *Nuclear Instruments and Methods* 233[B5](2): 272–273.

Schotterer, U and Oeschger, H 1980 Low-level liquid scintillation counting in an underground laboratory. *In* Stuiver, M and Kra, RS, eds, Proceedings of the 10th International ^{14}C Conference. *Radiocarbon* 22(2): 505–511.

Soini, E 1975a Stabilization of photomultipliers in liquid scintillation counters. *Review of Scientific Instruments* 46: 980–984.

_____1975b Rejection of optical cross-talk in photomultiplier tubes in liquid scintillation counters. *Wallac Report*, Turku, Finland: 1–9.

_____1977 Chemiluminescence monitoring and rejection in liquid scintillation counting. *Wallac Report*, Turku, Finland: 1–8.

Stipp, JJ, Knauer, GA and Godell, HG 1966 Florida State university radiocarbon dates I. *Radiocarbon* 8: 46–53.

Tamers, MA 1965 Routine carbon-14 dating using liquid scintillation techniques. *In* Chatters, RM and Olson, EA, eds, *Proceedings of the 6th International Conference on Radiocarbon and Tritium Dating*. USAAEC, CONF-650652: 53–67.

_____1975 Chemical yield optimisation of benzene synthesis for radiocarbon dating. *International Journal of Applied Radiation and Isotopes* 26: 676–682.

Valenta, R 1987 *Reduced background scintillation counting*. US Patent no. 4,651,006.

Wang, CH 1970 Quench compensation by means of gain restoration. *In* Bransome, ED, Jr, ed, *The Current Status of Liquid Scintillation Counting*. New York, Grune and Stratton: 305–312.

15

THE HISTORY OF AMS, ITS ADVANTAGES OVER DECAY COUNTING: APPLICATIONS AND PROSPECTS

H E GOVE

INTRODUCTION

Accelerator mass spectrometry (AMS), almost from its inception, involved the use of existing tandem Van de Graaff electrostatic accelerators, normally employed in nuclear physics research, and later, small tandem accelerators specifically designed for AMS, to directly detect long-lived cosmogenic radioisotopes in the presence of vastly larger quantities of their stable isotopes. Some early work was carried out using cyclotrons and even combinations of accelerators capable of accelerating heavy ions to energies of hundreds of MeV per nucleon but, except for special cases, tandem electrostatic accelerators are now the ones of choice for reasons that will be touched on below.

Although the original impetus was to detect radiocarbon because of the exciting consequences that it would have in the field of artifact dating, it was soon realized that many other cosmogenic radioisotopes could also be detected with unparalleled sensitivity by AMS. These isotopes include ^{10}Be, ^{26}Al, ^{36}Cl, ^{41}Ca and ^{129}I as well as trace amounts of stable isotopes in matrices of more abundant elements. A plenitude of applications in hydrology, geoscience, materials science, biomedicine, sedimentology, environmental sciences and many other fields emerged as soon as the AMS detection capabilities of the appropriate isotopes were demonstrated. Presented here is an account of the historical development of AMS, the international conferences that have been held on radiocarbon dating and on AMS since 1977, the reasons for the success of AMS over decay counting, its applications in various fields of research and some predictions of its future.

THE HISTORICAL DEVELOPMENT OF ACCELERATOR MASS SPECTROMETRY

The first measurements of ^{14}C in natural organic material by AMS of milligram samples of carbon took place at the University of Rochester during the week of May 14, 1977, using the University's tandem Van de Graaff electrostatic accelerator (Purser *et al* 1977; Bennett *et al* 1977). It was carried out by a team of nuclear physicists from the General Ionex Corporation led by K H Purser, from the University of Toronto led by A E Litherland and from the University of Rochester led by H E Gove. The measurements demonstrated that negative nitrogen ions were unstable – a fact that had been anticipated by physicists using tandem accelerators in their nuclear research. This meant that the interference between ^{14}C and its stable and vastly more abundant isobar, ^{14}N, was eliminated. Most importantly, the measurements included the direct detection of ^{14}C in a homely sample of contemporary barbecue charcoal. It represented a revolution in the field of carbon dating invented by Willard Libby (Libby, Anderson & Arnold 1949), at the time almost three decades old, principally because now it could be applied to samples too small or too precious to be dated by the Libby decay counting technique. Remarkably enough, about three weeks after this first measurement at Rochester of ^{14}C on a natural organic sample, Nelson, Korteling and Stott (1977) from Simon Fraser University, stimulated by Muller's suggestion, employed the tandem accelerator at McMaster University to detect ^{14}C in a sample of AD 1880–1890 wood. Purportedly neither the Rochester nor Simon Fraser group was aware of the other group's efforts at the time.

The advantages of atom counting by mass spectrometry over decay counting for long-lived radioisotopes had been recognized some eight years before. Oeschger *et al* (1970) advocated the development of mass spectrometric methods for radiocarbon as early as 1969, and in that same year, Anbar and coworkers mounted a valiant but unsuccessful effort to do so using low energy mass 29 negative molecules of ^{14}C injected into a combination of einzel lenses, Wien filters and electrostatic and magnetic deflectors (Schnitzer *et al* 1974). The attempt failed because chemical techniques were inadequate to eliminate trace impurities of interfering mass 29 negative molecules and $^{29}Si^-$ itself, and the effort was abandoned (Anbar 1978).

As far as can be determined, the first use of AMS occurred in 1939, when Alvarez and Cornog (1939) detected the rare isotope of helium, ^{3}He, using a 60-inch cyclotron. The high energies employed made the identification of these mass 3, atomic number 2 nuclei straightforward by well-known nuclear techniques.

The concept of using AMS to directly detect ^{14}C with a cyclotron was reintroduced by Muller in 1977 following a search for integrally charged quarks (Muller *et al* 1977). It was less than completely successful in detecting ^{14}C in contemporary organic samples because positive ions were employed, and the minuscule numbers of positive ^{14}C ions were overwhelmed by the flood of positive ^{14}N ions. The application of well-known range-separation techniques, but ingeniously employed by the Berkeley team (Muller, Stephenson & Mast 1978) were insufficiently robust to stem the flood.

It is of interest to note that Muller's mentor, Luis Alvarez, invented the tandem electrostatic accelerator employing negative ions in 1951 (Alvarez 1951). Had such an accelerator been employed by the Berkeley group, they would have had the same success as the Rochester group. The latter's efforts were independent of those at Berkeley, and were carried out virtually simultaneously. It is another remarkable example of the well-known phenomenon that, when the time is ripe for an advance to be made, it is often carried out by two or more groups independently and simultaneously. In reality, the time was over-ripe. The advance could have been made some 19 years before when the first tandem Van de Graaff accelerator was commissioned at the Atomic Energy of Canada's Chalk River Laboratories (Gove *et al* 1958).

Around the same time as the Rochester and Berkeley work, and again quite independently, a single-ended electrostatic accelerator was used by a group at Argonne to search for quarks of 1/3 charge (Schiffer *et al* 1978), and a group at Brookhaven used a tandem Van de Graaff to search for superheavy elements (Schwarzchild, Thieberger & Cumming 1978).

An absolutely key element in the success of tandem electrostatic accelerators for the detection of ions of rare atoms was the elimination of the interference from molecular ions of the same mass by their breakup in the terminal stripper foil or gas in the process of converting the singly charged negative molecular ions to multiply charged positive ions. This was pointed out by Purser in 1976 in a US patent for a system to detect ozone-destroying halides in the atmosphere (Purser 1977). In 1977, Purser *et al* (1977) and Bennett *et al* (1977) showed at Rochester that if three or more electrons are removed from a neutral mass 14 molecule like $^{12}CH_2$, the molecule disassociated in a Coulomb explosion, and the resultant fragments were swept aside before reaching the final detector.

Stimulated by Muller's suggestion and quite independent of the Rochester work, Raisbeck and Yiou at the René Bernas Institute used the Grenoble cyclotron to detect ^{10}Be (Raisbeck *et al* 1978) and Farwell *et al* (1980) at the University of Washington initiated an AMS program for the detection of ^{10}Be and ^{14}C using an FN tandem Van de Graaff accelerator at the Nuclear Physics Laboratory at that institution in Seattle, Washington. When the latter group began their program in June 1977, they were not only unaware of the Rochester work, but also that of Nelson *et al* at McMaster University and Raisbeck *et al* at Grenoble. Both the René Bernas and the University of Washington groups continue to make important contributions in ^{10}Be and ^{14}C measurements, respectively.

The initial work by the Rochester consortium (Purser *et al* 1977; Bennett *et al* 1977) was shortly followed by their comparison of the ages of samples supplied by Meyer Rubin of the US Geological Survey (USGS) and previously measured by him using the Libby decay counting method and measured at Rochester by AMS. The agreement was excellent (Bennett *et al* 1978), but it is worth stressing that the measurements were not made blind – it was too early in the development of AMS for that.

In an act that some considered daring, but admirable nonetheless, Muller (1978) and the Berkeley group performed two blind ^{14}C dating measurements in collaboration with Rainer Berger, that resulted in only a 50% success. Muller signalled each measurement with a letter to me giving his result before he learned the answer from Berger, and soliciting wishes for success. His daring attempt was defeated by the positive ion cyclotron, which was just not the instrument for carbon dating. In later years, a graduate student of Muller's was persuaded to design a small negative ion cyclotron for the purpose, so far with less than complete success (Bertsche *et al* 1987).

Subsequent work on AMS was carried out at Oxford to demonstrate that ^{14}C could be measured with completely acceptable sensitivity using small tandem accelerators with terminal voltages around 2 MV. All that was required was a negative ion energy high enough to have a reasonable probability of producing charge 3+ ions in the terminal stripper to ensure the elimination of mass 14 molecules. A 2 MV terminal tandem provided that energy (Doucas *et al* 1978). This was followed by the design of a small tandem accelerator, the "Tandetron" (Purser, Liebert & Russo 1980), and nine accelerators of this design have been installed in laboratories throughout the world (Suter 1990).

The next major advance in AMS was the detection of the cosmogenic radioisotope, ^{36}Cl, at Rochester. At a workshop on the dating of old groundwater held at the University of Arizona in March 1978, Bentley (1978) concluded that a measurement of this chlorine isotope provided the greatest potential for establishing groundwater ages. The long half-life (301,000 years) and the low natural abundance (recent rainwater samples have a ^{36}Cl to stable chlorine ratio of 10^{-12} to 10^{-13}) make the detection of ^{36}Cl in natural samples by decay counting difficult. The first measurements were carried out at Rochester less than a month later, on samples enriched over the natural level by a factor of 10 or so, and reported at the first conference on AMS held in Rochester in April 1978 (Naylor *et al* 1978). Several months later, ^{36}Cl was measured in natural water samples, and the background level for the ^{36}Cl to stable chlorine ratio was established to be below 3×10^{-15} (Elmore *et al* 1979).

The main problem AMS encounters in the detection of ^{36}Cl is the existence of the stable isobar, ^{36}S. Sulphur is a very common element, and even though its mass 36 isotope has a low abundance (0.02%), careful chemistry is required in sample preparation to reduce the sulphur to a level low enough to make viable the ^{36}S counting rate in the final dE/dx - E detectors. Even so, to achieve a sufficient separation provided by the dE/dx signal between Z = 16 and 17, a mass 36 energy of the order of 100 MeV is needed. This requires tandem terminal voltages in the 10 MV region. There are presently only a limited number of tandems worldwide with AMS research programs capable of such voltages.

The other major cosmogenic radioisotope measured by AMS is ^{129}I. It has a number of applications in studies of the marine environment such as the dating of sediments, tracing of slow water movement and dating/tracing of hydrocarbons and several others. It has an even longer half-life than ^{36}Cl (1.57×10^7 years), and the need to measure the ratio of $^{129}I/^{127}I$ to as low as 10^{-14} again precludes the possibility of using decay counting. It was first measured in natural samples at Rochester (Elmore *et al* 1980).

INTERNATIONAL CONFERENCES ON RADIOCARBON AND ACCELERATOR MASS SPECTROMETRY

Following the pattern set by the radiocarbon community, which has held International Radiocarbon Conferences at the sites of major decay-counting radiocarbon laboratories throughout the world every three years since Libby's seminal work, the AMS community does likewise at sites of AMS facilities starting at Rochester (Anbar 1978; Muller 1978; Naylor *et al* 1978). The second was held at Argonne National Laboratory in 1981 (Kutschera 1981), the third at Zurich in 1984 (Wölfli, Polach & Anderson 1984), the fourth (which celebrated the Anno Decimo of AMS) at Niagara-on-the-Lake, Canada in 1987 (Gove, Litherland & Elmore 1987) and the fifth in Paris, France in 1990 (Yiou & Raisbeck 1990).

The tenth International Radiocarbon Conference, held in Bern - Heidelberg in August 1979, was the first that had a session on AMS (Stuiver & Kra 1980). Seven papers were presented in that category with the lead-off paper presented by Gove *et al* (1980) for the General Ionex - Toronto - Rochester consortium with Meyer Rubin of the USGS as a co-author. The second by Purser, Liebert and Russo (1980) described the small tandem accelerator systems under construction at General Ionex Corporation for the Universities of Arizona, Toronto and Oxford. The remaining 5 papers described work underway at Argonne (Kutschera *et al* 1980), Oxford (UK) (Hedges *et al* 1980), Chalk River

(Canada) (Andrews *et al* 1980), Harwell (UK) (Shea *et al* 1980) and the University of Washington (Farwell *et al* 1980).

All subsequent International Radiocarbon Conferences, the 11th in Seattle (Stuiver & Kra 1983), the 12th in Trondheim (Norway) (Stuiver & Kra 1986) and the 13th in Dubrovnik (Yugoslavia) (Long & Kra 1989) have included AMS sessions as did the 14th conference, held in Tucson, Arizona in May 1991 which, like Seattle, is the site of both a decay-counting and an AMS laboratory.

It is interesting to speculate on how long it will be before ^{14}C dating by AMS will supplant that by decay counting? Recently, the decision was made by one of the premier radiocarbon laboratories in Europe, the Centre for Isotope Research in Groningen, The Netherlands, to add an AMS unit to their decay-counting facility. Of course, a similar decision had already been made at a leading radiocarbon decay-counting laboratory in the USA, namely the one at the University of Arizona, soon after the power of AMS was demonstrated to add an AMS unit. These decisions suggest that, before too many years, the International Radiocarbon Conferences may be devoted virtually exclusively to measurements by AMS.

THE ADVANTAGES OF ACCELERATOR MASS SPECTROMETRY OVER DECAY COUNTING

The main advantage of AMS over decay counting can be measured either in the ratio of the time it would take to collect 100 counts, *ie*, achieve a 10% statistical accuracy using the same amount of sample ordinarily employed in AMS in a 100% efficient decay-counting detector to the time taken in an AMS measurement. The equivalent comparison is the ratio of the sample size needed in decay counting to produce 100 counts in the same time as with AMS. Both ratios are the same, and depend on specific values of certain AMS parameters, such as the ion source current of the stable isotope and the product of the accelerator efficiency, and the terminal stripper yield for the chosen charge state.

Typical AMS parameters for this comparison have been given by Elmore and Phillips (1987) for ^{10}Be, ^{14}C, ^{26}Al, ^{36}Cl and ^{129}I. The ratios are, respectively, 6×10^7, 2×10^5, 3×10^6, 1.5×10^6 and 3×10^7. These are impressively large numbers. In the case of ^{14}C, they mean a measurement can be made of the date of an artifact 34,000 years old with a statistical accuracy of 10% (± 830 years) in 7 minutes using a 0.25 mg sample, whereas a sample of 50 g would be required for decay counting to obtain the same accuracy in the same 7 minutes. A modern sample could be measured to 1% (± 83 years) in an hour by AMS.

As can be seen from the above ratios, the AMS advantage for the longer-lived radioisotopes is even greater.

In the very early days of AMS, it was thought that it would be readily possible to measure ^{14}C dates back 100,000 years. The radiocarbon background observed when petroleum-derived graphite was measured, yielding a graphite "age" between 45 and 65 thousand years, was thought to come from the residual ^{14}C produced by large tandem accelerators, such as the one at Rochester. We believed that residual ^{14}C came from years of neutron production in nuclear physics research. However, small tandems such as the ones at Arizona and Toronto, which are too low in energy to produce neutrons, still suffer from the same background whose origin is not fully understood. However, for AMS measurements of very old samples, it is known that sample processing can add some contemporary carbon contamination, and corrections for this are well understood. The age limit for decay counting of unenriched samples ranges from 35,000 to 60,000 years and, in this respect, it is quite comparable to AMS.

In short, it is the small sample size and the short measuring times combined with precisions that now match decay counting that gives AMS a competitive edge over decay counting for carbon dating. Even the larger capital costs may not prevent AMS from dominating the field in the years ahead. It is these capital costs and the greater complexity of AMS hardware that have so far prevented AMS from supplanting decay-counting facilities.

At the Fifth AMS Symposium in Paris (Yiou & Raisbeck 1990), Suter summarized the status of AMS and, in particular, he listed the accelerator facilities that were devoted part or full time to AMS. They totaled 40. Of these, 9 were tandems designed specifically for AMS with terminal voltages around 2 MV. Another 22 or so were tandems adapted for AMS in the 6 to 14 MV terminal voltage range. Many of these have research programs shared between nuclear physics and AMS. The remainder are single-ended Van de Graaff accelerators, some tandem LINAC combinations and cyclotrons numbering about 9, which are used for special applications, or are under development in the hope that they will provide competition for tandems.

A tenth small tandem of terminal voltage around 2 MV dedicated to ^{14}C dating by AMS has recently been installed in the Woods Hole Oceanographic Institute. It will be used for the production-line measurements of ^{14}C in ocean water and deep-sea sediment samples.

SOME INTERESTING APPLICATIONS OF AMS

In the 13 years since the first AMS measurements were made, there are a number that the author thinks qualify for the "most interesting" category, recognizing that the rating is bound to be somewhat subjective.

In the ^{14}C category, these include the following:

A Viking Site in Newfoundland

Although it is certain in many people's minds that the Vikings discovered North America hundreds of years before Columbus, there had never been incontrovertible proof. At a Viking site at L'Anse aux Meadows in the northern tip of Newfoundland, characterized by the ruins of stone building structures, cooking firepits and artifacts almost certainly of Scandinavian origin, a fire pit clearly used for the smelting of iron was discovered.

Small inclusions of charcoal from the wood fuel used to melt the iron were found in the slag in this pit. This charcoal was carefully picked out bit by bit from the slag until a milligram or so was collected. A preliminary measurement of its age was carried out at Rochester by the General Ionex - Rochester - Toronto consortium and yielded a late 10th century date. Later, the material was dated much more carefully by the Isotrace facility at the University of Toronto. Their date was AD 984–1010 (Beukens 1986). The Viking Sagas say the area was settled by Lief Ericsson between AD 945 and 1007 (Beukens 1986). The fact that it was an iron-smelting firepit meant it was used by Europeans. Carbon from cooking firepits could have been of North American Indian origin for whom iron, at that time, was an unknown element.

The Initial Peopling of North America

The question of the date of the initial advent of humans to North America has been a matter of long-standing controversy in New World archaeological circles. Evidence from geological criteria, amino-acid racemization and uranium series analyses has suggested, from time to time, that some human bone samples discovered in North America may have dated back some 70,000 years. Such dates substantially preceding that of the last ice age have now been shown (Taylor *et al* 1985), by AMS radiocarbon measurements at the University of Arizona, to be wrong. The correct ages are generally around 5000 years. It was the ability of AMS to measure in bone the ^{14}C to stable carbon ratios in milligram samples of single amino acids, total amino acid components and/or multiple organic fractions of bone pretreated to remove non-indigenous organic components that provided the true bone ages (Taylor 1987).

The Shroud of Turin

Although of little scientific significance – certainly not compared with the date of the discovery of North America by the Vikings or the first advent of humans to the New World – the age of the Turin shroud has been of considerable interest to the general public for many years. The shroud, which bears a mysterious image – frontal and dorsal – of a crucified man, is widely believed to be the burial cloth of Christ. Its historical record, however, goes back in time only to ca AD 1353.

The actual age of the shroud's linen cloth could only be established by ^{14}C dating. Apparently, Libby offered to date the shroud after his decay counting method had been perfected, but his offer was declined because it would have required a handkerchief-size sample. The author and others renewed this offer in 1978, using AMS, which would reduce the sample size to that of a postage stamp.

Complex, if not Byzantine, negotiations followed involving the Vatican, its Pontifical Academy of Sciences, the Archdiocese of Turin, five AMS laboratories located at the University of Arizona, USA, CFR Gif-sur-Yvette, France, ETH Zurich, Switzerland, Oxford University, UK and the University of Rochester, USA, two laboratories using small proportional counters located at AERE Harwell, UK and Brookhaven National Laboratory, USA and, as coordinating institution, the British Museum, London, UK. All in all, a rather extraordinarily prestigious list of scientific and religious institutions to which can be added the National Science Foundation in the USA which provided funding to support the travel of the three US participants to a workshop held in Turin to establish a dating protocol.

Ten years later at 9:50 AM 6 May 1988, the first AMS ^{14}C measurement on the Turin shroud was carried out at the University of Arizona, using one quarter of the total 2 cm^2 sample they had received – a 1/2 cm^2 piece of the shroud cloth. In ten minutes the answer was known – the shroud was only about 650 years old! Subsequent measurements carried out at Zurich and Oxford confirmed the Arizona result. The flax from which the Turin shroud's linen was formed was harvested in AD 1325 with an uncertainty of plus or minus 33 years (Damon *et al* 1989). Its age was consistent with the shroud's historic date – arguably the least captivating result. The original protocol for dating the shroud (Gove 1987b), a report on progress in ^{14}C dating the shroud (Gove 1989), and a final assessment on the dating result (Gove 1990) have been provided by the author. The result was a public triumph for AMS, but a disappointment for those who hoped or believed it was the burial cloth of Christ.

In the ^{36}Cl category, one notes the following important developments:

^{36}Cl as a Measure of Groundwater Age

At the Arizona workshop on the dating of old groundwater (Bentley 1978) mentioned above, it was concluded that a measurement of ^{36}Cl in water samples would be an indicator of their ages. To test this, water samples were collected along the flow of a hydrologically well-understood aquifer in the Great Artesian Basin area of Australia. A reasonably good one-to-one agreement was obtained between the ^{36}Cl age and the hydrological age of the water over a period of one million years (Bentley *et al* 1986).

The ^{36}Cl Bomb Pulse

What was not foreseen at the Arizona groundwater dating workshop was the use of ^{36}Cl produced in the nuclear weapons tests carried out in the 1950s in the Pacific Ocean to measure water flow rates over the past 40 years. Tritium has been used this way for many years but, because of its 12-year half-life, it is becoming a progressively weaker signal.

These weapons tests injected ^{36}Cl produced by neutrons interacting with the chlorine in the seawater into the biosphere at a level two orders of magnitude above the pre- and postbomb test ambient levels. The pulse has a width at half maximum of about six years. It was first measured in Greenland ice cores (Elmore *et al* 1982), where the time scale was determined by counting the seasonal variations in the $^{18}O/^{16}O$ ratio and in groundwater (Bentley *et al* 1982). It has subsequently been widely used to measure groundwater flow rates in order to determine the suitability of potential sites for the storage of low- and high-level nuclear waste, as well as for other purposes.

There are a number of examples of interesting and important measurements of other radioisotopes and stable elements, including ^{10}Be, ^{26}Al, ^{129}I and the two stable isotopes, ^{186}Os and ^{187}Os, which will not be included here.

SOME PREDICTIONS FOR THE FUTURE OF AMS

In the 13 years since the genesis of AMS, the technique has seen applications in many areas of science. These have been catalogued in numerous review articles (Elmore & Phillips 1987; Gove *et al* 1979; Purser Litherland & Gove 1979; Litherland 1980; Gove *et al* 1980; Gove 1985, 1987a; Kutschera & Paul 1990) which, in addition, describe in some detail the principles governing AMS measurements by tandem electrostatic accelerators – a subject that has not been dealt with in this review.

To predict the future of a technique so replete in examples of its present applicability is both daunting and perhaps even audacious. There are, however, three areas that seem obvious and ones that are presently in their infancy. They are biomedicine, condensed matter and the use of cesium ion source microprobes. Two other developments are mentioned by Beukens (1992), namely the work on CO_2 gas sources for radiocarbon detection by AMS, and the construction of the second generation Tandetron AMS system of the type presently being installed at the Woods Hole Oceanographic Institution, Massachusetts, and recently ordered by the radiocarbon laboratory in Groningen, The Netherlands.

Applications of AMS to Biomedical Research

A number of radionuclides that could make useful tracers in medical research are readily detectable by AMS at extremely low concentrations. They include ^{3}H, ^{14}C, ^{26}Al, ^{36}Cl, ^{41}Ca and ^{129}I. Another radionuclide that may be of interest is ^{79}Se, but no effort has yet been devoted to its detection by AMS.

Of these, ^{14}C is the most universally used biochemical tracer, and is the one for which the capability of detection by AMS is most developed. It is also the one that is most easily, accurately and sensitively measured by tandem accelerators having terminal potentials of the order of one million volts. The current status of the field was recently discussed (Purser & Gove 1990; Davis 1992; Jope & Jope 1992).

^{14}C can be incorporated into most organic molecules and, further, the site of the ^{14}C atom so introduced can be accurately specified by the production procedure. The ^{14}C can be located at sites within the molecule that are not subject to exchange. ^{14}C does not substantially alter the chemical behavior of the molecule. In the past, it was necessary to have a dozen or so ^{14}C per molecule to reduce the amount of material ingested, especially if the material were toxic.

The idea of applying AMS detection of ^{14}C in medical research was first discussed by Kielson and Waterhouse in 1978. They noted, at that time, that for the past 35 years or so, tracer kinetic methods had been widely employed in medical research in studies of human metabolism. It was their feeling that improved techniques that could be employed by the use of AMS in such studies would not only have diagnostic importance, but would be of value to pharmacological studies where tracer kinetics are commonly employed.

Surprisingly enough, nine years later, in 1987, no liaison between AMS and the biomedical field had yet formed to exploit the power of AMS in medical research. In a paper published that year, Elmore (1987) discussed a possible application of AMS in the biological sciences, but the first actual application of AMS to the field was reported by J S Felton *et al* (1990) and J S Vogel *et al*

(1990) at the Fifth International Conference on Accelerator Mass Spectrometry in Paris.

If biomedical research using AMS is to be more widely performed, smaller and cheaper AMS facilities will be required. The general parameters of such a facility for the detection of ^{14}C in biological samples have been described (Purser & Gove 1990).

Applications of AMS in the Field of Condensed Matter

An early example of the power and sensitivity of AMS to detect a wide range of trace impurities in semiconductors was described by Anthony and Donahue (1987). About the same time and independently, the depth profiling of chlorine and nitrogen in ultrapure silicon wafers was demonstrated using a combination of neutron activation and AMS (Hossain *et al* 1987). The application of AMS to electronic and silver halide imaging research is continuing (Gove *et al* 1990). There is little doubt that this AMS/condensed matter interaction will flourish in the future.

Cesium Microprobes for AMS Negative Ion Sources

If cesium beams in AMS negative ion sources of sufficient intensities and with diameters approaching 50 microns or less can be developed, they could be used to scan polished surfaces of geological and other samples to determine the distribution of important elements in tiny inclusions in those samples. Very preliminary work in this area was carried out in the early 1980s at the University of Rochester (Gove 1985), but the technique is now being developed on the AMS facility at the University of Toronto (Wilson, Kilius & Rucklidge, ms in preparation).

REFERENCES

Alvarez, LW 1951 Energy doubling in dc accelerators. *Review of Scientific Instruments* 22: 705–706.

Alvarez, LW and Cornog, R 1939 He^3 in helium. *Physical Review* 56: 379.

Anbar, M 1978 The limitations of mass spectrometric radiocarbon dating using CN^- ions. *In* Gove, HE, ed, *Proceedings of the 1st Conference on Radiocarbon Dating with Accelerators*. University of Rochester: 152–155.

Andrews, HR, Ball, GC, Brown, RM, Davies, WG, Imahori, Y and Milton, JCD 1980 Progress in radiocarbon dating with the Chalk River MP tandem accelerator. *In* Stuiver, M and Kra, RS, eds, Proceedings of the 10th International ^{14}C Conference. *Radiocarbon* 22(3): 822–829.

Anthony, JM and Donahue, DJ 1987 Accelerator mass spectrometry solutions to semiconductor problems. *In* Gove, HE, Litherland, AE and Elmore, D, eds, Proceedings of the 4th International Symposium on Accelerator Mass Spectrometry. *Nuclear Instruments and Methods* B29: 77–82.

Bennett, CL, Beukens, RP, Clover, MR, Elmore, D, Gove, HE, Kilius, L, Litherland, AE and Purser, KH 1978 Radiocarbon dating with electrostatic accelerators: Dating of milligram samples. *Science* 201: 345–347.

Bennett, CL, Beukens, RP, Clover, MR, Gove, HE, Liebert, RB, Litherland, AE, Purser, KH and Sondheim, WE 1977 Radiocarbon dating using electrostatic accelerators: negative ions provide the key. *Science* 198: 508–510.

Bentley, HW 1978 Some comments on the use of chlorine-36 for dating very old ground water. *In* Davis, SN, ed, *Workshop on Dating Old Ground Water*. Tucson, University of Arizona Press: 102–111.

Bentley, HW, Phillips, FM, Davis, SN, Gifford, S, Elmore, D, Tubbs, LE and Gove, HE 1982 Thermonuclear ^{36}Cl pulse in natural water. *Nature* 300: 737–740.

Bentley, HW, Phillips, FM, Davis, SN, Habermehl, MA, Airey, PL, Calf, GE, Elmore, D, Gove, HE and Torgeson, T 1986 Chlorine 36 dating of very old groundwater 1. The Great Artesian Basin, Australia. *Water Resources Research* 22: 1991–2001 .

Bertsche, KJ, Friedman, PJ, Morris, DE, Muller, RA and Welsh, JJ 1987 Status of the Berkeley small cyclotron AMS project. *In* Gove, HE, Litherland, AE and Elmore, D, eds, Proceedings of the 4th International Conference on Accelerator Mass Spectrometry. *Nuclear Instruments and Methods* B29: 105–109.

Beukens, RP 1986 *Isotrace Laboratory Newsletter*. University of Toronto 3(1): 3.

_____ 1992 Radiocarbon accelerator mass spectrometry: Background, precision and accuracy. *In* Taylor, RE, Long, A and Kra, RS, eds, *Radiocarbon After Four Decades: An Interdisciplinary Perspective.* New York, Springer-Verlag, this volume.

Damon, PE, Donahue, DJ, Gore, BH, Hatheway, AL, Jull, AJT, Linick, TW, Sercel, PJ, Toolin, LJ, Bronk, CR, Hall, ET, Hedges, REM, Housley, R, Law, IA, Perry, C, Bonani, G, Trumbore, S, Wölfli, W, Ambers, JC, Bowman, SGE, Leese, MN and Tite, MS 1989 Radiocarbon dating the shroud of Turin. *Nature* 337: 611–615.

Davis, JC 1992 New biomedical applications of radiocarbon. *In* Taylor, RE, Long, A and Kra, RS, eds, *Radiocarbon After Four Decades: An Interdisciplinary Perspective.* New York, Springer-Verlag, this volume.

Doucas, G, Garmon, EF, Hyder, HRM, Sinclair, D, Hedges, REM and White, NR 1978 Detection of ^{14}C using a small van de Graaff accelerator. *Nature* 276: 253–255.

Elmore, D 1987 Ultrasensitive radioisotope, stable-isotope, and trace-element analysis in the biological sciences using tandem accelerator mass spectrometry. *Biological Trace Element Research* 12: 231–245.

Elmore, D, Fulton, BR, Clover, MR, Marsden, JR, Gove, HE, Naylor, H, Purser, KH, Kilius, LR, Beukens, RP and Litherland, AE 1979 Analysis of ^{36}CL in environmental water samples using an electrostatic accelerator. *Nature* 277: 22-25.

Elmore, D, Gove, HE, Ferraro, R, Kilius, LR, Lee, HW, Chang, H, Beukens, RP, Litherland, AE, Russo, CJ, Purser, KH, Murrell, MT and Finkel, RC 1980 Determination of ^{129}I using tandem accelerator mass spectrometry. *Nature* 286: 138–140.

Elmore, D and Phillips, FM 1987 Accelerator mass spectrometry for measurement of long-lived radioisotopes. *Science* 236: 543–550.

Elmore, D, Tubbs, LE, Newman, D, Ma, XZ, Finkel, R, Nishiizumi, K, Beer, J, Oeschger H and Andrée, M 1982 ^{36}Cl bomb pulse measured in a shallow ice core from Dye 3, Greenland. *Nature* 300: 735-737.

Farwell, GW, Schaad, TP, Schmidt, FH, Tsang, M-YB, Grootes, PM and Stuiver, M 1980 Radiometric dating with the University of Washington Tandem Van de Graaff accelerator. *In* Stuiver, M and Kra, RS, eds, Proceedings of the 10th International ^{14}C Conference. *Radiocarbon* 22(3): 838-849.

Felton, JS, Turteltaub, KW, Vogel, JS, Balhorn, R, Gledhill, BL, Southon, JR, Caffee, MW, Finkel, RC, Nelson, DE, Proctor, ID and Davis, JC 1990 Accelerator mass spectrometry in the biomedical sciences: applications in low exposure biomedical and environmental dosimetry. *In* Yiou, F and Raisbeck, GM, eds, Proceedings of the 5th International Conference on Accelerator Mass Spectrometry. *Nuclear Instruments and Methods* B52 (3,4): 517–523.

Gove, HE 1985 Accelerator-based ultrasensitive mass-spectrometry. *In* Bromley, DA, ed, *Treatise on Heavy Ion Science.* New York, Plenum Press: 431–463.

_____1987a Tandem-accelerator mass-spectrometry measurements of ^{36}Cl, ^{129}I and osmium isotopes in diverse natural samples. *Philosophical Transactions of the Royal Society of London* A323: 103–119.

_____1987b Turin workshop on radiocarbon dating the Turin shroud. *Nuclear Instruments and Methods* B29: 193–195.

_____1989 Progress in radiocarbon dating the shroud of Turin. *In* Long, A and Kra, RS, eds, Proceedings of the 13th International ^{14}C Conference. *Radiocarbon* 31 (3): 965–969.

_____1990 Dating the Turin shroud - an assessment. *Radiocarbon* 32(1): 87–92.

Gove, HE, Elmore, D, Ferraro, R, Beukens, RP, Chang, KH, Kilius, LR, Lee, HW, Litherland, AE and Purser, KH 1980 Radioisotope detection with tandem electrostatic accelerators. *Nuclear Instruments and Methods* 168: 425–433.

Gove, HE, Fulton, BR, Elmore, D, Litherland, AE, Beukens, RP, Purser, KH and Naylor, H 1979 Radioisotope detection with tandem electrostatic accelerators. *IEEE Transactions in Nuclear Science* 26: 1414–1421.

Gove, HE, Kuehner, JA, Litherland, AE, Almqvist, E, Bromley, DA, Ferguson, AJ, Rose, PH, Bastide, RP, Brooks, N and Connor, RJ 1958 Neutron threshold measurements using the Chalk River Tandem Van de Graaff accelerator. *Physics Review Letters* 1: 251–253.

Gove, HE, Kubik, PW, Sharma, P, Datar, S, Fehn, U, Hossain, TZ, Koffer, J, Lavine, JP and Lee, S-T 1990 Applications of AMS to electronic and silver halide imaging research. *In* Yiou, F and Raisbeck, GM, eds, Proceedings of the 5th International Conference on Accelerator Mass Spectrometry. *Nuclear Instruments and Methods* B52(3,4): 502–506.

Gove, HE, Litherland, AE and Elmore, D, eds, 1987 Proceedings of the 4th International Symposium on Accelerator Mass Spectrometry. *Nuclear Instruments and Methods* B29.

Hedges, REM, White, NR, Wand, JO and Hall, ET 1980 Radiocarbon dating by ion counting: Proposals and progress. *In*

Stuiver, M and Kra, RS, eds, Proceedings of the 10th International ^{14}C Conference. *Radiocarbon* 22(3): 816–821

Hossain, TZ, Elmore, D, Gove, HE, Hemmick, TK, Kubik, PW and Jiang, S (ms) Neutron activation analysis/accelerator mass spectrometry measurements of nitrogen and chlorine in silicon. Paper presented at Workshop on Applications of Nuclear Physics Techniques to Condensed Matter Physics. American Physical Society, Rutgers, New Jersey, October 14, 1987.

Jope, M and Jope, M 1992 Radiocarbon in the biological sciences. *In* Taylor, RE, Long, A and Kra, RS, eds, *Radiocarbon After Four Decades: An Interdisciplinary Perspective.* New York, Springer-Verlag, this volume.

Kielson, J and Waterhouse, C 1978 Possible impact of the new spectrometric techniques on ^{14}C tracer kinetic studies in medicine. *In* Gove, HE, ed, *Proceedings of the 1st Conference on Radiocarbon Dating with Accelerators.* University of Rochester: 391–397.

Kutschera, W, ed 1981 *Proceedings of the Symposium on Accelerator Mass Spectrometry.* ANL/PHY 81-1. Springfield, National Technical Information Service: 1–15.

Kutschera, W, Henning, W, Paul, M, Stephenson, CJ and Yntema, JL 1980 Radioisotope detection with the Argonne FN tandem accelerator. *In* Stuiver, M and Kra, RS, eds, Proceedings of the 10th International ^{14}C Conference. *Radiocarbon* 22(3): 807–815.

Kutschera, W and Paul, M 1990 Accelerator mass spectrometry in nuclear physics and astrophysics. *Annual Review of Nuclear and Particle Science* 40: 411–438.

Libby, WF, Anderson, EC and Arnold, JR 1949 Age determination by radiocarbon contents: World wide assay of natural radiocarbon. *Science* 109: 227–228.

Litherland, AE 1980 Ultrasensitive mass spectrometry with accelerators. *Annual Reviews of Nuclear and Particle Science* 30: 437–473.

Long, A and Kra, RS, eds 1989 Proceedings of the 13th International ^{14}C Conference. *Radiocarbon* 31(3): 229–1082.

Muller, RA 1978 Radioisotope dating with the LBL 88" cyclotron. *In* Gove, HE, ed, *Proceedings of the 1st Conference on Radiocarbon Dating with Accelerators.* University of Rochester: 33–37.

Muller, RA, Alvarez, LW, Holley, WR and Stephenson, EJ 1977 Quarks with unit charge: A search for anomalous hydrogen. *Science* 196: 521–523.

Muller, RA, Stephenson, EJ and Mast, TJ 1978 Radioisotope dating with an accelerator: A blind measurement. *Science* 201: 347–348.

Naylor, H, Elmore, D, Clover, MR, Kilius, LR, Beukens, RP, Fulton, BR, Gove, HE, Litherland, AE and Purser, KH 1978 Determination of ^{36}CL isotopic ratios. *In* Gove, HE, ed, *Proceedings of the 1st Conference on Radiocarbon Dating With Accelerators.* University of Rochester: 360–371.

Nelson, DE, Korteling, RG and Stott, WR 1977 Carbon-14: Direct detection at natural concentrations. *Science* 198: 507–508.

Oeschger, H, Houtermans, J, Loosli, H and Wahlen, M 1970 The constancy of cosmic radiation from isotope studies in meteorites and on the Earth. *In* Olsson, IU, ed, *Radiocarbon Variations and Absolute Chronology.* Proceedings of the 12th Nobel Symposium. New York, John Wiley & Sons: 471–498.

Purser, KH 1977 US Patent 4037100.

Purser, KH and Gove, HE 1990 A new instrument for ultra-sensitive ^{14}C tracers. *Transactions of the American Nuclear Society* 62: 10–11.

Purser, KH, Liebert, RB, Litherland, AE, Beukens, RP, Gove, HE, Bennett, CL, Clover, MR and Sondheim, WE 1977 An attempt to detect stable N^- ions from a sputter ion source and some implications of the results for the design of tandems for ultra-sensitive carbon analysis. *Revue de Physique Appliquée* 12: 1487–1492.

Purser, KH, Liebert, RB and Russo, CJ 1980 MACS: An accelerator-based radioisotope measuring system. *In* Stuiver, M and Kra, RS, eds, Proceedings of the 10th International ^{14}C Conference. *Radiocarbon* 22 (3): 794–805.

Purser, KH, Litherland, AE and Gove, HE 1979 Ultra-sensitive particle identification systems based upon electrostatic accelerators. *Nuclear Instruments and Methods* 162: 637–656.

Raisbeck, GM, Yiou, F, Fruneau, M and Loiseaux, JM 1978 Beryllium-10 mass spectrometry with a cyclotron. *Science* 202: 215–217.

Schiffer, JP, Renner, TR, Gemmell, DS and Mooring, FJP 1978 Search for +1/3 e fractional charges in Nb, W and Fe metal. *Physical Review* D17: 2241–2244.

Schnitzer, R, Aberth, WA, Brown, HL and Anbar, M 1974 *Proceedings of the 22nd Annual Conference*. Philadelphia, ASMS: 64 p.

Schwarzchild, AZ, Thieberger, P and Cumming, JB 1978 A search for super-heavy elements in nature – the tandem Van de Graaff as a high sensitivity mass spectrometer. *Bulletin of the American Physical Society* 22: 94.

Shea, JH, Conlon, TW, Asher, J and Read, PM 1980 Direct Detection of ^{14}C at the Harwell tandem. *In* Stuiver, M and Kra, RS, eds, Proceedings of the 10th International ^{14}C conference. *Radiocarbon* 22(3): 830–837.

Stuiver, M and Kra, RS, eds 1980 Proceedings of the 10th International ^{14}C Conference. *Radiocarbon* 22(3): 565–1016.

_____1983 Proceedings of the 11th International ^{14}C Conference. *Radiocarbon* 25(2): 171–796.

_____1986 Proceedings of the 12th International ^{14}C Conference. *Radiocarbon* 28(2A & 2B): 175–1030.

Suter, M 1990 Accelerator mass spectrometry: state of the art in 1990. *In* Yiou, F and Raisbeck, GM, eds, Proceedings of the 5th International Conference on Accelerator Mass Spectrometry. *Nuclear Instruments and Methods* B52(3,4): 211–223.

Taylor, RE 1987 AMS ^{14}C dating of critical bone samples: Proposed protocol and criteria for evaluation. *In* Gove, HE, Litherland, AE and Elmore, D, eds, Proceedings of the 4th International Symposium on Accelerator Mass Spectrometry. *Nuclear Instruments and Methods* B29: 159–163.

Taylor, RE, Payen, LA, Prior, CA, Slota, PJ, Jr, Gillespie, R, Gowlett, JAJ, Hedges, REB, Jull, AJT, Zabel, TH, Donahue, DJ and Berger, R 1985 Major revisions in the Pleistocene age assignments for North American human skeletons by C-14 accelerator mass spectrometry: None older than 11,000 C-14 years BP. *American Antiquity* 50: 136–140.

Vogel, JS, Turteltaub, KW, Felton, JS, Gledhill, BL, Nelson, DE, Southon, JR, Proctor, ID and Davis, JC 1990 Application of AMS to the biomedical sciences. *In* Yiou, F and Raisbeck, GM, eds, Proceedings of the 5th International Conference on Accelerator Mass Spectrometry. *Nuclear Instruments and Methods* B52: 524–530.

Wilson, GW, Kilius, LR and Rucklidge, JC (ms) *In-situ*, parts-per-billion analysis of precious metals in polished mineral samples and sulphide "standards" by accelerator mass spectrometry. Manuscript submitted to *Geochimica et Cosmochimica Acta*.

Wölfli, W, Polach, HA and Anderson, HH, eds, 1984 Proceedings of the 3rd International Symposium on Accelerator Mass Spectrometry. *Nuclear Instruments and Methods* B5: 91–448.

Yiou, F and Raisbeck, GM, eds 1990 Proceedings of the 5th International Symposium on Accelerator Mass Spectrometry. *Nuclear Instruments and Methods* B52 (3,4): 211–630.

RADIOCARBON ACCELERATOR MASS SPECTROMETRY: BACKGROUND, PRECISION AND ACCURACY

ROELF P BEUKENS

INTRODUCTION

Since 1977, accelerator mass spectrometry (AMS) has made important contributions to radiocarbon dating, and many new fields of research were opened up. The obvious advantages of AMS dating with respect to background, precision and accuracy are now well known. The actual limitations and consequences of AMS dating will be discussed here.

AMS can detect approximately 1% of all ^{14}C present in a sample. This efficiency is 100 to 1000 times greater than the efficiency achieved by decay counting, and allows AMS to use milligram-size instead of gram-size samples. This small sample capability has captured most of the attention in AMS dating, as it makes possible the dating of a large number of samples that previously could not be analyzed reliably. Arguably of greater importance has been that AMS presents the opportunity for better selection of samples and sample material. For example, the ability to date macrofossils instead of bulk sediment, seeds or food remains in pottery instead of associated charcoal, has yielded more reliable and meaningful results.

Although work is still continuing on the development of small cyclotrons, all currently operating AMS facilities use a linear tandem accelerator. Figure 16.1 shows a typical layout of such a tandem accelerator mass spectrometer. The negative ions are created in a negative ion sputter source from solid graphite targets. As nitrogen does not form negative ions, no atomic N^- is present at this point. The only nitrogen left is in the form of NH^- molecular ions. Standard mass spectromagnetic elements select the mass 12, 13 and 14 for sequential or simultaneous injection into the accelerator. At this moment, many AMS laboratories cannot accelerate the large $^{12}C^-$ currents and only inject mass 13 and 14. The mass 14 beam mainly consists of $^{12}CH_2^-$ and $^{13}CH^-$ ions, and of course, some $^{14}C^-$. These beams are accelerated to the central electrode of the accelera-

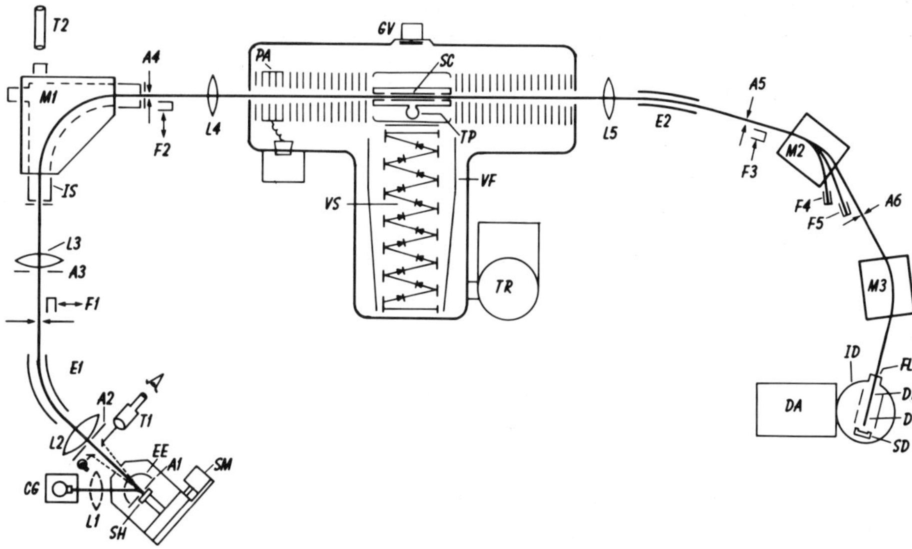

Fig 16.1. Schematic layout of the IsoTrace Tandetron AMS facility. No steerers are shown. Ion source: CG-cesium gun; L1-split einzel lens; SH-sample holder; SM-sample motion; EE-electric extraction; A2–phase space defining aperture; L2–einzel lens. Low-Energy-Mass Analysis: E1-electric analyzer; A3-phase space defining aperture; L3-einzel lens; M1-injection magnet; IS-electric isotope selector ("bouncer"). Accelerator: L4-matching einzel lens; PA-acceptance matching lens; SG-argon stripper gas cell; TP-recycling turbo pump; GV-rotating voltmeter used for stabilizing the terminal voltage; VS, TR and VF-accelerator power supply. High-Energy-Mass Analysis: L5-electric quadrupole lens; E2-energy and charge state defining electric analyzer; M2 and M3-magnetic analyzers; F4 and F5-^{12}C and ^{13}C-faraday cups; ID-ionization detector.

tor, which is usually at 2 MV for Tandetrons and 6 MV for Van de Graaff accelerators. The ions pass through the stripper gas cell or foils in the central electrode of the accelerator. Collisions with the stripper material remove several electrons, turning the negative carbon ions into positively charged ions. The accelerator functions as a molecular dissociator. If enough electrons are knocked off, hydride molecules will become unstable and fly apart as a result of the Coulomb forces. The remaining positive ions are accelerated again, and the correct energy and most prolific charge state is selected by a high-energy mass spectrometer. The most prolific charge state of the resulting carbon ions is +3 for 2 MV and +4 for 6 MV accelerators. At this point, the ^{12}C and ^{13}C currents are measured. The ^{14}C is then detected by a gas and/or solid-state ionization detector, which allows the ^{14}C ions to be identified by their energy and specific-energy loss, and separated from any background ions that might still be present.

In this chapter, emphasis will be placed mainly on the problems of background, precision and accuracy, as viewed from the perspective of the IsoTrace laboratory.

TABLE 16.1. Summary of Background Measurements at the IsoTrace Laboratory

Contamination source	Measured contribution
ME/q^2 ambiguities	Limits dating to 60,000 BP
E/q ambiguities	≤ 0.0004 pMC
Electronic noise	≤ 0.0004 pMC
Source contamination	~ 0.0025 pMC
Sample processing	0.08 pMC

BACKGROUNDS

The absence in AMS of cosmic radiation background eliminates the most important background component of decay counting; thus, older samples can be dated. As in other mass spectrometers, backgrounds from various sources do exist in AMS, and the knowledge and understanding of these backgrounds are essential for the reliable analysis of older samples. Several AMS laboratories have quoted measurements on old materials, but very little information exists on the level of environmental contamination of these materials. Some laboratories have tried to estimate the fixed and sample-size-dependent contributions to the background by performing contamination studies as a function of sample size. A more logical approach to the background problem, however, is to identify the individual contributions and measure them one by one. Such an analysis was done at the IsoTrace laboratory; Table 16.1 shows a summary of the results which are specific for the IsoTrace laboratory and do not necessarily apply to other laboratories.

Mass Spectrometric Backgrounds

Mass spectrometric background consists of atoms other than ^{14}C that mimic ^{14}C as a result of the finite resolution of mass spectrometric analysis. Figure 16.2 shows a magnetically analyzed spectrum of negative ions emitted by the negative ion sputter source from a carbon target. At mass 14, the $^{14}C^-$ is at least nine orders of magnitude less abundant than the interfering ions. The latter can be separated into the resonant interference, *ie*, ions that peak at mass 14, and the continuum interference, *ie*, ions that do not peak at mass 14 and are the result of tails of peaks at other masses. The resonant interferences are $^{13}CH^-$, $^{12}CH_2^-$, $^7Li_2^-$ and the hot molecule, $^{14}NH_2^-$, which decays in flight to $^{14}NH^-$. The most important continuum interferences are the $^{12}C^-$ and $^{13}C^-$ high-energy sputter tails characteristic of atomic ion sputtering of solids (Doucas 1977). Multiple charge exchanges in the accelerator create an additional source of continuum interference. Mass spectrometric selection provides information on the quantities, E/q and M/q of an ion, where E is the energy, M the mass and q the charge of

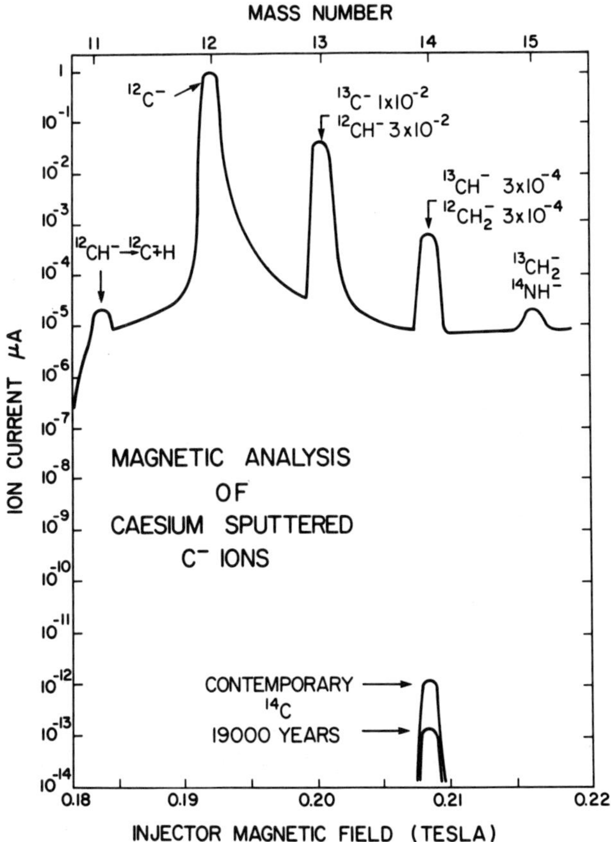

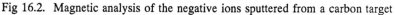

Fig 16.2. Magnetic analysis of the negative ions sputtered from a carbon target

the ion. These quantities can be determined in four ways:

1. Magnetic selection $(\mathcal{B}\rho)^2 = 2(M/q)(E/q)$
2. Electric selection $\mathcal{E}\rho = 2(E/q)$
3. Cyclotron selection $1/f = (2\pi/\mathcal{B})(M/q)$
4. Velocity selection $v^2 = 2(E/q)/(M/q)$.

In reality, cyclotron selection is not used on linear tandem accelerators, whereas velocity selection is obtained by a combined magnetic and electric $\mathcal{B} \times \mathcal{E}$ filter. Thus, all selections can be described by magnetic and electric analysis. The finite resolution of this analysis gives rise to interferences due to E/q and ME/q^2 ambiguities. The best place to look for them is in the particle spectrum obtained by the gas ionization detector. Figure 16.3 shows spectra of both a contemporary and an old sample, with three carbon peaks. All ions other than carbon have been removed from these spectra by means of the specific-energy-loss

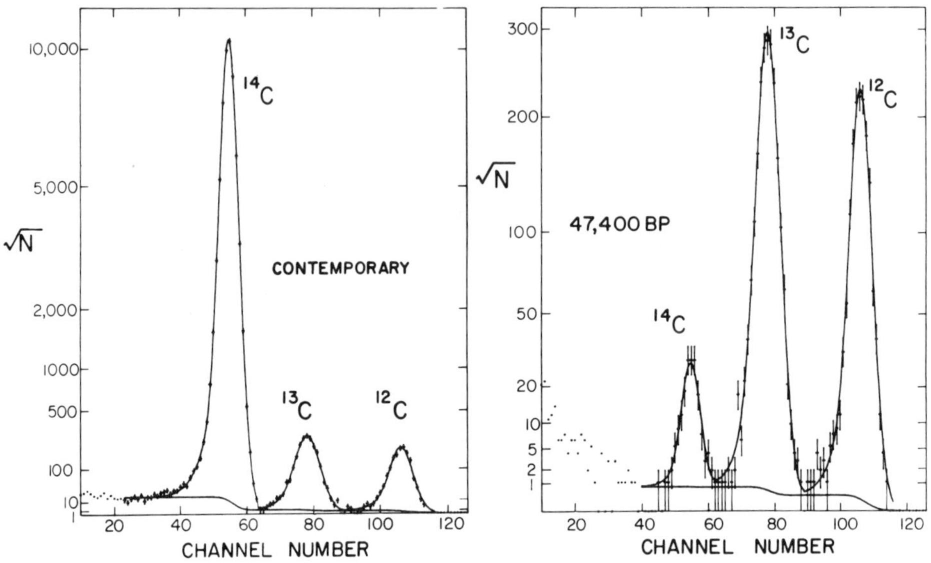

Fig 16.3. "E-final" (total energy – energy loss) spectra of a contemporary and an old sample. The vertical scale is the square root of the counts. The solid line represents the least-squares fit to the data of a standard spectral shape.

analysis. The difference between the specific-energy loss of the carbon isotopes is too small to allow effective removal of the ^{12}C and ^{13}C interferences from these spectra. The lowest energy peak is mainly due to ^{14}C particles, but can also include ^{12}C and ^{13}C E/q ambiguities. The two higher energy peaks are due to the ^{13}C and ^{12}C ME/q^2 ambiguities. This can easily be confirmed by injecting the $^{12}C^-$ beam into the accelerator while collecting the particle spectrum. This gives rise to two peaks, the E/q ambiguity at the same energy as the ^{14}C and the ME/q^2 ambiguity. A similar result can be obtained by injecting the $^{13}C^-$ beam. The E/q ambiguity is due to those particles selected by the electric analysis and partially passed by the magnetic analysis through small angle scattering on the residual gas in the vacuum of the magnetic analyzers. Similarly, the ME/q^2 ambiguity is due to those particles selected by the magnetic analysis and partially passed by the electric analysis through small angle scattering on the residual gas of the vacuum.

ME/q^2 ambiguities are not really a background, as the ion detector is energy dispersive, and the ^{14}C peak and the ^{13}C and ^{12}C ambiguity peaks are well separated. However, charge recombination in the detector creates a tail from these peaks which underlies the ^{14}C peak. This is an unavoidable complication, present in all ionization detectors. Standard spectrum analysis techniques, developed by nuclear physics to cope with this problem, can be used to remove the counts due to these tails as shown in Figure 16.3. The reduced statistical precision, which results from this analysis, limits the age of samples that can be analyzed at the IsoTrace laboratory to 60,000 BP. The simplest way to reduce or eliminate the ME/q^2 ambiguity is to install additional electric analysis in front of the particle detector.

The E/q ambiguity is a real background, because the ^{13}C and ^{12}C counts are indistinguishable from the ^{14}C counts. The main source of this ambiguity is the sputter tail of the $^{12}C^-$ peak (Fig 16.2). The existence of this continuum interference can be verified by varying the fields of the low-energy mass analysis around the mass 14 peak (Fig 16.4). A background was observed, underlying the mass 14 resonance, at a level of 0.008 pMC or 75,000 BP, which was due to the E/q ambiguity. The installation of a low-energy electric analyzer at the IsoTrace laboratory (Fig 16.1) reduced the $^{12}C^-$ sputter tail by a factor 100. This has also reduced the effect of the E/q ambiguity to $\leq$ 0.0004 pMC. Of course, the E/q ambiguity also can be reduced by additional magnetic analysis.

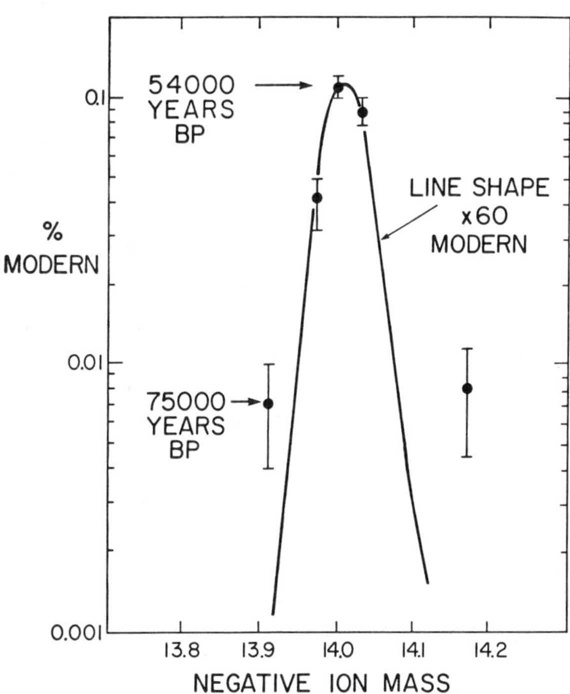

Fig 16.4. Measurement of the E/q ambiguities from the $^{12}C^-$ sputter tail. The shape of the mass 14 resonance was determined by the analysis of a 6000 pMC sample.

Electronic Noise

Electronic noise is mainly due to electric and microphonic noise in the particle detector and associated electronics, and can be observed in the low-energy channels of the particle spectrum for the old sample (Fig 16.3). Fortunately, this low-energy phenomenon is well separated from the ^{14}C peak in the energy dispersive particle spectrum. At IsoTrace, the contribution of this electronic noise was measured to be ≤ 0.0004 pMC by collecting a spectrum for several days without injecting any particles into the accelerator.

Source Contamination

The early results of 60,000 BP for old graphite samples (Bennett *et al* 1977) demonstrated the very low contamination and memory effects of the sputter ion source, commonly used in AMS radiocarbon dating.

Sample cross-contamination results when ^{14}C, sputtered from one sample, deposits in the vicinity of another sample, and is resputtered when this sample is being analyzed. It should be possible to avoid cross-contamination completely by proper ion-source design.

Contamination through ion-source memory effect is probably unavoidable, but regular cleaning of the ion-source components can keep this kind of contamination in check. Of all sputtered ^{14}C, only 10% or less is turned into negative ions, and the remaining ^{14}C atoms are deposited somewhere in the ion-source region. Eventually, the ^{14}C in the source will reflect the average of all previously analyzed samples, and the source is said to have a "memory" of these samples. The actual methods by which this ^{14}C is reintroduced into the system at a later time are not well understood, but the presence of this kind of contamination has been confirmed by unfortunate accidents with highly enriched samples at several AMS facilities. Such an accident occurred in 1985 at the IsoTrace laboratory, when a laser enrichment laboratory submitted a "background" sample, which, in fact, turned out to have a ^{14}C content of ≥ 20,000,000 pMC. All background samples run after this accident showed substantial contamination due to the memory effect. After standard ion-source cleaning, these background samples yielded again the same low levels as were obtained before the accident, demonstrating that standard source cleaning is effective in keeping the memory effect low.

A third type of contamination is from the hydrocarbons in the residual gas of the vacuum in the ion source. The main origins of these hydrocarbons are pump oil, "O"-ring and "finger" grease. Other possible sources include the interaction of atmospheric CO_2 or carbon from the above-mentioned memory effect with activated metal surfaces in the ion source in the presence of water vapor.

Measurements have yielded an apparent age of 13,000 to 7000 BP for these hydrocarbons, confirming the mixed old and modern origins of this type of contamination.

At the IsoTrace laboratory, we measured the ^{14}C generated from a blank aluminum target pellet to determine the contribution of all ion-source contaminations to an "infinitely old" sample, which was approximately 0.0025 pMC.

Contamination in Sample Processing

The total contribution from all sources of contamination discussed thus far is small enough to allow the contamination contributed by sample processing to be determined. In principle, this contamination should be negligible, as no extraneous carbon, such as in iron or oil on lithium pellets, needs to be introduced into the sample processing, and every step or chemical used can be checked independently. The simplest way to determine this contamination is to analyze a combusted or hydrolyzed ^{14}C-free sample (Table 16.2). In fact, very old samples are not necessarily ^{14}C-free, as contamination of such samples is very possible. As a result, many different samples had to be analyzed to determine their actual ^{14}C content. In AMS, a graphitization step is used in the processing, as most laboratories use solid graphite sputter sources. The contribution of the graphitization process was measured in commercially available CO_2, obtained by the combustion of natural gas ("cylinder" CO_2). The sum of graphitization and acid hydrolysis was measured on a marble sample, provided by the International Atomic Energy Agency in Vienna, and the sum of graphitization and combustion was measured on an anthracite sample, supplied by the Center for Isotope Research of the University of Groningen. The results of these three samples are statistically identical, implying that the acid hydrolysis and the combustion processes are contamination free. In the absence of cosmic-radiation background, the results in Table 16.2 look very low, but comparisons with some decay-counting laboratories indicate that their backgrounds due to sample processing are as low or lower. Although we have not been able to demonstrate the ideal "zero" contamination by sample processing,

TABLE 16.2 Sample Preparation Related Background Measurements at IsoTrace

Sample treatment	Material	^{14}C content (pMC)	Apparent age (yr BP)
Graphitization	CO_2 from natural gas	0.077 ± 0.005	57,630 ± 540
Acid hydrolysis	Marble	0.076 ± 0.009	57,690 ± 920
Combustion	Anthracite	0.081 ± 0.019	57,190 ± 1860

TABLE 16.3. High-precision intercomparison with the Quaternary Center (QRC), University of Washington

QRC results (yr BP)	IsoTrace results (yr BP)
4132 ± 18	4157 ± 21
6973 ± 20	7019 ± 20
≥ 55,000	53,100 ± 600

the limits are low enough to concentrate on the most important part of radiocarbon dating, the sample pretreatment, where the datable material is separated from the inevitable contamination. More than anything else, this determines the reliability of radiocarbon dating.

PRECISION AND ACCURACY

A 1.6 μA C^- beam from a contemporary sample contains 12 $^{14}C^-$ ions per second. As most ion sources can easily produce 5 to 50 μA, a precision of better than 1% can be obtained in a few hours. A precision estimate by itself is not significant, however, if it does not represent the real variance in the data. This is usually tested by analyzing a sample many times and averaging the results. In AMS dating, a precision should be calculated by combining the errors from counting statistics and current measurements. Beukens, Gurfinkel and Lee (1986) have shown that a precision calculated in this way accounts for all the variance in the data and that no other sources of errors should be present. The ^{14}C dating accuracy of several high-precision decay-counting laboratories involved in dendrocalibration has been demonstrated by the intercomparison of over a thousand samples. Table 16.3 summarizes a similar high-precision intercomparison between one of these laboratories and the IsoTrace Laboratory (Beukens 1990). The IsoTrace results were corrected for sample fractionation as well as sputter fractionation to a base of $\delta^{13}C = -25‰$, as described by Beukens, Gurfinkel and Lee (1986). The agreement is satisfactory, as it demonstrates that AMS can be as accurate as decay counting. Of course, a data base of a few examples is very small compared to the over 1000 decay-counting intercomparisons. AMS still has a long way to go before it can claim the same reliability as has been demonstrated by some of the decay-counting laboratories.

ACKNOWLEDGMENTS

The support of the Natural Sciences and Engineering Research Council of Canada through an Infrastructure Grant is gratefully acknowledged. I also wish to thank the sample preparators and the support staff from the IsoTrace Laboratory for their dedicated work. The intercomparison samples were kindly provided by Minze Stuiver.

REFERENCES

Bennett, CL, Beukens, RP, Clover, MR, Gove, HE, Liebert, RB, Litherland, AE, Purser, KH and Sondheim, WE 1977 Radiocarbon dating with electrostatic accelerators: Negative ions provide the key. *Science* 198: 508–510.

Beukens, RP 1990 High precision intercomparison at IsoTrace. *In* Scott, EM, Long, A and Kra, RS, eds, Proceedings of the International Workshop on Intercomparison of Radiocarbon Laboratories. *Radiocarbon* 32(3): 335–339.

Beukens, RP, Gurfinkel, DM and Lee, HW 1986 Progress at the IsoTrace radiocarbon facility. *In* Stuiver, M and Kra, RS, eds, Proceedings of the 12th International ^{14}C Conference. *Radiocarbon* 28(2A): 229–236.

Doucas, G 1977 Energy distribution of negative ions sputtered from caesiated surfaces. *International Journal of Mass Spectrometry and Ion Physics* 25: 71.

HYDROLOGY

PREFACE

WILLEM G MOOK

This section reviews the state of the art of applying radiocarbon to groundwater dating.

Radiocarbon dating of groundwater, which started with the original work of the Heidelberg Radiocarbon Laboratory, back in the late 1950s, is certainly one of the most, if not the most complicated and often questionable application of radiocarbon dating. The reason for this is to be found in the aqueous geochemistry of carbon in the unsaturated and saturated zones. Part of the carbon and radiocarbon content of dissolved inorganic carbon in groundwater is of inorganic and part is of organic origin.

One approach, which was adopted by the original researchers, is to study the carbon chemistry as well as the stable carbon isotopic composition, $^{13}C/^{12}C$, of groundwater. The first simple models were subsequently extended to sophisticated dissolution/isotopic exchange models for the unsaturated zone, and to models that trace the further chemical evolution of the groundwater in the saturated zone. In several published studies, this approach has been successful.

A second approach is concerned with combining radiocarbon data on groundwater samples in a dynamic modeling procedure. This phenomonological viewpoint provides direct information on groundwater movement, which is, in many cases, the most essential aspect, rather than on ages.

A third and more recent approach involves dating dissolved organic, instead of inorganic, constituents. The following, in my view, excellent chapters by some active scientists in this field discuss the fundamental and applied aspects of recent developments and the state of the art of radiocarbon applications in hydrogeological studies.

With its shortcomings on the one hand, and successes on the other, radiocarbon does not guarantee definite answers to hydrogeological questions. From a scientific point of view, the problem of the aqueous occurrence of radiocarbon underground probably is the most intriguing.

17

CHEMICAL AND ISOTOPIC CONSTRAINTS ON ^{14}C DATING OF GROUNDWATER

JEAN-CHARLES FONTES

INTRODUCTION

The conversion of the ^{14}C activity of the total dissolved inorganic carbon (TDIC) in terms of groundwater residence time is an interpretation requiring the discussion of several constraints. This chapter is a review of some geochemical conditions that must be considered for groundwater dating.

In the case of dating solid carbonates, it is not completely precluded that future analytical developments would permit the direct measurement of radiogenic ^{14}N in carbonate crystal, allowing the application of a parent-daughter method to derive "absolute" ^{14}C ages. The impossibility of distinguishing ^{14}N, the daughter product of ^{14}C, from the common dissolved nitrogen in groundwater, makes the ^{14}C clock strictly dependent on knowledge of the initial activity of the aqueous carbon. The initial activity A_0 of the TDIC is defined as its ^{14}C content after all chemical and isotopic processes have taken place and before any decay (Wigley 1975). Time value will refer to the decrease of A_0 by decay.

The main questions addressed below for the interpretation of the data are:

1. What proportion of the TDIC derives directly from atmospheric CO_2, via the soil CO_2, at the time of infiltration?
2. To what extent do isotope exchange reactions modify the ^{14}C content of the TDIC?

However, the discussion of the following important problems is beyond the scope of this review, and will not be considered here:

1. Possible delay in the atmosphere-soil-aquifer passage which would lead to a significant aging
2. Possible extra sources of active carbon in the aquifer itself
3. Validity of the extrapolation of present conditions of recharge/discharge processes of the aquifer to the past.

Further symbols and notations used in this paper are

g – soil CO_2; c – solid carbonates
A – ^{14}C activity (in % modern); δ – ^{13}C content in ‰ PDB
a, b and T – H_2CO_3, HCO_3^- and TDIC concentrations, respectively
$\varepsilon_{i\text{-}j}$ – stable isotope enrichment factor between species i and j ($\varepsilon_{ij} \approx \delta_i - \delta_j \approx -\varepsilon_{ji}$)

Temperature dependences (T in K) of various ε values are given by Mook (1974, 1980)

$\varepsilon_{ag} = -0.373 \times 10^3T^{-1} + 0.19$; $\varepsilon_{gb} = -9.483 \times 10^3T^{-1} + 23.89$; $\varepsilon_{cb} = -4.232 \times 10^3T^{-1} + 15.10$ (for calcite).

MODIFICATION OF ATMOSPHERIC ACTIVITY IN TDIC

Dilution

Data on the geographical distribution of atmospheric ^{14}C before thermonuclear testing are available from accelerator mass spectrometry (AMS) measurements on gaseous inclusions in ice (van de Wal *et al* 1990; Wilson & Donahue 1990), but mainly from ^{14}C analyses of tree rings (Stuiver & Kra 1986). Although temporal variations of up to a few per mil occurred regularly, it is likely that the ^{14}C concentration of the atmospheric reservoir has always been uniform at a given time. The global homogenization of the thermonuclear peak of 1963 documented the rather thorough tropospheric mixing of CO_2 (Levin, Münnich & Weiss 1980; Nydal & Lövseth 1983; Levin *et al* 1989).

Atmospheric $^{14}CO_2$ may reach the water table in several ways, including dissolution of CO_2 from root respiration, decay of organic matter and direct diffusion through the unsaturated zone, in the case of high pH waters. Under normal geochemical conditions, the main source is soil CO_2, which dissolves into the complex form of $H_2CO_3^0$ and aqueous CO_2, both generally grouped as H_2CO_3. Except at high pH, H_2CO_3 is present in groundwater in sufficient amounts to theoretically allow for micromeasurement by AMS. However, H_2CO_3 is not easily extracted from water without displacing the bicarbonate equilibrium. Further, in solutions that have been evolving through carbonate precipitation, part of the H_2CO_3 is derived from the bicarbonate according to the reaction opposite to that of dissolution

$$2HCO_3^- + Ca^{2+} = CaCO_3 + H_2CO_3 \ . \tag{17.1}$$

In this case, the problem of evaluating A_o of H_2CO_3 is the same as for the TDIC.

After its formation, the carbonic acid may dissolve solid carbonates (eg, reaction 1 from right to left). The active carbon from the soil is thus mixed with another source of carbon which may be old. Theoretically, initial activity can be corrected for this dilution from the chemical balance (Geyh & Wendt 1965; Tamers 1967) or from the isotopic balance (Vogel & Ehhalt 1963; Ingerson & Pearson 1964; Geyh & Wendt 1965), which both account for the mixing between these two sources of carbon. These balances may be expressed by the following rigorous equations (Fontes & Garnier 1979)

$$A_o = A_T = [(a + 0.5b)\ A_g + 0.5b\ A_c]/(a + b)\ , \qquad (17.2)$$

for the chemical balance (disregarding the presence of CO_3^{2-} for pH values close to neutral), and

$$A_o = A_T = [(A_g - A_c)(\delta_T - \delta_c)/(\delta_g - \delta_c)] + A_c\ , \qquad (17.3)$$

for the stable isotope balance.

For purposes of simplicity, these approaches are generally known as the Tamers (Eq 17.2) and Pearson models (Eq 17.3).

The assumptions underlying these treatments are similar. The aqueous carbon derives from simple mixing between two sources: the gaseous CO_2 of the soil zone and the solid carbonate, which is dissolved. No other loss of carbon and of carbon isotopes is possible, even after a long flow time except by radioactive decay. Implications are: 1) the solution is undersaturated with calcite or any other carbonate species; 2) the dissolution of CO_2 (gas) in the soil zone occurs without any isotope exchange with bicarbonate; 3) the solid carbonate of the matrix can only dissolve and does not exchange isotopes with the TDIC. This is often considered to be the case if the dissolution occurs below the water table where the excess of H_2CO_3 may dissolve some carbonate from the matrix. Although the TDIC may thus increase along the flow paths, these conditions are often improperly referred to as "closed-system conditions." The situation is even more complex if saturation is reached through calcium sulfate dissolution and/or dedolomitization (Plummer *et al* 1990) and a gas phase may appear (Deines, Langmuir & Harmon 1974).

The two models described above would produce overcorrected results and low A_o values if isotopic exchange occurred between the TDIC and the gas phase. Conversely, the models would give undercorrected and high A_o values if isotopic exchange takes place between the matrix carbonate and the dissolved species of inorganic carbon. However, Tamers' and Pearson's models often give reasonable and rather similar corrections (see Fontes & Garnier 1979; Fontes 1983, 1985, for a discussion) when the partial pressure of soil CO_2 is rather low, giving rise to low alkalinity values in the recharge zone ($\leq$ 2 to 3 mmol.kg^{-1} of HCO_3^- with a pH close to neutral).

Plummer (1977) proposed a complete chemical and isotopic approach in which the possible reactions of dissolution and precipitation are reconstructed step-by-step along the flow path. The influence of these reactions on the balance of aqueous carbon and of ^{13}C and ^{14}C is then evaluated (Plummer 1977; Wigley, Plummer & Pearson 1978). Plummer *et al* (1990) have recently developed an extension of this approach including redox reactions for sulfur species. Reardon and Fritz (1978) also coupled the chemical balance with a stable isotope balance so that the evolved values of groundwater alkalinity, pH and stable isotope composition could be related to acceptable initial conditions of temperature, pCO_2, pH and $\delta^{13}C$ of the gas phase in the recharge zone. Both these models allow for equilibrium of the TDIC with the gaseous CO_2 in the soil zone before the TDIC reaches the water table.

Isotope Exchange Reactions

The pure mixing models require modification to account for the possibility of exchange reaction during or after mixing occurs. The isotopic reaction

$$^{14}CO_2 + H^{12}CO_3^- = {}^{12}CO_2 + H^{14}CO_3^- \, , \qquad (17.4)$$

was considered from the early studies (Münnich 1957; Münnich & Vogel 1959; Vogel & Ehhalt 1963) as responsible for ^{14}C contents significantly higher than those indicated by the single dissolution process in modern groundwater. A value of 85 ± 5% for the initial activity of the TDIC was proposed as a standard value (Vogel & Ehhalt 1963) to account for the sum of mixing + exchange processes. This correction parameter produced acceptable results and is still extensively used. It is remarkable that the pioneering studies of Münnich and co-workers encompassed also the possibility of an isotope exchange with the solid phase (see Münnich & Roether 1963), for example, according to the reaction

$$H^{14}CO_3^- + Ca^{12}CO_3 = H^{12}CO_3^- + Ca^{14}CO_3 \; . \qquad (17.5)$$

MODELS FOR EXCHANGE MIXING PROCESSES

Concepts and Model Formalization

In a step-by-step approach, Mook (1972, 1976, 1980) considers partial or complete isotopic equilibration in the soil zone between the gas phase and all the carbon species involved in the reaction, including part of the dissolving carbonate. This treatment provides consistent results in systems where solid carbonates are not dominant (Mook 1976). Wallick (1976) considers the exchange in the soil zone, assuming that equilibrium is reached between H_2CO_3, HCO_3^- and the soil CO_2 before mixing occurs with old carbon from calcite dissolution. The latter is indicated by the Ca^{2+} balance corrected for the contributions of gypsum dissolution and base exchange.

The model (F & G) proposed by Fontes & Garnier (1976, 1977, 1979; Garnier & Fontes 1980; Fontes 1983, 1985) describes a complete exchange of a fraction of the TDIC with either the gas phase (Eq 17.4) or the solid (carbonate) phase (Eq 17.5). The amount of solid carbonate that is dissolved is taken, at any pH value, as half of the total carbonate alkalinity.

The final mathematical expressions for these models are somewhat similar in form. They include a mixing term based on the chemical balance (Tamers' correction of Eq 17.2) and an additive term, k, which may be considered to account for the isotope exchange relative to the Tamers model

$$A_o(\text{corrected}) = A_o(\text{Tamers}) + k \ . \qquad (17.6)$$

Expressions of k are

$$k(\text{Mook}) \approx 0.5(A_g - A_c)[\delta_T(a+b)-(a+0.5b)\delta_g-0.5b\ \delta_c]/[0.5(\delta_g-\delta_c)-\varepsilon_{gb}](a+b) \qquad (17.7)$$

$$k(\text{F \& G}) \approx (A_g - A_c)\ [\delta_T\ (a+b) - (a+0.5b)\delta_g - 0.5b\ \delta_c]/[\delta_g - \varepsilon - \delta_c]\ (a+b) \qquad (17.8)$$

where $\varepsilon_{gb} \approx \delta_g - \delta_b$ is the isotope enrichment factor between CO_2 (gas) and HCO_3^- (Mook, Bommerson & Staverman 1974; Mook 1980; Lesniak & Sakai 1990). As the exchange occurs through the aqueous phase, the F & G model is applied using ε_{gb} for ε in a first run. A positive value of k in this run means that the exchange takes place with the gas phase (Eq 17.4) and the calculation is then made. If k is negative, an exchange with the solid phase (Eq 17.5) is suggested. In that case, Equation (17.8) still applies (Garnier & Fontes 1980), substituting ε_{gb} by the isotope enrichment factor $\varepsilon_{bc} \approx \delta_b - \delta_c$ between the bicarbonate and the solid carbonate (taking $\varepsilon_{bc} \approx -\varepsilon_{cb}$ as derived by Mook (1980), or adopting the value $\varepsilon_{cb} = +0.9‰$ given by Rubinson and Clayton (1969)).

An extreme situation for the F & G model is such that the carbonic acid and the bicarbonate directly derived from the gas phase are in equilibrium with the latter. The term $(a + 0.5b)\delta_g$ from Equation (17.8) becomes

$$(a + 0.5b)\delta_g = 0.5b(\delta_g - \varepsilon_{gb}) + a(\delta_g + \varepsilon_{ag}) \ , \qquad (17.9)$$

where $\varepsilon_{ag} \approx \delta_a - \delta_g$ accounts for the fractionation between H_2CO_3 and the CO_2 gas (Vogel, Grootes & Mook 1970).

The term $(a + 0.5b)A_g$ from Equation (17.2) becomes

$$(a + 0.5b)A_g = A_g (1 + \varepsilon^{14}{}_{ag})a + 0.5b (1 + \varepsilon^{14}{}_{gb})A_g \qquad (17.10)$$

where ε^{14}_{ij} is the radiocarbon enrichment factor between species i and j, with (Saliège & Fontes 1984)

$$\varepsilon^{14}_{ij} \approx 2.3\varepsilon^{13}_{ij}, \qquad (17.11)$$

when expressed in decimal fraction (and not in per mil). This amendment referred to as F & G equilibrium (F & G equil) leads to higher values for A_o than the other form (F & G), which allows for a simple dissolution of an aliquot of the gas phase.

The following should be noted: 1) one would have expected the correction to be expressed in the form of a multiplying factor, rather than by the additive term k; 2) k is a ratio for which the denominator may be reduced, leading to large and sometimes over-corrected values (Fontes & Garnier 1979); 3) conditions of complete isotope exchange in the "open-system conditions" (see below) cannot be derived from the models, despite the undoubted existence of a physical and chemical continuum between "open-" and "closed-system" conditions. As discussed by Fontes & Garnier (1979) and Fontes (1983, 1985), these shortcomings complicate the use of any of the correction models.

An attempt to simplify the concept of the exchange-mixing model was introduced in a study made by the IAEA (Salem *et al* 1980) who, like Mook (1972), proposed that Pearson's mixing model may be considered as two steps: 1) equilibrium of HCO_3^- with the gas phase followed by 2) dissolution of solid carbonate. In its complete form, the final equation is

$$A_o = [\delta_T - \delta_c)(A_g - A_c) + (\delta_g - \varepsilon_{gb} - \delta_c)A_c]/(\delta_g - \varepsilon_{gb} - \delta_c) \ . \qquad (17.12)$$

This model implies that all chemical reactions may be accounted for by the stable isotope content of the end members (pure mixing), one of them being completely exchanged with the gas phase. As ε_{gb} is < 0, this treatment gives higher A_o values than Pearson's mixing model. For instance, for $A_c = 0$ (dissolution of dead carbonate), $A_g = 97\%$ (pre-bomb soil CO_2, see below), $\delta_c = 0‰$ (dissolution of marine carbonate), $\delta_g = -20‰$ (C_3 vegetation cover), $\varepsilon_{gb} = -7.9‰$ (25°C, Mook, Bommerson & Staverman 1974) and $\delta_T = -12‰$, the

IAEA model gives A_o = 96.2%, whereas Pearson's model gives A_o = 58.2% and an apparent age younger by 4150 years.

Evans *et al* (1979) propose an interesting approach to account for the dissolution-precipitation processes that occur in the presence of an infinite reservoir of carbonate. Here precipitation is considered to cause fractionation $\delta_c = \delta_b + \varepsilon_{cb}$, whereas dissolution does not.

The integration of a step-by-step process leads to a correction factor

$$A_g/A_o \approx m = [(\delta_c - \varepsilon_{cb} - \delta_g)/(\delta_c - \varepsilon_{cb} - \delta_T)]^{(1 + 10^{-3}\,\varepsilon_{cb})} \;. \qquad (17.13)$$

As ε_{cb} is small ($|\varepsilon_{cb}| \leq 1‰$ in the temperature range 0–25°C, Mook (1980)), Equation (17.13) reduces to

$$A_o \approx A_g\,[(\delta_T + \varepsilon_{cb} - \delta_c)/(\delta_g + \varepsilon_{cb} - \delta_c)] \;. \qquad (17.14)$$

Thus, in its approximate form, this model is very close to the Pearson mixing model[1], and produces similar A_o values despite the fact that the two derivations start from quite different concepts.

Eichinger (1983) proposes also a correction for the partial isotope exchange with carbonates of the aquifer matrix. The Tamers' value of A_o (Eq 17.2 with A_c = o) is corrected according to

$$A_o = A_o\,(\text{Tamers}) \times B \;. \qquad (17.15)$$

The correction factor B is the fraction of the TDIC that is not exchanged with the rock

$$B = (\delta_T - \delta_e)/(\delta_i - \delta_e) \qquad (17.16)$$

where δ_i is the ^{13}C content of the TDIC after completion of the dissolution process and before any exchange

$$\delta_i = [a\delta_a + 0.5b\,(\delta_a + \delta_c)]/(a + b), \qquad (17.17)$$

and where δ_e is the ^{13}C content of the TDIC in equilibrium with the solid carbonate evaluated through the appropriate isotope enrichment factors

$$\delta_e = [a(\delta_c - \varepsilon_{cb} + \varepsilon_{gb} + \varepsilon_{ag}) + b(\delta_c - \varepsilon_{cb})]/(a + b) \;. \qquad (17.18)$$

[1]A more complete derivation of the Evans *et al* (1979) approach would be even more comparable to Pearson's equation as: $A_o \approx [A_g - A_c]\,[(\delta_T + \varepsilon_{cb} - \delta_c)/(\delta_g + \varepsilon_{cb} - \delta_c)] + A_c$.

This correction is interesting because of its mathematical form which employs a multiplying factor. However, it is designed for isotope exchange reactions with solid carbonates only, and the fraction B may become larger than one if an isotope exchange occurred with the gas phase.

Field Performance of Models

Validity tests of the models are not easy to perform in aquifers because of the various possible origins of the flow paths. The unsaturated zone may theoretically provide more favorable conditions of cylindrical piston flow with known boundary conditions. Despite the complications introduced by the co-existence of three phases containing carbon, a study was made in the unsaturated zone of the chalk from Champagne (Moulin 1990; Dever & Fontes 1990). This carbonate environment was selected for its homogeneity and its very high specific surface (average pore diameter 0.2 μm) favorable to exchange processes. The depth of the tritium peak shows that it recharges in a pure piston flow with a rate of 0.75 ma^{-1} for a porosity of 0.42. Concentrations, ^{13}C and ^{14}C (AMS) of the TDIC were determined on samples of soil solutions obtained from suction cups at various depths between the surface and 20 m (average depth of the water table). The soil atmosphere was also sampled at the root zone at the same depths for isotope and partial pressure measurements. Natural vegetation has covered the site for several decades, and steady (seasonal) conditions may be assumed for CO_2 release. The average ^{13}C content of the gas samples collected at −1 m and −2 m is −21.95‰ *vs* PDB. We adopted a value of −22‰ for model calculations.

Samples of the chalk were obtained to a depth of 10.5 m. The ^{13}C content is very constant ($\delta^{13}C$ = +2.20‰ *vs* PDB, s = 0.15, n = 11) and very close to the value of +2.35 found for British chalk (Smith *et al* 1976), which is equivalent to that of Champagne. The ^{14}C activity values along the profile range from 1 to 3% modern (Dever 1985), indicating that isotope exchange and/or precipitation of secondary carbonates occurs.

Previous studies (Dever *et al* 1982) showed no significant delay between photosynthesis and release of CO_2 in the soil zone. We thus disregarded the possible generation of extra CO_2 by further oxidation of dissolved organic matter or of organic particulates in the unsaturated zone. The ^{14}C content of the soil gas at the estimated infiltration time, can thus be evaluated for each sample depth, from the recent ^{14}C variations of the atmospheric CO_2 (Levin, Münnich & Weiss 1980; Nydal & Lövseth 1983).

Calculations (Tables 17.1, 17.2, Fig 17.1) show that the values of the initial activity of the TDIC are underestimated by the mixing models (Tamers, Pearson,

TABLE 17.1. Unsaturated zone of the chalk of Champagne (Moulin 1990): variables and parameters for calculations of A_o according to the models

Depth m	Temp °C	H_2CO_3 $mmol.kg^{-1}$	HCO_3^- $mmol.kg^{-1}$	$\delta^{13}C$(TDIC) ‰PDB	$A^{14}C$(TDIC) % modern	$A^{14}C(CO_2$ gas) % modern*	$\delta^{13}C(CO_2$ gas) ‰ PDB**
02	8.7	0.18	3.10	−11.55	117.5	127	−20.6
05	10.2	0.19	3.35	−10.66	96.9	130	−19.8
08	9.6	0.21	4.65	−8.76	88.7	135	−18.0
11	9.6	0.19	3.20	−8.53	90.0	145	−17.6
14	8.8	0.21	2.95	−10.16	85.8	155	−19.1
16	9.2	0.52	2.80	−10.09	73.2	170	−18.1
20	9.7	0.71	2.95	−10.90	67.1	140	−18.5

*^{14}C activity of the gas phase at the time of infiltration is evaluated from the atmospheric distribution of $^{14}CO_2$ and from the depth reached by the samples in a piston-flow movement (see text).
**^{13}C content of the gas phase in equilibrium with the TDIC is calculated from Equation (17.19).

TABLE 17.2. Unsaturated zone of the chalk of Champagne. Calculated values of the initial activity of the TDIC*

Depth (m)	A_{TDIC}	F & G equil	AIEA	Tamers	Pearson	F & G	Eichinger	Evans
02	117.5	117.7	120.4	68.4	73.5	77.0	70.5	71.9
05	96.9	112.6	115.4	69.9	70.5	70.9	67.2	68.8
08	88.7	100.4	102.6	71.9	62.8	62.7	59.0	60.8
11	90.0	104.0	107.9	77.9	66.0	65.9	61.7	63.9
14	85.8	127.3	132.3	84.1	80.6	80.6	76.4	78.8
16	73.2	126.9	144.3	99.6	87.8	87.8	81.3	85.9
20	67.1	108.5	126.5	84.8	77.2	77.1	71.5	75.5

*Calculations made for an average temperature of 9.4°C with constant values of δ_c = +2.20‰, A_c = 3% for ^{13}C and ^{14}C contents of the chalk matrix, and a value of δ_g −22‰ for the ^{13}C content of the soil CO_2.

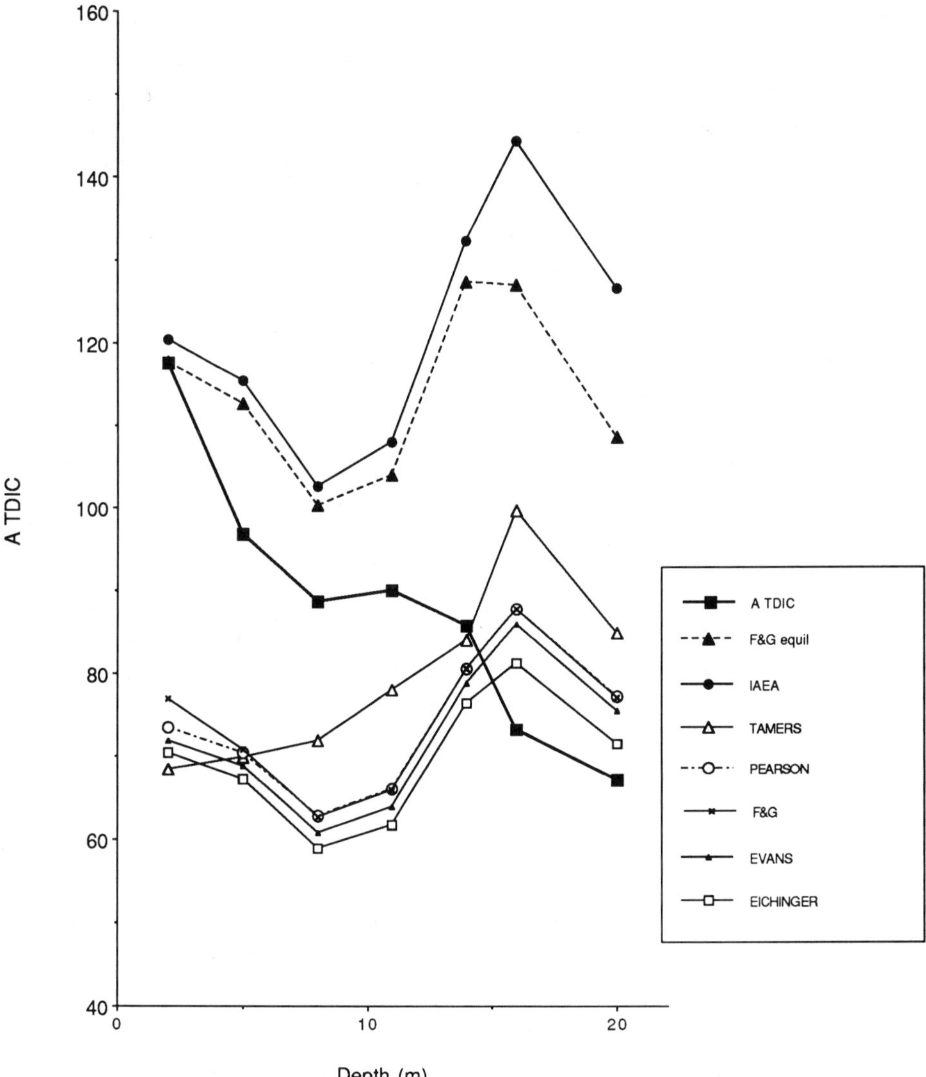

Fig 17.1. Evolution of the calculated values of the activity of the TDIC according to the various models and comparisons with the measured values in samples of pore fluid from the unsaturated zone of the chalk.

Evans) in the upper part of the profile and overestimated in its lower parts. The F & G model and the Eichinger model behave similarly, the latter producing the lowest values for A_0. This consistency reflects the influence of isotope exchange reactions with the solid phase. The Mook model, which is designed to account for exchange with CO_2, cannot be used in this context. Although agreement between calculated and measured values is poor, one must note that the models describe the general trend of the observed activity values for the TDIC to a depth of 8 m and below 16 m. A parametric adjustment could be sought to obtain better agreement with measured values. However, another phenomenon occurs between 8 and 16 m, which overwhelms the exchange-mixing process, *ie*, a second stage of matrix dissolution by the CO_2 released by the precipitation (reaction 1). In that case, the initial conditions adopted for A_0 no longer apply. A treatment of this transient state will be given elsewhere together with the general discussion of these results.

The IAEA and F & G (equil) models are able to describe the first stages of the process where the exchange occurs with the gas phase, but they produce excessively high values for the ^{14}C activity of TDIC at depths where the exchange with the matrix becomes progressively predominant. In the vicinity of the profile, the ^{14}C activity of the groundwater TDIC (A_T = 62.5% modern, δ_T = −10.5‰ *vs* PDB) approximates that calculated from the Eichinger and F & G pure mixing models using a ^{14}C of 120% for the gas phase. Although it is not possible to check experimentally the latter because of the complex flow near the water table, it may be concluded that these approaches reasonably evaluate the ^{14}C activity of the groundwater from carbonate-dominated systems. However, no model using the initial conditions of the root zone can yet adequately estimate the ^{14}C activity of TDIC during transport through the unsaturated zone.

OPEN SYSTEMS

Soil CO_2-Dominated Systems

If the CO_2 production in the soil zone is high (partial pressure 10^{-1} to 10^{-2}), the gas phase controls all isotopic reactions involving inorganic carbon. This is often defined as "open-system" conditions (Wigley 1975) by reference to the chemical situation in which the partial pressure of CO_2 controls the chain of equilibria. The isotope content of the components of the carbon system will differ from that of the CO_2 (gas) by the enrichment factors of the various reactions between gas, H_2CO_3 (ε_{ag}) and HCO_3^- (ε_{bg}). Under conditions of high CO_2 partial pressure, the species CO_3^{2-} can be disregarded.

For ^{13}C, the mass-balance equation is

$$\delta_{TDIC} = \delta_T = \Sigma m_i \, \delta_i / \Sigma m_i \qquad (17.19)$$

where m and δ are the molal concentration and ^{13}C content, respectively, of species i (H_2CO_3 and HCO_3^-) with $\delta_i \approx \delta_g - \varepsilon_i^{13}$, where ε_i^{13} refers to the ^{13}C enrichment factor between CO_2 gas and species i (ε_{gb}^{13} and $\varepsilon_{ga}^{13} \approx -\varepsilon_{ag}$).

For ^{14}C activities, Equation (17.19) is

$$A = \Sigma m_i A_i / \Sigma m_i \tag{17.20}$$

with

$$A_i = A_{CO_{2(g)}}/(1 + \varepsilon_i^{14}) \sim A_g\,(1 - \varepsilon_i^{14}) \ . \tag{17.21}$$

From Equation (17.18), and using the temperature dependence of ε_i given by Mook, Bommerson & Staverman (1974), we can calculate the stable isotope content of the CO_2 (gas) in equilibrium with the TDIC. If the calculated values are compatible with expected values for soil gases, Equations (17.21) and (17.11) can be used to estimate A_{TDIC} before any decay, *ie*, A_o.

Numerous data are now available for the stable isotope content of soil gases under different types of vegetation and climate (eg, Galimov 1966; Rightmire & Hanshaw 1973; Rightmire 1978; Fritz *et al* 1978; Reardon, Allison & Fritz 1979; Dörr & Münnich 1980; Dever *et al* 1982; Parada, Long & Davis 1983). As observed by Dörr and Münnich (1980), diffusion processes from the root zone give rise to an enrichment of soil gases by about 4.4‰ in ^{13}C compared to the organic matter from which they derived. Typical values for soil covered with Calvin-Benson type (C_3 type) vegetation are −20 to −22‰. For tropical soils, $\delta^{13}C$ values of soil CO_2 are often dependent on the importance of the cultivation of millet or sugar cane (Hatch – Slack, C_4 plants) and on the contribution of grass species that commonly follow C_4 cycles in arid and semi-arid regions (Winter, Troughton & Card 1976; Vogel, Fuls & Ellis 1978). Values ranging from −10 to −17‰ are common. Plants such as Euphorbiacae, obeying CAM photosynthetic cycles in semi-arid regions, lead to ^{13}C contents of soil gases close to −18‰, as in the Canary Islands (Gasparini *et al* 1990). Each study should thus require a site-specific discussion of the vegetation cover at the time of infiltration, which is precisely the unknown factor that studies aim to evaluate. However, ^{14}C concentrations higher than 100% provide evidence for the control of the equilibria by the gas phase in recently recharged groundwater systems. Such is the case in alluvial deposits from the Kalahari Desert (Mazor *et al* 1974); in sandstone from eastern Senegal (Castany *et al* 1974), in crystalline rocks from northeast Brazil (Salati, Menezes Leal & Mendes Campos 1974) and from Niger (Ousmane *et al* 1983), in volcanic rock from Gran Canaria (Gasparini *et al* 1990), and in numerous other locations where carbonates are absent in the recharge zone.

In cases where the ^{14}C and ^{13}C concentrations of the TDIC demonstrate that CO_2 determines their values, it is reasonable to extrapolate these conditions to samples with similar ^{13}C concentrations but with ^{14}C activities lower than 100%. In that case, the decrease in ^{14}C content, with respect to the value indicated by Equations (17.20) and (17.21), is attributed to radioactive decay. The value, A_g, is that of the soil atmosphere. Due to the difficulty in collecting a sufficient amount of sample for conventional ^{14}C analysis, few data are available on the ^{14}C content of soil CO_2 (Kunkler 1969; Reardon, Mozeto & Fritz 1980; Dever *et al* 1982; Haas *et al* 1983). With AMS measurements, data will become more numerous in the future. An approximate value for the ^{14}C content of the soil CO_2 may be derived from its ^{13}C content. The total ^{14}C fractionation (photosynthesis + decarboxylation + diffusion) during the transit between the free atmosphere and the soil can be derived from Equations (17.11) and (17.21)

$$A_g \approx A_{atm}\ [1 - 2.3(\delta_{atm} - \delta_g)/(10^3)]\ . \qquad (17.22)$$

Calculation gives an activity value of about 97% for prebomb soil CO_2 with ^{13}C content of −20‰, assuming an activity of 100% and a ^{13}C content of about −6 to −7‰ (Craig & Keeling 1963; Broecker & Peng 1982; Salomons & Mook 1986) for the atmospheric CO_2. However, due to fluctuations in ^{14}C cosmic production and to past variations in the partial pressure of CO_2 in the atmosphere (Neftel *et al* 1982), the time constancy of A_{atm} is probably a crude assumption. Nevertheless, open-system conditions are the most favorable for groundwater dating.

Systems dominated by soil CO_2 are easy to identify when they are carbonate-free. However, a complete equilibrium with the gas phase may even occur in groundwater systems where part of the TDIC derives from old carbonate dissolution, as will be discussed now.

Limits of the "Hard-Water Effect" Concept

If properly corrected for sulphate dissolution and other effects (Plummer 1977; Reardon & Fritz 1978; Fontes & Garnier 1977, 1979), the presence of Ca^{2+} (and Mg^{2+}) in the water is an index of carbonate dissolution. Such dissolution is often automatically considered as an index of ^{14}C dilution and reported as "hard-water effect" (see Mook 1980 for a discussion). This is not always the case, as is demonstrated by the example discussed below of a small karstic spring in southwest France (Fleyfel 1979).

The spring drains the shallow parts of karstified limestones with some dolomite, covered by soil and active vegetation cover. Carbonate dissolution is shown by the high Ca^{2+} and Mg^{2+} concentrations (Table 17.3), and the water is constantly supersaturated with calcite (Fleyfel & Bakalowicz 1980). The $\delta^{13}C$ values calculated for CO_2 in equilibrium with the TDIC (Equation 17.19) range between −20.3 and −21.5‰ *vs* PDB (Table 17.3), in good agreement with the

TABLE 17.3. Karstic spring (Moulis, Southwest France; Fleyfel (1979))

Date	T °C	H_2CO_3 mmol.kg^{-1}	HCO_3^- mmol.kg^{-1}	$Ca^{2+} + Mg^{2+}$ mmol.kg^{-1}	$\delta^{13}C$(TDIC) ‰ PDB	$A^{14}C$(TDIC) % modern	$\delta^{13}C(CO_2g)$* ‰ PDB
09-04-76	8.8	0.17	5.85	3.01	−12.02	121.8 ± 1.4	−21.5
13-04-77	8.7	0.18	5.38	2.81	−11.97	122.2 ± 1.9	−21.4
01-06-77	9.5	0.22	4.73	2.49	−12.17	126.9 ± 1.0	−21.4
09-12-77	9.5	0.42	5.42	2.99	−12.19	82.3 ± 1.0	−21.1
31-01-78	8.8	0.29	5.12	2.82	−12.44	120.0 ± 1.4	−21.6
05-03-78	8.7	0.43	4.36	2.39	−11.57	127.1 ± 2.0	−20.4
18-04-78	8.8	0.26	4.42	2.43	−11.87	104.0 ± 4.0	−21.0
02-02-78	9.3	0.25	4.18	2.32	−11.60	123.3 ± 1.8	−20.7
03-02-78	9.2	0.20	4.28	2.29	−11.71	106.4 ± 4.0	−20.9
03-02-78	9.2	0.26	4.32	2.30	−11.59	127.6 ± 3.2	−20.7
04-02-78	9.2	0.26	3.66	2.03	−11.55	118.0 ± 2.0	−20.5
01-05-78	8.6	0.31	4.52	2.49	−11.60	114.1 ± 1.0	−20.7
01-05-78	8.8	0.25	4.54	2.51	−11.32	106.5 ± 3.1	−20.5
01-05-78	8.9	0.23	4.50	2.53	−11.51	112.5 ± 1.6	−20.7
01-05-78	8.8	0.29	4.56	2.54	−11.73	109.8 ± 1.9	−20.8
02-05-78	8.8	0.28	4.60	2.51	−11.54	123.7 ± 4.2	−20.7
02-05-78	8.9	0.31	4.64	2.60	−11.81	118.1 ± 3.2	−20.9
03-05-78	8.9	0.30	4.66	2.60	−11.82	131.8 ± 4.3	−20.9

*^{13}C content of the gas phase in equilibrium with the TDIC (open-system conditions) is calculated from Equation (17.19).

measurements of CO_2 in the soil zone −20.8 to −23.1‰ *vs* PDB. Over a period of some months, the ^{14}C activity of the TDIC ranged from 82 to 131% modern (Table 17.3). These fluctuations are not age effects, as tritium was present in all water samples (Fleyfel 1979).

The highest values of ^{14}C activity observed for the TDIC are compatible with those inferred for soil CO_2 by applying Equations (17.20) and (17.21) to the average ^{14}C content of the atmosphere given by Nydal and Lövseth (1983) for the corresponding years. Direct sampling of the soil CO_2 in November 1978 gave the following isotope contents: A_g ^{14}C = 128.4 ± 3.7%, $\delta^{13}C$ = −22.7‰ *vs* PDB. Introducing these values together with $\delta CO_{2(atm)} \sim -8$‰ *vs* PDB (Salomons & Mook 1986), to Equation (17.21), gives a calculated value of 133% for the activity of atmospheric CO_2. This agrees with measurements for the years 1977–1978. For high values of ^{14}C activity of TDIC, high Ca^{2+}–Mg^{2+} concentrations do not indicate ^{14}C dilution by dead carbonate ("hard-water effect").

The problem is more complex with lower ^{14}C values of TDIC but still ≥100% (eg, 104%, sample from 18-04-78, Table 17.3). These samples show no major variations in chemical composition, which may suggest a greater dilution of ^{14}C by more dissolution of calcite. Calculated values of ^{13}C content of CO_2 in equilibrium with the TDIC (Table 17.3) do not show any deviation from the other samples.

The absence of an age effect implies that some chemical or isotopic dilution has occurred, which is not detectable in the heavy isotope content. This may be explained by dissolution of secondary calcite.

Although not very well known, the isotope fractionation between dissolved HCO_3^- and solid calcite is probably very small ($\varepsilon_{bc} \approx -0.15$‰ at 10°C, Mook (1980)). Thus, precipitation of a secondary calcite would not substantially modify the ^{13}C content of the remaining TDIC, and the newly deposited calcite under open-system conditions would have a ^{13}C content close to that of the TDIC. Two samples of secondary calcite deposited at the outlet of the fractures show a ^{13}C content of −8.30‰ and a ^{14}C activity of 95 and 96%, respectively. Both ^{13}C and ^{14}C concentrations suggest that precipitation occurred under conditions close to those of an open system. The various models of mixing and exchange may be applied to the TDIC with the following parameters: a = 0.26m mol.kg^{-1}, b = 4.42m mol.kg^{-1}, δ_c = 8.3%, A_c = 95%, A_g = 128%, ε_{gb} at 8.8°C = −9.7‰, δ_g = −23.1‰ (the lowest observed value in the soil) and δ_T = −11.87‰. The F & G model gives an A_o value of 103%, in very good agreement with the measured value of 104%. The simple mixing of the Pearson model (Eq 17.3) leads to the same result, whereas the Evans and Eichinger model gives values of A_0 that appear too low (30.3 and 41.9%, respectively). However, the system cannot be explained by a simple dilution, as the Tamers model gives a rather high value of 112.4% for A_0.

These results confirm that secondary calcite is present and may be redissolved from the fracture system or in the soil zone under conditions of partial isotope exchange. The presence of such secondary calcite deposited under "open-system conditions," but clearly aged, could also account for the lowest observed A_0 value of 82%. Old secondary calcite could be redissolved by pulses of aggressive water corresponding to variations in the ratio of the CO_2 production in soils *versus* its removal by infiltrating waters.

From these results, it appears that the concept of "hard-water effect" may be misleading if used without a critical discussion of the chemistry and isotope concentrations of the carbonate system, including secondary calcite.

Conclusions and Summary

Detailed processes of dilution and isotope exchange that affect the activity of the TDIC during infiltration through a carbonate soil zone cannot be simply accounted for by application of the existing models. However, an overall estimate of A_0 at the entrance into the aquifer is reasonably provided by the approaches of Pearson, Evans, and Fontes and Garnier. The very interesting behavior of the Pearson mixing model is probably due to the fact that incomplete isotope exchange processes may be approximated by a simple mixing model between two end-members. One end-member would be the non-exchanged system of simple mixing; the other end-member would be the completely exchanged TDIC. For mathematical reasons, the Evans model reduces to the Pearson model, for all practical purposes. In the carbonate-dominated environment, the Tamers model appears to give high values of A_0 because it does not take any exchange reaction into account. The Eichinger model seems to provide slightly lower values of the initial activity, possibly because it superimposes two types of dilution (one chemical and one isotopic).

As expected, none of the models that account for the gaseous exchange in the soil zone can be used in such an environment. The conclusion is that no systematic use of the model can be made without a detailed discussion of local environmental conditions. ^{14}C correction of A_0 is also site-specific. "Open-system" conditions represent the ideal case for time estimates, provided that it can be shown that no age distortion has been introduced by dissolution of secondary calcite precipitated under open-system conditions. This process is hardly noticeable in the ^{13}C content of the TDIC, and may lead to apparent aging of the TDIC.

Further, open-system conditions represent a considerable limitation for the "hard-water effect" concept. Evidence for dissolution of old carbonate (Ca^{2+} and Mg^{2+}) does not necessarily signify a decrease in ^{14}C activity of the TDIC, if isotope equilibrium was reached with the gas phase after dissolution occurred.

ACKNOWLEDGMENTS

This work was encouraged by W G Mook. Careful reviews and useful comments were provided by W G Mook, L N Plummer and Fred Robertson. The editorial assistance of R Kra and Ph Barker was appreciated.

REFERENCES

Broecker, WS and Peng, TH 1982 *Tracers in the Sea.* Palisades, New York, Eldigio Press.

Castany, G, Marcé, A, Margat, J, Moussu, H, Guillaume, Y and Evin, J 1974 Etude par les isotopes du mileu du régime des eaux souterraines dans les aquifères de grandes dimensions. In *Isotope Techniques in Groundwater Hydrology* I. Vienna, AIEA: 243–258.

Craig, H and Keeling, CD 1963 The effects of atmospheric N_2O on the measured composition of atmospheric CO_2. *Geochimica et Cosmochimica Acta* 27: 549–551.

Deines, P, Langmuir, D and Harmon, RS 1974 Stable carbon isotope ratios and the existence of a gas phase in the evolution of carbonate groundwaters. *Geochimica et Cosmochimica Acta* 38: 1147–1164.

Dever, L (ms) 1985 Approches chimiques et isotopiques des interactions fluides-matrice en zone non saturée carbonatée. Thesis (Doctorat ès Sciences), Université Paris-Sud, Orsay: 179 p.

Dever, L, Durand, R, Fontes, JCh and Vachier, P 1982 Géochimie et teneurs isotopiques de systèmes saisonniers de dissolution de la calcite dans un sol sur craie. *Geochimica et Cosmochimica Acta* 46: 1947–1956.

Dever, L and Fontes, JCh 1990 Isotopic changes of the total dissolved inorganic carbon as tracers of liquid-solid interactions in the unsaturated zone. In *Geochronology, Cosmochronology and Isotope Geology.* 7th International Conference, Canberra (abstract).

Dörr, H and Münnich, KO 1980 Carbon-14 and carbon-13 in soil CO_2. *In* Stuiver, M and Kra, RS, eds, Proceedings of the 10th International ^{14}C Conference. *Radiocarbon* 22(3): 909–918.

Eichinger, L 1983 A contribution to the interpretation of ^{14}C groundwater ages considering the example of a partially confined sandstone aquifer. *In* Stuiver, M and Kra, RS, eds, Proceedings of the 11th International ^{14}C Conference. *Radiocarbon* 25(2): 347–356.

Evans, GV, Otlet, RL, Downing, A, Monkhouse, RA and Rae, G 1979 Some problems in the interpretation of isotope measurements in United Kingdom aquifers. In *Isotope Hydrology* II. Vienna, IAEA: 679–708.

Fleyfel, M (ms) 1979 Etude hydrologique, géochimique et isotopique des modalités de minéralisation et de transfert du carbone dans la zone d'infiltration d'un aquifère karstique. Thesis (Docteur ingénieur), Université Paris 6: 221 p.

Fleyfel, M and Bakalowicz, M 1980 Etude géochimique et isotopique du carbone minéral dans un aquifère karstique. In *Cristallisation, Déformation, Dissolution des Carbonates.* Université Bordeaux III: 281–245.

Fontes, JCh 1983 Dating of groundwater. *In* Guidebook on Nuclear Techniques in Hydrology. *IAEA Technical Reports Series* 91: 285–317.

_____1985 Some considerations on groundwater dating using environmental isotopes. *In* Keynote Papers, IAH Congress, 18th, part 1. *Hydrogeology in the Service of Man*: 118–154.

Fontes, JCh and Garnier, JM 1976 Correction des activités apparentes en ^{14}C du carbone dissous: estimation de la vitesse des eaux des nappes captives. *In* Réunion Annuelle Sciences de la Terre. Paris, *Société Géologique de France* 77: 4 p.

_____1977 Determination of the initial ^{14}C activity of the total dissolved carbon. Age estimation of waters in confined aquifer. *In* Proceedings of the 2nd International Symposium on Water Rock Interactions. *Sciences Géologiques*: 363–376.

_____1979 Determination of the initial ^{14}C activity of the total dissolved carbon: A review of the existing models and a new approach. *Water Resources Research* 15: 399–413.

Fritz, P, Reardon, EJ, Barker, J, Brown, M, Cherry, A, Killey, WD and McNaughton, D 1978 The carbon isotope geochemistry of a small groundwater system in northeastern Ontario. *Water Resources Research* 14: 1059–1067.

Galimov, EM 1966 Carbon isotopes in soil CO_2. *Geochemistry International* 3: 889–898.

Garnier, JM and Fontes, JCh 1980 Hydrochimie, géochimie des isotopes du milieu et conditions de circulation dans la nappe captive des sables astiens (Hérault). *Revue Bureau de Recherches Géologiques et Minières* 2, Series 3: 199–214.

Gasparini, A, Custodio, E, Fontes, JCh, Jimenez, J and Nuñez, JA 1990 Exemple d'étude géochimique et isotopique de circulations aquifères en terrain volcanique sous climat semi-aride (Amurga, Gran Canaria, Iles Canaries). *Journal of Hydrology* 114: 61–91.

Geyh, MA and Wendt, I 1965 Results of water sample dating by means of the model of Münnich and Vogel. *In* Chatters, RM and Olson, EA, eds, *Proceedings of the 6th International Conference on Radiocarbon and Tritium Dating.* Clearinghouse for Federal and Technical Information, National Bureau of Standards, Washington, DC: 597–603.

Haas, H, Fisher, DW, Thorthenson, C and Weeks, EP 1983 $^{13}CO_2$ and $^{14}CO_2$ measurements on soil atmosphere sampled in the western Great Plains of the US. *In* Stuiver, M and Kra, RS, eds, Proceedings of the 11th International ^{14}C Conference. *Radiocarbon* 25(2): 301–314.

Ingerson, E and Pearson, FJ 1964 Estimation of age and rate of motion of groundwater by the ^{14}C-method. *In* Miyake, Y and Koyama, T, eds, *Recent Researches in the Field of Hydrosphere, Atmosphere and Nuclear Geochemistry.* Tokyo, Maruzen: 263–283.

Kunkler, JL 1969 The source of carbon dioxide in the zone of aeration of the Bandelier Tuff, near Los Alamos, New Mexico. *USGS Professional Paper* 650-13: B185–B188.

Lesniak, PM and Sakai, H 1990 Carbon isotope fractionation between dissolved carbonate (CO_3^{2-}) and CO_2 (g) at 25° and 40°C. *Earth and Planetary Science Letters* 95: 297–301.

Levin, I, Hesshaimer, V, Glöcker, R, Kromer, B, Münnich, KO and Francey, R 1989 Radiocarbon in atmospheric CO_2: Global distribution and trends. *In* Extended Abstracts, 3rd International CO_2 Conference. *WMO Report* 59: 55–60.

Levin, I, Münnich, KO and Weiss, W 1980 The effect of anthropogenic CO_2 and ^{14}C sources on the distribution of ^{14}C in the atmosphere. *In* Stuiver, M and Kra, RS, eds, Proceedings of the 10th International ^{14}C Conference. *Radiocarbon* 22(2): 379–391.

Mazor, E, Verhagen, BT, Sellschop, JPF, Robins, NS and Hutton, LG 1974 Kalahari groundwaters: their hydrogen, carbon and oxygen isotopes. In *Isotope Techniques in Groundwater Hydrology* 1. Vienna, AIEA: 203–223.

Mook, WG 1972 On the reconstruction of the initial ^{14}C content of groundwater from the chemical and isotopic composition. *In* Rafter, TA and Grant Taylor, T, eds, *Proceedings of the 8th International ^{14}C Conference*. Wellington, Royal Society of New Zealand: 342–352.

_____1976 The dissolution-exchange model for dating groundwater with ^{14}C. In *Interpretation of Environmental Isotope and Hydrochemical Data in Groundwater Hydrology*. Vienna, IAEA: 213–225.

_____1980 Carbon-14 in hydrogeological studies. *In* Fritz, P and Fontes, JCh, eds, *Handbook of Environmental Isotopes Geochemistry* 1. Amsterdam, The Netherlands, Elsevier: 50–74.

Mook, WG, Bommerson, JC and Staverman, WH 1974 Carbon isotope fractionation between dissolved bicarbonate and gaseous carbon dioxide. *Earth and Planetary Science Letters* 22: 169–176.

Moulin, M (ms) 1990 Genèse et évolution des minéraux secondaires en zone non saturée carbonatée: Etude géochimique et isotopique de la zone non saturée de la craie de Champagne. PhD thesis, Université Paris-Sud, Orsay: 220 p.

Münnich, KO 1957 Messung des ^{14}C-Gehaltes von hartem Grundwasser. *Naturwissenschaften* 44: 32–34.

Münnich, KO and Roether, W 1963 A comparison of carbon-14 and tritium ages of groundwater. In *Radioisotopes in Hydrology*. Vienna, AIEA: 97–404.

Münnich, KO and Vogel, JC 1959 ^{14}C Altersbestimmung von Süsswasser-Kalkablagerungen. *Naturwissenschaften* 46: 168.

Neftel, A, Oeschger, H, Schwander, J, Stauffer, B and Zumbrunn, R 1982 Ice core sample measurements give atmospheric CO_2 content during the past 40,000 yr. *Nature* 295: 220–223.

Nydal, R and Lövseth, K 1983 Tracing bomb ^{14}C in the atmosphere, 1962–1980. *Journal of Geophysical Research* 88(C6): 3621–3642.

Ousmane, B, Fontes, JCh, Aranyossy, JF and Joseph, A 1983 Hydrologie isotopique et hydrochimie des aquifères discontinus de la bande sahélienne et de l'Aïr (Niger). In *The Use of Isotopic Techniques in Water Resources Development*. Vienna, IAEA: 367–395.

Parada, CB, Long, A and Davis, SN 1983 Stable isotopic composition of soil carbon dioxide in the Tucson Basin, Arizona, USA. *Isotope Geosciences* 1: 219–236.

Plummer, LN 1977 Defining reactions and mass transfer in part of the Floridian aquifer. *Water Resources Research* 13: 801–812.

Plummer, LN, Busby, JF, Lee, RW and Hanshaw, BB 1990 Geochemical modeling of the Madison aquifer in parts of Montana, Wyoming and South Dakota. *Water Resources Research* 26: 1981–2014.

Reardon, EJ, Allison, GB and Fritz, P 1979 Seasonal chemical and isotopic variations of soil CO_2 at Trout Creek, Ontario. *Journal of Hydrology* 43: 355–371.

Reardon, EJ and Fritz, P 1978 Computer modeling of groundwater ^{13}C and ^{14}C isotope compositions. *Journal of Hydrology* 36: 201–224.

Reardon, EJ, Mozeto, AA and Fritz, P 1980 Recharge in northern clime calcareous sandy soils: soil water chemical and carbon-14 evolution. *Geochimica et Cosmochimica Acta* 44: 1723–1735.

Rightmire, CT 1978 Seasonal variation in P_{CO_2} and ^{13}C content of soil atmosphere. *Water Resources Research* 14: 691–692.

Rightmire, CT and Hanshaw, BB 1973 Relationship between the carbon isotope composition of soil CO_2 and dissolved carbonate species in groundwater. *Water Resources Research* 9: 958–967.

Rubinson, M and Clayton, RN 1969 Carbon-13 fractionation between aragonite and calcite. *Geochimica et Cosmochimica Acta*

33: 997–1022.

Salati, E, Menezes Leal, J and Mendes Campos, M 1974 Environmental isotopes used in a hydrogeological study of northeastern Brazil. In *Isotope Techniques in Groundwater Hydrology* 1. Vienna, AIEA: 259–282.

Salem, O, Visser, JH, Dray, M and Gonfiantini, R 1980 Groundwater flow patterns in the western Lybian Arab Jamahiriaya. In *Arid-Zone Hydrology: Investigations with Isotope Techniques.* Vienna, IAEA: 165–179.

Saliège, JF and Fontes, JCh 1984 Essai de détermination expérimentale du fractionnement des isotopes ^{13}C et ^{14}C du carbone au cours de processus naturels. *International Journal of Applied Radiation and Isotopes* 35(1): 55–62.

Salomons, W and Mook, WG 1986 Isotope geochemistry of carbonates in the weathering zone. *In* Fritz, P and Fontes, JCh, eds, *Handbook of Environmental Isotope Geochemistry* 2. Amsterdam, The Netherlands, Elsevier: 239–269.

Smith, DB, Downing, RA, Monkhouse, RA, Otlet, RL and Pearson, FJ 1976 The age of groundwater in the chalk of the London basin. *Water Resources Research* 12: 392–404.

Stuiver, M and Kra, RS, eds, 1986 Calibration issue. Proceedings of the 12th International ^{14}C Conference. *Radiocarbon* 28(2B): 905–1030.

Tamers, MA 1967 Radiocarbon ages of groundwater in an arid zone unconfined aquifer. *In* Isotope Techniques in the Hydrological Cycle.*American Geophysical Union Monograph* 11: 143–152.

Vogel, JC and Ehhalt, D 1963 The use of carbon isotopes in groundwater studies. In *Radioisotopes in Hydrology.* Vienna, IAEA: 383–395.

Vogel, JC, Ehhalt, D and Roether, W 1963 A survey of the natural isotopes of water in South Africa. In *Radioisotopes in Hydrology.* Vienna, IAEA: 407–416.

Vogel, JC, Fuls, A and Ellis, RP 1978 The geographical distribution of Kranz grasses in South Africa. *South African Journal of Sciences* 74: 209–215.

Vogel, JC, Grootes, PM and Mook, WG 1970 Isotopic fractionation between gaseous and dissolved carbon dioxide. *Zeitschrift für Physik* 230: 225–238

Wal, van de, RSW, van der Borg, K, Oerter, H, Reek, N, de Jong, AFM and Oerlemans, J 1990 Progress in ice dating of ice at Utrecht. *In* Yiou, F and Raisbeck, GM, eds, Proceedings of the 5th International Conference on Accelerator Mass Spectrometry. *Nuclear Instruments and Methods* B52(3,4): 469–472.

Wallick, EI 1976 Isotopic and chemical considerations in radiocarbon dating of groundwater within the semi-arid Tucson Basin, Arizona. In *Interpretation of Environmental Isotope and Hydrochemical Data in Groundwater Hydrology.* Vienna, IAEA: 195–212.

Wigley, TML 1975 Carbon dating of groundwater from closed and open systems. *Water Resources Research* 11: 324–328.

Wigley, TM, Plummer, LN and Pearson, FJ 1978 Mass transfer and carbon isotope evolution in natural water systems. *Geochimica et Cosmochimica Acta* 42: 1117–1139.

Wilson, AT and Donahue, DJ 1990 AMS carbon-14 dating of ice: Progress and future prospects. *In* Yiou, F and Raisbeck, GM, eds, Proceedings of the 5th International Conference on Accelerator Mass Spectrometry. *Nuclear Instruments and Methods in Physics Research* B52 (3,4): 473–476.

Winter, K, Troughton, JH and Card, A 1976 $\delta^{13}C$ values of grass species collected in the northern Sahara Desert. *Oecologia* 25: 115–123.

18

EFFECTS OF PARAMETER UNCERTAINTY IN MODELING ^{14}C IN GROUNDWATER

F J PEARSON, JR

INTRODUCTION

^{14}C occurs in groundwater in dissolved carbonate ($CO_{2(aq)}$, HCO_3^- and CO_3^{-2}) and organic carbon. Carbonate contains most of the carbon in groundwater, and many measurements of its ^{14}C content have been made. Recently, as our ability to measure the ^{14}C content of very small samples has improved, studies of the ^{14}C content of dissolved organic carbon have also begun (Long *et al* 1992). This chapter will focus on ^{14}C in dissolved carbonate.

Groundwater carbonate differs in two ways from most other substances dated by ^{14}C. First, the quantity of interest is not the age of the dissolved carbonate itself, but the age of the groundwater from which the carbonate was extracted. Several models relating carbonate age to groundwater age are discussed briefly. Second, the ratio of ^{14}C to the total sample carbonate ($^{14}C/C_{tot}$) is affected by geochemical reactions as well as by ^{14}C decay. Fontes (1992) discusses models to adjust for geochemical changes in $^{14}C/C_{tot}$ ratios.

This chapter summarizes a general model for geochemical effects on the $^{14}C/C_{tot}$ ratios of groundwater, discusses the uncertainties in the parameters of this model, and describes the effects of those uncertainties on modeled groundwater ages.

Models of Groundwater Flow

Although the focus of this chapter is on uncertainties associated with geochemical models, in practical applications, we must not neglect uncertainties in the models that relate carbonate ages to groundwater ages. Models of groundwater flow can be discussed in terms of two extreme conceptual models, the piston flow model and the exponential or well-mixed reservoir model.

The piston flow model treats groundwater as if it were flowing in a pipe with no dispersive mixing in the direction of flow. Here, the ^{14}C age of dissolved carbonate is equivalent to that of the increment of water from which the sample was taken. The piston flow model is usually appropriate to regional flow systems, and it is widely used in ^{14}C studies, although its choice is rarely explicit. It is generally the model assumed when geochemical models are used to interpret the $^{14}C/C_{tot}$ ratios of groundwaters.

At the other extreme is the well-mixed reservoir model, which can be thought of as a system with infinite dispersion. The age of water in such a system equals the reservoir volume divided by its throughput. This age may be considerably older than the age corresponding to the ^{14}C content of the water. Mathematically, the exponential model of Siegenthaler (1972) is a well-mixed reservoir model. This model is most successfully applied to smaller flow systems and to the interpretation of relatively short-lived isotopes such as ^{3}H.

Probably the most realistic models fall between the two extreme conceptual models. Numerical solutions to the differential equations of solute transport, as used in many groundwater transport studies, are close to the piston flow end member. Mixing cell, or finite state models, range from the equivalent of well-mixed reservoir models to the equivalent of finite difference solutions to the differential transport equations as the number of cells increases. The papers in International Atomic Energy Agency (1986) are a useful summary of such models.

Serious errors in groundwater ages can also result from lack of consideration of mixing processes. The contamination of old groundwater by a small amount of ^{14}C-bearing younger water is an obvious example. Mixing effects can be detected by using several isotopes (eg, ^{3}H, ^{85}Kr, ^{39}Ar, as well as ^{14}C) on a given sample (eg, Loosli, Lehmann & Däppen 1991).

Models of Geochemical Changes in $^{14}C/C_{tot}$ Ratios

Most models of geochemical change in $^{14}C/C_{tot}$ ratios consider the same geochemical processes, so it is possible to write a general model that reduces to the conventional models under appropriate boundary conditions. The effects of uncertainties in the parameters of this general model readily translate into effects of uncertainties in the more widely used models.

The source of ^{14}C to groundwater carbonate is ^{14}C-bearing $CO_{2(g)}$ dissolved in the recharge zone. (CO_2 from the oxidation of ^{14}C-bearing dissolved organic carbon in groundwater is a second possible source. In actual application, the model used here implicitly includes this source of ^{14}C, so it is not explicitly discussed further.) Although it is sometimes possible to collect samples on which the ^{14}C, ^{13}C and C_{tot} contents of recharge water can be measured directly, the initial conditions must generally be inferred from measurements on the sample itself.

In the aquifer, ^{14}C is diluted by ^{14}C-free carbonate from the dissolution of mineral carbonate, or the oxidation of aquifer organic carbon. Enough samples may be available from some aquifer systems that the carbonate geochemical evolution and the amount of ^{14}C-free carbonate added can be found directly. However, it is generally necessary to infer these quantities from measurements on the sample itself and assumptions about the processes operating.

^{14}C can also be lost from the solution by processes other than radioactive decay. Isotope exchange between dissolved and mineral carbonate should have only a limited effect on groundwater ^{14}C because of the very slow rate of diffusion of ions within solids. However, incongruent dissolution of carbonate minerals – the dissolution of one carbonate mineral such as dolomite, or a high-Mg calcite accompanied by virtually simultaneous precipitation of another, such as low-Mg calcite – occurs in many aquifer systems and has the same effect on the $^{14}C/C_{tot}$ ratio as true isotope exchange. The extent to which this process occurs can be inferred from the relative changes in the concentrations and the ^{13}C contents of the dissolved carbonate.

To summarize: Uncertainties in adjustments to the ^{14}C content of groundwater result from uncertainties in the ^{14}C and ^{13}C contents and concentrations of the C_{tot} of the water at recharge, in the ^{13}C content and amount of ^{14}C-free carbonate dissolved in the aquifer, and in the amount of incongruent dissolution by which ^{14}C can be chemically removed from the groundwater.

UNCERTAINTY ANALYSIS

To analyze uncertainties in adjusted ^{14}C ages requires a mathematical ^{14}C adjustment model and a procedure for performing uncertainty and sensitivity analysis on it. These are discussed in the following subsection. The analysis also requires definition of the values of the parameters in the adjustment model and their uncertainty distributions. The second and third following subsections address these uncertainties.

Mathematical Model of ^{14}C Evolution

The model used is after Wigley, Plummer, and Pearson (1978). It is based on an isotope balance equation stating that the change in the ^{13}C or ^{14}C content of dissolved carbonate equals the sum of the amounts of the isotope entering the water from all carbon sources less the sum of the amounts leaving to all sinks. The equation is written:

$$d(R_S C_{tot}) = \sum_{i=1}^{N} d(R_{E_i} C_{E_i}) - \sum_{j=1}^{M} d(R_{L_j} C_{L_j}) \qquad (18.1)$$

In this equation, C_{tot} = total dissolved carbonate in M/kg H_2O, C_{E_i} = moles of carbonate entering the water from source i and C_{L_j} = moles of carbonate leaving the groundwater to sink j. R_S, R_E and R_L are the ratios of ^{13}C or ^{14}C to ^{12}C (or total C_{tot}) in solution and in substances entering and leaving the solution, respectively. It is assumed that carbonates leaving groundwater do so slowly enough to maintain isotopic equilibrium with the solution. Thus, $R_{L_j} = R_S \alpha_{(L_j - S)}$ where $\alpha_{(L_j - S)}$ is the fractionation factor between substance j leaving the solution, and the total dissolved carbonate.

A solution to Equation (18.1) for ^{13}C written using δ-notation is (Wigley, Plummer & Pearson 1978; Pearson 1991)

$$\beta(\delta^{13}C_S + 1000) - (\delta^{13}C_E + 1000) = [\beta\,(\delta^{13}C_S^0 + 1000) - (\delta^{13}C_E + 1000)]\ \Omega \qquad (18.2)$$

in which $\delta^{13}C_S^0$ and $\delta^{13}C_S$ (in ‰ relative to Vienna PDB) refer to the initial solution and the sample, and $\delta^{13}C_E$ to the carbonate dissolving from the aquifer. When $\sum C_E \neq \sum C_L$,

$$\Omega = \left[\frac{C_{tot}}{C_{tot}^0}\right]^{\phi} . \qquad (18.3)$$

β and ϕ in Equations (18.2) and (18.3) include the amounts of carbonate entering and leaving the solution and the fractionation factors between the dissolved carbonate and the substance leaving the solution (Wigley, Plummer & Pearson 1978; Pearson 1991). For a sample in which only carbonate dissolution has occurred, $\beta = 1$ and $\phi = -1$, and Equations (18.2) and (18.3) reduce to the ^{14}C adjustment equations using $\delta^{13}C$ and C_{tot}, respectively, proposed by Ingerson and Pearson (1964).

Carbonate dissolving from an aquifer is likely to be ^{14}C-free because of the age of the dissolving minerals. Thus, $R_E = 0$ and a solution to Equation (18.1) is

$$R_S = R_S^0 \cdot \Omega . \qquad (18.4)$$

In this equation, R_S^0 is the ^{14}C content in pMC (in % relative to the standard activity = 0.95 ^{14}C content NBS OXI) of the dissolved carbonate of the initial solution and R_S is that of the dissolved carbonate after the chemical evolution of the water but without considering radioactive decay of ^{14}C. The groundwater model age is calculated from R_M/R_S, where R_M is the measured ^{14}C content of the dissolved carbonate. Values of Ω are calculated from $\delta^{13}C$ data using Equation (18.2) and from the sample chemistry using Equation (18.3), and are used in Equation (18.4) to adjust the ^{14}C content of groundwater.

To use these equations requires $\delta^{13}C$ values and concentrations of the total sample carbonate, C_{tot}, the initial ^{14}C-bearing solution, C^0_{tot} and the concentrations of carbonate entering, C_E and leaving, C_L, the groundwater in the aquifer. The last two parameters are required to evaluate β and ϕ in Equations (18.2) and (18.3).

The chemical parameters, C_E and C_L can be developed from the sample chemistry using mass balance techniques. When many samples are available from a known flow system, carbonate sources and sinks can be worked out with considerable precision (eg, Plummer, Parkhurst & Thorstenson 1983; Plummer *et al* 1990). For this uncertainty analysis, an idealized chemical evolution model was used based on the chemical composition of the water. The model assumes that all sulfate is from a $CaSO_4$ mineral, all magnesium from (Ca, Mg) CO_3 (dolomite), anions other than carbonate and sulfate are from a "NaCl" mineral, and that Ca^{+2} for Na^+ exchange may occur. The quantity of $CaCO_3$ dissolved is determined from a Ca^{+2} balance, and the carbonate content of the initial solution, C^0_{tot}, from a carbonate balance (Fontes & Garnier 1979; Pearson 1991).

The ^{13}C content of some samples is too high to result from any combination of reasonable values of isotopic and chemical parameters. These waters are modeled by assuming incongruent dissolution of carbonate. The modeled amount of carbonate dissolved, C_E, is increased, balanced by an equivalent amount of carbonate precipitated, C_L, until the modeled and measured ^{13}C values agree.

The isotope model and mass balance equations were programmed into a commercial spreadsheet (Lotus 1-2-3®) running under MS-DOS®. The uncertainty analyses were carried out using the Lotus add-in program @RISK®. @RISK® permits the user to define spreadsheet parameter values using statistical distributions of many types. It then recalculates the spreadsheet a given number of times, choosing values for uncertain parameters from among their defined distributions, using Monte Carlo or Latin hypercube (stratified Monte Carlo) sampling. It retains the values of chosen output cells from each calculation, calculates their statistics and displays them graphically.

Uncertainty in Model Parameters

Values and representative statistical distributions of the uncertain parameters are based on the sources of ^{14}C-bearing dissolved carbonate, sources of diluting ^{14}C-free carbonate and ^{14}C sinks other than radioactive decay.

Sources of ^{14}C-Bearing Dissolved Carbonate. The dominant source of ^{14}C to most groundwaters is CO_2 in soil air. Most of this CO_2 is produced by plant respiration and decay, and dissolves in the water during recharge at time zero. Its ^{13}C content depends on the ^{13}C content of the plants in the recharge area and on the extent of isotope fractionation during dissolution.

The distribution of $\delta^{13}C$ values in various types of plants is reasonably well known. Deines (1980) has summarized a large number of $\delta^{13}C$ values for plants. His data for C_3 and C_4 plants can be described by normal distributions with means and standard deviations of -27.0 ± 2.1 and $-12.7 \pm 1.2‰$. In practice, it is likely that the type of plants dominant in the area of recharge to an aquifer of interest will be known. For the purpose of this uncertainty analysis, C_3 plants are assumed.

Isotope fractionation between gaseous and dissolved CO_2 may also affect the ^{13}C content of the dissolved carbonate (Mook 1976). The amount of fractionation depends on the conditions of dissolution and can range from zero, under closed-system conditions, to perhaps 10‰, under open-system conditions (Deines, Langmuir & Harmon 1974; Reardon & Fritz 1978). For this uncertainty analysis, a triangular distribution has been assumed with minimum and most probable values equal to 0‰, corresponding to closed-system dissolution, and a maximum and least probable value of +10‰, corresponding to completely open-system conditions.

The concentration of the initial dissolved CO_2 is one of the parameters generated by the chemical mass balance model, and its mean and standard deviation are determined by those of the water chemical data. Uncertainties in the concentrations of all dissolved constituents are assumed to be normally distributed with a standard deviation (σ) of 5% except for alkalinity and total dissolved carbonate, which have been modeled with standard deviations of 5 and 10%.

Sources of ^{14}C-Free Carbonate. Diluting carbonate comes from the dissolution of aquifer mineral carbonate and less commonly from oxidation products of aquifer organic carbon or CO_2 of petroleum, natural gas or deep mantle origin.

The ^{13}C content of dissolved mineral carbonate will be that of the minerals themselves, because no fractionation occurs during dissolution. Most carbonates are marine and will have $\delta^{13}C$ values around 0 to +2‰. Fresh or brackish water carbonate with $\delta^{13}C$ values as negative as −5 to −10‰ may be present in some aquifers. Aquifer geology should constrain the choice of mineral carbonate $\delta^{13}C$ values. A normal distribution with a mean of +1 and standard deviation of ± 2‰ is used in this uncertainty analysis. Also, there are assumed to be no non-mineral sources of ^{14}C-free carbonate.

The amount of ^{14}C-free carbonate dissolved is also important in the isotope model. Its mean value and uncertainty come from the chemical mass balance model.

^{14}C Sinks Other Than Radioactive Decay. Precipitation of carbonate from groundwater is the dominant process of this type. Pure isotope exchange between dissolved and solid carbonate could also remove ^{14}C from the water, but this process is too slow to affect ^{14}C except at high temperatures (> ca 100°C). Concurrent dissolution and precipitation of carbonate minerals is indistinguishable from pure exchange, and has been demonstrated petrographically.

Carbonate precipitation is commonly accompanied by additional carbonate mineral dissolution and can be recognized by enriched ^{13}C in the dissolved carbonate (typically $\delta^{13}C$ values more positive than ca −10 to −8‰). The amount of precipitation and additional dissolution is important in the isotope model and is constrained by the requirement that the chemical model yield water with the $\delta^{13}C$ value measured on the sample. The uncertainty in the amount of precipitation depends on the uncertainty of the water chemical parameters, the $\delta^{13}C$ values of the aquifer minerals and the total dissolved carbonate of the sample. The uncertainties in all these parameters except the sample $\delta^{13}C$ values have been previously defined. The sample $\delta^{13}C$ is assumed to have a standard deviation of 0.5‰. As the sample $\delta^{13}C$ approaches the $\delta^{13}C$ in the aquifer minerals, *ie*, as the dissolved carbonate approaches isotopic equilibrium with aquifer carbonate, the uncertainties in the calculations become very large.

In fracture flow systems in fissured or porous media, a diffusive loss of ^{14}C from flowing water in the fractures to stagnant water in the matrix may take place (Neretnieks 1981). This type of ^{14}C loss is in the realm of flow system modeling, and its effects are not included here.

Precipitation is assumed to occur at isotopic equilibrium with the sample. ^{14}C fractionation during this process is taken as twice the ^{13}C fractionation. The amount of fractionation is calculated from the water chemistry and is uncertain to the extent that the water chemical parameters are uncertain. The fractionation factors used are those of Mook (1986) and are not considered uncertain.

Results and conclusions

Three different types of waters have been modeled to illustrate the uncertainty analysis procedure. These and other sets of calculations lead to conclusions about the sensitivity of ^{14}C adjustments to various model parameters, and about levels of uncertainty to be expected when modeling ^{14}C ages of waters of various types. The three waters illustrated range from a simple calcium-bicarbonate water relatively depleted in ^{13}C through a water more enriched in ^{13}C to a highly evolved calcium-bicarbonate-sulfate water with a high ^{13}C content.

Example Calculations

The first example is from an aquifer that is dominantly silica and has little carbonate and almost no other soluble minerals. The water contains about 3.9 millimoles (mM) C_{tot} with $\delta^{13}C = -13.8‰$ and $^{14}C = 23.7 \pm 0.3$ pMC. The chemical model shows that C_{tot} of the initial solution is 2.1 mM and $\sum C_E$ is 1.8 mM. The chemistry and ^{13}C content of this sample constrain the model such that a given $\delta^{13}C$ value for the initial carbonate can correspond to only one value for the dissolving mineral carbonate. The range of $\delta^{13}C$ values of C_3 plants ($-27 \pm 2‰$) corresponds to the range of marine carbonate (*ca* $+1 \pm 2‰$). This dependency is included in the uncertainty analysis, and no fractionation during soil CO_2 dissolution was included.

Because of the strong internal constraints on the $\delta^{13}C$ of this sample, the water was modeled using σC_{tot} values of 10% and 5% to explore uncertainties in the results due to the uncertainty chosen for the chemical parameters. Figure 18.1 shows the two distributions of modeled ages. Both are skewed and have most

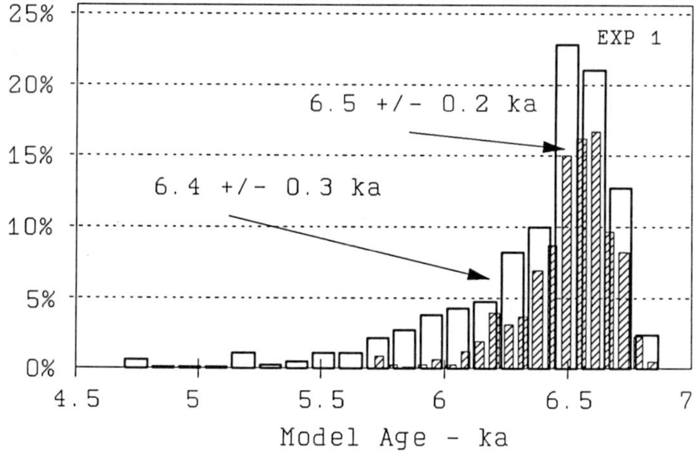

Fig 18.1. Distribution of model ages calculated using σC_{tot} = 10% and 5%: Example 1

probable values between 6500 and 6600 years. For reference, the age modeled using an initial ^{14}C content of 85 pMC is 10,600 years, and that calculated, using the simple correction factor of $\delta^{13}C/-25$, is 7000 years. The uncertainties of the modeled ages in this figure are rather small. The broader distribution, calculated using σC_{tot} = 10%, has σ = 300 years, and the narrower distribution, calculated with σC_{tot} = 5%, has σ = 200 years.

The water used in the second example is from a carbonate aquifer and contains about 6.6 mM C_{tot} with $\delta^{13}C$ = −8.9‰, and ^{14}C = 33.3 ± 0.4 pMC. Input parameters to model this water are not as constrained as were those of the first water. The chemical mass balance gives C^0_{tot} = 3.9 mM, and C_E = 2.7 mM. The most probable isotope values, $\delta^{13}C^0_{tot}$ = −27‰ and $\delta^{13}C_E$ = +1‰ lead to a calculated $\delta^{13}C$ of −15.4‰, far lighter than the measured −8.9‰. This measured $\delta^{13}C$ could be modeled with initial and mineral $\delta^{13}C$ values at the edge of their reasonable ranges, −15.9 and +1.0, or −17.3 and +3.0‰, respectively, for example. The measured ^{13}C can also be modeled with more central values for initial and mineral carbonates, if it is assumed that more carbonate is dissolved and reprecipitated than indicated by the chemical balance alone. This is equivalent to assuming that incongruent dissolution, or "isotope exchange" has acted to increase the ^{13}C content of the water and, concurrently, to reduce its ^{14}C content. To do uncertainty calculations for this water, an equation was developed empirically relating the amount of mineral carbonate dissolved and reprecipitated to the $\delta^{13}C$ values of the initial and mineral carbonate. The water was then modeled taking the $\delta^{13}C$ value of plants as −27 ± 2‰, the fractionation during CO_2 dissolution from 0 and 10‰, and the $\delta^{13}C$ of minerals as +1 ± 2‰.

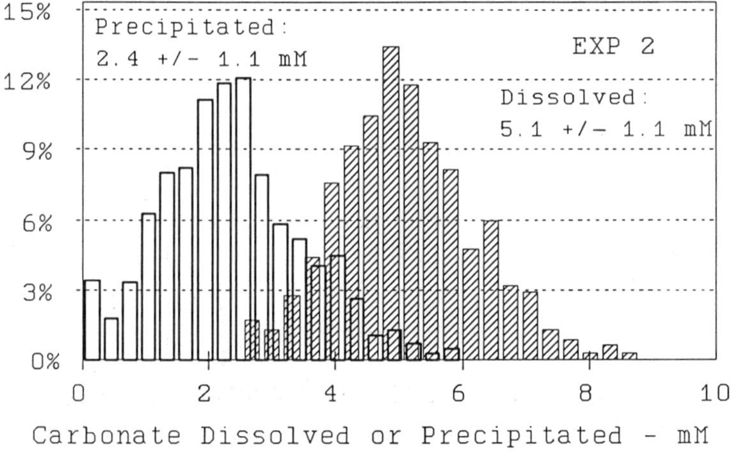

Fig 18.2. Distribution of amounts of dissolved and reprecipitated mineral carbonate: Example 2

Figure 18.2 shows the amounts of dissolved and precipitated mineral carbonate calculated using a σC_{tot} of 10%. The difference between the means of these distributions is 2.7 mM, the amount of mineral carbonate required by the chemical mass balance.

The modeled age of this sample is 800 ± 2000 years, as shown in Figure 18.3. This result is consistent with an age of 200 years determined from a ^{39}Ar measurement on this sample (Loosli, Lehmann & Däppen 1991). It is considerably lower than the age of 7700 years calculated using an initial ^{14}C content of 85%, but is not significantly different from the age of 600 years, which results from applying the simple correction factor $\delta^{13}C/-25$ (Pearson 1965).

The third example is highly evolved water from a $CaSO_4$-bearing carbonate aquifer with C_{tot} = 7.2 mM, $\delta^{13}C$ = −4.6‰, and ^{14}C = 13.28 ± 0.14 pMC. It was chosen to illustrate that as the dissolved carbonate approaches isotopic equilibrium with the rock, uncertainties in the model ages become extremely large.

Fractionation between dissolved carbonate and calcite depends on the distribution of dissolved carbonate species (the pH) and temperature, so its uncertainty should depend on the uncertainties chosen for the concentrations of the carbonate species. With σC_{tot} = 10%, σ of the fractionation is 0.2‰, whereas with σC_{tot} = 5%, σ of the fractionation is 0.1‰. The calculated $\delta^{13}C$ is −4.4‰, whereas $\delta^{13}C$ of the sample is −4.6‰. Thus, calculations made using mineral $\delta^{13}C$ values more negative than 0‰ will frequently lead to infinitely large correction factors. To make calculations possible, the range of mineral carbonate $\delta^{13}C$ was limited to between 0 and +3‰.

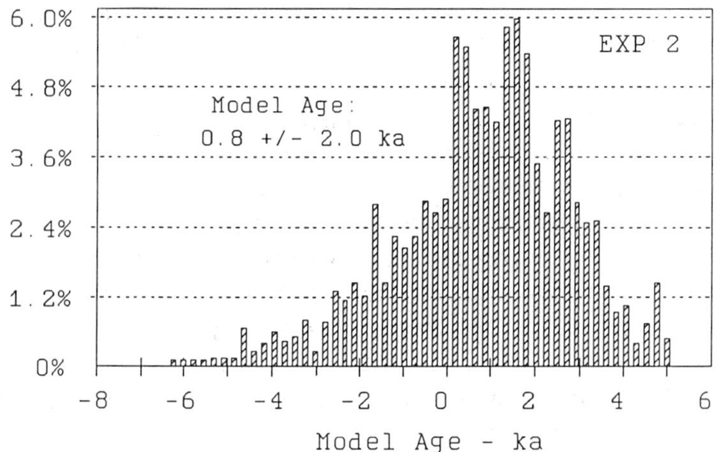

Fig 18.3. Distribution of model ages: Example 2

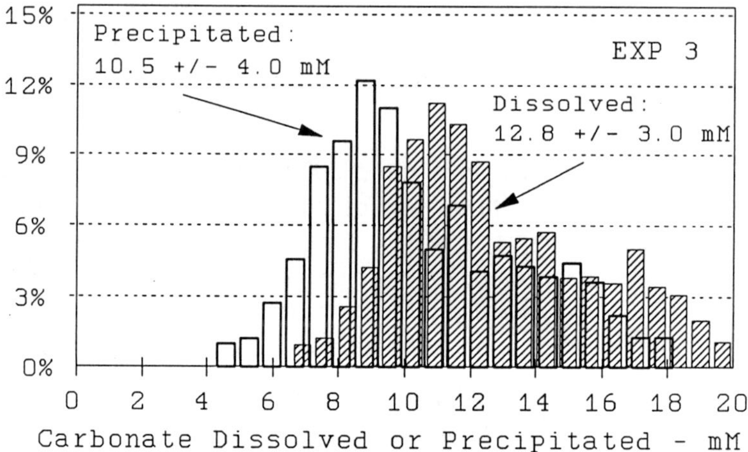

Fig 18.4. Distribution of amounts of dissolved and reprecipitated mineral carbonate: Example 3

Figure 18.4 compares the amounts of dissolved and reprecipitated carbonate. The difference between the mean values is 2.3 mM, the net amount of mineral carbonate precipitation required by the chemical mass balance, but the total amounts dissolved and reprecipitated are much larger than those of the previous example (Fig 18.2) because of the much higher ^{13}C content of this sample. Both distributions are skewed to high values because the sample is so nearly in isotopic equilibrium with mineral carbonate. Had mineral $\delta^{13}C$ values more negative than 0‰ been used in the modeling, the carbonate amounts would have extended to much higher values.

Figure 18.5 shows the distribution of model ages. Two sets of results are shown, one based on σC_{tot} = 10% and the other on σC_{tot} = 5%. The two overlap completely, showing that, in this case, the uncertainty in the chemical parameters does not affect the model ages. The modeled age is −1000 ± 4400 years, and the distributions have tails toward more negative ages. Negative ages reflect overcorrection for mineral carbonate dissolution and reprecipitation in certain ranges of the model parameters. In spite of this uncertainty, the conclusion is that the water is quite young, which is consistent with an age of 800 years determined from ^{39}Ar measurement on the sample (Loosli, Lehmann & Däppen 1991). For reference, the age calculated using an initial ^{14}C of 85 pMC is 15,000 years, whereas that calculated using a correction factor of $\delta^{13}C/-25$ is 2700 years. It is noteworthy that the latter age is within the 1σ range of ages calculated using the full model.

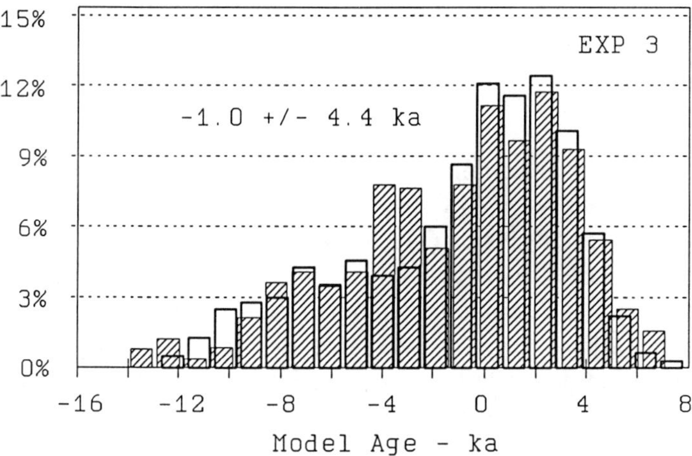

Fig 18.5. Distribution of modeled ages: Example 3

CONCLUSIONS

The examples given here, and additional calculations not illustrated, can be summarized as follows:

- Models considering the evolution of the chemical and stable carbon isotopic compositions of water can lead to reasonable ^{14}C ages for groundwaters.
- Uncertainties in modeled ^{14}C ages can result from uncertainties in the models themselves and in the parameters required for their use.
- Uncertain parameters include the sample chemical composition, the isotopic composition of initial dissolved and aquifer mineral carbonate, and the extent of water carbonate-rock reactions.
- Waters with $\delta^{13}C$ values more negative than around −10 to −12‰ yield model ^{14}C ages with uncertainties as small as several hundred years because the combination of sample chemistry and ^{13}C content limits the possible ranges of uncertain parameters.
- Waters with $\delta^{13}C$ values less negative than around −10 to −12‰ yield model ages with uncertainties of a few thousand years because such ^{13}C values can be produced by a range of initial and mineral $\delta^{13}C$ values and of amounts of minerals dissolved and reprecipitated.
- Waters with ^{13}C values less negative than around −6 to −4‰ approach isotopic equilibrium with the aquifer and may yield uncertainties as large as one half-life or more. In situations where rapid dissolution and reprecipita-

tion is possible – geothermal waters, for example – even young waters may have no measurable ^{14}C.

Acknowledgments

The need to consider uncertainties in ^{14}C model ages was impressed on me by my colleagues of the Working Group on the Isotope Hydrogeology of Northern Switzerland convened by Nagra (The Swiss National Cooperative for the Storage of Radioactive Waste), particularly Professor H H Loosli, whose concern was the intercomparison of results of a number of isotope techniques, and Dr H-J Schmassmann, whose concern was that the results be reasonable hydrogeologically. I am particularly grateful to Dr A Gautschi for encouraging the free-ranging discussion which led to this work.

References

Deines, P 1980 The isotopic composition of reduced organic carbon. *In* Fritz, P and Fontes, J Ch, eds, *Handbook of Environmental Isotope Geochemistry, Vol I, The Terrestrial Environment, A*. Amsterdam, Elsevier: 329–406.

Deines, P, Langmuir, D and Harmon, RS 1974 Stable carbon isotope ratios and the existence of a gas phase in the evolution of carbonate groundwaters. *Geochimica et Cosmochimica Acta* 38: 1147–1164.

Fontes, J-Ch 1992 Chemical and isotopic constraints on ^{14}C dating of groundwater. *In* Taylor, RE, Long, A and Kra, RS, eds, *Radiocarbon After Four Decades: An Interdisciplinary Perspective*. New York, Springer-Verlag, this volume.

Fontes, J Ch and Garnier, JM 1979 Determination of the initial ^{14}C activity of the total dissolved carbon: A review of the existing models and a new approach. *Water Resources Research* 15: 399–413.

Ingerson, E and Pearson, FJ, Jr 1964 Estimation of age and rate of motion of groundwater by the ^{14}C-method. *In* Miyake, Y and Koyama, T, eds, *Recent Researches in the Fields of Hydrosphere, Atmosphere and Nuclear Geochemistry*. Ken Sugawara Volume. Tokyo, Maruzen: 263–283.

International Atomic Energy Agency 1986 *Mathematical Models for Interpretation of Tracer Data in Groundwater Hydrology*. Vienna, IAEA-TECDOC-381, 234 p.

Long, A, Murphy, EM, Davis, SN and Kalin, RM 1992 Natural radiocarbon in dissolved organic carbon in groundwater. *In* Taylor, RE, Long, A and Kra, RS, eds, *Radiocarbon After Four Decades: An Interdisciplinary Perspective*. New York, Springer-Verlag, this volume.

Loosli, HH, Lehmann, BE and Däppen, G 1991 Dating by radionuclides. *In* Pearson, FJ, Jr, Balderer, W, Loosli, HH, Lehmann, BE, Matter, A, Peters, TJ, Schmassmann, H and Gautschi, A, eds, *Applied Isotope Hydrogeology: A Case Study in Northern Switzerland*. Amsterdam, Elsevier and Nagra Technical Report 88-01: 153–174.

Mook, WG 1976 The dissolution-exchange model for dating groundwater with ^{14}C. In *Interpretation of Environmental Isotope and Hydrochemical Data in Groundwater Hydrology*. Vienna, IAEA: 213–225.

_____1986 ^{13}C in atmospheric CO_2. *Netherlands Journal of Sea Research* 20: 211–223.

Neretnieks, I 1981 Age dating of groundwater in fissured rock: Influence of water volume in micropores. *Water Resources Research* 17: 421–422.

Pearson, FJ, Jr 1965 Use of C^{13}/C^{12} ratios to

correct radiocarbon ages of materials initially diluted by limestone. *In* Chatters, RM and Olsen, EA, eds, *Proceedings of the 6th International Conference on Radiocarbon and Tritium Dating*. Washington, US Atomic Energy Commission, CONF-650652: 357–366.

_____1991 Carbonate isotopes. *In* Pearson, FJ, Jr, Balderer, W, Loosli, HH, Lehmann, BE, Matter, A, Peters, TJ, Schmassmann, H and Gautschi, A, eds, *Applied Isotope Hydrogeology: A Case Study in Northern Switzerland*. Amsterdam, Elsevier and Nagra Technical Report 88-01: 175–237.

Plummer, LN, Busby, JF, Lee, RF and Hanshaw, BB 1990 Geochemical modelling of the Madison Aquifer in parts of Montana, Wyoming and South Dakota. *Water Resources Research* 26: 1981–2014.

Plummer, LN, Parkhurst, DL and Thorstenson, DC 1983 Development of reaction models for groundwater systems. *Geochimica et Cosmochimica Acta* 47: 665–686.

Reardon, EJ and Fritz, P 1978 Computer modelling of groundwater ^{13}C and ^{14}C isotope compositions. *Journal of Hydrology* 36: 201–224.

Siegenthaler, U 1972 Bestimmung der Verweildauer von Grundwasser im Boden mit radioaktiven Umweltisotopen (C-14, Tritium). *Gas-Wasser-Abwasser* 52: 283–290.

Wigley, TML, Plummer, LN and Pearson, FJ, Jr 1978 Mass transfer and carbon isotope evolution in natural water systems. *Geochimica et Cosmochimica Acta* 42: 1117–1139.

19

NUMERICAL MODELING WITH GROUNDWATER AGES

MEBUS A GEYH

HISTORICAL REVIEW

The use of ^{14}C analysis for dating groundwater with ages up to about 40,000 BP was introduced by Münnich (1957, 1968). Eriksson (1958) proved that hydrodynamic mixing processes usually must be taken into account. But the piston flow model assuming simple convective flow is still used for the interpretation of ^{14}C data (eg, Andres & Geyh 1970; Bath, Edmunds & Andrews 1979), despite its obvious oversimplification of the usual hydraulic situation.

As early as 1964, Ingerson and Pearson (1964) discussed the importance of the carbon chemistry for transforming ^{14}C data into actual groundwater ages. The content of both bicarbonate and dissolved carbon dioxide are the decisive parameters for the initial ^{14}C content, which must be known for this transformation of the conventional ^{14}C time scale into the groundwater ^{14}C time scale. This model assumes 100% modern carbon for soil CO_2, zero for solid carbonates and perfect stoichiometry for reactions involving carbon. The $\delta^{13}C$ values of total dissolved inorganic carbon (TDIC) compounds offer another possibility for correcting the carbon chemistry for initial ^{14}C content by adjusting for mixing of the carbon compounds (Mook 1976; Wigley 1977; Wigley, Plummer & Pearson 1978; Reardon & Fritz 1978; Fontes & Garnier 1979). All these cited papers dealing with the correction problem of groundwater ^{14}C dates have given the impression that hydrochemical processes rather than radioactive decay and hydrodynamic mixing are the decisive factors for the ^{14}C content in TDIC. The possibility of dating groundwater *via* ^{14}C was questioned.

These theoretical approaches were contrary to experience gathered in many hydrological studies of freshwater systems. We found that the hydrochemical correction for conventional ^{14}C ages of groundwater is rather constant in most cases after a few thousand ^{14}C years after groundwater recharge (Fig 19.1). This is demonstrated, for example, by the results of Phillips *et al* (1989). The mean values and their corresponding standard deviations calculated from the differences between the conventional ^{14}C dates and the corrected ages are com-

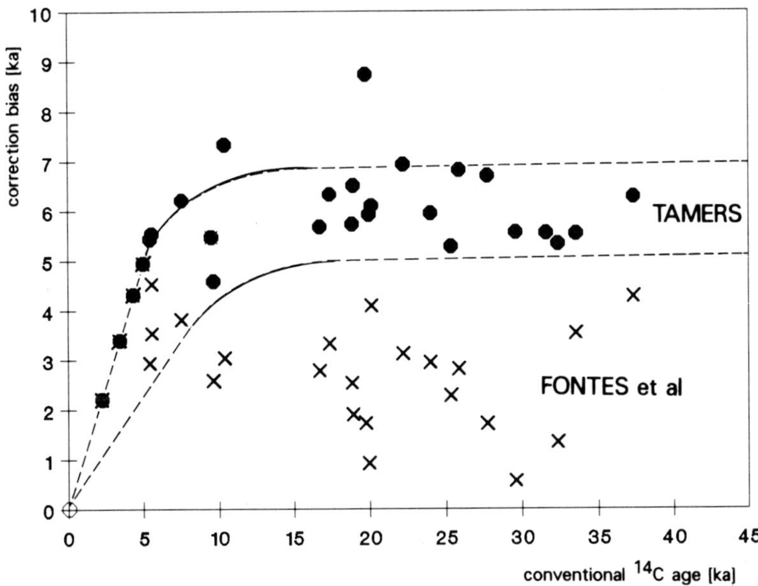

Fig 19.1. Difference between conventional and corrected ^{14}C ages of water samples studied by Phillips *et al* (1989). The simple hydrochemical correction by Tamers (1967) yields rather constant values after a few 1000 ^{14}C years. The correction bias based on δ^{13}C values and hydrochemical data developed by Fontes and Garnier (1979) shows much more scatter over the ^{14}C time scale and indicates possible overinterpretation of the hydrochemical and stable isotope data.

piled in Table 19.1. It is obvious that the standard deviations increase with the complexity of the correction models or the number of parameters involved. An exception is the reaction path model where sufficient data was not available. The most simple model, by Tamers (1967), yields a standard deviation of ± 590 years whereas the most complex one, by Mook (1976), yields ± 2030 years. The means of the various correction values cover a wide range (960–6120 yrs) although the results of the most advanced models are closer together (960–2690 yrs). An empirical estimate of the correction bias may be obtained by comparing the histogram of conventional ^{14}C dates of groundwater samples with the paleoclimatic record. In humid areas, for instance, it is well established that groundwater recharge was interrupted between about 20,000 and 12,000 BP (Andres & Geyh 1970; Bath, Edmunds & Andrews 1979). Paleohydrological information sometimes exists for semiarid areas (Geyh *et al* 1985). Another reasonable method for estimating the bias of the ^{14}C time scale of groundwater is based on empirical values determined for various geologically different catchment areas (Geyh 1972).

TABLE 19.1. Means and standard deviations of the differences between conventional ^{14}C groundwater ages and corrected ages using the results by Phillips *et al* (1989), who applied different hydrochemical correction models

Model	Correction bias (yr)	Standard Deviation
Vogel	1300	–
Tamers	6120	± 590
Reaction path	1650	± 760
Pearson	1140	± 1490
Fontes	2690	± 1800
Mook	960	± 2030

According to our studies and more recent ones, hydrodynamic processes resulting in complex convective mixing are usually more decisive for the interpretation of ^{14}C data than hydrochemical processes (eg, Geyh 1980; Phillips *et al* 1989). Multi-environmental isotope analyses often yield qualitative or quantitative hydrodynamic information rather than groundwater ages. The most recent improvement in the interpretation of ^{14}C data has been their application in conceptional and numerical mass transport modeling (Geyh & Backhaus 1979; Pearson, Noronha & Andrews 1983; Fröhlich *et al* 1987; Phillips *et al* 1989).

NUMERICAL MODELING - PRINCIPLES

Numerical flow modeling is a well-established method for estimating the amount of extractable groundwater, the resulting changes in the morphology of the groundwater table, and the flow pattern. The Darcy law is the basic equation

$$v_D = \frac{k \times \Delta}{d}$$

where v_D is the groundwater flow in m^3 through an area of 1 m^2 per time unit, k is the hydraulic conductivity in m per unit time, and Δ is the difference in the level of the water table over the distance d (hydraulic gradient). The field velocity is obtained from v_D by division by the void ratio n. Theoretically, numerical flow modeling can substitute groundwater dating via ^{14}C with all its correction problems. However, validation and calibration of such models is still a great problem. For instance, numerical flow modeling does not yield the present groundwater recharge rate if at least parts of the groundwater resource are fossil (Burdon 1977). In this case, the hydraulic gradient is a superposition of the one representing the present recharge-discharge steady-state and that of the discharge of the fossil water body.

Mass transport modeling improves the results of the numerical modeling by additional consideration of mass conservation. In the transport equation, the concentrations of dissolved constituents such as environmental isotopes are suitable parameters. If the groundwater is fresh and older than about 1000 years, ^{14}C is a suitable and rather conservative tracer. Evidence for this are similar $\delta^{13}C$ and bicarbonate values. Both result in a rather constant correction bias of conventional ^{14}C ages of groundwater (Table 19.1, Fig 19.1).

In the one-dimensional case, the equation for mass conservation (assuming no hydrochemical and physical reactions between the solid and dissolved constituents in the aquifer) is given by

$$\frac{\partial}{\partial t}(n \cdot C) = -\frac{\partial}{\partial x}\left(v_D \cdot C - n \cdot D \cdot \frac{\partial C}{\partial x}\right) - \lambda \cdot n \cdot C \pm I$$

where λ is the decay constant of the radioisotope and D is the dispersion coefficient. This equation describes the temporal change of concentration C due to convective solute transport and dispersion (terms 1 & 2, respectively, in the second parentheses), the first-order process of radioactive decay and the source/sink term $\pm$ I (Pearson, Noronha & Andrews 1983).

Mass transport modeling is not yet standard in numerical modeling because 1) computer time is considerably increased, 2) the data base is often insufficient and 3) the hydrochemical processes involving the monitored non-conservative tracers are often not yet well understood.

NUMERICAL MASS TRANSPORT MODELING

Radiocarbon as Age Indicator

Steady-State: Humid Regions. Tamers, Stipp & Weiner (1975) made one of the first attempts to model ^{14}C data for applied hydrogeologic studies. They measured and interpreted the ^{14}C content of the deep groundwater of the Biscayne aquifer in southern Florida to find safe extraction limits in the heavily exploited aquifer zones in order to prevent pollution from septic tanks in the region. The model calculations helped the authors to advise the management of the waterworks without disturbance and interruption of the water supply.

Another study involved an interaquifer mixing in a confined aquifer system north of Nuremberg, Germany (Geyh & Backhaus 1979; Geyh *et al* 1984). The groundwater, which flows to the northwest (Fig 19.2), is recharged along a fracture zone in the east and percolates into a deep sandstone aquifer. The thickness of the confining leaky aquitard decreases from about 200 m in the east to zero in the west. The hydraulic conductivity coefficients of the aquifer and the aqui-

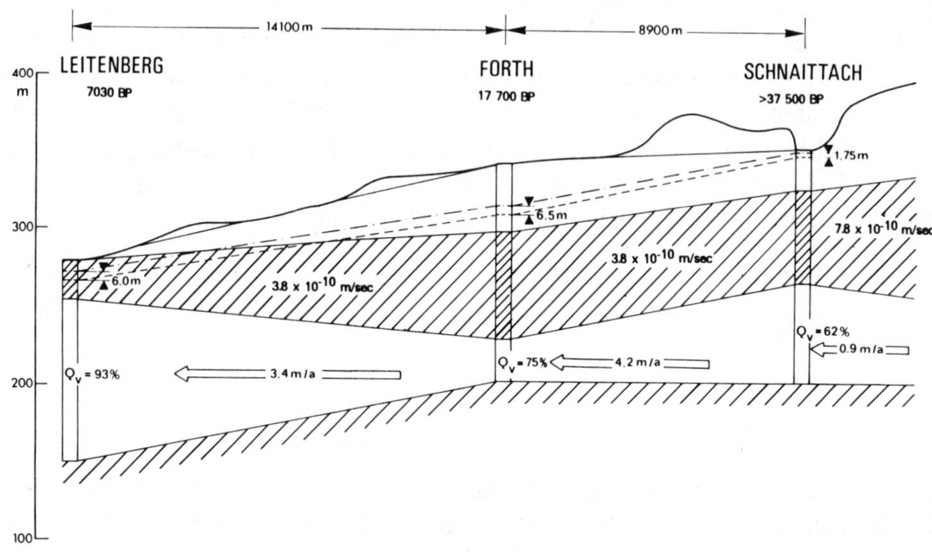

Fig 19.2. Hydrogeological section along the direction of groundwater flow in the study area, showing the conventional ^{14}C ages (BP) of the deep groundwater, the regional hydraulic conductivity (m/sec) of the aquitard, the percentage of vertical inflow of groundwater Q_v, and the tracer velocity (m/yr) (Geyh *et al* 1984)

tard differ by at least two orders of magnitude. Hence, the horizontal flow velocity is larger than the vertical one.

The conventional ^{14}C ages for TDIC in water samples from the confined aquifer unexpectedly decrease in the direction of groundwater flow, *ie*, the ^{14}C content increases (Fig 19.2). A conceptual one-dimensional transport model (Geyh *et al* 1984) was applied to explain this surprising finding. The following assumptions were made: 1) the porosity and hydraulic conductivity are constant throughout the study area; 2) water in the confined aquifer is rather well mixed due to pumping during sampling; 3) the mean residence time of the groundwater in the phreatic aquifer is near zero; 4) the initial ^{14}C content of groundwater in the unconfined and confined aquifers is the same; 5) secondary hydrochemical processes did not occur in the confined aquifer as the hydrochemical properties of all samples were very similar; and 6) steady-state recharge-discharge conditions exist.

Because the piezometric head of the unconfined aquifer in this example is always higher than that of the confined aquifer, water percolating through the aquitard is continuously mixed with the horizontally flowing groundwater, changing the isotopic and hydrochemical areal distribution. Near the catchment area, the admixed water from the aquitard is older than that entering directly from the catchment area (Fig 19.3). Hence, the apparent ^{14}C ages are slightly older than the actual travel time. Farther away in the direction of groundwater flow, is an area in which the conventional ^{14}C age of the groundwater agrees with the travel time. Still farther away, the input of ^{14}C atoms through the aquitard into the confined aquifer finally counterbalances loss by radioactive decay and the apparent ^{14}C ages remain constant, independent of the actual travel time. Even the flow direction can no longer be recognized for the one-dimensional case study (Geyh & Backhaus 1979; Fig 19.3).

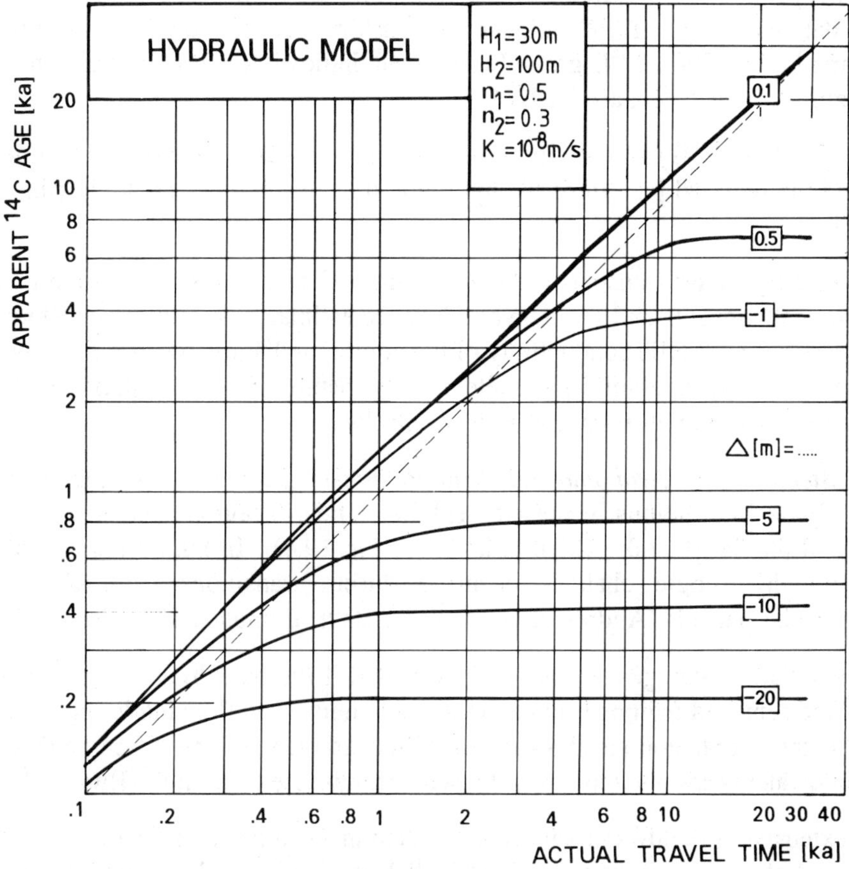

Fig 19.3. One-dimensional scheme for convective mixing of the groundwater in a confined aquifer (thickness H_2, total porosity n_2) with that entering from a leaky aquitard (thickness H_1, total porosity n_1, hydraulic conductivity k) and the resulting changes in apparent ^{14}C ages, compared to the actual travel time of the groundwater as a function of the difference (m) between the higher hydraulic head of the unconfined aquifer and the lower head of the confined aquifer (Geyh & Backhaus 1979). Near the catchment area, the apparent ^{14}C ages are older than the travel time; farther away, the two are equal. Still farther away, the apparent ^{14}C ages remain constant and independent of travel time.

In the study area, mass transport modeling of the ^{14}C data yielded the following hydrogeological information instead of water ages (Fig 19.2): 1) the regional k value for the aquitard (3.8×10^{-10} m/s, which agrees well with those less precisely determined by pump tests); and 2) a quantitative estimate of the mass balance (only 5–10% of the naturally discharged groundwater has its origin in the fractured zone) (Geyh *et al* 1984). Most of it comes from percolation through the aquitard.

These results are in contradiction to the previous hydrogeological concept but are independently supported by mass transport modeling with $\delta^{18}O$, rare gas and ^{39}Ar data (Geyh *et al* 1984). Linear relationships were found with respect to the apparent conventional ^{14}C groundwater ages, indicating two-component mixing (Mazor, Kaufman & Carmi 1973).

Phillips *et al* (1989) recently presented a similar successful study. They determined the regional pattern of the hydraulic transmissivity of an aquifer in New Mexico.

In summary, mass transport modeling using ^{14}C data helps to improve hydrogeological concepts and groundwater balance estimates, as well as to determine the regional hydraulic conductivity of the aquifer and aquitards or other relevant regionally valid hydraulic parameters. The results may be superior to those obtained by pumping tests that reflect only local conditions.

Non-Steady-State: Arid and Semiarid Regions. During the geological past, hydrological conditions in present arid and semiarid regions changed between dry and pluvial periods (Geyh *et al* 1985; Fig 19.4). In humid regions of the present, hydrologic changes occurred during transitions from glacial to interglacial periods (Andres & Geyh 1970; Bath, Edmunds & Andrews 1979).

Groundwater recharge rates are overestimated if the paleohydrology of present arid and semiarid regions is not considered in numerical flow modeling. In such cases, the actual non-steady-state hydraulic gradient of the "decaying" fossil groundwater is assumed to reflect present steady-state conditions (Burdon 1977).

An extensive groundwater survey conducted in 1976 north of Khartoum, Sudan and east of the Nile is an example (Fröhlich *et al* 1987). The groundwater table in the sandy-loamy Quaternary aquifer overlying the Nubian sandstone aquifer is about 35 m lower 35 km east of the Nile than at the Nile (Fig 19.5). For this gradient, numerical modeling yielded a horizontal bank infiltration rate of 17.5×10^6 m^3 yr^{-1} for the 50 km river bank in the study area. Taking the uncertainties of the method into account, this more or less counterbalances the pumping of 19×10^6 m^3 yr^{-1} water in 1976. The finite tritium values of dug well samples

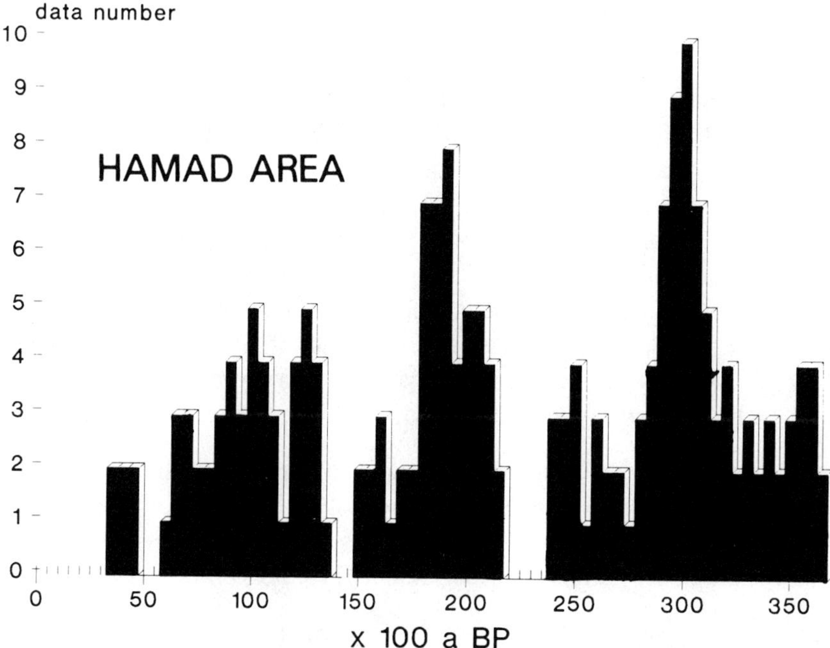

Fig 19.4. Hydrological scenario during the last 30,000 years in the present-day arid Hamad area in Iraq, Jordan, Syria and Saudi Arabia (Geyh *et al* 1985). Distinct periods of groundwater recharge are reflected by a histogram of the conventional ^{14}C dates for groundwater samples. The correction bias may amount to 2500 years. The histogram of conventional ^{14}C dates of stalagmite samples from Palestine looks similar.

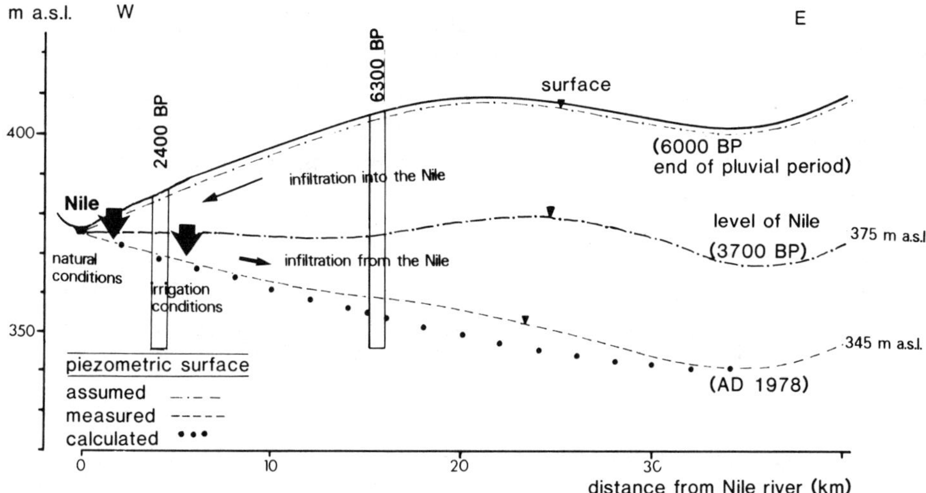

Fig 19.5. Hydrogeological section of the area east of the Nile River north of Khartoum, measured and calculated piezometric level as well as conventional ^{14}C ages of the groundwater. The gradient of the groundwater table gave the impression of intensive infiltration of Nile water. Numerical mass transport modeling of the ^{14}C data showed that the hydraulic gradient is mainly that of relict groundwater from a paleopluvial period about 6000 BP. Present-day seepage is negligible (Fröhlich *et al* 1987).

up to 5 km from the Nile, corresponding to water ages less than 30 years, apparently supports these calculations.

In 1986, we repeated the isotope hydrological survey and included ^{14}C analyses of water samples from the few drilled wells. The results were in obvious conflict with the hydraulic concept obtained from numerical modeling. According to the ^{14}C data, most of the groundwater recharged during pluvial periods about 4000–6000 years ago. A reinterpretation of the old data on electrical conductivity, nitrate and bicarbonate contents and stable isotope data of oxygen and hydrogen, as well as ^{3}H isotope data supports the concept that natural Nile water recharge takes 20 years to infiltrate and travel about 1.5 km from the Nile. The tritiated groundwater farther away indicates infiltration of Nile water from irrigation canals, particularly at distances of 1.5–4 km from the Nile.

A rather simple one-dimensional mass transport model for the ^{14}C data yielded a real bank infiltration rate of only $1.5 - 4 \times 10^{-6}$ m^3 yr^{-1} along the river bank. Hence, the actual groundwater availability is one magnitude of order smaller than that estimated by the numerical model (Fröhlich *et al* 1987).

Meanwhile, this concept has been indirectly confirmed by a drastic lowering of the groundwater table in the study area.

Radiocarbon as Quality Indicator

Non-Steady-State: Overexploited Groundwater Resources. During recent decades, even in humid areas, heavy usage of groundwater resources has disturbed the recharge/discharge water balance. Groundwater withdrawal and the resulting lowering of the groundwater table has been shown to increase groundwater percolation through aquitards separating adjacent aquifers. Changes of ^{14}C content, and often of chemical composition of samples from production wells are strong indications of overexploitation, *ie*, "groundwater mining" (Fig 19.6; Geyh & Backhaus 1979).

Under rather ideal conditions, conventional ^{14}C dates of groundwater show similar changes within the pumping area as a function of time. In such cases, changes in ^{14}C can help quantify the changes of hydrodynamic mixing induced by water abstraction. Fröhlich *et al* (1987) illustrated this in the Azraq area, Jordan.

Long-term trends in ^{14}C dates for heavily exploited aquifers showing complex stratification, however, are not yet understood. The considerable variation in only a few years makes it rather problematic to interpret ^{14}C data as groundwater ages (Geyh 1986). For example, ^{14}C data for samples from wells in the approximately 200-m-deep unconfined limestone aquifer below Mainz, Germany, has

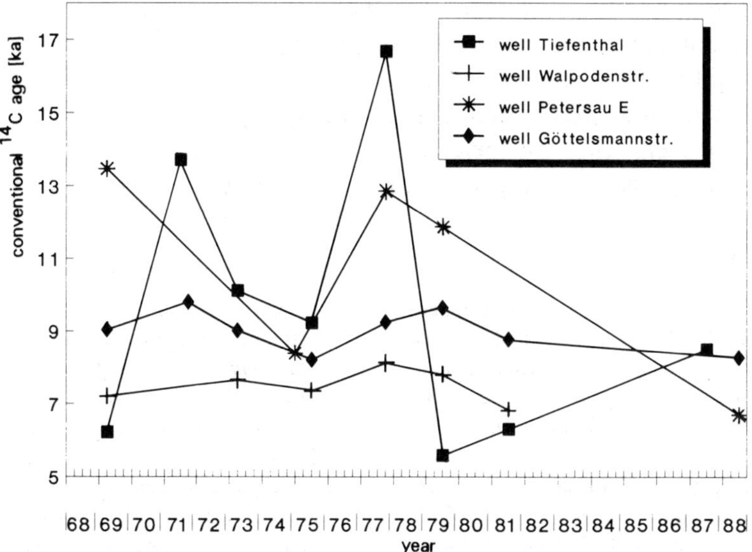

Fig 19.6. One of three representative types of trends in the temporal change of conventional ^{14}C groundwater ages for wells in the heavily exploited Mainz aquifer system. These trends are irregularly distributed among the wells. It is not yet known if ^{14}C data of groundwater collected from such disturbed groundwater resources can be hydraulically interpreted.

been monitored since 1968 (Geyh & Sonne 1983). Three temporal ^{14}C trends (Fig 19.6) do not show any local clustering. Analogous observations have been made in many other aquifers both unconfined and confined, fractured and sedimentary. This opens a wide field of future research questions:

1. What are the pumping parameters (eg, abstraction rate, total abstraction, depth of filter or pump) that are decisive for the observed ^{14}C changes?
2. What information can be derived from the similarity of the long-term ^{14}C trends? Can recommendations be made to optimize groundwater abstraction from single wells or at least for the aquifer as a whole?
3. Can the initial ^{14}C content be reconstructed from the trends?
4. Can ^{14}C data from overexploited aquifers be used for developing hydrogeological concepts? Are isotope hydrological studies at all useful under such conditions?

CONCLUSIONS

Many case studies already show that mass transport modeling with ^{14}C and other environmental isotope and hydrochemical data from groundwater samples collected from rather undisturbed groundwater resources are helpful for improving hydrogeological concepts, for estimating regional hydraulic field parameters and for calibrating numerical models. The situation is not always clear for heavily exploited aquifer systems. Further research is needed. Experience with freshwater systems supports the conclusion that hydrochemical effects are often of minor importance in comparison to the effects of convective mixing.

REFERENCES

Andres, G and Geyh, MA 1970 Isotopenphysikalische Untersuchungen über den Grundwasserhaushalt im überdeckten Sandsteinkeuper mit Hilfe von ^{14}C- und ^{3}H-Wasseranalysen. *Die Wasserwirtschaft* 8: 259–263.

Bath, AH, Edmunds, WM and Andrews, JN 1979 Palaeoclimatic trends deduced from the hydrochemistry of a Triassic sandstone aquifer, United Kingdom. *Isotope Hydrology 1978* (2). Vienna, IAEA: 545–566.

Burdon, DJ 1977 Flow of fossil groundwater. *Quarterly Journal of Engineering Geology* 10: 97–124.

Eriksson, E 1958 The possible use of tritium for estimating groundwater storage. *Tellus* 10: 472–479.

Fontes, JC and Garnier, JM 1979 Determination of the initial ^{14}C activity of the total dissolved carbon. A review of the existing models and a new approach. *Water Resources Research* 15: 399–413.

Fröhlich, K, Geyh, MA, Verhagen, BT and Wirth, K 1987 Isotopenhydrologische Methoden zur Begutachtung von Grundwasser in Trockengebieten. Entwicklung eines Instrumentariums für die Beurteilung gefährderter Vorkommen. *Forschungsberichte des Bundesministeriums für wirtschaftliche Zusammenarbeit* 85. Cologne, Welt-Forum: 179 p.

Geyh, MA 1972 Basic studies in hydrology and ^{14}C and ^{3}H measurements. *Proceedings of the 24th International Geology Congress* 11: 227-234.

____1980 Hydrogeologic interpretation of the ^{14}C content of groundwater – a status report. *Fisika* 12: 87–106.

____1986 Computer modeling of confined aquifer systems for interpretation of chemical and environmental isotope data. In *Mathematical Models for Interpretation of Tracer Data in Groundwater Hydrology* (IAEA-TECDOC-381). Vienna, IAEA: 165–179.

Geyh, MA and Backhaus, G 1979 Hydrodynamic aspects of carbon-14 groundwater dating. *Isotope Hydrology 1978* (2). Vienna, IAEA: 631–643.

Geyh, MA, Backhaus, G, Andres, G, Rudolph, J and Rath, HK 1984 Isotope study on the Keuper sandstone aquifer with a leaky cover layer. *Isotope Hydrology 1983*. Vienna, IAEA: 499–513.

Geyh, MA, Khouri, J, Rajab, R and Wagner, W 1985 Environmental isotope study in the Hamad region. *Geologisches Jahrbuch* C38: 3–15.

Geyh, MA and Sonne, V 1983 Monitoring of groundwater budget changes with isotope techniques in the NE Mainz Basin. *Proceedings of the International Conference on Groundwater Resources Plan D*: 357–365.

Ingerson, E and Pearson, FJ, Jr, 1964 Estimation of age and rate of motion of groundwater by the ^{14}C method. In *Recent Researches in the Fields of Hydrosphere, Atmosphere, and Nuclear Geochemistry*. Tokyo, Maruzen: 263–283.

Mazor, E, Kaufman, A and Carmi, I 1973 Hammat Geder (Israel): Geochemistry of a mixed thermal spring complex. *Journal of Hydrology* 18: 289–303.

Mook, WG 1976 The dissolution-exchange model for dating groundwater with ^{14}C. In *Interpretation of Environmental Isotope and Hydrochemical Data in Groundwater*

Hydrology. Vienna, IAEA: 213–225.

Münnich, KO 1957 Messung des ^{14}C-Gehaltes von hartem Grundwasser. *Naturwissenschaften* 34: 32–33.

_____1968 Isotopen-Datierung von Grundwasser. *Naturwissenschaften* 55: 158–163.

Pearson, FJ, Jr, Noronha, CJ and Andrews, RW 1983 Mathematical modeling of the distribution of natural ^{14}C, ^{234}U, and ^{238}U in a regional ground-water system. *In* Stuiver, M and Kra, RS, eds, Proceedings of the 11th International ^{14}C Conference. *Radiocarbon* 25 (2): 291–300.

Phillips, FM, Tansey, MK, Peeters, LA, Cheng, S and Long, A 1989 An isotopic investigation of groundwater in the Central San Juan Basin, New Mexico: Carbon 14 dating as a basis for numerical modeling. *Water Resources Research* 25: 2259–2273.

Reardon, EJ and Fritz, PE 1978 Computer modeling of groundwater ^{13}C and ^{14}C isotope compositions. *Journal of Hydrology* 36: 201–224.

Tamers, MA 1967 Surface-water infiltration and groundwater movement in arid zones of Venezuela. *Isotopes in Hydrology*: 339–353.

Tamers, MA, Stipp, JJ and Weiner, R 1975 Radiocarbon ages of groundwater as a basis for the determination of safe limits of aquifer exploitation. *Environmental Research* 9: 250–264.

Wigley, TML 1977 Carbon-14 dating of groundwater from closed and open systems. *Water Resources Research* 11: 324–328.

Wigley, TML, Plummer, LN and Pearson, FJ, Jr, 1978 Mass transfer and carbon isotope evolution in natural water systems. *Geochimica et Cosmochimica Acta* 42: 1117–1139.

20

NATURAL RADIOCARBON IN DISSOLVED ORGANIC CARBON IN GROUNDWATER

AUSTIN LONG, ELLYN M MURPHY, STANLEY N DAVIS and ROBERT M KALIN

INTRODUCTION

Groundwater dating is important to many studies of hydrologic systems, and radiocarbon dating is one of the most common radiometric techniques employed. Chapters 17, 18 and 19 treat theory and application of groundwater dating with dissolved inorganic carbon (DIC). A major source of uncertainty in radiocarbon dating of groundwater is the reconstruction of the geochemical processes in the aquifer involving carbon. In this chapter, we review published studies of the isotope geochemistry of dissolved organic carbon (DOC) in groundwater. We discuss studies of gases in the unsaturated zone, insofar as they impact the organic geochemistry of groundwater. Despite the relatively immature state of knowledge in these areas, importance of such studies to the overall understanding of carbon geochemistry in groundwater is clear. For example, some DOC components may oxidize to DIC within the aquifer, thus affecting the ^{14}C of the DIC. In many groundwater systems, the ^{14}C content of organic fractions will probably be an important link to understanding both the origin of DOC and DIC and the age relations among the soluble carbon species present. To date, the success of the application of radiocarbon dating in hydrogeologic studies has depended almost entirely on ^{14}C measurements of DIC in groundwater. As other chapters in this volume demonstrate, this application has evolved to highly sophisticated and computer-dependent modeling techniques, requiring as much information as possible about sources and sinks of dissolved carbon. However, as DOC is part of the groundwater carbon inventory, groundwater models for ^{14}C dating are incomplete without knowledge of the role of DOC. Additionally, the ^{14}C in certain fractions of DOC may yield independent estimates of age. In contrast to groundwater DIC ^{14}C studies, the hydrologic interpretation of ^{14}C measurements of DOC is still in its infancy. Reasons for this comparatively primitive stage of development are the difficulty of extracting and isolating DOC, the low carbon yield from extraction; hence, the requirement that ^{14}C be measured by accelerator mass spectrometry (AMS), and the complexity of the geochemistry of DOC in groundwater. In view of these handicaps, what compensating advantages make ^{14}C information on DOC worth pursuing?

Quantitative estimates of ages, or average subsurface residence times, of groundwater are extremely important in many situations. For example, ^{14}C ages of groundwater can reveal the times of major aquifer recharge and present rate of recharge. This is information valuable in evaluating the groundwater as a municipal or agricultural resource. Another application of groundwater dating is estimating the rate of water movement within polluted aquifers in order to estimate the dispersal rate of pollutants or contaminants. Perhaps the major justification for studying ^{14}C from DOC would be comparison with groundwater travel times inferred from ^{14}C in DIC. All methods of age estimation involve uncertainties, and for most studies, it is valuable to use multiple techniques that, as much as possible, have mutually independent uncertainties. In a groundwater system, the isotopic contents of the organic and inorganic carbon components are related. Some of the same processes that affect concentrations of ^{14}C in DOC also affect the concentration of ^{14}C in DIC. For instance, CO_2 produced from oxidation of organic matter most likely will alter the ^{14}C activity of the DIC. In addition, the lower pH associated with the higher concentrations of dissolved CO_2 will dissolve calcite if it is present in the aquifer, further altering the ^{14}C activity in the DIC (Pearson & Friedman 1970; Chapelle *et al* 1987). Under highly anaerobic conditions, methanogenic bacteria can consume CO_2 (Games, Hayes & Gunslaus 1978). No interpretation of groundwater ages from ^{14}C in DIC can be complete without knowledge of the geochemical behavior of the organic matter in the system.

Measurements of concentrations of carbon isotopes can also help distinguish between artificial and natural organic compounds in groundwater. This is possible because most artificial chemicals are synthesized from petroleum, which has undetectable amounts of ^{14}C, whereas natural organic compounds, if they contain carbon derived from photosynthetic organisms within the past 50,000 years, will have detectable ^{14}C. A more powerful potential technique for identifying intrusion of artificial compounds into the groundwater would combine a detailed chemical characterization of dissolved organic compounds with a study of the carbon isotopes of the individual compounds. This approach has the capability of distinguishing artificially derived from naturally derived components of DOC that are chemically similar. As isolation and identification techniques improve, it may become feasible to identify a large number of organic compounds from an aquifer, and to infer their most likely sources and pathways within the aquifer. In addition, the stability of most organic compounds is very sensitive to temperature, so that under ideal circumstances, it might be possible to identify limits on the maximum temperatures the groundwater has experienced during its subsurface migration. Such information would be valuable in evaluating potential geothermal resources as well as reconstructing patterns of regional groundwater circulation. This knowledge, when combined with isotopic information, should allow a much more detailed reconstruction of the history of groundwater than is possible at present. Computer models of groundwater flow, if tested and validated on the past performance of the aquifer, can more confidently project future performance of the aquifer, be it water resources, movement of contaminants or response of the system to global changes.

The first studies on age dating groundwater using ^{14}C in DOC have told us much about the geochemical behavior of DOC in groundwater. This review leans heavily on the pioneering work of Michael Thurman and his associates, who developed many of the techniques for separating DOC fractions that have been used by researchers at the University of Arizona (Thurman 1985a, b; Murphy, Long & Davis 1985; Murphy 1987; Murphy *et al* 1989a, b). In recent years, the hydrogeology research group at the University of Waterloo and researchers at the University of Maryland have contributed studies of ^{14}C in groundwater DOC (Wassenaar, Aravena & Fritz 1989; Purdy *et al* 1992).

POSSIBLE SOURCES OF DISSOLVED ORGANIC CARBON IN GROUNDWATER

For accurate groundwater dating by ^{14}C, an understanding of the nature and sources of organic carbon in groundwater is as important as an understanding of the sources and sinks of DIC. Organic carbon in subsurface systems may exist in either particulate or dissolved phases. Particulate organic carbon may derive from 1) lithified or fossiliferous plant remains, such as lignite or kerogen, 2) particle-associated humic substances that have been transported as DOC from the overlying soil and vadose zone, and 3) particle-associated organic carbon originating by oxidation of lignite or kerogen (Zachara, Cowan & Murphy 1990). Thurman (1985a) hypothesized that DOC in groundwater may derive from either 1) the interstitial soil waters of the vadose/recharge zone, dislodged from larger organic masses by microbiological activity, or 2) the dissolution or oxidation of kerogen embedded in the aquifer matrix. Assuming that the above five sources/processes are comprehensive, a combination of these sources will ultimately determine the nature, distribution and levels of the DOC.

The main factors affecting the distribution and concentration of DOC in groundwater are 1) the organic productivity of the overlying soils in the recharge/vadose zone, 2) the mass fraction of particulate organic carbon (f_{OC}) in the aquifer sediments, and 3) the pH of the groundwater (Murphy & Aiken 1991). As Thurman (1985b) showed, f_{OC} in the host rocks may have a profound effect on the concentration of DOC in groundwater. For instance, groundwaters associated with oil shales or oil-field brines typically have high DOC levels, reflecting the high organic carbon content in the host rock. The combination of basic pH and oil-shale deposits is largely responsible for the high DOC concentrations found in trona groundwaters. In contrast, the relatively low pH (pH<6) found in Atlantic Coastal Plain aquifers, on the east coast of the United States, (Purdy *et al* 1992) may slow the dissolution and mobilization of organic carbon from lignite deposits typically found in these sandy aquifer sediments. Although basalt flows are originally devoid of organic carbon, the presence of paleosol interbeds rich in organic carbon coupled with basic pH can lead to relatively high DOC in basalts (Wood & Low 1988). The overall DOC in groundwaters is also controlled by the respiration of heterotrophic micro-organisms, which could lead to a decrease in DOC along groundwater flow paths.

The literature generally classifies DOC according to physical properties of its isolatable components. Here we attempt to clarify nomenclature in common use. The DOC in groundwater is a diverse mixture of organic compounds, ranging from macromolecules, such as humic substances, to low-molecular-weight (LMW) compounds, such as simple organic acids (Thurman 1985a). The average molecular weight of fulvic acid, the dominant humic substance in groundwater, ranges from approximately 500 to 2000 atomic mass units (amu) (Thurman *et al* 1982; Aiken & Malcolm 1987). These averages are much lower than the average molecular weights of aquatic humic acids, which range from 500 to more than 10,000 amu (Thurman *et al* 1982). Organic geochemists generally classify larger, complex molecules in terms of functional groups and their chemical and physical behavior in water. Humic and fulvic acids are colored polyelectrolytic acids that are isolated from water by sorption onto XAD resins or weak-base ion exchange resins, or by some comparable procedure (Thurman 1985a). At pH 2, the humic acid precipitates and fulvic acid remains in solution. Humic and fulvic acids are often called hydrophobic acids because of their sorption to XAD resin at low pH. Murphy *et al* (1989a, b) referred to humic substances as the high-molecular-weight (HMW) fraction of the DOC in groundwater. Hydrophilic acids are those that pass through XAD resin columns at pH 2.

In the recharge zone of an aquifer, as the soil-derived DOC components move through the porous media, they react with the solid matrix, reducing the concentration of the DOC and changing the distribution of the components within the DOC. The concentration of the DOC in the soil interstitial waters decreases with depth in the soil profile, from average concentrations of ~20 mg/L in the O/A horizon (the uppermost zone in the profile, rich in plant fragments and organic matter) to ~1 mg/L in groundwater (Thurman 1985a;

Leenheer *et al* 1974). Figure 20.1, from Murphy and Aiken (1991), illustrates the decrease in DOC concentration and character with depth in soil profiles. Thurman (1985a) attributes changes in DOC concentration to selective microbial mineralization and sorption to soil particles.

Scharpenseel *et al* (1989) demonstrated the downward movement of organic matter in the soil horizon. They traced the migration of organic matter in undisturbed soil profiles by measuring the penetration of ^{14}C from thermonuclear bomb testing in the 1960s. This natural "bomb pulse" in the organic matter was traced to depths of 2 m in five soil orders. In all cases, the apparent mean residence time of the soil organic matter (based on ^{14}C content) increased with depth.

Siliceous sediments will have a chromatographic effect on the distribution of DOC in both the saturated and unsaturated zones (Fig 20.1). Cationic organic solutes are readily removed from the DOC by cation exchange on siliceous aquifer materials; this is a significant sorption mechanism. Organic solutes that may exist as cations in natural waters (*ie*, pH 6–8) include amino acids, polypeptides (amide groups) and phenolic compounds, such as catechol, cresol, alkyl phenol, vanillin, naphthol and syringaldehyde. Aquifer sediments retain neutral and LMW anionic organic compounds the least (Malcolm *et al* 1980). Neutral organic solutes are defined as those organic compounds possessing hydroxyl,

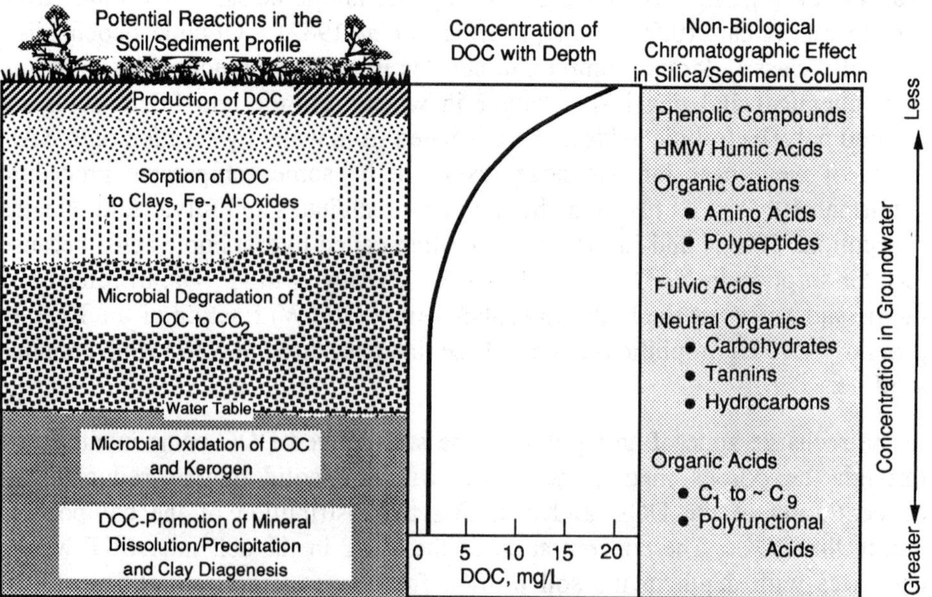

Fig 20.1. Idealized DOC profile in a soil zone, and processes controlling changes in DOC concentration and composition with depth (from Murphy and Aiken 1991)

ether, ketone, aldehyde, ester or lactone functional groups. The neutral organic solutes usually fall into the following categories: carbohydrates; tannins; and hydroxy and keto acids. Anionic organic solutes, such as volatile and nonvolatile fatty acids, form one of the largest categories of compounds in the DOC of natural waters.

In a study of forested watersheds in Adirondack Park, New York, Cronan and Aiken (1985) showed the fractionation of DOC in soil interstitial waters between the O/A horizon interface (20 cm) and the B horizon (50 cm). Although humic substances were the dominant DOC fraction in the O/A horizon interface (ca 50% of the DOC), they were a minor component of the DOC in the B horizon pore waters (~25%). The more mobile (less reactive) LMW anionic organic compounds were the dominant component of the DOC in the B horizon, accounting for over 50% of the DOC. In groundwater, the more mobile neutral and LMW anionic organic species may account for over 70% of the DOC due to this chromatographic effect, whereas humic substances are generally a minor component of the DOC (Murphy *et al* 1989a, b; Thurman 1979, 1985a, b).

Although no comprehensive study has been done on the transport and fractionation of DOC from the soil horizon to the saturated zone, a few general observations can be made. Groundwaters that are close to the recharge zone or groundwaters in shallow, unconfined aquifers will have a DOC composition and concentration influenced by soil or surface water organic matter. For example, the DOC in the Biscayne aquifer (Dade County, Florida, USA, ~13 mg/L DOC) originates from DOC in the Everglades, which ranges between 30 and 40 mg/L (Murphy 1987). The percentage of humic substances in the DOC of the Biscayne groundwaters, ~66%, is atypical of normal groundwaters, and reflects the unique Everglade origin (Thurman 1979). The DOC in the groundwaters of deep, confined aquifers or groundwaters that are distant from the recharge zone may be heavily influenced by kerogen or lignites in the aquifer matrix. In such cases, the isolated humic materials in the dissolved phase are a minor component of the total DOC (Thurman 1985a; Murphy *et al* 1989a), and may originate either partly or solely from organics in the buried sediments. Microorganisms in the subsurface sediments create a chemically dynamic environment, and microbial oxidation of sedimentary organic carbon can be a unique source of DOC to the groundwater. Thus, groundwater DOC contains humic substances microbiologically reprocessed from both soil-derived and kerogen-derived source material.

Groundwater dating by ^{14}C analysis of the DOC should work quite well in shallow, unpolluted waters, where the primary source of DOC is the soil horizon in the recharge zone. The ^{14}C activity in organic matter from an active soil indicates a range from ultramodern (containing ^{14}C produced by thermonuclear bombs) to the equivalent of 10,000 years (Stevenson 1982; Scharpenseel *et al* 1989). These ^{14}C activities were measured on soil-bound humic substances, which are probably much older and certainly less mobile (less prone to move within the aquifer) than the DOC in soil interstitial waters. The DOC in soil interstitial waters comes primarily from plant-root exudates (Rovira & McDougall 1967) and, like the DOC isolated from surface waters, probably contains a modern carbon signature. The ultramodern ^{14}C values that Hedges *et al* (1986) found in all fractions of particulate and DOC collected from Amazon River water demonstrates that the transported carbon was only a few years old. If these measurements are representative of ^{14}C in DOC of groundwater recharge water, they suggest that the soil-derived fractions of DOC in groundwater that perform as good tracers and travel freely with the water through the aquifer, would be useful for radiocarbon dating the water.

Groundwater dating by ^{14}C analysis of the DOC should also work well in deep aquifers that meet the following criteria: 1) the f_{OC} in the solid phase is minimal and 2) the organic productivity of the soils in the overlying recharge zone is high. On the other hand, in deep aquifers where the groundwater residence times are long, detailed information on the lithology and characterization of the DOC is needed. Without this information, the origin of the DOC is impossible to ascertain.

TECHNIQUES OF ISOLATION AND MEASUREMENT

Purdy *et al* (1992) modified the photo-oxidation technique of Armstrong, Williams and Strickland (1966) to convert DOC to CO_2 while still dissolved in the water. Purdy *et al* used a 1200-watt ultraviolet lamp irradiating 2 L of water from which the DIC had been collected by acidification. They added $Na_2S_2O_8$ to expedite oxidation. This procedure produced CO_2 from the total DOC. The CO_2 was then converted to graphite for ^{14}C analysis by AMS. The advantages of this technique are 1) it involves no column chromatography, thus eliminating the opportunity for addition of organic matter from decomposing column resin, and 2) virtually all processing can be accomplished in the laboratory, rather than on-site. The disadvantage is that the technique does not allow separate ^{14}C analysis of different molecular weight fractions of DOC.

Murphy (1987) and Murphy *et al* (1989a, b) described the procedures used in the field to isolate LMW and HMW compounds. Water was passed through a series of columns packed with pre-cleaned resin materials. The LMW compounds (C_1 to ca C_{10}, up to ca 140 amu) were collected on Silicalite, a silica-based molecular sieve, and the HMW compounds (mostly fulvic acid, 500 to

950 amu) were collected on columns of XAD-8 resin. Because there is a gap representing the intermediate molecular weights, these two columns recover only about 30 to 60% of the total DOC. The DOC collected on Silicalite is termed "hydrophilic," and that collected on the XAD-8 is "hydrophobic." Figure 20.2 illustrates Murphy's (1987) field sampling apparatus. To obtain each 200-μg sample of organic matter needed for both chemical characterization and AMS ^{14}C analysis, 100 to 1500 L of water must pass through the columns. Depending on the DOC content, amount of carbon required, and the collection apparatus employed, sample collection could take from a few hours to three days.

Column bleed (release of resin decomposition products) from organic-based resins such as XAD-8, can potentially alter the ^{14}C activity of the DOC sample; however, column bleed is not a significant problem for XAD-8 resins because the DOC bleed was <1% of the DOC, and our ^{14}C analysis of the resin revealed no detectable ^{14}C. Because it is composed of silicon and oxygen, Silicalite column bleed cannot affect the DOC. Silicalite is stable up to 1000°C. Thus, the organic compounds adsorbed on Silicalite are recovered, and the column is regenerated by thermal desorption at this temperature. DOC collected on XAD columns are eluted with NaOH solution.

Chemical fractions destined for isotopic analysis are burned to CO_2 either in a sealed silica glass tube with CuO oxidant or directly from the heated Silicalite molecular sieve in an oxygen stream. The stable isotope mass spectrometer consumes only a small portion of the CO_2 during $\delta^{13}C$ determination. The remainder of the CO_2 is reduced to graphite (Klouda *et al* 1984) for ^{14}C determination by AMS (Donahue *et al* 1987).

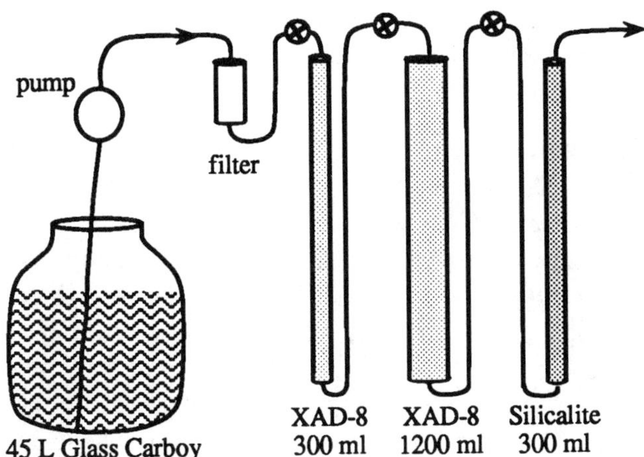

Fig 20.2. Sampling apparatus schematic

Important in the interpretation of carbon isotope values in DOC is the chemical characterization of the ^{14}C "dated" components in a sample. The characterization procedures usually involve chemical breakdown by hydrolysis, then gas chromatography/mass spectrometry. This consumes most of the DOC collected. Chemical characterization can, for example, detect the presence of natural or anthropogenic petroleum compounds. Although petroleum-based compounds would contain no detectable ^{14}C, if unrecognized, they could affect the interpretation of the ^{14}C data by diluting the non-petroleum DOC ^{14}C signal. Characterization also identifies additional parameters that can be monitored as indicators of evolving DOC downgradient in an aquifer and can signal addition of DOC from other sources.

Examples of Studies

The literature reveals only four sites that have been studied in sufficient detail to provide insight into the usefulness of ^{14}C in groundwater DOC for geochemical dating. These are the Milk River aquifer in Alberta, Canada, the Stripa iron mine in Sweden, the Rodney Site in Ontario, Canada and the Atlantic Coastal Plain aquifer in Maryland, USA.

Milk River Aquifer

The Milk River sandstone aquifer underlies 12,000 km^2 of southern Alberta, Canada and 2200 km^2 of northern Montana (Murphy *et al* 1989b). Groundwater flows generally northward. This artesian aquifer is dominantly quartz sandstone, grading downward to a sandy shale, and a confining shale lies above. Of 16 water samples collected and processed in their study, Murphy *et al* (1989b) analyzed 6 for ^{14}C and $\delta^{13}C$ in three fractions: DIC; hydrophobic DOC (HMW); and hydrophilic DOC (LMW) (see Table 20.1). Included also in Table 20.1 are the carbon isotope data in water samples from two wells near the recharge area (Wassenaar, Aravena & Fritz 1989).

Murphy *et al* (1989b) noted that Wells 8 and 16 both have anomalously high ^{14}C and $\delta^{13}C$ values. The authors noted that the percent Modern Carbon (pMC) value of 157 in Well 16 is likely contamination with organic matter contemporary with the well-drilling rather than with the sampling. Murphy *et al* suggest that straw may have been added to the hole to facilitate drilling (a not uncommon practice), and that this straw has fermented in the well. The identification of high amounts of methanol (a fermentation product) in the hydrophilic fraction, and the fact that this well water is electrochemically reducing, support this hypothesis. The major chemical components in Well 8 indicate that its groundwater is mixed with a younger component infiltrating from the overlying glacial till. Well 8 is not artesian. The hydrophilic components from

TABLE 20.1. Milk River Aquifer Carbon Isotope Data

Well no.	DIC ^{14}C pMC	DIC $\delta^{13}C$(‰)	Hydrophobic DOC ^{14}C pMC	Hydrophobic DOC $\delta^{13}C$(‰)	Hydrophilic DOC ^{14}C pMC	Hydrophilic DOC $\delta^{13}C$(‰)	Reference
MR-85	9.2	−12.5	30.6		46.8		Wassenaar *et al* (1989)
MR-52	1.7	−12.4	6.5		7.4		
4	0.2	+5.5	0.7	−24.3	0.8	−35.3	Murphy *et al* (1989)
5	28.1	−11.8	20.0	−25.8	0.4	−40.0	
7	18.6	−6.8	37.6	−26.0	1.5	−40.2	*ibid*
8	3.0	−12.6	7.2	−25.7	63.5	−20.2	*ibid*
11	0.0	−16.4	7.3	−25.8	0.6	−36.5	*ibid*
16	0.1	−0.8	4.3	−27.1	157.0	−19.6	*ibid*

Wells 4, 5, 7 and 11 are low in ^{14}C, although Wells 5 and 7 have relatively high ^{14}C in their DIC and hydrophobic DOC. This suggests diverse groundwater in general, and specifically that kerogen is a dominant source for the hydrophilic DOC in Wells 4, 5, 7 and 11 at least.

Wassenaar, Aravena and Fritz (1989) showed that, for Wells MR-85 and MR-52, which represent a downgradient flow sequence near the recharge area, the uncorrected ^{14}C data for each of the three measured carbon fractions indicate close to the same flow time between wells, from 12,460 to 14,760 years, based on simple ^{14}C decay. Such agreement among three data sets may be fortuitous, but it increases confidence in the calculated rate. Similarly, Murphy *et al* (1989b) found close correspondence between model-corrected ^{14}C values in DIC and ^{14}C values in hydrophobic DOC from Wells 5, 7 and 8 (Table 20.1). These data support both the assumption that the hydrophobic fraction originates primarily from the soil zone and the use of ^{14}C in DOC as an independent technique in the estimation of groundwater ages.

In a more recent analysis of the Milk River aquifer data, Wassenaar *et al* (1991) speculated that the hydrophobic DOC (aquatic fulvic acids) entered the flow system with a ^{14}C activity of about 65 pMC resulting from the mixing of young and old organic carbon in the vadose zone. Similarly, the modeled initial DIC CO_2 would have to have a ^{14}C activity of 35 pMC to be consistent with the age determined by hydraulic data and the interpretation of ^{18}O data. Under these modeling constraints, the CO_2 gas entering the flow system would be largely derived from the oxidation of kerogen or lignite in the vadose zone, which would negate the common assumption that the initial ^{14}C activity is 80 to 100 pMC in the recharge zone. The hydrophilic DOC, however, entered the flow system with a ^{14}C activity of 100 pMC, suggesting that this fraction of the DOC was unaffected by processes in the vadose zone.

Using the PHREEQE/CSOTOP model, the corrected DIC age difference between MR-85 and MR-52 was 12,800 years, identical to the composite age difference for the hydrophobic DOC. Wassenaar *et al* (1991) concluded that the hydrophobic DOC behaved as a conservative tracer along this flow path. The composite age difference for the hydrophilic DOC was 15,200 years, suggesting that this fraction of the DOC has been diluted by dead organic carbon along the flow path.

Stripa Mine

Several papers in the August 1989 issue of *Geochimica et Cosmochimica Acta* (note particularly the paper by Fritz *et al* 1989) describe recent detailed studies of the geology, hydrology and geochemistry of the Stripa iron mine in Sweden. The reason for the wide interest in this site is that because it is granite, it may be characteristic of deep-seated crystalline rocks in terms of their groundwater flow properties and ability to retain materials dangerous to the environment. Murphy *et al* (1989a) and Murphy (1987) reported seven ^{14}C measurements on hydrophilic and hydrophobic fractions from two boreholes in the granite (Fig 20.3). The water samples were obtained from 700 to 1200 m below the surface.

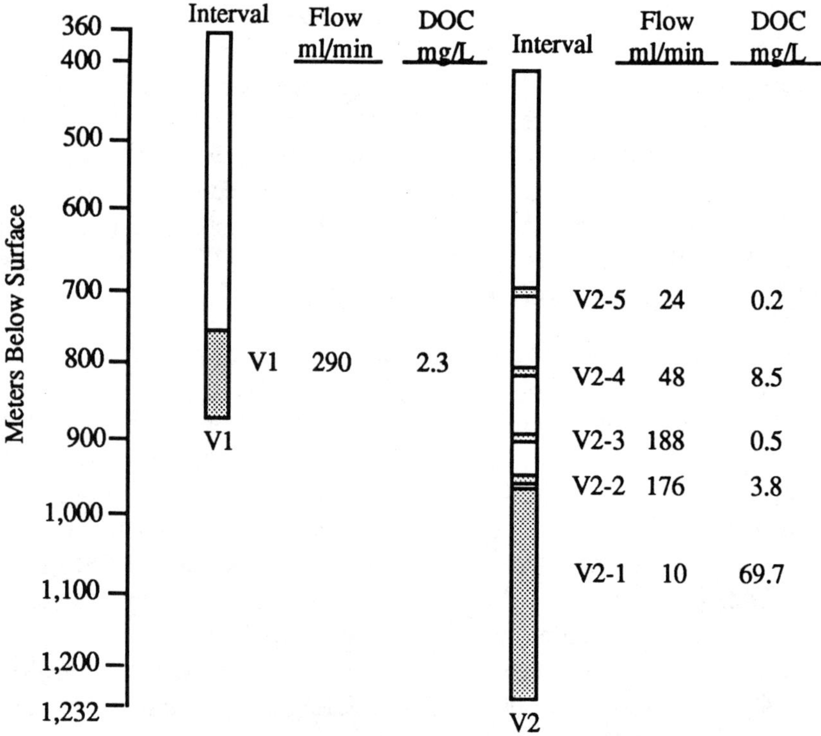

Fig 20.3. Sampling intervals at the Stripa site

TABLE 20.2. Stripa Mine Carbon Isotope Data

Sample (depth, m)	DIC ^{14}C pMC	DIC $\delta^{13}C$(‰)	Hydrophobic DOC ^{14}C pMC	Hydrophobic DOC $\delta^{13}C$(‰)	Hydrophilic DOC ^{14}C pMC	Hydrophilic DOC $\delta^{13}C$(‰)	Comment (Murphy)
V 1	14.9	−24.4			3.0	−34.6	Different DIC sample dates
(758–863)	19.7	−26.0					
V 2-1	87.6	−21.9	28.3	−32.7	73.9	−26.8	Different DIC sample dates
(967–1230)	59.0	−20.5					
V 2-2, 2-3	33.3	−27.5	40.6	−26.6	0.8	−29.3	V2-2 & V2-3 sampled separately
(898–903)	37.9						

Most analyses of DOC from Stripa were paired with ^{14}C analyses on DIC. In all but one case, the ^{14}C activity of the inorganic component exceeded that of the organic component. The carbon isotope values are consistent with the assumption that the DIC derived from oxidation of DOC (Table 20.2). Some isotopic values obtained may show the effects from the plasticizer in the Nylon tubing, n-butylbenzenesulphonamide (NBBS), which appears in the hydrophobic-neutral DOC fraction and lowers its ^{14}C activity. Teflon tubing is evidently non-contaminating, but only 1 of the 7 sampling ports was connected to the surface by Teflon tubing (*viz*, V2-5).

These data generally support the hypotheses that the hydrophobic DOC fraction derives from decay of young organic matter in soils, and that molecules from solid organic matter (in this case, possibly asphaltenes in fractures) dominate the hydrophilic fraction. The levels of ^{14}C in both DIC and DOC a kilometer below the surface indicate rapid flow of at least some of the groundwater in fractured granite. The rapid flow was probably induced by the long-term dewatering of the mine.

Rodney, Ontario

Wassenaar, Aravena and Fritz (1989) studied the Rodney site in Ontario, Canada, near Lake Erie. The water table is about 1 m below the surface of the site, and the aquifer is a shallow, unconfined sand, 3 to 4 m thick. Nitrates and herbicides are applied in farming at the site. Anaerobic bacteria denitrify the nitrate and reduce any sulfate within the saturated zone, using DOC as an energy source. Water samples from the saturated zone came from a single cluster of six piezometers at equally spaced depths 25 to 150 cm below the water table. Water samples 60 to 100 L in volume, collected in 20-L glass carboys, were stored at 4°C before processing. The procedures used to isolate DOC were similar to those used by Murphy (1987) (see Fig 20.4, modified from Wasserman *et al* 1989).

The DOC concentrations in the saturated zone ranged from 1 to 12 mg/L (4.6 mg/L was the average). The relative proportions of the hydrophilic and hydrophobic fractions of DOC varied widely with depth, with the hydrophobic fraction dominant at the upper and lower sample ports and the hydrophilic fraction dominant at intermediate depths. At the highest sampling port, methyl acetate dominated the hydrophilic components, with toluene, hydroxylamine and chloroform as minor components. At the lowest sampling port, the hydrophilic fraction had no dominant component but contained minor amounts of *n*-hexane, cyclohexane, cyclopentane, xylenes and stearic and palmitic acid. All four ^{14}C values on DOC were significantly different from each other and from ^{14}C values on solid organic matter, shown within squares in Fig 20.4. No ^{14}C data on DIC are available at this time.

The hydraulic situation at the Rodney site precludes simple radioactive decay as an explanation for differences in ^{14}C activity in the various DOC components analyzed. Wassenaar, Aravena and Fritz (1989) suggested three possible explanations for the distribution of ^{14}C activities of the hydrophobic component at the sampling point: 1) the hydrophobic fraction of the DOC is a mixture of soil-derived (high ^{14}C) and aquifer-derived (low ^{14}C) carbon, and the aquifer-derived component is greater in the lower levels; 2) the hydrophobic DOC contains a younger component that is more labile, or easily processed microbiologically than the older component, and the lower hydrophobic fraction experienced more

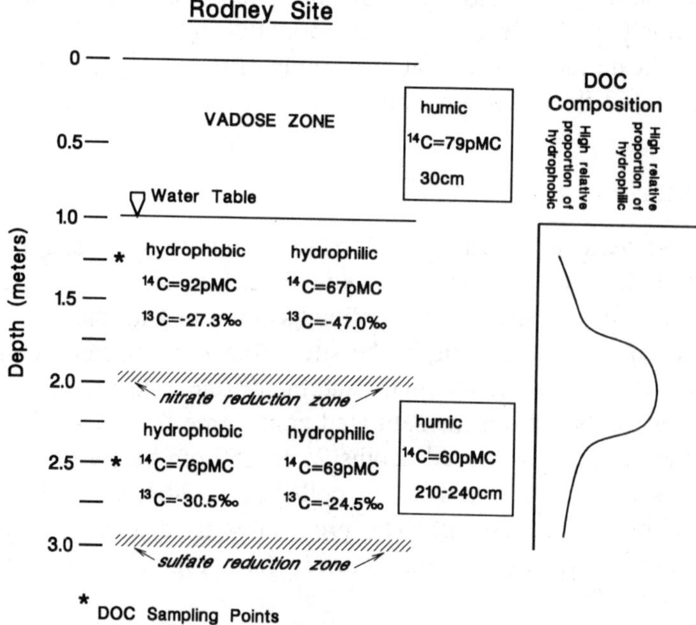

Fig 20.4. Summary of carbon isotope data from the Rodney site

effective microbiological activity than the upper hydrophobic fraction; 3) differences in ^{14}C activity may simply reflect seasonal differences in release of DIC from the soil zone. These explanations are not mutually exclusive. Available data do not rule out any of the three. Stable isotope analyses on the water support a seasonal stratification of water, and by implication, of the DIC. The $\delta^{18}O$ of −11‰ at 2.5 m is similar to that of winter/spring recharge, and the $\delta^{18}O$ of −8‰ at 1.25 m is similar to that of summer/fall recharge.

The ^{14}C concentrations in the hydrophilic fractions at the two radiocarbon sampling levels are similar and lower than in the hydrophobic fractions. Identification of the specific molecular compounds, listed above, reveals products of microbial fermentation. The low ^{14}C suggests that a significant portion of the hydrophilic fraction derives from solid (aquifer-derived) organic matter.

Atlantic Coastal Plain

Purdy *et al* (1992) compared ^{14}C concentrations in DOC and DIC from the two major aquifers in the Atlantic Coastal Plain. Water was sampled at four well sites in Maryland near the upper Chesapeake Bay. Some wells penetrated one, some both aquifers. They report ^{14}C measurements on total DIC and total DOC, summarized here in Table 20.3. Increasing site numbers indicate downgradient flow, though not necessarily a single flow path. The Purdy *et al* DIC ^{14}C analysis procedure, described above in the techniques section, measures ^{14}C in the total DOC from the sample. Water from Site 1 contained high (thermonuclear-produced) levels of ^{36}Cl and ^{3}H (tritium), indicating that it had been recharged since the late 1950s. Thermonuclear ^{14}C, though likely present at Site 1, is evidently obscured by carbon from sources other than plants grown since the late 1950s. The ^{14}C in both DIC and DOC decreases progressively in water from Sites 1 to 4. The ^{14}C in DIC decreases more than that in the DOC. Water from Sites 2 and 3 is not uniquely from the same aquifer as Sites 1 and 4, and not in the same flowpath. Purdy *et al* calculated 19,000- and 12,000-year ages for the water at Site 4 based on DIC and DOC, respectively. They obtained these dates simply by subtracting the uncorrected ^{14}C dates at Site 1 from the uncorrected ^{14}C dates at Site 4. They calculated yet another date for Site 4 by assuming that the water dissolved additional carbonate flowing from Site 1 to

TABLE 20.3. Atlantic Coastal Plain Sites (after Purdy *et al* 1992)

Site (well) no.	Formations penetrated	Total DOC (mg/L)	^{14}C activity in DOC (pMC)	Total DIC (mmoles/L)	^{14}C activity in DIC (pMC)
1	Aquia	0.57 ± 0.01	40.0 ± 0.5	2.40 ± 0.08	34.0 ± 0.4
2	Aquia & Magothy	1.06 ± 0.02	28.0 ± 0.4	4.10 ± 0.08	25.0 ± 0.4
3	Magothy	1.03 ± 0.02	13.0 ± 0.9	2.90 ± 0.06	7.0 ± 0.2
4	Aquia	0.62 ± 0.01	9.0 ± 0.2	3.00 ± 0.60	0.3 ± 0.2

Site 4. This gave a 17,400-year age for Site 4 water. Oxygen and hydrogen isotopic data, as well as chloride contents, are consistent with a glacial age for water at Site 4 (Purdy *et al* 1992). Only after more detailed studies reveal sources and sinks of DIC components and of DOC in the Atlantic Coastal Plain, will it be possible to define the age of water at Site 4.

Each of the studies involving ^{14}C measurements on groundwater reported here explores new territory. The common goal was to understand more about the sources, nature and processes involving dissolved organic carbon in groundwater. In some cases, the purpose was to evaluate the potential of ^{14}C in DOC as a means for dating groundwater. These studies have taught us that each study attempted to select a site representative of a larger class. However, each site had unique characteristics. Generalizations are nevertheless possible. DOC has different components that probably derive from different sources. The different components (sources) have different levels of ^{14}C. Water can lose or gain DOC as it moves through the aquifer. Human activity can affect ^{14}C levels in DOC. ^{14}C measurements, stable isotope analyses and chemical characterization of DOC, when combined with a detailed knowledge of the hydrogeology and geochemistry of groundwater systems, constitute a viable approach to understanding the organic geochemistry of groundwater.

CO_2 IN THE UNSATURATED ZONE: A RELATED PARAMETER

Understanding the isotope geochemistry of CO_2 in the unsaturated zone is important for the interpretation of geochemical and isotopic models that correct groundwater ^{14}C ages. The unsaturated zone is the source of much of the DIC in recharge waters. ^{14}C dating of groundwater using DIC requires knowledge of the ^{14}C and $\delta^{13}C$ values in newly recharged "zero-age" groundwater, as well as a knowledge of geochemical and isotopic processes which affect the carbon budget within the aquifer (see Chapters 17, 18 and 19).

Partial pressures of CO_2 in the unsaturated zone are generally at least 1 to 2 orders of magnitude greater than atmospheric values, even in semi-arid regions. Elevated levels of CO_2 in the unsaturated zone control carbonate geochemistry during groundwater recharge. Carbon dioxide in the unsaturated zone can derive from plant respiration, microbiological degradation of organic matter and from diffusion or advection of atmospheric gases into the unsaturated zone. The contribution of CO_2 from these sources is variable (Galimov 1966; Lerman 1972; Rightmire & Hanshaw 1973; Atkinson 1977; Fritz *et al* 1978; Rightmire 1978; Reardon, Allison & Fritz 1979; Dörr & Münnich 1980, 1986; Parada, Long & Davis 1983; Cerling 1984, 1991; Solomon & Cerling 1987; Quade, Cerling & Bowman 1989). Suggested sources of CO_2 at depth in the unsaturated zone include diffusion from the soil zone, oxidation of organic matter and CO_2 degassing from groundwaters (Haas *et al* 1983; Thorstenson *et al* 1983; Wood & Petraitis 1984; Turin 1986). Recent studies by Suchomel, Kreamer and Long (1990) demonstrated by AMS ^{14}C measurements that microbiological oxidation of petroleum-based contaminants can also be a source of CO_2 in the unsaturated zone.

EXAMPLES OF STUDIES

The literature reveals many studies that determined or modeled the isotopic composition of CO_2 in the unsaturated zone. Following are three examples representing 1) distribution and modeling of ^{14}C and $\delta^{13}C$, 2) variations of $\delta^{13}C$ of soil CO_2 due to the proportion of biomass using the C_4 photosynthetic pathway, and 3) the contribution to CO_2 in the unsaturated zone from oxidation of petroleum-based contaminants.

Western Great Plains, USA

The companion studies of Haas *et al* (1983) and Thorstenson *et al* (1983) measured P_{CO_2}, $\delta^{13}C$ and ^{14}C values of CO_2 in the unsaturated zone at eight sites in the Great Plains of North Dakota and Texas. Distinct seasonal variations in P_{CO_2} and $\delta^{13}C$ reflected the influence of biogenic CO_2 at all but one site, Site 4, which contained a P_{CO_2} of up to 19% and a stable isotopic signature from the oxidation of lignite in the unsaturated zone. The variation in the ^{14}C content of soil CO_2 was small, <2 pMC, at all but Site 4, at which the oxidation of lignite varied the ^{14}C content up to 20 pMC. Thorstenson *et al* (1983) used the data from Haas *et al* (1983) to develop a model of the diffusion, based on a modified form of Fick's Second Law, of $^{12}CO_2$, $^{13}CO_2$ and $^{14}CO_2$ from the soil zone to depth in the unsaturated zone. The novel aspect to this work is the treatment of $^{12}CO_2$, $^{13}CO_2$ and $^{14}CO_2$ as separate entities that diffuse in response to their own concentration gradients, regardless of the distribution of the other species. They found that the model could predict P_{CO_2} and $^{14}CO_2$ values over seasonal variations in the shallow unsaturated zone, but did not account for the decrease in ^{14}C activity near the water table where the ^{14}C activity of unsaturated zone CO_2 corresponded well with the ^{14}C activity of the groundwater.

Modeling of Carbon Isotopes in Soils

Cerling (1984, 1991) developed a model of the ^{13}C content in soil CO_2 and soil carbonate which includes both the effect of diffusion and the proportion of biomass with C_3 and C_4 photosynthetic pathways. The results of the modeling suggest that $\delta^{13}C$ of soil carbonates are most affected by the depth in the soil profile, the fraction of C_4 biomass and the P_{CO_2} of the atmosphere. This model predicts the $\delta^{13}C$ of the carbonates in the soil zone and soil CO_2 based on environmental parameters. This model is used in these studies for paleoclimatic reconstruction with measurements of soil carbonates, but also has applications for the determination of the initial $\delta^{13}C$ and ^{14}C values of groundwater.

Phoenix, Arizona, USA

Suchomel, Kreamer and Long (1990) tested the feasibility of using measurements of soil $^{14}CO_2$ and $^{13}CO_2$ as a tracer of organic contamination. The unsaturated zone in the study area was highly contaminated with organic and inorganic solvents. The authors selected two sampling sites in the unsaturated zone, one under an asphalt parking lot at the site of contamination and the other at an uncontaminated site in the vicinity. At the uncontaminated site, ^{14}C and $\delta^{13}C$ were 110 pMC and −18.5‰, respectively, and the P_{CO_2} ranged between 1.45 and 3%. These results are consistent with an expected CO_2 source of root respiration and oxidation of organic matter. At the contaminated site, $^{14}C = 8$ pMC, $\delta^{13}C = -24‰$ and the P_{CO_2} exceeded 15%. Apparent biochemical oxidation of the ^{14}C-free and light ^{13}C organic contaminants has occurred at the contaminated site. These results support the use of soil $^{14}CO_2$ and $^{13}CO_2$ measurements as a tracer of organic contamination.

Directions for Future Work

Field and Laboratory Techniques

To date, natural organic compounds in groundwater have been collected and processed only as fractions comprising numerous compounds with similar broad characteristics. The literature contains no isotopic analyses on individual compounds. Owing to the very low natural concentrations of the various compounds present in the water, the next major stage in research evolution awaits development of methods of collection, concentration and separation of individual organic species. This new methodology must be free of sources of external contamination, and, at least in part, must be field operable. Many potentially interesting compounds lend themselves to known methods of chromatographic separation, but no one has tested all the steps needed to concentrate sufficient organic compounds in the field, and subsequently to process all the concentrates in the laboratory.

Types of Water to be Sampled

In order to understand better the range of organic compounds present in groundwater, a vast number must be studied. The work Thurman and others began must be expanded to include most of the major types of hydrochemical systems. Examples include arid region aquifers that are fully oxygenated throughout, hydrothermal systems, large artesian basin aquifers that contain very old groundwater, circulation systems containing chemically reducing water, groundwater systems containing brackish water, and many more. Based on these broad studies, a sampling and analytical program should then be planned to select compounds for detailed isotopic analyses. The chemical characteristics should yield information on their origin and the processes that acted on the dissolved organic components. The isotopic analyses would yield information not only on origin and processes, but also on the times involved in the various processes.

Unsaturated Zone CO_2

Stable and radioactive isotopes of carbon are commonly used as tools in understanding the geochemistry of systems, or for verifying hydrologic models. Estimates of organic carbon as a source of unsaturated zone CO_2 and as a subsequent source of DIC in water, combined with mineral equilibrium geochemistry, are critical for mass balance geochemical evolution models. Study in this area is needed because of limited data on the stable and radioactive isotopic composition of unsaturated zone CO_2 from "undisturbed" hydrologic systems.

CONCLUSIONS

The combination of organic geochemistry of dissolved organic matter in groundwater with gas analysis in the unsaturated zone and stable isotope and ^{14}C analyses has demonstrated applications to groundwater dating and pollution studies. This is an area of investigation still in the juvenile stage, that, with sustained further effort, will bring additional insight into our understanding of some fundamental processes in groundwater chemistry about which we know very little. ^{14}C data on certain DOC fractions appear to reinforce dates derived from DIC, thus improving confidence in the groundwater ages. Carbon isotopes also provide evidence for both increasing the DIC by respiration and for decreasing the DIC by fermentation. The next step in ^{14}C studies of DOC should probably be the study of the separate species present, such as individual amino acids, separated from groundwater that has been collected from physically and chemically well-characterized systems. This may be the key to identification of contributions to DOC from different sources, which is so critical to the interpretation of ^{14}C in DOC in terms of the age of groundwater.

ACKNOWLEDGMENTS

Preparation of this review was supported by the Subsurface Science Program of the Ecological Research Division (ERD) of the US Department of Energy (DOE). Pacific Northwest Laboratory is operated for DOE by Battelle Memorial

Institute. This review is also a product of the Geohydrology Program, Department of Geosciences, The University of Arizona.

The authors wish to thank Meyer Rubin, George Burr and E M Thurman for their helpful comments and discussions. Technical editors at Pacific Northwest Laboratories and in the *RADIOCARBON* office were most helpful in finalizing the manuscript.

REFERENCES

Aiken, GR and Malcolm, RL 1987 Molecular weight of aquatic fulvic acids by vapor pressure osmometry. *Geochimica et Cosmochimica Acta* 51: 2177–2184.

Atkinson, TC 1977 Carbon dioxide in the atmosphere of the unsaturated zone: An important control of groundwater hardness in limestones. *Journal of Hydrology* 35: 111–123.

Armstrong, FAJ, Williams, PM and Strickland, JDH 1966 Photo-oxidation of organic matter in sea water by ultra-violet radiation, analytical and other applications. *Nature* 211(5048): 481–483.

Cerling, TE 1984 The stable isotopic composition of modern soil carbonate and its relationship to climate. *Earth and Planetary Science Letters* 71: 229–240.

_____1991 Carbon dioxide in the atmosphere: Evidence from Cenozoic and Mesozoic paleosols. *American Journal of Science* 291: 377–400.

Chappelle, FH, Zelibor, JL, Grimes, DJ and Knobel, LL 1987 Bacteria in deep coastal plain sediment of Maryland: A possible source of CO_2 to groundwater. *Water Resources Research* 23(8) 1625–1632.

Cronan, CS and Aiken, GR 1985 Chemistry and transport of soluble humic substances in forested watersheds of the Adirondack Park, New York. *Geochimica et Cosmochimica Acta* 49: 1697–1705.

Donahue, DJ, Jull, AJT, Linick, TW, Hatheway, A, Toolin, LJ and Gore, B 1987 Some results from the Arizona TAMS Facility, AMS ages of athletic, artistic and animal artifacts. *Nuclear Instruments and Methods* B29: 169–172.

Dörr, H and Münnich, KO 1980 Carbon-14 and carbon-13 in soil CO_2. *In* Stuiver, M and Kra, RS, eds, Proceedings of the 10th International ^{14}C Conference. *Radiocarbon* 22(3): 909–918.

_____1986 Annual variations of the ^{14}C content of soil CO_2. *In* Stuiver, M and Kra, RS, eds, Proceedings of the 12th International ^{14}C Conference. *Radiocarbon* 28(2A): 338–345.

Fritz, P, Fontes, J-Ch, Frape, SK, Louvat, D, Michelot, J-L and Balderer, W 1989 The isotope geochemistry of carbon in groundwater at Stripa. *Geochimica et Cosmochimica Acta* 53: 1765–1775.

Fritz, P, Reardon, EJ, Barker, J, Brown, RM, Cherry, JA, Killey, RW and McNaughton, D 1978 The carbon isotope geochemistry of a small groundwater system in Northeastern Ontario. *Water Resources Research* 14: 1059–1067.

Galimov, EM 1966 Carbon Isotopes of soil CO_2. *Geochimica International* 3: 889–897.

Games, LM, Hayes, JM and Gunslaus, RP 1978 Methane-producing bacteria: Natural fractions of the stable carbon isotopes. *Geochimica et Cosmochimica Acta* 42: 1295–1297.

Haas, H, Fisher, DW, Thorstenson, DC and Weeks, EP 1983 $^{13}CO_2$ and $^{14}CO_2$ Measurements on soil atmosphere samples in the sub-surface unsaturated zone in the western great plains of the US. *In* Stuiver, M and Kra, RS, eds, Proceedings of the 11th International ^{14}C Conference. *Radiocarbon* 25(2): 301–314.

Hedges, JI, Ertel, JR, Quay, PD, Grootes,

PM, Richey, JE, Devol, AH, Farwell, GW, Schmidt, FW and Salati, E 1986 Organic carbon-14 in the Amazon River system. *Science* 231: 1129–1131.

Klouda, GA, Currie, LA, Donahue, DJ, Jull, AJT and Zabel, TH 1984 Accelerator mass spectrometry sample preparation: methods for ^{14}C in 50 to 1000 microgram samples. *Nuclear Instruments and Methods* B5: 265–271.

Leenheer, JA, Malcolm, RL, McKinley, PW and Eccles, LA 1974 Occurrence of dissolved organic carbon in selected groundwater samples in the United States. *Journal of Research US Geologic Survey* 2(3): 3611–369.

Lerman, JC 1972 Soil-CO_2 and groundwater: carbon isotope composition. *In* Rafter, TA and Grant-Taylor, T, eds, *Proceedings of the 8th International Conference on Radiocarbon Dating*. Wellington, Royal Society of New Zealand: D93–D106.

Malcolm, RL, Aiken, GR, Thurman, EM and Avery, PA 1980 Hydrophilic organic solutes as tracers in groundwater recharge studies. *In* Baker, RA, ed, *Contaminants and Sediments* 2. Ann Arbor, Michigan, Ann Arbor Science Publishers, Inc.

Murphy, EM 1987 Carbon-14 measurements and characterization of dissolved organic carbon in groundwater. PhD Thesis, University of Arizona: 180 p.

Murphy, EM and Aiken, GR 1991 Natural organic carbon in groundwater. *In* Palmer, CD, ed, *The Geochemistry of Groundwater*, in press.

Murphy, EM, Davis, SN, Long, A, Donahue, DJ and Jull, AJT 1989 ^{14}C in fractions of dissolved organic carbon in groundwater. *Nature* 337: 153–155.

_____1989 Characterization and isotopic composition of organic and inorganic carbon in the Milk River aquifer. *Water Resources Research* 25: 1893–1905.

Murphy, EM, Long, A and Davis, SN 1985 Detection of ^{14}C in natural trace organics recovered from groundwater. *American Nuclear Society Transactions* 50: 191–192.

Murphy, EM, Long, A and Thurman, EM 1984 The separation and concentration of dissolved organic carbon in groundwater for carbon-14 analysis (abstract). *American Geophysical Union, Transactions* 65: 889.

Parada, CB, Long, A and Davis, SN 1983 Stable-isotopic composition of soil carbon dioxide in the Tucson Basin, Arizona, USA. *Isotope Geoscience* 1: 219–236.

Pearson, FJ and Friedman, II 1970 Sources of dissolved carbonate in an aquifer free of carbonate minerals. *Water Resources Research* 6(6): 1775–1781.

Purdy, CB, Burr, G, Rubin, M, Helz, GR and Mignery, AC 1992 Dissolved organic and inorganic ^{14}C concentrations and ages for coastal plain aquifers in southern Maryland. *Radiocarbon* 34, in press.

Quade, J, Cerling, TE and Bowman, JR 1989 Systematic variations in the carbon and oxygen isotopic composition of pedogenic carbon along elevation transects in the southern Great Basin, United States. *Geologic Society of America Bulletin* 101: 464–475.

Reardon, EJ, Allison, GB and Fritz, P 1979 Seasonal chemical and isotopic variations of soil CO_2 at Trout Creek, Ontario. *Journal of Hydrology* 43: 355–371.

Rightmire, CT 1978 Seasonal variation in P_{CO2} and ^{13}C content of soil atmosphere. *Water Resources Research* 14(4): 691–692.

Rightmire CT and Hanshaw BB 1979 Relationship between the carbon isotope composition of coil CO_2 and dissolved carbonate species in groundwater. *Water Resources Research* 14: 958–967.

Rovira, AD and McDougall, BM 1967 Microbiological and biochemical aspects of the rhizosphere. *In* McLaren, AD and Petersen, GH, eds, *Soil Biochemistry*: 417–463.

Scharpenseel, HW, Becker-Heidmann, P, Neue, HU and Tsutsuki, K 1989 Bomb-carbon, ^{14}C-dating and ^{13}C-measurements as tracers of organic matter dynamics as well as of morphogenetic and turbation processes. *The Science of the Total Environment* 81/82: 99–110.

Solomon, DK and Cerling, TE 1987 The annual carbon dioxide cycle in a montane soil: Observations, modeling and implications for weathering. *Water Resources Research* 23(12): 2257–2265.

Stevenson, FJ 1982 *Humus Chemistry, Genesis, Composition, Reactions*. New York, John Wiley & Sons: 443 p.

Suchomel, KH, Kreamer, DK and Long, A 1990 Production and transport of carbon dioxide in a contaminated vadose zone: A stable and radioactive carbon isotope study. *Environmental Science and Technology* 24: 1824–1831.

Thorstenson, DC, Weeks, EP, Haas, H and Fisher, DW 1983 Distribution of gaseous $^{12}CO_2$, $^{13}CO_2$ and $^{14}CO_2$ in the sub-soil unsaturated zone of the western US Great Plains. *In* Stuiver, M and Kra, RS, eds, Proceedings of the 11th International ^{14}C Conference. *Radiocarbon* 25(2): 315–346.

Thurman, EM 1979 Isolation, Characterization, and Geochemical Significance of Humic Substances from Groundwater. PhD Dissertation, University of Colorado.

_____1985a *Organic Geochemistry of Natural Waters*. Dordrecht, The Netherlands, Martinus Nijhoff: 497 p.

_____1985b Humic substances in groundwater. *In* Aiken, GR, McKnight, DM, Wershaw, RL and MacCarthy, P, eds, *Humic Substances in Soil, Sediment and Water*. New York, John Wiley & Sons.

Thurman, EM, Wershaw, RL, Malcolm, RL and Pinckney, DJ 1982 Molecular size of aquatic humic substances. *Organic Geochemistry* 4: 27–35.

Turin, HJ 1986 Carbon Dioxide and oxygen profiles in the unsaturated zone of the Tucson basin. Unpublished MS thesis, University of Arizona.

Wassenaar, L, Aravena, R, Hendry, J and Fritz, P 1991 Radiocarbon in dissolved organic carbon, a possible groundwater dating method: Case studies from western Canada. *Water Resources Research* 27(8): 1975–1986.

Wassenaar, L, Aravena, R and Fritz, P 1989 The geochemistry and evolution of natural organic solutes in groundwater. *In* Long, A and Kra, RS, eds, Proceedings of the 13th International ^{14}C Conference. *Radiocarbon* 31(3): 865–876.

Wood, WW and Low, WH 1988 Solute geochemistry of the Snake River Plain regional aquifer system, Idaho and eastern Oregon. *US Geological Survey Professional Paper* 1408D: 79 p.

Wood, WW and Petraitis, MJ 1984 Origin and distribution of carbon dioxide in the unsaturated zone of the southern high plains of Texas. *Water Resources Research* 20(9): 1193–1208.

Zachara, JM, Cowan, CE and Murphy, EM 1990 Influence of dissolved and solid associated organic substances on the sorption of metallic contaminants by layer silicates and subsurface materials. In *Proceedings of the International Conference on Metals in Soils, Plants and Animals*, April 30–May 2, 1990, Orlando, FL.

OLD WORLD ARCHAEOLOGY

PREFACE

FRED WENDORF

The following chapters discuss the impact of radiocarbon dating on our understanding of human cultural developments in three important areas of the Old World. In each case, the introduction of ^{14}C dating radically altered our view of the local prehistory and led to significant changes in the kinds of questions that archaeologists sought to study and how they did those studies. However, this new dating method did not lead to increased emphasis on chronology; on the contrary, the technique provided relatively independent and precise dates, and thus, freed archaeologists from what had been a sterile preoccupation with questions of relative dating.

The three chapters adopt different perspectives. Fred Wendorf discusses how ^{14}C dating overturned the traditional cultural sequences in both the Nile Valley and the Maghreb, and how these new chronologies have altered our view of North African prehistory and its relationship to similar developments in Europe and Southwest Asia. Donald Henry stresses the impact of ^{14}C dating or archaeological theory and research design. Peter Robertshaw, on the other hand, emphasizes the role of ^{14}C dating in the study of the later prehistory of sub-Saharan Africa, particularly of developments in the Iron Age. He sees a number of problems in the use of the technique, and suggests that the results have not always been as useful as had been anticipated. Despite such occasional difficulties, it is evident that, in all of these areas, radiocarbon dating has now become commonplace and archaeologists can no longer imagine research without it.

21

THE IMPACT OF RADIOCARBON DATING ON NORTH AFRICAN ARCHAEOLOGY

FRED WENDORF

INTRODUCTION

The first archaeological sample dated by ^{14}C came from North Africa. It was from Sakkara, a ceremonial center not far from modern Cairo. The dated sample was a piece of acacia wood from the tomb of Zoser, an Egyptian pharaoh of the Third Dynasty (Sample C-1); three other Egyptian samples were among the first hundred ^{14}C dates produced. After such a beginning, one would have expected ^{14}C dating to play a dominant role in the archaeology of North Africa; however, its application and impact have been uneven, particularly for the pharaonic periods in Egypt. For prehistory throughout North Africa, ^{14}C dating produced a true revolution in our ideas about the origin and development of almost every known cultural complex. The dating technique also profoundly changed our concept of cultural relationships within North Africa, and between North Africa and other areas. To appreciate the impact of ^{14}C dating on North African prehistory, one only has to read any of the syntheses written before the widespread application of ^{14}C dating (Gobert & Vaufrey 1932; Balout 1955; Vaufrey 1955). Virtually none of the chronological relationships so confidently proposed before 1960, from the Middle Paleolithic to the Neolithic, has survived the onset of absolute dating provided by ^{14}C.

Part of this problem is peculiar to North Africa. Most of the area is extreme desert, and erosion has destroyed much of the original context of many archaeological sites. Those that are preserved tend to occur in widely separated patches, and there are rarely opportunities for direct stratigraphic or other correlations between these patches. North Africa has also suffered from an excess of what can only be called "Eurocentricism"; *ie*, the archaeology has tended to be interpreted from the perspective of European archaeology. This was aggravated by the attitude that "everything worthwhile began in Europe."

Even in the Nile Valley, which had abundant remains of a civilization recognizably earlier than anything in Europe, our understanding of pre-pharaonic archaeology was complicated by the fact that the Nile behaved very differently from rivers in Europe, that the sediments in which the archaeology occurred were almost entirely derived from central and eastern Africa, and they reflected a unique environment.

In the absence of absolute dates, various theories were proposed to correlate climatic events in tropical Africa with those of more northerly latitudes, particularly Egypt and the Sahara. One of these was the "pluvial theory" (Nilsson 1931; Charlesworth 1957: 1112–1139). Simply stated, the theory suggested that during periods of glacial maximum, the storm tracks of the northern latitudes would be pushed southward, there would be less evaporation because of lower temperatures, and thus, the mid-latitude deserts, such as the Sahara, would be wetter. This theory was so firmly held by most Pleistocene geologists and prehistorians before ^{14}C dating that it became dogma. Only in the last few years has this dogma begun to crumble, in the face of overwhelming ^{14}C dated evidence from the Sahara and adjacent oceans (Thiede, Suess & Müller 1982; Wendorf & Schild 1980) that the Sahara was hyperarid during the last glacial maximum. The climatic model to explain this phenomenon is still not fully developed, and some scholars still believe that the Sahara was wet 20,000 years ago (Tillet 1983; Petit-Maire 1988).

RADIOCARBON DATING IN EGYPT

Although the first ^{14}C dated sample came from Egypt, and Libby's 1955 date list of 949 samples includes 18 from North Africa, all of these were from Egypt and Sudan (Fig 21.1). The first four were for validation, to check if there was a consistent correlation between the ^{14}C date and the established historical age. There was a considerable element of luck in the results of these first samples. The concordance with the historical ages was sufficiently close that Libby could see no obvious flaws in his theoretical model, and he was encouraged to proceed with the dating project. Even the first sample (C-1) yielded an average date (3979 BP ± 350) that was almost two standard deviations younger than the known historical age (4650 BP ± 75) (Libby 1955: 77). Later, as more samples of pharaonic age were dated, Egyptologists began to note a disturbing lack of agreement between the ^{14}C age determinations and the historical ages. By the early 1960s, these discrepancies were so evident that several Egyptologists suggested that ^{14}C dating was neither sufficiently precise nor consistent to be useful in reconstructing Egyptian history (Hayes 1964: 192; Smith 1964).

Part of the problem could be attributed to the kinds of samples that were sometimes selected for dating, particularly in the early years. Some of the dates were determined on inappropriate materials, some had long been in museum collections and thus possibly contaminated, and others were not closely tied to

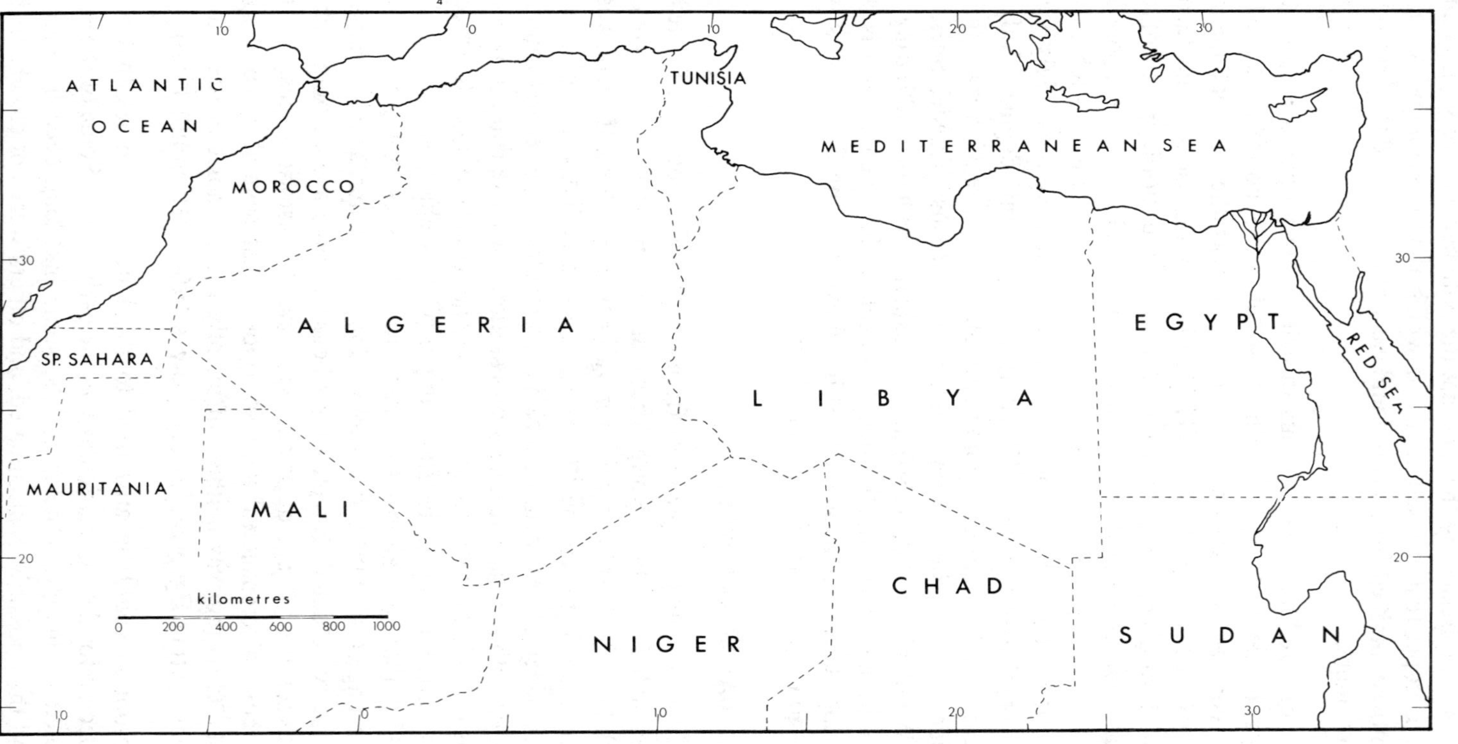

Fig 21.1. Map of North Africa

the historical event supposedly being dated. Nevertheless, there remained a disturbing number of ^{14}C dates of unquestionable context and association that differed significantly from the historical age.

These problems began to be resolved when it was discovered, through ^{14}C measurements of dated tree rings, that the amount of ^{14}C in the atmosphere had varied through time, and that these variations were of such magnitude that ^{14}C dates could differ significantly from the true age of the dated specimens. Suess (1970) soon proposed a calibration curve that was promptly applied to the available Egyptian ^{14}C dates (Long 1970). Nevertheless, serious doubts still remained about ^{14}C dating and the Egyptian chronology (Clark & Renfrew 1973; Clark 1978).

It was not until 1987 that a convincing correlation was made of calibrated ^{14}C dates and critical events in Egyptian history (Hassan & Robinson 1987). There may now be changes in Egyptology, like those which occurred in North American archaeology after ^{14}C dating had freed scholars from their preoccupation with chronology and permitted the current interest in cultural processes. Many social, political and economic changes that occurred during the development of Egyptian civilization remain to be explored, if Egyptologists can shift their attention from king lists and the archaeological remains of the ruling elite.

Not all of the conflicts between ^{14}C dating and the Egyptian historical chronology have yet been resolved, but it is now evident that the technique has enormous potential to contribute to a better understanding of the origin and development of Egyptian civilization. For example, a series of 64 consistent ^{14}C dates from ten pyramids and other monuments at Giza, Sakkara and Abu Roash (Haas *et al* 1987) suggests that these structures are 300–400 years older than had been thought, and that some of them might have been built during the Predynastic. These dates have not been received with enthusiasm, because they suggest either that our models of the development of complex society in Egypt are inadequate, or that the beginning of the Old Kingdom occurred at least 300 years earlier than now believed. The alternative explanations offered thus far are inadequate (Hassan & Robinson 1987: 129; *cf* Close 1988: 153).

Egyptian archaeological materials earlier than the Predynastic, particularly those of the Late Paleolithic, received little attention before 1960. There was, however, some important work on the Neolithic, particularly in the Fayum (Caton-Thompson & Gardner 1934), along the Nile (Brunton & Caton-Thompson 1928) and in the Delta (Junker 1929; see summaries in Kaiser (1957) and Hassan (1988)). The difficulties encountered in placing the finds in a chronological sequence are well illustrated by the work in the Fayum, where the relative positions of the Terminal Paleolithic ("Fayum B") and Neolithic ("Fayum A") were reversed.

The lack of interest in pre-Neolithic development in the Nile Valley was a result of several factors. First, and perhaps most important, extensive surveys of the Nile silts had yielded only scattered artifacts (Sandford & Arkell 1929, 1933, 1939; Sandford 1934; Caton-Thompson 1946). *In situ* living sites were known only in one area, the Kom Ombo Plain, just north of Aswan (Vignard 1923). It was widely assumed that most of the Paleolithic record of Egypt was buried below the modern Nile floodplain. Second, the few sites known at Kom Ombo, although apparently associated with the latest Pleistocene silts, frequently contained evidence of Levallois technology, a technique generally thought to be characteristic of the Middle Paleolithic. This led to the popular belief that Egypt and the Nile Valley had been culturally retarded during the Late Paleolithic (Movius 1953: 175; Huzayyin 1941: 169). Most prehistorians were concerned with the origins of the Upper Paleolithic and the beginnings of food production, problems to which Egypt seemed quite irrelevant.

It is also likely that political events in Egypt after World War II discouraged many prehistorians who might have worked in Egypt. Had there been any encouragement from local authorities, surely some scholar would have reasoned that one of the great waterways of the world must have a complex prehistoric record, that sites preserving that record must exist somewhere, and that a proper search would locate them. This is precisely what happened between 1962 and 1966, when Egypt and Sudan requested international help in the rescue of the archaeological remains from inundation by the lake behind the New High Dam at Aswan. Systematic surveys soon disclosed hundreds of *in-situ* prehistoric sites throughout the reservoir area and beyond (Wendorf 1968; Irwin, Wheat & Irwin 1968; Smith 1966). The excavation of many of these sites and the associated ^{14}C chronology still provide the basic framework for the prehistory of the Nile Valley (Wendorf & Schild 1976; Wendorf, Schild & Close 1989).

In many respects, it is fortunate that few showed interest in the prehistory of the Nile Valley before the Nubian campaign. In the Maghreb, numerous Paleolithic sites had been excavated before 1950 and a complex cultural sequence established. When ^{14}C dating became available, it was soon evident that this sequence was almost completely in error, and many adjustments had to made, some of which were undoubtedly painful to established scholars in that area. In the Nile Valley, on the other hand, those of us who were fortunate enough to be involved in the Aswan reservoir project worked without any preconceived notions about the chronology of the sites we were studying. The ^{14}C dates and the stratigraphy of the associated Nilotic deposits provided the chronological framework. The Nile sedimentary sequence was extremely difficult to decipher (we are only now beginning to understand its behavior during the Late Pleistocene), and thus, ^{14}C dates were the primary chronological tool. Unfortunately, the ^{14}C determinations were not always reliable. Laboratory problems that resulted in a few erroneous dates led us to place a Middle Paleolithic complex (the Khormusan) in the Late Paleolithic, thus reinforcing the previous concept of Nilotic conservatism (Wendorf, Schild & Haas 1979). These erroneous dates also led to a misinterpretation of the sedimentary chronology. Despite these problems, undoubtedly, without ^{14}C dating, our understanding of the sequence of prehistoric complexes would be very different. An extensive series of ^{14}C dates are now associated with Paleolithic materials along the Nile. The excavations that yielded the dated samples have documented a rich Paleolithic heritage. Often different from sites of the Levant and Europe, the Paleolithic of the Nile was fully equivalent in both cultural complexity and technological development.

Even so, several yet unexplained hiatus remain in the sequence. The data indicate the presence of distinct, but interrelated, Late Paleolithic complexes (most of which stressed the production of bladelets in their lithic technologies) from about 21,000 to 11,000 years ago, when the number of known sites declines abruptly, possibly because of a major change in the behavior of the Nile. Only three sites are known before 21,000 BP. Those show greater emphasis on blades (rather than bladelets), and could be classified as Upper, rather than Late, Paleolithic. The earliest of these sites is a shaft-and-gallery mine with several ^{14}C dates between 35,000 and 30,000 BP (Vermeersch *et al* 1982). The other two sites have TL dates of 24,700 BP ± 2500 (OXTL-253) and 21,590 BP ± 1520 (Oxford, 161-C-1) (Paulissen, Vermeersch & Van Neer 1985; Wendorf & Schild 1975: 138). These sites have begun to fill one major break in the Nilotic sequence, but a significant hiatus still remains between the latest-known Middle Paleolithic, estimated to date between 45,000 and 55,000 years ago (from several TL dates), and the earliest Upper Paleolithic. No sites are seen as transitional between the Middle and Upper Paleolithic.

RADIOCARBON DATING IN NORTHWESTERN AFRICA

Considerable archaeological research had been done in the Maghreb (Fig 21.1) and elsewhere in northwestern Africa before the advent of ^{14}C dating. Researchers believed that the Late Pleistocene and early Holocene archaeological sequence consisted of a Middle Paleolithic entity known as the Aterian, regarded as the chronological equivalent of the Upper Paleolithic in Europe, and two late Final Pleistocene and early Holocene complexes, the Iberomaurusian and Capsian. The stratigraphic primacy of the Aterian was well established at several sites, where it had been found below Iberomaurusian assemblages. There was, however, considerable disagreement over the chronological and cultural relationships between the two later entities.

The Iberomaurusian and Capsian never occurred in the same areas. The Iberomaurusian is a bladelet industry found only along the coast and the adjacent mountains; sites of the Capsian, on the other hand, are confined to the interior plateaus. Because of this geographic separation, the chronological relationship between these two complexes could not be determined through stratigraphic studies, nor were there significant differences in associated faunas that could not be explained by variations in the local environments. Skeletal studies also revealed differences in associated human physical type: the Iberomaurusian population was a heavily muscled, robust type now known as Mechta-Afalou, whereas the Capsian skeletons were more gracile Mediterraneans.

The problem was further complicated by the definition of several subdivisions of the Capsian. A Typical Capsian was regarded as the oldest because of the use of large blades, some of which were retouched into backed pieces, end-scrapers and, in particular, elegant burins. The Upper Capsian was believed to be later than the Typical Capsian because its sites often contained high frequencies of geometrics. Some stratigraphic work confirmed this chronological relationship: at four sites, both types of Capsian occurred, the Typical always below the Upper. Finally, a Neolithic of Capsian Tradition, was regarded as the most recent because of the presence of pottery. A scatter of Epipaleolithic sites were neither Typical nor Upper Capsian; perhaps because they did not form a cohesive unit, these miscellaneous sites were considered of little importance in the prehistory of northwestern Africa.

Some researchers believed the Iberomaurusian was later than the Typical Capsian and a facies of the Upper Capsian (Vaufrey 1955: 288). Others (Alimen 1957: 64; Balout 1955: 379) suggested that the Iberomaurusian began before the Typical Capsian and lasted on throughout the period of the Typical and Upper Capsian. Thus, they were regarded as two distinct social groups of different human physical type, exploiting adjacent areas with little or no interaction.

Radiocarbon dating came late to Northwest Africa. No samples from this area were in Libby's original date list and, as late as 1960, only one date was published. By 1963, there were still only 16 dates, which rapidly increased to 120 by 1968, and to 316 from 134 archaeological sites by 1973 (Camps, Delibrias & Thommeret 1973: 66). This spectacular growth in ^{14}C age determinations has continued until today (Close 1980, 1984, 1988). A recent summary of ^{14}C dates listed 865 from Algeria, Mali, Mauretania, Morocco, Tunisia and the Western Sahara; almost half of these are from Algeria (Shaw 1989: 20–21).

By 1973, it was evident from the ^{14}C dates that the sequence of archaeological complexes in the Maghreb was very different from what was previously believed. The dates showed that the Iberomaurusian began before 20,000 BP and lasted until around 10,000 BP. Most of the miscellaneous Epipaleolithic industries fell near the end of, or just after, the Iberomaurusian; their role in the prehistory of the Maghreb still remains obscure. The Typical and Upper Capsian were found to be contemporary (the Typical Capsian is now often regarded as only a regional facies of the Capsian). Both began shortly after 10,000 BP and ended around 6500 BP, whereas the Neolithic of Capsian Tradition may have begun as early as 8000 BP, suggesting that parts of the aceramic Capsian Tradition survived for some time after the introduction of pottery in this area.

The new ^{14}C dates also had a profound effect on how prehistorians view the Middle Paleolithic in the Maghreb. Throughout the Maghreb and southward across the Sahara are numerous Aterian sites characterized by typical Middle Paleolithic technology and tool assemblages that include bifacial foliates and tanged or pedunculated forms. Although known to be older than the Iberomaurusian and the Capsian, the Aterian was regarded by most archaeologists as the local equivalent of the European Upper Paleolithic. This was in part because of the resemblance of the bifacial foliates to similar pieces in the European Solutrean, and also because neither the Capsian nor the Iberomaurusian was believed to be earlier than Early Holocene or possibly the very last part of the Late Pleistocene. A few scholars, however, regarded at least some of the Aterian as much older because, in several localities, it appeared to occur in or just above beach features of the Last Interglacial.

Although there have been several ^{14}C age determinations on the Aterian, its age is still controversial. Many of the dates are infinite, thus supporting the view of the Aterian as pre-Upper Paleolithic (Camps 1955). Other samples have yielded finite dates; most of these can be rejected as probably contaminated or because they were on carbonates or other inappropriate materials. But several dates around 30,000 BP or slightly older on charcoal are difficult to reject on present evidence.

Unfortunately, very little archaeological research has been carried out in northwestern Africa since the mid-1970s. By that time, most of the chronological questions had been answered and the archaeological complexes had been described in minute detail; this provided a solid data base rarely equaled elsewhere in Africa. Had political events not intervened, archaeological studies in northwestern Africa would almost certainly have turned toward the investigation of problems relating to prehistoric food and raw material economies, and the social systems that structured the defined archaeological entities. A sea change might have occurred in archaeological goals, not unlike the reevaluation that occurred in North American archaeology in the 1960s. Fortunately, most of the key sites are still there, and perhaps some day we will see a resurgence of interest in the archaeology of this area. Because of the numerous ^{14}C age determinations available, future research can proceed from chronology and focus on more complex questions that relate to how prehistoric societies functioned.

RADIOCARBON DATING IN THE SAHARA

Almost no detailed archaeological research occurred in the Sahara (Fig 21.1) before the availability of ^{14}C dating. There were occasional exploration projects into the desert, some of which included archaeologists or individuals interested in archaeology. Most of these projects returned with large surface collections that clearly indicated a significant human prehistoric presence. However, almost no excavations were undertaken, and stratigraphic observations were rare, so that estimates of the ages of the sites from which these collections came were based on typological comparisons with adjacent areas. We knew the desert had been occupied, but little else.

The one notable exception was the work of Caton-Thompson (1952) at Kharga Oasis in the Western Desert in Egypt. Although undertaken in 1930–31, the Kharga project is still an outstanding example of interdisciplinary research. Working mostly in spring vents and wadi deposits, Caton-Thompson recorded a long sequence of archaeological assemblages, ranging from Late Acheulean to several varieties of Middle Paleolithic and, finally, to Neolithic. All were tied to a complex sequence of environmental fluctuations. Although the assemblages were not dated, Caton-Thompson attempted to correlate the pluvial events with similar events in East Africa and with the European glacial chronology, as it was then understood.

Apart from Kharga, systematic archaeological research in the Sahara began in the 1960s along with ^{14}C dating (Camps 1968). The application of the technique has normally been a great help, but occasionally it has been a hindrance. One of the most enduring problems in the Sahara has been the effort to date the widespread Aterian and the wet episode(s) with which it was associated. That this has been problematic is largely because inappropriate carbonate and snail shell samples were used for dating. Almost all of these samples yielded dates between 20,000 and 40,000 BP. Because they were the only age estimates available, they were accepted as the true ages of the Aterian (Conrad 1969,

1972; Servant & Servant-Vildary 1972; Hugot 1966; see discussion in Wendorf & Schild 1980: 231–234). It has even been suggested that the Aterian survived until the beginning of the Neolithic (Aumassip 1987: 236, 245). This is a major discrepancy when seen from the perspective of the Nile Valley and Cyrenaica, where good ^{14}C evidence exists for the Upper Paleolithic before 30,000 years ago, and for the Late Paleolithic by 21,000 years ago (McBurney 1967; Paulissen & Vermeersch 1987; Wendorf, Schild & Close 1989: 768–824). It is highly unlikely that these advanced complexes could have existed for several thousand years alongside the Middle Paleolithic Aterian without some trace of cultural interaction. Nevertheless, throughout the Sahara there is no trace of either the Upper or Late Paleolithic, although people would surely have been present had there been sufficient rainfall.

The problem of the age of the Aterian and the associated wet episodes in the Sahara is only now nearing resolution. The Nile Valley during the period between about 60,000 and 12,500 BP reveals no traces of wadi sediments interfingering with Nile sediments or of significant slopewash from the escarpments bordering the Valley (Wendorf, Schild & Close 1989: 774–777). This indicates that Egypt was hyperarid and probably even drier than today. The application of other dating techniques, including Uranium series, ESR, TL and ostrich eggshell epimerization, at two localities in Egypt (Wendorf, Close & Schild 1987; Kowalski *et al* 1989) and another in Libya (Petit-Maire 1982) has shown that some of the Aterian wet episodes date to the Last Interglacial and earlier (ca 175,000 to 70,000 BP). The Egyptian determinations were made on exactly the same sites and sediments that have yielded ^{14}C dates between 20,000 and 30,000 BP on carbonates and snail shells (Haas & Haynes 1980; Pachur *et al* 1987).

The major contribution of ^{14}C dating to Saharan prehistory has been toward our understanding of Early Holocene cultural and climatic events. In many parts of the Sahara now are well-dated sequences of Holocene archaeological complexes that are closely tied to wet and dry climatic cycles (Barich 1987; Kuper 1988; Petit-Maire & Riser 1983; Roset 1987; Neumann 1989; Wendorf & Schild 1980; Wendorf, Schild & Close 1984). In these areas, the available data are now beginning to approach the detail and complexity known in areas with much longer histories of archaeological research.

^{14}C dating has also demonstrated that ceramics occur throughout the southern Sahara at a much earlier date than in either the Egyptian Nile Valley or in northwestern Africa. Several ^{14}C dates older than 9000 BP are associated with well-made pottery, with distinct and often complex design motifs, known as Saharo-Sudanese or Early Khartoum ware. These ceramics occur at several sites in the Egyptian Sahara (Wendorf, Schild & Close 1984), Algeria (Camps, Delibrias & Thommeret 1973) and Niger (Roset 1987). The association of the dates with the ceramics is controversial, and will remain so until actual potsherds are directly dated. However, if the pottery is more than 9000 years old, as now seems likely, then pottery-making was independently invented somewhere in the Southern Sahara or Sahelian zone long before ceramics were known anywhere in southwestern Asia. It is generally thought that pottery use implies some degree of sedentism, which would have enormous social and economic implications for the nature of societies that functioned in the Early Holocene Sahara.

Another, still controversial, aspect of the ^{14}C-dated archaeological sites in the Sahara is the discovery of bones of cattle, thought to be domestic, in some of the same Early Holocene sites that yielded the early ceramics. Several sites in the Eastern Sahara with associated cattle remains yielded ^{14}C dates older than 9000 BP, and many more fall between 9000 and 8000 BP. These finds may represent an early stage in the development of the distinctive African pattern of cattle husbandry, wherein the animals are an important source of wealth and status, and are used primarily for milk and blood rather than meat (Wendorf, Schild & Close 1984). We anticipate that the age of these cattle remains will soon be established by directly dating amino acids extracted from the bones, using accelerator mass spectrometry (AMS).

NEW RADIOCARBON DATING IN NORTHERN AFRICA

The development of AMS ^{14}C dating can be expected to have a major impact on the prehistory of North Africa. It has already resolved a major problem by directly dating plant remains from Late Paleolithic sites at Wadi Kubbaniya in southern Egypt (Wendorf, Schild & Close 1989); analyses showed that grains of wheat and barley, thought to have been associated with a hearth dating around 18,000 BP, were intrusive, but that tubers of nut-grass and club-rush were properly associated with the occupation. AMS dating also has shown that the sedimentary sequence at Selima Oasis in northern Sudan is more complex than previously believed (Haynes, personal communication 1989).

While the potential of AMS dating has been most evident in the resolution of problems where context or association is in doubt (and will undoubtedly help resolve the problems of early cattle and ceramics noted above), this new technology also has enormous potential for dating sites where the preservation of charcoal and other organic materials is minimal; this is a common situation in much of North Africa, particularly in the Sahara. It is also likely that the near future will see the use of accelerator dating to assist in the identification and interpretation of settlements where multiple occupations occur.

REFERENCES

Alimen, H 1957 *The Prehistory of Africa.* London, Hutchinson.

Aumassip, G 1987 Neolithic of the Basin of the Great Eastern Erg. *In* Close, AE, ed, *Prehistory of Arid North Africa: Essays in Honor of Fred Wendorf.* Dallas, Southern Methodist University Press: 235–258.

Balout, L 1955 *Préhistoire de l'Afrique du Nord.* Paris, Arts et Métiers Graphiques.

Barich, BE 1987 Archaeology and Environment in the Libyan Sahara. Oxford, *BAR International Series* 368.

Brunton, G and Caton-Thompson, G 1928 *The Badarian Civilization and Prehistoric Remains Near Badari.* London, Quaritch.

Camps, G 1955 Le gisement atérien de Camp Franchet d'Espéray. *Libyca* 3: 17–56.

_____1968 Amekni, Néolithique ancien du Hoggar. *Mémoires du CRAPE* 10. Paris, Arts et Métiers Graphiques.

Camps, G, Delibrias G and Thommeret, J 1973 Chronologie des civilisations préhistoriques du nord de l'Afrique d'après le radiocarbone. *Libyca* 21: 65–89.

Caton-Thompson, G 1946 The Levalloisian industries of Egypt. *Proceedings of the Prehistoric Society* 12: 57–120.

_____1952 *Kharga Oasis in Prehistory.* London, Athlone Press.

Caton-Thompson, G and Gardner, EW 1934 *The Desert Fayum.* London, Royal Anthropological Institute.

Charlesworth, JK 1957 *The Quaternary Era with Special Reference to Glaciation.* London, Edward Arnold (2 vols).

Clark, RM 1978 Bristlecone pine and ancient Egypt: A re-appraisal. *Archaeometry* 20: 5–17.

Clark, RM and Renfrew, C 1973 Tree-ring calibration of radiocarbon dates and the chronology of ancient Egypt. *Nature* 243: 265–270.

Close, AE 1980 Current research and recent radiocarbon dates from Northern Africa. *Journal of African History* 21: 145–167.

_____1984 Current research and recent radiocarbon dates from Northern Africa, II. *Journal of African History* 25: 1–24.

_____1988 Current research and recent radiocarbon dates from Northern Africa, III. *Journal of African History* 29: 145–176.

Conrad, G 1969 *L'évolution continentale post-Hercynienne du Sahara algérien.* Paris, CNRS.

_____1972 Les fluctuations climatiques récentes dans l'est du Sahara occidental algérien. *Actes du Congrès Panafricain de Préhistoire, VI Session, Dakar, 1967*: 343–349.

Gobert, EG and Vaufrey, R 1932 Deux gisements extrêmes d'Ibéromaurisien. *L'Anthropologie* 42: 449–490.

Haas, H, Devine, J, Wenke, R, Lehner, M and Wölfli, W 1987 Radiocarbon chronology and the historical calendar in Egypt. *In* Aurenche, D, Evin, J and Hours, F, eds, Chronologies in the Near East. Oxford, *BAR International Series* 379: 586–606.

Haas, H and Haynes, CV 1980 Discussion of radiocarbon dates from the Western Desert. *In* Wendorf, F and Schild, R, eds, *Prehistory of the Eastern Sahara.* New York, Academic Press: 373–378.

Hassan, FA 1988 The Predynastic of Egypt. *Journal of World Prehistory* 2: 135–185.

Hassan, F and Robinson, SW 1987 High-precision radiocarbon chronometry of Ancient Egypt and comparisons with Nubia, Palestine and Mesopotamia. *Antiquity* 61: 119–135.

Hayes, WC 1964 *Chronology, Egypt – To End of Twentieth Dynasty.* Cambridge, Cambridge University Press.

Hugot, HJ 1966 Limites méridionales de l'Atérien. In *Actas del V Congreso Panafricano de Prehistoria y de Estudio de Cuaternario, 2, Santa Cruz de Tenerife*: 95–108.

Huzayyin, SA 1941 The place of Egypt in prehistory. *Mémoires de l'Institut d'Egypte* 49. Cairo, Institut d'Egypte.

Irwin, HT, Wheat, B and Irwin, LF 1968 University of Colorado investigations of

Paleolithic and Epipaleolithic sites in the Sudan, Africa. *University of Utah Papers in Anthropology* 90. Salt Lake City, University of Utah Press.

Junker, H 1929 Vorlaufiger Bericht über die Grabung der Akadamie der Wissenschaften in Wien auf der Neolithischen Siedlung von Merimde-Benisalâme (Westdelta). *Anzeiger der Akadamie der Wissenschaften in Wien, Philosophische-historische Klasse,* 16–18: 156–250.

Kaiser, W 1957 Zur inneren Chronologie der Nagadakultur. *Archaeologia Geographica* 6: 69–77.

Kowalski, K, Van Neer, W, Bochenski, Z, Mlynarski, M, Rzebik-Kowalska, B, Szyndlar, Z, Gautier, A, Schild, R, Close, AE and Wendorf, F 1989 A last interglacial fauna from the eastern Sahara. *Quaternary Research* 32: 335–341.

Kuper, R 1988 Neuere Forschungen zur Besiedlungsgeschichte der Öst-Sahara. *Archäologisches Korrespondenzblatt* 18: 127–42.

Libby, WF 1955 *Radiocarbon Dating,* 2nd edition. Chicago, University of Chicago Press.

Long, RD 1970 Ancient Egyptian chronology, radiocarbon dating and calibration. *Zeitschrift für Agyptische Sprache und Altertumskünde* 103: 30–48.

McBurney, CBM 1967 *The Haua Fteah (Cyrenaica) and the Stone Age of the South-east Mediterranean.* Cambridge, Cambridge University Press.

Movius, HL 1953 Old World prehistory. *In* Kroeber, AL, ed, *Anthropology Today.* Chicago, University of Chicago Press: 163–192.

Neumann, K 1989 Vegetationsgeschichte der Östsahara im Holozän: Holzkohlen aus prähistorischen Fundstellen. *In* Kuper, R, ed, *Forschungen zur Unweltgeschichte der Östsahara.* Köln, Heinrich-Barth-Institut: 13–181.

Nilsson, E 1931 Quaternary glaciations and pluvial lakes in British East Africa. *Geografiska Annaler* 13: 249–348.

Pachur, H-J, Röper, H-P, Kröpelin, S and Goschin, M 1987 Late Quaternary hydrography of the eastern Sahara. *Berliner Geowissenschaftliche Abhandlungen* 75: 331–384.

Paulissen, E and Vermeersch, PM 1987 Earth, man and climate in the Egyptian Nile Valley during the Pleistocene. *In* Close, AE, ed, *Prehistory of Arid North Africa: Essays in Honor of Fred Wendorf.* Dallas, Southern Methodist University Press: 29–67.

Paulissen, E, Vermeersch, PM and Van Neer, W 1985 Progress report on the Late Paleolithic Shuwikhat sites (Qena, Upper Egypt). *Nyame Akuma* 26: 7–14.

Petit-Maire, N, ed 1982 *Le Shati, Lac Pléistocene du Fèzzan (Libye).* Marseille, CNRS.

_____1988 Climatic change and man in the Sahara. *In* Bower, J and Lubell, D, eds, Prehistoric Cultures and Environments in the Late Quaternary of Africa. Oxford, *BAR International Series*: 19–42.

Petit-Maire, N and Riser, J, eds, 1983 *Sahara ou Sahel? Quaternaire récent du Bassin de Taoudenni (Mali).* Marseille, CNRS.

Roset, J-P 1987 Paleoclimatic and cultural conditions of Neolithic development in the Early Holocene of northern Niger (Air and Tenere). *In* Close, AE, ed, *Prehistory of Arid North Africa: Essays in Honor of Fred Wendorf.* Dallas, Southern Methodist University Press: 211–234.

Sandford, KS 1934 Paleolithic Man and the Nile Valley in Upper and Middle Egypt. *University of Chicago Oriental Institute Publication* 18. Chicago, University of Chicago Press.

Sandford, KS and Arkell, WJ 1929 Paleolithic Man and the Nile-Fayum Divide. *University of Chicago Oriental Institute Publication* 10. Chicago, University of Chicago Press.

_____1933 Paleolithic Man and the Nile Valley in Nubia and Upper Egypt. *University of Chicago Oriental Institute Publication* 17. Chicago, University of Chicago Press.

_____1939 Paleolithic Man and the Nile

Valley in Lower Egypt. *University of Chicago Oriental Institute Publication* 46. Chicago, University of Chicago Press.

Servant, M and Servant-Vildary, S 1972 Nouvelles données pour une interprétation paléoclimatique de séries continentales du bassin tchadien (Pléistocène récent, Holocène). *In* van Zinderen Bakker, EM, ed, *Palaeoecology of Africa* 6. Cape Town, Balkema: 87–92.

Shaw, T 1989 African archaeology: looking back and looking forward. *The African Archaeological Review* 7: 3–31.

Smith, HS 1964 Egypt and C14 dating. *Antiquity* 38: 32–37

Smith, PEL 1966 The Late Paleolithic of Northeast Africa in the light of recent research. *American Anthropologist* 68(2): 326–355.

Suess, HE 1970 Bristlecone-pine calibration of the radiocarbon time-scale 5200 BC to the present. *In* Olsson, IU, ed, *Radiocarbon Variations and Absolute Chronology*. Proceedings of the 12th Nobel Symposium. New York, John Wiley & Sons: 303–311.

Thiede, J, Suess, E and Müller, PJ 1982 Late Quaternary fluxes of major sediment components to the sea floor at the Northwest African continental slope. *In* von Rad, U, Hinz, K, Sarntheim, M and Seibold, E, eds, *Geology of the Northwest African Continental Margin*. Berlin, Springer-Verlag: 605–631.

Tillet, T 1983 *Le Paléolithique du Bassin Tchadien Septentrional (Niger-Tchad)*. Paris, CNRS.

Vaufrey, R 1955 Préhistoire de l'Afrique, Tome I, Le Maghreb. *Publications de l'Institut des Hautes Etudes de Tunis* 4. Paris, Masson.

Vermeersch, PM, Otte, M, Gilot, E, Paulissen, E, Gijselings, G and Drappier, D 1982 Blade technology in the Egyptian Nile Valley. *Science* 216: 626–628.

Vignard, E 1923 Une nouvelle industrie lithique, le Sébilien. *Bulletin de l'Institut Français d'Archéologie Orientale* 22: 1–756.

Wendorf, F, ed 1968 *The Prehistory of Nubia*. Dallas, Fort Burgwin Research Center and Southern Methodist University Press (2 vols + atlas).

Wendorf, F, Close, AE and Schild, R 1987 Recent work on the Middle Paleolithic of the Eastern Sahara. *The African Archaeological Review* 5: 49–63.

Wendorf, F and Schild, R 1975 The Paleolithic of the Lower Nile Valley. *In* Wendorf, F and Marks, AE, eds, *Problems in Prehistory: North Africa and the Levant*. Dallas, Southern Methodist University Press: 127–169.

_____1976 *Prehistory of the Nile Valley*. New York, Academic Press.

_____1980 *Prehistory of the Eastern Sahara*. New York, Academic Press.

Wendorf, F, Schild, R and Close, AE, eds, 1984 *Cattle-Keepers of the Eastern Sahara: The Neolithic of Bir Kiseiba*. Dallas, Department of Anthropology, Institute for the Study of Earth and Man, Southern Methodist University.

_____1989 *The Prehistory of Wadi Kubbaniya*. Dallas, Southern Methodist University Press (3 vols).

Wendorf, F, Schild, R and Haas, H 1979 A new radiocarbon chronology for prehistoric sites in Nubia. *Journal of Field Archaeology* 6: 219–223.

22

THE IMPACT OF RADIOCARBON DATING ON NEAR EASTERN PREHISTORY

DONALD O HENRY

INTRODUCTION

As within other scientific inquiries, archaeology embraces a strong interplay between technology and theory. Of the many technological advances in modern archaeology, ^{14}C dating has had perhaps the greatest influence on shaping theoretical developments within Near Eastern prehistory. Whereas this influence has been most obvious in the domain of major cultural-historic questions (eg, the emergence of social complexity), a more significant theoretical consequence of the technique rests in the way Near Eastern prehistorians have come to explain variability in the prehistoric record.

Coincident with the introduction of ^{14}C dating in the 1950s, was a shift in the focus of research away from the excavation of deeply stratified, single sites to areal investigations. Areal projects typically involved excavations of a wide variety of site types distributed within and between major prehistoric periods. Without ^{14}C dating, such investigations would have been incapable of fitting prehistoric occupations often lacking stratigraphic links into common chronological sequences. This change in spatial focus from site-specific to area-wide investigations also paralleled a rather fundamental paradigmatic shift in the way prehistorians viewed variability. Prior to the emergence of areal investigations, artifact and assemblage variability was explained primarily as a result of diachronic or evolutionary change. Areal investigations, on the other hand, came to pay much more attention to synchronic, functional factors and to adopt ecological explanations of assemblage variability.

BACKGROUND AND STATISTICS

Beginning in the mid-1950s, prehistorians working in the Near East began using ^{14}C dating. By 1965, a sufficient number of dates had accumulated to justify the publication of a compendium (Watson 1965) with similar syntheses and date lists following at fairly regular intervals (Henry & Servello 1974; Weinstein

TABLE 22.1. A time series showing the distributions of radiocarbon dates for major prehistoric periods in the Near East. Note the differences in geographic coverages for the date lists.

Period	Publication Date and Geographic Area				
	1965* Near East	1974** Near East	1984† Levant	1989‡ Levant	1990§ Arid zone
Middle Paleolithic	-	49	49	-	-
Upper Paleolithic	-	29	29	-	-
Epipaleolithic	1	64(19)ǁ	55	102	-
Neolithic	66	120	93	-	61
Chalcolithic	17	-	19	-	14

*Watson (1965)
**Henry and Servello (1974)
†Weinstein (1984)
‡Henry (1989)
§Zarins (1990)
ǁFor Levant only

1984; Zarins 1990). A review of these comprehensive syntheses, along with others covering specific time-frames, provides a rough measure of the rate at which dates are accumulating (Table 1). The uneven coverage of the date lists relative to temporal sweep and geography hinders most direct comparisons, but between 1974 and 1989, a general picture emerges of the dramatic rise in the numbers of dates from the Epipaleolithic of the Levant. Between 1974 and 1984, the reservoir of dates increased by over 500%.

Clearly, the striking growth in dating results from Near Eastern prehistoric deposits over the last 40 years denotes the technique's expanding importance as a research tool. What is somewhat surprising is that few of these dates resulted from research efforts specifically devoted to chronological problems, an approach often taken with other radiometric techniques (TL, U-series, amino acid racemization). When viewed collectively, however, the ^{14}C dates generated by numerous discrete research efforts significantly influenced our view of the Near Eastern prehistoric record in several specific areas.

IMPACT ON CULTURAL-HISTORIC QUESTIONS

Within the Near East, cultural-historic questions relating to certain transitional intervals in human development appear to have been shaped most strongly by ^{14}C dating. The most significant of these include the Middle to Upper Paleolithic transition, the emergence of social complexity in the late Pleistocene and the beginnings of food production in the early Holocene.

Although not directly a diachronic issue, the problem of the human occupation of the arid zone of the Near East is another important cultural-historic domain that has been greatly influenced by ^{14}C dating (Zarins 1990). Prior to the advent of the ^{14}C technique, prehistorians were restricted to exploring deeply stratified cave and rockshelter deposits for the discovery of chronological sequences. In the Near East and especially the Levant, these sheltered deposits are primarily restricted to the better-watered Mediterranean environmental zone with few such settings located in the adjacent steppe and desert zones (Cauvin & Sanlaville 1981: 582). Thus, concomitant with the widespread use of ^{14}C dating in the late 1960s, exploration of the arid zone was initiated (Marks 1971, 1976) with numerous projects following in the 1970s (Bar-Yosef & Phillips 1977; Henry 1982; Garrard & Stanley Price 1977; Cauvin 1981; Hanihara & Akazawa 1979).

The Middle to Upper Paleolithic Transition

One of the huge cultural-historic problems within Near Eastern prehistory centers on cultural and biologic continuity during the early part of the Last Glacial (Clark & Lindly 1990). It may seem somewhat surprising that ^{14}C dating has had much to do with this issue given that most of the interval falls beyond the range of the technique. However, recent ^{14}C dates from late Levantine Mousterian occupations at Tabun Cave (Weinstein 1984; Jelinek 1982) and from the open-air Negev desert site of Boker Tachtit (Weinstein 1984; Marks 1983) have significantly altered our understanding of the Middle Paleolithic record (Fig 22.1).

The point dates from Tabun Cave Unit I fall between 45,800 and 51,000 BP, whereas those from a firepit at Boker Tachtit Level 1 range from 44,930 to 47,280 BP. The assemblage recovered from the Unit I horizon at Tabun Cave belongs to the Levantine Mousterian B type (flake-dominated) industry which was previously thought to end the long 50,000–70,000-year Mousterian sequence in the Levant. Although the Level 1 assemblage from Boker Tachtit occupied a similar end-of-sequence position, as suggested by the dates as well as from Upper Paleolithic assemblages recovered from overlying strata, it represents a D type industry. Traditionally, the D type industry has been viewed as initiating the long Levantine Mousterian sequence, giving way to C and B type industries some 60–70,000 BP. The discovery and dating of Boker Tachtit has prompted a rethinking of this scheme.

Given the presence of earlier (90,000–60,000 BP) D type occupations in the arid zone (Negev and southern Jordan), coupled with the absence of evidence for C and B type industries from the same region, researchers have come to argue for development along two evolutionary paths during the Middle Paleolithic (Marks 1988). Whereas the traditional developmental scheme may have prevailed in the

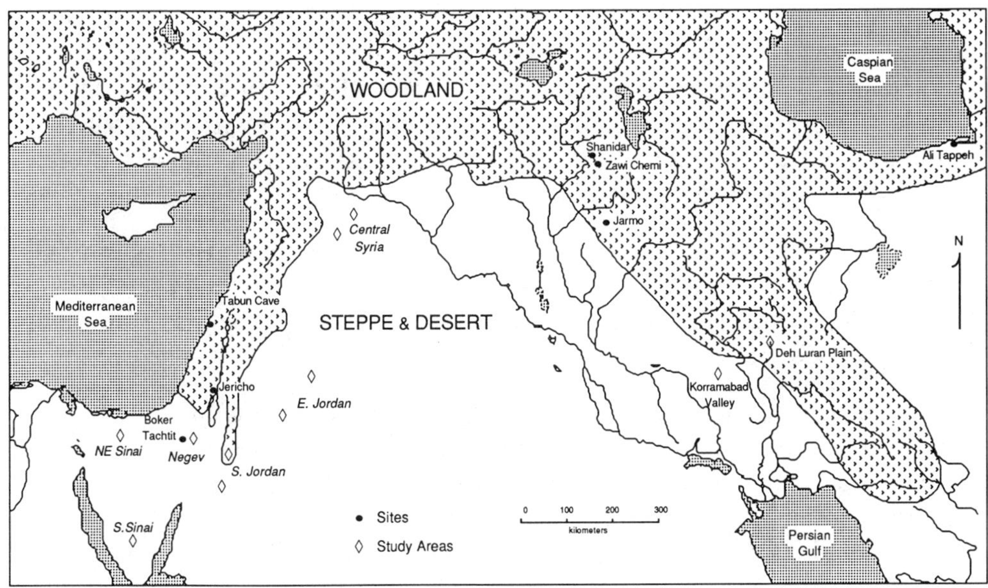

Fig 22.1. Map showing the distribution of sites and study areas relative to woodland and steppe/desert environmental zones

Mediterranean zone (although even this has recently been challenged), it seems clear that the D type industry remained relatively unchanged from the beginning to the end of the Middle Paleolithic sequence within the arid zone of the southern Levant.

The continuity seen in the cultural record of the early Last Glacial in the southern Levant is also significant to the current debate surrounding the origin of morphologically modern humans. If, as argued by some "replacement" advocates, morphologically modern humans evolved first in Africa and expanded into the Near East some 80–100,000 BP, one would expect such a population expansion to have left a "cultural track" and disrupted the cultural succession of the region. In that such a cultural track or disruption is not apparent, an expansion out of Africa must have occurred before approximately 100,000 BP, or modern humans must have evolved multiregionally in various parts of the Old World.

Emergence of Social Complexity

Near the end of the Pleistocene, populations in the Near East experienced fundamental changes in their economic, demographic and social structures. Large sedentary and socially ranked groups dependent upon the intensive collection of wild cereals replaced small, mobile egalitarian foraging groups over much of the region beginning around 12,500 BP. Although this was a widespread phenomenon, it is best understood within the Levant where it was represented by the Natufian archaeological culture (Henry 1983, 1989). In that the Natufian appeared transitional between long-held hunting and gathering patterns and the first steps of food production, considerable attention has been focused on its beginnings, as well as its evolution to the early Neolithic.

Over 100 ^{14}C dates from more than 40 sites within the interval have enabled prehistorians to correlate diverse paleoenvironmental, economic and demographic data in the formulation of explanatory models. Such correlations would not have been possible without these temporal controls.

The explanatory models address three major problems: 1) defining the factors that gave rise to the Natufian; 2) identifying the factors responsible for a shift from the intensive collection of wild cereals to their cultivation; and 3) tracing the relationships of various discrete societies that occupied the region about 13,000 to 10,000 years ago.

Bracketed between 12,500 and 13,000 BP, a sharp rise in the earth's temperature is thought to have triggered an expansion of warmth-loving cereal grasses out of their low elevational Pleistocene refuges into the hill zone of the Near East. This colonization of the uplands brought wild cereals into their optimum habitat – the Fertile Crescent – and extended their harvest period over several months, depending upon elevation. Cereals then provided a new, exceedingly attractive resource for terminal Pleistocene foragers.

Faunal, macrobotanic, palynologic, geomorphic and archaeologic evidence drawn from natural and cultural deposits in the Levant confirm such an expansion during this warm, moist episode. And it is at this time that we see the first intensive exploitation of wild cereals, coupled with the emergence of large villages, rapid population growth and social complexity. Establishing the synchroneity of these natural and cultural events is dependent upon ^{14}C dating.

A detailed chronology of the Natufian is equally important to understanding the society's dissolution and adoption of agriculture. Based upon correlations between certain kinds of retouch found on lithic artifacts, ^{14}C dates and stratigraphic evidence, a precise seriation scheme was developed for the Natufian (Bar-Yosef & Valla 1979). When data for the sizes and distributions of sites are ordered using these chronologic controls, a clear trend of population growth and expansion is denoted (Henry 1989). The results of dental and osteologic analyses of Natufian skeletal populations also point to a progressive reliance upon cereals, growing nutritional stress and the use of female infanticide as an attempt to control a rising population (Henry 1989). By 10,000 BP, the Natufian adaptive system had failed under mounting population pressure and environmental deterioration. Those Natufian groups situated in marginal arid areas were forced to return to mobile foraging, whereas communities adjacent to dependable water supplies in more verdant settings began cultivating cereals in order to supplement wild stands. In the absence of stratigraphic and direct artifact links, these synchronous alternative responses are confirmed solely through ^{14}C dates.

Origin of Agriculture

Some of the first ^{14}C dates for Near Eastern archaeology are tied to the question of the origin of agriculture. Samples from the Pre-Pottery Neolithic A horizon of Jericho, which yielded grains of domesticated barley (Hopf 1969), were dated to approximately 10,000 BP. Some scholars continue to argue for a much earlier occurrence of plant cultivation (Moore 1985), but the plant remains from these deposits are represented only by wild species (Henry 1989; Crabtree & Campana 1990). Recent AMS dates have confirmed the presence of wild species in the period immediately prior to 10,000 BP (Gowlett 1987), but rejected notions of such antiquity for domestic cereals (Legge 1986; Wendorf 1987). There is a certain irony here, in that most scholars fully expected the origin of agriculture to be pushed back in time with the accumulation of more ^{14}C dates, given that more precise chronologies accompanying ^{14}C dating have almost universally increased the time depth of other major prehistoric milestones. This has not taken place, however. With the greater number of dates and, perhaps more importantly, with the ability through AMS to date directly evidence (eg, seed and bone) attributable to food production, ^{14}C dating has not only failed to push back the antiquity of agriculture, but it has been the principal means by which claims of early agriculture have been refuted.

Impact on Explanations of Assemblage Variability

As mentioned earlier, it was during the 1950s that areal investigations were introduced to the study of Near Eastern prehistory. The shift in focus from the intensive excavation of deep, stratified tells and sheltered sites to the extensive investigation of a wide variety of site types over a large area was in part stimulated by the innovation of ^{14}C dating. Concomitant with the shift from single-site to areal investigations, prehistorians adopted a new paradigm for examining variability in the prehistoric record. Synchronic functional and ecologic perspectives emerged as explanatory alternatives to diachronic, evolutionary approaches. Given the broader theoretical movements in prehistoric archaeology, the technical innovation of ^{14}C dating was not the sole reason for the paradigmatic shift in Near Eastern prehistory. However, this new dating technique undoubtedly accelerated the adoption of a functional paradigm.

The Iraq-Jarmo project was the first example of an ecologically oriented areal investigation in the Near East. Undertaken in the early 1950s by Braidwood and his colleagues, the investigation centered on the problem of the first appearance of settled village-farming communities (Braidwood & Howe 1960). In emphasizing the environmental factors that may have influenced the transition from foraging to farming, the project involved a team of natural scientists and adopted a regional approach. And these facets of the research program relied upon the then new innovation of ^{14}C dating for chronologic control.

Comparisons of cultural and paleoenvironmental evidence drawn from within and outside the Iraq-Jarmo Project area were stressed:

> Regardless of dating problems, these archaeological efforts in the Jordan Valley and those in Iraqi-Kurdistan will prove to be interdependent and related. Regardless also of the question of specific ecological niches, both efforts lie within the general area that we feel may be crucial to the great transition (Braidwood & Howe 1960: 6).

^{14}C dating was the principal avenue by which these comparisons were to be brought into a common chronology. In 1955, various members of the research team collected samples for dating within the project area and further afield in Jordan, Lebanon, Syria, southern Turkey and even Egypt (Braidwood & Howe 1960: 23). ^{14}C dates from natural settings associated with highland lake pollen cores were also of considerable importance to the project in connecting paleo-environmental and cultural sequences (Wright 1968).

Also in northern Iraq, research during the late 1950s and early 1960s at the nearby sites of Shanidar Cave and Zawi Chemi Shanidar was much influenced by the results of ^{14}C dating (RL Solecki 1969, 1981; RS Solecki 1964). Radiometrically synchronous dates of ca 10,800 BP from the deposits of the cave and open site, coupled with their differences in elevations and local environments, prompted an adaptive explanation for the differences in artifact patterns seen between the two sites. R L Solecki (1969) suggested that seasonally governed movements of groups between different elevational belts accompanied by different activities accounted for the observed variations in artifact patterns.

During the 1960s, areal research efforts conducted in southwestern Iran continued the emphasis upon cultural ecologic interpretations of the prehistoric record (Hole, Flannery & Neely 1969; Hole & Flannery 1967). Although ^{14}C dates were important to developing an absolute chronology for the late Pleistocene Khorramabad Valley sites as well as the early Holocene village sites of the Deh Luran Plain, the sequences and inter site correlations were based mainly upon stratigraphic successions and artifact cross-reference dating. In that the investigated sites consisted of deeply stratified caves, rockshelters and early village communities, traditional time-correlation approaches were feasible.

McBurney's (1968) excavation of the Cave of Ali Tappeh on the Caspian coast deserves special mention when reviewing the impact of ^{14}C dating on Near Eastern prehistory. Although the project centered on the excavation of this deeply stratified site, through intensive dating of the deposit and comparisons with dated sequences for other Caspian coast caves, McBurney was able to trace the local responses to terminal Pleistocene climatic changes. By correlating (through interpolation and regression analysis) ^{14}C ages with depth, a rather precise chronology was developed for the deposit of Ali Tappeh. Against this chronology faunal successions were compared to nearby sites and eventually to the terminal Pleistocene climatic sequence of Europe. McBurney's use of ^{14}C dates in building a local chronology for the examination of human ecology on a regional – and even global – scale was a model for future researchers.

By the late 1960s, ^{14}C dating had become a common prehistoric research tool in the Near East. Skepticism concerning the accuracy of dates was rare. The research designs of most projects anticipated and, to some extent, depended upon the production of dates. It was at this time that areal investigations were launched that emphasized synchronic intersite variability as a means of reconstructing prehistoric procurement-settlement systems. Typified by Marks' (1971) pioneering effort in the highland Negev, these projects were accompanied by intensive, systematic, on-foot surveys that resulted in the discovery of tremendous numbers of sites of all varieties. Of particular significance here is that the diversity of sites and their constituent artifact assemblages were emphasized over the thickness or artifact richness of deposits. This new research bias was precipitated by an overarching theoretical movement that stressed questions of prehistoric human ecology and adaptation. And it was thought that these questions would be better addressed by collecting limited data from a great number and wide variety of prehistoric habitations than by retrieving a great amount of evidence from a single reference point.

During the 1970s and 1980s, areal investigations were confined principally to the Levant and its margins. Such efforts were initiated in northeastern Sinai (Bar-Yosef & Phillips 1977), southern Sinai (Phillips 1987), the Negev (Goring-Morris 1985), southern Jordan (Henry 1982; Clark *et al* 1988), eastern Jordan (Garrard & Stanley Price 1977; Garrard, Byrd & Betts 1986), and central Syria (Cauvin 1981; Hanihara & Akazawa 1979). ^{14}C dating was commonly used for chronological comparisons within and between these projects, and this resulted in a marked growth in the numbers of dates.

TABLE 22.2. Environmental Distributions for Paleolithic Industries of the Levant

Paleolithic period	Arid zone	Mediterranean zone
Middle	D Type	D,C,B types
Upper	Ahmarian	Levantine Aurignacian
Epipaleolithic	Mushabian	Geometric Kebaran
Neolithic	Harifian	Pre-Pottery Neolithic A

With the underlying adaptive, ecologic paradigm guiding these efforts, it is not surprising that newly discovered archaeologic complexes were interpreted as representing alternative adaptive modes to those related to traditionally defined complexes. These adaptive dichotomies, proposed for almost all of the major prehistoric periods, were based upon assumed differences in selective pressures between the better-watered wooded Mediterranean zone and the arid reaches of the Levant (Table 22.2).

Although Binford (1982) has argued that the recognition of such "parallel phyla" is a natural consequence of adding new analytic procedures and measures of variability to the traditional Bordean system of analysis, this was not entirely the case in the Levant. Programs of analysis stressing technologic over typologic attributes have been commonly used since the late 1960s. The distinguishing characteristics of those complexes rooted in the Mediterranean or the arid zones were based as much or more on technologic as typologic attributes. ^{14}C dating perhaps had more to do with the emergent perspective of regional adaptive differences than did newly adopted analytic procedures. Intrastratified contexts that offered opportunities for comparing the chronologies of dichotomous complexes were rare or non-existent. Thus, without ^{14}C dates that confirmed synchroneity, discussions of alternative prehistoric adaptations to different environmental zones within the Levant would have been little more than speculation.

SUMMARY

As a dating innovation, the ^{14}C technique is traditionally viewed as having had tremendous impact upon our understanding of the prehistoric record especially with regard to cultural-historic questions. In Near Eastern prehistory, the ability of archaeologists to construct precise chronologies with the aid of ^{14}C dating has done much to reshape our cultural-historic reconstructions, particularly over transitional intervals. But ^{14}C dating may have had an even greater effect on Near Eastern prehistory in the realm of theory and specifically on the way artifact/assemblage variability is explained. In providing another means of gaining chronologic controls beyond the excavation of deep stratified deposits, the technique opened up the possibility for archaeologists to begin examining artifact and site diversity from a more regional perspective. At the same time, an adaptive or functional paradigm emerged as an alternative to the previously held evolutionary approach to explaining variation in the prehistoric record.

REFERENCES

Bar-Yosef, O and Phillips, JL, eds 1977 *Prehistoric Investigations in Gebel Maghara, Northern Sinai.* Monographs of the Institute of Archaeology. Jerusalem, Hebrew University.

Bar-Yosef, O and Valla, F 1979 L'évolution du Natoufien. *Paléorient* 5: 145–152.

Binford, SL 1982 The archaeology of place. *Journal of Anthropological Archaeology* 1(1): 5–31.

Braidwood, R and Howe, B, eds 1960 Prehistoric investigation in Iraqi Kurdistan. *Studies in Ancient Oriental Civilization* 31.

Cauvin, M-C 1981 L'Epipaléolithique du Levant. *In* Cauvin, J and Sanlaville, P, eds, *Préhistoire du Levant.* Paris, Éditions du CNRS: 439–444.

Cauvin, J and Sanlaville, P, eds 1981 *Préhistoire du Levant.* Paris, Éditions du CNRS.

Clark, G and Lindly, J 1990 Modern human origins and the Levant and western Asia. *American Anthropologist* 91(4): 962–985.

Clark, GA, Lindly, J, Donaldson, M, Garrard, A, Coinman, N, Schuldenrein, J, Kish, S and Olszewski, D 1988 Paleolithic archaeology in the southern Levant. *Annual of the Department of Antiquities of Jordan* 31: 19–78.

Crabtree, P and Campana, D 1990 The late Natufian site of Salibiya I in the Jordan Valley: Preliminary investigations. *Anthro Quest: News of Human Origins, Behaviors and Survival* 42: 20–22.

Garrard, A, Byrd, B and Betts, A 1986 Prehistoric environment and settlement in the Azraq Basin: An interim report on the 1984 excavation season. *Levant* 18: 5–24.

Garrard, A and Stanley Price, NP 1977 A survey of the prehistoric sites in the Azraq Basin. *Paléorient* 3: 109–123.

Goring-Morris, A (ms) 1985 Terminal Pleistocene hunter/gatherers in the Negev and Sinai. PhD dissertation, Hebrew University.

Gowlett, JAJ 1987 The archaeology of radiocarbon accelerator dating. *Journal of World Prehistory* 1(2): 127–170.

Hanihara, K and Akazawa, T, eds 1979 *Paleolithic Site of Douara Cave and Paleogeography of the Palmyra Basin in Syria.* Tokyo, University of Tokyo Press.

Henry, DO 1982 The prehistory of southern Jordan and relationships with the Levant. *Journal of Field Archaeology* 9(4): 417–444.

_____1983 Adaptive evolution within the Epipaleolithic of the Near East. *Advances in World Archaeology* (2): 99–160.

_____1989 Correlations between reduction patterns and settlement patterns. *In* Henry, DO and Odell, GH, eds Alternative approaches to lithic analysis. *Archaeological Papers of the American Anthropological Association* 1: 139–156.

Henry, DO and Servello, AF 1974 Compendium of carbon-14 determinations derived from Near Eastern prehistoric deposits. *Paléorient* 2: 19–44.

Hole, F and Flannery, KV 1967 The prehistory of southwestern Iran: A preliminary report. *Proceedings of the Prehistoric Society* 33: 147–206.

Hole, F, Flannery, K and Neely, J, eds 1969 *Prehistory and Human Ecology of the Deh Luran Plain: An Early Village Sequence from Khuzistan, Iran.* Memoirs of the Museum of Anthropology. Ann Arbor, University of Michigan.

Hopf, M 1969 Plant remains and early farming in Jericho. *In* Ucko, PJ and FW Dimbleby, eds, *The Domestication and Exploitation of Plants and Animals.* London, Duckworth: 355–357.

Jelinek, A 1982 The Tabun cave and Paleolithic man in the Levant. *Science* 216: 1364–1375.

Legge, AJ 1986 Archaeological results from accelerator dating. *In* Gowlett, JAJ and Hedges, REM, eds, Seeds of discontent: Accelerator dates on some charred plant remains from the Kebaran and Natufian cultures. *Oxford University Committee for Archaeology Monograph* 11: 13–21.

Marks, AE 1971 Settlement and intrasite variability in the central Negev, Israel. *American Anthropologist* 73(5): 1237–44.

_____1976 Site D5, a Geometric Kebaran "A" occupation in the Nahal Zin. *In* Marks, AE, ed, *Prehistory and Paleoenvironments in the Central Negev, Israel* 1: Dallas, Texas, Southern Methodist University Press: 227–293.

_____1983 The sites of Boker Tachtit and Boker: A brief introduction. *In* Marks, AE, ed, *Prehistory and Paleoenvironments in the Central Negev, Israel* 3. Dallas, Texas, Southern Methodist University Press: 15–38.

_____1988 The curation of stone tools during the Upper Pleistocene. *In* Dibble, HL and Montet-White, A, eds, *Upper Pleistocene Prehistory of Western Asia.* Philadelphia, The University Museum: 275–285.

McBurney, CBM 1968 The cave of Ali Tappeh and the Epi-Paleolithic in NE Iran. *Proceedings of the Prehistoric Society* 23: 385–418.

Moore, AMT 1985 The development of Neolithic societies in the Near East. *Advances in World Archaeology.* New York, Academic Press: 1–70.

Phillips, J 1987 Upper Paleolithic hunter-gatherers in the Wadi Feiran, Southern Sinai. *In* Soffer, O, ed, *The Pleistocene Old World Regional Perspectives.* New York, Plenum Press: 164–181.

Solecki, RL 1969 Milling tools and the Epi-Paleolithic in the Near East. *In* Etudes sur le Quaternaire dans le monade. *Eighth International Congress of the Quaternary*: 984–994.

_____1981 An early village site at Zawi Chemi Shanidar. *Bibliotheca Mesopotamia* 13.

Solecki, RS 1964 Shanidar cave: A Late Pleistocene site in Northern Iraq. *Sixth International Congress of the Quaternary.* Lodz, Pabnstwowe Wydawnictwe Naukowe Oddzial w Kodzi: 413–423.

Watson, PJ 1965 The chronology of North Syria and North Mesopotamia from 10,000 BC to 2,000 BC. *In* Ehrich, RW, ed, *Chronologies in Old World Archaeology.* Chicago, The University of Chicago Press: 61–90.

Weinstein, J 1984 Radiocarbon dating in the Southern Levant. *Radiocarbon* 26(3): 297–366.

Wendorf, F 1987 The advantages of AMS to field archaeologists. *Nuclear Instruments and Methods* B29: 155–158.

Wright, HE, Jr 1968 Natural environment of early food production north of Mesopotamia. *Science* 20(4): 334–339.

Zarins, J 1990 Archaeological and chronological problems within the greater Southwest Asian arid zone, 8500–1850 BC. *In* Ehrich, RW, ed, *Chronologies in Old World Archaeology, Third Edition.* Chicago, The University of Chicago Press: 93–132.

23

RADIOCARBON DATING AND THE PREHISTORY OF SUB-SAHARAN AFRICA

PETER ROBERTSHAW

INTRODUCTION

The invention of ^{14}C dating was of enormous consequence for archaeology. In due course it provided an absolute chronological framework for later African prehistory that was to upset several notions about the relationship between African and European prehistory and history, which had served as ideological foundations for colonialism. I begin by reviewing some of these early developments; then I briefly examine aspects of the history of ^{14}C dating of the Stone Age followed by a more detailed discussion of the Iron Age. Archaeologists who study the Iron Age and the interface with history are particularly concerned with the issues of precision and calibration. Other current concerns include the absence of active dating laboratories in black Africa and the requirement of many more ^{14}C dates if progress in understanding prehistory is to go beyond the establishment of regional cultural sequences. The final section of the chapter addresses aspects of the future of ^{14}C dating in sub-Saharan Africa.

BEGINNINGS

Libby (1951) published the first ^{14}C date for an archaeological site in sub-Saharan Africa. The date was obtained from carbonized wood associated with a Lupemban stone blade recovered by the archaeologist Janmart from the site of Mufo in Portuguese Angola. The date of 11,189 ± 490 (C-580) was in accord with the geomorphological evidence, and was supported by a date of 14,503 ± 560 (C-581) for another sample of wood found lower in the excavation. The next Chicago date list (Libby 1952) contained several dates for sub-Saharan sites, ranging from the Neolithic site of Shaheinab near Khartoum to Great Zimbabwe. These dates included a number of samples from the Rhodesias (now Zambia and Zimbabwe), most of which came from excavations by Desmond Clark. Hallam Movius, known for his work on the Upper Palaeolithic of western Europe, seems to have acted as a clearing agency at Harvard for the submission of African samples to the Chicago laboratory during the

1950s. However, perusal of the early date lists in *Science* and *Radiocarbon* demonstrates that very few dates were run on archaeological samples from sub-Saharan Africa until the mid-1960s.

The small community of archaeologists active in the region was certainly aware of the potential of ^{14}C dating (see, eg, Cole 1954: 284–285), but with 1 or 2 exceptions, had little knowledge of how to proceed. Peter Shinnie recently remarked that when he excavated the earthworks site of Bigo in Uganda in 1957, few labs were accepting samples for dating and that he "certainly did not know how to set about getting such samples, nor how to raise the necessary funds to pay for laboratory dating" (Shinnie 1990: 227). Sonia Cole's remark made in 1963 that East Africa was "sadly lacking in carbon-14 dates" (Cole 1963: 216) was applicable to the whole of sub-Saharan Africa.

The ^{14}C dates obtained during the 1950s and early 1960s provided the first tentative indications of the absolute chronology of later prehistory. In some instances, the results appeared to confirm existing ideas about the antiquity of various technological traditions; for example, most archaeologists seemed comfortable with the dates from Kalambo Falls that indicated the transition from the Acheulean to the Sangoan occurred about 60,000 years ago. But there were also some surprises, notably the 21,000 BP date on shell associated with bone harpoons at Ishango in Central Africa, the mid-first millennium BC dates for iron-working at Nok in Nigeria, and the 7th and 8th centuries AD dates on a wooden lintel found in the Great Enclosure at Great Zimbabwe (see below).

Sending charcoal samples to laboratories for ^{14}C dating became standard practice for archaeologists in sub-Saharan Africa in the mid-1960s. From then until the end of the 1970s, the number of samples submitted for dating increased every year, keeping pace with both the growing number of active archaeologists and the commissioning of new laboratories. The emergence of ^{14}C as the primary dating method for the later prehistory of Africa came at an auspicious moment. The rise of African nationalism and the granting of independence to many countries in the late 1950s and early 1960s prompted an upsurge of interest in Africa's precolonial past, which fostered the discipline of African history and an awareness of the importance of archaeological investigation of the Iron Age. ^{14}C dating had already proven the antiquity of iron metallurgy in West Africa and was soon to provide unexpectedly early dates for the beginnings of the Iron Age over much of the subcontinent. It is important to understand that not only did ^{14}C dating provide archaeologists with dates, it also spared them from the necessity of having to excavate deep-sequence sites to construct relative chronologies; thus, archaeologists were able to devote their energies to the single-component sites characteristic of the Iron Age. As Desmond Clark remarked,

> . . . radiocarbon dating . . . [has] . . . released the prehistorian so that he may now try to interpret what a homogeneous assemblage of artifacts in sealed context can say about the activities of the human group that made and used them (Clark 1990: 191).

The close connection between ^{14}C dating and the study of African history is clearly borne out by the role that the *Journal of African History* has played in disseminating the results of ^{14}C tests. In 1961, the second volume of this journal published the first of what has become a long series of articles summarizing recent archaeological research on the later time periods together with a listing of new ^{14}C dates (Fagan 1961). In preparing this chapter, I have made considerable use of these articles and their appended lists of dates. Figures 23.1 and 23.2 show the number of dates reported in each *Journal of African History* article for several regions of sub-Saharan Africa. Hence, they serve as a simple summary of the history of ^{14}C dating. Some interesting patterns are apparent in the data. However, a few words of caution are needed: the date lists are almost certainly not comprehensive, as the compilers of the lists met with varied success in attempting to track down new dates that had been processed since the preceding list was published. The earlier lists excluded dates older than about 1000 BC, but this cut-off point was increasingly ignored by compilers during the 1970s; hence, the earlier lists may be "missing" some dates, but I suspect the number of such dates is insignificant for the purposes of this presentation.[1] Date lists have appeared at irregular intervals; I have therefore attempted to make the lists comparable by averaging the number of "new" dates per year by dividing the total number of dates in a list by the number of years that elapsed since the previous list was published.

Bearing in mind the above caveats, the figures appear to show an increase in the number of dates processed each year until the end of the 1970s, with a possible spurt in growth in the early 1970s. This trend correlates with increases both in the number of dating laboratories and in the number of archaeologists active in sub-Saharan Africa; it comes as no surprise. A less obvious but equally predictable trend is towards processing more dates per site. What does, perhaps, come as a surprise is the decline in the number of dates processed each year that seems to set in during the 1980s; I postpone discussion of this observation until the last part of the chapter.

[1]This sampling bias applies particularly to South Africa where the majority of research in the 1960s and early 1970s was directed towards the Stone Age.

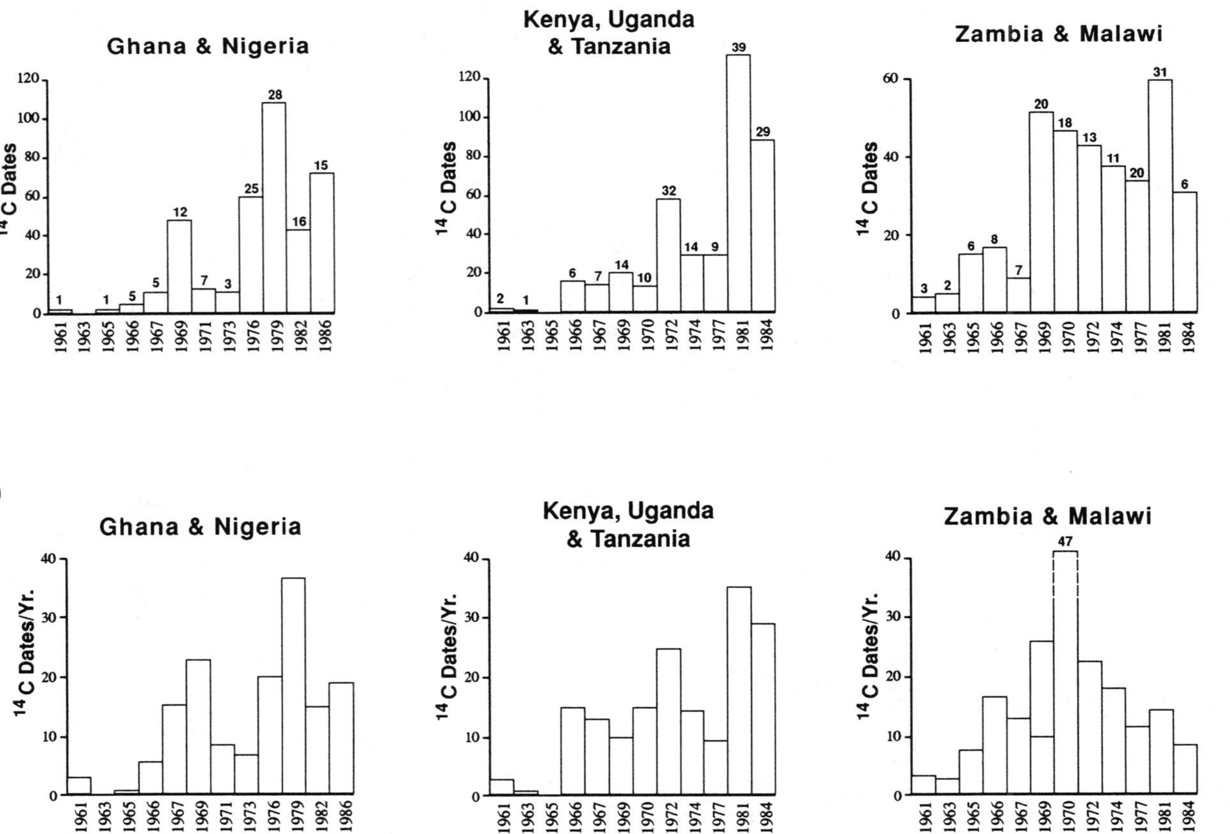

Fig 23.1. Frequencies of ^{14}C dates for various regions of Africa published in issues of the *Journal of African History*. A) Number of dates reported in each year that a date list was published; the numbers on the bars refer to the number of sites for which dates were reported. B) Average of the number of "new" dates per year (obtained by dividing the total number of dates in a list by the number of years that elapsed since the previous list was published).

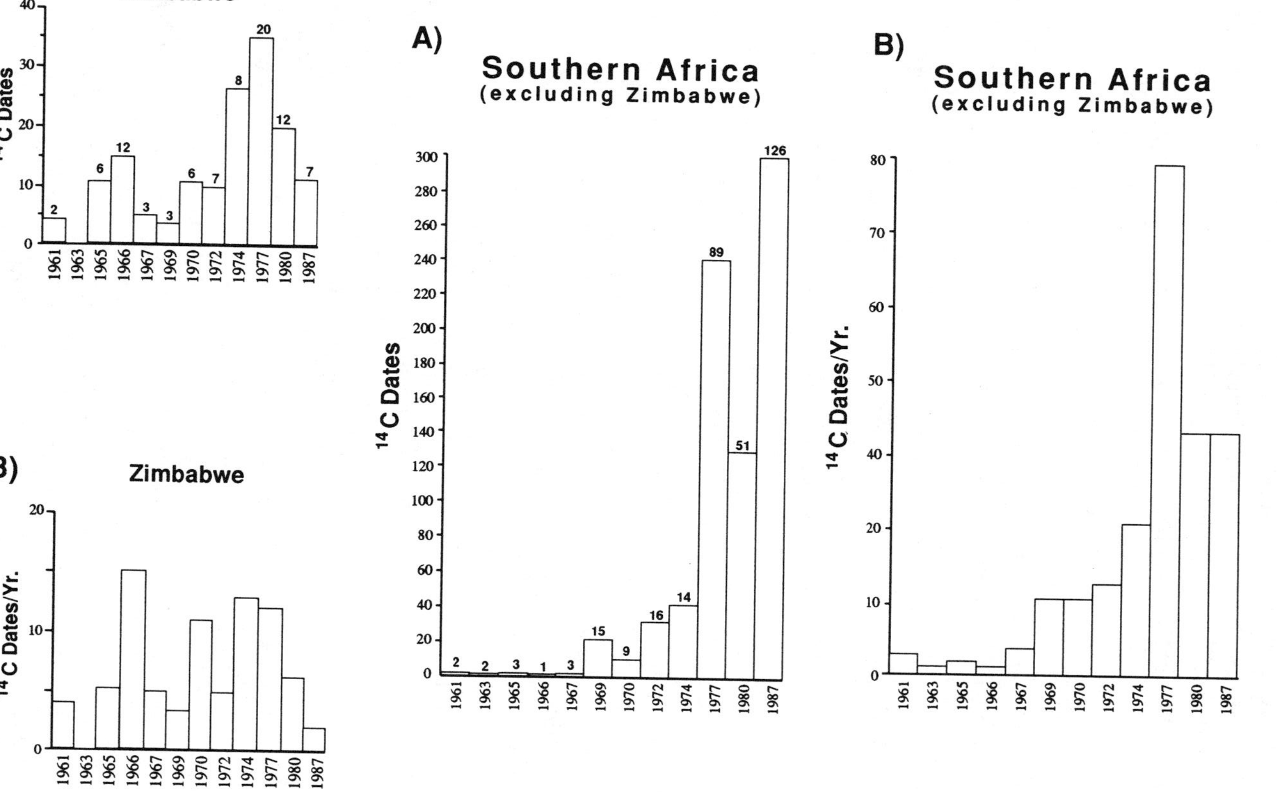

Fig 23.2. Frequencies of ^{14}C dates for Zimbabwe and the remainder of southern Africa published in issues of the *Journal of African History*. A) Number of dates reported in each year that a date list was published; the numbers on the bars refer to the number of sites for which dates were reported. B) Average of the number of "new" dates per year (obtained by dividing the total number of dates in a list by the number of years that elapsed since the previous list was published).

A significant contribution to the acceleration of ^{14}C dating in sub-Saharan Africa from the mid-1960s was made by dating laboratories within the region. The Gulbenkian Radiocarbon Dating Laboratory in Salisbury (now Harare, Zimbabwe) came into operation in October 1962. A laboratory was established in Pretoria in 1969, the same year that ^{14}C dates from the Dakar (Senegal) laboratory were first published in the *Journal of African History*. The University of the Witwatersrand (Johannesburg) began to report dates on archaeological samples in 1987. However, the laboratories in South Africa are the only ones still producing dates on a regular basis; Pretoria has processed more than 4000 dates. The Salisbury laboratory suffered numerous setbacks in its brief history, its counter proving particularly irksome. The laboratory produced a whole suite of dates in the early 1970s that were either directly refuted by other laboratories or considered too young by the archaeologists who submitted the samples (Soper 1974: 176). By 1978, the laboratory was closed. The absence of active dating laboratories in black Africa and the difficulty of obtaining hard currency to pay for dating are among the reasons for the decline in the use of radiocarbon dating by African archaeologists in the 1980s.

RADIOCARBON DATING OF THE STONE AGE

Although much of the African Stone Age is beyond the range of even the most optimistic projections of the capabilities of AMS dating, the application of ^{14}C to problems of Stone Age chronology has been especially rewarding, despite some early hiccoughs. Initial interest centered on establishing a firm chronology that would permit correlations to be made between African and European industrial traditions, as well as climatic events, and hence, to discover whether Africa was "peripheral or paramount" (*cf*, Clark 1975). African archaeologists became painfully aware of the need for an absolute chronology when the Pluvial Theory (correlating glacial periods in higher latitudes with wetter periods in tropical Africa discerned through geomorphological studies) was discredited (Flint 1959). It seemed that ^{14}C dating would fill the vacuum. The first results appeared to indicate that Africa was peripheral for later human biological and cultural evolution. Dates from Kalambo Falls in Zambia suggested that the Acheulean tool kit was replaced by the Sangoan only about 60,000 years ago, whereas the dating of South African sites appeared to confirm that the African Middle Stone Age (MSA) was roughly contemporary with the Upper Paleolithic of Eurasia. Since it was generally assumed, during the 1960s and before, that there was a correlation between Upper Paleolithic technology and anatomically modern humans, it was implied that the late survival of MSA flake technology in sub-Saharan Africa was to be linked to the late penetration of modern humans into "remote" regions occupied by archaic *Homo sapiens* (Clark 1990: 201). In a similar vein, the microlithic Later Stone Age (LSA) was reckoned to be fully developed only in the second millennium BC (Clark 1959: 188).

The LSA chronology collapsed as many more dates were processed in the late 1960s, and a radically revised chronology for the Stone Age was published in 1972 (Beaumont & Vogel 1972; Vogel & Beaumont 1972; see also Carter & Vogel 1974). The new dates, which were concentrated in South Africa, revealed that the LSA could be divided into "Early" and "Later" components; the former was considered to be roughly equivalent, on the basis of typology, to the Western European Upper Palaeolithic, and the latter to the Mesolithic. Dates from several rockshelters in southern Africa placed the MSA/LSA interface at greater than 37,000 years ago, consistent with the idea that the "Early LSA" was chronologically, as well as typologically, equivalent to the Upper Palaeolithic. More recent work (summarized by Deacon 1984: 238; 1990: 175–180) indicates that the MSA/LSA transition falls some time after 40,000 and before 25,000 BP. Beaumont and Vogel (1972) also divided the MSA into "Final MSA" and "Full MSA," with the latter falling beyond the range of ^{14}C dating, at that time. "Full" MSA stone artifact assemblages were shown to be associated with remains of anatomically modern humans. Thus, Vogel and Beaumont (1972: 51) argued that "it seems likely that *Homo sapiens sapiens* already occupied much of Africa by the advent of the 'Final MSA', that is before 50,000 BC at the latest, and there are indications that his origins may reach back to the beginnings of the 'Full MSA' at probably greater than 100,000 BC." These suggestions have received support both from more recent archaeological and geomorphological research in South Africa (Beaumont, De Villiers & Vogel 1978; Butzer, Beaumont & Vogel 1978; Singer & Wymer 1982; Deacon & Geleijnse 1988) and from studies of mitochondrial DNA (Cann, Stoneking & Wilson 1987). Considerable attention has also focused on the Howieson's Poort Industry, once thought to represent an intermediate stage between the MSA and the LSA (eg, Beaumont & Vogel 1972). It is now recognized as an integral part of the MSA (Volman 1984). The backed tools which characterize this industry may represent some of the earliest evidence anywhere in the world for composite tools. In addition to composite tools, southern Africa has also laid claim to early rock art; painted slabs from the site of Apollo 11 in Namibia are associated with charcoal dated to about 26,000 BP and overlain by deposits with dates of approximately 19,000 BP (Wendt 1976). ^{14}C dates are also associated with both paintings and engravings from various South African sites that span the last 10,000 years (Thackeray 1983).

The accumulation of dates for the Stone Age not only placed the origins of blade and microlithic technologies in Africa at least as early as in Europe, but also permitted an examination of changing regional settlement patterns in southern Africa through the last 10,000 years. Plots of the available ^{14}C dates revealed no sites in the drier areas of the inland plateau with occupation falling between 9500 and 4600 BP, nor were any sites assigned to the Smithfield industrial complex dated to this interval (Deacon 1974). Deacon's paper is a landmark study in the application of ^{14}C dating to African archaeology, for it was one of the first attempts to assess the patterning within a large number of ^{14}C dates.[2] Although admitting the possibility of sampling errors, Deacon argued cogently that the absence of sites in the interior could be correlated with paleoenvironmental evidence for a hyperarid episode in this already dry region. Her analysis demonstrated that the model of two parallel and separately evolving traditions, the Wilton and the Smithfield, could not be sustained. This model had been the foundation of Later Stone Age archaeology in southern Africa since its formulation by Goodwin and Van Riet Lowe (1929). Deacon's research also showed that the so-called Wilton assemblages had been identified on the basis of a temporal type – the segment or crescent – dating to between about 7000 and 3000 BP and occurring in the context of a single evolving toolmaking tradition (see also Parkington 1984). Although recent research has begun to fill the "gap" in the radiocarbon chronology of the interior (Wadley & Verhagen 1986), Deacon's revision of the classificatory framework has not been seriously challenged. Indeed, it is the explanation rather than the existence of chronological and spatial variation in stone tool assemblages in southern Africa that has become a major focus of research (see, eg, Parkington 1980).

Our knowledge of stone tool traditions outside southern Africa is not as good. ^{14}C dates have helped in the construction of regional sequences, but better documented artifact assemblages, as well as many more dates, are required before archaeologists will be able to aspire to the levels of complexity that characterize discussions of chronological patterning in the South African Stone Age.

RADIOCARBON DATING OF THE IRON AGE

As with dating the Stone Age, some of the first ^{14}C dates for the Iron Age in sub-Saharan Africa were "red herrings." In 1955, Roger Summers reported the first ^{14}C dates for Great Zimbabwe – a site whose interpretation has generated more controversy than any other in Africa because of competing political and ideological viewpoints (see Hall 1990). These dates were obtained on two wooden lintels carrying a wall above a drain in the Great Enclosure; samples from these lintels processed at Chicago and the short-lived London laboratory gave results around the mid-first millennium AD. As these dates were not in

[2]A comparable analysis of ^{14}C dates from the Sahara was published the same year (Geyh & Jäkel 1974).

agreement with conventional archaeological wisdom, Summers (1955) suggested that the wood may have been taken from an earlier feature and re-used, whereas those with a political axe to grind cited the dates as evidence for pre-Bantu or exotic origins for the site. The controversy was eventually resolved by redating the lintels; three dates processed at Pretoria all fell around the 14th century AD. It seems that the early dates on the lintels were simply wrong, perhaps in part as a result of the use of the solid carbon method (Huffman & Vogel 1979).

In 1960, 11 richly adorned human skeletons were recovered from Ingombe Ilede on the north bank of the Zambezi River. As with Great Zimbabwe, the ^{14}C dates turned out to be much earlier than expected; four dates fell in the 7th to 11th centuries AD (Fagan 1965). However, in this case, the error lay in the collection of the charcoal samples rather than in the laboratory analysis, for none of the samples was in indisputable association with any of the burials. Further excavations at the site in 1968 revealed that the dates must relate to an earlier occupation, since charcoal from a midden with pottery like that found in the burials was dated to about AD 1400 (Phillipson & Fagan 1969).

Some of the surprises provided by ^{14}C dates for Iron Age sites have been validated, in contrast to the dating of Great Zimbabwe and Ingombe Ilede. For instance, the ceramic sequence for southern Africa established by Schofield (1948) was overturned and Mapungubwe, previously thought to have been an offshoot of the Zimbabwe state, was shown to antedate it, resulting in a new understanding of the origins of the state in southern Africa (Hall & Vogel 1980; Huffman 1986). In West Africa, considerable excitement and discussion were generated by the radiocarbon dating of charcoal associated with the superb bronzes excavated at Igbo-Ukwu in Nigeria (Shaw 1975). The ninth-century dates indicated that these items were made before the rise of the trans-Saharan trade in which Islamic merchants played a pivotal role. The excavator, well aware of Ingombe Ilede, argued for the unlikelihood of incorrect association between the charcoal samples and the artifacts at Igbo-Ukwu. However, these dates "no longer seem anomalous in light of growing evidence for extensive pre-Arab trade in West Africa" (McIntosh & McIntosh 1986: 433–434).

Apart from the interest generated by the ^{14}C dating of spectacular sites, the major contribution of ^{14}C dating to Iron Age archaeology in sub-Saharan Africa has been in attempting to unravel the complexities of the expansion of Bantu-speaking peoples across the subcontinent. Prior to the 1960s, the primary sources of evidence about the origins and spread of the Bantu were linguistic. In 1965, the British Institute in Eastern Africa initiated the Bantu Studies Research Project (BSRP) in order to add an archaeological dimension to the linguistic reconstruction. In a seminal paper, Roland Oliver (1966: 371) argued that the crucial contribution of archaeology would lie in the provision of ^{14}C dates so that the chronology and directions of Bantu expansion could be firmly established. As Oliver hoped, the number of ^{14}C dates proliferated, and their interpretation has vexed archaeologists ever since. By 1975, about 400 published ^{14}C dates for the Iron Age of Bantu Africa provoked Phillipson (1975) to attempt an analysis of these dates based on Ottaway's (1973) use of dispersion diagrams. Grouping the dates into ceramic traditions, Phillipson argued that the interquartile range of each group of dates signified the floruit of each tradition. He interpreted the resulting geographical-temporal zoning of the traditions as confirmation of his hypothesis of two streams of Early Iron Age penetration of Bantu speakers into southern Africa. Interestingly, Phillipson (1975: 341) dismissed calibration as "at present of little practical relevance to the historian of the Iron Age in sub-Saharan Africa." Phillipson's analysis of the ^{14}C dates and the hypothesis to which it related generated considerable interest; for example, his views were published in *Scientific American* (Phillipson 1977). However, his work also came under attack. Huffman (1979) raised a number of objections: 1) floruits may reflect archaeological excavation priorities, not the actual duration of the culture; 2) the complete context of a date is not considered by Phillipson, who treats the dates as simple mathematical entities disassociated from the human events that they are supposed to date; 3) linked dates (those dating the same event) are not treated differently from separate dates; 4) aberrant dates – those that differ from other ^{14}C dates of the same human event by more than their standard deviations – are not distinguished from wrong dates – those that are not associated with the human event they purport to date.

While Phillipson's analysis is a reflection of the tremendous importance that ^{14}C dating has had for archaeological understanding of Bantu expansion, Huffman's critique marked the beginning of a new awareness among archaeologists in sub-Saharan Africa of the problems and potential pitfalls in the interpretation of ^{14}C dates for reconstructing prehistory. This awareness is vital for archaeologists whose knowledge of absolute chronology rests almost exclusively on ^{14}C dating. Only in West Africa has another dating technique – thermoluminescence (TL) – been used consistently to date Iron Age sites.

The importance of critical evaluation of ^{14}C dates by archaeologists is illustrated by the debate concerning the antiquity of pastoralism in East Africa. Claims have been made for the presence of domestic animals at various sites perhaps as early as 7000 BP (Bower & Nelson 1978; Nelson & Kimengich 1984; Bower & Chadderdon 1986). Such claims are based on three assumptions: 1) the animal bones have been identified correctly; 2) the bones are contextually associated with the samples used for ^{14}C dating; 3) the ^{14}C dates are accurate.

Discussion has centered mostly upon the third of these assumptions because the dates have been obtained from samples of bone apatite. An analysis of the congruence between paired apatite and collagen dates from East Africa raised serious doubts about the reliability of apatite dates (Collett & Robertshaw 1983). The problem of the association between the sample of charcoal or other material dated and the event or object to which the date is supposed to refer has been important in discussions of the chronology of the Iron Age in recent years, and, in particular, of the validity of the first millennium BC dates for iron technology in the interlacustrine region of East Africa.

CURRENT CONCERNS

^{14}C dating has made an enormous contribution to the later prehistory of sub-Saharan Africa, not only through the dating of archaeological sites, but also through the dating of paleoenvironmental evidence. However, the role of ^{14}C dating in African archaeology is changing, and the future value of the technique is uncertain. The 1980s saw a reduction in the number of ^{14}C dates processed each year (Figs 23.1, 23.2). This factor may, in part, correlate with a shift in research emphasis among archaeologists primarily interested in the Stone Age to the earlier periods, such as the MSA, where radiocarbon dating is not usually applicable but where other dating methods (electron-spin resonance (ESR), TL, etc) are beginning to provide good chronological control. It is also quite probable that there has been a reduction in the amount of archaeological research undertaken (no statistics are available), which, in turn, might be linked to political and economic instability in many African countries, as well as to economic constraints outside Africa. Moreover, as ^{14}C dating of charcoal samples from archaeological sites has become a routine procedure, few research laboratories will commit their resources to such samples. Thus, archaeologists are compelled to send their samples to commercial laboratories. Indigenous African archaeologists, now far more numerous than in earlier decades, are often unable to pay for ^{14}C dates, since they have little or no access to hard currency. Also, there are no active dating laboratories in Africa, except for those in South Africa, which most Africans could not or would not choose to patronize.

Current perceptions of ^{14}C dating among Africanist archaeologists are varied. On the one hand, some argue that ^{14}C dating of archaeological sites has established the chronological framework for the later prehistory of sub-Saharan Africa and that future dating programs will have an important role to play in testing patterns of settlement over time and space (Parkington & Hall 1987). Others, however, take a less optimistic view:

> . . . it is ironic that the advent of high-precision dating has virtually removed all hope that conventional ^{14}C dates can be anything but a relatively low precision tool for interpreting chronology. . . . It is now clear that for most conventional ^{14}C dates, we must settle for a real calendar date range of no better than 300–400 years, . . .

using the recommended two-sigma limits (McIntosh & McIntosh 1986: 416). Even the optimists point out drawbacks to the reliance on ^{14}C dates, noting that the concern among archaeologists to use ^{14}C to demonstrate the antiquity of various developments, such as iron-working, has diverted attention away from study of the social interactions that followed such developments (Parkington & Hall 1987: 15).

Some archaeologists have detected an over-reliance on ^{14}C dating with damaging consequences. McIntosh and McIntosh (1988) argue that archaeology in West Africa was "a child of the radiocarbon age," as a result of which chronology has become the major goal of much archaeology in the region. Partly because of this, West Africanists have developed a "fear of or inability to do pottery analysis and description . . . that threatens our ability to say anything meaningful about change through time or interaction across space" (McIntosh & McIntosh 1988: 90). This problem is not confined to West Africa; many archaeologists are reluctant to invest their energies in constructing material culture sequences through the time-consuming business of analyzing the pottery and stone artifacts recovered in excavations, especially when it seems simpler, assuming funds are available, to send a couple of charcoal samples to a ^{14}C laboratory. Yet, such sequences combined with stratigraphic evidence may be "the only way to ascertain what the temporal relationship of events dated by two or more C-14 dates might actually be" (McIntosh & McIntosh 1988: 91; see also Posnansky & McIntosh 1976: 162). Also prevalent among archaeologists is a tendency both to accept dates acritically and to consider them much more precise than they actually are. Too few archaeologists heed Sutton's (1982: 293) comment that "laboratory results should be seen not as dates in themselves but as evidence to be assessed." A naive attitude to dates is exemplified by the readiness of archaeologists to dismiss ^{14}C dates that do not fit their preconceived ideas about prehistoric chronology as the result of laboratory errors. In fact, most discrepancies are due to incorrect associations made by the archaeologist; wrong, rather than aberrant, dates are common in African archaeology (Huffman 1979: 235).

High-precision calibration of the ^{14}C time scale is arguably one of the two major achievements in the field of ^{14}C dating in recent years – the other being the development of accelerator mass spectrometric (AMS) dating. African prehistorians can no longer ignore calibration; yet the results of calibration have been disappointing, in a sense. While calibrated dates are more accurate, they are usually far less precise than the uncalibrated results, especially when the physicists' injunction to calibrate at two sigma is obeyed. The kinks in the calibration curve for the last 500 years are particularly irksome for African archaeologists, for this is the period when archaeology meets history in much of sub-Saharan Africa. Radiocarbon dates of events remembered in oral tradition are virtually worthless for linking archaeological and historic chronologies. Also distressing is the dramatic flattening of the calibration curve between 800 and 400 BC, when metallurgy makes its appearance in sub-Saharan Africa. McIntosh and McIntosh (1988: 107) bemoan "the stunning imprecision of calibrated dates for early metallurgy in West Africa." More generally, Africanists are concerned about the validity of applying a 30-year correction, based on the dating of tree rings from the Cape (Vogel *et al* 1986), to charcoal samples from the Southern Hemisphere prior to calibration. How applicable is this correction to charcoal samples from other regions of sub-Saharan Africa, particularly equatorial Africa?

THE FUTURE OF RADIOCARBON DATING FOR THE ARCHAEOLOGY OF SUB-SAHARAN AFRICA

The role of ^{14}C dating in African archaeology is changing. Its major contribution has been to establish regional chronologies for later prehistoric cultures, technologies and economies. Much of this task has been completed, particularly in southern Africa where the bulk of archaeological research has been conducted. Further refinements in chronology will come not so much from more ^{14}C dates, but from the development of other complementary dating methods. A return to archaeological basics – the analysis of lithic and ceramic assemblages – will be crucial for the construction of more detailed local sequences than is possible from ^{14}C dating alone, and for the interpretation of the spatial and temporal variation in prehistoric cultures.

This does not mean that ^{14}C dating is becoming redundant for African archaeology, but that its role needs to be redefined. Many more dates will be needed, primarily to test observed patterns in regional archaeological data. As Parkington and Hall (1987: 17) recently stressed, "the intensity of ^{14}C dates from specific regions, that have been linked to specifically designed research problems, is still too low to allow firm conclusions about more subtle patterns in prehistoric behavior." African archaeologists are on the verge of transcending culture-stratigraphic sequences to discern, as Flannery (1967) expressed it, "the system behind the Indian and the artifact." We will need tight chronological control to understand the workings of this "system." Given the imprecision of a single calibrated ^{14}C date, suites of dates that can be statistically combined will be required even from single-component sites.

The scientists in the laboratories should welcome the challenge of high-precision dating of sites (± 20 or even 10 years for sites younger than 2000 BP, and ± 40 or 50 years for sites back to 10,000 BP). If such precision is to be achieved, a calibration curve that is demonstrably accurate for sub-Saharan Africa will be required. Moreover, those who process the ^{14}C dates must be more closely involved in archaeological research design to ensure good sampling strategies, good communication between laboratories and archaeologists, and reasonable turnaround times for the dating of samples. Far too often, dating results, which might serve to re-orient fieldwork, become available only after opportunities and funding for fieldwork are exhausted. Active collaboration between laboratory scientists and archaeologists will be difficult, especially since there are no active dating laboratories in black Africa, but such collaboration will also be the key to successful grantsmanship.

The contribution of AMS dating to the archaeology of sub-Saharan Africa has been rather limited up till now. The technique has been used mostly to evaluate evidence for early agriculture and pastoralism, for example, dating of the finger millet seeds found at Gobedra in Ethiopia and the sheep hair from Wonderwerk Cave in South Africa (Gowlett *et al* 1987). The potential of AMS dating is, however, much greater. The transition from the Middle to the Later Stone Age may be dated more accurately and more precisely than it is at present, resulting in, among other things, a better understanding of the origins of microlithic technology. While it looks as if the origins of anatomically modern humans lie beyond the range of AMS dating, application of the method to later human skeletal material will be crucial to the elucidation of later Pleistocene human evolution. AMS dating will also assist in the resolution of the problems surrounding the dating of bone from tropical regions (Stafford *et al* 1987). The ability to date very small samples may eliminate some of the archaeological problems of finding samples for dating in close association with the objects to be dated, particularly in the case of sites where features such as hearths have not been discovered.

A range of rather basic issues concerning the prehistory of sub-Saharan Africa remains to be resolved by ^{14}C dating, using both conventional and AMS techniques. These include the distribution of settlement at the Last Glacial Maximum, the antiquity of foraging in the tropical forests and the origins and spread of both agriculture and metallurgy. Many more dates of much higher precision will be needed if archaeologists are to go beyond these basic issues to explore the details of prehistory and to confront aspects of the archaeological record relevant to the theoretical concerns of the discipline. In this regard, it is clear that the future of ^{14}C dating in sub-Saharan Africa lies in close collaboration between archaeologists and laboratory scientists.

ACKNOWLEDGMENTS

I thank various colleagues who attended the Society of Africanist Archaeologists conference in Florida for their remarks on ^{14}C dating issues. Russell Barber kindly commented upon a preliminary version of this chapter.

REFERENCES

Beaumont, PB, De Villiers, H and Vogel, JC 1978 Modern man in sub-Saharan Africa prior to 49,000 years BP: A review and evaluation with particular reference to Border Cave. *South African Journal of Science* 74: 409–419.

Beaumont, PB and Vogel, JC 1972 On a new radiocarbon chronology for Africa south of the Equator. *African Studies* 31: 65–89 and 155–182.

Bower, J and Chadderdon, TJ 1986 Further excavations of Pastoral Neolithic sites in Serengeti. *Azania* 21: 129–133.

Bower, JRF and Nelson, CM 1978 Early pottery and pastoral cultures of the Central Rift Valley, Kenya. *Man* 13: 554–566.

Butzer, KW, Beaumont, PB and Vogel, JC 1978 Lithostratigraphy of Border Cave, Kwazulu, South Africa: A Middle Stone Age sequence beginning c 195,000 BP. *Journal of Archaeological Science* 5: 317–341.

Cann, RL, Stoneking, M and Wilson, AC 1987 Mitochondrial DNA and human evolution. *Nature* 325: 31–36.

Carter, PL and Vogel, JC 1974 The dating of industrial assemblages from stratified sites in eastern Lesotho. *Man* 9: 557–570.

Clark, JD 1959 *The Prehistory of Southern Africa*. Harmondsworth, England, Penguin Books: 341 p.

_____1975 Africa in prehistory: peripheral or paramount? *Man* 10: 175–198.

_____1990 A personal memoir. *In* Robertshaw, P, ed, *A History of African Archaeology*. Portsmouth, New Hampshire, Heinemann: 189–204.

Cole, S 1954 *The Prehistory of East Africa*. Harmondsworth, England, Penguin Books: 301 p.

Cole, S 1963 *The Prehistory of East Africa* (revised edition). New York, Macmillan: 382 p.

Collett, D and Robertshaw, P 1983 Problems in the interpretation of radiocarbon dates: the Pastoral Neolithic of East Africa. *African Archaeological Review* 1: 57–74.

Deacon, HJ and Geleijnse, VB 1988 The stratigraphy and sedimentology of the main site sequence, Klasies River, South Africa. *South African Archaeological Bulletin* 43: 5–14.

Deacon, J 1974 Patterning in the radiocarbon dates for the Wilton/Smithfield complex in southern Africa. *South African Archaeological Bulletin* 29: 3–18.

_____1984 Later Stone Age people and their descendants in southern Africa. *In* Klein, RG, ed, *Southern African Prehistory and Paleoenvironments*. Rotterdam, AA Balkema: 221–328.

_____1990 Changes in the archaeological record in South Africa at 18,000 BP. *In* Gamble, C and Soffer, O, eds, *The World at 18,000 BP: Volume 2: Low Latitudes*. London, Unwin Hyman: 170–188.

Fagan, BM 1961 Radiocarbon dates for sub-Saharan Africa (from *c* 1000 BC) – I. *Journal of African History* 2(1): 137–139.

_____1965 Radiocarbon dates for sub-Saharan Africa (from *c* 1000 BC) – III. *Journal of African History* 6(1): 107–116.

Flannery, KV 1967 Culture history v culture process: a debate in American archaeology. *Scientific American* 217: 119–122.

Flint, RF 1959 On the basis of Pleistocene correlation in East Africa. *Geological Magazine* 96: 265–284.

Geyh, MA and Jäkel, D 1974 Late Glacial and Holocene climatic history of the

Sahara Desert derived from a statistical assay of ^{14}C dates. *Palaeogeography, Palaeoclimatology, Palaeoecology* 15: 205–208.

Goodwin, AJH and Van Riet Lowe, C 1929 The Stone Age cultures of South Africa. *Annals of the South African Museum* 27: 1–289.

Gowlett, JAJ, Hedges, REM, Law, IA and Perry, C 1987 Radiocarbon dates from the Oxford AMS system: Archaeometry datelist 5. *Archaeometry* 29(1): 125–155.

Hall, M 1990 'Hidden history': Iron Age archaeology in southern Africa. *In* Robertshaw, P, ed, *A History of African Archaeology*. Portsmouth, New Hampshire, Heinemann: 59–77.

Hall, M and Vogel, JC 1980 Some recent radiocarbon dates from southern Africa. *Journal of African History* 21: 431–455.

Huffman, TN 1979 African origins. *South African Journal of Science* 75(5): 233–237.

_____1986 Iron Age settlement patterns and the origins of class distinction in Southern Africa. *In* Wendorf, F and Close, A, eds, *Advances in World Archaeology* 5. New York, Academic Press: 291–338.

Huffman, TN and Vogel, JC 1979 The controversial lintels from Great Zimbabwe. *Antiquity* 53: 55–57.

Libby, WF 1951 Radiocarbon dates, II. *Science* 114: 291–296.

_____1952 Chicago radiocarbon dates, III. *Science* 116: 673–681.

McIntosh, SK and McIntosh, RJ 1986 Recent archaeological research and dates from West Africa. *Journal of African History* 27: 413–442.

_____1988 From stone to metal: New perspectives on the later prehistory of West Africa. *Journal of World Prehistory* 2(1): 89–133.

Nelson, CM and Kimengich, J 1984 Early phases of pastoral adaptation in the Central Highlands of Kenya. *In* Krzyzaniak, L and Kobusiewicz, M, eds, *Origin and Early Development of Food-Producing Cultures in North-Eastern Africa*. Poznan, Polish Academy of Sciences: 481–487.

Oliver, R 1966 The problem of the Bantu expansion. *Journal of African History* 7: 361–376.

Ottaway, B 1973 Dispersion diagrams: a new approach to the display of carbon-14 dates. *Archaeometry* 15(1): 5–12.

Parkington, J 1980 Time and place: some observations on spatial and temporal patterning in the Later Stone Age sequence in Southern Africa. *South African Archaeological Bulletin* 35: 73–83.

_____1984 Changing views of the Later Stone Age of South Africa. *In* Wendorf, F and Close, A, eds, *Advances in World Archaeology* 3. New York, Academic Press: 89–142.

Parkington, J and Hall, M 1987 Patterning in recent radiocarbon dates from Southern Africa as a reflection of prehistoric settlement and interaction. *Journal of African History* 28: 1–25.

Phillipson, DW 1975 Chronology of the Iron Age in Bantu Africa. *Journal of African History* 16(3): 321–342.

_____1977 The spread of the Bantu language. *Scientific American* 236: 106–114.

Phillipson, DW and Fagan, BM 1969 The date of the Ingombe Ilede burials. *Journal of African History* 10(2): 199–204.

Posnansky, M and McIntosh, RJ 1976 New radiocarbon dates for northern and western Africa. *Journal of African History* 17(2): 161–195.

Schofield, JF 1948 *Primitive Pottery*. Cape Town, South African Archaeological Society.

Shaw, T 1975 Those Igbo-Ukwu radiocarbon dates: facts, fictions and probabilities. *Journal of African History* 16(4): 503–517.

Shinnie, PL 1990 A personal memoir. *In* Robertshaw, P, ed, *A History of African Archaeology*. Portsmouth, New Hampshire, Heinemann: 221–235.

Singer, R and Wymer, J 1982 *The Middle Stone Age at Klasies River Mouth in South Africa*. Chicago, University of Chicago Press: 234 p.

Soper, R 1974 New radiocarbon dates for

eastern and southern Africa. *Journal of African History* 15(2): 175–192.

Stafford, TW, Jull, AJT, Brendel, K, Duhamel, RC and Donahue, D 1987 Study of bone radiocarbon dating accuracy at the University of Arizona NSF accelerator facility for radio-isotope analysis. *Radiocarbon* 29(1): 24–44.

Summers, R 1955 The dating of the Zimbabwe Ruins. *Antiquity* 29: 107–111.

Sutton, JEG 1982 Archaeology in West Africa: a review of recent work and a further list of radiocarbon dates. *Journal of African History* 23: 291–313.

Thackeray, AI 1983 Dating the rock art of southern Africa. *South African Archaeological Society Goodwin Series* 4: 21–26.

Vogel, JC and Beaumont, PB 1972 Revised radiocarbon chronology for the Stone Age in South Africa. *Nature* 237: 50–51.

Vogel, JC, Fuls, A, Visser, E and Becker, B 1986 Radiocarbon fluctuations during the third millennium BC. *In* Stuiver, M and Kra, RS, eds, Proceedings of the 12th International ^{14}C Conference. Radiocarbon 28(2B): 935–938.

Volman, TP 1984 Early prehistory of southern Africa. *In* Klein, RG, ed, *Southern African Prehistory and Paleoenvironments.* Rotterdam, AA Balkema: 169–220.

Wadley, L and Verhagen, B Th 1986 Filling the gap: a new Holocene chronology for the Transvaal Stone Age. *South African Journal of Science* 82(5): 271.

Wendt, WE 1976 'Art mobilier' from the Apollo 11 Cave, South West Africa: Africa's oldest dated works of art. *South African Archaeological Bulletin* 31: 5–11.

NEW WORLD ARCHAEOLOGY

PREFACE

R E TAYLOR

The impact of radiocarbon dating on the conduct of archaeological research has been of particular significance in the development of New World prehistoric studies over the last four decades. Several observers (Willey & Phillips 1958: 44; Dean 1978: 226) noted that ^{14}C-based age estimates provided a means of deriving chronological relationships independent of assumptions about cultural processes and unrelated to the manipulation of archaeologically derived data. In the United States, the rise of the "New Archaeology" in the 1970s took place within this context. As Lewis Binford has reflected, ^{14}C chronology "changed the activities of archaeologists . . .they direct(ed) their methodological investments toward theory building rather than towards chronology building" (quoted in Gittins 1984: 238). It has also been suggested that the advent of ^{14}C dating led to a noticeable improvement in archaeological field methods (Johnson 1965: 764) and was, at least in part, responsible for the increasing attention given to statistical approaches in the evaluation of archaeological data (Thomas 1978: 323).

The four chapters in this section reflect the wide-ranging influence of ^{14}C data on the conduct of New World archaeological studies over the last forty years. Each topic is one in which Libby expressed a considerable interest and, at times, a specific point of view. Vance Haynes deals with the contributions of ^{14}C to the geochronology of the peopling of the New World. In his view, the most important lesson of the last forty years is that the accuracy of ^{14}C age estimates can only be verified by a combination of proper stratigraphic controls, in conjunction with concordance with data from other lines of inquiry. R E Taylor reviews the torturous history of the ^{14}C dating of bone, noting the early difficulties and continuing debates, particularly in the changing context of concerns about the validity of ^{14}C values on North American Paleoindian skeletal materials. He also discusses the role of AMS technology in developing a preparative methodology which would accurately identify organics indigenous to a bone irrespective of its diagenetic history. S L Fedick and K A Taube emphasize the role that ^{14}C has played in gradually resolving several competing chronological constructs for the Classic Maya region of Mesoamerica based on *calendric and ceramic correlations. Radiocarbon data have been the key elements in defining the temporal range of the pre-Classic, and thus, current understanding of the development of Maya civilization in the context of changing environments, subsistence, settlements patterns and demography. The final chapter by Rainer Berger focuses specific attention on the diverse contributions to archaeological research carried out by Libby's laboratory at the University of California, Los Angeles (UCLA). About two-thirds of the ^{14}C dates measured at his laboratory at the University of Chicago were obtained on archaeological samples, and this pattern continued at UCLA. Berger notes Libby's interest in the question of when humans entered the New World, which provided the stimulus for the Tule Springs project, one of the largest "big science" interdisciplinary projects in New World archaeology.*

References

Dean, JS 1978 Independent dating in archaeological analysis. In *Schiffer, MB, ed,* Advances in Archaeological Method and Theory *1. New York, Academic Press: 223–265.*

Gittins, GO 1984 Radiocarbon chronometry and archaeological thought. Unpublished doctoral dissertation, University of California, Los Angeles.

Johnson, F 1965 The impact of radiocarbon dating upon archaeology. In *Chatters, RM and Olson, EA, eds,* Proceedings of the Sixth International Conference on Radiocarbon and Tritium Dating. *Springfield, Virginia, Clearinghouse for Federal Scientific and Technical Information: 762–780.*

Thomas, DH 1978 The awful truth about statistics in archaeology. American Antiquity *43: 231–244.*

Willey, GR and Phillips, P 1958 Method and Theory in American Archaeology. *Chicago, University of Chicago Press.*

24

CONTRIBUTIONS OF RADIOCARBON DATING TO THE GEOCHRONOLOGY OF THE PEOPLING OF THE NEW WORLD

C VANCE HAYNES, JR

INTRODUCTION

Before radiocarbon dating, early man sites were dated geologically, either by association of artifacts with remains of extinct animals or with geologic deposits that could be related to glacial events (Oakley 1964: 51–57). This primitive geochronology began in the first half of the 19th century with the recognition by Boucher de Perthes of handaxes in terrace deposits of the Somme River near Abbeville in northern France. This was during the formative years of geology and Darwinian evolution, so he can be excused for attributing the terrace deposits to the biblical flood, thus dating the handaxes as antediluvial. It was English geologist, Hugh Falconer, who in 1858 correctly read the terraces as derived from glacial-age rivers, thus assigning the Abbevillean artifacts to the Pleistocene Epoch (Macgowan & Hester 1962: 63–64).

Evidence for early man in the New World began as early as in the Old World with the discovery of artifacts with mastodon bones in Missouri in 1838, and a few years later, of a human pelvis with fossilization equivalent to that of extinct animal bones in Mississippi River terrace deposits near Natchez (Wormington 1957: 226). Stone artifacts were also found in terrace gravels thought to be Pleistocene in New Jersey and Ohio. These controversial finds were eventually disproved, but they adequately convey the problems with dating early man sites before radiocarbon dating (Meltzer 1983).

Geological dating depended upon estimates for the times of Pleistocene glaciations. These were crude at best and resulted from the recognition of four main glacial periods within the Pleistocene Epoch. River terraces in glacial areas were then associated with glaciation, so that the highest terrace (eg, T-4) represented the first glaciation, the next lowest (eg, T-3) the next glacial, and so on down to the Holocene floodplain (T-0). Further refinement was the recognition of cold and warm climate faunas whereby associated cultural remains could be assigned to either glacials or interglacials.

Quantitative estimates for the lengths of glaciations in years were little more than intelligent guesses based upon sedimentation rates or astronomical theory (Zeuner 1958: 134–142) until the establishment of the glacial varve chronology in Sweden by DeGeer in the early part of the 20th century (Nilsson 1983: 43–45, 52–53). Recognizing the annual nature of alternating laminae of fine and coarse sediments derived from glacial outwash and laid down in quiet waters, DeGeer dated the deglaciation of Scandinavia by counting varved clays deposited in the Baltic during final deglaciation. His protégé, Ernst Antevs, then attempted to establish a varve chronology for the last or Wisconsin glaciation of North America by counting varves in glacial lake deposits in New England and eastern Canada. His results were hampered by several gaps in the record that required "guesstimates," one of which was based upon the rate of recession of Niagara Falls upon its emergence from under glacial ice. Antevs then correlated the four Wisconsin substages to the Scandinavian sequence by equating the last Wisconsin stage (W-IV) with the Salpausselkä II moraine in Finland (Antevs 1931: 313–324).

Geological Dating of Folsom Man

The best example of the complexities and tenuousness of dating early man sites in America before radiocarbon dating is the dating of the Folsom occupation of the Lindenmeier site in northern Colorado by geologist Kirk Bryan and his then student Louis L Ray (1940). In 1927, the small cattle-shipping town of Folsom in northeastern New Mexico became famous for the nearby discovery of stone projectile points in direct association with bones of an extinct form of bison. This discovery of what became known as Folsom points with buried skeletons of *Bison antiquus* clearly placed early man (now called Paleoindian) in the New World at least as early as the end of the Pleistocene (Wormington 1957: 23–29). Bryan (1937) was the first geologist to report on the geology of the Folsom site. However, the geological situation there did not lend itself to correlation to the glacial chronology.

The discovery, a few years later, of another Folsom site, a campsite, on the Lindenmeier Ranch in the piedmont of the Colorado Rocky Mountains provided a better opportunity for geological dating because the drainages of the region were derived from glaciated valleys in the Rocky Mountain Front Range. By studying and mapping glacial deposits, Ray was able to define four Wisconsin glacial moraines in the upper Cache La Poudre River drainage that he correlated with outwash terraces farther downstream. He and Bryan then correlated these with alluvial terraces of the Colorado piedmont (Bryan & Ray 1940). However, the Lindenmeier site is confined to an isolated valley and deeply buried in alluvium that could not be readily correlated to the terrace sequence. In order to make this critical connection, they assumed that the Folsom occupation at Lindenmeier was contemporary with the Folsom-like Clovis occupation at Dent in a terrace of the South Platte River. This was correlated with the Kersey terrace of the Cache La Poudre which, in turn, they correlated to the Corral Creek or Wisconsin III substage of Rocky Mountain glaciation.

In order to quantify the age of the sites, Ray and Bryan correlated their Rocky Mountain glacial sequence with Antevs' North American varve chronology for Wisconsin glaciation, but only after reinterpreting his various rates of recession between Long Island, New York to Cochrane, Ontario. They concluded that the Folsom occupation of the region occurred between 25,000 and 10,000 years ago, but leaned toward the older end of the range. Antevs (1941: 41) disagreed with their correlation of the Corral Creek moraine to Wisconsin III and considered it equivalent to Wisconsin IV, thus arriving at an age of 10,000 years, or a little more, for the Folsom occupation (Haynes 1990).

RADIOCARBON DATING OF FOLSOM SITES

With the advent of radiocarbon dating, the application of the new technology to early man in America became of great interest to archaeologists as well as Willard Libby. The radiocarbon dating of Folsom man was a prime objective. The first attempt was on a sample of charcoal from the type site. In 1933, paleontologist, Harold J Cook, one of the initial investigators of the site, collected charcoal, presumably for paleoecological information, from a hearth believed to be of Folsom age. The resulting date of 4283 ± 250 (C-377) was considered "surprisingly young" (Arnold & Libby 1950). Cook's subsequent investigation of the site revealed the hearth to be in a secondary channel fill inset against older alluvium. An attempt to date the Lindenmeier Folsom occupation had similar results, yielding a young date of 5020 ± 300 (C-451) (Roberts 1951). These false starts clearly indicated a need for proper evaluation of site stratigraphy before submitting samples for radiocarbon dating. Kirk Bryan died 21 August 1950. Had he examined the sedimentary contexts of the hearths, I have little doubt that he would have recognized their post-Folsom age.

The first apparently reasonable radiocarbon date for Folsom came from the Lubbock Lake site in Texas where burned bison bone yielded a date of 9883 ± 350 (C-558) (Libby 1955). However, as bone, used as a material for radiocarbon dating, gained a bad reputation (Haynes 1967a: 163–168), researchers looked upon the Lubbock Folsom date as one needing verification by radiocarbon dating charcoal in direct association. This came in 1960 with a determination made on charcoal from the Folsom level at Lindenmeier at 10,780

± 135 (I-141) (Haynes & Agogino 1960). The standard deviation in the initial publication was ± 375 years due to calculation based upon preliminary counting data. Subsequent publication of the final results in *RADIOCARBON* provided the more precise sigma without changing the mean value (Walton, Trautman & Friend 1961).

Recent re-evaluation of the Lubbock Lake geochronology has shown a post-Folsom age for the burned bone originally attributed to the Folsom occupation, thus making the Lindenmeier Folsom date the first reliable radiocarbon date obtained for the Folsom complex (Holliday & Johnson 1986).

In 1973, renewed investigations at the type Folsom site produced a better understanding of the stratigraphic context of the Folsom level as well as bone and charcoal for radiocarbon dating (Anderson & Haynes 1978). The distal end of a bison long bone yielded enough purified collagen to provide a radiocarbon date of 10,260 ± 110 (SMU-179) (Hassan 1976), but only 0.33 g of charcoal was obtained. This could have been dated at the time, but I did not do so because of the large standard deviation that would have resulted. The sample was saved in anticipation of improved counting technology that would allow more precise dating of small samples.

Confident that such improvement would surely come, I did not anticipate the revolutionary technology of accelerator mass spectrometry (AMS). This ideal technique for dating single lumps or flecks of charcoal was applied to the type Folsom sample in 1987 by selecting five discrete lumps for individual analysis as well as a composite made by combining a small chip off each lump. The six analyses resulted in a statistical average of 10,890 ± 50 (AA-1708–1712, AA-1213) (Table 24.1), the most precise radiocarbon date yet obtained for the Folsom complex (Haynes *et al*, in press). However, this result raises serious questions regarding the significant difference from the aforementioned bone date (Fig 24.1). The charcoal consisted of two types of wood, pine and juniper, neither of which probably had inherent ages in excess of the standard deviation. The apparently young value for the collagen date, as well as data to be discussed later, revives my distrust of bone dates.

Radiocarbon Dating of Clovis Sites

In 1932 and 1933, fluted projectile points, much like Folsom points but more robust, were found in direct association with bones of at least 12 mammoths at the railroad siding of Dent on the South Platte River north of Denver, Colorado (Wormington 1957: 43–44). At first, this was thought to be a variant of the Folsom complex where the larger, stouter Folsomoid projectile points were used to kill mammoths and the more delicate Folsom points were used for smaller

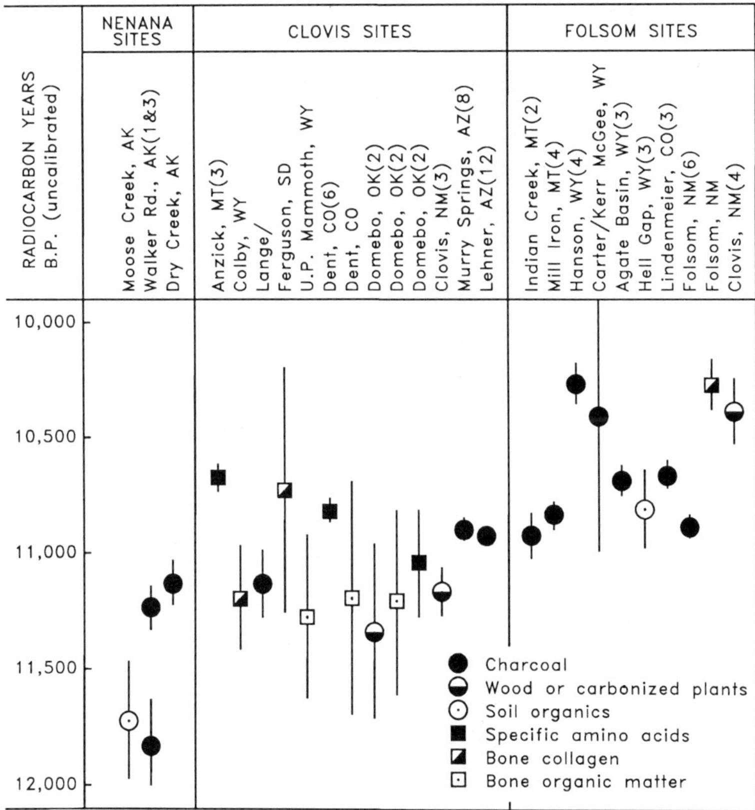

Fig 24.1. Comparison of Nenana, Clovis, Folsom and related radiocarbon dates on various materials, arranged in order of decreasing latitude. Numbers in parentheses are the number of dates averaged in Table 24.1. Averages do not necessarily apply to a single occupation or event.

game such as bison. However, subsequent finds of the Dent-style fluted point stratigraphically below Folsom in Blackwater Draw near Clovis, New Mexico, eventually demonstrated that Clovis points, as they then became known, were ancestral to Folsom points (Sellards 1952: 28–31, 54–58). The stratigraphy at the type Clovis site indicated that an erosional hiatus separated the Clovis occupational horizon from the Folsom (Evans 1951), raising the question of how much time separated the two cultural complexes. In an early assessment of radiocarbon dating as applied to archaeology, Johnson, Arnold and Flint (1957) concluded that Folsom ranged in age from 10,000 to 12,000 years ago and that Clovis "may have existed from 12,000 to 16,000 years ago" with little basis given for the older end of the Clovis range.

The first charcoal radiocarbon dates for Clovis came from the Lehner site in southwestern Arizona (Haury, Sayles & Wasley 1959). Charcoal from two sep-

TABLE 24.1. Selected Clovis, Folsom and Related Radiocarbon Dates

Site	Material	Date (yr BP)	Sample no.
	Nenana sites		
Moose Creek, Alaska	Soil organics	11,730 ± 250	GX-6281
Walker Road, Alaska	Charcoal	11,010 ± 230	AA-1683
	Charcoal	11,170 ± 180	AA-1681
	Charcoal	11,300 ± 120	AA-2264
	Average of 3*	11,230 ± 90	
	Charcoal	11,820 ± 200	Beta-11254
Dry Creek, Alaska	Charcoal	11,120 ± 85	SI-2880
	Clovis Sites		
Anzick site, Montana	Aspartic acid	10,240 ± 120	AA-2978
	Glutamic acid	10,820 ± 100	AA-2979
	Hydroxyproline	10,710 ± 100	AA-2980
	Glycine	10,940 ± 90	AA-2981
	Alanine	10,370 ± 130	AA-2982
	Average of 5	10,680 ± 50	
Colby site, Wyoming	Collagen	11,200 ± 220	RL-392
	Apatite	10,864 ± 141	SMU-254
	Collagen	8719 ± 392	SMU-278
Lange/Ferguson, South Dakota	Charcoal	11,140 ± 140	AA-905
	Bone	10,730 ± 530	I-13104
UP Mammoth site, Wyoming	Tusk organics	11,280 ± 350	I-449
Dent site, Colorado	XAD hydrolysate	10,980 ± 90	AA-2941
	Aspartic acid	10,660 ± 170	AA-2942
	Glutamic acid	10,800 ± 110	AA-2943
	Hydroxyproline	10,600 ± 90	AA-2945
	Glycine	10,710 ± 90	AA-2946
	Alanine	10,670 ± 120	AA-2947
	Average of 6	10,810 ± 40	
	Bone organics	11,200 ± 500	I-622
Domebo site, Oklahoma	Elm wood	11,045 ± 647	SM-695
	Elm wood	11,490 ± 450	AA-823
	Average of 2	11,340 ± 370	
	Bone A	11,220 ± 500	SI-172
	Bone B	11,200 ± 600	SI-175
	Average of 2	11,210 ± 390	
	XAD purified	11,480 ± 450	AA-825
	Pro-Hypro	10,860 ± 450	AA-811
	XAD hydrolysate	10,810 ± 420	AA-805
	Average of 3	11,040 ± 250	

TABLE 24.1 (continued)

Site	Material	Date (yr BP)	Sample no.
Blackwater Draw,	Carbonized plants	11,630 ± 350	A-491
New Mexico	Carbonized plants	11,170 ± 110	A-481
Clovis level	Carbonized plants	11,040 ± 240	A-490
	Average of 3	11,170 ± 100	
Murray Springs site,	Charcoal	11,190 ± 180	SMU-18
Arizona	Charcoal	11,150 ± 450	A-805
	Charcoal	11,080 ± 180	Tx-1413
	Charcoal	10,930 ± 170	Tx-1462
	Charcoal	10,890 ± 180	SMU-27
	Charcoal	10,840 ± 70	SMU-41
	Charcoal	10,840 ± 140	SMU-42
	Charcoal	10,710 ± 160	Tx-1459
	Average of 8	10,900 ± 50	
Lehner site,	Charcoal (F_1)	10,770 ± 140	SMU-168
Arizona	Charcoal (F_1)	10,940 ± 100	A-378
	Charcoal (F_1)	11,080 ± 200	SMU-181
	Charcoal (F_2/D,Z)	10,950 ± 110	SMU-194
	Charcoal (F_2/D,Z)	10,860 ± 280	SMU-164
	Charcoal (F_2/D,Z)	10,950 ± 90	SMU-290
	Charcoal (F_2/D,Z)	11,080 ± 230	SMU-196
	Charcoal (F_2/F_0)	10,620 ± 300	SMU-347
	Charcoal (F_2/F_0)	10,700 ± 150	SMU-297
	Charcoal (F_2/F_0)	10,710 ± 90	SMU-340
	Charcoal (F_2/F_0)	11,170 ± 200	SMU-264
	Charcoal (F_2/F_0)	11,470 ± 110	SMU-308
	Average of 12	10,930 ± 40	
	Folsom sites		
Indian Creek site,	Charcoal	10,980 ± 150	Beta-4619
Montana	Charcoal	10,630 ± 280	Beta-13666
	Average of 2	10,930 ± 100	
Mill Iron site,	Charcoal	11,340 ± 120	Beta-13026
Montana	Charcoal	11,320 ± 130	Beta-16179
	Charcoal	11,360 ± 130	Beta-20111
	Charcoal	11,570 ± 170	NZA-625
	Charcoal	11,560 ± 920	NZA-624
	Average of 5	11,360 ± 70	
	Charcoal	11,010 ± 140	Beta-16178
	Charcoal	10,760 ± 130	Beta-20110
	Charcoal	10,770 ± 85	AA-3669
	Charcoal	10,990 ± 170	NZA-623
	Average of 4	10,840 ± 60	
Hanson site,	Charcoal	10,225 ± 125	Beta-31072
Wyoming	Charcoal	10,300 ± 150	Beta-22514-ETH-3229
	Charcoal	10,700 ± 670	RL-374
	Charcoal	10,080 ± 330	RL-558
	Average of 4	10,260 ± 90	
Carter/Kerr McGee site, Wyoming	Charcoal	10,400 ± 600	RL-917

TABLE 24.1 (continued)

Site	Material	Date (yr BP)	Sample no.
Agate Basin, Wyoming	Charcoal	10,780 ± 120	SI-3733
Folsom level	Charcoal	10,665 ± 85	SI-3732
	Charcoal	10,375 ± 700	I-472
	Average of 3	10,690 ± 70	
Hell Gap, Wyoming	Soil organics**	10,930 ± 200	A-503
Folsom-Goshen-	Soil organics**	10,690 ± 500	A-504
Midland level	Soil organics**	10,290 ± 500	A-502
	Average of 3	10,820 ± 170	
Lindenmeier site,	Charcoal	10,780 ± 135	I-141
Colorado	Charcoal	10,560 ± 100	TO-337
	Charcoal	10,500 ± 80	TO-342
	Average of 3	10,660 ± 60	
Folsom site,	Composite	10,780 ± 100	AA-1213
New Mexico	Charcoal	11,060 ± 100	AA-1708
	Charcoal	10,890 ± 150	AA-1710
	Charcoal	10,760 ± 140	AA-1709
	Charcoal	10,850 ± 190	AA-1711
	Charcoal	10,910 ± 100	AA-1712
	Average of 6	10,890 ± 50	
Blackwater Draw,	Collagen	10,260 ± 110	SMU-179
New Mexico	Carbonized plants	10,250 ± 320	A-380-379
Folsom level	Carbonized plants	10,490 ± 900	A-386
	Carbonized plants	10,170 ± 250	A-488
	Carbonized plants	10,490 ± 200	A-492
	Average of 4	10,380 ± 140	

*Averaging follows Long and Rippeteau (1974)
**Corrected for fractionation of standard (Haynes *et al* 1967)

arate hearths first produced five dates with a wide spread of 5000 years between 7022 ± 450 (A-32) and 12,000 ± 450 (A-406) (Wise & Shutler 1958), thus prompting Haury to state that "No aboriginal fire was kept burning that long!" With the application of the carbon dioxide method using proportional counters, it was immediately obvious to Paul Damon, then the new director of the Arizona Radiocarbon Laboratory, that the problem with the Lehner Clovis dates was contamination *via* the old solid carbon counters (Shutler & Damon 1959). Application of carbon dioxide dating by both the Michigan laboratory (Crane & Griffin 1959) and Copenhagen (Tauber 1960) yielded dates of 11,290 ± 500 (M-811) and 11,180 ± 140 (K-554), respectively, clearly older than Folsom.

Seeing a need for more Clovis dates, George Agogino and I visited the Dent site in 1960 and collected fragments of mammoth bone from the backdirt of the original excavations. After pretreatment following the pyrolysis technique of May (1955), Isotopes, Inc provided a date of 7200 ± 200 (I-473), obviously too young (Trautman & Willis 1966). Assuming that this sample contained shellac as a preservative, a second attempt was made after extracting the preservative using the solvent extraction technique of de Vries and Oakley (1959) for radiocarbon dating the Piltdown skull. The result for Dent of 11,200 ± 500 (I-622) was in excellent agreement with the Lehner site dates.

The discovery of the Union Pacific (UP) mammoth site in Wyoming in 1960 provided another opportunity to date an archaeologically associated mammoth (Irwin, Irwin & Agogino 1962). Unfortunately, none of the associated stone artifacts were diagnostic Clovis points, but the two bifacial knives and a graver were not out of place in known Clovis assemblages. The radiocarbon date on the very well-preserved ivory of 11,280 ± 350 (I-449) is consistent with a Clovis age. As with the Dent mammoth bone, I pretreated the sample by pyrolysis, a technique no longer used because it fails to eliminate humic acid contaminants. However, in the case of the UP mammoth, the ivory was so white and well preserved that typically dark brown humic acids were probably not present in significant quantity.

A year after the UP mammoth discovery, another Clovis site was reported, this time in central Oklahoma (Leonhardy 1966). A radiocarbon date on the acid-soluble organic fraction of bone of 11,220 ± 500 (SI-172) is essentially the same as one on the humic acid fraction of 11,200 ± 600 (SI-175), suggesting that much, if not all, of the humic acids were derived from the degradation of bone organic matter rather than from extraneous sources. A tree stump approximately correlated with the bone- and artifact-bearing unit dated 11,045 ± 647 (SI-695).

The first radiocarbon dates on the Clovis level of the type site resulted from new archaeological investigations in 1962–63 (Warnica 1966; Haynes & Agogino 1966). Three dates, 11,630 ± 350 (A-491), 11,170 ± 110 (A-481), and 11,040 ± 240 (A-490) are statistically the same (Damon, Haynes & Long 1964: 101; Haynes & Agogino 1966: 14), yielding a weighted average of 11,170 ± 100 (Table 24.1). The remarkable time constraint of Clovis dates between 11,000 and 11,300 BP, as well as the lack of any evidence of Clovis progenitors in underlying strata, begged for explanation. With a new assessment that the Two Creeks interstadial ended by 11,850 ± 100 years ago instead of 11,400 (Broecker & Farrand 1963), I was prompted to suggest that the ice-free corridor through Canada proposed by Johnston (1933) may have opened during Two Creeks time, thus allowing Clovis progenitors to reach North America south of Canada soon thereafter (Haynes 1964).

The discovery, in 1966, of the Murray Springs Clovis site, 11 miles north of the Lehner site, and new excavations at the latter in 1974–75, resulted in the most precise dates yet obtained for Clovis. The statistical average of eight charcoal dates from the Clovis occupation surface at Murray Springs is 10,900 ± 50 whereas 12 charcoal samples from Lehner average 10,930 ± 40 (Table 24.1). A reassessment of all Clovis and Folsom non-bone radiocarbon dates reveals a remarkably tight constraint on Clovis to three centuries between 11,200 and 10,900 BP and a 700-year spread of Folsom dates from 10,900 to 10,200 BP (Fig 24.1). From these data, it is clear that the transition from Clovis to Folsom took place within the standard deviation of radiocarbon dating (Haynes 1991).

Other Folsom and Related Radiocarbon Dates

Table 24.1 and Figure 24.1 show radiocarbon dates for six other Folsom sites. The Folsom level at the type Clovis site is in organic diatomaceous earth overlying the Clovis level, as discussed below. The carbonized plant remains in both of the dated occupation levels were probably contemporary with the occupations, and the stratigraphic consistency reinforces this assessment.

All of the dates from the basal occupation level (Folsom, Goshen, Agate Basin) at the Hell Gap site are on bulk sediments containing finely divided charcoal, probably cultural, plus soil organic matter of lesser reliability (Haynes, Damon & Grey 1966). Charcoal for AMS dating will be collected in the near future.

The Mill Iron site (Frison 1988), neither Clovis nor Folsom, is characterized by unfluted projectile points (Plainview, Midland, Goshen) probably closely related to either Clovis or Folsom. Radiocarbon dates show two clusters, one at 11,360 ± 70 and the other at 10,840 ± 60 (Table 24.1), suggesting two periods of occupation. A single bone tool made of mammoth bone may go with the earlier cluster. This brings to mind the unidentified occupation level under the Folsom level at Agate Basin (Haynes 1987) and raises the question of a possible relation. However, the occurrence of Cretaceous lignite in the local bedrock at Mill Iron, as well as a charcoal date of 23,720 ± 220 (AA-3668) from the cultural horizon, suggests that, because of the likelihood of contamination by "dead" carbon, the younger cluster is more likely correct and applies to a single occupation (Fig 24.1).

The overlap of standard deviations indicates that probably the only foreseeable way we will be able to determine the relative age of the diversity of Paleoindian complexes appearing in the few centuries following Clovis will be through stratigraphic succession at multicomponent sites.

Clovis and Folsom Radiocarbon Dates on Bone Amino Acids

Without reviewing the entire history of radiocarbon dating of bone, I will note here that what has been logically anticipated as the ideal substance for radiocarbon dating, bone protein or collagen, is a great disappointment despite a variety of promising techniques (Taylor 1987: 53–61). With the isolation of specific collagen amino acids and AMS dating, Stafford (1991) has produced new dates for both the Dent and Domebo Clovis sites. From these data, I have selectively averaged values that are on the theoretically best quality fractions (Stafford, personal communication) and that are within one sigma of each other (Table 24.1) for plotting in Figure 24.1. It is readily apparent that the amino-acid dates are significantly younger than previous radiocarbon dates (Stafford *et al* 1988). These, together with a third amino-acid date from human skull fragments from the Anzick site (Stafford 1991) significantly overlap with Folsom radiocarbon dates (Table 24.1, Fig 24.1), a situation that is not consistent with the stratigraphy at the multicomponent sites of Clovis, Hell Gap, and Agate Basin. The Clovis amino acid dates, along with the collagen date from the Folsom type site, mentioned earlier, still leave uncertainties in the reliability of bone dates compared to charcoal or wood dates (Fig 24.1).

RADIOCARBON DATING AND PRE-CLOVIS SITES

So far, I have considered only the stratigraphically consistent, well-established, and well-dated evidence for the Paleoindian occupation of the New World. The present evidence for pre-Clovis inhabitants of America is based mainly on radiocarbon dates and remains controversial (Dincauze 1984; Lynch 1990). In the 40 years since the practical application of radiocarbon dating, numerous potential pre-Clovis sites have been reported and subsequently either disproved or ignored because of inadequate evidence or lack of substantiation through comparable finds (Haynes 1967b, 1969, 1971; Payen 1982).

THE TULE SPRINGS EXPERIENCE

A classic example of a pre-Clovis radiocarbon-dated site that failed to hold up under close scrutiny is Tule Springs, Nevada. One of Libby's Chicago lab dates (C-914) indicated that an obsidian flake found in deposits containing extinct fauna in the Las Vegas valley was greater than 23,800 years old (Libby 1955). This prompted new excavations at the site by the Southwest Museum of Los Angeles in 1955 and 1956. The investigation resulted in additional artifact finds, hearth features and a new radiocarbon date of >28,000 (L-533B) (Olson & Broecker 1961). Results reported by Harrington and Simpson (1961) convinced many American archaeologists that man had entered the New World before the main substage of Wisconsin glaciation (Krieger 1962).

In February 1962, Willard Libby invited several prominent scientists interested in early man in America to decide on a site of critical importance to American archaeology. He offered his radiocarbon dating laboratory be used in solving important questions regarding early man (Shutler 1967: 3–6). It was agreed that

Tule Springs should be the site because a few archaeologists still doubted the validity of the claims of Harrington and Simpson (Wormington 1957: 197–198).

Libby's laboratory assistant, R Ervin Taylor, was assigned to be the courier between his new laboratory at UCLA and the site. Erv would take back samples to UCLA in one trip, return results a week or so later and take back a new batch of samples to Libby's lab for dating by Gordon J Fergusson. Fortunate in being chosen to do the Quaternary geology of the site, I had little doubt about the validity of the claims of the Southwest Museum. Their report, taken at face value, was quite convincing. I believe most others on the project felt the same way. Because of the industrial connections of Herschel C Smith, a prominent contractor with an avid interest in early man, industry provided an incredible array of heavy earth-moving equipment free of charge, as well as numerous other services including detailed aerial photo coverage and contour maps with an interval of one foot.

The first order of business was to convince Hersch Smith that the overburden should not be stripped from the site because it contained the strata that were crucial to understanding the age of the artifacts and hearth features. The local operators union put up the labor for nothing more than workmen's compensation. However, union operators are trained to make every minute count of the customer's time. Going slow was not in their psyches. With this in mind, I selected a grid line at the extreme northwestern end of the mapped area for a stratigraphic trench that could be used to keep the equipment busy whenever finds in other trenches and areas required the slow, tedious excavation techniques of the archaeologists. The result was that Trench K was extended across Tule Springs Wash and into the alluvial fan for a total length of 3600 ft and a maximum depth in one area of 40 ft (Shutler 1967: Figs 5, 6).

The second order of business was to convince Libby that no carbon sample should be taken for radiocarbon dating until its relation to the stratigraphic framework was known. Having just learned radiocarbon dating from Paul Damon and Austin Long, I brought pretreatment materials to the site, and soon discovered that the "charcoal" samples were completely soluble in base solution and therefore not charcoal. Examination of the stratigraphic context soon revealed that the naturally oxidized wood was associated with bowl-shaped sediments at the base of spring-fed ponds (Haynes 1967b: 15–104). Thus, the lenticular concentrations looked like hearths. The underlying feeder conduits in some cases contained jumbled masses of bones of extinct vertebrates, including Pleistocene mammoth, bison, horse and camel (Mawby 1967: 105–129). The "bone tools" of the earlier excavations were very likely the natural products of animals, stuck in the spring conduit, trampling and breaking the bones of previous victims. Also, what had been described as fire-reddened earth turned out to be post-depositional stains of hydrated iron oxides deposited by groundwater. The overall result of the 1962–63 Tule Springs investigations was the realization that previous claims for the presence of early man there before 12,000 BP could not be substantiated (Shutler 1967: 304–308).

RECENT RADIOCARBON DATING OF PRE-CLOVIS SITES

Today, some archaeologists strongly support two sites, Meadowcroft rockshelter in Pennsylvania (Adovasio, Boldurian & Carlisle 1988) and Monte Verde in Chile (Dillehay 1986), as representing pre-Clovis peopling of the New World. However, neither site has remained unchallenged. The older dates for human occupation at Meadowcroft (Stratum IIa) are stated to be on fine-grained charcoal collected from thin, irregular lenses interpreted as hearths. However, pretreatment revealed some of the samples to be so soluble in base solution that the reaction had to be cut short to ensure enough residue for radiocarbon dating (Haynes 1977). As with Tule Springs, this raises the question of whether or not the samples are really charcoal and the features really hearths. At my suggestion, the dating of the humate fraction, usually discarded because of younger contaminants from soil-derived humic acids, produced a date several times *older* than the residue. This abnormal situation remains unexplained (Haynes 1980, 1987) and, along with other questions (Mead 1980; Kelley 1987; Tankersley, Munson & Smith 1984; Dincauze 1981), leaves the validity of the pre-Clovis occupation at Meadowcroft open to further verification, preferably by independent investigation.

The Monte Verde site in Chile has yielded several radiocarbon dates on three separate materials, *ie*, wood, charcoal and bone, all lying between 11,800 and 13,600 BP (Collins & Dillehay 1986; Pino Quivira & Dillehay 1988), thus providing a convincing case for a 12,000 BP age of the strata if the older date results from burning old wood. Otherwise a spread of about 1800 years requires explanation. The archaeological evidence at this age consists of an incredible variety of artifacts, features, and perishable materials that make up an assemblage nothing short of bizarre, ranging from stone, bone and wood artifacts so crude as to be highly controversial, to those that are rather sophisticated. While this situation by itself does not discredit the site, it raises questions about the interpretations (Lynch 1990) that need independent verification. Like Tule Springs, Monte Verde may be another case where the implications of pre-Clovis radiocarbon dates have caused an overemphasis of "possibilism" (Lynch 1990; Dincauze 1984).

On the other hand, if the well-made bifacial projectile points from Monte Verde are of the age claimed, it lends some support to Bryan's (1978) claims for the

14,000 BP age of the El Jobo complex at Taima Taima, Venezuela, because of similarities between El Jobo projectile points and the two specimens from Monte Verde. Until the Monte Verde and Meadowcroft claims are substantiated by either independent investigations or replication of other sites with similar artifact assemblages and radiocarbon dates, pre-Clovis peopling of the New World south of Canada will not be considered established beyond a reasonable doubt.

CLOVIS ORIGIN

Recent archaeological investigations in the Nenana Valley of central Alaska have demonstrated the presence there, at least as early as Clovis time (Fig 24.1), of a cultural complex that is Clovis-like in most respects except for the absence of the diagnostic Clovis point (Powers & Hoffecker 1989; Goebel & Powers 1990). This, the Nenana Complex, occurs at three sites: Dry Creek, Moose Creek and Walker Road from which there are six radiocarbon dates between 11,000 and 11,800 BP (Table 24.1). At Walker Road, the concentration of artifacts within a circular area with a shallow basin, unlined hearth in the center suggests a tent-like shelter (Goebel, personal communication). For the moment, it appears that Clovis progenitors passed from the Nenana Valley to North America south of Canada about Two Creeks time and developed fluting somewhere along the way (Haynes 1987).

Potential pre-Clovis sites such as Taima Taima and Monte Verde, if valid, may not have been Clovis progenitors. They may have been pre-Clovis migrants unrelated to Clovis and possibly related to the early stemmed projectile point complexes of the Great Basin (Bryan 1988). However, Jared Diamond (1990a, b) has emphasized the improbability of humans entering an uninhabited continent and not multiplying or, more improbably, dying out. While this does not preclude the possibilities implied by Monte Verde and Meadowcroft, it makes their potential verification all the more important to science.

CONCLUSION AND LIMITATIONS

In the final analysis of the contributions of radiocarbon dating to the geochronology of the peopling of the New World, it has undoubtedly been almost as great as the discovery of early man in its own right. Without it, we might still have sorted out sequences of Paleoindian development and, thus, have the relative ages of some cultural complexes, but precise quantitative chronologies would not have been possible. Neither would we be able to correlate Paleoindian developments with glaciation and environmental changes on a precise basis. On the other hand, neither should it be used as an end in itself in regard to dating cultural changes.

A major flaw in the application of radiocarbon by non-specialists in the method is inadequate awareness of its limitations. It is as important to realize that radiocarbon dates can be wrong as it is to realize their benefits. All too often a radiocarbon determination is reported with no clear indication of what was dated. For properly evaluating a bone date, for example, it is essential to know what was analyzed - carbonate, apatite, soluble or insoluble organic matter, collagen or a specific amino acid. When the term "charcoal" is stated as the material dated, this should imply that the sample was truly insoluble in basic solution during pretreatment. For samples that yield significant percentages of soluble organic matter, it is useful to analyze both fractions. Large discrepancies in the dates raise questions as to the origins of the organic fractions, Meadowcroft being a case in point. Also useful is the quality rating system for materials dated as emphasized by Mead and Meltzer (1984).

The nuclear basis for the radiocarbon method is so sound that the method is commonly used as the standard for evaluating or calibrating other dating techniques. However, the most important lesson to be learned from the past 40 years of radiocarbon dating in geoarchaeology is that its accuracy can only be verified by a combination of proper stratigraphic control in conjunction with reassuring repetition and compatibility with the results of other disciplines.

Stratigraphic investigations at Paleoindian sites over the past 60 years have shown a remarkable degree of consistency in the position of artifacts with respect to the alluvial deposits of low-order tributary streams (Haynes 1968, 1984, 1991). The earliest Paleoindian levels (Clovis, Folsom, Agate Basin, etc) are at or near the base of the alluvial sections. Therefore, I would regard radiocarbon ages between 12,000 and 10,000 BP occurring high in an alluvial section as abnormal and one that required closer scrutiny. Such a situation is by no means impossible. Early Holocene overbank deposition on a Pleistocene terrace could be the cause, but this is so rare that it would require special attention. Sites along the South Platte River in Colorado are a case in point (Holliday 1987; Haynes 1990).

We have not likely reached the end of improvements to the method, and the next few decades may see significant improvements in the extension of the range of radiocarbon dating as well as calibration, as a better understanding of ^{14}C production and variation develops. Also, the small-sample capability of AMS dating means that more attention will be paid to dating specific molecules rather than bulk samples of soils, sediments, shells and bone.

The first graphic summary of Paleoindian radiocarbon dates showed an abrupt rise in radiocarbon-dated sites at 10,400 BP with only two earlier sites, both questionable (Libby 1961). Although not mentioned, one of these is undoubtedly Tule Springs. Jelinek's (1962) histogram shows no abrupt rise, but his graphs are based upon numbers of radiocarbon dates rather than number of sites so dated. A later assessment showed the abrupt rise to occur approximately 1000 years earlier than Libby's, mainly due to Clovis dates (Haynes 1967c), and several pre-Clovis sites listed as of questionable validity have since been dropped from further consideration by most of the archaeological profession (Payen 1982; Dincauze 1984). Today, the abrupt appearance of Clovis in North America persists at 11,300 BP but with a new crop of pre-Clovis candidates. It will be interesting to see if these prevail or disappear into the background like electronic noise on an oscilloscope screen.

The proper application of radiocarbon dating to problems of the initial peopling of the New World will eventually result in either replication of pre-Clovis dates with solid associations or a continuum of questionable sites and dates. There probably will never be a clear, cut-and-dry break between positive evidence for early man in America and the negative evidence for pre-early man. Nature provides too many opportunities for misinterpretation (Haynes 1988).

ACKNOWLEDGMENTS

This research was supported by the National Science Foundation (Grants EAR-8607479 and EAR-8820395), the National Geographic Society, and the University of Arizona Social and Behavioral Sciences Research Institute. Appreciation is expressed to George C Frison and Thomas W Stafford for making unpublished radiocarbon dates available and to Doris Sample for expert word processing. I dedicate this paper to the memory of long-time friend and colleague, Cynthia Irwin-Williams.

REFERENCES

Adovasio, JM, Boldurian, AT and Carlisle, RC 1988 Who are those guys?: Some biased thoughts on the initial peopling of the New World. *In* Carlisle, RC, ed, Americans before Columbus: Ice-age origins. University of Pittsburgh, *Ethnology Monographs* 12: 45–62.

Anderson, AB and Haynes, CV, Jr 1978 How old is Capulin Mountain: Correlation between Capulin Mountain volcanic flows and the Folsom type site, northeastern, New Mexico. *In* Shelton, N, ed, *Proceedings of the 1st Conference on Scientific Research in the National Park Service*: 893–899.

Antevs, E 1931 Late glacial clay chronology of North America. *Smithsonian Report for 1931*: 313–324.

_____1941 Age of the Cochise cultural stages. *Medallion Papers* 29: 31–56.

Arnold, JR and Libby, WF 1950 *Radiocarbon Dates (September 1, 1950)*. Chicago. University of Chicago, Institute for Nuclear Studies.

Broecker, WS and Farrand, WR 1963 Radiocarbon age of the Two Creeks Forest Beds, Wisconsin. *Geological Society of America Bulletin* 74: 795–802.

Bryan, AL 1978 An El Jobo mastodon kill at Taima-taima, Venezuela. *Science* 200: 1275–1277.

_____1988 The relationship of the stemmed point and fluted point traditions in the Great Basin. *In* Willig, JA, Aikens, CM and Fagen, JL, eds, *Nevada State Museum Anthropological Papers* 21: 53–74.

Bryan, K 1937 Geology of the Folsom deposits in New Mexico and Colorado. *In* MacCurely, GG, ed, *Early Man*. Philadelphia, JB Lippincott: 139–152.

Bryan, K and Ray, LL 1940 Geologic antiquity of the Lindenmeier site in Colorado. *Smithsonian Miscellaneous Collections* 99(2): 1–76.

Collins, MB and Dillehay, TD 1986 The implications of the lithic assemblage from Monte Verde for early man studies. *In* Bryan, AL, ed, *New Evidence for the Pleistocene Peopling of the Americas.* Orono, University of Maine: 339–355.

Crane, HR and Griffin, JB 1959 University of Michigan radiocarbon dates IV. *American Journal of Science Radiocarbon Supplement* 1: 173–198.

Damon, PE, Haynes, CV, Jr and Long, A 1964 Arizona radiocarbon dates V. *Radiocarbon* 6: 91–107.

Diamond, J 1990a The latest on the earliest. *Discover* 11(1): 50.

_____1990b Clovis claims: Reply to Dillehay. *Discover* 11(5): 8.

Dillehay, TD 1986 The cultural relationships of Monte Verde: A Late Pleistocene settlement site in the sub-Atlantic forest of south-central Chile. *In* Bryan, AL, ed, *New Evidence for the Pleistocene Peopling of the Americas.* Orono, University of Maine: 319–337.

_____1990 Clovis claims: reply to Diamond. *Discover* 11(5): 8.

Dincauze, DF 1981 The Meadowcroft papers (Adovasio, Gunn, Donahue, Stuckenrath, Gilday, Lord, Volman, Haynes, Mead). *Quaternary Review of Archaeology* 2: 3–4.

_____1984 An archaeological evaluation of the case for pre-Clovis occupations. *In* Wendorf, F and Close, A, eds, *Advances in World Archaeology* 3: 275–323. New York, Academic Press.

Evans, GL 1951 Prehistoric wells in eastern New Mexico. *American Antiquity* 17: 1–8.

Frison, GC 1988 Paleoindian subsistence and settlement during post-Clovis times on the northwestern plains, the adjacent mountain ranges, and intermontane basins. *In* Carlisle, RC, ed, Americans Before Columbus: Ice-age origins. University of Pittsburgh, *Ethnology Monographs* 12: 83–106.

Goebel, T and Powers R 1990 The Nenana complex of Alaska and Clovis origins. Society for American Archaeology, 55th annual meeting, *Programs and Abstracts*: 94.

Harrington, MR and Simpson, RD 1961 Tule

Springs, Nevada with other evidences of Pleistocene man in North America. *Southwest Museum Papers* 18: 146 p.

Hassan, AA 1976 Geochemical and mineralogical studies on bone material and their implications for radiocarbon dating. PhD dissertation, Southern Methodist University. Ann Arbor, Michigan, University Microfilms.

Haury, EW, Sayles, EB and Wasley, WW 1959 The Lehner mammoth site, southeastern Arizona. *American Antiquity* 25: 2–30.

Haynes, CV, Jr 1964 Fluted projectile points: their age and dispersion. *Science* 145: 1408–1413.

_____1967a Bone organic matter and radiocarbon dating. In *Radioactive Dating and Methods of Low-Level Counting.* Vienna, IAEA: 163–168.

_____1967b Quaternary geology of the Tule Springs area, Clark County, Nevada. *In* Wormington, HM and Ellis, D, eds, Pleistocene studies in southern Nevada. *Nevada State Museum Anthropology Papers* 13: 411 p.

_____1967c Carbon-14 dates and early man in the New World. *In* Martin, PS and Wright, HE eds, Pleistocene extinctions: The search for a cause. *International Association for Quaternary Research* 6. New Haven, Connecticut, Yale University Press.

_____1968 Geochronology of late-Quaternary alluvium. *In* Morrison, RB, ed, *Means of Correlation of Quaternary Successions* 8. University of Utah Press.

_____1969 The earliest Americans. *Science* 166: 709–715.

_____1971 Time, environment, and early man. *Arctic Anthropology* 8: 3–14.

_____1977 When and from where did man arrive in northeastern North America: A discussion. *In* Newman, WS and Salwen, B, eds, Amerinds and their paleoenvironment in northeastern North America. *Transactions of the New York Academy of Sciences* 228: 165–166.

_____1980 Paleoindian charcoal from Meadowcroft Rock Shelter: Is contamination a problem? *American Antiquity* 45: 583–587.

Haynes, CV, Jr 1984 Stratigraphy and late Pleistocene extinction in the United States. *In* Martin, PS and Klein, RG, eds, *Quaternary Extinctions: A Prehistoric Revolution.* Tucson, University of Arizona Press: 345–353.

_____1987 Clovis origin update. *The Kiva* 52: 83–93.

_____1988 Geofacts or fancy. *Natural History* 97: 4–11.

_____1990 The Antevs-Bryan years and the legacy for Paleoindian geochronology. *In* Laporte, LF ed, Establishment of a geologic framework for paleoanthropology. *Geological Society of America Special Paper* 242: 55–68.

_____1991 Archaeological and paleohydrological evidence for a terminal Pleistocene drought in North America and its bearing on Pleistocene extinction. *Quaternary Research* 35: 438–450.

Haynes, V, Jr and Agogino, GA 1960 Geological significance of a new radiocarbon date from the Lindenmeier site, Colorado. *Denver Museum of Natural History Proceedings* 9: 23 p.

_____1966 Prehistoric springs and geochronology of the Clovis site, New Mexico. *American Antiquity* 31(6): 812–821.

Haynes, CV, Jr, Beukens, RP, Jull, AJT and Davis, OK, in press, New radiocarbon dates for some old Folsom sites using accelerator technology. *In* Stanford, D and Day, JS, eds, *Ice Age Hunters of the Rockies.* Denver Museum of Natural History.

Haynes, CV, Jr, Damon, PE and Grey, DC 1966 Arizona radiocarbon dates VI. *Radiocarbon* 8: 1–21.

Haynes, CV, Jr, Grey, DC, Damon, PE and Bennett, R 1967 Arizona radiocarbon dates VII. *Radiocarbon* 9: 1–14.

Holliday, VT 1987 Geoarchaeology and late Quaternary geomorphology of the middle South Platte River, northeastern Colorado. *Geoarchaeology* 2: 317–329.

Holliday, VT and Johnson, E 1986 Re-evaluation of the first radiocarbon age for the Folsom culture. *American Antiquity* 51: 332–338.

Irwin, C, Irwin, HT and Agogino, GA 1962 Ice age man vs. mammoth in Wyoming. *National Geographic* 121(6): 828–837.

Jelinek, AJ 1962 An index of radiocarbon dates associated with cultural material. *Current Anthropology* 3: 451–475.

Johnson, F, Arnold, JR and Flint, RF 1957 radiocarbon dating. *Science* 124: 240–242.

Johnston, WA 1933 Quaternary geology of North America in relation to the migration of man. *In* Jenness, D, ed, *American Aborigines*. Toronto, University Toronto Press: 11–45.

Kelly, RL 1987 A comment on the pre-Clovis deposits at Meadowcroft Rock Shelter. *Quaternary Research* 27: 332–334.

Krieger, AD 1962 The earliest cultures in the western United States. *American Antiquity* 28(2): 138–143.

Leonhardy, FC 1966 Domebo: A Paleoindian mammoth kill in the prairie-plains. *Contributions of the Museum of the Great Plains* 1: 53 p.

Libby, WF 1955 *Radiocarbon Dating*, second edition. Chicago, University of Chicago Press: 175 p.

_____1961 Radiocarbon dating: The method is of increasing use to the archaeologist, the geologist, the meteorologist, and the oceanographer. *Science* 133: 621–629.

Long, A and Rippeteau, B 1974 Testing contemporaneity and averaging radiocarbon dates. *American Antiquity* 39: 205–215.

Lynch, TF 1990 Glacial-age man in South America: A critical review. *American Antiquity* 55: 12–36.

Macgowan, K and Hester, JA, Jr 1962 *Early Man in the New World.* Garden City, New York, Doubleday: 333 p.

Mawby, JE 1967 Fossil vertebrates of the Tule Springs site, Nevada. *In* Wormington, HM and Ellis, D, eds, Pleistocene studies in southern Nevada. *Nevada State Museum Anthropology Papers* 13: 411 p.

May, I 1955 Isolation of organic carbon from bones for C^{14} dating. *Science* 121: 508–509.

Mead, JI 1980 Is it really that old? A comment about the Meadowcroft "Overview." *American Antiquity* 45: 579–582.

Mead, JI and Meltzer, DJ 1984 North American Late Quaternary extinctions and the radiocarbon record. *In* Martin, PS and Klein, RG, eds, *Quaternary Extinctions: A Prehistoric Revolution*. Tucson, University of Arizona Press: 440–450.

Meltzer, DJ 1983 The antiquity of man and the development of American archaeology. *In* Schiffer, M, ed, *Advances in Archaeological Method and Theory* 6.

Nilsson, T 1983 *The Pleistocene.* Dordrecht, The Netherlands, D Reidel Publishing Company: 651 p.

Oakley, KP 1964 *Frameworks for Dating Fossil Man.* Chicago, Aldine: 335 p.

Olson, EA and Broecker, WS 1961 Lamont natural radiocarbon measurements VI. *Radiocarbon* 3: 141–175.

Payen, LA 1982 The pre-Clovis of North America: Temporal and artifactual evidence. PhD dissertation, University of California, Riverside: 491 p.

Pino Quivira, M and Dillehay, TD 1988 Monte Verde, south-central Chile: Stratigraphy, climate change and human settlement. *Geoarchaeology* 3: 177–191.

Powers, WR and Hoffecker, JF 1989 Late Pleistocene settlement in the Nenana Valley, central Alaska. *American Antiquity* 54: 263–287.

Ray, LL 1940 Glacial chronology of the southern Rocky Mountains. *Geological Society of America Bulletin* 51: 1851–1918.

Roberts, FHH 1951 Radiocarbon dates and early man. *In* Johnson, F, assembler, Radiocarbon dating. *Society for American Archaeology Memoir* 8: 20–21. Also in *American Antiquity* 17(2): 1.

Sellards, EH 1952 *Early Man in America.* Austin, University of Texas Press: 211 p.

Shutler, R, Jr 1967 Archaeology of Tule Springs. *In* Wormington, HM and Ellis, D, eds, Pleistocene studies in southern Nevada. *Nevada State Museum Anthro-*

pological Papers 13: 411 p.

Shutler, R, Jr and Damon, PE 1959 University of Arizona radiocarbon dates II. *American Journal of Science Radiocarbon Supplement* 1: 59–62.

Stafford, TW, Jr, Brendel, K and Duhamel, RC 1988 Radiocarbon, ^{13}C and ^{15}N analysis of fossil bone: Removal of humates with XAD-2 resin. *Geochimica et Cosmochimica Acta* 52: 2257–2267.

Stafford, TW, Hare, PE, Currie, L, Jull, AJT and Donahue, DJ 1990 Accuracy of North American human skeleton ages. *Quaternary Research* 34: 111–120.

_____1991 Accelerator radiocarbon dating at the molecular level. *Journal of Archaeological Sciences* 18: 35–72.

Tankersley, KB, Munson, CA and Smith, D (ms) 1984 Coal contamination: Possibilities, probabilities, and occurrences. Paper presented at the 49th annual meeting of the Society for American Archaeology, Portland, Oregon.

Tauber, H 1960 Copenhagen radiocarbon dates IV. *American Journal of Science Radiocarbon Supplement* 2: 12–25.

Taylor, RE 1987 *Radiocarbon Dating: An Archaeological Perspective.* Orlando, Florida, Academic Press: 370 p.

Trautman, MA and Willis, EH 1966 Isotopes, Inc. radiocarbon measurements V. *Radiocarbon* 8: 161–203.

Vries, H de and Oakley, KP 1959 Radiocarbon dating of the Piltdown skull and jaw. *Nature* 184: 224–226.

Walton, A, Trautman, MA and Friend, JP 1961 Isotopes, Inc. radiocarbon measurements I. *Radiocarbon* 3: 58.

Warnica, JM 1966 New discoveries at the Clovis site. *American Antiquity* 31: 345–357.

Wise, EN and Shutler, R, Jr, 1958 University of Arizona radiocarbon dates (I). *Science* 127: 72–73.

Wormington, HM 1957 Ancient man in North America, fourth edition. *Denver Museum Natural History* 4: 322 p.

Zeuner, FE 1958 *Dating the Past.* Darien, Connecticut, Hafner Publishing Co: 565 p.

25

RADIOCARBON DATING OF BONE: TO COLLAGEN AND BEYOND

R E TAYLOR

INTRODUCTION

Discussions concerning the reliability of ^{14}C-based age determinations on bone have occurred throughout all four decades of radiocarbon research. The accuracy of bone ^{14}C determinations was questioned by Libby even before the first bone ^{14}C analysis was undertaken. Despite the amount of attention given to the exclusion of contamination by isolation and purification of specific chemical and, most recently, molecular fractions of bone, a tradition of skepticism concerning the general reliability of bone ^{14}C values remains (eg, Brown 1988: 225). Concerns about the accuracy of ^{14}C values obtained on seriously collagen-degraded bones (eg, Gillespie 1989; Stafford *et al* 1990, 1991) maintain the negative connotations associated with this sample type.

From the point of view of the archaeologist or paleoanthropologist, this is an unfortunate situation, since bone material is present in many sites where other organics are not present or, if present, have a questionable relationship with an actual or purported cultural expression. Bone is also often an important source of data for a wide range of studies aimed at reconstructing paleoenvironmental constraints on past human behavior (Price 1989). More specifically, bone has been used in attempts to reconstruct the evolving dietary patterns of human populations (Schoeninger 1989). Direct dating of material, from which so much biocultural and human ecological data can potentially be extracted, could measurably contribute to reducing the many ambiguities and uncertainties associated with our understanding of the evolution of past human behavior based on archaeological and paleoanthropological evidence (Taylor 1982, 1987a).

INITIAL VIEWS AND EXPERIENCES

Libby at Chicago

The experience of Libby and his collaborators at Chicago in the late 1940s and early 1950s yielded a hierarchy of reliable sample types. At the top of the list of recommended materials was charcoal and "charred organic materials such as heavily burned bone." In second place was "well-preserved wood," followed by "grasses, cloth and peat," "well-preserved antler" and, at the bottom, "well-preserved shell" (Libby 1952: 43). The carbon derived from "charred bone" would include that derived from the bone itself but, depending on the degree of burning, would generally be dominated by carbonized skin, hair and other residual fatty and connective tissue adhering to the bone. Only two bones were measured by the Chicago laboratory; both were bison and, in each case, the sample material was characterized as "burned." The first was from Sage Creek, Wyoming and yielded an age of 6876 ± 250 yrs BP (C-302: Libby 1952: 122). This sample would be redated a decade later using a method designed to extract collagen; the revised age was 8750 ± 120 (UCLA-697A: Berger, Horney & Libby 1964; see also Table 2). The other bone measured by the Chicago laboratory was a sample of bison bone initially thought to be associated with the Folsom level at Lubbock Lake, Texas. It yielded an age of 9883 ± 350 ^{14}C yrs BP (C-558: Libby 1952: 82).

The ^{14}C value on the Lubbock Lake bone sample had been obtained because of a previous anomalous ^{14}C determination. The first sample submitted from Folsom, New Mexico – a charcoal sample from a hearth – yielded a "surprisingly young" age of 4283 ± 250 years (C-377: Arnold & Libby 1950: 10, 1951: 116). C-377 was later interpreted as belonging to a "secondary channel" rather than the Folsom level (Libby 1952: 82). More recently, the context of the bone sample used for C-558 has also undergone scrutiny. Geologic evidence combined with additional ^{14}C determinations have been used to argue that C-558 may not, in fact, have come from the Folsom levels at Lubbock Lake (Holliday & Johnson 1986; *cf* Haas, Holiday & Stuckenrath 1986).

Libby's view up to the early 1970s was that bone itself would be a "very poor prospect" because of its relatively low, largely inorganic, carbon content, its very porous structure and the "potential for alteration" – *ie*, the expectation of isotopic exchange of the carbonates in the bone with groundwater and soil carbonates. While he conceded that it was "barely conceivable" that bone measurements might yield accurate results, he suggested that because the quantities required were so large and other more acceptable sample materials associated with the bone were available, "it (did) not seem to be an urgent matter to pursue" (Libby 1952: 43, 1956: 44). No ^{14}C dates on unburned bone appeared in any of the Chicago date lists.

Other Early Investigations

Other pioneering laboratories, however, did undertake ^{14}C measurements on unburned bone samples. They confirmed Libby's essentially *a priori* view that bone as such would yield inconsistent results when compared with other sample types (eg, de Vries & Barendsen 1954; Ralph 1959: 56; Olson & Broecker 1961: 142). Reviewing the literature in the early 1960s, Olson (1963: 61–65) documented that ^{14}C measurements on bone were most often rejected when compared with the acceptance of ^{14}C values obtained on charcoal, wood, peat, antler and shell. Figure 25.1 illustrates the dismal view held by early investigators of the validity of bone ^{14}C dates in comparison to other sample types.

The principal cause of the initial anomalous results was the use of the total or whole-bone matrix which included the inorganic carbon constituent of the bone. Inorganic carbonates from a bone buried in a soil profile can be derived from either primary carbonates associated with the indigenous apatite structure – calcium carbonate incorporated in calcium phosphate crystals and other amorphous, carbon-containing inorganic materials – or from secondary (diagenetic) carbonates, which are transported into the bone matrix from the groundwater and soil environment by chemical exchange and/or through dissolution and reprecipitation processes. The consensus that quickly developed was that "bone carbonate is worthless" (Olson & Broecker 1961: 142). In at least one case, the "bone carbonate" ^{14}C values obtained on bone from an archaeological context, which had been measured after the 1954 initiation of atmospheric testing of thermonuclear weapons, were reported to contain "bomb" ^{14}C, indicating that external contamination of the carbonate fraction could be severe (Grant-Taylor 1971).

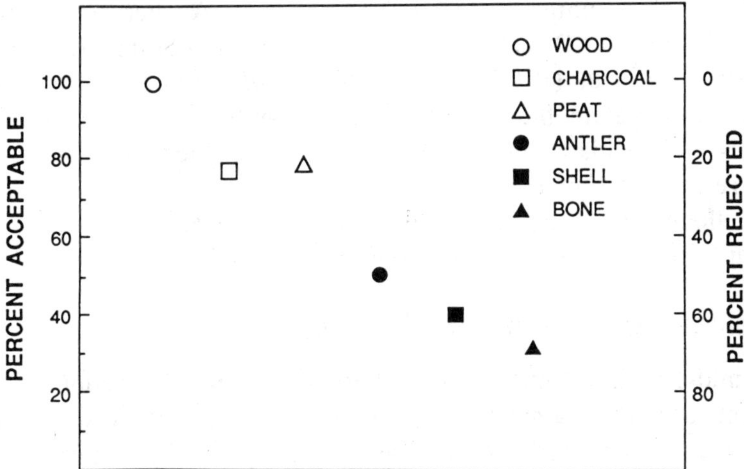

Fig 25.1. Acceptance and rejection rates of different sample types in the early history of ^{14}C dating based on data compiled by Olson (1963)

TABLE 25.1. Comparisons of ^{14}C Determinations on Inorganic and Organic Components of Bone Samples

Source	Inorganic fraction	Organic fraction
Old Crow, Yukon	$27,000 ^{+3000}_{-2000}$*	1350 ± 150**
12-Mile Creek, Kansas	10,435 ± 260[†]	10,245 ± 335[‡]
CA-SJo-112, California	880 ± 90[§]	2875 ± 70[‖]

*GX-1640: Irving & Harrington (1973); **RIDDL-145: Nelson *et al* (1986)
[†]GX-5812A: Rogers & Martin (1984); [‡]GX-5812: Rogers & Martin (1984)
[§]Average of two analyses, UCR-449A and UCR-450A: Taylor & Slota (1979).
[‖]Average of four analyses, UCR-449B, UCR-449C, UCR-450B and UCR-450C: Taylor & Slota (1979)

Subsequent studies have shown that ^{14}C values obtained on a total carbonate fraction can be significantly older, essentially the same age, or younger than an organic fraction from the same bone. Table 25.1 illustrates the lack of consistency in the ^{14}C ages exhibited by the total inorganic and organic fractions in the same sample. An example of a significantly inflated inorganic fraction ^{14}C age is that obtained on a presumed tool made from a caribou tibia recovered from the Old Crow Basin, Yukon Territory, Canada. The original ^{14}C determination was made on a sample of CO_2 released from the bone by treatment with acid (Irving & Harrington 1973). Although this fraction was characterized as "bone mineral apatite," a more accurate designation would probably be "total inorganic carbonate fraction." A later accelerator mass spectrometry (AMS) ^{14}C analysis on an organic fraction of this bone yielded an age of 1350 ± 150 BP (RIDDL-145: Nelson *et al* 1986). Studies to develop techniques to separate the *in situ* primary apatite fraction from diagenetic carbonates were initiated in the late 1960s (eg, Haynes 1968; Hassan 1976; Hassan, Termine & Haynes 1977; Haas & Banewics 1980). While such studies continue, there have been continuing difficulties in accomplishing this on a consistent basis (Haas, personal communication, 1990).

TOWARD COLLAGEN: USE OF ORGANIC FRACTIONS

By the mid-1950s, attention had shifted toward the extraction of organic constituents of bone as a means of obtaining more accurate ^{14}C determinations. An early pretreatment methodology was designed primarily to minimize loss of the residual organics: the bone was first pyrolysized at 340°C in a nitrogen atmosphere – converting the bone to "black char" – and then treated with HCl to remove the carbonates (May 1955). A procedure designed to isolate and purify residual collagen in fossil bone was described by Sinex and Faris (1959), who demonstrated the procedure by extracting a gelatin-like precipitate from an antler excavated from a late Upper Paleolithic cultural context, using an EDTA solution and dialysis. Previous ^{14}C measurements already carried out on this sample by the Yale Laboratory (Barendsen, Deevey & Gralenski 1957: 911) had identified significant discrepancies among different fractions; a total carbonate (acid-soluble) fraction yielded an age of 7060 ± 400 BP (Y-158-1), whereas the age of the organic (acid-insoluble) fraction was 10,760 ± 250 BP (Y-158-2). Using molecular weight and the hydroxyproline concentration as criteria, Sinex and Faris identified the organic preparation as gelatin formed from collagen. Unfortunately, they did not obtain a ^{14}C measurement on their extract.

Chemical Fractions

Several chemical pretreatment strategies used to isolate and purify bone organic components were outlined in the early 1960s by several investigators. Three methods outlined by Berger, Horney & Libby (1964) included: 1) weak acid demineralization to remove the inorganic carbonates and recovery of the acid-insoluble fraction ("collagen") by filtration for dating; 2) total acid hydrolysis, precipitation of the mineral component and evaporation of the supernatant to dryness to provide a total organic fraction; and finally, 3) the use of EDTA and dialysis, which essentially involves a molecular weight separation, followed by washing in water to recover and purify the organic product (*cf* Sellstedt, Engstrand & Gejvall 1966). Table 25.2 summarizes the results of the application of the first method listed above to four bone samples. In two of the four cases, the bone collagen ^{14}C values were somewhat younger than the associated organics. The Sage Creek bison sample had been previously dated by the Chicago laboratory and a significantly younger age was obtained. The younger age of C-302 was later attributed to the use of "alcohol not uniformly 'dead' used in the preparation of the very early samples before ethylene as a quenching agent was introduced" (Berger, Horney & Libby 1964: 1000).

TABLE 25. 2. Comparison Between Collagen Fractions and Other Associated Organics Reported in Berger, Horney & Libby (1964)

Sample	Bone collagen	Associated organics
Human ("Whiskey Lil"): Chimney Cave, Nevada	2500 ± 80 (UCLA-689)	Skin: 2510 ± 80 (UCLA-690) Mat: 2590 ± 80 (UCLA-692)
Human: Santa Rosa Island, California	3970 ± 100 (UCLA-140)	Charcoal: 4260 ± 85 (UCLA-140)
Human: India	1800 ± 100 (UCLA-684)	Charcoal: 2270 ± 110 (TF-90)
Bison: Sage Creek, Wyoming	8750 ± 120 (UCLA-697A)*	Charred bone organics: 8840 ± 140 (UCLA-697B)

*Previously dated by the Chicago Laboratory, 6876 ± 250 (C-302: Libby 1955: 122)

The first of the Berger, Horney & Libby methods was used – after extraction of the bone in acetone to remove possible preservatives – to obtain a ^{14}C age of 3310 ± 150 BP (BM-86) on the acid-insoluble fraction of a humerus of the Galley Hill skeleton. When discovered in the 19th century, this skeleton was thought to be Pleistocene in age. However, fluorine values obtained in 1949 suggested that the skeleton was a post-Pleistocene burial that had intruded into Pleistocene sediments (Barker & Mackey 1961: 41; Oakley 1963).

Over the next decade, researchers began to examine biochemical indices that might be useful in characterizing collagen to provide a quantitative baseline against which degraded bone could be compared. These indices or "fingerprints" came to include a distinctive nitrogen/carbon ratio, a characteristic pattern in the relative concentrations of constituent amino acids, and the presence and relative concentration of the "collagen amino acid," hydroxyproline.

While the isolation of the acid-insoluble organic fraction was adopted by several other laboratories and investigators, it was specifically noted that, in some cases, this fraction yielded ^{14}C values younger than that of the associated organic material which, in most cases, was charcoal. In one set of data assembled by Tamers and Pearson (1965), the amount of the deviation tended to increase with the age of the sample with the acid-insoluble fractions exhibiting younger ages up to about 3000 years. The observation of the typically younger ages being obtained was echoed by Haynes (1967), who suggested that the anomalies might be traced to the presence of soil humic compounds in the bone organic carbon extracts. However, Haynes questioned Tamers and Pearson's conclusion that humic-acid contamination would *always* result in bone dates that were too young. Haynes examined four different acid/base solubility fractions – including extracted humic and fulvic acids – in four fossil and modern bones of presumed known age and observed up to a 3600-year variation in the ^{14}C ages of the extracted fractions. In the preparation of his acid-insoluble fractions, Haynes used *cold* 1N HCl under reduced pressure, a modification of the first approach of Berger, Horney & Libby (1964), which had been suggested by Krueger (1965) to increase sample yields of the acid-insoluble fraction ("collagen").

The use of ion-exchange chromatography to isolate a total amino-acid fraction of bone for ^{14}C dating was first undertaken in a series of bone samples by Ho, Marcus and Berger (1969) to solve a specific problem – the ^{14}C dating of bones of *Smilodon californicus* (popularly known as the saber-toothed "cat"), which had been impregnated with petroleum tar from the Rancho La Brea tar pits in Los Angeles, California. Following pretreatment of the samples with ether to remove as much of the surface-adhering tar as possible, the bones were first decalcified in excess HCl, hydrolyzed, passed onto an ion-exchange column and washed with distilled H_2O. The column was then eluted with NH_4OH to recover the amino acids. The relatively large sample sizes available, combined with the excellent preservation of the collagen in the bones resulting from the unique circumstance of the exclusion of water and bacterial contamination afforded by tar penetration, made these ^{14}C measurements feasible to pursue. Two samples of wood and a total amino-acid fraction of a *Smilodon californicus* collected from the same pit and level at Rancho La Brea yielded statistically identical ^{14}C values at about 14,500 ^{14}C yrs BP.

Longin Technique

The solubility of collagen in acidic hot water was used by Longin (1971) as the basis of a method, which came to be widely employed beginning in the 1970s, to isolate and purify residual collagen in fossil bone samples. The approach was developed to solve what Longin stated to be two problems with previous techniques: they could leave contaminants in the final product or, to accomplish complete purification, the pretreatment procedure was so severe that a significant portion of the residual collagen was lost. In the Longin technique, HCl was used to eliminate the carbonates. The collagen was then extracted in the form of gelatin by heating with water at a pH of 3 at about 90°C for several hours. According to Longin, with this technique, "only collagen is present in the gelatin formed; the impurities remain in the residue and can be eliminated by centrifugation." He suggested that other organic contaminants such as humic acids – "which are almost insoluble in the acid medium" – would remain with the residue (Longin 1971: 242).

Longin reported very good concordance between the ^{14}C values obtained on the gelatin fraction of eight bone samples associated with charcoal or carbonized bone organics ("burned bone") samples (Evin *et al* 1971). The ages obtained ranged from 4000 to about 24,000 years. In only one case was a statistically significant difference registered – the oldest gelatin sample was about 1000 years younger than the charcoal control. The ages of humics extracted from 2 of the 3 charcoal and carbonized organic ("burned bone") control sample pairs cited by Longin were younger than those obtained on charcoal samples from which the humics had been extracted by base (NaOH) treatment. In one case, the age of humic extract (10,900 ± 400) was about 6000 years younger than the bone gelatin extract and NaOH-treated burned bone sample. In the second example, there was about a 1000-year difference.

By the mid-1970s, the application of various approaches to the chemical pretreatment of bone and the publication of results had built up a sufficient corpus of data so that reviews and evaluations of the problems and relative merits of the various proposed strategies could be undertaken. Olsson *et al* (1974) noted the long-standing lack of confidence in using any inorganic constituent because of the inability of any generally applicable method to isolate the indigenous apatite structure and exclude diagenetic carbonates. The Uppsala group also noted the general rejection of the use of any fraction solubilized by acid or EDTA treatment, but, at the same time, noted the need to restrict the amount of treatment with acid – either by reducing the length of treatment or adjusting the strength of the solution – to maximize the amount of insoluble material remaining. They cautioned against the use of the term "collagen" to refer simply to a fraction from which the inorganic constituents had been removed. Reflecting on the experience of the Australian National University laboratory, Polach (1971) had also previously questioned using the term "collagen" to refer to an acid-insoluble residue.

Uppsala and de Vries Principle

Following from a comment of Hessel de Vries that it would be highly improbable that contamination would cause the same error in different fractions, the suggested treatment of bone developed at the Uppsala laboratory involved first physical cleaning and demineralization in an excess of acid, followed by the preparation of two components – an acid-soluble and acid-insoluble fraction – both further purified by dialysis and extraction in hot water. Although proposing this procedure for a majority of bones, the Uppsala group noted explicitly that the pretreatment method should be chosen to fit the chemical composition of the bone, *ie*, the amount of residual collagen and the nature of potential contaminants (Olsson *et al* 1974; El-Daoushy *et al* 1978).

Several laboratories applied the Uppsala/de Vries principle by examining the distribution of ^{14}C ages in different organic fractions and implicitly or explicitly adopting the view that concordance in the ^{14}C values obtained on different fractions of the same bone provided strong evidence of the essential accuracy of the age inferences. Based on the Uppsala laboratory experience, this approach was adopted in the late 1970s by the University of California, Riverside group in examining bone samples. A procedure to extract four organic fractions from each bone sample was developed. After careful inspection, physical cleaning under magnification, and ultrasonic cleaning in a weak acid followed by the removal of inorganic carbonates, the procedure called for the extraction of: 1) a total acid-soluble fraction (*never* referred to as "collagen"); 2) a total base-soluble fraction; 3) a total acid-insoluble fraction after gelatin conversion with base-soluble fraction removed; 4) a total acid-soluble fraction after gelatin conversion with base-soluble fraction removed (Taylor 1983). A summary of the protocol is presented in Figure 25.2. Unfor-

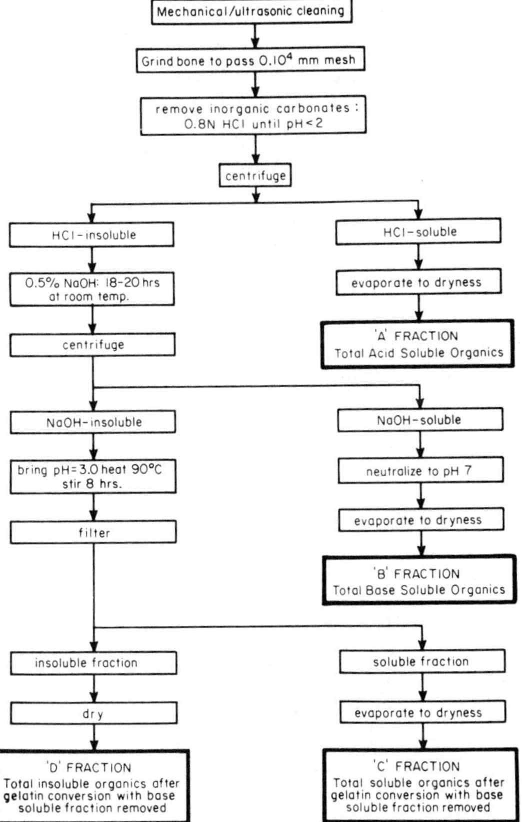

Fig 25.2. Procedure to prepare bone differential acid-, base- and water-solubility fractions for ^{14}C measurement (taken from Taylor 1987b: 60)

tunately, ideal conditions were rarely met since the organic yields from a number of the fractions in the vast majority of bones – given the amount of bone from archaeological contexts typically available for processing – were often insufficient to obtain meaningful ^{14}C analysis with decay counting.

RESULTS OF THE USE OF AMS TECHNOLOGY

As we have noted, one of the initial analytical problems involved in the critical examination of ^{14}C determinations on bone was that the yields of organic product from many samples – especially those from tropical environments and/or of Pleistocene age – are relatively low. With conventional ^{14}C decay counting technology, these low yields often raised difficulties, sometimes requiring adjustments in pretreatment procedures to reduce sample loss or even making it impractical to pursue ^{14}C measurements and often impeding an intensive investigation of the biogeochemistry of a bone sample using ^{14}C to ensure that indigenous bone fractions were identified and extracted. We have commented already that the first published approach to the dating of an organic fraction of bone was designed, in large part, to reduce the loss of the residual organics during pretreatment. The advent of accelerator (or atomic/ultra-high sensitivity) mass spectrometry essentially eliminated such problems and has permitted meaningful routine analyses to be carried out on sample sizes several orders of magnitude below that possible with conventional decay counting instrumentation.

Although the conceptual basis of AMS technology was first set forth in 1977 (Muller 1977; Nelson, Korteling & Scott 1977; Bennett *et al* 1977), it was not until about 1983 that reasonably routine operational status was achieved for ^{14}C analysis by several laboratories (Kutschera 1983). The technical features of the various types of AMS systems used for ^{14}C measurements have been extensively reviewed (eg, Litherland 1984, 1987; Hedges 1987; Linick *et al* 1989; Beukens 1992; Gove 1992). Taylor (1987a: 94) has listed the early general references along with comments on results and implications of the use of AMS ^{14}C technology in the measurement of archaeologically related samples (*cf* Hester 1987). Despite the initial expectation that AMS technology could be used to extend the ^{14}C time scale (eg, Muller 1977), the principal contribution of AMS ^{14}C analysis – particularly as it relates to bone ^{14}C studies – continues to be, first and foremost, its ability to obtain ^{14}C measurements on milligram – and, with some additional effort, submilligram – amounts of carbon. A less important capability of AMS technology is the ability to obtain meaningful ^{14}C measurements with counting times measured in minutes and hours rather than days and sometimes weeks. Currently, most ^{14}C analyses carried out on bone are conducted by investigators attached to AMS facilities (Hedges 1992; Hedges & van Klinken 1992; Long *et al* 1989).

Over the last decade, the use of AMS technology, combined with the application of a range of analytical and preparative biochemical techniques, has intensified the examination of the distribution of ^{14}C ages exhibited in various organic products contained in fossil bone. Studies carried out by several groups of investigators have provided a foundation on which to build a framework for understanding the interrelationship of various biogeochemical diagenetic processes, the sources and characteristics of contamination and the pattern of ^{14}C ages found in fossil bones.

Several reviews of the AMS ^{14}C data obtained on fossil bone over the last decade have reflected on the potential sources and effects of contamination in these samples. They have proposed, in the context of the AMS-derived data, various criteria to evaluate the likelihood that a given ^{14}C analysis would yield an accurate age estimate. For example, Taylor (1987c) proposed a protocol which – in addition to tests to positively detect, and, if present, remove organic preservative(s) applied (often without documentation) to bone – suggested, as a minimum requirement, the isolation of 1) a total HCl-insoluble, 2) a total amino-acid, and 3) a humic/fulvic-acid fraction. The basis of this suggestion was that the ^{14}C analysis on a total HCl-insoluble fraction would provide an age estimate on the "generic organics" in the bone whereas the humic- or fulvic-acid fraction ^{14}C value would permit a positive determination of the age of what is typically the principle organic contaminant. The degree of concordance of the total HCl-insoluble and total amino-acid fractions and the age separation between these two fractions (if any) and the humic/fulvic fraction would provide a quantitative basis on which to evaluate the distribution of ^{14}C activities in the major organic constituents of a bone.

It was also proposed that additional consideration should be given to the separation of one or more individual amino acids – the most likely being glycine. The isolation of hydroxyproline had been suggested as an optimum fraction of bone ^{14}C dating, because this amino acid has been thought to have a limited distribution other than as a constituent of collagen (Wand 1981; Stafford *et al* 1982; *cf* Taylor 1987b: 56) However, from a practical point of view, the use of hydroxyproline is severely limited even with the use of AMS technology due to the fact that it is not present in detectable amounts in many fossil bones. In addition, this amino acid, along with many others, has been detected in natural waters, sometimes in significant concentrations (Thurman 1985: 151–155; Long *et al* 1989).

The increasing number of ^{14}C determinations on bone using AMS technology has revealed to several research groups a pattern with important implications for procedures developed to guide the processing of bone samples for ^{14}C analysis. Based on the experiences of the Oxford AMS laboratory, Hedges and Law (1989) note that with bone, the issues and problems associated with sample pretreatment protocols will depend specifically on the amount of recoverable collagen contained in a bone specimen. The University of Arizona AMS group has noted that "the reality emerges that bone specimens for ^{14}C dating exist in a continuum of quality, ranging from well-preserved . . . to completely devoid of original amino acids, and consequently undatable" (Long *et al* 1989: 232). They have advanced a series of explicit acceptance/rejection criteria to determine which bones can and cannot be expected to yield accurate ^{14}C age estimates. These criteria include the proportion of the original collagen remaining, the degree to which the bone exhibits a collagen-like amino-acid pattern and the presence of anomalous concentrations of certain amino acids. As one means of quantifying the degree of collagen degradation, they have used a "glycine depletion ratio" (GDR) scale in which the ratio of glycine contained in a sample of modern bone (GDR = 1.0) is compared with that of a fossil bone sample (Long *et al* 1989: 234–235).

For bone samples containing significant quantities of well-preserved collagen, it is now generally agreed that appropriate physical and chemical pretreatment can, in most cases, effectively isolate and purify the residual collagen. At the Oxford AMS Laboratory, initial pretreatment procedures involve the isolation of a demineralized HCl-insoluble residue followed by the preparation of various fractions using base (NaOH) extraction, charcoal absorption and/or ion exchange and elution (Gillespie, Hedges & Wand 1984). Improvements to avoid contamination with amino-sugar reaction products have involved the initial extraction of collagen as gelatin after demineralization, following from the procedure of Longin (1971) but also including a base extraction step. This procedure has been recently semiautomated at the Oxford AMS facility by using flow cells, through which appropriate solutions are pumped in proper sequence under computer control (Law & Hedges 1989).

In addition, for reasonably well-preserved bone samples, various types of chromatographic methods have been used to remove humate compounds (eg, Stafford *et al* 1987; Stafford, Brendel & Duhamel 1988; Gillespie, Hedges & Humm 1986). Currently, for such bones, the Arizona AMS group routinely extracts the total amino-acid fraction (Long *et al* 1989: 234; *cf* van Klinken & Mook 1990). Another AMS group has recently modified the Longin method of collagen extraction by adding an ultrafiltration step designed to exclude low molecular weight species. Brown *et al* (1988) assumed that most of the exogenous contamination will be contained in this fraction. Another approach developed to purify collagen for stable isotope analyses involves the use of collagenase, which preferentially isolates peptides of known length from the surviving collagen fragments (DeNiro & Weiner 1988b).

In contrast to several techniques currently employed to purify collagen, one of the most serious challenges currently confronting investigators is that of obtaining reliable ^{14}C determinations on chemically and microbiologically degraded bone – those samples from which the collagen has been almost entirely removed from the bone matrix. A review of four different bone collagen preparation procedures at the University of Toronto AMS facility concluded that none of these techniques can eliminate contaminants in seriously degraded bone (Gurfinkel 1987). The Oxford Radiocarbon Accelerator Unit does not regard current methods of pretreating bone for ^{14}C dating "as sufficiently reliable for dating when bones have lost >95% of their protein." They note that "such a policy removes the possibility of dating 90% of the bone (with probable selection bias of the remaining 10%) recovered from sites within 40% of the equator" (Hedges & Law 1989). Although studies are currently underway by several groups, there is currently no consensus as to biogeochemical methods that can be routinely used in bones exhibiting very low or trace amounts of collagen to distinguish indigenous non-collagen proteins, amino acids, peptides and other products of collagen diagenesis from external contamination.

Recent measurements undertaken by Stafford *et al* (1987, 1990, 1991), examining the range of ^{14}C values exhibited in bones of varying degrees of preservation and denaturation of the collagen component have highlighted problems in the ^{14}C dating of collagen-deficient bone. Using a relatively well-preserved bone – *ie*, a bone exhibiting a collagen-like amino-acid profile – of a mammoth (Domebo *Mammuthus* sp) associated with a wood sample dated at 11,490 ± 450 ^{14}C years (AA-823), these researchers obtained ^{14}C measurements on a wide spectrum of organic fractions prepared by a variety of preparative methods. These fractions ranged from untreated gelatin, HCl-insoluble residues with and without gelatinization, ion-exchange-purified components, individual and combined amino acids, and fulvic acids. In the Stafford *et al* (1987) report, 9 fractions yielded ^{14}C ages within two sigma of the wood value, whereas 5 fractions exhibited still younger ages. With the exception of one organic fraction, which was prepared using a solvent extraction technique and fulvic acids, the youngest organic fraction was about 2000 (± 500) years younger than the actual age of the bone. The principal contaminant in the bone was identified as humic compounds (fulvic acids), which yielded ^{14}C ages of about 5000 ^{14}C years.

Stafford *et al* (1990) also obtained AMS ^{14}C measurements on a comparable series of fractions – 25 analyses on 14 different chemical fractions – from another mammoth (Escapule *Mammuthus* sp) bone exhibiting a non-collagen amino-acid composition pattern. The actual age of the mammoth was inferred to be approximately 11,000 BP on several lines of evidence. The ^{14}C age of individual amino acids extracted from the Escapule mammoth ranged from about 3000 to 4500 BP. The youngest ^{14}C value – 2270 ± 360 (AA-2660) – was obtained on a combined aliquot containing four amino acids: aspartic acid; glutamic acid; serine; and threonine. Extracted fulvic acid yielded ages ranging from about 7200–8800 BP. The oldest ^{14}C value obtained on the Escapule mammoth – 12,280 ± 110 BP (AA-2972) – was obtained on a fraction described as "nonhydrolyzable organic matter (humins)."

Based on data from these and three other bone samples (including human bones from two portions of the Del Mar skeleton), Stafford *et al* (1990: 114–118) suggested that ^{14}C analyses on organic fractions of non-collagenous bone samples – including total and individual amino acids – will significantly underestimate the actual age of the bone. In their view, these ages "will be a measure of the bones mean ^{14}C activity, not its true geologic age." They argue that the only means of evaluating ^{14}C dates on samples of trace amounts of protein material and a non-collagenous amino-acid pattern is to obtain ^{14}C determinations on the "fulvic, humic acid, and humin fractions" and compare them with ^{14}C measurements on individual and total amino acids. According to them, even if the fulvic/humic-acid fraction(s) and amino-acid extract ages are concordant, the concordant age should still be considered as a minimum age, which still could be thousands of years too young.

There now appears to be a general consensus among investigators concerning the reliability of bone ^{14}C values: 1) that where appropriate biochemical purification procedures are employed, accurate ^{14}C age estimates can be obtained on bones retaining significant amounts of intact collagen; and 2) that bones seriously depleted in their original protein (mostly collagen) content can yield seriously anomalous ^{14}C values.

Beyond Collagen: The Use of Non-Collagen Proteins

Noting the difficulty of isolating autochthonous organics in collagen-deficient bone for various purposes – eg, paleodietary inferences based on bone ^{13}C and ^{15}N data (De Niro & Weiner 1988a) – several researchers have suggested the use of non-collagen components to examine their usefulness for bone where collagen is seriously degraded and present in low or trace amounts. For example, Long *et al* (1989: 238) have suggested that "phosphoproteins may be protected from degradation because of their bonding to apatite structures in bone." Gillespie (1989: 240) noted the existence of osteocalcin, osteonectin and other phosphoproteins, proteoglycans, and glycoproteins as well as blood (eg, Nelson *et al* 1986) proteins such as hemoglobin, serum albumin and the immunoglobins as a potential source of material for ^{14}C dating of collagen-depleted bones (*cf* Masters 1987).

The first suite of ^{14}C measurements of a non-collagen protein in fossil bone – osteocalcin – has been reported by Ajie *et al* (1990). Also known in the biomedical literature as "Gla-containing protein," and "Bone Gla protein" (Termine 1988), osteocalcin was discovered in the laboratory of Peter Hauschka of the Harvard School of Dental Medicine in the early 1970s during a search for the source of the calcium-binding, vitamin-K-dependent amino acid, gamma-carboxyglutamic acid or GLA (Hauschka, Lian & Gallop 1975; Hauschka 1977; Hauschka & Gallop 1977).

The first published proposal to use osteocalcin for the dating of fossil bone was made by Hauschka (1980). He pointed to the following characteristics of this protein: 1) it is relatively abundant, comprising about 1% of the total bone protein and some 10%–20% of all non-collagenous proteins in bone; 2) it appears to bind tightly to hydroxyapatite, the major mineral component of bone. In this bound form, this protein has an excellent chance of being protected from biochemical degradation; 3) GLA has not been detected in many species of bacteria, plants and invertebrates. If this limited distribution is confirmed by additional study, major sources of diagenetic contamination would be eliminated; and 4) several techniques to isolate and purify osteocalcin have been developed.

Table 25.3 presents compositional and biogeochemical data in a series of fossil bone samples with ^{14}C ages ranging from a few thousand to in excess of 20,000 years, listed in order of their decreasing collagen content. From this suite of bone samples, collagen in the form of gelatin was prepared using procedures described by DeNiro & Epstein (1981) and Schoeninger & DeNiro (1984). The methods of extraction and purification of osteocalcin are outlined in Gundberg *et al* (1984) and Poser *et al* (1980). Although there is significant variation in collagen/gelatin content in these fossil bones, osteocalcin concentration varies in a very narrow range. This is interpreted as supporting the initial inference of the biochemical stability of the hydroxyapatite environment in which the osteocalcin is deposited.

In the first seven of the samples listed in Table 25.3, the C/N, HYP concentration, and GLY/GLU values of the collagen fractions indicate that the collagen extract in the form of gelatin has undergone minimal denaturation even though its concentration in the bone is significantly reduced. However, for the bone

TABLE 25.3. Biogeochemical and ^{14}C Data on Gelatin and Osteocalcin Fractions of Fossil Human and Moa Bone*

	Gelatin fraction					Osteocalcin fraction		
Bone type	%wt	C/N**	HYP†	GLY/ GLU‡	^{14}C age (yrs BP)	%wt	C/N**	^{14}C age (yrs BP)
Human	11.9	3.1	85	4.7	3845 ± 60	.24	4.2	3855 ± 93
Human	11.6	3.1	86	4.6	7260 ± 90	.25	4.3	9090 ± 120
Moa	11.0	3.4	86	4.7	15,000 ± 200	.31	3.8	15,275 ± 230
Moa	10.5	3.4	86	4.7	18,650 ± 250	.31	3.8	18,780 ± 260
Moa	9.9	3.4	85	4.7	10,490 ± 88	.31	3.8	11,050 ± 110
Human	4.4	3.1	85	4.7	3630 ± 100	.25	4.2	3661 ± 100
Moa	2.2	5.7	86	4.6	24,510 ± 250	.31	3.8	24,965 ± 220
Human	0.2	11.1	59	3.4	1720 ± 550	.25	4.2	3260 ± 120

*Adapted from Ajie *et al* (1990) and reprinted by permission of the authors and Elsevier Science Publishers. **Elemental carbon/nitrogen ratios; †Hydroxyproline composition in residues/1000; ‡Glycine/glutamic acid ratios

Table 25.4. Major Non-Collagen Proteins*

Protein	Other names used
Osteonectin	"Secreted protein, acidic and rich in cysteine" (SPARC)
Osteocalcin	"Gla-containing proteins," "Bone Gla protein" (BGP), "Matrix Gla protein"
Proteoglycans	"Ground substance" and "mucopolysaccharides"
Osteopontin	"Sialoprotein I"
Sialoprotein[II]	"Bone sialoprotein"

*>1% by weight: based on Termine (1988)

sample exhibiting less than 1% residual collagen, the HYP concentration has been significantly reduced, the C/N and GLY/GLU ratios diverge significantly from the modern collagen pattern, and there are significant discordances in the collagen and osteocalcin fractions. Future studies are planned that will include an examination of ^{14}C measurements on total amino-acid and osteocalcin extracts of known-age bones exhibiting both collagen and non-collagen amino-acid profiles, along with a demonstration of the purity of the osteocalcin extracts. In the next section, the results of ^{14}C analysis on osteocalcin extracts from a series of human skeletons will be presented.

It should be noted that osteocalcin is only one of a number of non-collagen organic components of bone which have been found to exist in amounts greater than 1% by weight in modern bone (Table 25.4). Each component merits investigation to determine the degree to which a given non-collagen protein extract retains isotopic integrity in situations where moderate-to-severe diagenetic effects have eliminated collagen from the bone matrix.

CASE STUDY: RADIOCARBON DATING OF NEW WORLD HUMAN SKELETONS

Radiocarbon age determinations on human bone from sites in the western hemisphere have assumed a unique place in the torturous history of one of the most contentious and long-standing debates in American archaeology – the timing of the initial peopling of the New World. A relatively large corpus of ^{14}C determinations places the archaeologically well-documented Paleoindian Clovis tradition in North America between about 10,900 and 11,300 ^{14}C years BP (Haynes 1992). Of the more than 100 major sites in North America with purported evidence of a "pre-Clovis" occupation, only a very small percentage remain under active consideration. For the remaining alleged pre-Clovis sites in North (Payen 1982; Dincauze 1984) and South (Lynch 1990) America, either the cultural character of the material remains or the adequacy of the geochronological data associated with the remains – or both – have been questioned (Stanford 1982; Owen 1984; Irving 1985; Bonnichsen *et al* 1987).

Because of the controversy surrounding any proposed pre-Clovis occurrence, there would seem to be an obvious advantage to directly date human bone. Any question concerning the human attribution of the material dated and its associations would be totally satisfied when dating is carried out on a bone sample clearly identified, on widely accepted morphological grounds, as *Homo sapiens* (Taylor 1991).

California Paleoindians

The 1970s saw the application of several of the chemical pretreatment procedures, which had been developed over the previous ten-year period, to human bone samples associated with contexts of purported late Pleistocene age from New World localities. Two of these samples – the Laguna (south coastal southern California) and Los Angeles (Baldwin Hills section of western Los Angeles) human skeletons – figured prominently in discussions on the timing of the peopling of the New World during this decade. These data have been summarized under [A] in Table 25.5 along with results of the application of two other Quaternary dating methods – amino-acid racemization (AAR) and uranium series (U series) – applied to other alleged Pleistocene-age New World human bone samples (Taylor *et al* 1985; Stafford *et al* 1990, 1991) – and, in one instance, to an artifact fabricated from bone recovered from Old Crow in the Yukon – both purporting to document late Pleistocene occurrences of human populations in the New World (Nelson *et al* 1986).

The published ^{14}C ages assigned to the Laguna skull (17,150 ± 1470 BP: UCLA-1233A) and Laguna long-bone fragments (>14,800 ^{14}C BP: UCLA-1233B) were obtained on acid-insoluble fractions after NaOH treatment following initial extractions with ether (Berger & Libby 1969). The age of >23,000 BP (UCLA-1430) had been obtained on an amino-acid fraction from a portion of the skull of the Los Angeles skeleton. The amino-acid extract had been obtained using the same procedure applied to the *Smilodon* bones from the La Brea tar pit samples (Berger *et al* 1971). The finite ^{14}C age obtained on the Laguna bone had also been used as the calibration sample for determining an aspartic-acid racemization rate constant (k_{asp}) for the application of the AAR method to a series of human bone from California. These included the Del Mar skeleton from the San Diego region of coastal southern California and, adjusted to a different temperature regime, the Sunnyvale human skeleton from the San Francisco Bay region of California (Bada, Schroeder & Carter 1974; Bada & Helfman 1975). The inferred AAR ages on these human skeletons ranged from over 40,000 to about 70,000 years.

TABLE 25.5. Summary of revisions in age estimates on bone of purported Pleistocene age based on ^{14}C determinations. Data for Arizona AMS measurements on Del Mar skeleton taken from Stafford *et al* (1990: Table 3), for Laguna from Berger (1991), for Natchez from Cotter (1991) and for Tepexpan from Stafford *et al* (1991: 52). Other data taken from Taylor *et al* (1985) and Brooks *et al* (1991; Table 1) where original references are listed. Except for the Old Crow sample, all bones have been identified as human skeletal material.

Skeleton(s)/ artifact	[A] Original estimate		[B] Revised estimate	
	Basis	Age	^{14}C age	Laboratories
Sunnyvale	AAR	70,000	3600–4850	UCR/Arizona AMS
	U-series	8300/9000	6300*	UCSD (Scripps)/Oxford AMS
Haverty	AAR	>50,000	4050–5350**	UCR
[Angeles Mesa]			5200**	GX (Geochron)
			7900–10,500**	UCLA
			2730–4630[†]	UCR/LLNL-CAMS AMS
			4600–13,500[‡]	UCR/LLNL-CAMS AMS
			5250[†]	DSIR, New Zealand AMS
			15,900[‡]	DSIR, New Zealand AMS
Del Mar	AAR	41,000–48,000	4900*	UCSD (Scripps)/Oxford AMS
	U-series	11,000/11,300		
			4830	Arizona AMS
			1150–5060*	Arizona AMS
Los Angeles	^{14}C	>23,000*	3560	UCR/Arizona AMS
[Baldwin Hills]	AAR	26,000		
Taber	Geologic	22,000–60,000	3550	Chalk River AMS
Yuha	^{14}C	22,000[§]	1650–3850	Arizona AMS
	AAR	23,000		
	U-series	5800		
Old Crow[‖]	^{14}C	23,000	1350	Simon Fraser/McMaster AMS
Laguna	^{14}C	7100[#]	5100	UCSD (Scripps)/Oxford AMS
		17,150[¶]		
		>14,800[@]		
Natchez	Geologic	"Pleistocene"	5580	Arizona AMS
Tepexpan	Geologic	"Late Pleistocene"	1980	Arizona AMS

*Amino-acid fraction; **Acid-insoluble fraction, decay counting; [†]Collagen (gelatin) fraction; [‡]Osteocalcin fraction; [§]Diagenetic carbonate; [‖]Artifact fabricated from bone; [#]Laguna skull, "first run" (Berger 1992); [¶]Laguna skull, "second run" (Berger 1992); [@]Laguna long-bone fragment.

The immediate reaction both to the ^{14}C- and AAR-based age estimates on these samples by academic American prehistorians – although few found their way into print – ranged from puzzlement to rejection (eg, Taylor & Payen 1979; Meighan 1978, 1983). Willey (1978: 528) probably expressed a widespread view among professional archaeologists at the time when he commented that " . . . the collagen dating of human skeletons is suggestive – but no more – of very early habitation in the area [California]." The principal problem for many archaeologists was that most of the skeletal samples lacked archaeological context. Because of the general absence of artifacts or definitive stratigraphic or contextual data associated with the purported Pleistocene-age human skeletons, the position that probably best characterized the view of most archaeologists on both these and other purported "pre-Clovis" materials was agnosticism. The evidence was generally considered inconclusive.

Radiocarbon Applications

The first major archaeological issue that used AMS-based ^{14}C age determinations addressed the validity of the Pleistocene age inferences obtained by AAR and U-series measurements on human bone from several California localities. The Sunnyvale skeleton was chosen for initial examination because a major, unambiguous temporal discrepancy existed between the AAR-inferred age and ages inferred on several other lines of evidence. These included the ^{14}C-dated geological context, the archaeological context of the feature from which the skeleton had been recovered and considerations based on skeletal morphology. The latter indicated that, from an anatomical perspective, the skeleton was fully modern and indistinguishable from a population of skeletons recovered from well-documented late Holocene archaeological sites in the San Francisco Bay region. With the exception of the AAR-inferred age, all the other evidence pointed to an early-to-middle Holocene age (Gerow 1981). If the 70,000 BP age was even approximately correct, the Sunnyvale hominid would have been one of the oldest directly dated *Homo sapiens sapiens* skeletons, not just in the western hemisphere, but in the whole world. It would also have been necessary to explain why a human population had undergone essentially no morphological change over a 70,000-year period.

Four ^{14}C measurements – one by decay counting and three by AMS direct counting – were obtained on three organic fractions from post-cranial Sunnyvale bone. These fractions included an "A," "B" and "D" organic fraction defined according to the procedure outlined in Figure 25.2. As summarized in Table 25.5, the AMS ^{14}C values ranged from 3600 ± 600 (UCR-1437A/AA-50) to 4850 ± 400 BP (UCR-1437B/AA-52). A decay-counting determination on UCR-1437A yielded an age of 4390 ± 150 BP (Taylor 1983; Taylor *et al* 1983). A total amino-acid fraction of a Sunnyvale bone prepared by another investigator and measured by the Oxford AMS laboratory, produced an age of 6300 ± 400 BP (OxA-187: Bada *et al* 1984).

Several observations can be made on this application of AMS ^{14}C analysis to Sunnyvale bone samples. First, the range in all of ^{14}C ages is about 2700 (± 500) years. Such a range of values could be interpreted to indicate that the procedures employed yielded unreliable ^{14}C age estimates. Such an observation would need to be evaluated within the context of the specific question that these experiments were intended to address. The question that generated the first set of measurements was simple and specific: Was the Sunnyvale skeleton of early terminal Pleistocene age (*ie*, 40,000–70,000 BP) as suggested by the AAR data, or was it early-to-middle Holocene (*ie*, 4500–10,000 BP), as suggested by all of the other data? A range of 2700 or even 5000 years in the results was sufficient to provide an appropriate answer in the context of the initial question.

Similar comments can be advanced for all but one of the ^{14}C values obtained on the human skeletal and bone artifact samples listed in Table 25.5. The fundamental archaeological and paleoanthropological question again was: Are the human skeletons Pleistocene or Holocene specimens? Based on these data, the conclusion of Taylor *et al* (1985) was that all such human bone samples in the New World directly dated by ^{14}C appear to be of Holocene age. This conclusion continues to be supported by an AMS-based ^{14}C age determination obtained on a portion of the pelvis of "Natchez Man," a specimen found in 1846 near Natchez, Mississippi. Because of the reported association with late Pleistocene megafauna, the Natchez pelvis had previously been considered by some to be possibly of Pleistocene age. The ^{14}C age of the specimen, which exhibits a good glycine depletion ratio as defined by the Arizona AMS laboratory, is 5580 ± 80 ^{14}C yrs BP (AA-4051: Cotter 1991).

Haverty Human Skeletons

A suite of conventional and AMS-based ^{14}C values obtained on a group of human skeletons from the Haverty [Angeles Mesa] site in the western portion of the City of Los Angeles, California currently constitutes the only exception to the conclusion that all human skeletons in the western hemisphere directly dated by ^{14}C appear to be of Holocene age. The Haverty skeletons were recovered in 1924 and consist of at least 8 individuals – 3 males, 3 females and 2 sub-adults of indeterminate sex. Several of these individuals are represented by nearly complete cranial and post-cranial materials, an extremely rare occurrence in pre-contact Native American remains. The number of individuals along with the combination of adolescents and adults, both male and female, and the reported presence of artifacts, suggest a burial locale and thus a similar (± 100 yr) age. Table 25.5 includes a summary of a suite of 17 ^{14}C values (6 by conventional decay counting and 11 by direct/AMS analysis) on 5 of the Haverty skeletons obtained on different bone fractions by 3 decay-counting laboratories and 3 AMS facilities (Brooks *et al* 1991).

Attention was directed to the Haverty skeletal collection in the mid-1970s as a result of a suggestion that at least one of the skeletons exhibited an age of >50,000 yrs. This age was inferred on the basis of a relatively high (>0.5) D/L_{asp} ratio reported on a sample of Haverty bone (Taylor 1983: 651). This assignment of age was contradicted by the initial series of ^{14}C values obtained by two laboratories (UCR and Geochron) employing decay counting on various organic fractions of bones from two of the skeletons. The ages obtained initially ranged from 4050 ± 100 (UCR-1568A) on Haverty Individual 4 to 5350 ± 150 (UCR-1349A) on Haverty Individual 1. Subsequently, measurements carried out by the UCLA laboratory yielded significantly older ages. However, the relatively large statistical errors associated with both of these values – 10,500 ± 2000 (UCLA-1924) and 7900 ± 1440 (UCLA-1924B) – made it difficult to compare with the previously obtained values.

The range in ages called into question the assumption that all of the skeletons represented essentially a single temporal unit. Also, the possibility existed that a portion of the observed range in ages might be, in part, attributable to variations in the degree of collagen preservation. Nitrogen and amino-acid composition analysis carried out on the ^{14}C-dated skeletons did indicate significant variation in the degree of preservation of the collagen component in different Haverty skeletons. These variations are also reflected in the D/L_{asp} ratios exhibited in the Haverty bones with increasing D/L_{asp} ratios correlated with decreasing nitrogen concentrations. This effect is now well documented at several other sites (Taylor *et al* 1989).

Bones from four of the Haverty skeletons were used to provide a preliminary examination of the effects of variability in collagen preservation on the accuracy of ^{14}C values obtained from osteocalcin extracts from fossil bone. Results from these analyses are listed in Table 25.6. Very good concordance in the ^{14}C ages of the gelatin and osteocalcin fractions, and agreement with the initial UCR and Geochron ^{14}C values was obtained in the case of Haverty 1 – the bones of which exhibited a collagen-like amino-acid profile, nitrogen concentration and C/N ratio. However, as the gelatin yield decreased and the nitrogen, GLY/GLU and C/N ratios reflected progressive denaturation of the collagen, there was a tendency for the amount of discordance in the gelatin/osteocalcin ^{14}C ages to increase with one of the osteocalcin ^{14}C values indicating an age of almost 16,000 BP. If the osteocalcin ^{14}C determinations on Haverty Skeletons 4 and 5

TABLE 25.6. Comparisons between collagen/gelatin and osteocalcin ^{14}C data on human bones from the Haverty skeletal collection (based on Ajie *et al* 1990 and Brooks *et al* 1991)

Haverty skeleton	Gelatin fraction N_a*	GLY/GLU**	wt%†	C/N‡	^{14}C age	Osteocalcin fraction ^{14}C age
1	1.1	4.06	12.9	3.3	5260 ± 520 (UCR-3082/CAMS-442)	5540 ± 230§ (UCR-3083/CAMS-439)
2	<0.01	2.06	1.3	8.7	2730 ± 190 (UCR-3084/CAMS-445)	4630 ± 260 (UCR-3087/CAMS-438)
4	0.02	2.23	1.2	9.9	3870 ± 350 (UCR-3086/CAMS-440)	12,600 ± 460 (UCR-3088/CAMS-440) 13,500 ± 220‖ (UCR-3090/CAMS-434)
			0.8	18.3	5250 ± 90 (HA-104A/DSIR)	15,900 ± 250 (HA-104B/DSIR)
5	0.38	3.91	0.8	18.3	4710 ± 190 (UCR-3089/CAMS-441)	11,960 ± 500 (UCR-3085/CAMS-437)

*Amino-acid nitrogen content; **GLY = Glycine; GLU = Glutamic acid; †Weight percent of gelatin extracted from bone sample; ‡Carbon/nitrogen ratio. The C/N ratio of modern bone is about 3.0 (Hare & von Endt 1990); §Decay counting ^{14}C analysis on Haverty Skeleton 1: 5350 ± 150 (UCR-1349A: total acid-soluble fraction) and 5280 ± 180 (UCR-1349D: total acid-insoluble fraction after gelatin conversion with base-soluble fraction removed). ‖Duplicate ^{14}C analysis from split of CO_2 from UCR-2088/CAMS-433).

reflect actual ages, these skeletons would represent the oldest human remains in the western hemisphere.

At the present time, evidence seems to be insufficient to determine which of several possible explanations might account for the discordance in the ^{14}C values of the gelatin/osteocalcin fractions obtained on three of the skeletons. The view that all of the Haverty skeletons are essentially the same age may be in error. It is also possible that extraction procedures employed introduce variable amounts of contamination and/or cross-contamination occurs between the osteocalcin, other proteins, and most seriously, from soil organic contaminants. Studies to examine these questions are currently underway.

CONCLUSION

All investigators agree that bone can potentially be a very difficult sample type with which to work – requiring great attention to detail in sample pretreatment and preparation. The capabilities of AMS technology have provided researchers with a very powerful tool, making it possible to develop a detailed ^{14}C "map" of various types of carbon-containing compounds in a bone sample. An important goal of such studies will be to develop a preparative methodology which would accurately identify organics indigenous to a bone, irrespective of its diagenetic history. Currently, there is a general consensus on the ability to obtain relatively accurate ^{14}C age estimates on bone that still retains a significant amount of intact collagen. The future of bone ^{14}C research will be to demonstrate a capability to obtain accurate ^{14}C age determinations on bone samples characterized by trace amounts of extractable collagen.

ACKNOWLEDGMENTS

Studies of the UCR Radiocarbon Laboratory reported in this chapter were supported by the National Science Foundation (Anthropology/Archaeometry Program) and the Accelerator Mass Spectrometry Laboratory, Lawrence Livermore National Laboratory (LLNL), with additional resources provided by UCR Chancellor, Rosemary S J Schraer. Comments of P E Hare on an earlier draft of this chapter are very much appreciated as are the very helpful editorial suggestions of Renee Kra. Thomas W Stafford, Jr pointed out the reference to Thurman (1985) and made other suggestions. This is contribution 90-26 of the Institute of Geophysics and Planetary Physics, University of California, Riverside.

REFERENCES

Ajie, H, Kaplan, IR, Slota, PJ, Jr and Taylor, RE 1990 AMS radiocarbon dating of bone osteocalcin. *In* Yiou, F and Raisbeck, GM, eds, Proceedings of the 5th International Symposium on Accelerator Mass Spectrometry. *Nuclear Instruments and Methods* B52(3,4): 433–437.

Arnold, JR and Libby, WF 1950 *Radiocarbon Dates (September 1, 1950)*. Chicago: University of Chicago, Institute for Nuclear Studies.

_____1951 Radiocarbon dates. *Science* 113: 111–120.

Bada, JL, Gillespie, R, Gowlett, JAJ and Hedges, REM 1984 Accelerator mass spectrometry radiocarbon ages of amino acid extracts from Californian paleoindian skeletons. *Nature* 312: 442–444.

Bada, JL and Helfman, PM 1975 Amino acid racemization dating of fossil bones. *World Archaeology* 7: 160–183.

Bada, JL, Schroeder, R and Carter, GF 1974 New evidence for the antiquity of man in North America deduced from aspartic acid racemization. *Science* 184: 791–793.

Barendsen, GW, Deevey, ES and Gralenski, LJ 1957 Yale natural radiocarbon measurements III. *Science* 126: 908–919.

Barker, H and Mackey, J 1961 British Museum natural radiocarbon measurements III. *Radiocarbon* 3: 39–45.

Bennett, CL, Beukens, RP, Clover, MR, Gove, HE, Liebert, RB, Litherland, AE, Purser, KK and Sondheim, WE 1977 Radiocarbon dating using accelerators: Negative ions provide the key. *Science* 198: 508–509.

Berger, R 1992 Libby's UCLA radiocarbon laboratory: Contributions to archaeology. *In* Taylor, RE, Long, A and Kra, RS, eds, *Radiocarbon After Four Decades: An Interdisciplinary Perspective*. New York, Springer-Verlag, this volume.

Berger, R, Horney, AG and Libby, WF 1964 Radiocarbon dating of bone and shell from their organic components. *Science* 144: 999–1001.

Berger, R and Libby, WF 1969 UCLA radiocarbon dates IX. *Radiocarbon* 11(1): 194–209.

Berger, R, Protsch, R, Reynolds, R, Rozaire, R and Sackett, JR 1971 New radiocarbon dates based on bone collagen of California Paleoindians. *Contributions of the University of California Archaeological Research Facility* 12: 43–49.

Beukens, RP 1992 Accelerator mass spectrometry: Background, precision and accuracy. *In* Taylor, RE, Long, A and Kra, RS, eds, *Radiocarbon After Four Decades: An Interdisciplinary Perspective.* New York, Springer-Verlag, this volume.

Bonnichsen, R, Stafford, D and Fastook, JL 1987 Environmental change and development history of human adaptive patterns: the Paleoindian case. *In* Ruddiman, WF and Wright, HE, Jr, eds, *North America and Adjacent Oceans During the Last Deglaciation.* Boulder, Geological Society of America: 403–424.

Brooks, S, Brooks, RH, Kennedy, GE, Austin, J, Firby, JR, Payen, LA, Prior, CA, Slota, PJ, Jr, and Taylor, RE 1991 The Haverty human skeletons: Morphological, depositional and geochronological characteristics. *Journal of California and Great Basin Anthropology*, in press.

Brown, FH 1988 Geochronometry. *In* Tattersall, I, Delson, E and Van Couvering, J, eds, *Encyclopedia of Human Evolution and Prehistory.* New York, Garland Publishing: 222–225.

Brown, TA, Nelson, DE, Vogel, JS and Southon, JR 1988 Improved collagen extraction method by modified Longin method. *Radiocarbon* 30(2): 171–177.

Cotter, JL 1991 Update on Natchez Man. *American Antiquity* 56: 36–39.

DeNiro, MJ and Epstein, S 1981 Influence of diet on the distribution of nitrogen isotopes in animals. *Geochimica et Cosmochimica Acta* 45: 341–351.

DeNiro, MJ and Weiner, S 1988a Chemical, enzymatic and spectroscopic characteriztion of "collagen" and teeth organic fraction from prehistoric bones. *Geochimica et Cosmochimica Acta* 52: 2197–2206.

_____1988b Use of collagenase to purify collagen from prehistoric bones for stable isotope analysis. *Geochimica et Cosmochimica Acta* 52: 2425–2432.

Dincauze, DF 1984 An archaeological evaluation of the case of Pre-Clovis occupations. *In* Wendorf, F and Close, A, eds, *Advances in World Archaeology.* New York, Academic Press, 3, 275–323.

El-Daoushy, MFAF, Olsson, IU and Oro, FH 1978 The EDTA and HCl methods of pretreating bones. *Geologiska Föreningens i Stockholm Förhandlingar* 100: 213–219.

Evin, J, Longin, R, Marien, G and Pachiaudi, Ch 1971 Lyon natural radiocarbon measurements II. *Radiocarbon* 13(1): 52–73.

Gerow, BA 1981 Amino acid dating and early man in the New World: A rebuttal. *Society for California Archaeology Occasional Papers* 3: 1–11.

Gillespie, R 1989 Fundamentals of bone degradation chemistry: Collagen is not "the way." *In* Long, A and Kra, RS, eds, Proceedings of the 13th International ^{14}C Conference. *Radiocarbon* 31(3): 239–246.

Gillespie, R and Hedges, REM 1983 Sample chemistry for the Oxford high energy mass spectrometer. *In* Stuiver, M and Kra, RS, eds, Proceedings of the 11th International ^{14}C Conference. *Radiocarbon* 25(2): 771–774.

Gillespie, R, Hedges, REM and Humm, MJ 1986 Routine AMS dating of bone and shell proteins. *In* Stuiver, M and Kra, RS, eds, Proceedings of the 12th International ^{14}C Conference. *Radiocarbon* 28(2A): 451–456.

Gillespie, R, Hedges, REM and Wand, JO 1984 Radiocarbon dating of bone by accelerator mass spectrometry. *Journal of Archaeological Sciences* 11: 165–170.

Gove, HE 1992 The history of AMS, its advantages over decay counting: Applications and prospects. *In* Taylor, RE, Long, A and Kra, RS, eds, *Radiocarbon After Four Decades: An Interdisciplinary Perspective.* New York, Springer-Verlag, this volume.

Grant-Taylor, TL 1971 Contamination of radiocarbon samples. *Radiocarbon Users*

Conference. Department of Scientific and Industrial Research, Wellington, New Zealand.

Gundberg, CM, Hauschka, PV, Lian, JB and Gallop, PM 1984 Osteocalcin: Isolation, characterization and detection. *Methods in Enzymology* 107: 516–544.

Gurfinkel, DM 1987 Comparative study of the radiocarbon dating of different bone collagen preparations. *Radiocarbon* 29(1): 45–52.

Haas, H and Banewics, JJ 1980 Radiocarbon dating of bone apatite using thermal release of CO_2. *In* Stuiver, M and Kra, RS, eds, Proceedings of the 10th International ^{14}C Conference. *Radiocarbon* 22(2): 537–544.

Haas, H, Holliday, V and Stuckenrath, R 1986 Dating of Holocene stratigraphy with soluble and insoluble organic fractions at the Lubbock Lake archaeological site, Texas: An ideal case study. *In* Stuiver, M and Kra, RS, eds, Proceedings of the 12th International ^{14}C conference. *Radiocarbon* 28(2A): 473–485.

Hare, PE and von Endt, D 1990 Variable preservation of organic matter in fossil bone. *1989 Yearbook of the Carnegie Institution of Washington*. Washington DC, Carnegie Institution of Washington.

Hassan, AA 1976 (ms) Geochemical and mineralogical studies on bone material and their implications for radiocarbon dating. PhD dissertation. Southern Methodist University, Dallas, Texas.

Hassan, AA, Termine, JD and Haynes, CV, Jr 1977 Mineralogical studies on bone apatite and their implications for radiocarbon dating. *Radiocarbon* 19(3): 364–384.

Hauschka, PV 1977 Quantitative determination of gamma-carboxyglutamic acid in proteins. *Analytical Biochemistry* 80: 212–223.

_____1980 Osteocalcin: A specific protein of bone with potential for fossil dating. *In* Hare, PE, Hoering, TC and King, K, eds, *Biogeochemistry of Amino Acids*. New York, Wiley Interscience: 75– 82.

Hauschka, PV and Gallop, PM 1977 Purification and calcium-binding properties of osteocalcin, the gamma carboxy glutamate-containing protein of bone. *In* Wasserman, RH, ed, *Calcium Binding Proteins and Calcium Functions*. Amsterdam, Elsevier/ North Holland: 338–347.

Hauschka, PV, Lian, JB, Cole, DEC and Gundberg, CM 1989 Osteocalcin and matrix Gla protein: Vitamin K-dependent proteins in bone. *Physiological Review* 69: 990–1047.

Hauschka, PV, Lian, JB and Gallop, PMN 1975 Direct identification of the calcium-binding amino acid gamma-carboxy-glutamate in mineralized tissue. *Proceedings of the National Academy of Science* (USA) 72: 3925–3929.

Haynes, CV, Jr 1967 Bone organic matter and radiocarbon dating. In *Radiocarbon Dating and Methods of Low-level Counting*. Vienna, IAEA, 163–167.

_____1968 Radiocarbon: Analysis of inorganic carbon of fossil bone and enamel. *Science* 161: 687–688.

_____1992 Contributions of radiocarbon dating to the geochronology of the peopling of the New World. *In* Taylor, RE, Long, A and Kra, RS, eds, *Radiocarbon After Four Decades: An Interdisciplinary Perspective*. New York, Springer-Verlag, this volume.

Hedges, REM 1987 Radiocarbon dating by accelerator mass spectrometry: Some recent results and applications. *Philosophical Transactions of the Royal Society of London* A323: 57–73.

_____1992 Sample treatment strategies in radiocarbon dating. *In* Taylor, RE, Long, A and Kra, RS, eds, *Radiocarbon After Four Decades: An Interdisciplinary Perspective*. New York, Springer-Verlag, this volume.

Hedges, REM and Law, IA 1989 The radiocarbon dating of bone. *Applied Geochemistry* 4: 249–253.

Hedges, REM and van Klinken, GJ 1992 A review of current approaches in the pretreatment of bone for radiocarbon dating by AMS. *In* Long, A and Kra, RS, eds, Proceedings of the 14th International

^{14}C Conference. *Radiocarbon* 34, in press.

Hester, JJ 1987 The significance of accelerator dating in archaeological method and theory. *Journal of Field Archaeology* 14: 445–451.

Ho, TY, Marcus, LG and Berger, R 1969 Radiocarbon dating of petroleum-impregnated bone from tar pits at Rancho La Brea, California. *Science* 164: 1051–1052.

Holliday, VT and Johnson, E 1986 Re-evaluation of the first radiocarbon age for the Folsom culture. *American Antiquity* 51: 332–338.

Irving, WN 1985 Context and Chronology of Early Man in the Americas. *Annual Review of Anthropology* 14: 529–555.

Irving, WN and Harrington, CR 1973 Upper Pleistocene radiocarbon-dated artifacts from the Northern Yukon. *Science* 179: 335–340.

Klinken, GJ van and Mook, WG 1990 Preparative high-performance liquid chromatographic separation of individual amino acids derived from fossil bone collagen. *Radiocarbon* 32(2): 155–164.

Krueger, HW 1965 The preservation and dating of collagen in ancient bones. *In* Chatters, RM and Olson, EA, eds, *Proceedings of the 6th International Conference on ^{14}C and Tritium Dating*. Springfield, Clearing House for Federal Science and Technical Information: 332–337.

Kutschera, W 1983 Accelerator mass spectrometry: From nuclear physics to dating. *In* Stuiver, M and Kra, RS, eds, Proceedings of the 11th International ^{14}C Conference. *Radiocarbon* 25(2): 677–691.

Law, IA and Hedges, REM 1989 A semi-automated bone pretreatment system and the pretreatment of older and contaminated samples. *In* Stuiver, M and Kra, RS, eds, Proceedings of the 13th International ^{14}C Conference. *Radiocarbon* 31(3): 247–253.

Libby, WF 1952 *Radiocarbon Dating*. Chicago, University of Chicago Press.

_____1955 *Radiocarbon Dating*, 2nd edition. Chicago, University of Chicago Press.

Linick, TW, Damon, PE, Donahue, DJ and Jull, AJT 1989 Accelerator mass spectrometry: The new revolution in radiocarbon dating. *Quaternary International* 1: 1–6.

Litherland, AE 1984 Accelerator mass spectrometry. *Nuclear Instruments and Methods* 233(B5): 100–108.

_____1987 Fundamentals of accelerator mass spectrometry. *Philosophical Transactions of the Royal Society of London* A323: 5–19.

Long, A, Wilson, AT, Ernst, RD, Gore, BH and Hare, PE 1989 AMS radiocarbon dating of bones at Arizona. *In* Long, A and Kra, RS, eds, Proceedings of the 13th International ^{14}C Conference. *Radiocarbon* 31(3): 231–238.

Longin, R 1971 New method of collagen extraction for radiocarbon dating. *Nature* 230: 241–242.

Lynch, TF 1990 Glacial-age man in South America: A critical review. *American Antiquity* 55: 12–36.

Masters, RM 1987 Preferential preservation of noncollagenous protein during bone diagenesis: Implications for chronometric and stable isotopic measurements. *Geochimica et Cosmochimica Acta* 51: 3209–3214.

May, I 1955 Isolation of organic carbon from bones for ^{14}C dating. *Science* 121: 508–509.

Meighan, CW 1978 California. *In* Taylor, RE and Meighan, CW, eds, *Chronologies in New World Archaeology*. New York, Academic Press: 223–240.

_____1983 Early man in the New World. *In* Masters, PM and Flemming, NC, eds, *Quaternary Coastlines and Marine Archaeology: Towards a Prehistory of Landbridges and Continental Shelves*. New York, Academic Press: 441–462.

Muller, RA 1977 Radioisotope dating with a cyclotron. *Science* 196: 489–494.

Nelson, DE, Korteling RG and Scott, WR 1977 Carbon-14: Direct detection at natural concentrations. *Science* 198: 507–508.

Nelson, DE, Morland, RE, Vogel, JS, Southon, JR and Harrington, CR 1986 New radiocarbon dates on artifacts from the northern Yukon Territory: Holocene not upper Pleistocene in age. *Science* 232: 749–751.

Oakley, KP 1963 Dating skeletal material. *Science* 140: 488.

Olson, EA (ms) 1963 The problem of sample contamination in radiocarbon dating. Unpublished PhD dissertation, Columbia University, New York.

Olson, EA and Broecker, WS 1961 Lamont natural radiocarbon measurements VII. *Radiocarbon* 3: 141–175.

Olsson, IU, El-Daoushy, MFAF, Abd-El-Mageed, AI and Klasson, M 1974 A comparison of different methods of pretreatment of bones. *Geologiska Föreningens i Stockholm Förhandlingar* 96: 171–181.

Owen, RC 1984 The Americas: The case against an ice-age human population. *In* Smith, FH and Spencer, F, eds, *The Origins of Modern Humans: A World Survey of the Fossil Evidence*. New York, Alan R Liss, Inc: 517–563.

Payen, LA (ms) 1982 The pre-Clovis of North America: Temporal and artifactual evidence. Unpublished PhD dissertation, University of California, Riverside.

Polach, H 1971 Radiocarbon dating of bone organic and inorganic matter. In *Radiocarbon Users Conference*. Wellington, New Zealand, Department of Scientific and Industrial Research: 165–181.

Poser, JW, Esch, F, Ling, NC and Price, PA 1980 Isolation and sequence of the vitamin K-dependent protein from human bone. *Journal of Biological Chemistry* 225: 8685–8691.

Price, TD, ed 1989 *The Chemistry of Prehistoric Human Bone*. Cambridge, Cambridge University Press.

Ralph, EK 1959 University of Pennsylvania radiocarbon dates III. *American Journal of Science Radiocarbon Supplement* 1: 45–58.

Rogers, RA and Martin, LD 1984 The 12 Mile Creek site: A reinvestigation. *American Antiquity* 49: 757–764.

Schoeninger, MF 1989 Reconstructing prehistoric human diet. *In* Price, TD, ed, *The Chemistry of Prehistoric Human Bone*. Cambridge, Cambridge University Press: 38–67.

Schoeninger, MF and DeNiro, MJ 1984 Nitrogen and carbon isotopic composition of bone collagen from marine and terrestrial animals. *Geochimica et Cosmochimica Acta* 48: 625–639.

Sellstedt, H, Engstrand, L and Gejvall, N G 1966 New application of radiocarbon dating to the dating of collagen residue in bones. *Nature* 212: 572–574.

Sinex, FB and Faris, B 1959 Isolation of gelatin from ancient bones. *Science* 129: 969.

Stafford, TW, Jr, Brendel, K and Duhamel, RC 1988 Radiocarbon, ^{13}C and ^{15}N analysis of fossil bone: Removal of humates with XAD-2 resin. *Geochimica et Cosmochimica Acta* 52: 2197–2206.

Stafford, TW, Jr, Duhamel, RC, Haynes, CV, Jr and Brendel, K 1982 The isolation of proline and hydroxyproline from fossil bone. *Life Science* 31: 931–938.

Stafford, TW, Jr, Hare, PE, Currie, L, Jull, AJT and Donahue, DJ 1990 Accuracy of North American human skeleton ages. *Quaternary Research* 34: 111–120.

_____1991 Accelerator radiocarbon dating at the molecular level. *Journal of Archaeological Sciences* 18: 35–72.

Stafford, TW, Jr, Jull, AJT, Brendel, K, Duhamel, RC and Donahue, D 1987 Study of bone radiocarbon dating accuracy at the University of Arizona NSF accelerator facility for radioisotope analysis. *Radiocarbon* 29(1): 24–44.

Stanford, DJ 1982 A critical review of archaeological evidence relating to the antiquity of human occupation of the New World. *Smithsonian Contributions to Anthropology* 30: 202–218.

Tamers, MA and Pearson, FJ, Jr 1965 Validity of radiocarbon dates on bone. *Nature* 208: 1053–1055.

Taylor, RE 1982 Problems in the radiocarbon dating of bone. *In* Currie, LA, ed, *Nuclear and Chemical Dating Techniques: Inter-*

preting the Environmental Record. Washington DC, American Chemical Society, 453–473.

_____1983 Non-concordance of radiocarbon and amino acid racemization age estimates on human bone: Implications for the dating of the earliest Homo sapiens in the New World. *In* Stuiver, M and Kra, RS, eds, Proceedings of the 11th International ^{14}C Conference. *Radiocarbon* 25(1): 647–654.

_____1987a Dating techniques in archaeology and paleoanthropology. *Analytical Chemistry* 59: 317A–331A.

_____1987b *Radiocarbon Dating: An Archaeological Perspective.* New York, Academic Press.

_____1987c AMS ^{14}C dating of critical bone samples: Proposed protocol and criteria for evaluation. *Nuclear Instruments and Methods* B29: 159–163.

_____1991 Frameworks for dating the Late Pleistocene peopling of the Americas. *In* Dillehay, TD and Meltzer, DJ, eds, *The First Americans: Search and Research.* Boca Raton, Florida, CRC Press, in press.

Taylor, RE, Ennis, PJ, Slota, PJ, Jr and Payen, LA 1989 Non-age-related variations in aspartic acid racemization in bone from a radiocarbon-dated late Holocene archaeological site. *In* Long, A and Kra, RS, eds, Proceedings of the 13th International ^{14}C conference. *Radiocarbon* 31(3): 1048–1056.

Taylor, RE and Payen LA 1979 The role of archaeometry in American archaeology: approaches to the evaluation of the antiquity of Homo sapiens in California. *In* Schiffer, M, ed, *Advances in Archaeological Method and Theory 2.* New York, Academic Press: 239–283.

Taylor, RE, Payen LA, Gerow, B, Donahue, DJ, Zabel, TH, Jull, AJT and Damon, PE 1983 Middle Holocene age of the Sunnyvale human skeleton. *Science* 220: 1271–1273.

Taylor, RE, Payen, LA, Prior, CA, Slota, PJ Jr, Gillespie, R, Gowlett, JAJ, Hedges, REB, Jull, AJT, Zabel, TH, Donahue, DJ and Berger, R 1985 Major revisions in the Pleistocene age assignments for North American human skeletons by ^{14}C accelerator mass spectrometry: None older than 11,000 ^{14}C years BP. *American Antiquity* 50: 136–140.

Taylor, RE and Slota, PJ, Jr 1979 Fraction studies on marine shell and bone samples for radiocarbon analysis. *In* Berger, R and Suess, HE, eds, *Radiocarbon Dating.* Proceedings of the 9th International ^{14}C Conference. Berkeley, University of California Press, 422–431.

Termine, JD 1988 Non-collagen proteins in bone. In *Cell and Molecular Biology of Vertebrate Hard Tissues.* New York, John Wiley & Sons.

Thurman, EM 1985 *Organic Geochemistry of Natural Waters.* Norwell, Massachusetts, Kluwer: 151–155.

Vries, H de and Barendsen, GW 1954 Measurements of age by the carbon-14 technique. *Nature* 174: 1138–1141.

Wand, JO (ms) 1981 Microsample preparation for radiocarbon dating. Doctoral dissertation, Oxford University.

Willey, GR 1978 A summary scan. *In* Taylor, RE and Meighan, CW, eds *Chronologies in New World Archaeology.* New York, Academic Press: 513–563.

26

THE ROLE OF RADIOCARBON DATING IN MAYA ARCHAEOLOGY: FOUR DECADES OF RESEARCH

SCOTT L FEDICK and KARL A TAUBE

INTRODUCTION

Since its inception in the 1950s, ^{14}C dating has had a profound role in our understanding of the ancient Maya of southern Mexico and Central America, both in terms of culture history and developmental processes. In the case of the Classic Maya period, from AD 300–900, ^{14}C dating has primarily served to augment two other forms of interrelated dating: ceramic chronology and the Long Count system of calendric notation. In this sense, the advent of ^{14}C dating has been less of a revolution than a gradual resolution of several competing chronologic constructs for the Classic Maya.

The archaeology of the Maya region does not begin and end with the Classic Maya. ^{14}C dating has extended the temporal range far beyond the Classic period system of Long Count dating and even the appearance of ceramics in the Maya region. Thus, ^{14}C dating has been invaluable to our understanding of the development of complex society in the Maya region. ^{14}C dating also has been of great use in areas where the Long Count system was not commonly employed, such as the northern Maya lowlands, the highlands and the coastal plain and piedmont. In contrast to the Long Count, ^{14}C dating is by no means limited to elite contexts, and thus, has been of central importance in our understanding of Maya subsistence, settlement patterns and demography. In this chapter, we can touch on only some of the major topics of Maya research that have been explored through ^{14}C dating. For an in-depth treatment of Maya chronology, we refer the reader to Gareth Lowe's (1978) excellent overview. The vast majority of ^{14}C dates discussed in this chapter were published originally as uncalibrated dates converted to BC/AD convention. We use the published uncalibrated dates unless otherwise indicated, and retain the BC/AD convention to facilitate the placement of reported dates within defined cultural periods of Maya development.

RADIOCARBON DATING AND THE CORRELATION PROBLEM

It is impossible to describe the role of ^{14}C dating in Maya archaeology without mentioning the Long Count, which remains our most detailed means of absolute dating in the Maya area. In terms of complexity and calendric precision, the Long Count system is unique in the ancient New World. Appearing on stone monuments and portable objects throughout the Maya lowlands, Long Count dates have served as a crucial means of dating Classic Maya remains. The Long Count is a continuous count of days from a base line in remote antiquity. Composed of five units of time, the Baktun, Katun, Tun, Uinal and Kin, the Long Count does not repeat until the completion of 13 Baktuns, *ie*, 16,872,000 days, or 5125.6 solar years.

Although of great importance during the Classic period, the Long Count does not appear on Postclassic (after ca AD 900) Maya monuments. At the time of the Spanish Conquest, the Yucatec of the northern Maya lowlands used a much abbreviated form in which the largest unit of time, the approximately 400-year Baktun, was suppressed. Instead, the roughly 20-year Katun served as the primary unit of calendric notation. Due to the permutations of the calendar, a particularly numbered Katun, such as Katun 13 Ahau, would repeat roughly every 256.25 years. Thus, the so-called Short Count is of limited efficacy in matters of great antiquity, but is very valuable for dating Classic Long Count texts.

During the early Colonial era, Short Count dates denoting Tun and Katun endings and other historical events were correlated with the Christian calendar (Morley 1920; Thompson 1935). An especially important event was the ending of a Katun 13 Ahau in the Julian year of 1539 (Thompson 1935: 59). With the assumption that the Katun count remained unbroken since the Classic period, the major problem was how many roughly 256.25-year periods occurred between the Classic period Long Count inscriptions and the historical events of the 16th century, or, in terms of the Classic Maya, what Long Count notation corresponded to the Katun 13 Ahau ending of 1539. One correlation, proposed by Herbert Spinden (1924), argues that the Katun 13 Ahau ending of 1539 corresponds to 12.9.0.0.0, the completion of 12 Baktuns and 9 Katuns in the present great cycle. The Goodman-Martinez-Thompson correlation, or GMT, places the 1539 event at 11.16.0.0.0. The difference between these two correlations is one round of 13 Katuns, or about 256.25 years. In terms of the Christian calendar, the Spinden correlation places Long Count dates nearly 260 years before the GMT. Although the Spinden and GMT are the two most widely cited correlations, they are by no means the only ones. Over the years, a great many distinct correlations have been proposed. Until the correlation problem was resolved, the Long Count remained an extremely precise but entirely floating chronology.

Following the early concern with the recording and interpretation of stone monuments, Maya archaeologists began the more laborious task of archaeologic excavation. It was soon recognized that with seriation and cross-dating, pottery provided an extremely useful means of relative dating. George Vaillant (1935) was the first to provide a comprehensive ceramic sequence for the Maya area. Vaillant distinguished six ceramic phases for the ancient Maya, which closely correspond to the periods now known as the Preclassic, Early Classic, Late Classic, Terminal Classic, Early Postclassic and Late Postclassic. Vaillant was acutely aware of the value of cross-dating Maya vessels with ceramic chronologies of highland Mexico, although his attempts at cross-dating and absolute chronology were flawed by his conviction that Teotihuacan was destroyed in the 11th or 12th centuries (Vaillant 1935: 121). It is now recognized that this event occurred during the eighth century (Millon 1981). Although the absolute dating of the Maya ceramic phases by Vaillant was not entirely accurate, his relative dating of particular wares with periods has been proven to be generally correct.

It was quickly recognized in studies of Maya ceramic chronology that the Long Count could provide a means of correlating ceramic phases with absolute time. Thus, Vaillant (1935) attempted to correlate Maya ceramic sequences with the Long Count system. However, Vaillant was limited by how little was then known of the ancient Maya. According to Vaillant (1935: 136–137), the GMT correlation was problematic for one major reason: it suggested that the southern Maya area was depopulated following the "Maya Great Period," *ie*, the Late Classic period (ca 600–900 BC). We now realize, however, that this did occur at the end of the Classic period (ca AD 900), and that the depopulation of the Peten was a major event of the Classic Maya collapse (Culbert 1973).

In his Maya ceramic studies, Vaillant (1927, 1935; Merwin & Vaillant 1932: 75–82) made much use of ceramics from Long Count stela dedicatory caches at Copan. However, the association of pottery phases with Long Count monuments is best known from the Carnegie Institution of Washington excavations at Uaxactun. In several studies, researchers noted the stratigraphic association of pottery types with architectural phases and Long Count dates at Uaxactun (Ricketson 1928; Ricketson & Ricketson 1937; Smith 1955). Because of these investigations, the ceramic phases of Early Classic (ca AD 300–600) Tzakol and Late Classic (ca AD 600–900) Tepeu have been correlated with the Long Count calendar (Smith 1955: 3–4). These two Uaxactun phases, along with the still earlier Mamom and Chicanel, have served as the basis for ceramic chronology in the central Maya lowlands.

By the 1950s, the Long Count was correlated with many aspects of Classic Maya culture, including architecture, ceramic phases and elite art. In spite of its importance, the correlation of the Long Count with the Christian calendar was by no means resolved. Although the GMT correlation was generally favored, the Spinden and other correlations were still cited as equally possible solutions (see Thompson 1960: 306–309). The advent of ^{14}C dating provided an excellent opportunity for resolving this issue, especially since it involved a broad time scale of 260 years or more. It is not surprising that one of the first applications of ^{14}C dating in the Maya region was the resolution of the correlation problem.

The most satisfactory solution to the correlation problem would be the direct radiometric dating of an object bearing a Long Count date. With their relatively clear dedicatory dates, the carved *sapodilla* wood lintels of Late Classic Tikal provided an excellent means to test the various proposed correlations. J L Kulp of Lamont Geological Observatory, Columbia University, tested the lintel from Tikal Structure 10, now in the American Museum of Natural History in New York (Kulp, Feely & Tryon 1951). Soon afterwards, Willard Libby (1955) tested two beams from Lintels 2 and 3 of Tikal Temple IV, which had been sent to Basel, Switzerland in the 19th century. Libby obtained two samples, which were each measured over a span of six weeks. According to Libby (1955: 131), the dating of the Basel samples was to test the sensitivity of the ^{14}C method by comparing the ^{14}C dates to a known calendric system. However, Libby did not specify which Long Count correlation value he favored; thus, rather than serving as a test for the sensitivity of ^{14}C, the dates obtained by Libby had more of a direct bearing on the validity of the various proposed correlations.

The early ^{14}C dates from Structure 10 and Temple IV supported not the widely favored GMT correlation, but rather the Spinden correlation. The Structure 10 lintel, with a proposed dedicatory Long Count date of 9.15.10.0.0, provided ^{14}C dates of AD 350 ± 200 and AD 550 ± 150, both within a one-sigma range for the Spinden correlation date of August 30, AD 481 (Kulp, Feely & Tryon 1951: 565). Lintel 3 of Tikal Temple IV bears a clear Long Count dedicatory date of 9.15.10.0.0, the same date reconstructed for the Structure 10 Lintel. Libby (1955: 132) noted that the ^{14}C dates obtained from the Lintels 2 and 3 samples, AD 469 ± 120 and AD 433 ± 170 also fell within the Spinden date of AD 481.

The initial ^{14}C support for the Spinden correlation was quickly challenged by Mayanist Linton Satterthwaite (1956), who noted that the lintels from Tikal Structure 10 and Temple IV were reduced for transportation, and the ^{14}C samples obtained did not reflect the latest growth for the wooden beams. In fact, in their brief reports, neither Kulp (Kulp, Feely & Tryon 1951: 566) nor Libby (1955: 131–132) make any mention of the relation of the wood samples to the exterior or interior of the original tree. Satterthwaite argued that, due to post-sample growth error, the ^{14}C dates obtained by Kulp and Libby did not accurately reflect the dedication dates of the wooden lintels.

In 1955, the University Museum of the University of Pennsylvania began the massive Tikal Project, which continued active fieldwork at the site until 1970. One of the first topics addressed by the Tikal Project was the ^{14}C dating of the wooden lintels from Temple 1, Temple IV and Structure 10. In a thorough ^{14}C analysis by Satterthwaite and Ralph (1960), 17 counting runs were made of samples from 6 beams from Temple 1, and 33 runs from 10 beams from Temple IV. Unlike those obtained by Kulp and Libby, the samples were taken *in situ* at Tikal, with great attention paid to obtain samples corresponding to the most recent exterior growth. Nonetheless, the rectangular lintel beams could only have been pared down from large circular logs. To test for possible post-sample growth error, the dated samples were also derived from the smaller cylindrical vault beams, which correspond closely to the natural diameter and form of a trunk or limb, and at times, portions of exterior bark were preserved. Satterthwaite and Ralph (1960: 166) note that "the post-sample growth error for the vault beams must be close to zero."

The series of ^{14}C dates obtained by the Tikal Project from Temples I and IV were internally consistent. The ^{14}C dates of Temple 1, with a probable dedicatory date between the roughly 17-year span of 9.13.3.0.0 to 9.14.0.0.0, are before those of Temple IV, with a probable dedicatory date in the roughly 10-year span between 9.15.10.0.0 to 9.16.0.0.0. In both cases, the Temple I and Temple IV dates corresponded to the GMT correlation. The sample dates from 5 of the 6 lintel and vault beams from Temple I were all within a standard deviation of the GMT correlated years of AD 695–712. The one exception was a single vault beam that produced a ^{14}C date corresponding to the earlier Spinden correlation. Satterthwaite and Ralph (1960: 177) note that this particular beam was of a more brittle and less gummy wood, and suggest that it was reused from an earlier structure. The GMT correlation also fell within a one-sigma range for the ten *in situ* samples from lintel and vault beams of Temple IV.

The Structure 10 lintel was subsequently dated at Lamont in 1958 (Broecker, Olson & Bird 1959), and the date obtained, AD 456 ± 60, was within a standard deviation range of the previous results. However, Satterthwaite and Ralph (1960: 175) obtained a later date for the Lamont sample; both dates averaged to

AD 606 ± 57. A slightly more recent date was also obtained by Satterthwaite and Ralph (1960: 175) for the Basel portion of Lintel 3 of Temple IV. Thus, the possibility arose that University of Pennsylvania Radiocarbon Laboratory was producing consistently later dates. In order to avoid a possible bias toward more recent dates, Satterthwaite sent a series of lintel and vault beam samples from Temple IV and Structure 10 to be dated by Libby at the UCLA Radiocarbon Laboratory. Fergusson and Libby (1963: 13–14) noted that their Temple IV dates corresponded to the University of Pennsylvania results, which also supported the GMT correlation. Five UCLA dates were obtained from Structure 10 lintels and beams, supplying an average date of AD 625 ± 25. Fergusson and Libby (1963) noted that the Structure 10 dates fall between the GMT and Spinden correlations for the reconstructed Long Count date of 9.15.10.0.0 3 Ahau 3 Mol. Satterthwaite and Ralph (1960: 176) suggest that the early dates for the Structure 10 lintel derive from post-sample growth error. Although Satterthwaite (1967: 228) subsequently argued that the damaged date of the Structure 10 lintel may have actually been an earlier Long Count date of 9.4.10.0.0, Jones notes that this is not a viable option (Jones & Satterthwaite 1982: 104), and the previously reconstructed date of 9.15.10.0.0 3 Ahau 3 Mol is correct.

The radiometric analysis of the carved Tikal lintels by the Tikal Project provided strong material support for the GMT correlation, and this has not been contradicted by the extensive subsequent array of ^{14}C dates obtained from Classic Maya remains. Although the calibration of ^{14}C years to actual Christian years places the Tikal ^{14}C dates some 60 to 90 years later (Stuiver & Becker 1986: Table 3); this placement still favors the GMT among the most likely correlations. Recent advances in independent fields of inquiry, such as Maya archaeoastronomy (Lounsbury 1982), have provided further support for the GMT correlation. It is noteworthy that the only questions regarding the methodology and results of the Tikal research have been raised by individuals who favored other correlations (see Satterthwaite 1967; Chase 1986). Any correlation which differs greatly from the GMT must now take into account the impressive series of dates from Tikal.[1]

[1]Aside from Tikal, wooden lintels with carved inscriptions are known from other sites in the Maya area. Stephens (1962, vol 1: 117) briefly mentions a large carved lintel from the House of the Governor at Uxmal. Unfortunately, this important but otherwise unrecorded piece was destroyed during the 1842 fire at Catherwood's rotunda in New York. Temple VI of the recently rediscovered site of Tzibanche, Quintana Roo, contains a carved wooden lintel (Harrison 1972). The poorly preserved text has a date reconstructed as 9.6.0.0.0 9 Ahau 3 Uayeb, or 9.15.2.0.0 9 Ahau 3 Yax. Harrison (1972: 500) appears to favor the former date. A carved lintel stylistically

THE OLMEC PROBLEM

^{14}C dating has complemented the Long Count system of dating in another major research problem, the chronologic relationship of the Olmec to the Maya. During the first half of the 20th century, the age of the Olmec prompted considerable debate. A number of Mesoamerican researchers, such as Alfonso Caso (1942), Matthew Stirling (1940) and Miguel Covarrubias (1943) considered the Olmec to be early, essentially the "mother culture" of Mesoamerican civilization. However, Maya scholars, most notably J Eric Thompson (1941) and Sylvanus Morley (1946: 40–42), viewed the Maya to be at least as early as the Olmec. During the 1938 and 1939 fieldwork at Tres Zapotes, Stirling discovered a fragmentary stela bearing a face in Olmecoid style. Of special interest was the opposite side of the monument, which displayed the lower portion of a Long Count date reconstructed by Stirling (1940) as 7.16.6.16.18 6 Etz'nab 1 Uo. With the GMT correlation, this date corresponds to 31 BC, almost 350 years before the Leyden Plate, then the earliest known Maya Long Count date. Perhaps it is not surprising that two of the leading Mayanists, Thompson (1941, 1954: 50) and Morley (1946: 41), rejected Stirling's reconstruction of Stela C. According to Thompson (1941: 47), the Olmec style "certainly survived until shortly before the arrival of Cortes, and . . . appears to have reached its apex at the period of Mexican influence at Chichen Itza." In other words, to Thompson, the Olmec were a Postclassic phenomenon.

One of the difficulties in correlating the Olmec with the Maya was the absence of Olmec material in stratified contexts in the lowland Maya region. Vague similarities between La Venta ceramics and wares from the Peten caused Drucker (1947: 6) to equate the La Venta Olmec to the Early Classic Tzakol phase of Uaxactun. However, with the publication of the first ^{14}C dates from La Venta (Drucker, Heizer & Squier 1957, 1959), it was clear that the La Venta Olmec were considerably earlier. The seven ^{14}C dates from La Venta Groups A and C were from 1154–574 BC, with a maximum 1-sigma range of 1434–274 BC (Drucker, Heizer & Squier 1957: 265). Two samples from post-abandonment contexts supplied an average ^{14}C date of 333 ± 195 BC (*ibid*: 267). According to the authors, the ^{14}C dates suggest that La Venta was occupied from 800–400 BC.

similar to the Tikal examples was recently shown in a major exhibition of Maya art (Clancy *et al* 1985: No. 74). However, the carving on the exhibited portion of the lintel is not genuine.

M D Coe (1974) discussed a small and finely carved late Classic Maya wooden box with a hieroglyphic text containing a series of dates. The latest date refers to a Long count position of 9.12.9.7.12 9 Eb 15 Ceh. In view of the recent advances in AMS ^{14}C dating, a date could easily be obtained with minimal damage to the object.

Although the first ^{14}C dates from La Venta implied that the Olmec were early, Drucker, Heizer and Squier were reluctant to consider the Olmec as the first great civilization in Mesoamerica. Instead, they were quick to point out early ^{14}C dates from the highland Maya site of Kaminaljuyu, which suggested some contemporaneity with La Venta. Drucker, Heizer and Squier (1957: 260) tentatively suggested that the Olmec site of San Lorenzo was later than La Venta.[2] Drucker, Heizer and Squier (1957) suggest that the Olmec may be contemporaneous to the beginnings of the Classic Maya. However, excavations of San Lorenzo by Michael Coe determined that San Lorenzo is actually a pre-La Venta site (Coe & Diehl 1980). Of the 12 ^{14}C samples from San Lorenzo phase contexts, the period of Olmec florescence at San Lorenzo, 10 were internally consistent with dates ranging from 1150 ± 140 to 920 ± 140 BC (Coe & Diehl 1980: 395–396). The authors place the San Lorenzo phase at 1150–900 BC. Although it is conceivable that comparable remains will be found buried under later structures in the Maya region, no known contemporaneous Maya sites approach Early Formative San Lorenzo in either scale or complexity.

Occupation History of the Maya Area

In moving beyond problems of chronologic control and placement of different cultures in relation to the Maya, ^{14}C dating has helped to establish the initial occupation of the region and early developmental history of Maya culture. Numerous Paleoindian sites of the New World have been securely ^{14}C dated to as early as 12,000 years ago (see Meltzer 1989), but no such sites have been reported within the Maya lowlands. The only evidence for Paleoindian presence in the Maya lowlands consists of isolated fluted-lanceolate and fishtail projectile points which have been cross-dated to the late Paleoindian period, about 9000–7500 BC (Hester, Kelly & Ligabue 1981; Zeitlin 1984).

^{14}C dating has been applied to an excavated campsite in the Guatemalan highlands, providing the earliest independently dated site in the Maya area (Gruhn & Bryan 1977). This site, Los Tapiales, has been interpreted as a Paleoindian campsite containing four hearths and approximately 100 lithic artifacts, including a variety of cutting and scraping tools and a single fluted projectile point base. Gruhn & Bryan (1977) obtained a series of nine ^{14}C dates from Los Tapiales, derived from the hearths, about 2780–5870 BC, and considered them too young

[2]Following the publication of the San Lorenzo dates, Berger, Graham and Heizer (1967) reassessed the ^{14}C dating of La Venta and provided additional ^{14}C dates. The authors concluded that the Olmec period at La Venta does not date from 800–400 BC, but rather from 1000–600 BC, and suggested that La Venta is coeval with San Lorenzo. This revision of La Venta chronology has not received widespread acceptance.

to be associated with a fluted projectile point. The earlier ^{14}C dates, 9000–8000 BC, derived from scattered charcoal within the stratigraphic units, are argued to be more acceptable (Gruhn & Bryan 1977: 240–243). Further evaluation of this important site would seem in order.

The Archaic occupation of the greater Maya area has been temporally anchored by ^{14}C dating of sites in both highlands and lowlands. The rockshelter site of Ocozocoautla, situated in the Central Depression of the Chiapas highlands of Mexico, provides us with a view of Archaic period (ca 7500–2000 BC) life in the Maya periphery (MacNeish & Peterson 1962). A series of ^{14}C dates traces the use of the shelter and changes in subsistence patterns from about 6770 ± 400 BC to AD 90 ± 200, with less well-defined use extending into the Postclassic. Pottery and maize pollen are first present in an occupation layer that was ^{14}C dated to 1320 ± 200 BC.

Within the peripheral Maya area of the Chiapas Pacific Littoral of Mexico, we have well-documented evidence for Archaic occupation associated with Chantuto shell-midden sites (Voorhies 1976). A series of 16 ^{14}C assays establish the occupation sequence at these sites, beginning in the Late Archaic period, from around 3000–2000 BC (Voorhies 1976: 41–44). Additional, as yet unpublished, ^{14}C dates from excavations conducted since 1988 at Chantuto sites support this occupation date range (Michaels & Voorhies 1989; Voorhies & Michaels 1990). Other researchers have ^{14}C dated evidence for Late Archaic presence in the Gulf Coast periphery of the Maya lowlands along the coast of Veracruz (eg, Wilkerson 1975).

Recent investigations by the Belize Archaic Archaeological Reconnaissance (BAAR) project represent the only concerted effort to define Archaic settlement history within the Maya lowlands (Zeitlin 1984). The BAAR project recorded about 150 sites with possible preceramic components presumably dating from about 9000–2000 BC, but the chronology of these sites was complicated by poor stratigraphic sequences and a paucity of usable samples for ^{14}C dating. Two charcoal samples recovered from the Bentz Landing site (BAAR 6) on Progresso Lagoon in northern Belize yielded ^{14}C dates of 1280 ± 85 and 1325 ± 85 BC (Zeitlin 1984: 364). These samples were found in association with a small amount of lithic debitage and Preclassic ceramics. The stratum below these samples contained stone tools of the late Archaic Melinda complex, cross-dated to between 4000 and 3000 BC (Zeitlin 1984). Additional excavation of the identified Belize sites is clearly needed to produce a sound absolute chronology for the lowland Maya Archaic.

Reconstructions of early sedentary villages and agricultural development in the lowland interior have been subject to radical revisions over the last 40 years.

These revisions have resulted from the complementary contributions of excavation and ^{14}C dating, regional settlement pattern investigations and ecological research. Early notions of prehistoric Maya settlement and agriculture were based on the premise that the tropical lowlands were a hostile, agriculturally marginal environment, capable of supporting only a sparse population dependent on extensive, long fallow, slash-and-burn (swidden) cultivation of maize (Morley 1946; Turner 1978).

Betty Meggers (1954) went so far as to state that the environmental limitations of the tropical lowlands would not have allowed for the *in situ* growth of Classic Maya civilization. According to Meggers, a developed culture must have been introduced into the lowlands from elsewhere, only to briefly flourish before entering a long period of decline under conditions of environmental degradation. At the time of Meggers' writing, evidence for Preclassic occupation in the Maya lowlands was scant, as well as difficult to date due to the lack of absolute chronologic control. Meggers' diffusionist scenario of lowland colonization at about AD 320 was rendered obsolete by a number of excavation projects which documented a long Preclassic ceramic and architectural developmental sequence anchored by ^{14}C dating to about 900 BC at the Pasion River sites of Seibal and Altar de Sacrificios (Adams 1971; Berger *et al* 1974; Willey, Culbert & Adams 1967; Willey *et al* 1975).

While an early presence in the central lowlands was established by ^{14}C dating, questions concerning the agricultural adaptation of formative Maya farmers remained. The discovery of raised and drained fields in riverine-associated swamps of the western and eastern lowlands (Siemens & Puleston 1972; Siemens 1983) suggested evidence of early Maya intensive agriculture in wetland settings. Excavations of supposed raised fields along the Rio Hondo of northern Belize resulted in the recovery of a trimmed wooden post yielding a ^{14}C date of 1110 ± 230 BC (Puleston 1977). More recent excavations along the Rio Hondo by Bloom, Pohl and Stein (1985) produced a sample of charred maize remains with a ^{14}C date of 670 ± 190 BC. These data suggest that the Maya practiced wetland cultivation of maize in the area at least by the transition between the Early and Middle Preclassic periods, about 1000 BC (Bloom, Pohl & Stein 1985: 26; Fedick & Ford 1990).

Most scenarios of early Maya development were pushed backward in time, by more than 1500 years, by a series of ^{14}C dates originating from the site of Cuello, in the eastern riverine area of northern Belize. Research directed at Cuello by Norman Hammond during 1975 and 1976 produced a series of 10 internally consistent ^{14}C dates for Early Preclassic settlement at the site beginning about 2600 BC, based on the MASCA calibration (Hammond 1977;

Hammond *et al* 1976). The earliest occupation of Cuello, termed the Swasey phase, was associated with a well-developed Swasey ceramic complex, burials and architecture (Hammond *et al* 1979). Hammond argued that the ^{14}C chronology of Cuello provided clear evidence for independent development of Maya culture within the lowlands, even suggesting that the early Maya may have acted as an influence on emergent Olmec society: a dramatic role reversal in terms of perceived cultural developments (Hammond 1977: 121).

Acceptance of the Swasey phase of northern Belize as representative of the earliest Maya is affirmed in most recent textbooks (Hammond 1982; Henderson 1981; Morley, Brainerd & Sharer 1983; Weaver 1981). Some however, expressed strong doubts concerning the supposedly early beginnings and long duration of the Swasey phase (M D Coe 1987: 37; Marcus 1983: 459–460; *cf* Pring 1979). These skeptics found it difficult to accept the diverse, sophisticated Swasey ceramics as rudimentary in any sense, particularly when they showed little resemblance to previously known Early Formative or Pre-classic ceramics from other parts of Mesoamerica and rather clear resemblances to Eb, Xe, and Mammon ceramics of the Maya Middle Preclassic. Also difficult to accept was the notion of a 1000-year phase apparently lacking in ceramic change.

New ^{14}C dates from numerous samples excavated at Cuello since 1976 have prompted a very recent reevaluation of the Swasey phase by Andrews and Hammond (1990). The 19 calibrated dates resulting from the post 1976 excavations of Swasey and Bladen phase contexts fall between 900 BC and 400 BC, firmly within the Middle Preclassic. This reevaluation aligns the early Cuello materials with similar Middle Preclassic developments elsewhere in the Maya lowlands. Explanation of the very early dates obtained from the samples excavated during 1975 and 1976 remains problematic, although the researchers suggest that at least some of the early dates may be the result of sample amalgamation and inclusion of redeposited exogenous charcoal that antedated the Swasey-phase village occupation. Once again, a new series of ^{14}C dates will necessitate the rewriting of our textbooks.

Conclusions and Prospectus

In conclusion, it can be said that ^{14}C dating has contributed significantly to our understanding of Maya cultural development within the context of a rich variety of research perspectives. ^{14}C dating has provided direct support for the GMT correlation for the Long Count, by far the most detailed absolute chronologic system in New World prehistory. Through ^{14}C dating, we now know that Olmec civilization antedated that of the Classic Maya. ^{14}C dating has extended our comprehension of Maya prehistory through the Preclassic and Archaic periods, even to the first documentation of Paleoindian peoples in the Maya region. The waxing and waning of Cuello as the definitive, securely dated, earliest Maya site has taught us to be wary.

In addition to the topics discussed in this chapter, a few other research domains, problems and prospects associated with ^{14}C dating deserve mention. Numerous researchers have sought to test hypotheses that evaluate aspects of Maya development and decline as response to climatic change and natural disaster (Dahlin 1979, 1983; Folan *et al* 1983; Folan & Hyde 1985; Gunn & Adams 1981; Sharer 1974, 1984; Sheets 1971, 1979, 1983, 1986). The role of ^{14}C dating in establishing the absolute chronologic sequence of climate change and volcanic eruptions has contributed to attempts at understanding culture process within an environmental context (Denton & Karlen 1973; Sanchez & Kutzbach 1974; Wendland & Bryson 1974; Grove 1979).

One aspect of ecologic research in which ^{14}C has been attempted and subsequently precluded is the dating of pollen-bearing lake sediments from the Maya lowlands. Researchers of the Historic Ecology Project working in the central Peten lakes region of Guatemala report that 29 ^{14}C dates, measured by three laboratories, of lake sediment cores were rejected due to uncertainties of carbonate error in hard-water lakes, redeposition of soil carbonate of various ages and inadequate amounts of carbon for dating in the sediments (Rice, Rice & Deevey 1985: 95; Vaughan, Deevey & Garrett-Jones 1985: 74). The only exception to these problems was a date obtained from a sample of woody tissue of terrestrial origin recovered near the bottom of a lake sediment core (Ogden & Hart 1977). The Historical Ecology Project researchers rejected previously published ^{14}C dates for a sediment core from Laguna de Petenxil, Guatemala, which has often been cited as yielding pollen evidence for forest disturbance and maize cultivation by about 2000 BC (Tsukada 1966; Cowgill *et al* 1966; *cf* Vaughn, Deevey & Garrett-Jones 1985: 74). In precluding ^{14}C as a means of dating lake sediments, researchers have been forced to rely on correlating pollen sequences from the cores, with human activities believed to have caused the pollen events.

^{14}C dating has not always supplied clear solutions to chronologic problems in the Maya area, as further exemplified by the so-called "Cotzumalhuapa problem" of highland Guatemala (Parsons 1967; Thompson 1948). Whereas current ^{14}C data suggest that Cotzumalhuapa dates to the 5th to 7th centuries AD (Parsons 1967: Fig 15), the style, epigraphy and iconographic content of carved monuments suggests a later date corresponding to the Terminal Classic. The improvement of absolute chronologic control continues to be a vital issue in Maya studies.

For the Terminal Classic and Postclassic chronology of the northern Maya lowlands of Yucatan, ^{14}C dating remains second to ceramic seriation. The chronologic relationship between two major Terminal Classic and Early Postclassic ceramic spheres, Cehpech and Sotuta, is still poorly understood (Andrews & Sabloff 1986; Ball 1978; Lincoln 1986; Smith 1971). This is a pressing problem, since it directly concerns the political and economic role of Chichen Itza in northern Yucatan. The problem partly lies in the surprisingly limited number of ^{14}C dates from the Puuc region and Chichen Itza, and the limited number of excavations being performed at Chichen Itza and other Sotuta-related sites. The recent excavations at Isla Cerritos are an important and noteworthy exception (Andrews *et al* 1988). Until more excavation and ^{14}C dating is performed in Yucatan, we will have to rely upon ceramic chronology and the very tenuous and frequently misleading short count dates appearing in ethnohistoric accounts.

One of the most promising chronologic developments in Maya research is accelerator mass spectrometry (AMS) dating, which allows for accurate measurements from relatively small samples of carbon. Thus, it would now be possible to date the four pre-Hispanic Maya codices by using only minute amounts of paper. At present, only the Grolier Codex has been dated by ^{14}C (MD Coe 1973: 150), and it is likely that a more accurate date could now be obtained with AMS technology. With AMS ^{14}C dating, it is now possible to obtain direct measurements from pottery (Gabasio *et al* 1986). The direct dating of pottery could resolve many of the issues relating to Cehpech and Sotuta ceramic spheres and the overlap problem of the Late Classic and Early Postclassic Yucatan.

Still another potential application of ^{14}C dating, particularly AMS, is the direct dating of plant remains recovered by flotation. Whereas in the past, it was generally assumed that the potential for recovering plant remains from a tropical setting was very limited, recent flotation work in the Maya lowlands has proven otherwise (Cliff & Crane 1989; Crane 1986; Miksicek 1983; Miksicek *et al* 1981). Thus, flotation recovery of plant remains could prove to be not only a valuable source for datable samples, but also would have a direct bearing on chronologic questions related to plant use, dietary change and the construction of agricultural engineering features such as raised fields and terraces.

^{14}C dating also will be an important means of evaluating other absolute dating techniques, such as obsidian hydration dating (Hammond 1989), as they are applied experimentally in the Maya area. Although ^{14}C dating has resolved many of the major chronologic issues of Maya archaeology, it is clear that ^{14}C will continue to play a vital role in problems of ancient Maya culture history and social process.

REFERENCES

Adams, REW 1971 The ceramics of Altar de Sacrificios. *Papers of the Peabody Museum of Archaeology and Ethnology, Harvard University* 63(1).

Andrews, A, Gallareta Negron, T, Robles Castellanos, F, Cobos Palma, R and Cervera Rivero, P 1988 Isla Cerritos: an Itza trading port on the north coast of Yucatan, Mexico. *National Geographic Research* 4(2): 196–207.

Andrews, EW, V and Hammond, N 1990 Redefinition of the Swasey phase at Cuello, Belize. *American Antiquity* 55(3): 570–584.

Andrews, EW, V and Sabloff, JA 1986 Classic to Postclassic: A summary discussion. *In* Sabloff, JA and Andrews, EW, V, eds, *Late Lowland Maya Civilization Classic to Postclassic*. Albuquerque, University of New Mexico Press: 433–456.

Ball, J 1978 Archaeological pottery of the Yucatan-Campeche coast. *In* Studies in the archaeology of coastal Yucatan and Campeche, Mexico. *Middle American Research Institute Publication* 46: 69–140.

Berger, R, De Atley, S, Protsch, R and Willey, GR 1974 Radiocarbon chronology for Seibal, Guatemala. *Nature* 252(5483): 472–473.

Berger, R, Graham, JA and Heizer, RF 1967 A reconsideration of the age of the La Venta Site. *Contributions of the University of California Archaeological Research Facility* 3: 1–24.

Bloom, PR, Pohl, M and Stein, J 1985 Analysis of sedimentation and agriculture along the Rio Hondo, northern Belize. *In* Pohl, M, ed, Prehistoric lowland Maya environment and subsistence economy. *Papers of the Peabody Museum of Archaeology and Ethnology, Harvard University* 77: 21–33.

Broecker, WS, Olson, EA and Bird, J 1959 Radiocarbon measurements on samples of known age. *Nature* 183(4675): 1582–1584.

Caso, A 1942 Definicion y extension del complejo 'Olmeca.' In *Mayas y Olmecas, segundo reunion de mesa redonda sobre problemas antropologicos de Mexico y Centro America*. Tuxtla Gutierrez, Sociedad Mexicana de Antropologia: 43–46

Chase, AF 1986 Time depth or vacuum: the 11.3.0.0.0 correlation and the lowland Maya Postclassic. *In* Sabloff, JA and Andrews, EW, V, eds, *Late Lowland Maya Civilization: Classic to Postclassic*. Albuquerque, University of New Mexico Press: 99–140.

Clancy, FS, Coggins, CC, Culbert, TP, Gallenkamp, C and Johnson, RE 1985 *Maya: Treasures of an Ancient Civilization*. New York, Harry N Abrams: 240 p.

Cliff, MB and Crane, CJ 1989 Changing subsistence economy at a Late Preclassic Maya community. *In* McAnany, PA and Isaac, BL, eds, Prehistoric Maya economies of Belize. *Research in Economic Anthropology, Supplement* 4. Greenwich, Connecticut, JAI Press: 295–324.

Coe, MD 1973 *The Maya Scribe and His World*. New York, The Grolier Club:160 p.

_____1974 A carved wooden box from the Classic Maya civilization. *In* Greene Robertson, M, ed, *Primera Mesa Redonda de Palenque, Part II*. Pebble Beach, The Robert Louis Stevenson School: 143 p.

_____1987 *The Maya*, 4th edition. New York, Thames and Hudson: 200 p.

Coe, MD and Diehl, RA 1980 *In the Land of the Olmec: The Archaeology of San Lorenzo Tenochtitlan*, 2 vols. Austin and London, University of Texas Press.

Coe, WR 1967 *Tikal: A Handbook of the Ancient Maya Ruins*. Philadelphia, The University Museum, University of Pennsylvania: 123 p.

Covarrubias, M 1943 Tlatilco, archaic Mexican art and culture. *Dyn* 4–5: 40–46.

Cowgill, UM, Hutchinson, GE, Racek, AA, Goulden, CE, Patrick, R and Tsukada, M 1966 The history of Laguna de Petenxil: A small lake in northern Guatemala. *Memoirs of the Connecticut Academy of Arts and Sciences* 17: 1–126.

Crane, CJ 1986 Late Preclassic Maya agriculture, wild plant utilization, and land-use practices. *In* Robertson, RA and Freidel, DA, eds, *Archaeology at Cerros, Belize, Central America. An Interim Report*, vol 1. Dallas, Texas, Southern Methodist University Press: 147–165.

Culbert, TP 1973 The Maya downfall at Tikal. *In* Culbert, TP, ed, *The Classic Maya Collapse*. Albuquerque, University of New Mexico Press: 63–92.

Dahlin, BH 1979 Cropping cash in the Protoclassic: A cultural impact statement. *In* Hammond, N and Willey, G, eds, *Maya Archaeology and Ethnohistory*. Austin, University of Texas Press: 21–37.

_____1983 Climate and prehistory on the Yucatan peninsula. *Climate Change* 5(3): 245–263.

Denton, G and Karlen, W 1973 Holocene climatic variations: their pattern and possible cause. *Quaternary Research* 3(2): 155–205.

Drucker, P, 1947 Some implications of the ceramic complex of La Venta. *Smithsonian Institution, Miscellaneous Collections* 107 (8).

Drucker, P, Heizer, RF and Squier, RJ 1957 Radiocarbon dates from La Venta, Tabasco. *Science* 126(3263): 72–73.

_____1959 Excavations at La Venta, Tabasco, 1955. *Bureau of American Ethnology Bulletin* 170: 320 p.

Fedick, SL and Ford, A 1990 The prehistoric agricultural landscape of the central Maya lowlands: An examination of local variability in a regional context. *World Archaeology* 22(1): 18–33.

Fergusson, GJ and Libby, WF 1963 UCLA radiocarbon dates II. *Radiocarbon* 5: 1–22.

Folan, WJ, Gunn, J, Eaton, JD and Patch, RW 1983 Paleoclimatological patterning in southern Mesoamerica. *Journal of Field Archaeology* 10(4): 453–468.

Folan, WJ and Hyde, BH 1985 Climatic forecasting and recording among the ancient and historic Maya: An ethnohistoric approach to epistemological and paleoclimatological patterning. *In* Folan, WJ, ed, *Contributions to the Archaeology and Ethnohistory of Greater Mesoamerica*. Carbondale, Southern Illinois University Press: 15–48.

Gabasio, Evin, J, Arnal, GB and Andrieux, P 1986 Origins of carbon in potsherds. *In* Stuiver, M and Kra, RS, eds, Proceedings of the 12th International ^{14}C Conference. *Radiocarbon* 28(2A): 711–718.

Grove, JM 1979 The glacial history of the Holocene. *Progress in Physical Geography* 3: 1–54.

Gruhn, R and Bryan, AL 1977 Los Tapiales: A Paleo-Indian campsite in the Guatemalan highlands. *Proceedings of the American Philosophical Society* 121(3): 235–273.

Gunn, J and Adams, REW 1981 Climatic change, culture, and civilization in North America. *World Archaeology* 13(1): 87–100.

Hammond, N 1977 The earliest Maya. *Scientific American* 236(3): 116–133.

_____1982 *Ancient Maya Civilization*. New Brunswick, New Jersey, Rutgers University Press: 337 p.

_____1989 Obsidian hydration dating of Tecep phase occupation at Nohmul, Belize. *American Antiquity* 54(3): 513–521.

Hammond, N, Pring, D, Berger, R, Switsur, VR and Ward, AP 1976 Radiocarbon chronology for early Maya occupation at Cuello, Belize. *Nature* 260(5552): 579–581.

Hammond, N, Pring, D, Wilk, R, Donaghey, S, Saul, FP, Wing, ES, Miller, AV and Feldman, LH 1979 The earliest Maya? Definition of the Swasey phase. *American Antiquity* 44(1): 92–110.

Harrison, PD 1972 The lintels of Tzibanche, Quintana Roo. *Proceedings, 40th International Congress of Americanists*. Rome and Genoa: 495–501.

Henderson, JS 1981 *The World of the Ancient Maya*. Ithaca, New York, Cornell University Press: 271 p.

Hester, TR, Kelly, TC and Ligabue, G, 1981 A fluted Paleo-Indian projectile point from

Belize, Central America. *Working Papers 1, Colha Project*. San Antonio, Center for Archaeological Research, University of Texas.

Jones, C and Satterthwaite, L 1982 The Monuments and Inscriptions of Tikal: The Carved Monuments. *University Museum Monograph* 44, *Tikal Report* 33. Philadelphia, The University Museum, University of Pennsylvania.

Kulp, JL, Feely, HW and Tryon, LE 1951 Lamont natural radiocarbon measurements, I. *Science* 114(2970): 565–568.

Libby, WF 1955 *Radiocarbon Dating*, 2nd edition. Chicago, University of Chicago Press: 175 p.

Lincoln, CE 1986 The chronology of Chichen Itza: A review of the literature. *In* Sabloff, JA and Andrews, EW, V, eds, *Late Lowland Maya Civilization: Classic to Postclassic*. Albuquerque, University of New Mexico Press: 141–196.

Lounsbury, F 1982 Astronomical knowledge and its uses at Bonampak, Mexico. *In* Aveni, AF, ed, *Archaeoastronomy in the New World*. Cambridge, Cambridge University Press: 143–186.

Lowe, GW 1978 Eastern Mesoamerica. *In* Taylor, RE and Meighan, CW, eds, *Chronologies in New World Archaeology*. New York, Academic Press: 331–393.

MacNeish, RS and Peterson, FA 1962 The Santa Marta rock shelter, Ocozocoautla, Chiapas, Mexico. *Papers of the New World Archaeological Foundation* 14.

Marcus, J 1983 Lowland Maya archaeology at the crossroads. *American Antiquity* 48(3): 454–488.

Meggers, BJ 1954 Environmental limitations on the development of culture. *American Anthropologist* 56(5): 801–824.

Meltzer, DJ 1989 Why don't we know when the first people came to North America? *American Antiquity* 54(3): 471–490.

Merwin, RE and Vaillant, GC 1932 The ruins of Holmul, Guatemala. *Memoirs of the Peabody Museum of Archaeology and Ethnology* 3(2).

Michaels, GH and Voorhies, B (ms) 1989 Settlement systems and resource use by Preceramic estuarine dwellers in southern Mexico. Paper presented at Circum-Pacific Prehistory Conference, Seattle, Washington, August 4.

Miksicek, CH 1983 Macrofloral remains of the Pulltrouser area: settlements and fields. *In* Turner, BL, II, and Harrison, PD, eds, *Pulltrouser Swamp: Ancient Maya Habitat, Agriculture, and Settlement in Northern Belize*. Austin, University of Texas Press: 94–104.

Miksicek, CH, Bird, RM, Pickersgill, B, Donaghey, S, Cartwright, J and Hammond, N 1981 Preclassic lowland maize from Cuello, Belize. *Nature* 289(5793): 56–59.

Millon, R 1981 Teotihuacan: city, state, civilization. *In* Sabloff, JA, ed, *Archaeology: Supplement to the Handbook of Middle American Indians* 1. Austin, University of Texas Press: 198–243.

Morley, SG 1920 The inscriptions at Copan. *Carnegie Institution of Washington Publication* 219: 643 p.

_____ 1946 *The Ancient Maya*. Stanford, California, Stanford University Press: 520 p.

Morley, SG and Brainerd, GW; revised by Sharer, RJ 1983 *The Ancient Maya*, 4th edition. Stanford, California, Stanford University Press: 708 p.

Ogden, JG, III and Hart, WC 1977 Dalhousie University natural radiocarbon measurements II. *Radiocarbon* 19(3): 392–399.

Parsons, LA 1967 Bilbao: Guatemala. *Publications in Anthropology* 11. Milwaukee Public Museum: 197 p.

Pring, DC 1979 The Swasey ceramic complex of northern Belize: A definition and discussion. *Contributions of the University of California Archaeological Research Facility* 41: 215–229.

Puleston, DE 1977 The art and archaeology of hydraulic agriculture in the Maya lowlands. *In* Hammond, N, ed, *Social Process in Maya Prehistory*. London, Academic Press: 449–467.

Rice, DS, Rice, PM and Deevey, ES, Jr 1985 Paradise lost: Classic Maya impact on a lacustrine environment. *In* Pohl, M, ed,

Prehistoric lowland Maya environment and subsistence economy. *Papers of the Peabody Museum of Archaeology and Ethnology, Harvard University* 77: 91–05.

Ricketson, OG, Jr 1928 A stratification of remains at an early Maya site. *Proceedings of the National Academy of Sciences* 14(7): 505–508.

Ricketson, OG, Jr and Ricketson, EB 1937 Uaxactun, Guatemala: Group E, 1926–1931. *Carnegie Institution of Washington Publication* 477: 314 p.

Sanchez, WA and Kutzbach, JE 1974 Climate of the American tropics and subtropics in the 1960s and possible comparisons with climatic variations of the last millennium. *Quaternary Research* 4(2): 128–135.

Satterthwaite, L 1956 Radiocarbon dates and the correlation problem. *American Antiquity* 21(4): 416–419.

_____1967 Radiocarbon and Maya Long Count dating of 'Structure 10' (Str 5D-52, first story), Tikal. *Revista Mexicana de Estudios Antropologicos* 21: 225–249.

Satterthwaite, L and Ralph, EK 1960 New radiocarbon dates and the Maya correlation problem. *American Antiquity* 26(2): 165–184.

Sharer, RJ 1974 The prehistory of the southeastern Maya periphery. *Current Anthropology* 15(2): 165–187.

_____1984 Lower Central America as seen from Mesoamerica. *In* Lange, FW and Stone, DZ, eds, *The Archaeology of Lower Central America*. Albuquerque, University of New Mexico Press: 63–84.

Sheets, PD 1971 An ancient natural disaster. *Expedition* 14(1): 24–31.

_____1979 Environmental and cultural effects of the Ilopango eruption in Central America. *In* Sheets, P and Grayson, D, eds, *Volcanic Activity and Human Ecology*. New York, Academic Press: 525–564.

_____1983 Summary and conclusions. *In* Sheets, P, ed, *Archaeology and Vulcanism in Central America: The Zapotitan Valley of El Salvador*. Austin, University of Texas Press: 275–293.

Sheets, PD 1986 Natural hazards, natural disasters, and research in the Zapotitan Valley of El Salvador. *In* Urban, PA and Schortman, EM, eds, *The Southeast Maya Periphery*. Austin, University of Texas Press: 224– 238.

Siemens, AH 1983 Wetland agriculture in Pre-Hispanic Mesoamerica. *The Geographical Review* 73(2): 166–181.

Siemens, AH and Puleston, DE 1972 Ridged fields and associated features in southern Campeche: New perspectives on the lowland Maya. *American Antiquity* 37(2): 228–239.

Smith, RE 1955 Ceramic sequence at Uaxactun, Guatemala. *Middle American Research Institute Publication* 20, 2 vols: 214 p.

_____1971 The pottery of Mayapan including studies of ceramic material from Uxmal, Kabah, and Chichen Itza, 2 vols. *Papers of the Peabody Museum of Archaeology and Ethnology, Harvard University* 66.

Spinden, HJ 1924 The reduction of Maya dates. *Papers of the Peabody Museum of American Ethnology and Archaeology, Harvard University* 6(4): 286 p.

Stephens, JL 1962 *Incidents of Travel in Yucatan*, 2 vols. Norman, University of Oklahoma Press.

Stirling, M 1940 An initial series from Tres Zapotes, Veracruz, Mexico. *National Geographic Society Contributed Technical Papers, Mexican Archaeology Series* 1(1).

Stuiver, M and Becker, B 1986 High-precision decadal calibration of the radiocarbon time scale, AD 1950–2500 BC. *In* Stuiver, M and Kra, RS, eds, Proceedings of the 12th International ^{14}C Conference. *Radiocarbon* 28(2B): 863–910.

Thompson, JES 1935 Maya chronology: The correlation question. *Contributions to American Archaeology* 3(14): 51–104.

_____1941 Dating of certain inscriptions of non-Maya origin. *Carnegie Institution of Washington, Theoretical Approaches to Problems* 1: 85 p.

_____1948 An archaeological reconnaissance

in the Cotzumalhuapa region, Escuintla, Guatemala. *Carnegie Institution of Washington, Contributions to American Anthropology and History* 9(44): 56 p.

_____1954 *The Rise and Fall of Maya Civilization.* Norman, University of Oklahoma Press: 334 p.

_____1960 *Maya Hieroglyphic Writing: An Introduction.* Norman, University of Oklahoma Press: 347 p.

Tsukada, M 1966 The pollen sequence. *In* Cowgill, UM, Hutchinson, GE, Racek, AA, Goulden, CE, Patrick, R and Tsukada, M, eds, The history of Laguna de Petenxil: A small lake in northern Guatemala. *Memoirs of the Connecticut Academy of Arts and Sciences* 17: 63–66.

Turner, BL, II 1978 The development and demise of the swidden thesis. *In* Harrison, PD and Turner, BL, II, eds, *Pre-Hispanic Maya Agriculture.* Albuquerque, University of New Mexico Press: 13–22.

Vaillant, GC (ms) 1927 The chronological significance of Maya ceramics. PhD dissertation, Harvard University.

_____1935 Chronology and stratigraphy in the Maya area. *Maya Research* 2: 119-143.

Vaughn, HH, Deevey, ES, Jr and Garrett-Jones, SE 1985 Pollen stratigraphy of two cores from the Peten lake district, with an appendix on two deep-water cores. *In* Pohl, M, ed, Prehistoric lowland Maya environment and subsistence economy. *Papers of the Peabody Museum of Archaeology and Ethnology, Harvard University* 77: 73–89.

Voorhies, B 1976 The Chantuto people: An Archaic period society of the Chiapas littoral, Mexico. *Papers of the New World Archaeological Foundation* 41: 147 p.

Voorhies, B and Michaels, GH (ms) 1990 The Chantuto people revisited: New research on the Late Archaic period of coastal Chiapas, Mexico. Paper presented at the 55th Annual Meeting of the Society for American Archaeology, Las Vegas, Nevada, April 18–22.

Weaver, MP 1981 *The Aztecs, Maya, and Their Predecessors: Archaeology of Mesoamerica*, 2nd edition. New York, Academic Press: 597 p.

Wendland, WM and Bryson, RA 1974 Dating climatic episodes of the Holocene. *Quaternary Research* 4(1): 9–24.

Wilkerson, SJK 1975 Pre-agricultural village life: The Late Preceramic period in Veracruz. *In* Graham, JA, ed, Studies in Ancient Mesoamerica, II. *Contributions of the University of California Archaeological Research Facility* 27: 111–122.

Willey, GR, Culbert, TP and Adams, REW, eds, 1967 Maya lowland ceramics: A report from the 1965 Guatemala City conference. *American Antiquity* 32(3): 289–316.

Willey, GR, Smith, AL, Tourtellot, G, III and Graham, I 1975 Excavations at Seibal, Department of Peten, Guatemala. *Memoirs of the Peabody Museum of Archaeology and Ethnology, Harvard University* 13(1): 56 p.

Zeitlin, RN 1984 A summary on three seasons of field investigations into the Archaic period prehistory of lowland Belize. *American Anthropologist* 86(2): 358–369.

27

LIBBY'S UCLA RADIOCARBON LABORATORY: CONTRIBUTIONS TO ARCHAEOLOGY

RAINER BERGER

INTRODUCTION

Willard F Libby's research, leading to the discovery of radiocarbon dating, was principally carried out at the University of Chicago after World War II with his main collaborators, E C Anderson and J R Arnold. At the behest of Libby, A V Grosse and his collaborators at the Houdry Process Corporation demonstrated the existence of radiocarbon in nature by concentrating the isotope from a source of biogenic methane. A Committee on Carbon-14 was formed from members of the American Anthropological Association and the Geological Society of America to select a significant slate of samples for dating. Committee members were Frederick Johnson, Donald Collier, Richard Foster Flint and Froelich Rainey, who all assisted Libby with advice and dating priorities. The basic technique for measuring radiocarbon was solid carbon dating, which Libby described in detail in his book entitled "Radiocarbon Dating" and two updated editions. Typically, samples were counted for 48 hours to accommodate the large numbers submitted. All radiocarbon dates obtained prior to the fall of 1951 by Libby's original research team are listed in his publications (Libby 1952, 1955, 1965).

In 1954, Libby was appointed by President Eisenhower to a five-year term as the principal scientific member of the US Atomic Energy Commission. His activities there included the development of nuclear reactors, worldwide fallout studies, development of health physics and the training of atomic scientists at universities through research and support grants. Further, he was instrumental in organizing the International Atomic Energy Agency and funded the first Atoms for Peace Conference (Berger 1983).

With his return to academic life in 1959, Libby joined the Chemistry Department of the University of California, Los Angeles, and shortly thereafter became the director of the Institute of Geophysics and Planetary Physics of the statewide university. In this capacity, he recruited Gordon Fergusson from New Zealand, who worked, at the time, for the United Nations in New York, to start a new radiocarbon laboratory at UCLA. This facility used, and still uses today, the CO_2 gas proportional counting method, with accuracy up to about one order of magnitude greater than the original solid carbon method in use at the first radiocarbon laboratory in Chicago. Since 1960, the UCLA laboratory has dated a wide range of samples in archaeology, geochemistry and geophysics and other environmentally related fields. The first research assistant in the UCLA radiocarbon laboratory was R E Taylor, who now directs the University of California, Riverside Radiocarbon Laboratory. I arrived in 1963, to join Libby's team. In this contribution, I highlight the UCLA laboratory's work only in anthropology and archaeology, whereas a summary in other fields awaits future publication elsewhere.

DATING EARLY HUMANS IN THE NEW WORLD

In Libby's Nobel address delivered on 12 December 1960 and adapted in *Science* (Libby 1961), he showed a histogram relating the number of human occupation sites in America *versus* age (Fig 27.1). The histogram shows that the bulk of human occupation occurs after 11,500 radiocarbon years. Yet, it contained hints of an earlier presence in the 20,000-year range.

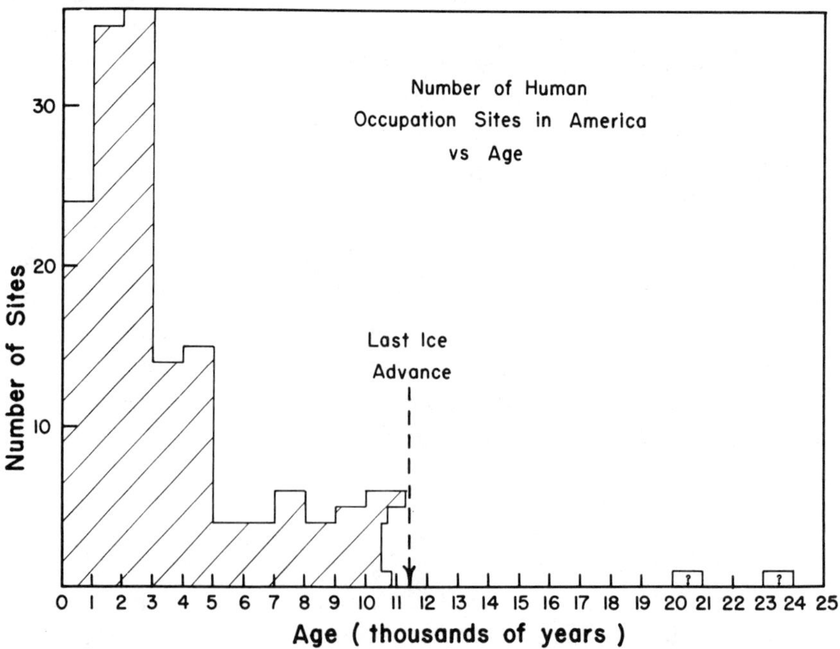

Fig 27.1. Number of human occupation sites in America *versus* their radiocarbon age in 1961 (after Libby 1961)

As a result of Libby's interest in the question when early man actually arrived in the New World, the Tule Springs Project came into being. On 28 February 1962, a meeting was held in Los Angeles attended by Wallace S Broecker, J Desmond Clark, G J Fergusson, R F Heizer, Carl Hubbs, W F Libby, C W Meighan, H B Nicholson, P C Orr, Charles Rozaire, Richard Shutler, Jr and Herschel C Smith. A number of potential early man sites were discussed, with Tule Springs, Nevada emerging as the best choice.

The then brand new UCLA Radiocarbon Laboratory and its members were made available for this multidisciplinary study at a location where earlier studies had suggested that humans had been present more than 28,000 years ago (Harrington & Simpson 1961). Details of the excavations and analyses are found in a summary publication, which contains reports by Shutler, Haynes, Mawby, Mehringer, Bradley and Deacon plus special reports by Brooks, Dove, Fitzwater, Stein, Taylor and Tuohy (Nevada State Museum 1967).

Working under the archaeological direction of Richard Shutler, Jr, the geological direction of C Vance Haynes, Jr and the dating expertise of Fergusson and Libby, a massive excavation project was undertaken northwest of Las Vegas, with R E Taylor serving as field-to-laboratory liaison. In fact, a truck-mounted field dating unit, capable of producing dates on site, was developed. However, the logistics of providing liquid nitrogen for processing samples in the desert proved to be too cumbersome, so specimens were flown to the UCLA laboratory for dating. In the end, the 1962–1963 investigations did not substantiate the previously obtained archaeological dates of 23,800 and 28,000 years for human occupation at Tule Springs in the path of the Las Vegas Wash. Rather, the evidence for man was found to be on the order of 10,000 to 12,000 years. Even though the archaeological results proved to be disappointing to Libby, as they were complicated by the highly mixed depositional environment of the seasonal desert wash, the coordinated use of heavy excavation equipment, coupled with an interdisciplinary approach, provided a well-documented history of the archaeology, geology, palynology and paleontology of this southern Nevada location. In order to appreciate the magnitude of the study, which was funded by the National Science Foundation and private sources, it should be noted that 200,000 tons of overburden were removed, two miles of trenches (12 feet wide and up to 30 feet deep) were cut and 58 radiocarbon dates determined (Fergusson & Libby 1964). Tule Springs illustrates Libby's "big science" approach to an interdisciplinary archaeological study searching for early humans in North America.

Another study reaching back into Early Human times was the dating of basketry from Falcon Hill, Nevada, where Lovelock cultural deposits have been found in very mixed and disturbed settings. After the early occupants of Falcon Hill excavated caches for storage purposes, burrowing by rodents left basketry fragments from many ages scattered throughout the deposits, regardless of depth. The excavations of this site were carried out by Richard Shutler, Jr and Donald Tuohy of the Nevada State Museum. Findings revealed a long sequence of basket-making, reaching back in time more than 9500 years (UCLA-675: 9540 ± 120; Berger, Fergusson & Libby 1965). According to Charles Rozaire, Los Angeles County Museum of Natural History, the earliest form of basketry was made by twining. First, the stitches leaned to the left, and several thousand years later, to the right. About 2500 years ago, coiled basketry was added to the repertoire, followed by the plaited weave of Lovelock Wicker. By directly dating different manufacturing styles, a basketry chronology was established, which can serve as a temporal control in excavations within the limitations explained above (Berger, Fergusson & Libby 1965).

The Santa Barbara Museum of Natural History, through a succession of curators, David Banks Rogers, Fay-Cooper Cole and P C Orr, extensively explored Santa Rosa Island, some 30 miles offshore in the Pacific (Orr 1968). Particularly fascinating were skeletal remains of dwarf mammoths eroding out of the ground, often in proximity to red-baked fire areas containing charcoal. This finding prompted an investigation into the nature of the fire areas, whether they were man-made (Orr & Berger 1966; Berger & Orr 1966; Wendorf 1982; Cushing *et al* 1984, 1986), and what their age was (Fergusson & Libby 1962; Berger, Fergusson & Libby 1965; Berger & Libby 1966). In brief, some appear to be due to forest fires burning tree stumps underground, whereas others are due to the oxidizing effects of groundwater. Areas burned by forest fires usually contain charcoal, ash and outlines of root systems extending outward from the main fire zone. However, some sites do not have the outlines of ancient tree stump fires nor the appearance of oxidized groundwater channels, as they are completely bowl-shaped. These features range from about 10,000 to more than 40,000 ^{14}C years ago, although the frequency is greatest in more recent time, and becomes progressively sparser with increasing antiquity.

One of the most enigmatic of these fire areas is the Woolley Mammoth site (Fig 27.2A) (Berger 1980), which combines a large fire area with skeletal material of a young dwarf mammoth and stone tools about the perimeter of the burned zone (Fig 27.2B). So far, only charcoal from the lower fire has been dated to greater than 40,000 years ago. Up to this point, the mammoth bones have not been analyzed, pending extension of the dating range of the University of California AMS facility to reach beyond 40,000 years. A recent thermoluminescence date for the fire-baked clays from this location yielded an age of around 43,000 years (UCLA-2100E). The object of chronometric testing is to establish

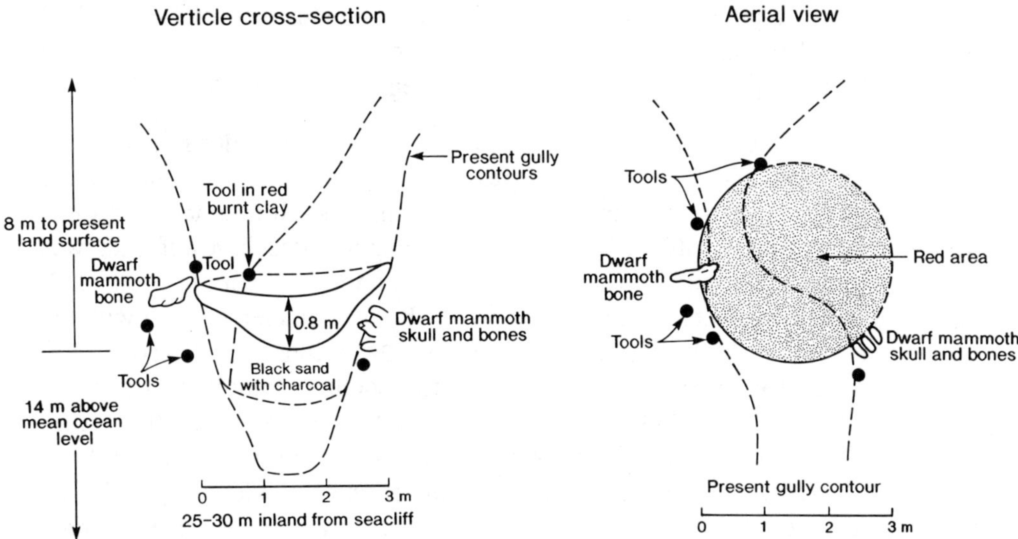

Fig 27.2. The Woolley mammoth site (after Berger 1980)

whether or not mammoth bones and fire are contemporaneous. Ultimately, the question needs to be addressed whether this location is an accidental assemblage or an archaeological site. Orr (1968) has discussed the appearance of mammoths on Santa Rosa Island. The most recent date for their extinction is as late as 8000 years ago (see below).

The application of radiocarbon dating in archaeology introduced new problems in sample selection. Whereas field associations between, say, charcoal, bone and stone tools were often true, at times they were accidental. Hence, it became more desirable to date objects directly to avoid uncertainties of association. Since bone is one of the hardest tissues, it survives long time spans, and is an excellent candidate for direct dating, especially if it is clearly human bone. During the early years of radiocarbon dating, bone was assayed using its carbonate fraction as a source of ^{14}C. Serious discrepancies in dates from the same stratigraphic level were unacceptable to archaeologists as discussed in a study on domestication by Berger and Protsch (1973) and Protsch and Berger (1973). One source of the difficulties was of a geochemical nature in that bone carbonate can exchange with groundwater carbonates to give various ages depending on the extent of carbonate exchange. To avoid this problem, Berger, Horney and Libby (1964) developed collagen dating, which bases a radiocarbon date on the organic fraction of bone, commonly called collagen.

Since then, collagen dating has undergone numerous modifications to make it more accurate. Berger, Fergusson and Libby (1965) reported the original measurements and applications, including a dwarf mammoth date of 8000 ± 250 (UCLA-705) from Santa Rosa Island, California, which appears to be the ultimate terminal phase for the extinction of megafauna in the New World.

At about this time, L S B Leakey started to give presentations in the Los Angeles area on his African discoveries, which also introduced him to the question of the antiquity of man in the New World. A skull fragment and some post-cranial material collected in 1933 from Laguna Beach, California was shown to him. He brought this to the UCLA laboratory's attention. Both skull and long-bone fragments were dated separately by the collagen method with the following results: skull, first run: 7100 ± 1300 BP, second run 17,150 ± 1470 BP (UCLA-1233A) based on a 4.3% fill of the counter; long-bone fragment, > 14,800 years (UCLA-1233B) (Berger & Libby 1969: 194–95). Another result, this time derived by accelerator mass spectrometry (AMS), was obtained from a fraction used for an earlier amino-acid racemization date, with an age of 5100 BP (Bada *et al* 1984; Taylor *et al* 1985). In fact, it would be desirable to redate the Laguna skull fragment in two separate laboratories, using AMS techniques and exactly the same pretreatment protocol, to avoid the introduction of carbon from chemicals of varying specific activity. As always, small samples are highly susceptible to contamination, which can then lead to very discordant dates.

Another challenging dating problem was the Pleistocene tar-pit bone deposits of Rancho La Brea about 10 km from UCLA. To remove the tar contamination, a chromatographic separation technique was developed, allowing for measurement of the pure bone amino-acid fraction (Berger & Libby 1968b; Ho, Marcus & Berger 1968; Marcus & Berger 1984). The results showed that the Rancho La Brea deposits are stratified and not highly mixed, as suspected earlier. Further, the first finite radiocarbon dates for the upper portion placed this important site into chronometric perspective. The only human skeleton ever found in the tar pits was dated to 9000 ± 80 years (Berger *et al* 1971).

When it became known that bones could be dated directly, a number of samples were submitted principally from western United States sites, which all resulted in Holocene dates (Taylor *et al* 1985), except one. This particular sample of Los Angeles Man was found in 1936 in the Los Angeles River bed. It was dated by the same procedure as the La Brea skeleton and calculated to be over 23,600 years old (Berger *et al* 1971). However, a date reported by Taylor *et al* (1985) states a revised age of 3560 BP. Again, the discrepancy between these dates may be due to different chemical protocols that can vary somewhat from laboratory to laboratory. Since most laboratories use the same radiocarbon standards, differences between counter or accelerator assays are less likely than problems generated by the use of different chemicals, especially those containing organic carbon. Thus, critically important samples should be dated by at least two laboratories using the same chemical methodology and carefully inter-calibrated equipment for radiocarbon assay.

In northern California, the Mostin site is a large early human locality, located along Kelsey Creek where it flows into Clear Lake. First discovered by Julian Mostin in 1973, it later was partially excavated by T S Kaufman (1980), who found burials and abundant obsidian artifacts. Radiocarbon dates of the site yielded ages of about 11,000 years ago on charcoal (UCLA-2165: 11,250 ± 240) and collagen (UCLA-2171: 10,470 ± 490) for a human burial containing three ground-stone artifacts. Some 70 burials have been lost to seasonal high-water stages of Kelsey Creek (J Mostin, personal communication). Taken as a whole, the Mostin site appears to be the oldest and largest cemetery area in the Americas. In contrast to the radiocarbon dates that support each other, obsidian measurements carried out so far at the Mostin site provide a discordant array of dates: a Borax Lake-source obsidian has a hydration maximum at the Mostin location of 7.4 microns, whereas other sites in the Clear Lake area at maximum depth show readings of 8–11 microns. The obsidian hydration process is dependent on local variations of chemical composition of the mineral itself, its temperature history and differences in groundwater chemistry. Understanding these regional variations in the Clear Lake region is currently the object of another dissertation project (J Parker, personal communication).

In retrospect, the various locales discussed here have provided evidence of varying quality for the existence of early humans. Using Oakley's (1964) chronologic evaluation scheme, it is possible to group archaeological finds in order of firmness of evidence. He subdivides dating categories into two main groups, the first called relative dating (R-1 through R-3), and the other, chronometric or absolute dating (A-1 through A-4). The criteria employed for defining the subgroups are listed below:

<u>Relative Dating</u>

R-1	Age related to the deposit where found;
R-2	Stage in local archaeologic or stratigraphic sequence;
R-3	Position of stage in wider stratigraphy or archaeologic context.

<u>Chronometric or Absolute Dating</u>

A-1	Direct dating of specimen (bone, etc);
A-2	Direct dating of source deposit (K/A dating of volcanic minerals in stratum);

A-3	Age of specimen inferred by correlation of source bed with deposit of known age;
A-4	Age inferred from theoretical considerations (eg, dates obtained by interpreting a local geologic sequence in terms of climatic fluctuations).

This classification assumes that the chronometric technique used yields correct results and that the associations in reality are what they appear to be. Laboratory and field errors may also enter this classification scheme not immediately addressed by the definitions. For practical purposes, suspect associations will be denoted by A?, and laboratory results that seem difficult to accept by L?.

In summary, the early man locations described here can be classified as follows in alphabetical order for quick comparison:

Falcon Hill	A-1
La Brea	A-1
Laguna	A-1/L?
Los Angeles Man	A-1/L?
Mostin	A-1/A-2
Woolley Mammoth	A-2/A?
Tule Springs	A-2/A?

From a rigorous chronometric point of view based on A-1 relationships, the earliest evidence of mankind in California or Nevada dates to the 10,000–12,000-year range, whereas there are hints of an earlier presence at a lesser level of confidence.

AMERICAN PREHISTORIC PROJECTS

The entire number of dating projects with which the UCLA laboratory was involved is too large to include here. Complete information on all the projects will be found in published UCLA date lists (Fergusson & Libby 1962, 1963, 1964; Berger, Fergusson & Libby 1965; Berger & Libby 1966, 1967, 1968a, b, 1969; Berger & Ericson 1983; Berger & Protsch 1989).

One of the earlier chronometric projects involved the Hopewell-Adena mounds in Ohio (Fergusson & Libby 1963). The principal collector of the samples was F J Soday of the Skelly Oil Co, Tulsa, Oklahoma. In general, all three Adena mounds were very close in age, ranging from 1825 to 2040 BP, whereas the

Hopewell mounds covered a broader time range, from 1000 to 2700 BP. The three Hopewell mounds were of quite dissimilar structure and artifact inventory, strengthening support for a longer occupation period. Additional dates are summarized in Berger, Fergusson and Libby (1965).

A second major project was the dating of Illinois mounds excavated by Joseph Tainter of Northwestern University, Evansville, Illinois and Jane Buikstra of the University of Chicago. Whereas the Hopewell-Adena study was based on charcoal samples, the Illinois project involved bone samples (Berger & Protsch 1989). The ages start at about 2000 BP and reach into Colonial times, indicating prolonged interment practices at these locations. For Buikstra's Helton site, no bone pathology was observed, as reported in Connor (1985).

In Mexico and Mesoamerica, dating projects focused on complex societies, from onset through florescence and decline. The first major project involved the Cuicuilco pyramid area just south of Mexico City. Excavations were directed by R F Heizer, University of California, Berkeley, who published a summary of the 1957 work together with J A Bennyhoff, University of Rochester (Heizer & Bennyhoff 1958). The radiocarbon dates, in turn, were published by Fergusson and Libby (1963, 1964). On the basis of these dates, a radiocarbon chronology was established: Ticoman I: 500–400 BC, Ticoman II: 400–300 BC, Ticoman III: 300–200 BC, Cuicuilco/Tezoyuca: 200–100 BC, Chimalhuacan: 100 BC to AD 1, Tzacualli: AD 1–100. Interestingly, one of the dates places the eruption of the Xitli volcano in the 5/6th century AD, based on a root carbonized by volcanic action.

One of the most important chronometric studies in Guatemala was the calendar comparison between the Mayan and current Roman calendric system. The UCLA laboratory actively worked on two Tikal projects, based on wood from lintel and vault beams that had been carefully selected to avoid post-sample growth errors. These samples were submitted by L Satterthwaite of the University Museum, University of Pennsylvania, to check on the veracity of previous Pennsylvania radiocarbon dates. Upon comparison, the UCLA and Pennsylvania dates differed by 34 ± 34 years and a second set by 19 ± 50 years. These dates were determined before radiocarbon laboratories routinely used tree-ring calibrated dates. For a second analysis, I obtained wood samples from the sap wood edge of the Tikal lintel carving in the Museum für Völkerkunde, Basel, Switzerland. The dates were corrected for ^{13}C fractionation and tree-ring calibrated (Suess 1965). Both support the Goodman-Thompson-Martinez (Thompson 1950) correlation similar to other results (Berger 1968; Fergusson & Libby 1963; Berger & Libby 1969).

At Teotihuacan, 25 miles northeast of Mexico City, a major chronologic study focused on the type site of the culture following the preclassic phase in central Mexico. This urbanized area of some nine square miles was on open ground without city walls. Large temples and palaces were in the middle of the city. Apparently, Teotihuacan was violently destroyed by fire and never rebuilt. Of a suite of 11 samples, only one dates to the estimated time of the fire: UCLA-613 at 1290 ± 80 BP, equivalent to AD 660–814 when calibrated. Ignacio Bernal and R F Heizer believe that the other dates represent re-use of timber as explained in more detail in Berger, Fergusson and Libby (1965).

The origin of the Olmec culture, especially *vis-a-vis* the Mayan, puzzled Mesoamericanists for years. UCLA participated in several seasons of excavations at La Venta, Tabasco and published the data in three date lists (Berger, Fergusson & Libby 1965; Berger & Libby 1966, 1968). According to the earliest dates, La Venta was built as a ceremonial center in 3050 BP, which, upon tree-ring calibration, becomes 1415 to 1255 BC (Stuiver & Kra 1986).

For years, the UCLA laboratory has been active in West Mexican archaeology, in the states of Colima, Jalisco and Nayarit, leading to the establishment of the first West Mexican chronology and cultural history. The samples, submitted by C W Meighan and his students, dated shaft tombs to around 2000 BP (Long & Taylor, 1966) as well as number of major sites, such as Amapa, Tizapan and Morett (Fergusson & Libby 1963; Berger, Fergusson & Libby 1965; Berger & Libby 1967. Our understanding of the chronology of the culture history of western Mexico is based largely on the corpus of UCLA ^{14}C dates (Meighan & Nicholson 1989).

European Medieval Buildings

This study series began in 1962 with an inquiry into the origin of cruck-framed houses in England to determine if buildings in existence today would reach back into Anglo-Saxon times of the 6th century. None were found to date that far into the past (Fergusson & Libby 1963). Work continued on determining the origin of the medieval bay system (Berger 1970) and early medieval structures in Ireland (Berger 1989). Most of the structures were dated using wood samples; some were also checked dendrochronologically and, in the absence of datable wood, by the radiocarbon assay of charcoal inclusions in mortar (Berger 1991).

Our overall interest began with defining architectural-historical chronology, and turned later towards an independent check of Suess' "wiggles," as reported first in Horn, Charles and Berger (1967) and summarized by Berger (1985). In essence, medieval buildings of known construction date or of unknown origin could be correctly dated by applying the "Suess Correction" (Berger 1970), which was later refined and extended by many laboratories, culminating in the *Radiocarbon* Calibration Issue (Stuiver & Kra 1986). This lists the latest conversions of radiocarbon years to tree-ring calibrated or calendric years with perhaps only one more issue refining the calibration relationships to be published in the future (Stuiver & Pearson 1992).

EGYPTIAN PHARAONIC CHRONOLOGY

Inasmuch as the radiocarbon production rate in the past has been modulated by geomagnetic and heliomagnetic field changes, it became of paramount importance to check the deviations first measured by Suess (1965) also in the distant past as far back as historically possible. To that effect, a collaboration was established between I E S Edwards, British Museum, W B Emery, University College, London, G T Martin, Corpus Christi College, Cambridge and the UCLA laboratory to obtain historically well-dated samples from Egypt. It was generally agreed that this chronology was the best defined in the world with the greatest antiquity. Berger (1985) confirmed Suess' stipulated variations, by historic-archaeologic means as well as extensively modifying radiocarbon dates to correspond to calendric ages. Initially, Libby was very skeptical of the need for calibration, but his views changed as more and more evidence came to light that quantitatively supported the deviations. Subsequently, he actually became very excited about the geophysical implications as discussed in this commemorative volume.

AFRICAN PREHISTORY

The UCLA radiocarbon laboratory has had a long history of collaboration with J Desmond Clark of the University of California, Berkeley, and dated numerous sites in Africa. Interestingly, in all these years of collaboration, Clark never made a single mistake in estimating the likely age of the samples he submitted. Most of the samples dated found their way into a monumental publication, "Atlas of African Prehistory" (Clark 1967). Other discussions of the later Pleistocene cultures of Africa are published separately (Clark 1965).

In conjunction with L S B Leakey, samples from the upper strata of Olduvai Gorge and the Omo River site were dated. Both were collagen dates from mammalian bone. Omo River gave 15,500 ± 300 (UCLA-1319) and Olduvai Gorge Bed 5: 10,100 ± 600 (UCLA-1321) (Berger & Libby 1969).

Reiner Protsch (1973) dated Sub-Saharan fossil hominid bones by the collagen method. All the measurements have been summarized in Berger and Protsch (1989). These 19 radiocarbon dates became the chronometric control of a study indicating that anatomically modern humans occur sympatrically, with, and perhaps even antedating in Europe, the Rhodesian group of Neanderthals. This makes a case for an offshoot group of Neanderthals, located outside the mainstream of human evolution, which leads from *Homo erectus* directly to *Homo sapiens*.

OCEANIA

Most of the radiocarbon investigations in the Pacific were carried out in the Philippines, especially on Palawan, and in Sarawak. The sites were Duyong, Manunggul, Niah and Tabon Caves; the principal excavators were R B Fox, National Museum, Manila, and R Shutler, Jr of Simon Fraser University. In brief, these dates show that man was present at Niah more than 40,000 years ago (Berger & Libby 1966), whereas the Palawan dates reach to more than 22,000 years ago. The latter dates were limited as finite dates by the size of the sample (Fergusson & Libby 1964).

Shutler collected another suite of dates in New Hebrides, where the oldest date of cultural material is about AD 1000. A greater time range applies to ceramics found on Guam by F Reinman, California State University, Long Beach. Carbon-bearing sherds were compared with associated charcoal (Taylor & Berger 1968), and there was reasonable agreement. In the end, C W Meighan thought that carbon dates on ceramics might be preferable to associated charcoal, because large sherds are less likely to move in the soil than small charcoal pieces.

Since the last UCLA date list (Berger & Protsch 1989), many more samples from the Pacific have been dated to ultimately characterize the migration of seafarers from the Asian mainland into the Pacific islands, such as New Zealand, Hawaii, Tahiti and neighboring Easter Island (Berger & Shutler, Pacific migrations, ms in preparation).

EPILOGUE

In 1976, Willard F Libby hosted the Ninth International Radiocarbon Conference, together with Hans E Suess at Los Angeles and La Jolla (Berger & Suess 1979). Only a few years later on 8 September 1980, pulmonary complications ended Willard F Libby's life at the UCLA Medical Center and his steady involvement in the UCLA radiocarbon laboratory. During his years at UCLA, his laboratory spawned new radiocarbon facilities at the Universities of California, Riverside and Irvine, at the J W Goethe University in Frankfurt, Germany and indirectly, the University of California AMS Center as well.

REFERENCES

Bada, JL, Gillespie, JA, Gowlett, JAJ and Hedges, REM 1984 Accelerator mass spectrometry radiocarbon ages of amino acid extracts from California paleo Indian skeletons. *Nature* 312: 442–444.

Berger, R 1968 Recent investigations toward the Maya calendar correlation problem. Stuttgart-Munchen, *Verhandlungen des XXXVIII International Amerikanistenkongresses*: 209–212.

_____ed 1970 *Scientific Methods in Medieval Archaeology*. Berkeley-Los Angeles, University of California Press: 459 p.

_____1980 The Woolley mammoth site, Santa Rosa Island, California. *In* Power, DM, ed, *The California Islands*. Santa Barbara, Santa Barbara Museum of Natural History: 73–78.

_____1983 Willard Frank Libby 1908–1980. *PACT* 8: 13–16.

_____1985 Suess' wiggles and deviations proven by historical and archaeological means. *Meteoritics* 20(2/2): 395–402.

_____1989 Early medieval Irish buildings: Radiocarbon dating of mortar. *PACT* 29: 415–422.

_____1991 Radiocarbon dating of early medieval Irish monuments. *Proceedings of the Royal Irish Academy*, in press.

Berger, R and Ericson, J 1983 UCLA radiocarbon dates X. *Radiocarbon* 25(1): 129–136.

Berger, R, Fergusson, GJ and Libby, WF 1965 UCLA radiocarbon dates IV. *Radiocarbon* 7: 336–371.

Berger, R, Horney, AG and Libby, WF 1964 Radiocarbon dating of bone and shell from their organic components. *Science* 144: 999–1001.

Berger, R and Libby, WF 1966 UCLA radiocarbon dates V. *Radiocarbon* 8: 467–497.

_____1967 UCLA radiocarbon dates VI. *Radiocarbon* 9: 477–504.

_____1968a UCLA radiocarbon dates VII. *Radiocarbon* 10(1): 149–160.

_____1968b UCLA radiocarbon dates VIII. *Radiocarbon* 10(2): 402–416.

_____1969 UCLA radiocarbon dates IX. *Radiocarbon* 11(1): 194–209.

Berger, R and Orr, PC 1966 The fire areas on Santa Rosa Island, II. *Proceedings of the National Academy of Sciences USA* 56: 1678–1682.

Berger, R and Protsch, R 1973 The domestication of plants and animals in Europe and the Near East. *In* Buccellati, G, ed, The Gelb volume: Approaches to the study of the ancient Near East. *Orientalia* 42: 214–227.

_____1989 UCLA radiocarbon dates XI. *Radiocarbon* 31(1): 55–67.

Berger, R, Protsch, R, Reynolds, R, Rozaire, C and Sackett, JR 1971 New radiocarbon dates based on bone collagen of California palaeoindians. *Contributions of the University of California Archaeological Research Facility, Berkeley* 12: 43–49.

Berger, R and Suess, HE, eds 1979 *Radiocarbon Dating*. Proceedings of the 9th International ^{14}C Conference. Berkeley/Los Angeles, University of California Press: 787 p.

Clark, JD 1965 The later Pleistocene cultures of Africa. *Science* 150: 833–836.

_____1967 *Atlas of African Prehistory*. Chicago, University of Chicago Press:

Connor, M 1985 Population structure and biological variation in the late Woodland of western central Illinois. PhD dissertation, University of Chicago.

Cushing, J, Daily, M, Noble, M, Roth, VL and Wenner, A 1984 Fossil mammoths from Santa Cruz Island, California. *Quaternary Research* 21: 376–384.

Cushing, J, Wenner, AM, Noble, E and Daily, M 1986 Groundwater hypothesis for the origin of fire areas on the Northern Channel Islands, California. *Quaternary Research* 26: 376–217.

Fergusson, GJ and Libby, WF 1962 UCLA radiocarbon dates I. *Radiocarbon* 4: 109–114.

_____1963 UCLA radiocarbon dates II. *Radiocarbon* 5: 1–22.

_____1964 UCLA radiocarbon dates III. *Radiocarbon* 6: 318–339.

Harrington, MR and Simpson, RD 1961 Tule Springs, Nevada, with other evidences of Pleistocene Man in North America. *Southwest Museum (Los Angeles) Papers* 18.

Heizer, RF and Bennyhoff, JA 1958 Archaeological investigations in Cuicuilco, Valley of Mexico. *Science* 127: 232–233.

Ho, T-Y, Marcus, LF and Berger, R 1968 Radiocarbon dating of petroleum impregnated bone from tar pits at Rancho La Brea, California. *Science* 164: 1051–1052.

Horn, W, Charles, FWB and Berger, R 1967 The cruck-built barn at Middle Littleton in Worcestershire, England. *Journal of the Society for Architectural Historians* XXV (4): 221–239.

Kaufman, TS 1980 Early prehistory of the Clear Lake area, Lake County, California. PhD dissertation, University of California, Los Angeles.

Libby, WF, 1952, *Radiocarbon Dating*. Chicago, University of Chicago Press: 124 p.

_____1955 *Radiocarbon Dating*, 2nd edition. Chicago, University of Chicago Press: 175 p.

_____1961 Radiocarbon dating. *Science* 133: 621–629.

_____1965 *Radiocarbon Dating*. Chicago, University of Chicago Press, Phoenix Science Series: 175 p.

Long, SV and Taylor, RE 1966 Chronology of a West Mexican shaft-tomb. *Nature* 212: 651–652.

Marcus, LF and Berger, R 1984 The significance of radiocarbon dates for Rancho La Brea. *In* Martin, PS and Klein, RG, eds, *Quaternary Extinctions*. Tucson, University of Arizona Press: 159–183.

Meighan, CW and Nicholson, HB 1989 The ceramic mortuary offerings of prehistoric West Mexico: An archaeological perspective. In *Sculpture of Ancient West Mexico*. Los Angeles County Museum of Art and University of New Mexico Press: 29–69.

Nevada State Museum 1967 Pleistocene studies in southern Nevada. This is the Tule Springs report containing contributions by Shutler, R, Jr, Haynes, CV, Mawby, JE, Mehringer, PJ, Jr, Bradley, WG and Deacon JE with special reports by Brooks, RH, Dove, C, Fitzwater, RJ, Stein, WT, Taylor, DW and Tuohy, DR. *Nevada State Museum Anthropological Papers* 13: 441 p.

Oakley, KP 1964 *Frameworks for Dating Fossil Man*. London, Weidenfeld and Nicholson: 355 p.

Orr, PC 1968 *Prehistory of Santa Rosa Island*. Santa Barbara, Santa Barbara Museum of Natural History: 253 p.

Orr, PC and Berger, R 1966 The fire areas on Santa Rosa Island I. *Proceedings of the National Academy of Sciences USA* 56: 1409–1416.

Protsch, R, 1973 The dating of upper Pleistocene subsaharan fossil hominids and their place in human evolution: with morphological and archaeological implications. PhD dissertation, University of California, Los Angeles.

Protsch, R and Berger, R 1973 Earliest radiocarbon dates for domesticated animals. *Science* 179: 235–239.

Stuiver, M and Kra, RS, eds, 1986 Calibration Issue. Proceedings of the 12th International ^{14}C Conference. *Radiocarbon* 28 (2B): 805–1030.

Stuiver, M and Pearson, GW 1992 Calibration of the radiocarbon time scale, 2500–5000 BC. *In* Taylor, RE, Long, A and Kra, RS, eds, *Radiocarbon After Four Decades: An Interdisciplinary Perspective*. New York, Springer-Verlag, this volume.

Suess, HE 1965 Secular variations of the cosmic-ray-produced carbon-14 in the atmosphere and their interpretations. *Journal of Geophysical Research* 70: 5937–5952.

Taylor, RE and Berger, R 1968 Radiocarbon dating of the organic portion of ceramic and wattle-and-daub house construction materials of low carbon content. *American Antiquity* 33(3): 363–366.

Taylor, RE, Payen, LA, Prior, CA, Slota, PJ, Jr, Gillespie, R, Gowlett, JAJ, Hedges, REM, Jull, AJT, Zabel, TH, Donahue, DJ and Berger, R 1985 Major revisions in the Pleistocene age assignments for North American Human skeletons by C-14 accelerator mass spectrometry: none older than 11,000 C-14 years BP. *American Antiquity* 50: 136–140.

Thompson, JE, Jr 1950 Maya hieroglyphic writing. *Carnegie Publication* 589. Washington, DC, Carnegie Institution of Washington: 347 p.

Wendorf, M 1982 Prehistoric manifestations of fire and fire areas of Santa Rosa Island, California. PhD dissertation, University of California-Berkeley.

EARTH SCIENCES

MEYER RUBIN

To the earth scientist of today, research without the benefits of radiocarbon dating is nearly incomprehensible. The temporal nature of the various earth science fields – including geology, climatology, oceanography, palynology and volcanology – requires dates to establish rates, corroborate theories, and generally satisfy the professional desire to construct as complete a picture as possible. The following chapters are good examples of the important role radiocarbon plays in three earth science fields.

Plafker, LaJoie and Rubin exemplify the use of radiocarbon dating by geologists in the study of tectonics and earthquakes. The authors use the time of onset and termination of buried peat layers and their rate of growth to deduce recurrence intervals and subsidence patterns for great 1964-type earthquakes. For short-term repetitive geologic events of this nature, no other dating technique matches radiocarbon's effectiveness.

Dorothy Peteet, an active researcher in palynology, presents an excellent summary of the field, covering its history of correlation, changes and advances, as affected by radiocarbon dating. She notes the impact AMS dating has had and will continue to have on palynology, and the concurrent climatic inferences that have become possible due to smaller sample size requirements.

Fairbanks, Charles and Wright use $^{230}Th/^{234}U$ dating by mass spectrometry to calibrate radiocarbon dates beyond tree-ring chronology for the Barbados sea-level record. Sea-level records and meltwater plumes, as measured by $\delta^{18}O$ in forams in ocean cores, are used to distinguish global meltwater pulses, which may amplify changes in seasonal insolation that apparently trigger the large-scale glacial cycles. Meltwater pulses are dated in planktonic foraminifera using AMS ^{14}C and calibrated with $^{230}Th/^{234}U$ dates on corals to document the timing of the disintegration of the different continental ice sheets.

For forty years, the earth sciences and radiocarbon dating have evolved together integrally. Future usage of radiocarbon dating will increasingly incorporate the benefits of AMS, with its small sample size, to explore conventional and as yet unforeseen applications in every field of earth science.

DETERMINING RECURRENCE INTERVALS OF GREAT SUBDUCTION ZONE EARTHQUAKES IN SOUTHERN ALASKA BY RADIOCARBON DATING

GEORGE PLAFKER, K R LAJOIE and MEYER RUBIN

INTRODUCTION

We outline here research in progress directed towards determining the seismotectonic cycle in the part of south-central Alaska that was deformed by vertical tectonic displacements during the great 1964 Alaska earthquake. Pre-1964 vertical displacements in this same area are recorded in the coastal stratigraphy and geomorphology. ^{14}C analysis of fossil peat, wood, salt marsh plants and shell material from shoreline sedimentary deposits provides the quantitative dating for this study. The history of vertical deformation yields recurrence intervals for 1964-type events and may ultimately provide a means of forecasting future events.

Vertical Displacements and Sea-Level Datum

The emergence or submergence of a strandline is the algebraic sum of 1) tectonic uplift or subsidence of the land, 2) eustatic sea-level change due to the increase in ocean volume from warming of the surface waters and melting of glaciers, 3) isostatic changes mainly due to postglacial rebound, and 4) sediment compaction. To determine the magnitude of the episodic late Holocene tectonic displacements, the other three factors must be known or assumed. Recent ^{14}C data from Barbados indicate that sea level has been rising at an average rate close to 1.5 mm yr^{-1} over the last 5000 years (Bard *et al* 1990). The isostatic component is much less certain, but a geophysical model that utilizes global ^{14}C-dated postglacial shoreline elevations, predicts an isostatic sea-level rise of about 0.5 ± 0.5 mm yr^{-1} in south-central Alaska (Peltier & Tushingham 1989). Compaction effects are likely to be negligible in thin-sediment sequences on or near bedrock, but they could be locally significant in areas of thick, saturated sediments subject to earthquake-induced liquefaction and lateral spreading such as in upper Cook Inlet or the Copper River delta. For the purposes of this chapter, we arbitrarily assume that for most localities, a reasonably close approximation to the tectonic uplift or subsidence is obtained by adding 1.5 ± 0.5 mm yr^{-1} to strandline submergence or emergence to correct for the eustatic and isostatic components.

Radiocarbon Methods

The peat and wood samples in this study were dated by ^{14}C using proportional gas counters. They were pretreated in the standard technique of acid-alkali-acid boilings in HCl and NaOH solutions. The counting was done in several counters, with sufficient time to give 10,000 counts per sample per counter for a one-sigma uncertainty of less than 80 years. A conservative multiplier of two has been used to allow for laboratory uncertainties in addition to the standard counting uncertainties. ^{14}C dates were calibrated to calendar ages using the tree-ring data of Stuiver and Pearson (1986) and the calibration program of Stuiver and Reimer (1986). The calendar ages in Table 28.1 are fair approximations and better estimates of the true age than the raw, uncalibrated ^{14}C dates, but are not absolute. In many cases, the intersection of the ^{14}C date with the calibration curve (reflecting the initial atmospheric ^{14}C concentration) gives two or more possible dates. Some of the calibrated ages in Table 28.1 are the arithmetic average of the range of these intersections. The uncertainty listed is this range, using the one-sigma uncertainties of the ^{14}C date and that of the calibration curve. While acknowledging the uncertainty of the ages of the organic matter, we believe they are sufficiently accurate for the tentative conclusions we make. These data, together with results of many more samples being dated by both gas counting and accelerator mass spectrometry, will be the basis for a more detailed study.

THE 1964 EARTHQUAKE

The 1964 Alaska earthquake (moment magnitude = 9.2) was the greatest seismotectonic event of the 20th century in terms of the known surface area affected by crustal deformation (Plafker 1969, 1972). Because an extensive coastal area was affected, the data on vertical crustal deformation are among the most complete ever obtained for a great earthquake. The earthquake occurred at the eastern end of the Aleutian arc (Fig 28.1) subduction zone where the convergence rate between the Pacific and North American plates is about 6.3 cm yr^{-1} in a N18°W direction (Minster & Jordan 1978).

Coseismic Displacements

Strandline and geodetic data revealed that vertical tectonic deformation during the 1964 Alaskan earthquake involved a segment of the eastern Aleutian arc 800 km long with an area of more than 140,000 km^2. Vertical deformation in this region consisted of a broad asymmetric downwarp as large as 2 m with uplift as

TABLE 28.1. Radiocarbon-dated samples at lower Alaganic Slough (60°24.4′N, 145°25.4′W) Copper River delta. See Figure 28.8 for stratigraphic section.

No.	Field no.	Lab no.*	Material dated	^{14}C age** (BP)	Calibrated age† (BP)	Sample depth‡ (cm)	Comments
1	88APR11A	W-6084	*Carex*	1640 ± 160	1539 ± 176	450–470	Represents high-tide zone
2	88APR11B	-6085	*Carex*	1610 ± 220	1524 ± 233	400–405	Represents high-tide zone
3	88APR11C	-6088	Peat	1500 ± 160	1329 ± 155	309–395	2.5–5.0 cm above base of peat layer 26 cm thick
4	88APR11D	-6089	Peat	1450 ± 120	1327 ± 106	375–380	15–16.5 cm above base of peat layer 26 cm thick
5	88APR11E	-6092	Peat	1330 ± 140	1308 ± 136	370–375	0–2.5 cm at top of peat layer 26 cm thick
6	88APR11F	-6094	Wood	1260 ± 140	1282 ± 129	370–375	Branch 3cm in diameter at top peat layer 26 cm thick
7	88APR11G	-6097	*Carex*	960 ± 120	1216 ± 138	350–355	Represents high-tide zone
8	88APR11H	-6098	*Carex*	830 ± 120	844 ± 122	270–275	Represents high-tide zone
9	88APR11I	-6102	Peat	350 ± 120	735 ± 103	250–255	2.5–5.0 cm above base of peat layer 51 cm thick
10	88APR11J	-6104	Peat	580 ± 120	423 ± 139	230–235	2.5–5.0 cm below top of peat layer 51 cm thick
11	88APR11K	-6105	Wood	320 ± 120	593 ± 79	230–235	Outer rings of tree stump
12	88APR11L	-6108	*Carex*	310 ± 120	378 ± 149	215–220	Represents high-tide zone
13	88APR11M	-6109	*Carex*	310 ± 120	377 ± 151	180–185	Represents high-tide zone
14	88APR11N	-6110	*Carex*	<200	<200	120–125	Represents high-tide zone
1	88APR11O	-6112	*Carex*	240 ± 120	297 ± 163	90–95	Represents high-tide zone
516	88APR11P	-6114	*Carex*	<200	<200	60–65	Represents high-tide zone

*Run at USGS laboratory in Reston, Virginia by proportional gas counting

**Error quoted is one standard deviation of the counting statistics times an error multiplier of two to allow for all laboratory uncertainties.

†Calendar age calculated from Stuiver and Reimer (1986). Error listed is the width of intersection of ^{14}C age plus its error with the calibration curve with its error. band. See text for additional information on methods.

‡Depths given are below ground surface at site.

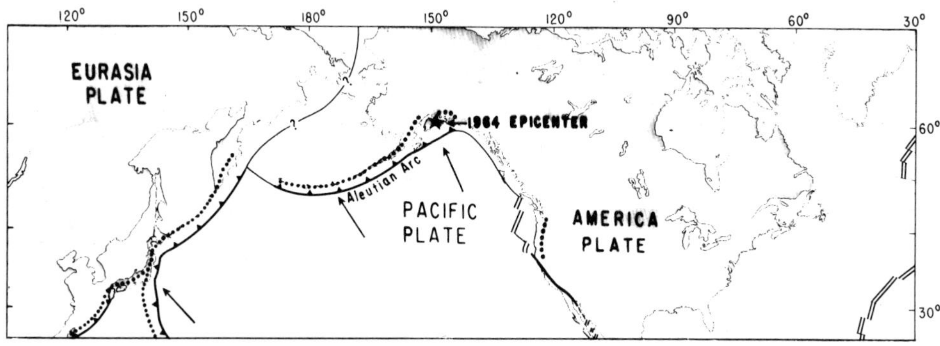

Fig 28.1. Setting of the 27 March 1964 earthquake relative to major tectonic elements around the north Pacific margin. Boundaries of lithospheric plates are indicated by barbed lines for the surface trace of subduction zones that dip beneath the arcs (barbs on upper plate), single lines for predominantly strike-slip faults and double lines for spreading oceanic rises. Dotted lines are volcanic arcs related to subduction. Arrows indicate the horizontal component of motion of the Pacific plate relative to the America and Eurasia plates (Minster & Jordan 1978).

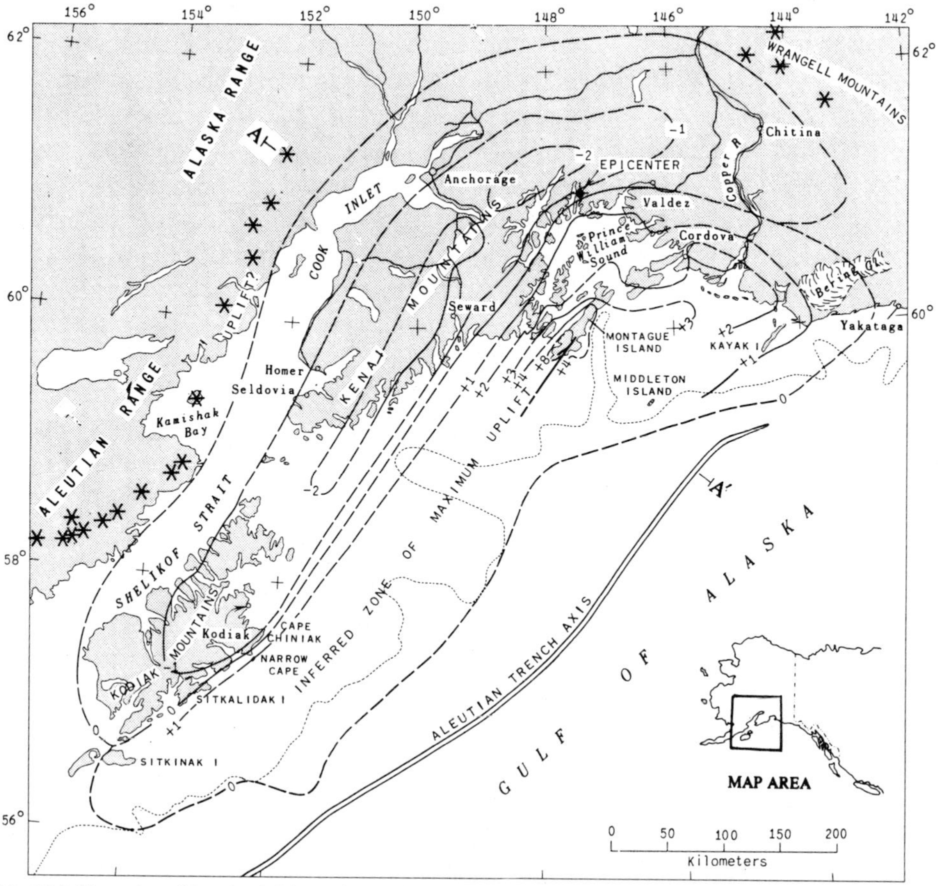

Fig 28.2. Tectonic uplift and subsidence that accompanied the 1964 Alaska earthquake (Plafker 1969). The land-level change, in meters, is shown by the contours, which are dashed where approximate or inferred. The outer edge of the continental shelf, at −200 m, is indicated by the dotted line; the Aleutian Trench axis is shown by the double solid line. Active or dormant eruptive centers of the volcanic arc are shown by stars. See Fig 28.3 for cross-section A-A′.

large as 11.3 m on its seaward side and 0.3 m on its landward side (Fig 28.2). Also, triangulation data showed that the region between Anchorage and the Gulf of Alaska coast shifted seaward at least 20 m (Plafker 1969, 1972). Movement on subordinate northwest-dipping reverse faults displaced the surface on Montague Island as much as 7.9 m dip-slip, and the faults extended offshore to the southwest. A train of destructive sea waves (tsunami) was generated by sudden upheaval of the sea floor on the continental shelf along the zone of major uplift and subsidiary faulting (Plafker 1969).

Rupture Mechanism

The 1964 earthquake resulted from a complex rupture along the plate interface (Aleutian megathrust) and on subsidiary faults in the upper plate (Plafker 1969, 1972). Surface displacements provide critical constraints on the rupture mechanism of this major seismotectonic event. Measured vertical and horizontal displacements indicate that the rupture released near-horizontal compressional elastic strain oriented roughly perpendicular to the trend of the eastern Aleutian arc (Fig 28.3A). The surface displacements and the mainshock and aftershock patterns are consistent with a simple elastic rebound model in which the margin of the North American plate is strained by underthrusting of the Pacific plate. Between the rupture events, the plates are locked together along the Aleutian megathrust resulting in regional horizontal compression and widespread subsidence (lower curve, Fig 28.3B). When the megathrust is loaded beyond its strength, rupture occurs, and the North American plate rebounds seaward along the main fault and along subsidiary faults that break through the upper plate (Fig 28.3A). The observed vertical surface deformation reflects the effects of elastic rebound and slip on the megathrust and on northward-dipping subsidiary faults in the upper plate (upper curve, Fig 28.3B). The net vertical displacement is the algebraic sum of coseismic and interseismic displacements for multiple earthquake cycles.

Paleoseismic Data

Documentary records of great earthquakes (moment magnitude ≥8), especially in North America, rarely span one full recurrence cycle at a given site. In some areas, the recurrence intervals, or duration of time between great seismotectonic events, can be deduced from the study of multiple strandlines that reflect repeated coseismic vertical displacements. The techniques of paleoseismology are primarily geologic (Wallace 1981), and in southern Alaska, they include analysis of stratigraphy, geomorphology and seismically-induced sedimentary structures.

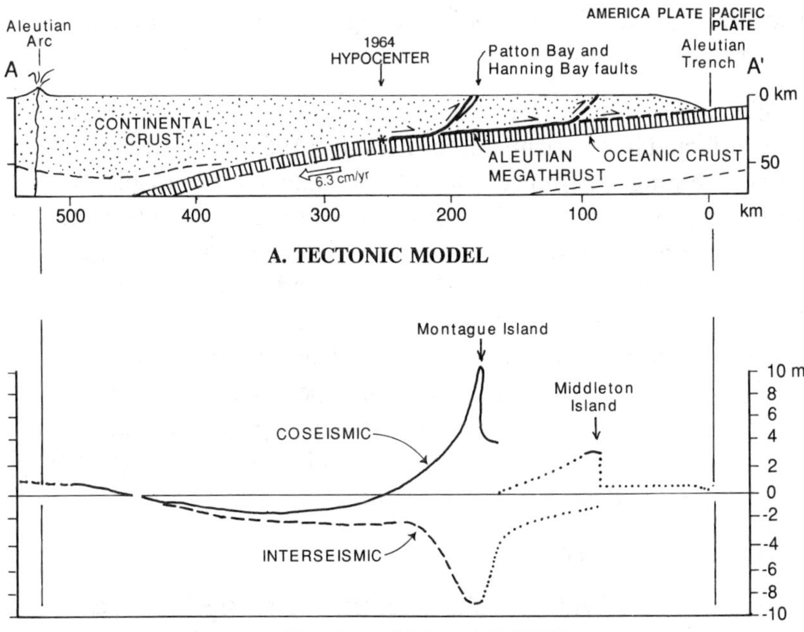

Fig 28.3. Schematic section and profiles along Line A-A′ in Fig 28.2. A. A simplified structural cross-section showing oceanic crust of the Pacific plate underthrust beneath continental crust of the America plate along the Aleutian megathrust. The star indicates the hypocenter of the 1964 earthquake; heavy black lines are the segments of the Aleutian megathrust and subsidiary faults that slipped during the earthquake (solid where known, dashed where inferred). B. Profiles of the 1964 coseismic displacements (upper line) and gradual interseismic displacements that preceded the 1964 earthquake for at least 800 years (lower line); dotted lines are inferred displacements beneath the continental shelf. The coseismic curve is based on the displaced 1964 shoreline, and the interseismic curve is based on ^{14}C-dated Holocene marine strandlines and submerged freshwater peat beds.

In the area affected by the 1964 Alaska earthquake, paleoseismologic data on previous great seismotectonic events come from deformed strandlines, the erosional and depositional expressions of former sea levels. The extensive coast affected by vertical displacements provides an excellent opportunity to determine both the ages and magnitudes of coseismic displacements, as well as the long-term history of cumulative vertical crustal movements (*ie,* over more than one earthquake cycle of strain accumulation and release). Much of the earthquake-affected region has also been subject to variable amounts of interseismic subsidence, so that it is the algebraic sum of the coseismic and interseismic displacements that determines whether a particular shoreline is one of long-term net uplift, net subsidence or stability (Fig 28.3). Because of the short historic record of less than 250 years in Alaska, and an even shorter instrumental record, geologic techniques provide the only means for deducing the long-term history of vertical tectonic displacements in the region and of the recurrence intervals of great earthquake events.

In the following sections, we consider paleoseismic data from different parts of the region affected by coseismic deformation in 1964 to illustrate the techniques used in determining recurrence intervals and interseismic displacements along shorelines with different combinations of coseismic and interseismic displacements. These are: 1) widespread pre-1964 interseismic subsidence along the Gulf of Alaska margin; 2) coseismic uplift and late Holocene net uplift at Middleton Island; 3) large coseismic uplift with apparent late Holocene net stability relative to sea level at Montague Island; 4) coseismic subsidence and net late Holocene subsidence or stability in upper Cook Inlet, and (5) coseismic uplift and net late Holocene subsidence in the Copper River delta area. New data that provide the best control for the dates of the last two 1964-type events at the Copper River delta are also presented.

Pre-1964 Regional Subsidence

Subaerial peat layers (with or without rooted tree stumps) that occur below the pre-1964 high-tide strandlines indicate interseismic subsidence within much of the region deformed by large coseismic vertical displacement (both uplift and subsidence) in 1964. Subsidence is known to have occurred in seaward parts of the Kodiak Islands, in the southern Kenai Mountains, in upper Cook Inlet, throughout the Prince William Sound and Copper River regions, and eastward to Kayak Island and Cape Suckling (locations on Fig 28.2). ^{14}C dates from buried peat and wood indicate that submergence took place gradually or in small increments for about 800 years at rates that range from 2.5–9.5 mm yr^{-1} (Plafker 1969 and unpublished data).

Coseismic and Long-Term Uplift at Middleton Island

Well-preserved Holocene marine terraces are known to occur in the area of 1964 coseismic deformation only at Middleton Island in the Gulf of Alaska (Fig 28.4). These Holocene terraces are identical in most respects to one formed on Middleton Island by 3.5 m coseismic uplift in 1964 except that the older ones have higher relict sea cliffs along their landward margins.

At Middleton Island, five pre-1964 Holocene terraces have been identified, and a sixth terrace was formed by coseismic uplift in 1964 (Plafker 1969, 1986; Plafker & Rubin 1978). The terraces are a steplike flight of gently sloping surfaces in Pleistocene bedrock separated by relict sea cliffs as much as 10 m high. Strandlines are about 3.5 (1964), 11, 20, 26, 36 and 46 m above present sea level (Fig 28.5). The time of uplift of each terrace can be approximated by dating driftwood or marine shells in the strandline deposits or *in-situ* shells on

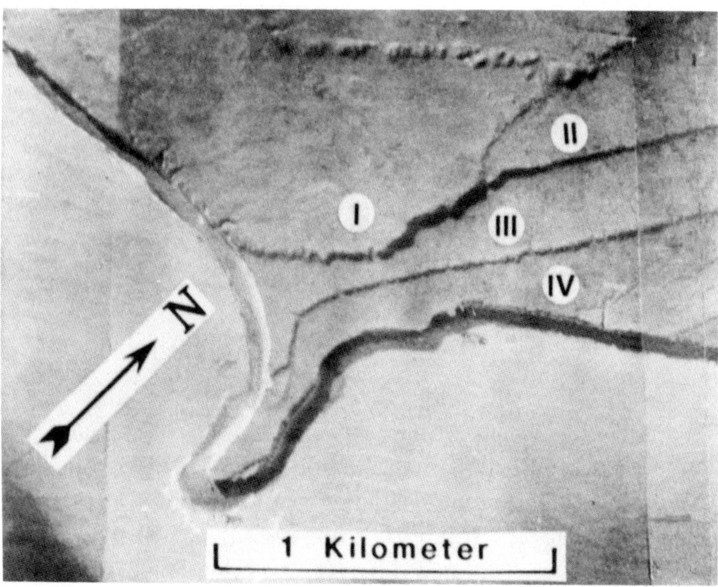

Fig 28.4. 1947 vertical aerial photo mosaic of the southeastern end of Middleton Island showing wave-cut platforms and sea cliffs of the four oldest of five pre-1964 marine terraces, each of which was formed by 5–7 m of coseismic uplift (see Fig 28.5 for elevations and ages). Terrace designations (I–IV) correspond to those in Figure 28.5.

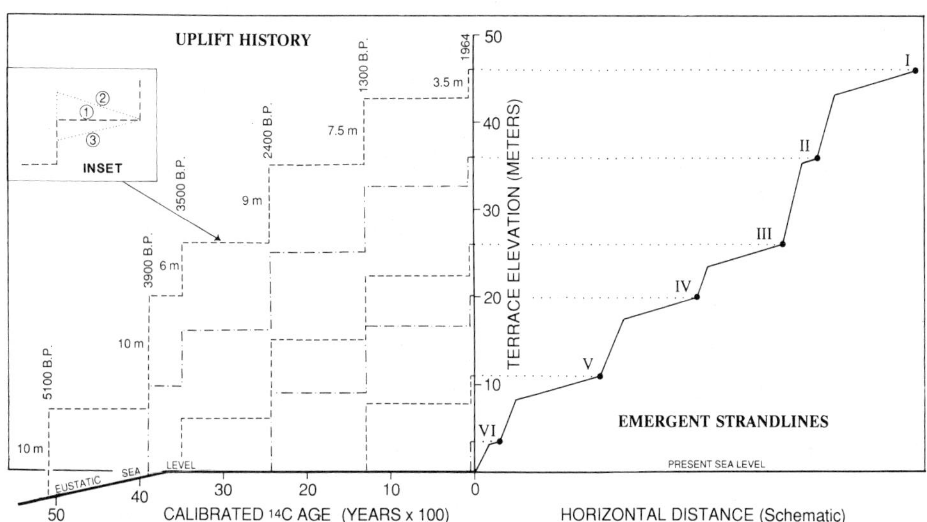

Fig 28.5. Model for origin of uplifted Holocene strandline terraces on Middleton Island. Schematic profile in the right part of diagram shows the present elevations of six emergent strandlines (I–VI). The net uplift history of each terrace is shown by the stepped lines in the left part of the diagram (Plafker 1987; note that sea-level curve shown is after Coleman & Smith 1964). Alternative interseismic histories shown in the inset are: 1) no change; 2) interseismic subsidence, and 3) interseismic uplift.

the rock platforms; a minimum age for the terrace uplift can be obtained by dating the oldest peat deposits formed on the surface after emergence. Corresponding calibrated ^{14}C ages (all ± ca 140 years) on peat and driftwood from the five older terraces at Middleton Island are about 1300, 2400, 3500, 3900 and 5100 years (Plafker & Rubin 1978). These data suggest that the last pre-1964 uplift occurred at about 1300 years, and that sudden uplifts have occurred at Middleton Island at intervals of around 400–1300 years (average 800 yrs) since the island emerged from the sea at about 5000 BP. Figure 28.5 models the uplift history for each of the Middleton Island terraces based on their ages and elevations. The interseismic displacements at Middleton Island are unknown; possibilities include relative stability, subsidence less than coseismic uplift or uplift (Fig 28.5, inset diagram).

The average (secular) uplift rate of Middleton Island based on emergent strandlines was 10–11 mm yr^{-1} over the past 5000 years (depending upon assumptions regarding eustatic sea-level rise), but the apparent rate over the past 1300 years has been only 2.7 mm yr^{-1} (3.5 m/1300 yrs), assuming constant sea level, or 4.2 mm (3.5 m + 1.95 m/1300 yrs), assuming a 1.5 mm yr^{-1} eustatic sea-level rise. Extrapolation of the average uplift rate suggests that more than half the strain accumulated since the formation of the second lowest strandline at 1300 years has yet to be released, assuming that the 1964 rupture was a characteristic event for this region. Consequently, if the Holocene trend continues, another coseismic uplift at least as large as that of 1964 should occur at Middleton Island in the near future (Plafker & Rubin 1978; Plafker 1987). Alternatively, the rupture process may be more random, as is suggested by apparent variations in terrace heights, in which case, all we can say is that strain will be released during an earthquake at some unspecified future time. Many more data of this type are needed, however, before forecasts, even on a time scale of centuries, become practicable. The Middleton Island data clearly demonstrate the potential of relatively inexpensive geomorphic and radiocarbon studies for significantly extending the history of great tectonic earthquakes in coastal regions.

Coseismic Uplift and Long-Term Stability at Montague Island

Stratigraphic data indicate that some coastal areas affected by coseismic uplift in 1964 experience gradual interseismic subsidence that approximately equals the coseismic displacement. In these areas, subaerial peat and forest horizons interbedded with intertidal sediments reflect successive uplift events, but there are no emergent marine surfaces that would indicate long-term net uplift. The best example is southern Montague Island, where the pre-1964 interseismic subsidence almost equals the 1964 coseismic uplift of about 11 m (Fig 28.3B), and a series of pre-1964 buried strandline deposits show about 7 m net uplift over a period of about 4500 years (Plafker, Lajoie & Rubin, unpublished data).

In this case, a plot of vertical ground displacement *vs* time would produce a sawtooth curve similar to Figure 28.9, but with approximately equal coseismic and interseismic displacements. Because long-term uplift has approximately kept pace with sea-level rise, the only emergent late Holocene marine terrace on Montague Island is the one formed in 1964.

Coseismic and Long-Term Subsidence

Evidence for prehistoric coseismic subsidence may be preserved as soil, peat and forest horizons that were suddenly lowered below extreme high tide and subsequently buried and preserved beneath transgressive marine sediments. Figure 28.6 shows hypothetical tectonic histories for five successive subaerial or high marsh layers that were submerged by coseismic subsidence. The stepped curves are time-displacement paths for each of these events, assuming tectonic stability and rapid progradation of sediment between events with sediment thickness equal to coseismic displacement. In this situation, the time of each earthquake is best approximated by the ^{14}C age of samples from the top of a peat layer, from the outer rings of trees rooted in peat or from fossil marsh grasses in sediment just above the peat. In order for successive layers to be formed and preserved, interseismic sedimentation (probably augmented by gradual tectonic uplift) must build up the ground surface high enough to form new layers of high marsh peat or subaerial peat (inset diagram, Fig 28.6). The actual path of relative emergence will vary from place to place depending upon sediment supply and the amount of the interseismic uplift component. In the upper Cook Inlet region, as much as 2.5 m of sedimentation over 75 km^2 occurred in less than 20 years after the 1964 subsidence of 2 m (Ovenshine, Lawson & Bartsch-Winkler 1976; Bartsch-Winkler 1988), so that the relative emergence path differs substantially from the idealized simple linear rise shown by the inset diagram of Figure 28.6.

At least two buried bog and forest horizons are locally well developed in slough bank exposures in the intertidal zone at Girdwood in upper Cook Inlet. Radiocarbon dates on wood indicate that the upper horizon at a pre-1964 depth of about 0.7 m is at least 700 years old, and the base of the sampled section at 4.5 m depth is about 2800 years old (Karlstrom, *in* Plafker 1969). At the head of Knik Arm, two peat layers interbedded with intertidal mud deposits were exposed in a trench for a gas pipeline at less than 2 m depth. ^{14}C ages for samples from the base of the peats are around 800 and 1300 BP, respectively (John Reeder, personal communication 1989). Studies of Holocene sections throughout the upper Cook Inlet region have identified at least 7 submerged organic layers as much as 4500 years old, to a depth of more than 12 m (Bartsch-Winkler & Schmoll, written communication 1990; Combellick 1986). Because of limited exposures of intertidal layers and the difficulty of interpreting

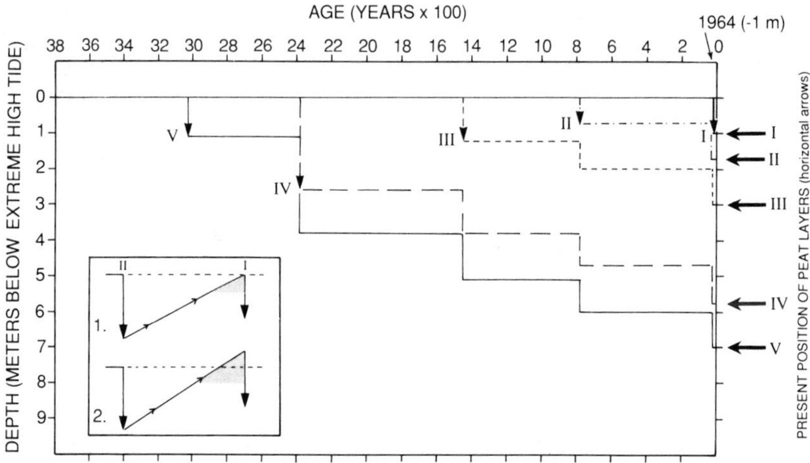

Fig 28.6. Model for origin of stratigraphic sequence containing peat, forest and soil layers (I–V) interbedded with marine sediment resulting from repeated episodes of coseismic subsidence. The vertical arrows indicate time and size of coseismic subsidence events and the subsequent stepped displacement history, assuming interseismic stability and sedimentation exactly equal to coseismic subsidence. The actual displacement history of a given layer, however, is the algebraic sum of coseismic subsidence, eustatic sea-level rise since the layer was formed, possible interseismic uplift and sedimentation. Consequently, the thickness of the sediment between organic layers is not likely to equal the coseismic subsidence. The inset diagram indicates the probable path of interseismic sediment accumulation and possible interseismic uplift that results in accumulation of organic layers (shaded) either within the high intertidal zone (1) or in the high intertidal zone and above sea level (2). In this model, the upper contact of the organic layers should be abrupt and may be overlain by coseismic sand-blow or tsunami deposits; the lower contact should be more gradational. Section shown from Bartsch-Winkler & Schmoll (unpublished data 1990).

subsurface samples, it is not clear how many of these dated organic layers, if any, are earthquake-related. Some, if not all, may have been submerged by the late Holocene sea-level rise and/or by isostatic subsidence. On-going studies by the US Geological Survey and the Alaska Geological Survey are directed towards distinguishing earthquake-related organic layers from those formed by other processes active in the region.

Coseismic Uplift and Long-Term Submergence

The Copper River delta area east of Cordova has an exceptionally complete record of multiple abrupt coseismic uplift events separated by gradual interseismic subsidence that exceeds the coseismic uplift, which was about 2 m in 1964. The Copper River delta consists of at least 180 m of prograding Holocene sediment deposited by the Copper River and its distributaries (Reimnitz 1966). The subaerial part of the delta has very low relief, and it is bordered by tide flats, as much as 1.6 km wide, separated from the open ocean by barrier sand islands. Freshwater peat-forming marsh (mostly sedge), brush (willow and alder) and conifer forest (spruce) cover the subaerial part of the delta, and salt-tolerant grasses of the *Carex/Potentilla* community occupy a broad area of the uppermost tidal flats. Because of the very low surface relief, a small amount of uplift causes large horizontal shifts in the position of the high-tide strandline, so that a wide coastal strip is abruptly converted from intertidal to subaerial marsh. In 1964, uplift of about 2 m resulted in seaward shifts of the strandline as much as 4 km. Much smaller uplift events should be preserved in the stratigraphic record.

The stratigraphic section consists of thin (10–30 cm) beds of freshwater peat with sharp basal contacts that grade upwards into thicker (1–3 m) beds of varved (?) intertidal silt containing the rooted remains of salt-tolerant sedge (Fig 28.7). Each peat-silt pair reflects a complete earthquake cycle, consisting of an abrupt emergence (coseismic uplift) above the highest-tide level, followed by gradual, though not necessarily uniform, submergence. The rate of submergence

Fig 28.7. Bank exposure in lower Alaganic Slough on the Copper River delta showing resistant ledges formed by a lower bed of freshwater peat 24 cm thick (foreground) and an upper bed of peat to 51 cm thick, that locally contains rooted stumps of brush and small trees (background). The peat beds have sharp contacts with underlying intertidal silt and sharp to locally gradational contacts with the overlying intertidal silt.

is the algebraic sum of tectonic and isostatic subsidence, sediment compaction and eustatic sea-level rise. In each cycle, peat accumulation begins some time after emergence and terminates upon gradual resubmergence into the intertidal zone; each uplift event occurred immediately after deposition of the uppermost horizon in each silt bed. The abundant supply of glacial sediment carried by the Copper River and local glacier-fed rivers (Reimnitz 1966) maintains the delta surface within about 1.5 m of the highest intertidal level where salt-tolerant sedge is able to survive. The sedge remains, which occur throughout the silt beds, indicate that silt accumulation roughly keeps pace with submergence. Net submergence, reflected primarily by the thickness of the section, is 2.8 mm yr^{-1} over the past 1450 years but varies from 2.2–5.9 mm yr^{-1} (Fig 28.8). Subtracting the assumed 1.5 mm yr^{-1}, eustatic sea-level rise yields a net subsidence rate of 1.3 mm yr^{-1} over 1450 years.

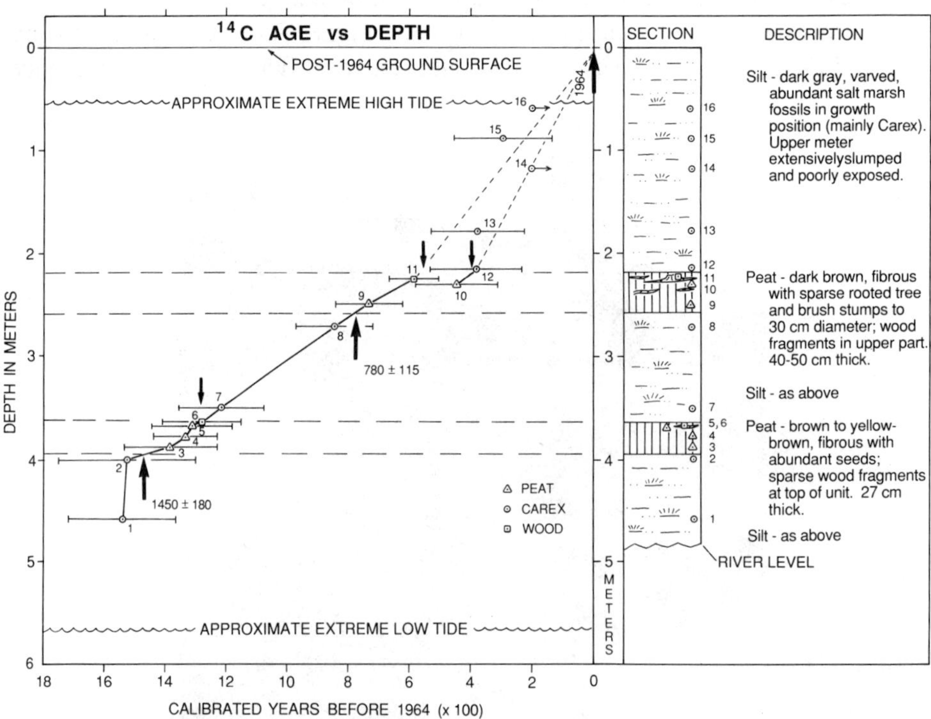

Fig 28.8. Stratigraphy and sedimentation history of interbedded subaerial peat and intertidal silt at lower Alaganic Slough, Copper River delta. Radiocarbon-dated samples of peat (◬), wood (⊡) and intertidal sedge in silt (⊙) correspond with those in Table 28.1. Large arrows at base of peat layers indicate interpolated ages of coseismic uplift that resulted in change from intertidal to subaerial deposition. The two coseismic uplift events at this site have calibrated ^{14}C ages of 780 and 1450 BP. Small arrows at top of peat layers indicate the approximate time when the peat layer was drowned by gradual interseismic submergence; two equally plausible submergence times are shown for the upper layer.

Two buried peat and forest horizons are widely exposed in the eroded banks of distributaries west of the Copper River (Fig 28.7), and at least three comparable deeper horizons have been sampled in hand-drilled auger holes. Figure 28.8 is a representative stratigraphic section and sedimentation history that includes the two upper pre-1964 organic horizons. In this section, the ^{14}C age of the fossil marsh plants in intertidal deposits immediately below the peat layers closely approximates the time just before an uplift event that abruptly raised the surface above the highest tide level. At one locality, a uniform medium-grained sand layer 11 cm thick under the peat is interpreted as a sand-blow deposit similar to those that formed on the delta during the 1964 earthquake. The ^{14}C age of the basal peat is a minimum for the uplift, because it dates the beginning of peat formation following coseismic emergence. The time of marine transgression due to interseismic subsidence is approximated by the ^{14}C age of the youngest peat, of the fossil marsh grass in intertidal sediment directly above the peat, or the outer rings of trees rooted in the peat.

At the site shown in Figure 28.8, the last two pre-1964 uplift events are at 780 and 1450 cal BP, and indicated intervals between events are 780 and 670 years. Preliminary data from hand-auger samples suggest at least two additional older events at about 2400 and 3000 cal BP, with corresponding intervals between events of 950 and 600 years. Because the calculated coseismic uplifts in the pre-1964 events were comparable in amount to uplift that accompanied the 1964 earth-quake (ca 2 m), it is likely that all the earthquakes that produced them were comparable in magnitude and mechanism.

Based on the data shown in Figure 28.8, the vertical displacement history in this area can be depicted diagrammatically as a sawtooth curve consisting of sudden coseismic uplift and gradual interseismic subsidence (Fig 28.9). In Figure 28.9, uplift events are comparable to the 1964 uplift of about 2 m, and interseismic submergence rates are assumed to be linear at about 6 mm yr^{-1} (the higher of the two pre-1964 rates shown in Fig 28.8). The two buried peat beds accumulated during subaerial exposure after coseismic uplift and before subsequent submergence below extreme high tide (Fig 28.9 B"-C", B′-C′); accumulation of intervening intertidal silt occurred during the latter part of the interseismic interval (Fig 28.9 C"-D", C′-D′). A plot of the displacement history requires information on 1) the actual interseismic submergence rates, 2) the pre-uplift depth of the intertidal deposits below extreme high tide, and 3) the elevation of the surface after uplift. Precise dating of closely-spaced samples of fossil plants in the intertidal deposits (possibly coupled with varve counts) may provide a more realistic subsidence curve, and enable us to discriminate between the linear subsidence path shown on Figure 28.9 and possible alternative non-linear paths shown on the inset. The present distribution of sedge communities in the inter-

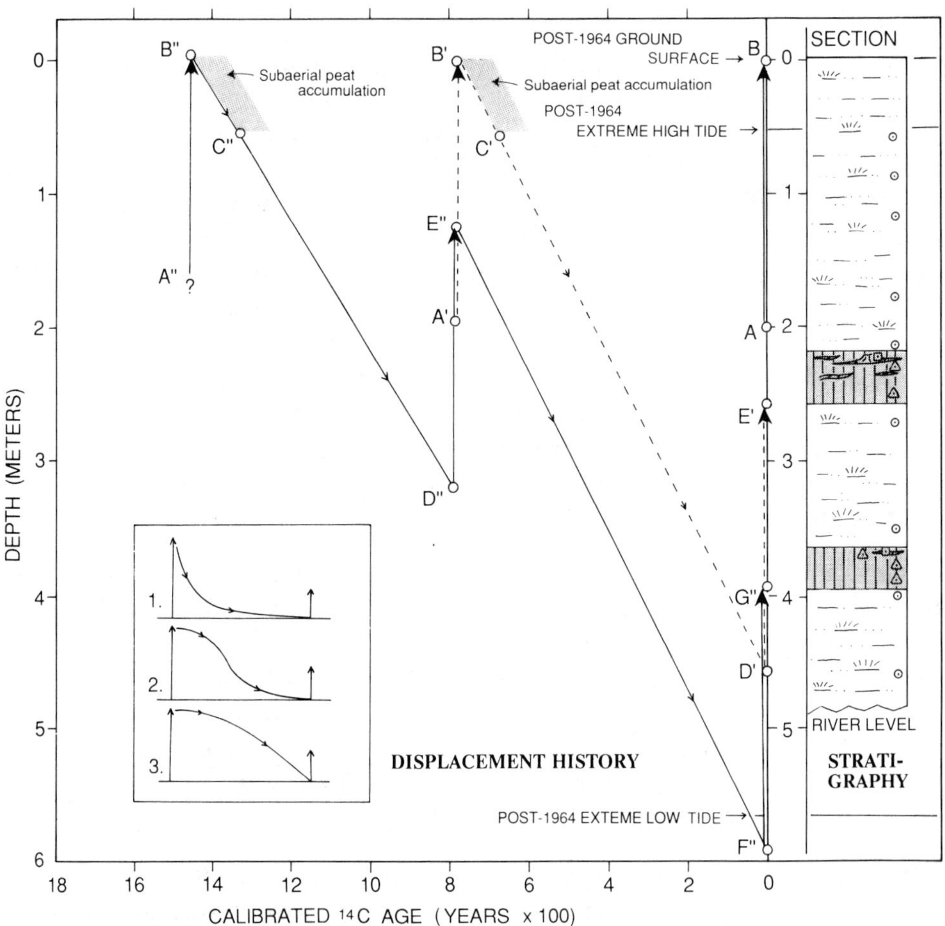

Fig 28.9. Stratigraphy and displacement history for the Copper River delta assuming coseismic uplift events comparable to the 1964 uplift of about 2 m and linear interseismic submergence at a rate of ca 6 mm yr^{-1}. Coseismic uplifts indicated by vertical lines (A-B, A'-B', A"-B"). Peat accumulation occurs during the interval from coseismic uplift to resubmergence below extreme high tide (post-B, B'-C', B"-C"), and accumulation of intertidal deposits occurs during the remainder of the interseismic interval (C'-D', C"-D"). The inset diagram shows three possible nonlinear interseismic subsidence paths.

tidal zone suggests that most of the silt in the sections accumulated within 1.7 m of extreme high tide (J F Thilenius, written communication 1990), and it may be possible to refine further the initial depths by detailed study of the flora in the present intertidal zone. If the interseismic subsidence rate and initial depth below high tide are known, the approximate amount of coseismic uplift for each event could be calculated, and the entire earthquake cycle in this area could be approximated.

Discussion of Earthquake Recurrence Data

It is apparent from all lines of geologic evidence that the recurrence intervals for 1964-type events commonly exceed 400 years, and can be as much as 1300 years. Consequently, the historic seismic record in Alaska is much too short to be useful in deducing recurrence of these great events or in estimating the recurrence times of future major earthquakes. Thus, predictions of an imminent great earthquake in the "Yakataga seismic gap" based on historic seismicity (McCann, Perez & Sykes 1980; Jacob 1984) are unjustified in an area where terrace data suggest a recurrence interval for the most recent great events of about 1400 years (Plafker 1986, 1987). In such regions, paleoseismic studies are the only technique presently available to us for obtaining recurrence data.

In the area affected by vertical displacements in 1964, only the Copper River delta and Middleton Island have reasonably well-dated records of multiple prehistoric great earthquakes. Ages we have deduced for previous earthquake events in the Copper River delta area (ca 665–895, 1270–1630, 2220–2580?, 2820–3180? BP) are not entirely synchronous with those determined from marine terraces at Middleton Island (ca 1150–1430, 2280–2560, 3360–3940, 3750–4030, 4850–5130 BP), even though both areas experienced coseismic vertical displacements in 1964. The youngest prehistoric event at the Copper River delta does not have a known corresponding terrace at Middleton Island suggesting: 1) a terrace was not formed at Middleton Island during that event; 2) a terrace was formed and was subsequently removed by erosion, or 3) a terrace was formed but has not been recognized. The 1270–1630 and 2220–2580? Copper River delta events have equivalent age terraces at Middleton Island, and it is likely that these represent events or sequences of events with synchronous vertical displacements in both areas. The 2820–3180? event on the Copper River delta does not overlap with the 3360–3940 event at Middleton Island by about 200 years. However, the age for this event in the Copper River delta is still too uncertain for any definite conclusions to be drawn regarding whether it correlates with an event on Middleton Island. Additional data from both areas may eventually allow more confident discrimination between synchronous and non-synchronous great earthquakes along this segment of the southern Alaska margin.

Conclusions

The tectonic displacements related to the March 27, 1964 Alaska earthquake and the paleoseismic history from strandlines in the same region demonstrate the complexity and tectonic mobility of the earth's crust at the eastern end of the Aleutian arc. These data indicate: 1) a seismic cycle in which interseismic subsidence may locally be equal to or larger than the coseismic uplift; 2) at least

800 years since the last 1964-type earthquake in this same region as indicated by the duration of pre-1964 gradual strandline subsidence; 3) recurrence intervals for great 1964-type events of about 600–950 years in the Copper River delta-Prince William Sound region and 400–1300 years in the Middleton Island-Icy Bay region. More and better data of this type, including precise radiometric dating of paleoshoreline deposits, should contribute to understanding the Holocene deformational processes, and may make it possible to forecast the relative susceptibility of specific localities along the coast to future great tectonic earthquakes.

ACKNOWLEDGMENTS

This paper has benefitted from technical reviews by Brian Atwater and Susan Bartsch-Winkler as well as by an anonymous but perceptive and critical reviewer.

References

Bard, E, Hamelin, B, Fairbanks, RG and Zindler, A 1990 Calibration of the ^{14}C timescale over the past 30,000 years using accelerator mass spectrometric U-Th ages from Barbados corals. *Nature* 345: 405–410.

Bartsch-Winkler, S 1988 Cycle of earthquake-induced aggradation and related tidal channel shifting, upper Turnagain Arm, Alaska, USA. *Sedimentology* 35: 621–628.

Coleman, JM and Smith, WG 1964 Late recent rise of sea level. *Geological Society of America Bulletin* 75:9 833–840.

Combellick, RA 1986 Chronology of late-Holocene earthquakes in southcentral Alaska–Evidence from buried organic soils in upper Turnagain Arm. *Geological Society of America Abstracts* 18: 569.

Jacob, KH 1984 Estimates of long-term probabilities for future great earthquakes in the Aleutians. *Geophysical Research Letters* 11: 295–298.

McCann WR, Perez, OJ and Sykes, LR 1980 Yakataga gap, Alaska: Seismic history and earthquake potential. *Science* 207: 1309–1314.

Minster, JB and Jordan, TH 1978 Present-day plate motions. *Journal of Geophysical Research* 83: 5331–5354.

Ovenshine, AT, Lawson, DE and Bartsch-Winkler, SR 1976 The Placer River silt–intertidal sedimentation caused by the Alaska earthquake of March 27, 1964. *US Geological Survey Journal of Research* 4: 151–162.

Peltier, WR and Tushingham, AM 1989 Global sea level rise and the greenhouse effect: Might they be connected. *Science* 244: 806–810.

Plafker, G 1969 Tectonics of the March 27, 1964, Alaska earthquake. *US Geological Survey Professional Paper* 543-I: 74 p.

_____1972 The Alaskan earthquake of 1964 and Chilean earthquake of 1960–Implications for arc tectonics. *Journal of Geophysical Research* 77(5): 901-925.

_____1986 Geologic studies related to earthquake potential and recurrence in the "Yakataga seismic gap." *US Geological Survey Open File Report* 86-92: 135-143.

_____1987 Application of marine-terrace data to paleoseismic studies. *In* Crone, AJ and Omdahl, EM, eds, Proceedings of conference XXXIX – Directions in paleoseismology. *US Geological Survey Open-File Report* 87-673: 146–156.

Plafker, G and Rubin, M 1978 Uplift history and earthquake recurrence as deduced from marine terraces on Middleton Island, Alaska. *In* Proceedings of Conference VI, Methodology for identifying seismic gaps and soon-to-break gaps. *US Geological Survey Open-File Report* 78-943: 687-721.

Reimnitz, E (ms) 1966 Late Quaternary history and sedimentation of the Copper River Delta and vicinity. PhD thesis, University of California, San Diego: 160 p.

Stuiver, M and Pearson, GW 1986 High-precision calibration of the radiocarbon time scale, AD 1950–500 BC. *In* Stuiver, M and Kra, RS, eds, Proceedings of the 12th International ^{14}C Conference. *Radiocarbon* 28(2A): 839–862.

Stuiver, M and Reimer PJ 1986 A computer program for radiocarbon age calibration. *In* Stuiver, M and Kra, RS, eds, Proceedings of the 12th International ^{14}C Conference. *Radiocarbon* 28(2B): 1022–1030.

Wallace, RE 1981 Active faults, paleoseismology, and earthquake hazards in the western United States. *In* Simpson, DW and Richards, PG, eds, Earthquake prediction – An international review. *American Geophysical Union, Maurice Ewing series* 4: 209–216.

29

MAJOR CONTRIBUTIONS OF RADIOCARBON DATING TO PALYNOLOGY: PAST AND FUTURE

DOROTHY M PETEET

INTRODUCTION

Before the advent of radiocarbon chronology, Quaternary palynologic records provided a biostratigraphic framework, which was used extensively as a chronostratigraphic guide, with implied synchrony of vegetational and climatic change from site to site. With the establishment of an absolute ^{14}C time scale, pollen zones in various geographic areas proved to be sometimes correlative but often time-transgressive. Absolute chronologies revealed errors in interpretation and promoted exploration of new dynamic patterns and comparisons.

The insights offered to the science of palynology through radiocarbon dating have been striking, and have affected the reconstruction of past vegetation and climate in numerous ways. This chapter briefly describes the early history of the use of palynology as a chronostratigraphic tool, reviews some of the major applications of radiocarbon chronology to palynology over the last 40 years, and highlights some of the problems facing Quaternary paleoecologists as we move into the next decade.

The most exciting recent contribution of ^{14}C dating to palynology arrived with the development of the accelerator mass spectrometry (AMS) technique. This method is important as a chronostratigraphic tool because 1) the actual biological indicators themselves can be dated, 2) it permits the dating of narrower stratigraphic intervals than bulk dates, and 3) dating of terrestrial macrofossils allows the dating of lakes with "hardwater" ^{14}C errors. Continued application of AMS ^{14}C dating to current and future palynological problems will ensure major contributions to this field.

BACKGROUND

Quaternary Palynology as a Chronostratigraphic Tool

Palynology, the study of fossil pollen and spores, developed as a tool for investigation of vegetational and climatic change in Scandinavia during the

1920s. Von Post (1916) presented the first pollen percentage diagrams, then Erdtman's (1921) paper enlared upon von Post's methods. Despite the awareness of numerous caveats in interpretation of the pollen stratigraphy, basic similarities throughout late-glacial and postglacial sequences suggested that intervals in sediment cores could be correlated regionally, and even globally across the North Atlantic. Thus began the use of Quaternary pollen diagrams for correlation purposes.

The pollen assemblage zone is defined as a biostratigraphic unit, a sequence within a diagram characterized by its flora (Faegri & Iversen 1975). However, confusion between biozones and corresponding time units has led to misuse of pollen assemblage zones. For example, Jessen's (1938) original division of Danish pollen diagrams was biostratigraphic, but the use of "Jessen's zones" as chronostratigraphic units outside of Denmark, though convenient, is not valid. The chronologic implication of biostratigraphic zones ended when radiocarbon dating provided an independent time scale.

The examples below illustrate the use of palynologic correlations as a means of providing geochronologic information. This chapter then reveals how radiocarbon dating has challenged, overturned, or modified these correlations.

Palynologic correlations include:

1. The Scandinavian Allerød-Younger Dryas climatic reversal with the midwestern North American Great Lakes Two Creeks/Valders glacial advance.
2. The Allerød/Younger Dryas event in Europe with late-glacial pollen stratigraphy in eastern North America (Deevey 1939; Leopold 1956).
3. Application of the Scandinavian postglacial Blytt-Sernander climatic sequence (Preboreal/Boreal/Atlantic/Subboreal/Subatlantic (Fig 29.1) to northeastern North America (Deevey 1939; Deevey & Flint 1957).
4. The use of Hypsithermal, a term for the Boreal through Atlantic time of maximum warmth in Europe, by Deevey and Flint (1957) for pollen zones in North America (Fig 29.1).
5. The European elm decline, thought initially to be synchronous everywhere on the continent.
6. The North American hemlock decline, considered synchronous throughout the Midwest and eastern United States (Fig 29.2).
7. The ragweed increase in northeastern North America, which, along with

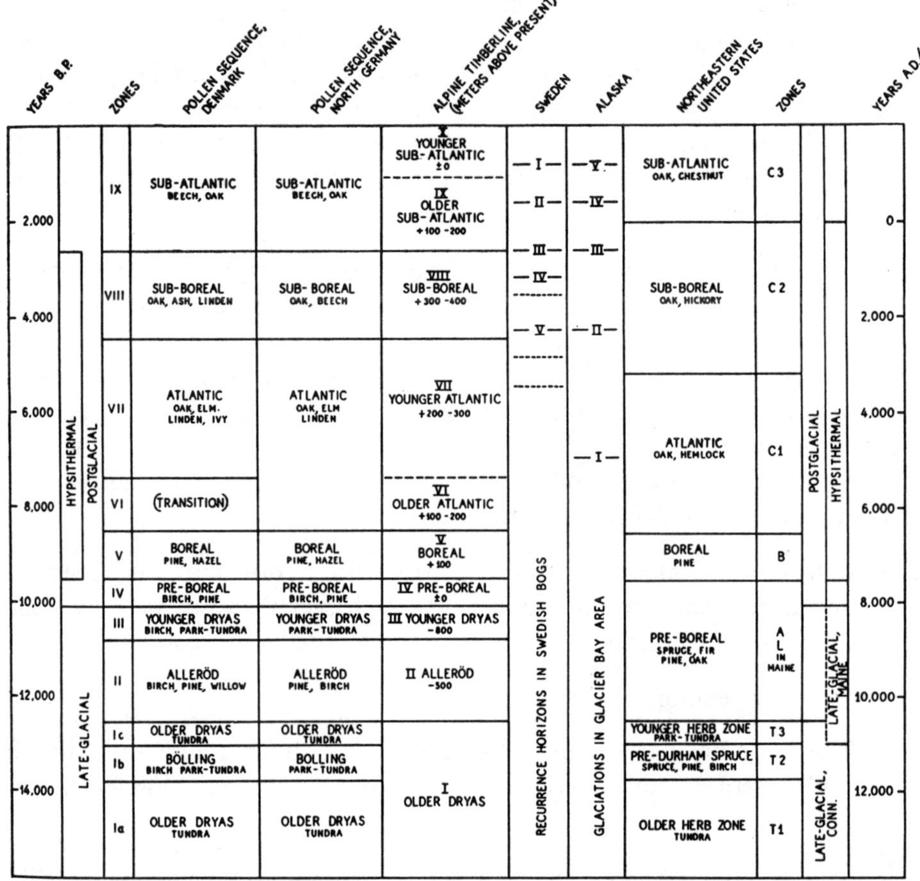

YEARS B.P.
ZONES
POLLEN SEQUENCE, DENMARK
POLLEN SEQUENCE, NORTH GERMANY
ALPINE TIMBERLINE, (METERS ABOVE PRESENT)
SWEDEN
ALASKA
NORTHEASTERN UNITED STATES
ZONES
YEARS A.D./BC.
2,000
4,000
6,000
8,000
10,000
12,000
14,000
HYPSITHERMAL
POSTGLACIAL
LATE-GLACIAL
IX
SUB-ATLANTIC
BEECH, OAK
VIII
SUB-BOREAL
OAK, ASH, LINDEN
OAK, BEECH
VII
ATLANTIC
OAK, ELM, LINDEN, IVY
OAK, ELM LINDEN
VI
(TRANSITION)
V
BOREAL
PINE, HAZEL
IV
PRE-BOREAL
BIRCH, PINE
III
YOUNGER DRYAS
BIRCH, PARK-TUNDRA
PARK-TUNDRA
II
ALLERÖD
BIRCH, PINE, WILLOW
PINE, BIRCH
Ic
OLDER DRYAS
TUNDRA
Ib
BÖLLING
BIRCH PARK-TUNDRA
PARK-TUNDRA
Ia
X YOUNGER SUB-ATLANTIC ±0
IX OLDER SUB-ATLANTIC +100 -200
VIII SUB-BOREAL +300 -400
VII YOUNGER ATLANTIC +200 -300
VI OLDER ATLANTIC +100 -200
V BOREAL +100
IV PRE-BOREAL ±0
III YOUNGER DRYAS -800
II ALLERÖD -500
I OLDER DRYAS
RECURRENCE HORIZONS IN SWEDISH BOGS
GLACIATIONS IN GLACIER BAY AREA
SUB-ATLANTIC
OAK, CHESTNUT
C3
SUB-BOREAL
OAK, HICKORY
C2
ATLANTIC
OAK, HEMLOCK
C1
BOREAL
PINE
B
PRE-BOREAL
SPRUCE, FIR PINE, OAK
A
L
IN MAINE
YOUNGER HERB ZONE
PARK-TUNDRA
T3
PRE-DURHAM SPRUCE
SPRUCE, PINE, BIRCH
T2
OLDER HERB ZONE
TUNDRA
T1
LATE-GLACIAL, MAINE
LATE-GLACIAL, CONN.
0
2,000
4,000
6,000
8,000
10,000
12,000

Fig 29.1. Deevey and Flint (1957) correlation of pollen zones in eastern North America with Blytt-Sernander scheme of Europe and radiocarbon chronology. (Reprinted from *Science* 125. © 1957 by the AAAS.)

the increase of other weedy species, has been used to date pollen profiles (RB Davis 1967; Webb 1973; Davis 1983).

Whether or not some of these pollen assemblage zones depicting marked increases or decreases in a particular flora are indeed correlative or time-transgressive from site to site is still open to question. Refinement of earlier bulk radiocarbon dating would test the regional correlations.

Traverse (1988) discusses a novel application of palynology as a chronostratigraphic tool to infer the timing of isostatic rebound (Hafsten 1956). Using the marine-lake mud contact in cores from numerous small lake basins in Scandinavia, Hafsten (1956) placed the pollen flora at each site with a pollen

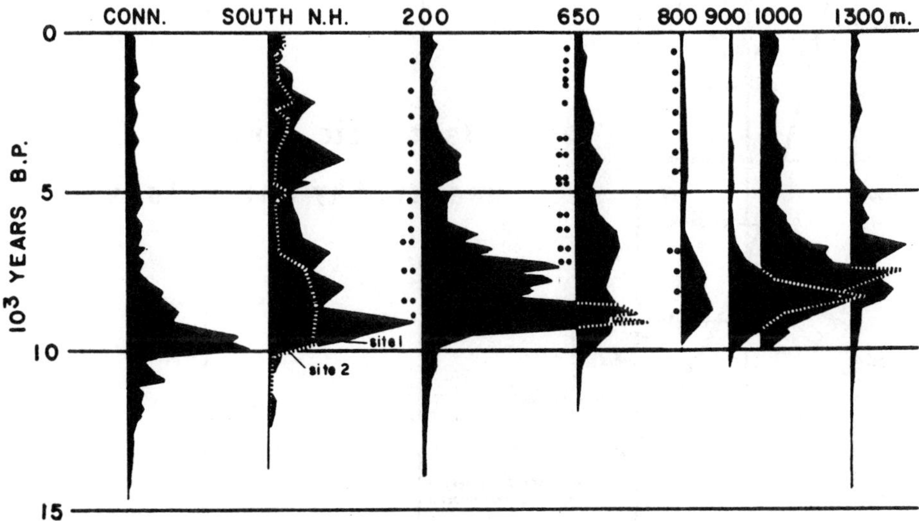

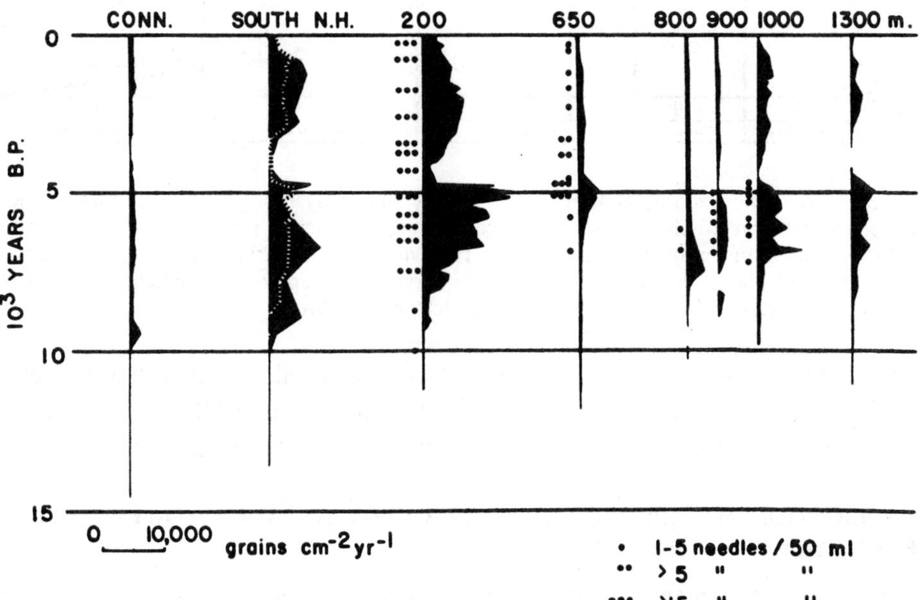

Fig 29.2. Hemlock pollen decline at Rogers Lake in Connecticut, two sites in southern New Hampshire (silhouette and dotted line), and a series of six sites at different elevations in the White Mountains of New Hampshire. Macrofossils were analyzed at the White Mountain sites only. • = 1–5 needles per sample; •• = 6–15 needles per sample; ••• = > 15 needles per sample (after Davis, Spear & Shane 1980). (Reprinted from *Quaternary Research* 14 by permission of the authors and Academic Press.)

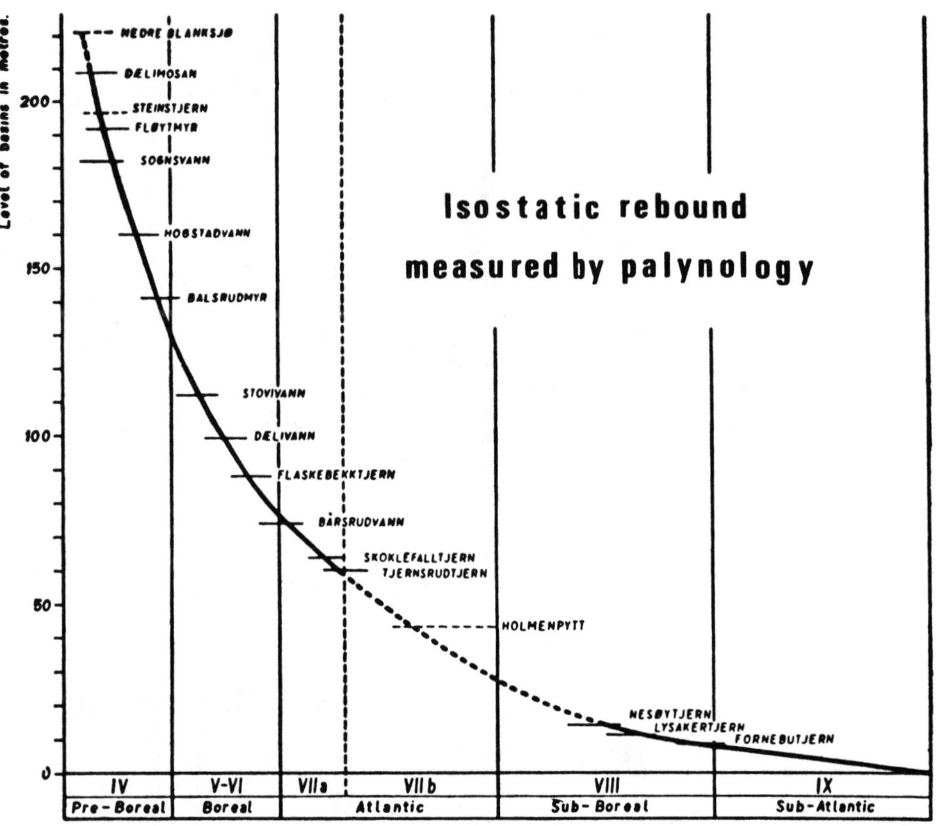

Fig 29.3. An application by Hafsten (1956) of palynologically based chronology that had been correlated with varved-clay studies before common application of radiocarbon dating. Names refer to small basins in southeastern Norway. Hafsten found the marine-lake contact in sediment cores and then placed the pollen flora of that interval in the appropriate pollen analysis zone. Present elevation of the basins is shown on the left (after Traverse 1988).

zone in the scheme of Jessen (1938). In this way he calculated isostatic rebound of 2.2 cm/yr from a basin at 220 m with a contact at 10,000 BP (Fig 29.3). Radiocarbon dating has not yet confirmed this correlation.

APPLICATION OF RADIOCARBON CHRONOLOGY TO PALYNOLOGY - THE LAST FORTY YEARS

The initial, fundamental contribution ^{14}C dating made to palynology was the establishment of the timing of major changes in the environment that characterized the last glacial-interglacial cycle. The initiation of ice advance and retreat and major changes in vegetation and climate were dated for localities throughout Europe (Godwin 1956; Mangerud 1980; Andersen 1981). The chronologic method was then applied to North America (Bryson *et al* 1969;

Dreimanis 1977; Denton & Hughes 1981; Porter 1981; Wright 1981; Prest 1984; Ruddiman & Wright 1987) and other extra-European geologic and palynologic sites. The uniformity of a time scale opened numerous paths for biologic, ecologic and climatic investigation, in particular because paleodata within these research fields could be independently compared. The ^{14}C chronology also paved the way for the linkup between palynology and stable isotopes, varves and tree rings.

The pattern of glacial-to-interglacial temperate vegetational sequences defined for Europe (Wjimstra 1969; van der Hammen, Wjimstra & Zagwijn 1971; Wjimstra & van der Hammen 1974; Woillard & Mook 1982) demonstrates dramatic and sometimes rapid climatic variability over the last climatic cycle. Radiocarbon-dated pollen stratigraphy in western North America also depicted an environmental sequence of stades and interstades that was more complex than had previously been recognized (Heusser 1972). Enrichment ^{14}C dating (Grootes 1977) has enabled a test of the correlation of the warm and cold events from site to site as far back as 70,000 BP (Woillard & Mook 1982).

The radiocarbon technique tested the following palynologic correlations:

Younger Dryas Equivalent in Midwestern North America

One of the first hypotheses under the scrutiny of radiocarbon dating was the synchrony of the European Allerød/Younger Dryas oscillation with an equivalent oscillation in North America. As more and more European pollen and glacial sites were ^{14}C dated, it became clear that the European Younger Dryas occurred at approximately 11,000–10,000 BP (Lowe & Walker 1980).

Early application of radiocarbon to the problem of possible correlation came with establishing the timing of the Two Creeks forest and overlying till of the Lake Michigan lobe (Valders, now Greatlakean oscillation). With the dating of Two Creeks forest, it became evident that the midwestern North American Two Creeks interstade occurred about 11,850 BP (Broecker & Farrand 1963). The overlying till (Greatlakean) was deposited possibly as much as 1000 years earlier than the till of the Younger Dryas glacial advance of Europe (Wright 1989).

Whether or not a Younger Dryas climatic equivalent is recorded in midwestern pollen records has been a controversial topic for a number of years. Wright (1989) summarized the controversy (Birks 1976; Saarnisto 1974) and the attempts to correlate local glacial advance with vegetational change. Shane (1987) recently reassessed the problem and interpreted palynologic changes from a site in Ohio as a regional Younger Dryas climatic equivalent. Further dating, in particular by AMS, should resolve the controversy.

Younger Dryas in Eastern North America

In the Maritime Provinces of eastern Canada, late-glacial pollen stratigraphy from several lake sites indicate a vegetational change at approximately 11,000–10,000 BP, which is also reflected in inorganic deposits sandwiched between organic layers (Mott *et al* 1986). The documentation of this shift generated a reevaluation of the hypothesis that this oscillation was also expressed in the vegetational history of southern New England (Peteet 1987).

Indications of a late-glacial climatic reversal in southern New England pollen stratigraphy were first described about 50 years ago and reinforced by subsequent investigations (Deevey 1939; Leopold 1956; Ogden 1959). However, later interpretation of the pollen stratigraphy, as well as a consideration of pollen influx from Rogers Lake, Connecticut, prompted a rejection of the Younger Dryas hypothesis for the eastern United States (Davis & Deevey 1964; MB Davis 1967). This conclusion was widely accepted (eg, Watts 1983; Gaudreau & Webb 1985; Davis & Jacobson 1986) until recently, when AMS-dated macrofossils showed the regional boreal shift to be between 11,000 and 10,000 years (Peteet *et al* 1990), apparently correlating with the Younger Dryas. The dated macrofossil species suggest that the palynologic shift is climatically driven.

Application of the Blytt-Sernander Scheme for North America

Initially, similarity of climatic stages in the interpretation of the pollen zones made the parallel with European pollen zones a seemingly simple correlation. However, as additional pollen sites were investigated, the diversity in landscape and vegetation in North America and the complexity in interpretation made this scheme difficult to accept. Uncertainty in the identification of spruce or pine pollen to species, varying migration rates, and the lack of detailed radiocarbon-dated stratigraphy ensures that this question is still open in parts of North America.

The Use of Hypsithermal, a Time of Maximum Holocene Warm/Dry

Many radiocarbon-dated chronologies at different latitudes have demonstrated that the Hypsithermal is a time-transgressive event. This Holocene warm/dry interval occurred between 8000 and 4000 BP in Minnesota (Wright 1976) and much of the midwest (Webb, Cushing & Wright 1983; Barnosky, Anderson & Bartlein 1987). However, it began and ended much earlier at high latitudes, and in particular, in Pacific northwestern North America, it occurred from 10,000–8000 BP (Ritchie, Cwynar & Spear 1983; Heusser, Heusser & Peteet 1985). From studies of sites in the White Mountains of New Hampshire, Davis, Spear and Shane (1980) conclude that the Hypsithermal lasted from 9000–5000 BP, roughly correlative with the midwestern region. Thus, radiocarbon chronology has dramatically changed our view of this regionally variable event, and provokes further questions concerning its origin.

European Elm Decline

The distinctive and widespread elm decline in pollen stratigraphy of the British Isles suggested that the phenomenon may have been synchronous throughout the islands (Faegri & Iversen 1975). However, as radiocarbon ages were derived from sediments recording the sharp negative values, it became clear that the timing of the elm decline differed from site to site (Birks 1973). The idea of selective utilization of elm as cattle fodder by Neolithic peoples, propounded by Troels-Smith (1960), has received support in recent years despite the lack of direct evidence (Walker & West 1970).

North American Hemlock Decline

The apparent rapidity and synchroneity with which hemlock pollen declined in widely scattered sites in eastern North America at approximately 4800 BP led palynologists to surmise that a forest pathogen was responsible (Davis 1976a; Davis, Spear & Shane 1980). AMS ^{14}C dating will enhance the precision with which we can determine the timing of this sharp pollen decline.

Eastern North American Ragweed Increase

The ragweed increase in eastern North American pollen stratigraphy is taken to indicate the decimation of forests by the arrival of Europeans. Thus used as a settlement indicator, the stratigraphic marker has been compared with historically documented land use changes that occurred approximately AD 1700 in southern Connecticut (Brugam 1978; Davis 1976b, 1983). Radiocarbon dating and Pb-210 analysis continue to define the timing of regional ragweed variability throughout eastern North America.

Other Important Applications of Radiocarbon Dating to Palynology

Timing of Deglaciation from Basal Lake and Bog Sediments. The development of radiocarbon dating provided data points for baseline maps of initial ice retreat from glaciated areas of both northern and southern hemispheres. Compilation of these data enabled the reconstruction of snapshot time intervals of ice extent (Denton & Hughes 1981; Prest 1984), which have been used for paleoclimatic reconstructions (CLIMAP 1981). In turn, these paleoclimatic scenarios have been used as boundary conditions for general climate model (GCM) experiments (Kutzbach & Guetter 1984; Manabe & Hahn 1977; Rind & Peteet 1985) in testing mechanisms and causes of climate change.

Isopolls and Isochrones. Isopolls are defined as lines of equal percentage representation of a particular pollen type (Szafer 1935). The use of isopolls was recently reintroduced by Birks, Deacon & Peglar (1975), Bernabo & Webb (1977) and Huntley & Birks (1983). Using existing radiocarbon chronologies along with extensive interpolation and extrapolation, synoptic maps were drawn to depict the time-sequential movement of vegetation. An example (Fig 29.4) shows isopolls of spruce, pine, oak and herb pollen depicted at 4000 BP (Bernabo & Webb 1977). On the basis of such isopolls, these and other authors have inferred northward migration rates of taxa and species, in particular following deglaciation. It is important to recognize the major assumptions and uncertainties in such inferrences: 1) pollen percentages reflect vegetation percentages; 2) ^{14}C dates on bulk sediments represent the age of the pollen; and 3) conventional (β-counting) ^{14}C dates often have poor time resolution.

Pollen Influx. The pollen influx method was developed in order to establish a measure of pollen change in contrast to pollen frequency. The calculation involves the number of grains accumulating on a unit of the sediment surface per unit time. Knowledge of the sediment accumulation rate and thus numerous radiocarbon dates per core are required for this method. Although pollen influx is useful for comparison to pollen percentage at any given site, Pennington (1973) and Davis, Moeller and Ford (1984) have shown that pollen influx rates over comparable time intervals may vary widely within a single lake for reasons such as sediment focusing. This renders the interpretation of pollen influx values problematic. However, influx is useful as a comparative measure against pollen percentage values.

Age of Peat Initiation. Well-dated palynostratigraphic sections of bogs and lakes extending throughout the late-glacial and Holocene result in a continual body of knowledge concerning age of initiation of peat formation on a landscape. From these records, climatic implications for initiation of peat formation have been developed, particularly in Scandinavia (Lundqvist 1962), the British Isles (Barber 1981) and most recently in western Canada (Ovendon 1990; Zoltai & Vitt 1990). Dramatic changes in peatland distribution across the western interior of Canada during the Holocene indicate a large response of the landscape to climatic changes, in particular, to the decrease in aridity after 6000 BP (Zoltai & Vitt 1990).

Lake-Level Changes. Radiocarbon-dated changes in sediment stratigraphy and biostratigraphy are used for studies of changes in lake levels (Butzer *et al* 1972; Street-Perrott *et al* 1989; Harrison 1989). With synoptic maps linking hydro-

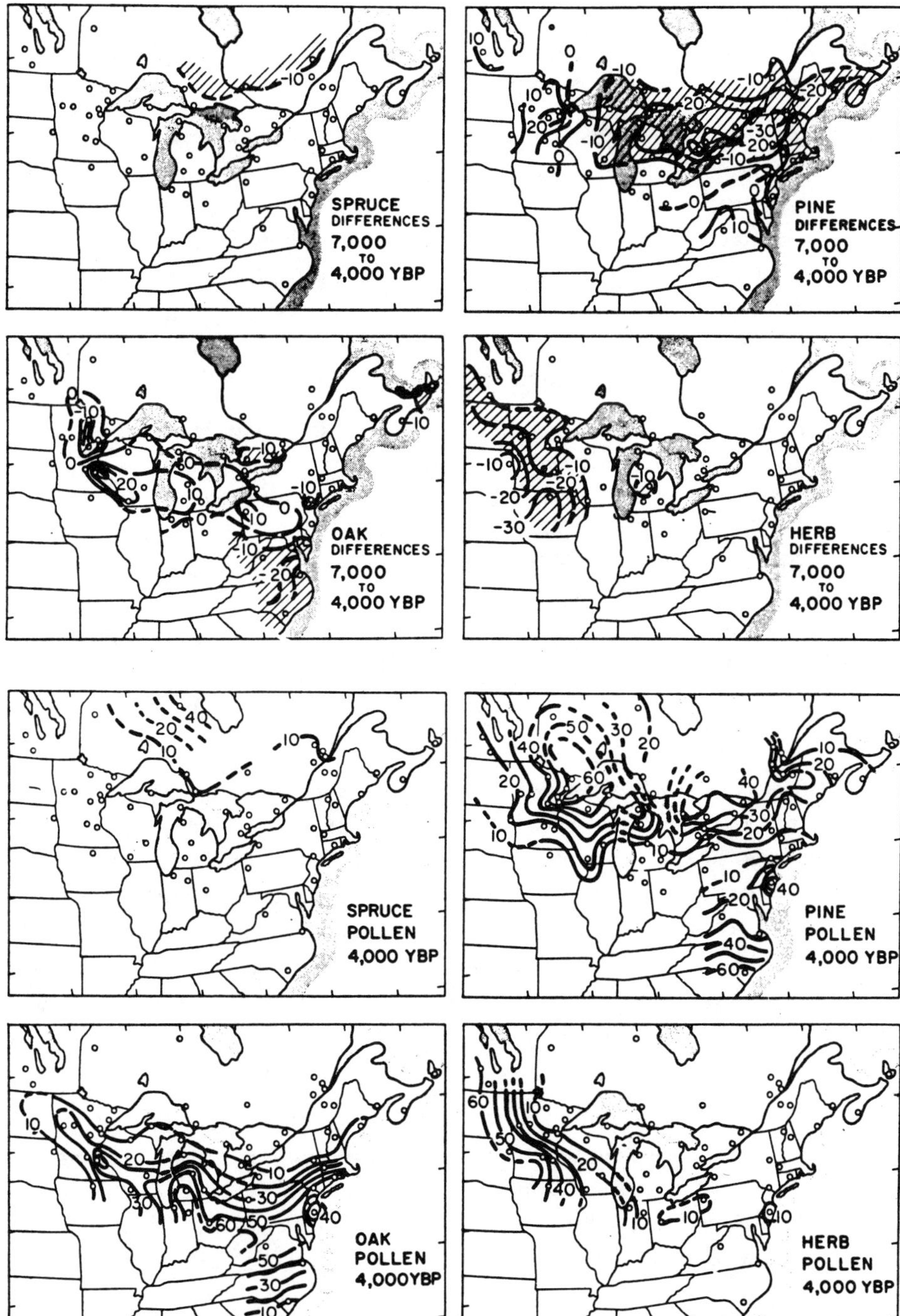

Fig 29.4. Top – Difference maps depicting patterns of change for pollen types – spruce (*Picea*), pine (*Pinus*), oak (*Quercus*) and herb (Graminae, Chenopodiineae, *Artemisia, Ambrosia*, Compositae and *Plantago*) – between 7000 and 4000 BP. Bottom – Isopolls of the same pollen types at 4000 BP. ○ = sample sites (after Bernabo & Webb 1977). (Reprinted from *Quaternary Research* 8 by permission of the authors and Academic Press.)

logic changes regionally, large-scale patterns of climate change are visible. As in other applications, the resolution and quality of the documented radiocarbon stratigraphy determine the usefulness of lake-level data.

Land-Ocean Correlation. Essential to understanding mechanisms and causes of climatic change is the comparison of timing of events on land, ocean and in ice cores. Mangerud *et al* (1984) successfully utilized a marker tephra and the radiocarbon method to correlate Younger Dryas events in Norway with the North Atlantic. Other palynologists seeking to establish the ocean-land climatic relationships have correlated continental lake pollen stratigraphy with marine oxygen isotope stratigraphy (Heusser & Shackleton 1979; Turon 1984), but radiocarbon stratigraphy is limited to 70,000 BP (Grootes 1978).

Rates and Patterns of Tree Migration. Using radiocarbon-dated pollen stratigraphy, palynologists have inferred rates of tree migration northward following the last ice-sheet retreat. A particular pollen percentage is assumed to indicate the establishment of the population on the landscape. Migration rates of trees are compared among eastern North America (Davis 1981), Europe (Huntley & Birks 1983) and western Canada (Cwynar & MacDonald 1987). Macrofossils are better indicators of presence than pollen percentages. Therefore, the refinement of this method using macro-fossils as population establishment indicators (Mott *et al* 1986; Peteet 1991) may change the interpretation of many of these migration rates.

Relationship of Vegetational Change to Archaeologic Sites and Major Faunal Changes and Extinctions. Godwin (1956), in one of the first applications of pollen analysis, linked the Iron Age to the beginning of the Subatlantic pollen zone of central Europe. As radiocarbon dating became a universal yardstick, connections between it and palynology have continued to flourish in the Americas (Delcourt *et al* 1986). One of the most interesting controversies in the New World concerns the relationship of late Quaternary mammalian extinctions, climate change and the arrival of humans (Mead & Meltzer 1984; Grayson 1987).

Fire History. ^{14}C-dated charcoal fragments, in conjunction with palynology, serve to define the timing of major regional alteration of vegetation in late Pleistocene history. Thus, radiocarbon chronology has been used to document the fire history of large geographic regions such as the New Jersey, USA Pine Barrens (Forman 1979) and the north central Amazon basin (Sanford *et al* 1985).

SIGNIFICANCE OF AMS ^{14}C DATING FOR PALYNOLOGY

The most recent technical advance in radiocarbon dating has had a major impact in palynology. In particular, the use of AMS ^{14}C dating of single, identifiable macrofossils in sediments (Andrée *et al* 1986; Lowe *et al* 1988) for high-resolution chronology development is changing our interpretation of the plant communities and climate during the late Pleistocene because the environmental indicators themselves, eg, *Dryas, Picea* can be dated. One example of a major change in interpretation is the documentation of the very early deglaciation of the Queen Charlotte Islands, western Canada (Mathewes *et al* 1985). Another is the discovery that mixtures of plant assemblages of the Tunica Hills, Louisiana region, have different ages. This alters the interpretation of the past climate of this important southern glacial site (Givens & Givens 1987). Jackson, Whitehead & Davis (1986) found similar age ranges in the paleovegetational record from the Midwest. AMS ^{14}C dating of a mixed assemblage of Late Pleistocene insect fossils from the Lamb Spring site, Colorado, has forced a reassessment of climate during deglaciation (Elias & Toolin 1990). AMS ages from macrofossils in the British Isles point to a Younger Dryas duration of only 600 years (Cwynar & Watts 1989). The application of the AMS technique for dating pollen residues (Brown *et al* 1989) will likely continue to modify our interpretation of the timing of vegetational change on the landscape.

UNSOLVED PROBLEMS AND FUTURE CONTRIBUTIONS

AMS ^{14}C dating has the potential to radically change interpretation of much of palynology. In particular, the dramatically improved resolution, precision and ability to date *in situ* recognizable plant parts eliminates many of the dating problems that have plagued the science for decades (Lowe, & Walker 1980). Application of AMS dating to many currently unresolved climatic questions, such as those mentioned below, should result in further landmark contributions to palynology and paleoclimatic research.

Deglaciation History

The timing of deglaciation in eastern North America is currently based on radiocarbon dating of bulk sediments. With improved AMS dating of macrofossils, major controversies over the timing of ice retreat in southern New England (Cotter *et al* 1984) may be resolved. Precise worldwide timing of the deglaciation would be extremely valuable, as land chronology could then be compared with global rates of sea-level rise during the last deglaciation (Fairbanks 1989). Additionally, this revised land chronology would allow comparison of climatic oscillations (*ie*, Younger Dryas) on a worldwide scale, aid in the establishment of rates of climate change and refine hypotheses concerning modes of operation of the ocean-atmosphere system (Broecker, Peteet & Rind 1985).

Synchroneity of N-S Hemisphere Glacial Maxima

Available radiocarbon dates on bulk samples in glaciated regions led Broecker and Denton (1990) to conclude that synchronous climate changes of similar severity characterized the last glacial-to-interglacial cycle. With AMS-improved chronologic resolution and precision, this synchroneity hypothesis may be tested. Establishment of this synchroneity worldwide would greatly aid our understanding of the links between ocean, ice, land and atmosphere.

Existence of a North American Ice-Free Corridor During the Last Glacial Maximum?

The existence of an ice-free corridor in western North America during the last glacial maximum has been controversial for many years. Recent dates on terrestrial macrofossils of less than 11,000 BP in the region (Fig 29.5) indicate

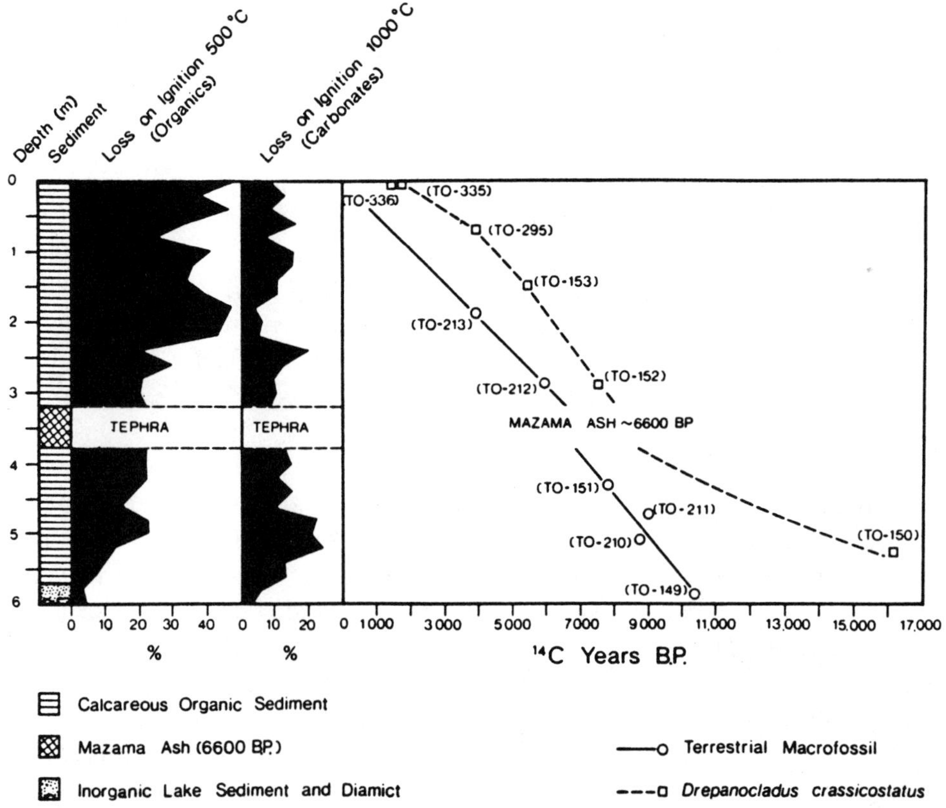

Fig 29.5. Sediment stratigraphy, loss-on-ignition results and radiocarbon age-depth relation for Toboggan Lake core, in the "ice-free corridor." Near-instantaneous deposition of Mazama ash is assumed (after MacDonald *et al* 1987). (Reprinted from *Geology* 15 by permission of the Geological Society of America.)

that glacial age dates on aquatic mosses are too old (MacDonald *et al* 1987). Additional research is needed to answer this question definitively.

Ancient Tropical Climate – Stability or Variability?

The question of the origin of enhanced species diversity during times and in areas of warmer climates has never been solved. Whereas diversity in past studies was attributed to stable climatic conditions, more recent tropical studies point to disturbance history (Campbell & Frailey 1984; Colinvaux *et al* 1985; Sanford *et al* 1985; Colinvaux, Olson & Liu 1988). A discrepancy still exists between ocean and land temperature estimates for the last glacial maximum (Rind & Peteet 1985). Detailed AMS radiocarbon and pollen and macrofossil stratigraphy may shed some light on the history of climate in regions of high biological diversity such as the Amazon Basin.

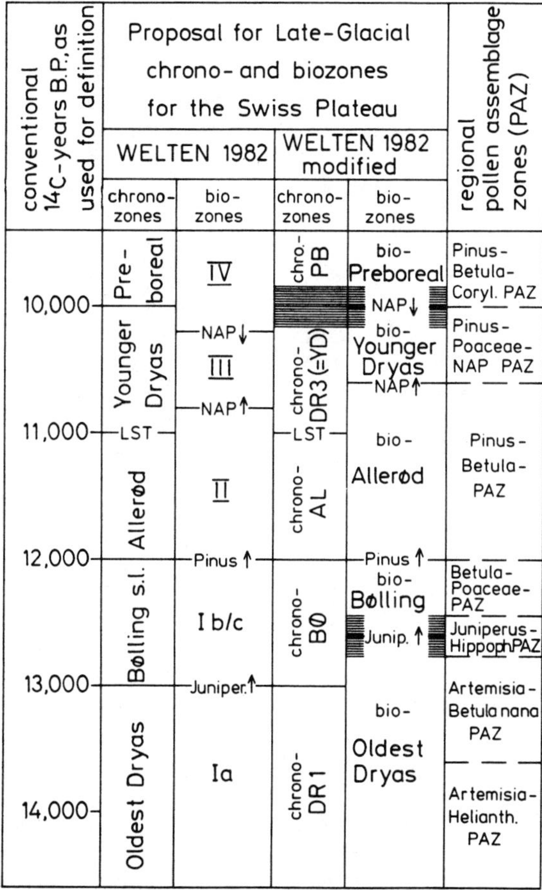

conventional 14C-years B.P., as used for definition
Proposal for Late-Glacial chrono- and biozones for the Swiss Plateau
WELTEN 1982
WELTEN 1982 modified
chrono-zones
bio-zones
chrono-zones
bio-zones
regional pollen assemblage zones (PAZ)
10,000
11,000
12,000
13,000
14,000
Pre-boreal
Younger Dryas
LST
Allerød
Bølling s.l.
Oldest Dryas
IV
NAP↓
III
NAP↑
II
Pinus ↑
I b/c
Juniper.↑
Ia
chro.-PB
chrono-DR3 (=YD)
LST
chrono-AL
chrono-BØ
chrono-DR1
bio-Preboreal
NAP↓
bio-Younger Dryas
NAP↑
bio-Allerød
Pinus ↑
bio-Bølling
Junip. ↑
bio-Oldest Dryas
Pinus-Betula-Coryl. PAZ
Pinus-Poaceae-NAP PAZ
Pinus-Betula-PAZ
Betula-Poaceae-PAZ
Juniperus-HippophPAZ
Artemisia-Betula nana PAZ
Artemisia-Helianth. PAZ

Fig 29.6. Proposed modification of the Late-Glacial chrono- and biozonation for the Swiss Plateau. Hatched areas mark plateaus of constant ^{14}C ages (after Ammann & Lotter 1989). (Reprinted from *Boreas* 18 by permission of Universitetsforlaget AS.)

Plateaus in Carbon 14

Ammann & Lotter (1989) established a detailed late-glacial radiocarbon stratigraphy for the Swiss Plateau, based on over 90 AMS ^{14}C dates on terrestrial macrofossils. This chronology focuses on a most interesting phenomenon, two plateaus of constant ^{14}C ages, occurring at 12,700 and 10,000 BP (Fig 29.6). Whether these plateaus are an artifact of the lake sedimentation history or represent changes in ^{14}C production remains to be tested. Additional lake and mire studies at widely scattered sites are needed to answer this question. Late-glacial terrestrial records can also be compared with the late-glacial AMS U-Th ages from corals (Bard *et al* 1990). If plateaus indeed exist in the chronology, the actual duration of the Younger Dryas climate event, generally believed to extend from 11,000–10,000 BP, may have been as long as 1700 years (Becker & Kromer 1986).

References

Ammann, B and Lotter, AF 1989 Late-glacial radiocarbon and palynostratigraphy on the Swiss Plateau. *Boreas* 18: 109–126.

Andersen, BG 1981 Late Weichselian ice sheets in Eurasia and Greenland. *In* Denton, GH and Hughes, TA, eds, *The Last Great Ice Sheets*. New York, Wiley-Interscience: 1–65.

Andrée, M, Oeschger, H, Siegenthaler, U, Riesen, T, Moell, M, Ammann, B and Tobolski, K 1986 ^{14}C dating of plant macrofossils in lake sediment. *In* Stuiver, M and Kra, RS, eds, Proceedings of the 12th International ^{14}C Conference. *Radiocarbon* 28(2A): 411–416.

Barber, KE, 1981 *Peat Stratigraphy and Climatic Change*. Rotterdam, Balkema: 219 p.

Bard, E, Hamelin, B, Fairbanks, RG and Zindler, A 1990 Calibration of the ^{14}C timescale over the past 30,000 years using mass spectrometric U-Th ages from Barbados corals. *Nature* 345: 405–410.

Barnosky, CW, Anderson, PM and Bartlein, PJ 1987 The northwestern US during deglaciation; vegetational history and paleoclimatic implications. *In* Ruddiman, WF and Wright, HE, Jr, eds, *North America and Adjacent Oceans During the Last Deglaciation*. Geological Society of America, The Geology of North America K-3: 289–321.

Becker, B and Kromer, B 1986 Extension of the Holocene dendrochronology by the Preboreal pine series, 8800 to 10,100 BP. *In* Stuiver, M and Kra, RS, eds, Proceedings of the 12th International ^{14}C Conference. *Radiocarbon* 28(2B): 961–967.

Bernabo, JC and Webb, T 1977 Changing patterns in the Holocene pollen record of northeastern North America: A mapped summary. *Quaternary Research* 8: 64–96.

Birks, HJB 1973 *Past and Present Vegetation of the Isle of Skye: A Paleoecological Study*. Cambridge, Cambridge University Press: 415 p.

_____1976 Late-Wisconsin vegetational history at Wolf Creek, central Minnesota. *Ecological Monographs* 46: 395–429.

Birks, HJB, Deacon, J and Peglar, S 1975 Pollen maps for the British Isles 5000 years ago. *Proceedings of the Royal Society* B 189: 87–105.

Broecker, WS and Denton, GH 1990 The role of ocean-atmosphere reorganizations in glacial cycles. *Geochimica et Cosmochimica Acta* 53: 2465–2501.

Broecker, WS and Farrand, WB 1963 Radiocarbon age of the Two Creeks forest bed, Wisconsin. *Geological Society of America Bulletin* 74: 795–802.

Broecker, WS, Peteet, DM and Rind, D 1985 Does the ocean-atmosphere system have more than one mode of operation? *Nature* 315: 21–25.

Brown, TA, Nelson, DE, Mathewes, RW, Vogel, JS and Southon, JS 1989 Radiocarbon dating of pollen by accelerator mass spectrometry. *Quaternary Research* 32: 205–212.

Brugham, RB 1978 Pollen indicators of land-use change in southern Connecticut. *Quaternary Research* 9: 349–62.

Bryson, RA, Wendland, WM, Ives, JD and Andrews, JT 1969 Radiocarbon isochrones on the disintegration of the Laurentide ice sheet. *Arctic and Alpine Research* 1: 1–14.

Butzer, KW, Isaac, GL, Richardson, JL and Washbourn-Kamau, C 1972 Radiocarbon dating of East African lake levels. *Science* 175: 1069–76.

Campbell, KE and Frailey, D 1984 Holocene flooding and species diversity in southwestern Amazonia. *Quaternary Research* 21: 369–375.

CLIMAP Project Members 1981 Seasonal reconstruction of the earth's surface at the last glacial maximum. *Geological Society of America Map & Chart Series* 36.

Colinvaux, PA, Miller, MC, Liu, K, Steinitz-Kannan, M and Frost, I 1985 Discovery of permanent Amazon lakes and hydraulic disturbance in the upper Amazon basin. *Nature* 313: 42–45.

Colinvaux, PA, Olson, K and Liu, K 1988 Late-glacial and Holocene pollen diagrams from two endorheic lakes of the inter-Andean plateau of Ecuador. *Review of Paleobotany & Palynology* 55: 83–99.

Cotter, JFP, Evenson, EB, Sirkin, L and Stuckenrath, R 1984 The interpretation of "bog-bottom" radiocarbon dates in glacial chronology. *In* Mahaney, W, ed, *Correlation of Quaternary Chronologies*. Toronto, Canada, Geobooks: 299–316.

Cwynar, LS and MacDonald, GM 1987 Geographical variation of lodgepole pine in relation to population history. *The American Naturalist* 129: 463–469.

Cwynar, LC and Watts, WA 1989 Accelerator-mass spectrometer ages for late-glacial events at Ballybetagh, Ireland. *Quaternary Research* 31: 377–380.

Davis, MB 1967 Pollen accumulation rates at Rogers Lake, Connecticut, during late- and postglacial time. *Review of Paleobotany & Palynology* 2: 219–230.

_____1976a Outbreak of forest pathogens in Quaternary history. *Proceedings of the 4th International Conference of Palynology*. Lucknow, India.

_____1976b Erosion rates and land use history in southern Michigan. *Environment Conservation* 3: 139–148.

_____1981 Quaternary history and the stability of deciduous forests. *In* West, DC, Shugart, HH and Botkin, DB, eds, *Forest Succession, Concepts and Application*. New York, Springer-Verlag: 132–177.

_____1983 Holocene vegetational history of the eastern United States. *In* Wright, HE, Jr, ed, *Late Quaternary Environments of the United States* 2. Minneapolis, University of Minnesota Press: 166–181.

Davis, MB and Deevey, ES 1964 Pollen accumulation rates: Estimates from late-glacial sediments of Rogers Lake, Connecticut. *Science* 145: 1293–1295.

Davis, MB, Moeller, RE and Ford, J 1984 Sediment focusing and pollen influx. *In* Haworth, EY and Lund, JWG, eds, *Lake Sediments and Environmental History*. Minneapolis, University of Minnesota Press: 261–294.

Davis, MB, Spear, RW and Shane, LCK 1980 Holocene climate of New England. *Quaternary Research* 14: 240–250.

Davis, RB 1967 Pollen studies of near surface sediments in Maine lakes. *In* Cushing, EJ and Wright, HE, Jr, eds, *Quaternary Paleoecology*. New Haven, Yale University Press: 143–173.

Davis, RB and Jacobson, GL 1986 Late-glacial and early Holocene landscapes in northern New England and adjacent areas of Canada. *Quaternary Research* 23: 341–368.

Deevey, ES 1939 Studies on Connecticut lake sediments. I. A postglacial climate chronology for southern New England. *American Journal of Science* 237: 691–724.

Deevey, ES and Flint, RF 1957 Postglacial

hypsithermal interval. *Science* 125: 182–184.

Delcourt, PA, Delcourt, HR, Cridlebaugh, PA and Chapman, J 1986 Holocene ethnobotanical and paleoecological record of human impact on vegetation in the Little Tennessee River Valley, Tennessee. *Quaternary Research* 25: 330–339.

Denton, G and Hughes, T, eds 1981 *The Last Great Ice Sheets*. New York, Wiley-Interscience: 484 p.

Dreimanis, A 1977 Late Wisconsin glacial retreat in the Great Lakes region, North America. *Annals of New York Academy of Science* 288: 70–89.

Elias, S and Toolin, LJ 1990 Accelerator dating of a mixed assemblage of Late Pleistocene insect fossils from the Lamb Spring Site, Colorado. *Quaternary Research* 3: 122–126.

Erdtman, OGE 1921 Pollen-analytische Untersuchungen von Torfmooren und marinen Sedimenten in Sudwest-Schwede. *Arkiv fur Botanik* 17: 10.

Faegri, K and Iversen, J 1975 *Textbook of Pollen Analysis*. New York, Hafner Press: 295 p.

Fairbanks, RG 1989 A 17,000-year glacio-eustatic sea level record: influence of glacial melting rates on the Younger Dryas event and deep-ocean circulation. *Nature* 342: 637–642.

Forman, RT 1979 *Pine Barrens: Ecosystem and Landscape*. New York, Academic Press: 601 p.

Gaudreau, D and Webb, T, III 1985 Late-Quaternary pollen stratigraphy and isochrone maps for the northeastern United States. *In* Holloway, RG and Bryant, VM, Jr, eds, *Pollen Records of Late Quaternary North American Sediments*. Dallas, Texas, American Association of Stratigraphic Palynologists: 245–280.

Givens, CR and Givens, FM 1987 Age and significance of fossil white spruce (*Picea glauca*) Tunica Hills, Louisiana-Mississippi. *Quaternary Research* 27: 283–296.

Godwin, H 1956 *The History of the British Flora*, 1st edition. Cambridge, Cambridge University Press.

Grayson, DK 1987 An analysis of the chronology of Late Pleistocene mammalian extinctions in North America. *Quaternary Research* 28: 281–289.

Grootes, P 1978 Carbon-14 time scale extended: Comparison of chronologies. *Science* 200: 11.

Hafsten, J 1956 Pollen-analytic investigations on the late Quaternary developments in the inner Oslofjord area. *University Bergen Arbok, Naturvidenskap* 8: 1–161.

Hammen, van der, T, Wijmstra, TA and Zagwijn, WH 1971 The floral record of the Late Cenozoic of Europe. *In* Turekian, KK, ed, *The Late Cenozoic Glacial Ages*. New Haven, Yale University Press: 391–424.

Harrison, S 1989 Lake levels and climatic change in eastern North America. *Climate Dynamics* 3: 157–167.

Heusser, CJ 1972 Palynology and phytogeographical significance of a late-Pleistocene refugium near Kalaloch, Washington. *Quaternary Research* 2: 189–201.

Heusser, CJ, Heusser, LE and Peteet, DM 1985 Late Quaternary climatic change on the American North Pacific coast. *Nature* 315: 485–487.

Heusser, LE and Shackleton, N 1979 Direct marine-continental correlation: 150,000-year oxygen isotope-pollen record from the North Pacific. *Science* 204: 837–840.

Huntley, B and Birks, HJB 1983 *An Atlas of Past and Present Pollen Maps of Europe 0–13,000 Years Ago*. Cambridge, Cambridge University Press: 667 p.

Jackson, ST, Whitehead, DR and Davis, OK 1986 Accelerator radiocarbon date indicates mid-Holocene age for hickory nut from Indiana late-glacial sediments. *Quaternary Research* 25: 257–259.

Jessen, K 1938 Some west Baltic pollen diagrams. *Quartar* 1: 124–139.

Kutzbach, JE and Guetter, PJ 1984 Sensitivity of late-glacial and Holocene climates to the combined effects of orbital parameter changes and lower boundary condition changes: "Snapshot" simulations with a general circulation model for 13, 9, and 6 ka BP. *Annals of Glaciology* 5: 85–87.

Leopold, EB 1956 Two late-glacial deposits in southern Connecticut. *Proceedings of the National Academy of Sciences* 42: 863–867.

Lowe, JJ, Lowe, S, Fowler, A, Hedges, REM and Austin, TJF 1988 Comparison of accelerator and radiometric radiocarbon measurements obtained from Late Devensian lateglacial lake from Llyn Gwernan, North Wales, UK. *Boreas* 17: 355–369.

Lowe, JJ and Walker, MJC 1980 Problems associated with radiocarbon dating the close of the Lateglacial period in the Rannoch Moor area, Scotland. *In* Lowe, JJ, Gray, JM and Robinson, JE, eds, *Studies in the Lateglacial of Northwest Europe.* Oxford, Pergamon Press: 123–137.

Lundqvist, B 1962 Geological radiocarbon datings from the Stockholm station. *Sveriges Geologiska Undersoekning* C 589: 3–23.

MacDonald, GM, Beukens, RP, Kieser, WE and Vitt, DH 1987 Comparative radiocarbon dating of terrestrial macrofossils and aquatic moss from the "ice-free corridor" of western Canada. *Geology* 15: 837–840.

Manabe, S and Hahn, DG 1977 Simulation of the tropical climate of an ice age. *Journal of Geophysical Research* 82: 3889–3911.

Mangerud, J 1980 Ice-front variations of different parts of the Scandanavian ice sheet 13,000 to 10,000 yr BP. *In* Lowe, JJ, Gray, JM and Robinson, JE, eds, *Studies in the Lateglacial of North-West Europe.* Oxford, Pergamon Press: 23–30.

Mangerud, J, Lie, SE, Furnes, H, Kristiansen, IL and Lomo, L 1984 A Younger Dryas ash bed in Western Norway and its possible correlation with tephra in cores from the Norwegian Sea and the North Atlantic. *Quaternary Research* 21: 85–104.

Mathewes, RW, Vogel, JS, Southon, JR and Nelson, DE 1985 Accelerator radiocarbon date confirms early deglaciation of the Queen Charlotte Islands. *Canadian Journal of Earth Science.* 22: 790–792.

Mead, JI and Meltzer, DJ 1984 North American Late Quaternary extinctions and the radiocarbon record. *In* Martin, PS and Klein, J, eds, *Quaternary Extinctions.* Tucson, University of Arizona Press: 440–450.

Mott, RJ, Matthews, JV, Jr, Grant, DR and Beke, GJ 1986 A late-glacial buried organic profile near Brooksdale, Nova Scotia. *Current Research, Part B. Geological Survey of Canada Paper* 86(1B): 289–294.

Ogden, JG, III 1959 A late-glacial pollen sequence from Martha's Vineyard, Massachusetts. *American Journal of Science* 257: 366–381.

Ovendon, L 1990 Peat accumulation in northern wetlands. *Quaternary Research* 33: 377–386.

Pennington, W 1973 Absolute pollen frequencies in the sediments of lakes of different morphometry. *In* Birks, HJB and West, RG eds, *Quaternary Plant Ecology*: 79–104.

Peteet, DM 1978 Younger Dryas in North America – Modeling, data analysis and re-evaluation. *In* Berger, W and Labeyrie, L, eds, *Abrupt Climatic Change: Evidence and Implications.* Dordrecht, The Netherlands, Reidel: 185–193.

_____1991 Postglacial migration history of lodgepole pine near Yakutat, Alaska. *Canadian Journal of Botany*, in press.

Peteet, DM, Vogel, JS, Nelson, DE, Southon, JR, Nickmann, RJ and Heusser, LE 1990 Younger Dryas climatic reversal in northeastern USA? AMS ages for an old problem. *Quaternary Research* 33: 219-230.

Porter, SC 1981 The Late Pleistocene. *In* Wright, HE, Jr, ed, *Late-Quaternary Environments of the United States.* Minneapolis, University of Minnesota Press: 407 p.

Post, L, von 1916 Om skogstradpollen i sydsvenska torfmossclagerfolyder. *Geologiska Föreningens i Stockholm Förhandlingar* 38: 384–390.

Prest, VK 1984 The Late Wisconsinan glacier complex. Map 1584. *In* Fulton, RJ, ed, Quaternary Stratigraphy of Canada – A

Canadian Contribution to IGCP Project 24. *Geological Survey of Canada Paper* 84-10.

Rind, D and Peteet, D 1985 Terrestrial conditions at the last glacial maximum and CLIMAP sea-surface temperature estimates: Are they consistent? *Quaternary Research* 24: 1–22.

Ritchie, JC, Cwynar, LC and Spear, RW 1983 Evidence from northwest Canada for an early Holocene Milankovitch thermal maximum. *Nature* 305: 126–128.

Ruddiman, WF and Wright, HE, Jr, eds, 1987 *North America and Adjacent Oceans During the Last Deglaciation.* Geological Society of America, The Geology of North America K-3.

Saarnisto, M 1974 The deglaciation history of the Lake Superior region and its climatic implications. *Quaternary Research* 4: 316–339.

Sanford, RL, Saldarriaga, J, Clark, KE, Uhl, C and Herrera, R 1985 Amazon rain-forest fires. *Science* 227: 53–55.

Shane, LCK 1987 Late-glacial vegetation and climatic history of the Allegheny Plateau and the till plains of Ohio and Indiana. *Boreas* 16: 1–20.

Street-Perrott, FA, Marchand, DS, Roberts, N and Harrison, SP 1989 Global lake-level variations from 18,000 to 0 years ago: A paleoclimatic analysis. *DOE Report* TR046.

Szafer, W 1935 The significance of isopollen lines for the investigation of geographical distribution of trees in the post-glacial period. *Bulletin International de l'Academie Polonaise des Sciences et des Lettres* B1: 235–239.

Traverse, A 1988 *Paleopalynology.* Boston, Unwin Hyman: 600 p.

Troels-Smith, J 1960 Ivy, mistletoe, and elm: Climatic indicators-fodder plants. *Danmarks Geologiske Undersoegelse* 4: 1–32.

Turon, J 1984 Direct land/sea correlations in the last interglacial complex. *Nature* 309: 673–676.

Walker, D and West, RG, eds, 1970 *Studies in the Vegetational History of the British Isles.* Cambridge, Cambridge University Press, 265 p.

Watts, WA 1983 Vegetational history of the eastern United States 25,000 to 10,000 years ago. *In* Wright, HE, Jr, ed, *Late Quaternary Environments of the United States* 2. Minneapolis, University of Minnesota Press: 166–181.

Webb, T, III 1973 A comparison of modern and presettlement pollen from southern Michigan USA. *Review of Paleobotany & Palynology* 6: 137–156.

Webb, T, III, Cushing, EJ and Wright, HE, Jr 1983 Holocene changes in the vegetation of the Midwest. *In* Wright, HE, Jr, ed, *Late Quaternary Environments of the United States* 2. Minneapolis, University of Minnesota Press: 142–165.

Wijmstra, TA 1969 Palynology of the first 30 metres of a 120 m deep section in northern Greece. *Acta Botanica Neerlandica* 18(4): 511–527.

Wijmstra, TA and Hammen, T, van der 1974 The last interglacial-glacial cycle: State of affairs of correlation between data obtained from the land and from the ocean. *Geologie en Mijnbouw* 53(6): 386–392.

Woillard, GM and Mook, WG 1982 Carbon-14 dates at Grande Pile: Correlation of land and sea chronologies. *Science* 215: 159–161.

Wright, HE, Jr 1976 The dynamic nature of Holocene vegetation – a problem in paleoclimatology, biogeography, and stratigraphic nomenclature. *Quaternary Research* 6: 581–596.

_____1981 The Holocene. *In* Wright, HE, Jr, ed, *Late-Quaternary Environments of the United States.* Minneapolis, University of Minnesota Press, 277 p.

_____1989 The Amphi-Atlantic distribution of the Younger Dryas paleoclimatic oscillation. *Quaternary Science Reviews* 8: 295–306.

Zoltai, SC and Vitt, DH 1990 Holocene climatic change and the distribution of peatlands in western interior Canada. *Quaternary Research* 33: 231–240.

30

ORIGIN OF GLOBAL MELTWATER PULSES

RICHARD G FAIRBANKS, CHRISTOPHER D CHARLES and JAMES D WRIGHT

INTRODUCTION

The fact that frequencies measured in climate records are the same as those predicted by the astronomical theory of climate change is undisputed (Hays, Imbrie & Shackleton 1976). However, the mechanisms by which these small changes in seasonal insolation are amplified into glacial cycles remain a fundamental mystery of the Earth's climate system. The Barbados postglacial sea-level record is sufficiently detailed to resolve, for the first time, the rates as well as the magnitude of continental ice melting (Fairbanks 1989, 1990) (Fig 30.1A). The Barbados meltwater discharge curve is not smooth but pulsed, with peaks at 12,000 ^{14}C years[1] and 9500 ^{14}C years (Fig 30.1B). Sea level rose more than 24 m during each of these pulses, with annual rates of sea-level rise exceeding 3 cm/yr. These enormous pulses must mark the ice-sheet response to a change in one or more of the climate amplifiers (eg, greenhouse gases and oceanic heat transports). The suspected amplifiers have different time constants and different regional sensitivities. Therefore, the discovery of both the pulsed deglaciation itself and the geographic origin of the pulses may help pinpoint the factors responsible for the timing of the large sea-level change associated with the last deglaciation, as well as the cause of previous "terminations" which recur every 100,000 ^{14}C years during the late Pleistocene Epoch (Broecker 1984).

Probably four ice sheets contribute to the shape of the global discharge curve: Laurentide; Fennoscandian; Barents; and Antarctic. Accelerator mass spectrometry (AMS) ^{14}C dating of meltwater plumes, as measured by $\delta^{18}O$ analyses of planktonic foraminifera, clearly documents the timing of the disintegration of the different continental ice sheets. Although $\delta^{18}O$ records of large meltwater plumes were discovered more than 15 years ago in continental margin sediments

[1]All ^{14}C ages are with respect to the 5568-year half-life and are uncorrected for secular changes in atmospheric ^{14}C. They are corrected for the estimated oceanic "reservoir age" of 400 years except where noted for the Southern Ocean.

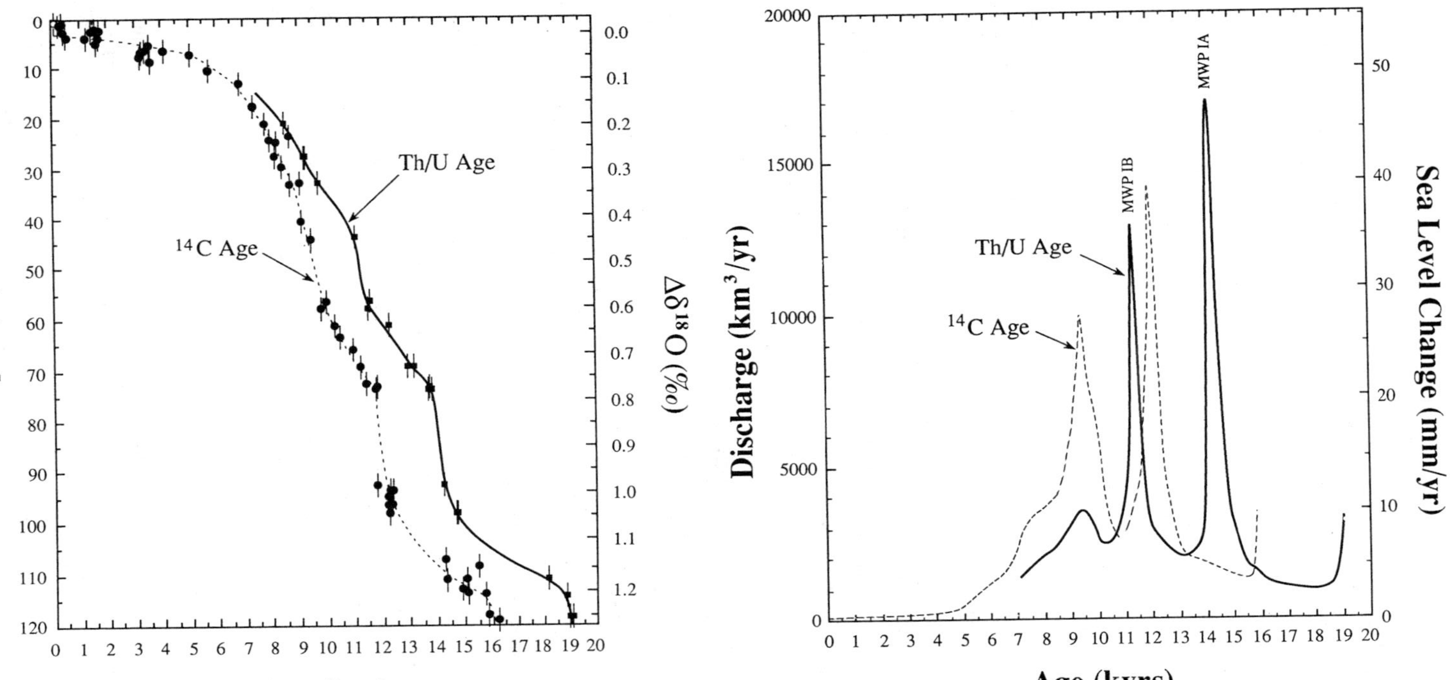

Fig 30.1A. Barbados sea-level curves based on ^{14}C-dated and $^{230}Th/^{234}U$-dated *A palmata* plotted with *A palmata* ^{14}C dates from four other Caribbean locations reported in Lighty *et al* (1982) (from Fairbanks 1990). The $^{230}Th/^{234}U$ sea-level curve is based on age determinations by mass spectrometry (squares) (Bard *et al* 1990a). Reprinted by permission of *Nature*. © 1990 MacMillan Magazines Ltd.

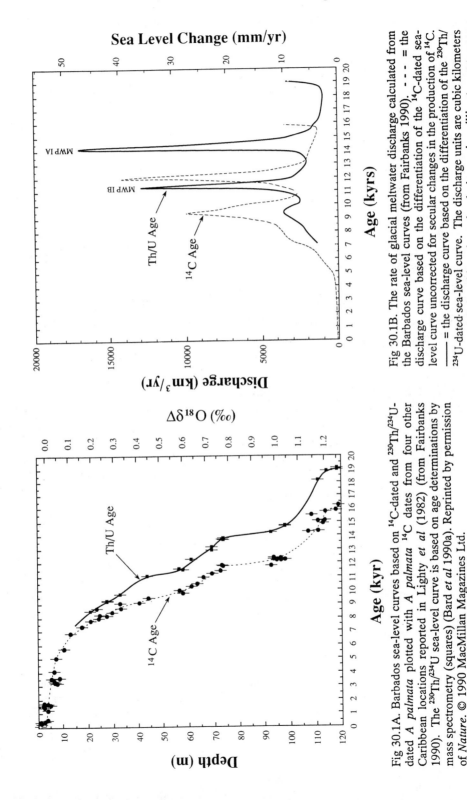

Fig 30.1B. The rate of glacial meltwater discharge calculated from the Barbados sea-level curves (from Fairbanks 1990). - - - = the discharge curve based on the differentiation of the ^{14}C-dated sea-level curve uncorrected for secular changes in the production of ^{14}C. —— = the discharge curve based on the differentiation of the $^{230}Th/^{234}U$-dated sea-level curve. The discharge units are cubic kilometers per year (left axis) and sea-level change in millimeters per year (right axis).

(Kennett & Shackleton 1975; Emiliani, Rooth & Stipp 1978; Leventer *et al* 1982), they were notoriously difficult to date because of terrestrial contamination. AMS ^{14}C dating of monospecific planktonic foraminifera from these high accumulation rate cores reveals meltwater discharge records in unprecedented detail (Broecker *et al* 1990b; Jones & Keigwin 1988). When compared with terrestrial evidence for glaciation, it is possible, for the first time, to document the sequential melting of the great continental ice sheets.

In this chapter, we examine the geographic origin of the two meltwater pulses by first documenting the timing of the meltwater plumes for the different continental ice sheets to determine which ice sheets were primarily responsible for the pulses. Analyses of the $\delta^{18}O$ of modern surface waters combined with ocean model experiments show that $\delta^{18}O$ oscillations that have the appearance of meltwater plumes can be created by ocean/climate processes. For example, changes in the production rates of North Atlantic Deep Water (NADW) may create $\delta^{18}O$ oscillations that are indistinguishable from meltwater plumes. Next, we examine the timing of several proposed amplifiers of climate change and, with the aid of General Circulation Models (GCMs), assess the regional sensitivity to these potential amplifiers. One difficulty in this approach is that comparing amplifiers of climate change to proxy records of climate change requires a standardized chronology. Most of the proxy climate records are reported in uncorrected radiocarbon years, whereas records of potential forcing, such as the CO_2 records from ice cores or astronomical calculations of local insolation, are measured or calculated in sidereal years. Here, it is important to note that the Barbados sea level record is measured in both ^{14}C years and $^{230}Th/^{234}U$ years (Bard *et al* 1990a, b; Fairbanks 1990). This new radiocarbon calibration is critical for postglacial climate research for several reasons. First, radiocarbon is not calibrated to the tree-ring chronology for the deglaciation. Second, abundant evidence from fossil trees and lake sediments indicates time intervals in the postglacial, spanning 300 to 400 years long, which had nearly constant radiocarbon ages. Third, climate researchers traditionally have assumed that sidereal and radiocarbon years are more or less interchangeable when comparing climate records (Prell 1984). The Barbados ^{14}C calibration (through $^{230}Th/^{234}U$) indicates that this assumption may be in error by as much as 2500 years (Bard *et al* 1990a, b).

COMPARISON OF GLOBAL MELTWATER DISCHARGE CURVE AND MELTWATER PULSES MEASURED IN THE CIRCUM-ATLANTIC

The meltwater discharge estimates derived from the Barbados sea-level curve must reflect the global discharge from the large continental ice sheets and has been adequately described elsewhere (Fairbanks 1989, 1990). An independent sea-level record from New Guinea, spanning the interval 7000 to 11,000 ^{14}C years, is concordant with the Barbados results (Chappell & Polach 1991). The discharge curve has been computed in radiocarbon and ^{230}Th/^{234}U years (Fig 30.1B), which we equate to sidereal years based on cross-calibration to the dendrocalibrated radiocarbon time scale (Stuiver *et al* 1986) and annual band counting in living corals (Bard *et al* 1990a, b; Edwards, Taylor & Wasserburg 1988). The two prominent peaks in meltwater discharge, termed meltwater pulse (MWP) IA and IB, are dated at 12,000 ^{14}C years (14,100 ^{230}Th/^{234}U yrs) and 9500 ^{14}C years (11,300 ^{230}Th/^{234}U yrs), respectively. The MWPs reached values of 17,000 km^3/yr for MWP-IA and 13,000 km^3/yr for MWP-IB.

Since the Barbados sea-level and discharge curves were derived, numerous meltwater plumes have been measured in the circum-North Atlantic and Gulf of Mexico and dated by AMS ^{14}C. The plumes are identified by anomalously low δ^{18}O values measured in high-accumulation-rate cores along the continental margins of the North Atlantic. Figure 30.2 illustrates the δ^{18}O records for six cores in the circum-Atlantic which are reported to be examples of meltwater plumes from the Barents, Fennoscandian and Laurentide ice sheets. There is a clear distinction between the timing of meltwater plumes from these three ice sheets. The results from two cores in the Norwegian Sea (V27-60 and PS21295-4) document a meltwater plume centered between 15,500 and 14,500 ^{14}C years (Jones & Keigwin 1988; Lehman *et al* 1991). This meltwater plume is believed to mark the early disintegration of the Barents ice sheet (Jones & Keigwin 1988). A δ^{18}O record from the North Sea (Troll 3.1), adjacent to the Fennoscandian ice sheet, documents a meltwater plume between 14,200 and 13,200 ^{14}C years (Lehman *et al* 1991). The most dramatic and best-dated MWP is found in the Gulf of Mexico, recording an MWP centered at 12,000 ^{14}C years (Broecker *et al* 1990b). In several of the Gulf of Mexico records and all MWP records outside of the Norwegian Sea, a dramatic δ^{18}O decrease occurs between 10,000 and 9000 BP. The shift is generally associated with warming in the North Atlantic, so it does not usually result in an "overshoot" of δ^{18}O values. Thus, the MWP after 10,000 ^{14}C years is difficult to define by available δ^{18}O records alone.

Fig 30.2. Oxygen isotope records of planktonic foraminifera showing pulsed and sometimes sequential melting of different continental ice sheets plotted on a map of observed free-air gravity anomalies over Canada and Fennoscandia (from Peltier 1990). The AMS-^{14}C-dated meltwater plumes are from Broecker *et al* (1990b) Gerg 87G13-31; Jones and Keigwin (1988) PS21295-4; Lehman *et al* (1991) V27-60, V28-14 and Troll 3.1 (shallow-water benthic foraminifera were analyzed in this core); Bard *et al* (1987) SU81-18; Labracherie *et al* (1989) MD84-551.

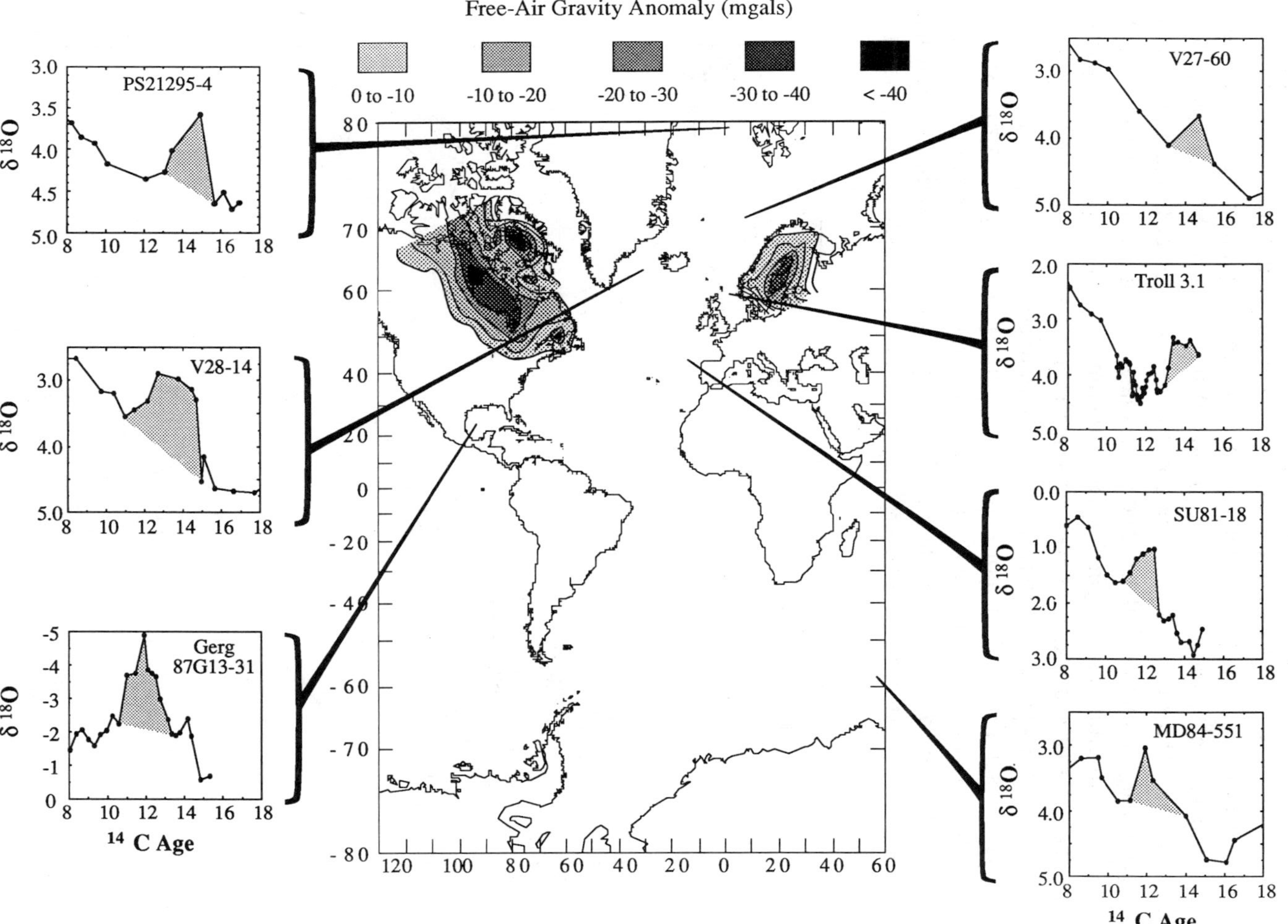
Free-Air Gravity Anomaly (mgals)
0 to -10
-10 to -20
-20 to -30
-30 to -40
< -40
PS21295-4
V28-14
Gerg
87G13-31
V27-60
Troll 3.1
SU81-18
MD84-551
δ 18O
14 C Age

It is important to note that small shifts in the isotherms and/or changes in salinity unrelated to MWPs can produce a $\delta^{18}O$ anomaly indistinguishable from glacial meltwater. Changes in the seasonal flux of a foraminiferal species can also produce a $\delta^{18}O$ transient. Examination of the predicted $\delta^{18}O$ of calcite for the modern surface ocean shows the sensitivity of this tracer to changes in the evaporation/precipitation balance and water mass mixing. It is well known that the $\delta^{18}O$ of foraminiferal calcite tests are predominantly a function of temperature and $\delta^{18}O$ of sea water, as well as some minor effects associated with the ecology and life cycles of different species. Figure 30.3A shows an example of the modern salinity field for northern hemisphere spring (March, April, May), which was used to compute the $\delta^{18}O$ of sea water for this season based on measurements of the $\delta^{18}O$-salinity relationship for all of the major salinity gradients (Fig 30.3B). A set of seven $\delta^{18}O$-salinity equations was used to compute the

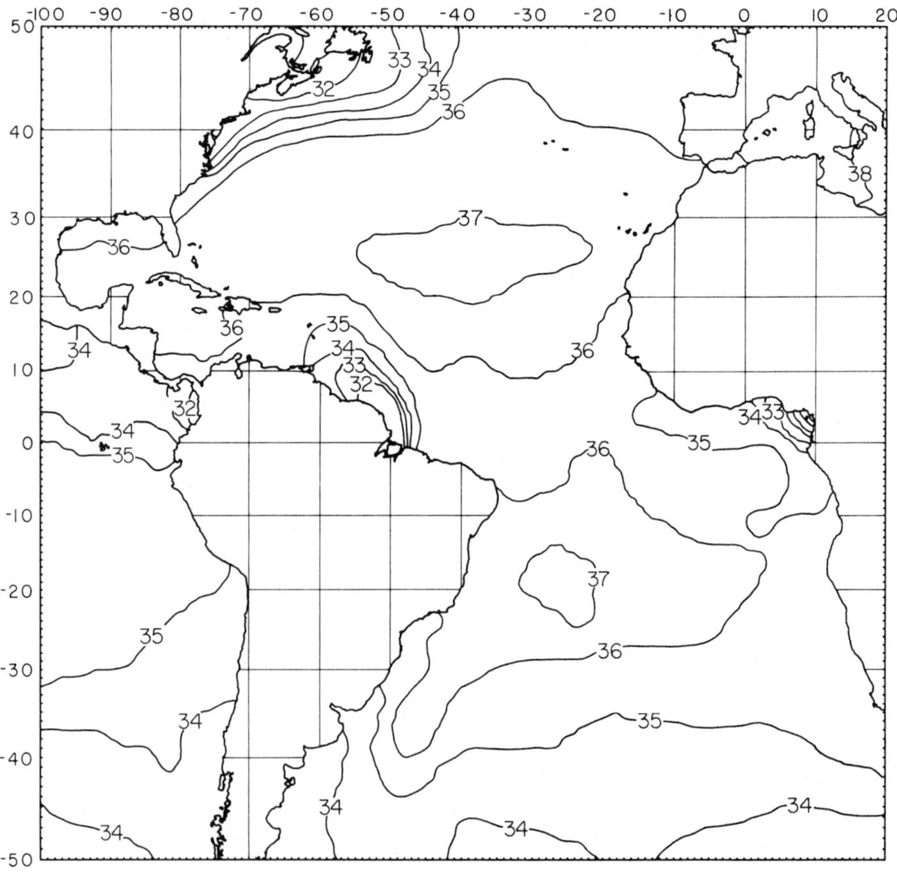

Fig 30.3A. Map of modern Atlantic surface water salinity averaged for the months of March, April and May (from Levitus 1982)

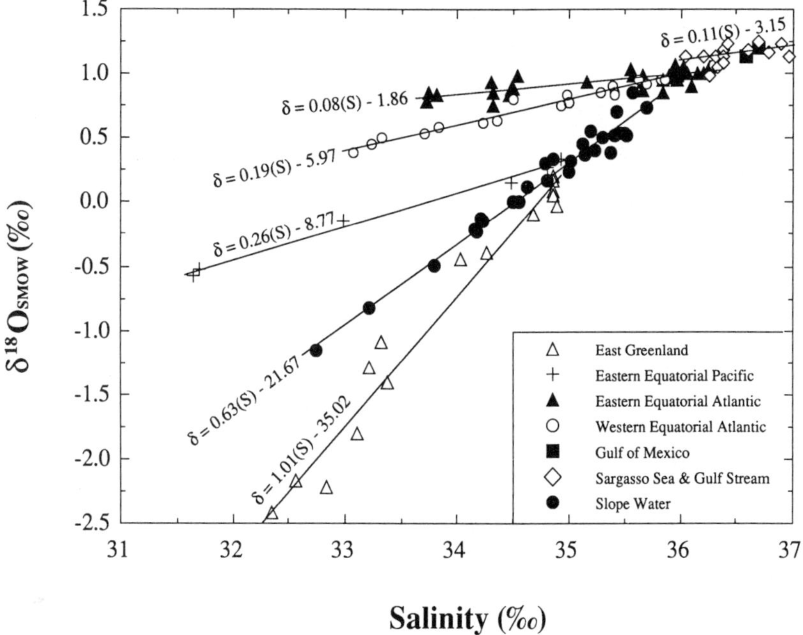

Fig 30.3B. Plots of $\delta^{18}O$ and salinity measurements across all major salinity gradients in the Atlantic and eastern tropical Pacific (Fairbanks, ms in preparation)

surface-water $\delta^{18}O$ on a 1° × 1° grid (Fairbanks, ms in preparation) for each season. Combining the estimated $\delta^{18}O$ of sea water with the measured monthly temperature from Levitus (1982) (Fig 30.3C), we computed the estimated $\delta^{18}O$ of calcite for each month. Figure 30.3D is the computed annual average $\delta^{18}O$ of calcite, and illustrates that it is easy to create $\delta^{18}O$ transients that are not MWPs. In the subpolar region, the effect of decreasing temperature on the $\delta^{18}O$ of calcite dominates the opposite effect of decreasing $\delta^{18}O$ of sea water poleward. In the polar waters, where surface water is freezing, the salinity has a greater influence on the isopleths of $\delta^{18}O_{calcite}$. This relationship adds some confidence to the meltwater origin of the Fram Strait core PS21295-4 (Fig 30.2). However, the best way to verify that a $\delta^{18}O$ anomaly is an MWP is to trace its increasing amplitude toward the source. The 4‰ $\delta^{18}O$ pulse from the Gulf of Mexico (Gerg 87G13-31) is the only MWP that is of indisputable origin. Although all of the other $\delta^{18}O$ anomalies in Figure 30.2 have been reported in the literature as MWPs, their relationship to meltwater awaits verification.

Climate Forcing Factors

According to the astronomical theory of climate change, the Pleistocene ice ages were caused primarily by changes in the seasonal distribution of incoming solar

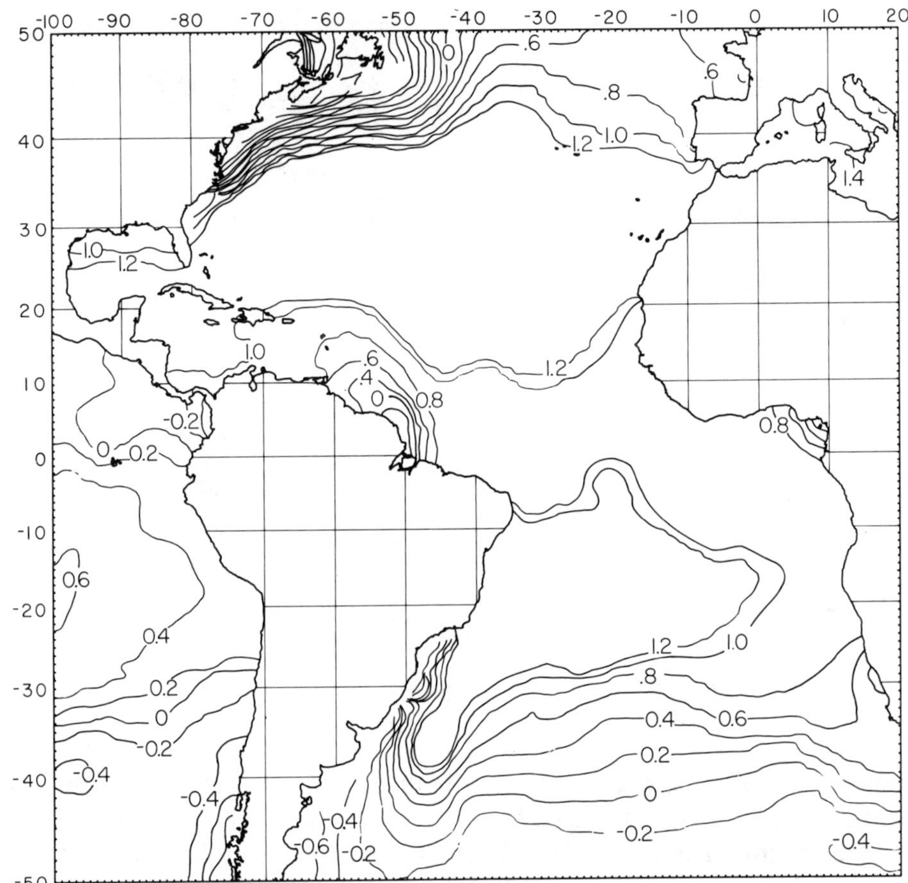

Fig 30.3C. The estimated $\delta^{18}O$ of modern Atlantic surface water averaged for the months of March, April and May. The map is based on the salinity *vs* $\delta^{18}O$ equations in Figure 30.3B applied to the salinity map in Figure 30.3A.

radiation associated with orbital variations (Milankovitch 1941). The northern hemisphere ice sheets accumulate in the 40°–80° N latitude range, whereas the Antarctic ice sheet is located poleward of 70° S. Insolation changes associated with the 41,000-year tilt cycle increase poleward in both hemispheres, while the 23,000-year precessional cycle dominates insolation changes at middle and low latitudes and is out of phase between hemispheres. Figure 30.4 shows the summer insolation at 50°, 60°, 70° and 80° N latitude, as calculated from the equations of Berger (1978). Summer insolation between 24,000 and 11,000 years increases by 11% at 50°N compared to 15% at 80°N. The relationship between insolation and a "filtered" sea-level curve is reasonably good for the last deglaciation, now that the sea-level curve is corrected to sidereal years. However, we know from comparison of insolation to the present-day sea level,

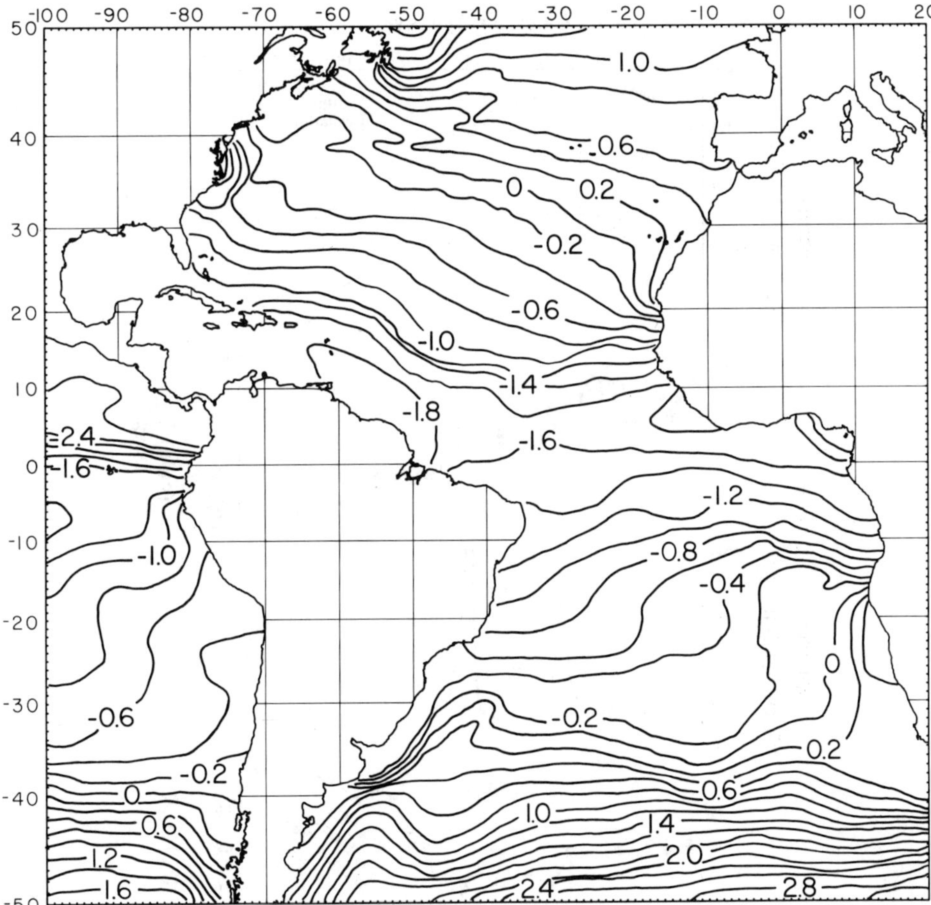

Fig 30.3D. The predicted annual average $\delta^{18}O$ of calcite is based on the "paleo-temperature" equation of Epstein *et al* (1953) using the 1° × 1° gridded temperature and $\delta^{18}O_{\text{sea water}}$ values.

as well as from longer records of sea level, that insolation alone does not correspond directly to ice volume or its sea-level equivalent. In the next section, we examine the timing and regions of influence of two suspected amplifiers of the insolation forcing function, in order to assess their possible roles in generating the meltwater discharge curve (Fig 30.1B).

One popular explanation for the non-linear response of climate change is variability in the production rate of NADW (Stommel 1961; Rooth 1982; Broecker, Peteet & Rind 1985; Bryan 1986; also see Broecker & Denton 1989, for a review). Several characteristics of the NADW thermohaline circulation provide the basis for this hypothesis. The process of NADW formation leads to excessive heat loss from the ocean to the atmosphere in the northern hemisphere

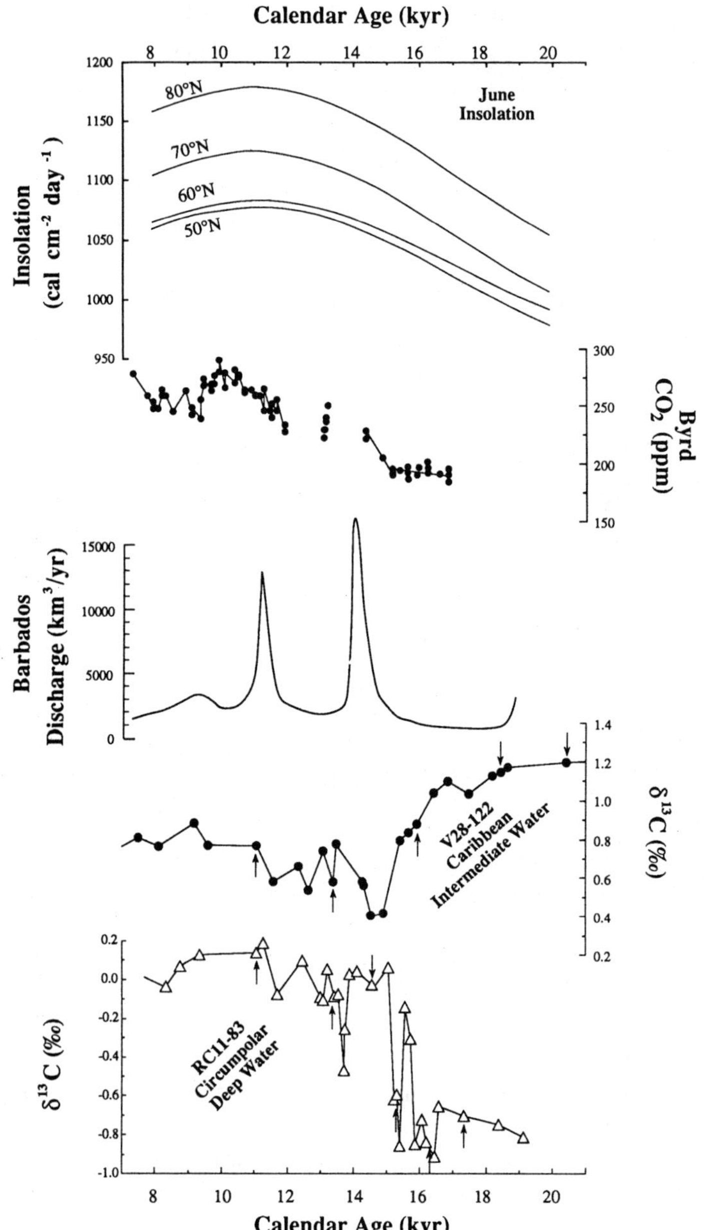

Fig 30.4. Record of atmospheric CO_2 from Byrd (Antarctica) polar ice core (Neftel *et al* 1988; Staffelbach *et al* 1991) compared to global meltwater discharge curves (Fairbanks 1990) and northern hemisphere summer insolation (Berger 1978), all reported in calendar or $^{230}Th/^{234}U$ years. Comparison of the Southern Ocean record of NADW flux (core RCl1-83 benthic foraminifera $\delta^{13}C$ record) with a mid-depth $\delta^{13}C$ time series of Atlantic intermediate water chemistry (core V28-122 benthic foraminifera $\delta^{13}C$ record) (Charles & Fairbanks, ms) adjusted to calendar years using the ^{14}C calibration of Bard *et al* (1990a, b), showing the coincidence of NADW production and deglaciation. AMS-^{14}C-dated levels are indicated by arrows.

winter. The heat released upon NADW formation is equivalent to 35% of the solar energy incident on the North Atlantic polar latitudes today, and therefore, changes in the production of NADW might explain large, rapid changes in European continental climate (Broecker & Denton 1989). NADW also provides a crucial source of heat and salt to other oceans, especially the Southern Ocean, where a convergence of oceanic heat is required to melt sea ice on an annual basis (Gordon 1986). Further, NADW variability would affect oceanic nutrient and CO_2 distributions on a global scale, and therefore, may indirectly change atmospheric CO_2 content. Finally, modeling studies suggest that it may be possible for the NADW cell to switch catastrophically between stable on and off modes (Bryan 1986; Manabe & Stouffer 1988; Maier-Reimer & Mikolajewicz 1989).

Changes in the flux of North Atlantic Deep Water as measured in the Southern Ocean

Although investigators of deep-sea sediment cores in the North Atlantic have tried to delineate the history of NADW production during the last deglaciation, the chronology of the records analyzed and the generalities of the results remain uncertain (Mix & Fairbanks 1985; Boyle & Keigwin 1987; Jansen & Veum 1990; Keigwin & Jones 1989). Oppo and Fairbanks (1987, 1990) and Charles and Fairbanks (ms) show that the Southern Ocean is one of the best regions in which to study the details of NADW variability and the consequent climatic effects. Taking advantage of the global-scale mixing of deep water that occurs in the Antarctic Circumpolar Current, Charles and Fairbanks (ms) used benthic foraminiferal $\delta^{13}C$ from a Southern Ocean core (RC11-83, 41°36′S; 09°48′E, 4718 m) with a high sedimentation rate to document the timing of NADW flux from the Atlantic Ocean during the last deglaciation (Fig 30.5). The deglacial time scale for RC11-83 is derived from 11 AMS ^{14}C dates on seven levels in the core (Charles & Fairbanks, ms). The AMS ^{14}C measurements were made at the University of Arizona NSF Accelerator facility for Radioisotope Analysis by A J T Jull and D J Donahue.

The RC11-83 benthic $\delta^{13}C$ record (Fig 30.4) indicates that deglacial changes in the nutrient chemistry of the deep Southern Ocean were rapid and large. An abrupt 1‰ shift from low values characteristic of glacial conditions to high values occurred from 12,600 to 12,200 ^{14}C years. This rapid rise in Southern Ocean $\delta^{13}C$ values was most likely associated with a change in status of NADW, switching from the "off" mode to the "on" mode. The RC11-83 record also shows significant oscillations superimposed on this basic NADW off-on switch. A prior oscillation occurred between 13,300 ^{14}C years and 12,800 ^{14}C years, and a low $\delta^{13}C$ event occurred at 11,800 ^{14}C years.

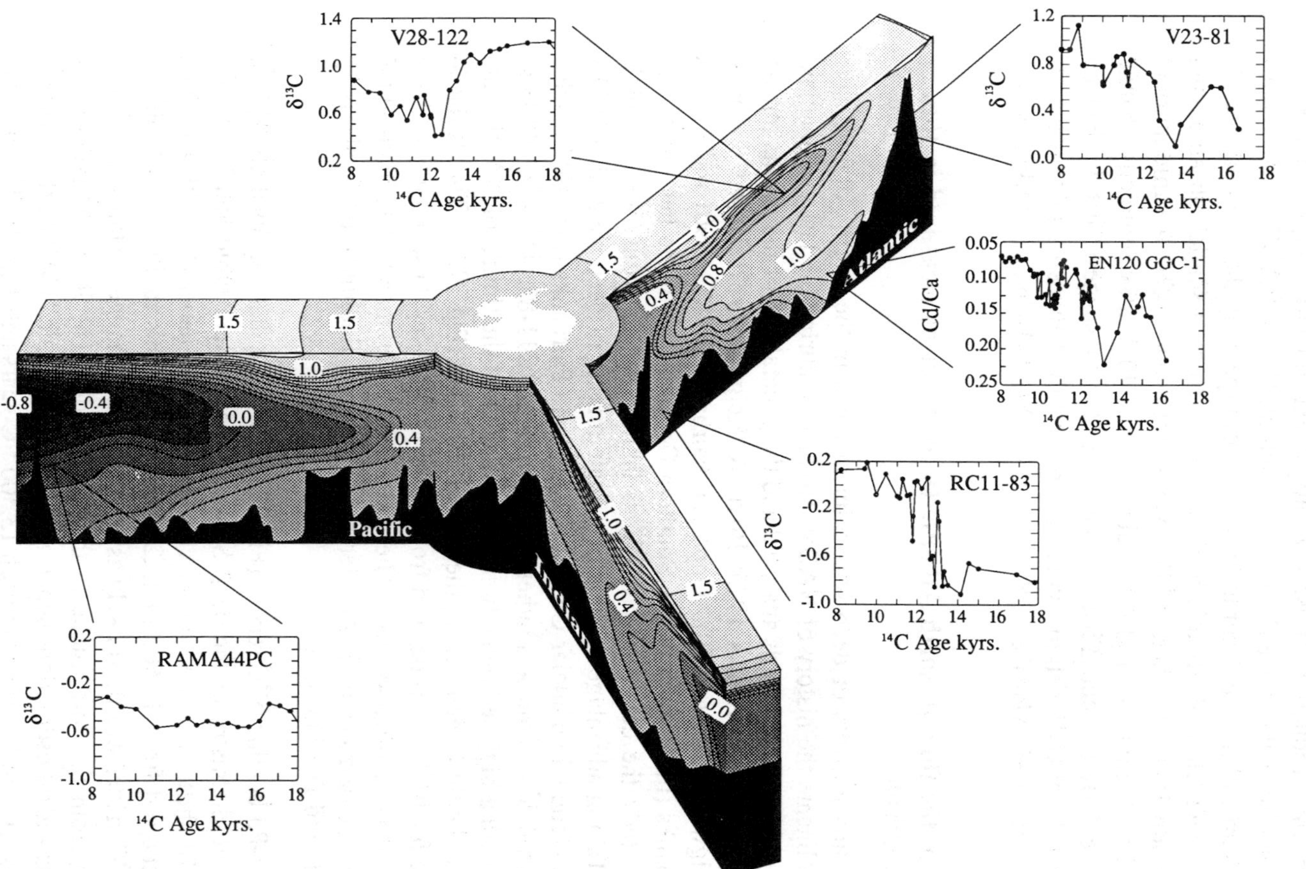

Fig 30.5. GEOSECS $\delta^{13}C$ data from the different ocean basins (Kroopnick 1985) with postglacial times series of $\delta^{13}C$ in sediments dated by AMS ^{14}C for the North Atlantic (Jansen & Veum 1990), intermediate-depth Atlantic (Oppo & Fairbanks), Southern Ocean (Charles & Fairbanks, ms) and Pacific Ocean (Keigwin 1987). Also shown for the North Atlantic are Cd/Ca in sediments dated by AMS (Boyle & Keigwin 1987).

Comparison with other detailed deglacial records from the deep North Atlantic shows significant differences for the interval from 16,000 to 13,500 ^{14}C years (Fig 30.5). The North Atlantic records show this region was first inundated with NADW between 17,000 and 15,000 ^{14}C years, followed by minimum NADW between 14,000 and 13,000 ^{14}C years, reaching or exceeding glacial δ^{13}C values (Boyle & Keigwin 1987; Jansen & Veum 1990). The striking transition from nutrient enrichment (low δ^{13}C) to nutrient depletion (high δ^{13}C), which is dated between 12,600 and 12,200 ^{14}C years in RC11-83, is apparent in all three records (Fig 30.5). Another step in the same direction, but with much lower amplitude, is apparent from about 10,000 to 9,000 ^{14}C years.

The δ^{13}C records that monitor mid-depth North Atlantic water provide further documentation of a rapid change in upper NADW nutrient chemistry from 14,000 to 12,400 ^{14}C years (Fig 30.5). In these records, the sense of change prior to 12,200 ^{14}C years is opposite to that of the deep Atlantic. Whereas relative nutrient depletion prevailed during glacial periods, a shift to more nutrient-rich conditions occurred at 12,700 ^{14}C years, synchronous with the timing of deep-ocean change. The inverse mid-depth and deep-Atlantic signals could both reflect the same phenomenon if, as in today's ocean, relatively nutrient-rich water from the Southern Ocean were drawn northward across the equator at mid-depth by the removal of surface and thermocline water to form southward-flowing NADW.

Where the various North Atlantic nutrient proxy records disagree, we suggest that RC11-83 should be taken as the more comprehensive measure of NADW flux variability for two reasons. First, deglacial sedimentation rates in RC11-83 are highest. Second, RC11-83 δ^{13}C values presumably monitor the flux of NADW relative to Pacific and Indian outflow, while mixing between NADW and Southern Ocean Water may vary independently of NADW flux in some North Atlantic locations. This issue is important for the interval from 17,000 to 15,000 ^{14}C years, because the North Atlantic records indicate a switch from high to low nutrients, suggesting that the initiation of NADW began 3000 years earlier than indicated by the Southern Ocean record. The RC11-83 record, on the other hand, shows more uniformly low δ^{13}C values until 13,000 ^{14}C years, indicating minimal NADW flux throughout the glacial period. This issue is also important for characterizing the Younger Dryas interval (11,000 to 10,000 ^{14}C years), the period during which, according to previous interpretations of a Bermuda Rise core (Boyle & Keigwin 1987), NADW production was temporarily reduced. The Southern Ocean record does not show an anomaly during this period, implying that the total flux of NADW may not have varied significantly. Thus, North Atlantic cores situated in the mixing zones between water masses may be too sensitive to small circulation changes to document accurately net changes in the NADW flux.

It is possible, however, that deviations from the simple mixing models for Circum-Polar Deep Water (CPDW) $\delta^{13}C$ can occur. For example, some evidence suggests that CPDW $\delta^{13}C$ was lower than in the deep Pacific during glacial periods, and this distribution of $\delta^{13}C$ cannot be easily explained by mixing alone unless Indian Ocean deep water compensated for the relatively nutrient-depleted Pacific water mixing in the circumpolar region. One could also invoke non-conservative effects that are unique to the Antarctic, such as proposed by Boyle (1990). A more serious problem is the observation that Cd/Ca ratios in benthic foraminifera from Southern Ocean cores do not increase nearly as much as the $\delta^{13}C$ might predict during glacial periods (Boyle 1991). We have no definitive answer for why the Southern Ocean $\delta^{13}C$ at times might be lower than the Pacific, and the discrepancy between $\delta^{13}C$ and Cd/Ca demands more research into the decoupling of these tracers from the nutrient elements in the Southern Ocean (eg, see Charles & Fairbanks 1990). Nevertheless, we make the simplest interpretation that seems justified from the modern distribution of $\delta^{13}C$ (Kroopnick 1985) and the core-top calibration of benthic foraminifera $\delta^{13}C$ in deep water (Belanger, Curry & Matthews 1981): a strong relative contribution of NADW is the only means for increasing CPDW $\delta^{13}C$ above Pacific (mean ocean) values.

Ice-core record of atmospheric CO_2 change

Antarctic ice-core specialists have emphasized the primary role of atmospheric CO_2 content in amplifying the glacial-interglacial cycles (Lorius *et al* 1990). The Antarctic CO_2 results are cited as an important independent calibration for the so-called CO_2 amplifier factor, which is incorporated in GCMs. Evaluating the magnitude of the CO_2 amplifier is at the heart of climate research today. Recently, the deglacial ice from Vostok (Antarctica) and Dye 3 (Greenland) ice cores have been reanalyzed for atmospheric CO_2, and additional data have been added to the Byrd (Antarctica) record in the critical deglacial interval (Barnola *et al* 1991; Staffelbach *et al* 1991). Re-evaluation of the air dating at Vostok indicates that the age difference between air and ice is about 6000 years during the coldest periods, rather than 4000 years as previously estimated (Barnola *et al* 1991). The long and variable closure age and poor sampling resolution of Vostok make the record unsuitable for comparison to the high-resolution ocean/climate records from the Southern Ocean, North Atlantic and Barbados. However, new data from Dye 3 and Byrd (Antarctica) are critical for this comparison (Fig 30.4).

Previous results from Dye 3 indicated that the CO_2 shift from glacial to postglacial occurred in approximately 650 years (Stauffer *et al* 1985). The new CO_2 measurements in this interval show anomalously high CO_2 concentrations and high variability between samples, characteristic of problems associated with melt layers (Staffelbach *et al* 1991). In their original paper, Stauffer *et al* (1985) argued for the validity of the rapid shift in CO_2 based on the fact that the shift occurred within an interval of anomalously low $\delta^{18}O$. This interval of low $\delta^{18}O$, also recognized in the Camp Century (Greenland) ice core, is similar to the interval interpreted as the Younger Dryas (which has been interpreted as an unusually cold interval) (Dansgaard, White & Johnson 1989). Interestingly, this is the same interval which Fairbanks (1990) reinterpreted to be a warm interval correlative with MWP-IA. It is now clear from the composite CO_2 data set for Dye 3 that the highest CO_2 values, exceeding 350 ppm, and the intervals of greatest variability occur in the two core sections with lowest $\delta^{18}O$ values (Staffelbach *et al* 1991). Regardless of the stratigraphic interpretation for this specific interval, the CO_2 results from Dye 3, and even CO_2 data sampled in intervals with low $\delta^{18}O$, are now suspect.

By default, we are left with only those CO_2 results from Byrd ice core, which, unfortunately, are dated by ice-flow models. New CO_2 data from the deglacial section of Byrd confirm a slow increase (Staffelbach *et al* 1991) from 1350 to 1000 m deep. According to the age model of Paterson and Hammer (1987) the glacial to postglacial change in CO_2 corresponds to the time interval from 20,000 to 10,750 BP. Based on their new CO_2 data for Dye 3 and Byrd, Staffelbach *et al* (1991) conclude that the glacial-to-postglacial rise in CO_2 (190–270 ppm) occurred slowly over a 10,000-year interval, and not rapidly over a 650-year interval, as they previously reported (Stauffer *et al* 1985) (Fig 30.4).

DISCUSSION

Increasing CO_2 and insolation are predicted to cause greatest climate changes in the polar regions. According to GCM simulations, the northern hemisphere polar region is most affected by an increase of atmospheric CO_2 (Fig 30.6). The difference in air temperatures for a $2 \times CO_2$ simulation is estimated to be as much as 4°C for North America for today's climate (Stouffer *et al* 1989). However, the CO_2 data from Byrd show only a 15% increase immediately prior to MWP-IA, thus implying extremely small thermal consequences for the Laurentide ice sheet.

Taken at face value, the AMS-^{14}C-dated meltwater plumes in the circum-Atlantic indicate sequential melting of the northern hemisphere ice sheets starting with the Barents, followed by the Fennoscandian and, finally, the Laurentide ice sheet (Fig 30.2). This north to south deglaciation is consistent with the ICE-3 recon-

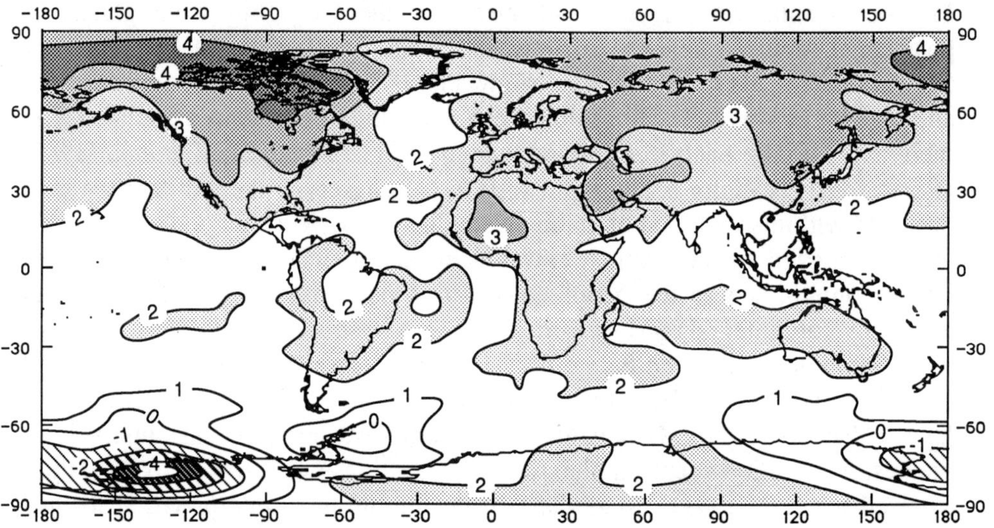

Fig 30.6. Map showing the predicted change in air temperature due to double CO_2 according to the model run of Stouffer, Manabe and Bryan (1989). Reprinted by permission of *Nature*. © 1989 MacMillan Magazines Ltd.

struction of Tushingham and Peltier (1991), based on regional sea-level curves and terrestrial reconstructions. The dating of the meltwater plumes in the Fram Strait (PS21295-4) and the Norwegian Sea (V27-60) suggests that the Barents ice sheet may have been responsible for the small initial rise of sea level between 16,000 and 15,000 ^{14}C years BP (Fig 30.1A) (Lehman *et al* 1991). It is clear from the meltwater plume records that the Laurentide ice sheet (Gerg13-31) was primarily responsible for the large sea-level increase between 12,300 and 11,800 ^{14}C years.

However, processes in the polar and subpolar regions may create apparent MWPs which are also tied to large-scale climate change (Broecker 1990a). For example, in the model simulation of Manabe and Stouffer (1988), in which they "turned-off" the thermohaline circulation, we have computed the estimated $\Delta\delta^{18}$O based on their temperature and salinity anomaly maps (Fig 30.7). We can see from Figure 30.7C that oscillations of the thermohaline circulation can create apparent MWPs in the polar and subpolar regions up to −0.75‰. This is the magnitude of δ^{18}O anomalies in the Norwegian/Greenland Sea, which were assumed to be MWPs based solely on the low δ^{18}O anomalies (Jones & Keigwin 1988; Lehman *et al* 1991). The implications of these simulations are that low δ^{18}O anomalies could be either a product of meltwater plumes or temporary shutdown of the thermohaline circulation or both. Interestingly, all high-resolution deep-water records from the North Atlantic, as well as the Southern Ocean record of RC11-83, show an abrupt "switch-on" of the thermohaline cir-

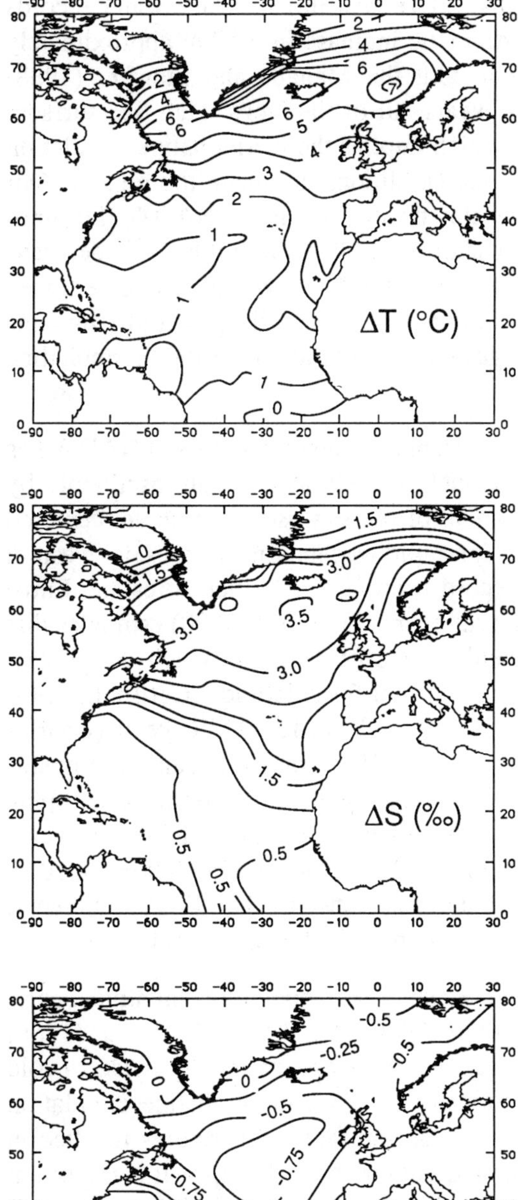

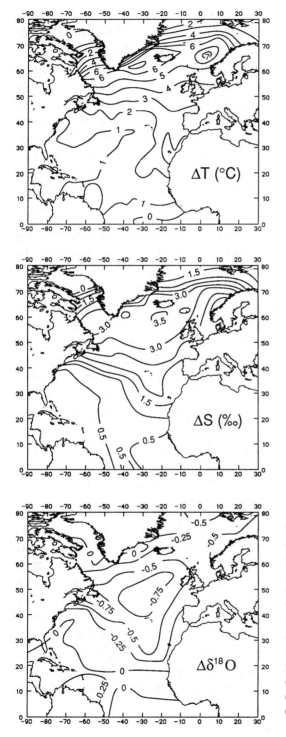

Fig 30.7. Maps showing the difference in North Atlantic surface-water temperature (A) and surface-water salinity (B) between model simulation with and without thermohaline circulation (from Manabe & Stauffer 1988). In the polar and subpolar latitudes, the $\delta^{18}O$ of sea water changes at a rate of −0.5‰ per unit salinity decrease compared to +0.22‰ for each degree temperature drop. The estimated difference in $\delta^{18}O$ shows that the effects of decreasing temperature and salinity tend to cancel out; however, in polar regions, the effects of decreasing salinity dominate (C).

culation at 13,000 ^{14}C years (Figs 30.4, 30.5). All three surface water records from the Norwegian Sea recorded a dramatic increase in $\delta^{18}O$ at approximately 13,000 ^{14}C years, which is consistent with the Manabe and Stouffer (1988) simulation (Fig 30.7). At 16,000 ^{14}C years, all of the Norwegian Sea surface-water records have high $\delta^{18}O$ values, which implies the thermohaline circulation is "switched-on" or possibly that sea ice conditions dominate the region. Our Southern Ocean core (RC11-83) indicates no net change in NADW flux until 13,000 ^{14}C years, which makes it difficult to explain the initial $\delta^{18}O$ decrease around 15,000 ^{14}C years. In summary, unless the $\delta^{18}O$ anomalies can be traced back to a point source with a large amplitude $\delta^{18}O$ signal, the possibility remains that they are artifacts of salinity changes due to oscillations of the thermohaline circulation.

This complex interplay between meltwater, marine precipitation and NADW has been developed into a conceptual model, known as the salt oscillator, by Broecker *et al* (1990a). The development of the salt oscillator model centered around the need for an explanation for rapid vegetation changes in western Europe and oscillations in the $\delta^{18}O$ record from Greenland ice cores and European lake cores during the Younger Dryas event. Broecker (1990) estimates that the annual production of NADW yields 4×10^{21} calories, which is approximately 10% of the atmospheric heat flux north of 50°N latitude (Carissimo, Oort & Vonder Haar 1985). The "switch-on" of NADW at 12,600 ^{14}C years (Charles & Fairbanks, ms) should have had immediate effects on European air temperatures according to GCM simulations. How this is transmitted to the catastrophic destruction of the Laurentide ice sheet between 12,300 and 11,800 ^{14}C years remains unresolved. A brief reduction in thermohaline circulation dated at 11,500 ^{14}C years (Fig 30.4) is a probable consequence of the MWP, centered at 12,000 ^{14}C years. This may be the best example of Broecker's "salt oscillator."

The influence of colder North Atlantic sea surface temperatures on atmospheric temperatures was simulated by Rind *et al* (1986) (Fig 30.8). Rind *et al* (1986) assigned winter sea surface temperatures that were approximately twice as cold as those simulated by Manabe and Stouffer's (1988) thermohaline circulation "off" mode. These extremely cold winter sea-surface temperatures resulted in winter air temperatures approximately 4°C cooler across Europe. It is important to note that the North American continent warmed slightly in this simulation, presumably due to a greater fraction of polar heat transport via the atmosphere to compensate for the loss in the oceanic heat transport. According to Stone (1978), the total meridional heat flux is primarily governed by physical factors other than the mode of transport (eg, ocean or atmosphere). Thus, a reduction in one mode will be at least partially compensated by the other.

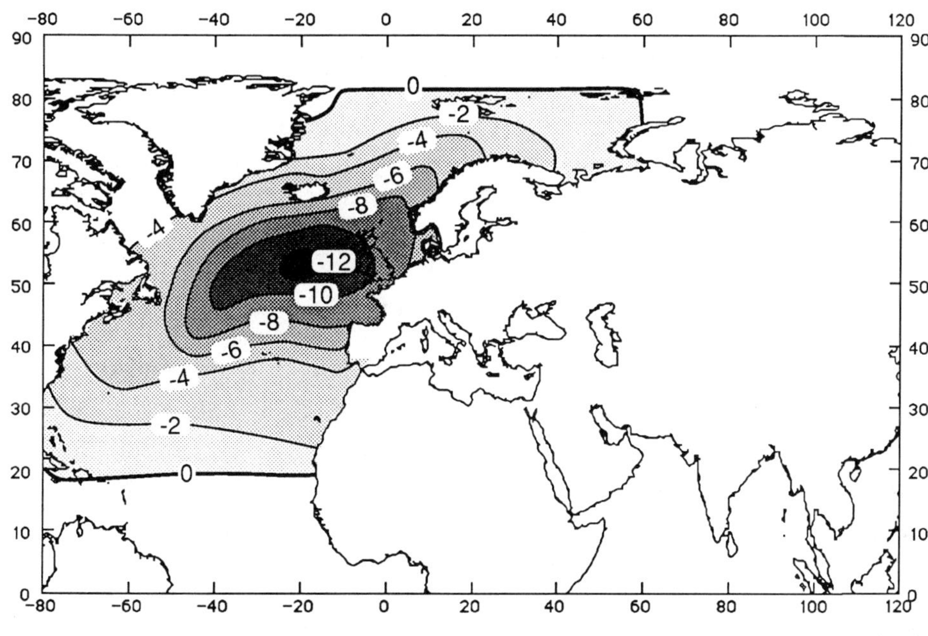

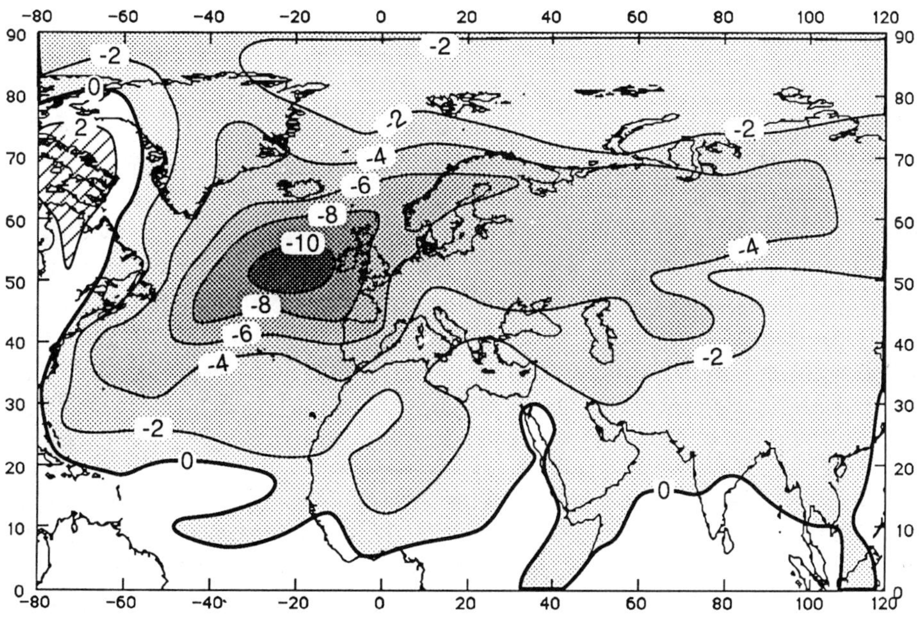

Fig 30.8. Maps showing influence of decreased winter sea-surface temperature in the North Atlantic on the winter air temperatures, as simulated by the Goddard Institute for Space Studies (GISS) atmospheric GCM (Rind *et al* 1986). Note that estimated air temperatures over Europe decrease while air temperatures over North America increase.

The simulation of Rind *et al* (1986) shows only a minor influence of NADW shut-down on the air temperatures over North America. More important, winter air temperatures are not generally considered critical for the massive ablation of ice sheets such as occurred in North America between 12,300 and 11,800 ^{14}C years. During the critical summer months, the northern North Atlantic Ocean is not an important source of heat to the North American continent. However, warm air masses are generally considered to be essential for the rapid ablation of an ice sheet (Oerlemans & van der Veen 1984). The strong winds associated with cyclones efficiently penetrate the temperature inversion layer over ice sheets, allowing warm air to erode the ice sheet. In summary, the thermohaline circulation status has direct effects on the winter air temperatures and vegetation history of Europe. However, the link to the Laurentide ice sheet must be indirect.

Thus, we must look to other indirect consequences of the "switch-on" of thermohaline circulation at 12,600 ^{14}C years on the ablation rates of the Laurentide ice sheet. It is important to note that the coincidence of the MWP computed from the Barbados sea-level curve and the measured meltwater plume from the Laurentide ice sheet demonstrates that the rise in sea level is due to a sudden change in the ablation rates, and not moisture supply rates to the Laurentide ice sheet.

One possible scenario that explains the coincidence of the "switch-on" of NADW and the catastrophic ablation of the Laurentide ice sheet is the simultaneous opening of the atmospheric corridor between the Laurentide ice sheet and Greenland (Fig 30.9). Today, the seasonally open water in the Labrador Sea and Baffin Bay, combined with the high elevation of Greenland to the east, causes cyclones to track northward to the Arctic, providing an important source of heat and moisture. During summer, the southeast coast of Greenland and Baffin Bay show the highest frequencies of cyclonic activity in the arctic (Reed & Kunkel 1960). The strong meridional flow of air greatly reduces the latitudinal climate differences along the west coast of Greenland (Putnins 1970; Vowinckel & Orvig 1970).

Today, the δ^{18}O difference between Dye 3 ice core in south Greenland and Camp Century in northwest Greenland is only 1.6‰, partially the result of the strong meridional flow of air. Fisher and Alt (1985) point out that the latitudinal gradient in δ^{18}O on the Greenland ice cap is an excellent measure of the meridional temperature gradients. Fairbanks (1990) applied this concept by comparing the difference in δ^{18}O between Dye 3 in south Greenland with Camp Century in northwest Greenland (Fig 30.10). During the Last Glacial Maximum, the difference between Dye 3 and Camp Century was 6‰, reflecting the zonal

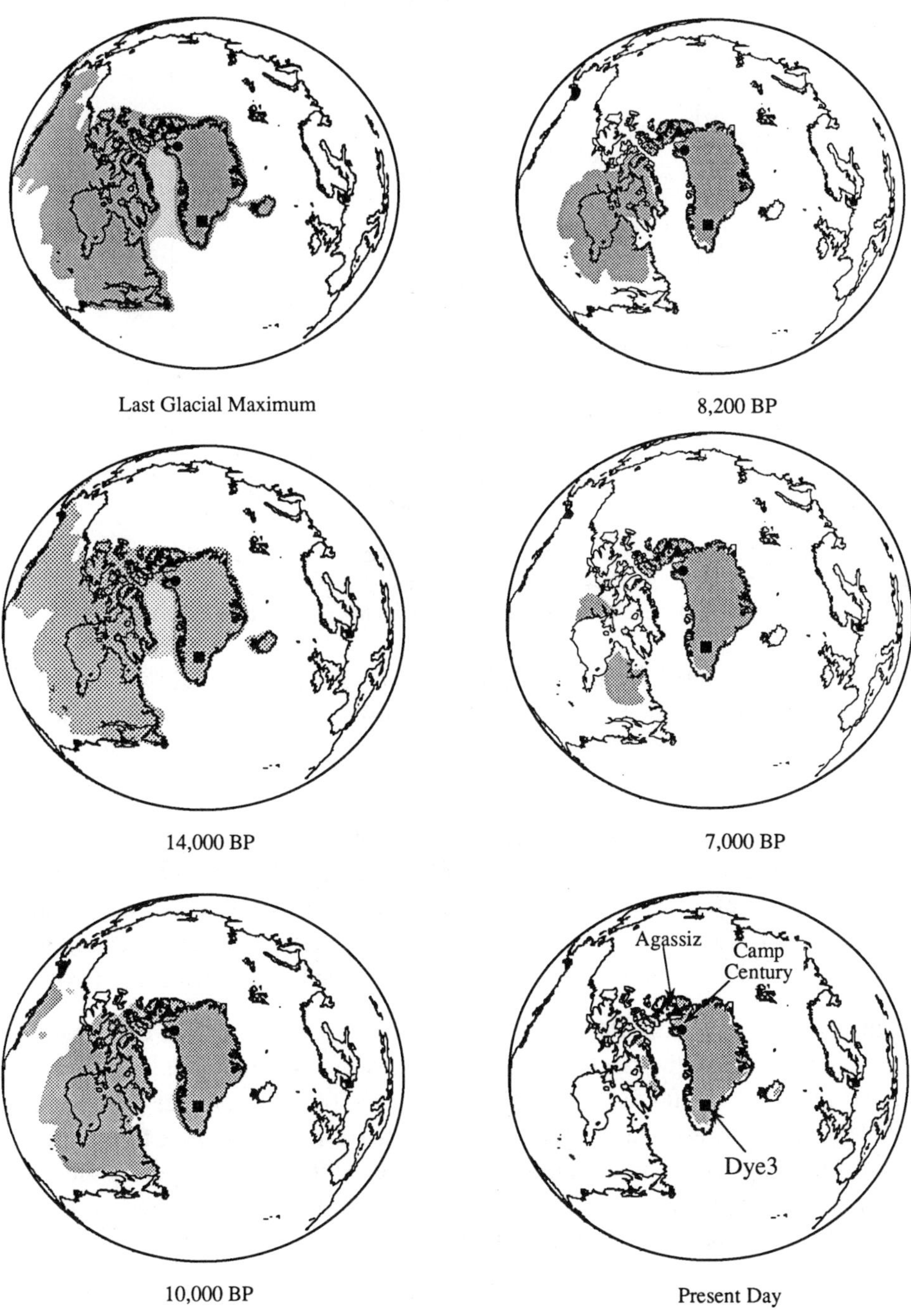

Fig 30.9. Maps of the deglaciation history of North America (modified after Denton & Hughes 1981).

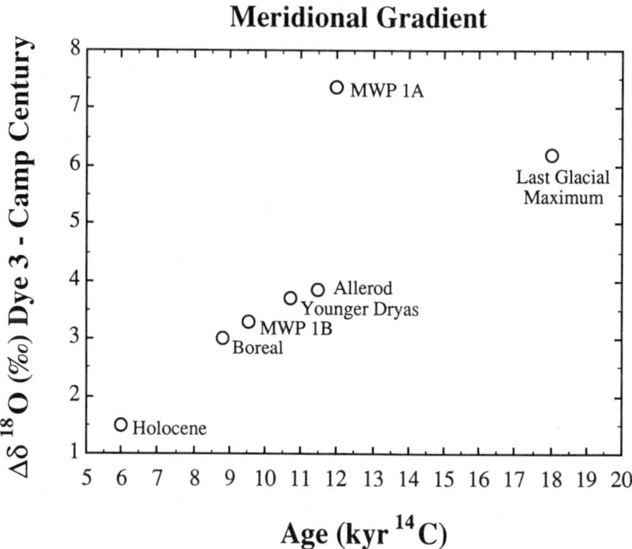

Fig 30.10. Meridional climate gradient along the west coast of Greenland estimated by the $\delta^{18}O$ difference between Dye 3 (southeast) and Camp Century (northwest) (from Fairbanks 1990). Assignment of ^{14}C chronozones is based upon the age model presented in Fairbanks (1990).

conditions documented by the CLIMAP group (1981). The $\Delta\delta^{18}O$ remained approximately 6‰ until it shifted suddenly to 4‰ during the Bølling/Allerød transition (Fairbanks 1990).

Although very little carbonate material is found in marine sediments of the Labrador Sea and Baffin Bay, Andrews *et al* (1990) successfully measured the stable isotope records from seven cores in this region, which are constrained by 35 ^{14}C dates. They found that the bulk of the sediment in their cores was deposited between 7000 and 11,500 ^{14}C years, and that there is a "conspicuous gap in ^{14}C dates between 13,300 ^{14}C years and 'old' basal dates" from their cores.

Therefore, evidence from the dated marine records support the ice core results, which indicate that the atmospheric corridor west of Greenland opened up approximately at 12,000 to 13,000 ^{14}C years. By 9500 ^{14}C years, ice-core results from Agassiz Ice Cap clearly indicate that the meridional heat flux through this corridor reached its maximum (Koerner & Fisher 1990), causing yearly melting of the ice cap (Fig 30.11). The melt-layer record on Agassiz ice cap also confirms that dramatic variations in the meridional heat flux along this corridor have occurred in the past.

Thus, the results from deep-sea and ice cores indicate that the massive disintegration of the Laurentide ice sheet at 12,000 ^{14}C years was coincident with a switch from zonal to meridional heat flux associated with the "switch-on" of the Atlantic thermohaline circulation (Broecker & Denton 1989) and the atmospheric

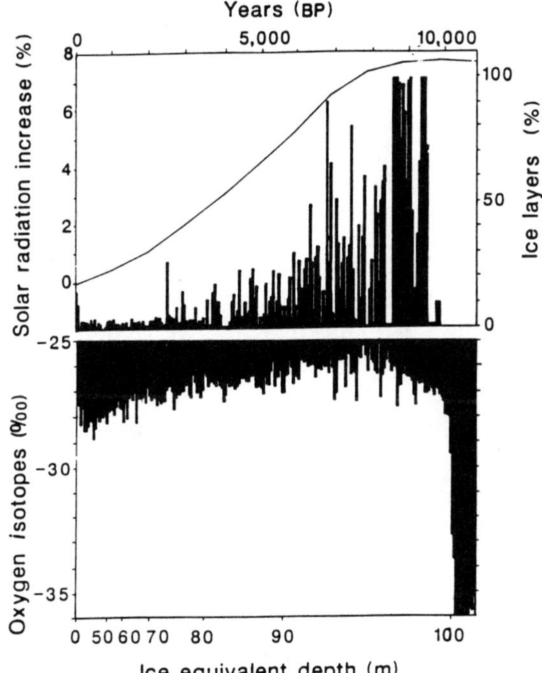

Fig 30.11. Percentage of annual melt layers that show melting averaged over 50-year intervals in ice core A-84 from the Agassiz Ice Cap at 1730 m elevation on Ellesmere Island, compared to local summer insolation record (from Koerner & Fisher 1990). Note intense melting peaked at approximately 9500 BP, according to the age model of Koerner and Fisher (1990). Work on Canadian ice caps showed that melt layers can be traced regionally between ice caps, and that melting correlates to late-summer sea-ice conditions in the region (Koerner 1989). Reprinted by permission of the author and *Science*. © 1989 AAAS.

heat pump *via* the Labrador Sea and Baffin Bay corridor. The proximity of the atmospheric corridor at the eastern edge of the Laurentide ice sheet, and the effectiveness of the atmospheric heat pump during critical summer months, makes this a likely trigger responsible for the catastrophic disintegration of the Laurentide ice sheet at 12,300 to 11,800 ^{14}C years.

CONCLUSIONS

Several new findings have impelled us to re-evaluate what is known about the amplifiers of glacial-to-postglacial climate change. These new results include:

1. The Barbados high-resolution sea-level record and global meltwater discharge curve (Fairbanks 1989, 1990);
2. The Barbados radiocarbon calibration for the deglaciation based on $^{230}Th/^{234}U$ dating by mass spectrometry (Bard *et al* 1990a);

3. New CO_2 data for Vostok, Dye 3 and Byrd ice cores (Barnola *et al* 1991; Staffelbach *et al* 1991);

4. AMS ^{14}C dating of NADW flux to the Southern Ocean (Charles & Fairbanks, ms), documenting the strength of the thermohaline circulation;

5. AMS-^{14}C-dated meltwater plumes in the circum-Atlantic ocean, which imply the sequential melting of the major ice sheets (Jones & Keigwin 1988; Broecker *et al* 1990; Lehman *et al* 1991);

6. New ^{14}C-dated marine cores in Baffin Bay, which indicate that seasonally open water was initiated approximately at 13,000 ^{14}C years (Andrews *et al* 1990).

Many amplifiers of postglacial climate change have been proposed. The timing and estimated magnitude of these amplifiers and their regional impact on climate are quite different. Compilation of AMS-^{14}C-dated meltwater plumes indicates that early deglaciation progressed from north to south. Using the Barbados radiocarbon calibration based on $^{230}Th/^{234}U$, we are able to place insolation, CO_2, ice-core climate records, deep-water reorganization and the Barbados meltwater curve on one time scale. The first major phase of deglaciation is coincident with the "switch-on" of the thermohaline circulation, as documented by AMS-^{14}C-dated carbon isotope records of deep-water circulation patterns. A probable consequence of the "switch-on" of the thermohaline circulation is seasonally open water in the Labrador Sea and Baffin Bay which in turn "switched-on" the "atmospheric heat pump" that operates along this corridor today. The implication for this study is that small changes in the oceanic heat flux in the Norwegian/Greenland Sea area may result in significant changes in the atmospheric heat transport through the much narrower corridor west of Greenland and immediately adjacent to the Laurentide ice sheet.

Although many aspects of the earth's energy budget can be researched using proxy records, we have virtually no tracers for many others. For example, insolation, water-mass and air-mass shifts, and atmospheric CO_2 and CH_4 are reasonably well known. However, we have no tracer for the most important greenhouse gas of all, water vapor. This chapter provides evidence for the causes of the first MWP dated at 12,000 ^{14}C years, but leaves explanations on the possible causes for the second MWP, dated at 9500 ^{14}C years, for another paper (Guilderson & Fairbanks, ms in preparation).

Acknowledgments

We would like to thank Julie Cole, Tom Guilderson, Mimi Katz and Wallace Broecker for helpful suggestions and comments on this manuscript. This research was supported by NSF grants OCE88-19438 and ATM90-01139 to RGF.

REFERENCES

Andrews, JT, Evans, LW, Williams, KM, Briggs, WM, Jull, AJT, Erlenkeuser, H and Hardy, I 1990 Cryosphere/ocean interactions at the margins of the Laurentide Ice Sheet during the Younger Dryas chron: SE Baffin Shelf, Northwest Territories, 5. *Paleoceanography*.

Bard, E, Arnold, M, Duprat, J, Moyes, J and Duplessy, J-C 1987 Retreat velocity of the North Atlantic polar front during the last deglaciation determined by accelerator mass spectrometry. *Nature* 328: 791–794.

Bard, E, Hamelin, B, Fairbanks, RG and Zindler, A 1990a Calibration of the ^{14}C timescale over the past 30,000 years using mass spectrometric U-Th ages from Barbados corals. *Nature* 345: 405–410.

Bard, E, Hamelin, B, Fairbanks, RG, Zindler, A, Mathieu, G and Arnold, M 1990b U/Th and ^{14}C ages of corals from Barbados and their use for calibrating the ^{14}C time scale beyond 9000 years BP. *In* Yiou, F and Raisbeck, GM, eds, Proceedings of the 5th International Symposium on Accelerator Mass Spectrometry. *Nuclear Instruments and Methods* B52: 461–468.

Barnola, J-M, Pimienta, P, Raynaud, D and Korotkevich, YS 1991 CO_2-climate relationship as deduced from the Vostok ice core: a re-evaluation based on new measurements and on a re-evaluation of the air dating. *Tellus* 43B: 83–90.

Belanger, PE, Curry, WB and Matthews, RK 1981 Core-top evaluation of benthic foraminiferal isotopic ratios for paleooceanographic interpretations. *Palaeogeography, Palaeoclimatology, Palaeoecology* 33: 205-220.

Berger, A 1978 Long-term variation of caloric insolation resulting from the Earth's orbital elements. *Quaternary Research* 9: 139–167.

Boyle, E 1990 Effect of depleted planktonic $^{13}C/^{12}C$ on bottom water during periods of enhanced relative Antarctic productivity. *EOS* 71: 1357–1358.

_____1991 Quaternary ocean paleochemistry. In *US National Report to International Union of Geodesy and Geophysics 1987–1990*: 634–638.

Boyle, E and Keigwin, L 1987 North Atlantic thermohaline circulation during the past 20,000 years linked to high-latitude surface temperature. *Nature* 330: 35–40.

Broecker, WS 1984 Terminations. *In* Berger, AL, Imbrie, J, Hays, J, Kukla, G and Saltzman, B, eds, *Milankovitch and Climate*. Dordrecht, The Netherlands, Reidel: 687–698.

_____1990 Salinity history of the northern Atlantic during the last deglaciation. *Paleoceanography* 5: 459–467.

Broecker, WS, Bond, G, Klas, M, Bonani, G and Wölfli, W 1990a A salt oscillator in the glacial Atlantic?, 1, The concept. *Paleoceanography* 5: 469–478.

Broecker, WS and Denton, GG 1989 The role of the ocean-atmosphere reorganizations in glacial cycles. *Geochimica et Cosmochimica Acta* 53: 2465–2501.

Broecker, WS, Klas, M, Clark, E, Trumbore, S, Bonani, G, Wölfli, W and Ivy, S 1990b Accelerator mass spectrometric radiocarbon measurements on foraminifera shells from deep-sea cores. *Radiocarbon* 32(2): 119–133.

Broecker, WS, Peteet, D and Rind, D 1985 Does the ocean-atmosphere have more than one stable mode of operation? *Nature* 315: 21–25.

Bryan, F 1986 High-latitude salinity effects and interhemispheric thermohaline circulations. *Nature* 323: 301–304.

Carissimo, BC, Oort, AH and Vonder Haar, TH 1985 Estimating the meridional energy transports in the atmosphere and ocean. *Journal of Physical Oceanography* 15: 82–91.

Chappell, J and Polach, H 1991 Post-glacial sea-level rise from a coral record at Huon Peninsula, Papua New Guinea. *Nature* 349: 147–149.

Charles, CD and Fairbanks, RG 1990 Glacial-interglacial changes in the isotopic gradients of Southern Ocean surface water. *In* Bleil, U and Thiede, J, eds, The Geologic History of Polar Oceans: Arctic *vs* Antarctic. *NATO ASI Series* 308. Dordrecht, The Netherlands, Kluwer Academic Publishers: 519–538.

____(ms) North Atlantic deep water and its climate effects over the last deglaciation: Evidence from Southern Ocean isotope records. Submitted to *Nature.*

CLIMAP Project Members 1981 *Geological Society of America Map and Chart Series* MC 36.

Dansgaard, W, White, JW and Johnson, SL 1989 The abrupt termination of the Younger Dryas climate event. *Nature* 339: 532–534.

Denton, GH and Hughes, DJ 1981. *The Last Great Ice Sheets.* New York, Wiley-Interscience.

Edwards, RL, Taylor, FW and Wasserburg, GJ 1988 Dating earthquakes with high-precision thorium-230 ages of very young corals. *Earth and Planetary Science Letters* 90: 371–381.

Emiliani, C, Rooth, C and Stipp, JJ 1978 The late Wisconsin flood into the Gulf of Mexico. *Earth and Planetary Science Letters* 41: 159–162.

Epstein, S, Buchsbaum, R, Lowenstam, HA and Urey, HC 1953 Revised carbonate-water isotopic temperature scale. *Geologic Society of America Bulletin* 64: 1315–1326.

Fairbanks, RG 1989 A 17,000-year glacio-eustatic sea level record: Influence of glacial melting rates on the Younger Dryas event and deep-ocean circulation. *Nature* 3421: 637–642.

____1990 The age and origin of the "Younger Dryas climate event" in Greenland ice cores. *Paleoceanography* 5: 937–948.

Fisher, D and Alt, BT 1985 A global oxygen isotope model – semi-empirical zonally averaged. *Annals of Glaciology* 7: 117–124.

Gordon, A 1986 Interocean exchange of thermocline water. *Journal of Geophysical Research* 91: 5037–5046.

Hays, JD, Imbrie, J and Shackleton, NJ 1976 Variations in the Earth's orbit, pacemaker of the Ice Ages. *Science* 194: 1121-1132.

Jansen, E and Veum, T 1990 Two-step deglaciation: timing and impact on North Atlantic Deep Water circulation. *Nature* 343: 612–616.

Jones, G and Keigwin, LD 1988 Evidence from Fram strait (78°N) for early deglaciation. *Nature* 336: 56–59.

Keigwin, LD 1987 North Pacific deep water formation during the latest glaciation, *Nature* 330: 362–364.

Keigwin, LD and Jones, GA 1989 Glacial-Holocene stratigraphy, chronology, and paleoceanographic observations on some North Atlantic sediment drifts. *Deep-Sea Research* 36: 845–867.

Kennett, JP and Shackleton, NJ 1975 Laurentide ice sheet meltwater recorded in Gulf of Mexico deep-sea cores. *Science* 188: 147–150.

Koerner, RM 1989 Ice core evidence for extensive melting of the Greenland ice sheet in the last interglacial. *Science* 244: 964–968.

Koerner, RM and Fisher, DA 1990 A record of Holocene summer climate from a Canadian high-Arctic ice core. *Nature* 343: 630–631.

Kroopnick, P 1985 The distribution of carbon-13 in the world oceans. *Deep Sea Research* 32: 57–84.

Labracherie, M, Labeyrie, LD, Duprat, J, Pichon, J, Bard, E, Arnold, M and Duplessy, JC 1989 The last deglaciation in the Southern Ocean. *Paleoceanography* 4: 629–638.

Lehman, SJ, Jones, GA, Keigwin, LD, Andersen, ES, Butenko, G and Ostmo, S-R 1991 Initiation of Fennoscandian ice-sheet retreat during the last deglaciation. *Nature* 349: 513–516.

Leventer, A, Williams, DF and Kennett, JP 1982 Dynamics of the Laurentide ice sheet during the last deglaciation: evidence from the Gulf of Mexico. *Earth and Planetary Science Letters* 59: 11–17.

Levitus, S 1982 Climatological atlas of the world ocean. *NOAA Professional Paper* 13. Washington, DC, US Government Printing Office: 173 p.

Lorius, C, Jouzel, J, Raynaud, D, Hansen, J and Le Treut, H 1990 The ice-core record: Climate sensitivity and future greenhouse warming. *Nature* 347: 139–145.

Maier-Reimer, E and Mikolajewicz, U 1989 Experiments with an OGCM on the cause of the Younger Dryas. *In* Ayala-Castanares, A, Wooster, W and Yanez-Arancibia, A, eds, *Oceanography 1988*. Mexico, UNAM Press: 87–100.

Manabe, S and Stouffer, RJ 1988 Two stable equilibria of a coupled ocean-atmosphere model. *Journal of Climate* 1: 841–866.

Milankovitch, M 1941 *Canon of Insolation and the Ice Age Problem*. Belgrade, Royal Serbian Academy. (Translation, Israel Program for Scientific Translation, Jerusalem, 1969).

Mix, AC and Fairbanks, RG 1985 North Atlantic surface-ocean control of Pleistocene deep-ocean circulation. *Earth and Planetary Science Letters* 73: 231–243.

Neftel, A, Oeschger, H, Staffelbach, T and Stauffer, B 1988 The CO_2 record in the Byrd ice core 50,000–5,000 years BP. *Nature 33*: 609–611.

Oerlemans, J and van der Veen, CJ 1984 *Ice Sheets and Climate*. Boston, D Reidel Company: 217 p.

Oppo, DW and Fairbanks, RG 1987 Variability in the deep and intermediate water circulation of the Atlantic Ocean during the past 25,000 years: Northern Hemisphere modulation of the Southern Ocean. *Earth and Planetary Science Letters* 86: 1–15.

_____1990 Atlantic Ocean thermohaline circulation of the last 150,000 years: Relationship to climate and atmospheric CO_2. *Paleoceanography* 5: 277–288.

Paterson, WSB and Hammer, C 1987 Ice core and other glaciological data. *In* Ruddiman, WF and Wright, HE, eds, *The Geology of North America K3, North America and Adjacent Oceans During the Last Deglaciation*. The Geological Society of America: 91–109.

Peltier, WR 1990 Glacial isostatic adjustment and relative sea level change. In *Studies in Geophysics*. Washington, DC, National Academy Press: 37–51.

Prell, W 1984 Variation of monsoonal upwelling: response to changing solar radiation. *In* Hansen, JE and Takahashi, T, eds, *Climate Processes and Climate Sensitivity*. Geophysics Monographs 29, Maurice Ewing Volume 5: 48–57.

Putnins, P 1970 The climate of Greenland. *In* Orvig, S, ed, *Climates of Polar Regions, World Survey of Climatology*. New York, Elsevier: 3–128.

Reed, RJ and Kunkel, BA 1960 The arctic circulation in summer. *Journal of Meteorology* 17: 489–506.

Rind, D, Peteet, D, Broecker, WS, McIntyre, A and Ruddiman, WF 1986 The impact of cold North Atlantic sea surface temperatures on climate: Implications to the Younger Dryas cooling (11–10K). *Climate Dynamics* 1: 3–33.

Rooth, C 1982 Hydrology and ocean circulation. *Progress in Oceanography* 1(1): 131–149.

Staffelbach, T, Stauffer, B, Sigg, A and Oeschger, H 1991 CO_2 measurements from polar ice cores: more data from different sites. *Tellus* 43B: 91–96.

Stauffer, B, Neftel, A, Oeschger, H and Schwander, J 1985 CO_2 concentration in air extracted from Greenland ice samples. *In* Greenland Ice Core: Geophysics, Geochemistry and the Environment. *Geophysics Monographs* 33: 85–89.

Stommel, H 1961 Thermohaline convection with two stable regimes of flow. *Tellus* 13: 224–230.

Stone, PH 1978 Constraints on dynamical transports of energy on a spherical planet. *Dynamics of Oceans and Atmosphere* 2: 123–139.

Stouffer, RJ, Manabe, S and Bryan, K 1989 Interhemispheric asymmetry in climate response to a gradual increase of atmospheric CO_2. *Nature* 342: 660–662.

Stuiver, M, Kromer, B, Becker, B and Ferguson, CW 1986 Radiocarbon age calibration back to 13,300 years BP and the ^{14}C age matching of the German oak and US bristlecone pine chronologies. *In* Stuiver, M and Kra, RS, eds, Proceedings of the 12th International ^{14}C Conference. *Radiocarbon* 28(2B): 969–979.

Tushingham, AM and Peltier, WR 1991 ICE-G: A new global model of late Pleistocene deglaciation based upon geophysical predictions of post-glacial relative sea level change. *Journal of Geophysical Research* 96: 4497–4523.

Vowinckel, E and Orvig, S 1970 The climate of the North Polar Basin. *In* Orvig, S, ed, *Climates of Polar Regions, World Survey of Climatology*. New York, Elsevier: 3–128.

ENVIRONMENTAL SCIENCES

PREFACE

LLOYD A CURRIE

It was fitting that Environmental Sciences was the concluding session of the conference, Four Decades of Radiocarbon Studies: An Interdisciplinary Perspective, *in that Bill Libby created the multidisciplinary "Environmental Doctor" program during the concluding years of his scientific career. In his introductory remarks to the Willard F Libby Collected Papers, Serge Korff noted that Bill "was a man of broad perspective, interested in the world's problems, and with the skill to approach solutions." Korff noted also that the UCLA Environmental Science and Engineering program constituted part of Libby's "remarkable triple career," which began with research leading to the Nobel Prize in Chemistry, followed by a period of public service as Commissioner of the Atomic Energy Commission, and concluded with his being Director of the Institute of Geophysics and Planetary Physics at UCLA. Throughout his career, Libby held a figurative and literal view of science as a global discipline. His graduate students learned quickly that an appreciation of such a view was a prerequisite for acceptance, reflected also in the diversity of topics in their concurrent thesis research. Libby's global vision, in which environmental science was intrinsically bound to such disciplines as isotope geophysics, stratospheric chemistry, meteorology and population dynamics, is perhaps best expressed in Libby's own description of "how a new fact of nature" (nuclear testing fallout) can increase our knowledge in many areas: "As we learn about the way the world-wide fallout particle, probably as tiny as a virus molecule, wends its way from the stratosphere through the tropopause into the troposphere and, within a few weeks, collides with a water droplet and thus is brought to the earth's surface by rain, we shall learn about the circulation of the atmosphere, about the way in which rain is formed, and about the questions which will naturally arise more and more frequently as the world's population increases world-wide pollution of the atmosphere not only with fission products but with the other by-products of our new technological age." (1956* Proceedings of the National Academy of Sciences *42: 945–962.)*

Libby's remarks form an eloquent introduction to this section. A common theme of the chapters by Ingeborg Levin, R L Otlet and L A Currie is the isotopic and chemical impact of human activities on the global atmosphere, as assessed by very sophisticated measurements of radiocarbon and other species. Ingeborg Levin and co-authors address the continuing global increase in the carbonaceous greenhouse gases, carbon dioxide and methane. Astute sampling of these gases at clean tropospheric sites in the northern and southern hemispheres, together with careful assay for ^{13}C *and* ^{14}C*, when combined with appropriate modeling, provides the basic data needed to apportion the several anthropogenic and natural sources of these gases. The isotopic* CO_2 *record, for example, demonstrated the influence of fossil-fuel-induced seasonality in the northern hemisphere, and gas exchange with circumpolar surface water in the southern hemisphere. Data on the isotopic budget of atmospheric methane showed that attention must not be limited to biospheric and fossil sources; increasingly,* $^{14}CH_4$ *emissions from nuclear power plants are having a significant impact. Important sampling strategies were proposed for quantitatively assessing that source term. The environmental impact of atmospheric* ^{14}C *emissions from the nuclear energy cycle, is at the heart of the chapter by R L Otlet, M J Fulker and A J Walker. These authors note that the* **same** *nuclear reaction –* $^{14}N(n,p)$ *– is responsible for the most important contribution of both cosmogenic and anthropogenic (nuclear fuel cycle)* ^{14}C*, the latter from nitrogen impurities. This chapter focuses primarily on the impact of nuclear facility release of* ^{14}C *on nearby populations. Atmospheric* CO_2 *was cleverly sampled by nature, in the hawthorn berry, which proved ideal as an integrating monitor. Studies of temporal and spatial patterns of* ^{14}C *specific activity around Sellafield demonstrated the importance of such measurements for estimating doses to human populations.*

Just as Libby's fallout particle drifted through the atmosphere in accord with the laws of physics and chemistry, so too, do carbonaceous particles introduced by acts of nature and humankind. L A Currie reviewed the history of atmospheric particulate carbon, from 13th century England to the present. On both regional and global scales, these particles also pose potential threats to health, visibility and climate. The role of radiocarbon has become centered in apportioning natural and anthropogenic sources, because it uniquely and quantitatively discriminates fossil from biospheric sources. AMS has enhanced such discrimination for individual classes of compounds. Understanding additional isotopic and chemical variables, together with ^{14}C*, enables us to interpret natural archival records and resolve the manifold sources contributing to the tropospheric carbonaceous particle budget.*

31

RADIOCARBON IN ATMOSPHERIC CARBON DIOXIDE AND METHANE: GLOBAL DISTRIBUTION AND TRENDS

INGEBORG LEVIN, RAINER BÖSINGER, GEORGES BONANI ROGER J FRANCEY, BERND KROMER, K O MÜNNICH MARTIN SUTER, NEIL B A TRIVETT and WILLY WÖLFLI

INTRODUCTION

For many years, there has been a growing concern in the field of atmospheric chemistry about anthropogenic and natural perturbations of the major atmospheric cycles of carbon, nitrogen and sulfur, and recently, oxygen. The concern is mainly due to the implications of these trace gases on global climate. In view of the atmospheric carbon cycle, the most abundant trace gases, carbon dioxide and methane, just recently became the subject of detailed ^{14}C investigations. These may play an important role in providing the supplementary and independent information needed to better evaluate the current observations.

The most important process of radiocarbon generation is natural *cosmic-ray production* in the upper atmosphere. ^{14}C is then finally oxidized to carbon dioxide and becomes part of the CO_2 cycle. Atmosphere-biosphere exchange and anaerobic decomposition of organic material introduce the ^{14}C signature into the atmospheric methane cycle. Thermonuclear bomb testing in the atmosphere produced significant pulses of ^{14}C into the carbon cycle. Since the last atomic bomb test in China in 1981, the *direct bomb ^{14}C input* to the atmosphere has decreased to nearly zero. Large quantities of bomb ^{14}C are, however, still stored in the atmosphere, the biosphere and the ocean. Bomb ^{14}C is subsequently released into the atmosphere through aerobic (CO_2) and anaerobic (CH_4) decomposition of organic material as well as biomass burning (CO_2 and CH_4). ^{14}C releases by *nuclear power plants* are becoming more and more important in recent years, and today have considerably disturbed the ^{14}C budget of atmospheric methane (Wahlen *et al* 1989). Finally, *fossil fuel* sources of CO_2 and CH_4 act as a "negative source" in the ^{14}C budget of both trace gases and, thus, may be quantified in their global budgets.

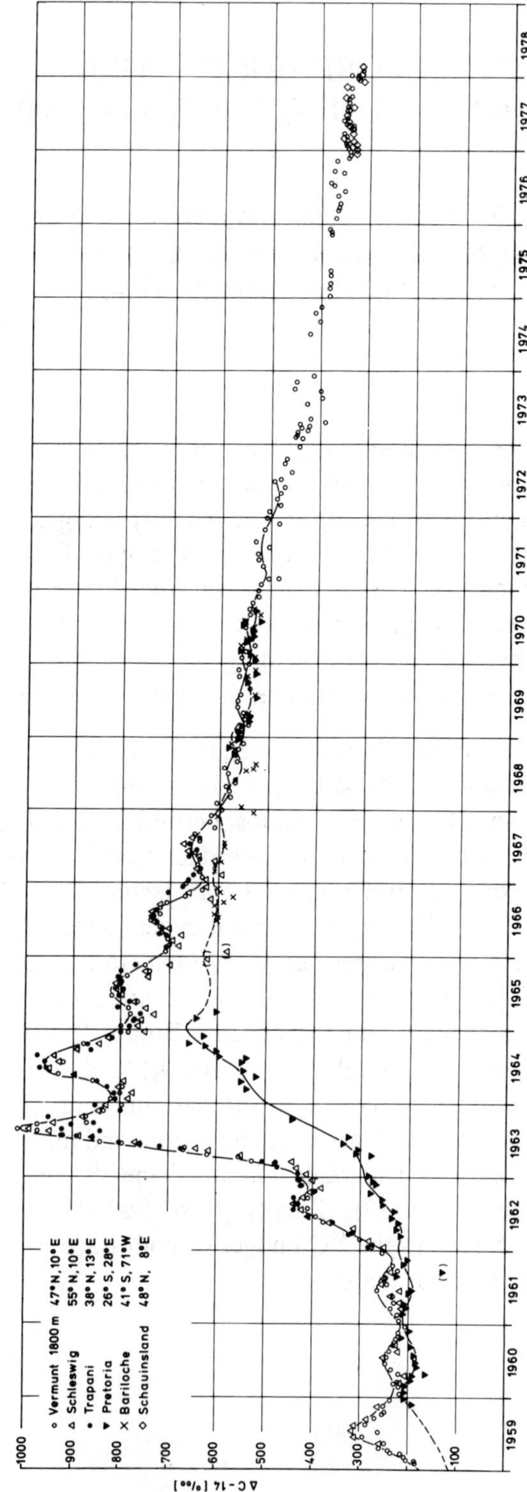

Fig 31.1. Long-term $^{14}CO_2$ observations at different sites in the northern and the southern hemispheres (Levin, Münnich & Weiss 1980)

Atmospheric $^{14}CO_2$ measurements have been made at several stations in the world since the beginning of atmospheric nuclear weapon testing (Nydal & Lövseth 1983; Levin *et al* 1985; Manning *et al* 1990). As an example, Heidelberg measurements from northern and southern hemisphere sites are presented in Figure 31.1. These data, dominated by the propagation of the atmospheric ^{14}C weapon test spike in the early 1960s, have provided invaluable information about the air-sea gas exchange rate of CO_2. They further provide an independent measure of the interhemispheric mixing rate and of the troposphere-stratosphere exchange rate.

We present here recent continuous observations of ^{14}C in atmospheric CO_2 as well as spot measurements of $^{14}CH_4$ from both hemispheres. The main features of current variabilities are discussed in relation to the sources described above.

SAMPLING SITES

We have collected continuous bi-weekly integrated atmospheric CO_2 samples at the following sites:

Alert (83°N, 62°W, 210 m asl). $^{14}CO_2$ collection started in 1987 at the WMO BAPMoN High Arctic Research Laboratory Alert operated by the Atmospheric Environment Service, Toronto, Canada. The station is located at the northeastern tip of Ellesmere Island and is thus far removed from major industrial regions. This site is well suited to monitor the background air of the northern part of the hemisphere (Trivett & Worthy 1989).

Schauinsland (48°N, 8°E, 1205 m asl). $^{14}CO_2$ collection started in 1977 at the WMO BAPMoN station Schauinsland operated by the Federal Environmental Agency (Umweltbundesamt) Berlin, FRG. For a more detailed description of the station site, see Levin (1987).

Jungfraujoch (47°N, 8°E, 3454 m asl). $^{14}CO_2$ collection started in 1986 at the High Alpine Research Station Jungfraujoch in the Swiss Alps. Because of its exposed position, most of the time, and particularly during the winter season, we collect air that is not directly affected by the surrounding ground-level sources.

Izaña (28°N, 16°W, 2376 m asl). $^{14}CO_2$ collection started in 1984 at the WMO BAPMoN station Izaña, Tenerife, Spain. At this high-altitude observatory in the North Atlantic trade-wind system, we collect air, which, most of the time is representative of the free troposphere at this latitude (Schmitt, Schreiber & Levin 1988).

Cape Grim (41°S, 145°E, 95 m asl). $^{14}CO_2$ collection at the Cape Grim Baseline Air Pollution Station started in 1987. The station is located near the top of a

95 m promontory on the western side of the northwestern tip of Tasmania, Australia. A detailed description of the site is given by Francey (1984).

Georg von Neumayer (71°S, 8°W, 16 m asl). $^{14}CO_2$ samples from the Georg von Neumayer station (GvN) have been measured since 1983. The German Antarctic GvN station is situated on the Ekström ice shelf close to the ice edge. $^{14}CO_2$ sampling is performed at some distance from the GvN station (1500 m) in a van. Local fossil fuel CO_2 contamination is minimized by locating the van upwind of the potential contaminating sources. In addition, air sampling is controlled by wind direction and condensation nucleus readings. At this coastal site winds are constantly from the east. Even during the winter season, the sea is not completely covered with ice. Large polynyas offshore the station provide access to surface ocean water, as indicated by high sea-salt levels all year round (Wagenbach *et al* 1988).

Spot samples for ^{14}C analyses in atmospheric methane were collected from 1987 to 1990 at the Schauinsland, Jungfraujoch, Izaña and Georg von Neumayer stations. Continuous sampling has also been made during 1988 and 1989 from ambient air at the Heidelberg Institute situated on the outskirts of Heidelberg (49°N, 8°E, 110 m asl).

In September 1988, a meridional profile of methane isotope samples was collected on the Atlantic Ocean (30°W) from 50°N to 30°S during a cruise of the *FS Polarstern*.

EXPERIMENTAL

$^{14}CO_2$

Levin, Münnich & Weiss (1980) and Schoch *et al* (1980) previously described sampling and laboratory procedures in great detail. In brief, CO_2 samples are collected continuously by quantitative absorption in sodium hydroxide solution. After shipment to the Heidelberg laboratory, CO_2 is extracted from the solution in a vacuum system by adding an inorganic acid. For ^{14}C analysis, the CO_2 gas is purified by activated charcoal and counted in a high-precision proportional counter system. The $^{13}C/^{12}C$ ratio is measured by mass spectrometry from small aliquots of the CO_2 gas.

All $^{14}CO_2$ activities are expressed as the per mil deviation ($\Delta^{14}C$(‰)) from 95% of the NBS oxalic acid activity corrected for decay (Stuiver & Polach 1977). As the sampling and laboratory procedures were made quantitatively, the $\Delta^{14}C$ values presented here are not corrected for fractionation of *individual* $\delta^{13}C$ values. To obtain a direct comparison with plant material, the values are corrected to $\delta^{13}C = -25$‰ using the mean value of all samples at one site for the observation period. The Alert data were corrected, using a value of $\delta^{13}C$ = −9.82‰, the Schauinsland data of $\delta^{13}C$ = −8.13‰, the Jungfraujoch data of $\delta^{13}C$ = −10.25‰ (at this site, the efficiency of the absorption was < 100%), the Izaña data of $\delta^{13}C$ = −8.29‰, the Cape Grim data of $\delta^{13}C$ = −8.32‰, and the GvN data of $\delta^{13}C$ = −8.13‰. The 1σ precision of a single $^{14}CO_2$ analysis is typically $\Delta^{14}C$ = ± 3‰, including the total error of the $\delta^{13}C$ measurement of less than 0.5‰.

$^{14}CH_4$

Conventional radiocarbon dating of atmospheric methane requires extraction of methane from an air volume of the order of 1000 m^3. The required amount of carbon for $^{14}C/^{12}C$ measurements by accelerator mass spectrometry (AMS) (Bonani *et al* 1987; Kromer *et al* 1987) does, however, only require an air sample volume of about 1–2 m^3 (1 mg carbon, *ie*, 2 cm^3 STP CH_4). A system to purify and to isolate CH_4 from atmospheric (1.7 ppmv) and source samples (up to several hundred ppmv) has been developed by Bösinger (1990): CH_4 is enriched in two steps (one in the field, one in the laboratory) over activated charcoal at −196°C and 0.4 bar after H_2O and CO_2 are removed by silica gel and molecular sieve. Under these conditions, CH_4 is quantitatively adsorbed on the charcoal while most of the air is pumped off. Desorption of CH_4 from the charcoal is achieved by flushing with helium at 300°C. After separation from remaining contaminants such as CO, CO_2 and non-methane-hydrocarbons by gas chromatography with pure synthetic air as carrier gas, the methane fraction is trapped in the sampling loop of a combustion cycle. Here, CH_4 is oxidized over platinum at 600°C and 900°C. The resulting CO_2 and H_2O are collected in cold traps at −78°C and −196°C. The overall yield for atmospheric samples is (92 ± 6)%. The CO_2 is used for conventional $^{13}C/^{12}C$ measurement and subsequently converted to graphite for AMS ^{14}C analysis. All $^{14}CH_4$ data are corrected for fractionation by means of parallel $^{13}C/^{12}C$ AMS analysis to $\delta^{13}C$ = −25‰. The ^{14}C activity is reported as pMC, the percent with respect to NBS oxalic-acid-based standard activity corrected for decay (Stuiver & Polach 1977). The total 1σ reproducibility for ^{14}C is ± 1.2 pMC. First results of an intercalibration with Martin Wahlen at the Wadsworth Center for Laboratories and Research, Albany, New York, show no systematic differences in absolute $^{14}CH_4$ results.

RESULTS AND DISCUSSION OF THE ATMOSPHERIC $^{14}CO_2$ DATA

Long-Term Trend and Meridional $^{14}CO_2$ Gradient

Figures 31.2 A–F show monthly mean ^{14}C concentrations calculated from bi-weekly data observed in atmospheric CO_2 from 1983 to 1989 at Alert, Schauinsland, Jungfraujoch, Izaña, Cape Grim and GvN. A steady but declining

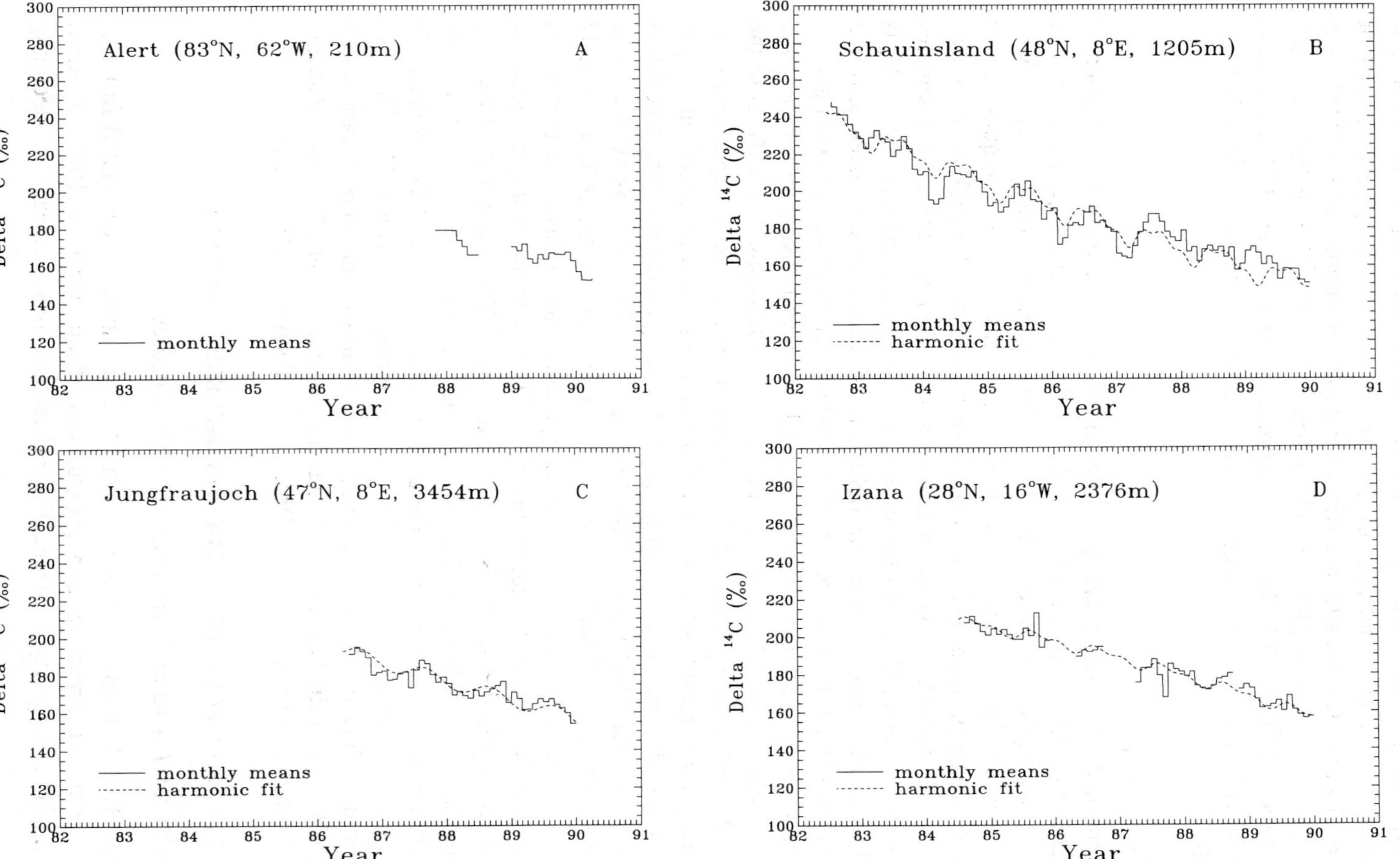

Fig 31.2. ^{14}C observations in atmospheric CO_2 at six stations between 83°N and 71°S. Plotted are monthly mean values calculated from bi-weekly samples. The dashed lines are calculated from the monthly data with a four-harmonic median analysis technique described by Trivett (1989).

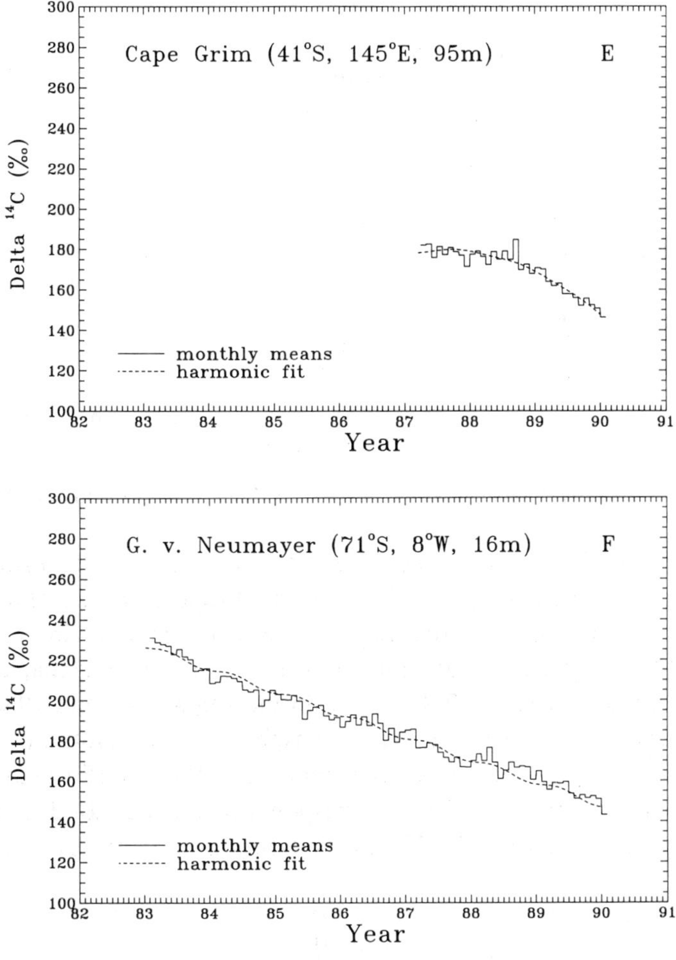

Fig 31.2 (continued)

decrease in $\Delta^{14}C$ is observed at all stations where long-term records are available. The mean $\Delta^{14}C$ decline calculated from the 1983 to 1985 Schauinsland and GvN records was (−14.1 ± 1.9)‰ yr^{-1}. From 1985 to 1989, the mean decrease at Schauinsland, GvN and Izaña was (−9.7 ± 1.0)‰ yr^{-1}. This long-term decrease is partly due to the bomb ^{14}C spike still equilibrating with the ocean surface water and to the ongoing input of ^{14}C-free fossil fuel CO_2 into the atmosphere. Short-term fluctuations in the decrease rate, and, particularly, the strong draw-down at Cape Grim in 1989, may be due to the year-to-year variability of the major $^{14}CO_2$ and CO_2 fluxes involved.

Significant differences in the yearly mean $^{14}CO_2$ level are observed for the individual sites: Figure 31.3 shows the yearly median ^{14}C values at five differ-

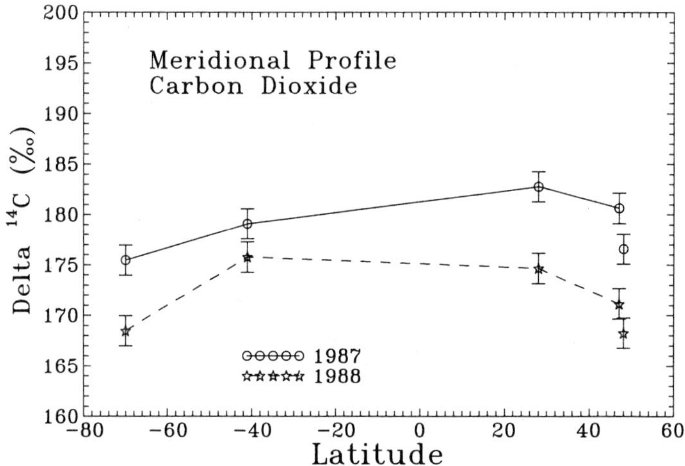

Fig 31.3. Observed yearly median atmospheric $^{14}CO_2$ activity at five stations in 1987 and 1988

ent stations in 1987 and 1988. The lower ^{14}C levels at the Schauinsland station compared to Jungfraujoch are attributed to a European-scale influence of the ^{14}C-free fossil fuel sources in this region (see below and Levin *et al* (1989)). Between 47°N (Jungfraujoch) and the maritime baseline site, Izaña, at 28°N, we observe a mean difference of 2.8 ± 1.0‰ in the years 1987 and 1988. As both sites are assumed to represent the free troposphere in their latitudinal belts, this difference can be interpreted as a fossil fuel CO_2 offset of 350 ppm × 2.8‰ = 1 ppm at 47°N, if compared to 28°N corresponding to about 20–30% of the total meridional gradient between 50°N and the South Pole (Conway *et al* 1988).

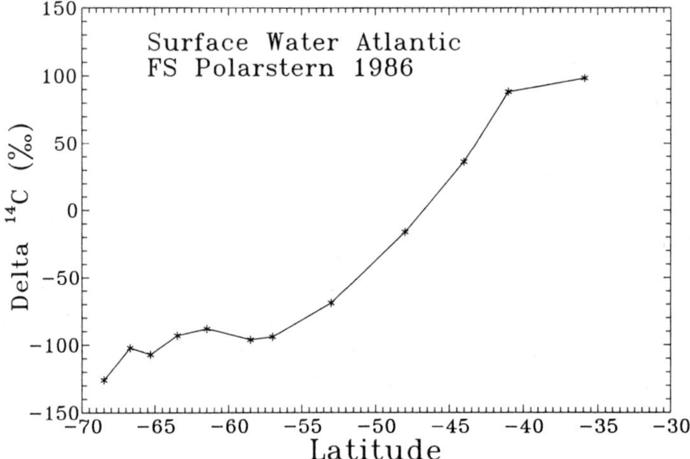

Fig 31.4. $\Delta^{14}C$ in surface water of the South Atlantic Ocean. The samples were taken during a *FS Polarstern* cruise in Winter 1986.

For the period 1985 to 1989, the mean ^{14}C difference between Baseline Izaña and the Antarctic GvN station is 6 ± 1.5‰. We observe a significant draw-down of the ^{14}C activity in mid-latitudes of the southern hemisphere at Cape Grim in 1987 as well. We attribute this to the CO_2 gas exchange with surface water, low in ^{14}C, of the circum-Antarctic upwelling region emerging at 40°S (Levin *et al* 1987) (*cf* Figure 31.4, ^{14}C measurements of surface water in the South Atlantic from the *FS Polarstern* cruise in 1986).

Seasonal Cycles

The monthly median ^{14}C observations at all stations with more than three years of data have been detrended and analyzed for a seasonal cycle with up to four harmonics using a technique described by Trivett (1989) (Fig 31.2). The median detrended seasonal cycles for the northern hemisphere stations and the GvN station calculated from Figure 31.2 adjusted to the observed median in 1987 are given in Figure 31.5. Plotted are the $\Delta^{14}C$ deviations from the yearly median ^{14}C level at Izaña. A marked seasonality of the $^{14}C/^{12}C$ ratio is observed at all three northern hemisphere stations. The European continental influence at Schauinsland is obvious and highest during the winter half year (*cf* also Levin *et al* (1989)). In the summer (May to October), the ^{14}C activity at Schauinsland does not differ very much from that at the more remote Jungfraujoch ($\delta\Delta^{14}C$ = −2.5‰, about 50% of the winter offset). Moreover, the seasonal pattern at Jungfraujoch and Izaña do compare rather well, indicating that the predominant sources are very common at both stations. The maximum ^{14}C activity at both

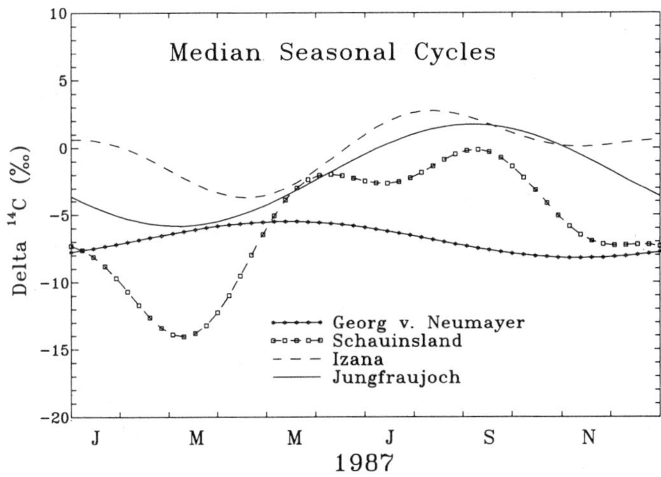

Fig 31.5. Median detrended seasonal cycles of $\Delta^{14}C$ in atmospheric CO_2. Plotted are the deviations from the Izaña yearly median value in 1987.

stations is observed in August/September; the minimum at Izaña in April/May with a peak-to-peak amplitude of $\Delta^{14}C$ = 6.5‰. The minimum ^{14}C activity at Jungfraujoch is observed already in March with a slightly higher peak-to-peak amplitude of $\Delta^{14}C$ = +8.0‰. Both effects may be explained by the larger fossil fuel source at northern mid-latitudes.

Three explanations for the $^{14}CO_2$ seasonality in northern hemisphere clean air (Izaña and Jungfraujoch) are possible:

1. The northern latitude fossil fuel source is seasonal and should be stronger than that quoted by Rotty (1987). At least in the European and Asian parts of the hemisphere – contributing about 2/3 of the global anthropogenic source – the difference of the fossil fuel CO_2 production between summer and winter reaches a factor of two (Levin *et al* 1989).
2. During photosynthesis, a strong fractionation occurs between the ^{14}C and the ^{12}C isotope (about 36‰). The spring and early summertime CO_2 concentration draw-down due to the biospheric uptake of CO_2 causes a $^{14}CO_2$ enrichment in the atmosphere. This effect causes a maximum that should occur in late summer, thus coinciding with the ^{14}C maximum observed at all northern hemisphere stations.
3. Finally, the seasonally varying input of natural ^{14}C from the stratosphere may partly drive the seasonal cycle of tropospheric $^{14}CO_2$.

At GvN we observe a seasonal pattern in $^{14}CO_2$, which is considerably smaller than in the northern hemisphere. At this site, the lowest ^{14}C concentrations are consistently observed during local spring, whereas the $^{14}C/^{12}C$ maximum occurs in early austral winter. At present, there is no unequivocal explanation for the $^{14}CO_2$ seasonal cycle observed. The modeling approach to this question is focusing on a possible seasonal cycle of gas exchange rate (parameterized by marine ^{222}Rn background (Heimann, Monfray & Polian 1990)), on stratospheric air mass intrusions (indicated by cosmogenic radioisotopes (Wagenbach *et al* 1988)), and possibly on a yet unknown biospheric contribution at this latitude.

RESULTS AND DISCUSSION OF THE ATMOSPHERIC $^{14}CH_4$ DATA

Comparison of isotopic ratio measurements and the concentration of atmospheric CH_4 at sites with different influences from the dominant methane sources, eg, continental European sites and "clean air" over the Atlantic, yields information on the CH_4 emissions in the source area. Air samples from Schauinsland and Jungfraujoch are significantly influenced by the predominantly continental methane sources, as there are emissions from ruminants, solid waste deposits, as well as natural and artificial wetlands and fossil fuel emissions. Although a considerable ^{14}C *depletion* over the continent due to ^{14}C-free fossil fuel CH_4 sources could be expected, the $^{14}CH_4$ level at Schauinsland has been observed to be up to 10 ± 1.5 pMC, and at Jungfraujoch up to 6 ± 1.5 pMC higher if compared to maritime "clean air" over the Atlantic (Izaña) (Fig 31.6A, B). This is due to $^{14}CH_4$ emissions from European pressurized water reactors (PWR).

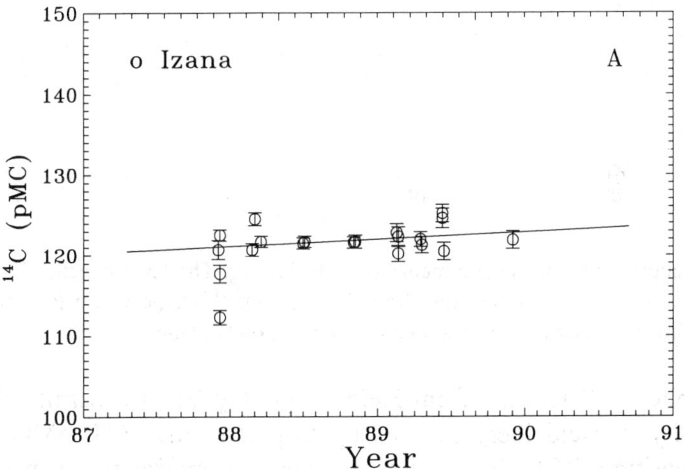

Fig 31.6A. ^{14}C activity in atmospheric methane at Izaña. The straight line is a least squares fit through the data. The slight increase in ^{14}C of 0.9 ± 0.6 pMC is attributed to increasing $^{14}CH_4$ releases from pressurized water reactors.

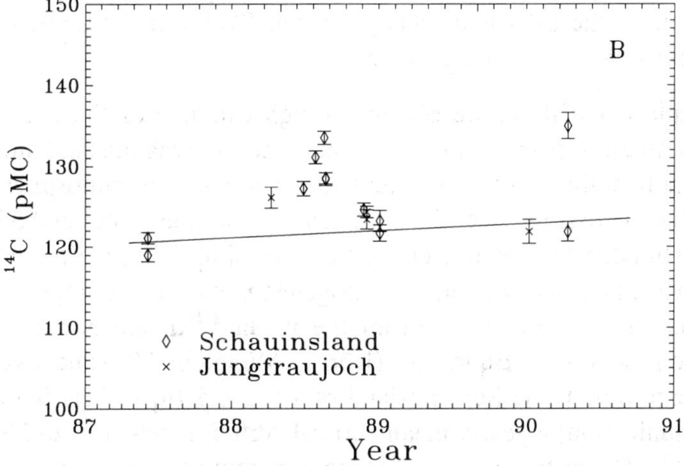

Fig 31.6B. ^{14}C activity in atmospheric methane at Jungfraujoch and Schauinsland. The solid line is the fitted curve through the Izaña data (A). Particularly at the Schauinsland station, European-scale $^{14}CH_4$ contaminations up to 10 pMC are observed.

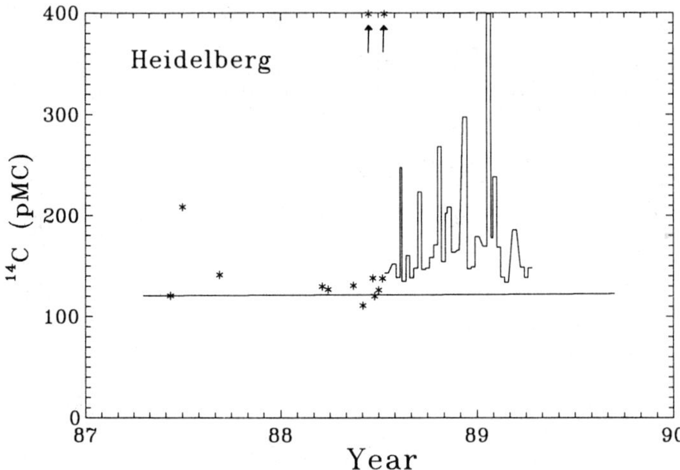

Fig 31.7. ^{14}C activity in atmospheric methane at Heidelberg. The level is generally higher than at the Atlantic site, Izaña (straight line from Fig 31.6A). $^{14}CH_4$ peaks up to 600 pMC were identified as direct contamination from nearby nuclear power plants.

The atmospheric $^{14}CH_4$ data from Heidelberg (Fig 31.7) confirm this finding: $^{14}CH_4$ activity in Heidelberg reaches up to greater than 600 pMC, about six times the maritime $^{14}CH_4$ level. The neighboring nuclear power plants, Biblis (35 km northwest of Heidelberg) and Philippsburg (25 km southwest of Heidelberg), are the sources of this contamination. With an atmospheric dispersion estimate, a mean $^{14}CH_4$ emission rate of $(4.0 \pm 1.7)\ 10^{11}$ Bq $GWe^{-1}\ yr^{-1}$ was calculated (Bösinger 1990). This is in good agreement with direct measurements of the off-gas discharge of two PWRs in the USA, resulting to 3.5×10^{11} Bq $GWe^{-1}\ yr^{-1}$ (Kunz 1985).

With a simple calculation, we obtain a rough estimate of the expected mean $^{14}CH_4$ contamination from this reactor source at the Schauinsland site. We use the natural radioactive noble gas ^{222}Radon as a measure of atmospheric dilution of near-surface emissions over the continent. Using the same method, Levin *et al* (1989) estimated the fossil fuel source. For simplicity, we assume that the nuclear power plant distribution is homogeneous over the European continent, which, of course, is not true. From the mean ^{222}Rn emanation rate of the European continent of 53 Bq $m^{-2}\ h^{-1}$ (Dörr & Münnich 1990) and a yearly mean ^{222}Rn concentration at the Schauinsland of 1.9 ± 0.5 Bq m^{-3} (H Sartorius, personal communication) a yearly mean vertical "dilution velocity" of 28 m h^{-1} can be calculated. Using this number for the apparent dispersion of reactor ^{14}C at the Schauinsland and a mean European $^{14}CH_4$ source from PWRs of 1.7×10^{13} Bq yr^{-1} or $6.4\ 10^{-4}$ Bq $m^{-2}\ h^{-1}$ in 1988 (Koelzer 1988; Kunz 1985) (this number includes estimated emissions from all European states except Scandinavia,

Bulgaria and the USSR) leads to a mean $^{14}CH_4$ offset at the Schauinsland site of 10 pMC. This is in the same order of magnitude as our observations.

This simple calculation illustrates that $^{14}CH_4$ effluents from PWRs are already a significant contaminant on the European scale, but also of relevance for the global atmospheric methane budget. Because of this serious contamination, the identification of fossil methane sources by $^{14}CH_4$ measurements, particularly in an industrialized region such as Heidelberg, is no longer unambiguous. However, the nuclear power plant $^{14}CH_4$ source – up to now restricted only to the northern hemisphere – provides a unique tracer for global $^{14}CH_4$ modeling. This is illustrated by the global distribution of $^{14}CH_4$: Figure 31.8 shows a meridional profile of ^{14}C in atmospheric methane over the Atlantic Ocean in September 1988. The data have been complemented by observations from GvN, Izaña and Jungfraujoch from the same period. A not yet significant difference of 1.8 ± 1.4 pMC between the northern and the southern hemispheres can be derived from the data. Using a two-box model of the atmosphere, Wahlen *et al* (1989) postulated a north-south gradient of 2.7 to 5.0 pMC for 1987, increasing in future years with an increasing ^{14}C source from PWRs in the northern hemisphere. The authors observed a north-south gradient of 3.3 ± 1.1 pMC for the end of 1987 in agreement with our findings.

Another consequence of the increasing $^{14}CH_4$ source from the nuclear industry is a worldwide increase of ^{14}C in atmospheric methane. At the maritime site, Izaña, during 1987 to 1989, we indeed observed a slight increase of 0.9 ± 0.6

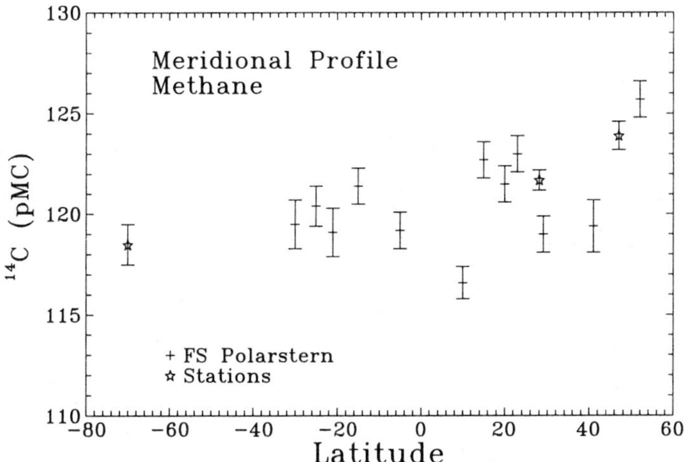

Fig 31.8. Meridional profile of ^{14}C in atmospheric methane. The samples were collected at 30°W over the Atlantic Ocean in Autumn 1988. The profile includes data from GvN, Izaña and Jungfraujoch from the same period. A north-south difference of 1.8 ± 1.4 pMC is calculated from the measurements.

pMC yr^{-1}. Because of the short record of observations, this is, however, not yet significant. With a simple one-box model, using a mean atmospheric residence time for methane of 10 years, including the bomb ^{14}C released through the biosphere, a fossil fuel source of 15% of the global methane release rate, and a global PWR source of 4.3×10^{13} Bq yr^{-1} in 1988, we calculate a global mean atmospheric $^{14}CH_4$ increase rate for 1988 of 3 pMC yr^{-1}. This is a factor of three higher than our observations.

Conclusions

High-precision $^{14}CO_2$ observations performed at continental and baseline sites provide considerable new and independent information on the carbon dioxide sources and sinks on a global scale. In combination with model simulations, these observations will provide a powerful tool for ongoing carbon cycle research.

The atmospheric $^{14}CH_4$ level is already considerably disturbed by ^{14}C emissions from nuclear power plants. A quantitative determination of the evolution of this contaminating source is necessary to successfully use radiocarbon for evaluating the global methane budget. One approach is continuous observations of $^{14}CH_4$ in background air of the northern and southern hemispheres, thus monitoring the postulated increasing interhemispheric gradient. In combination with a global transport model, this gradient is a direct measure of the northern hemispheric $^{14}CH_4$ source itself. Another possibility is continuous observations representative of relevant source regions, which, combined with continuous ^{222}Rn observations, allow for a direct estimate of the $^{14}CH_4$ flux.

Acknowledgments

This work was funded by the Umweltbundesamt, Berlin, under contract no. 104 02 627, and by the Bundesminister für Forschung und Technologie, Bonn, under contract nos. 0741112, 07 KF 022 and 07 KFT 11. We are very grateful to the personnel of the Alert, Cape Grim, Georg von Neumayer, Izaña, Jungfraujoch and Schauinsland stations for their careful work in collecting the numerous atmospheric $^{14}CO_2$ samples. Special thanks are due Dietmar Wagenbach for many helpful discussions.

REFERENCES

Bösinger, R (ms) 1990 Isotopenmessungen an atmosphärischem und quellnahem Methan. Doctoral thesis, University of Heidelberg.

Bonani, G, Beer, J, Hofmann, H, Synal, H, Suter, M, Wölfli, W, Pfleiderer, C, Kromer, B, Junghans, C and Münnich, KO 1987 Fractionation, precision and accuracy in ^{14}C and ^{13}C measurements. *In* Gove, HE, Litherland, AE and Elmore, D, eds, Proceedings of the 4th International Symposium on Accelerator Mass Spectrometry. *Nuclear Instruments and Methods* B29: 87–90.

Conway, TJ, Tans, PP, Waterman, LS, Thoning, KW and Masarie, KA 1988 Atmospheric carbon dioxide measurements in the remote troposphere. *Tellus* 40B: 81–115.

Dörr, H and Münnich, KO 1990 ^{222}Rn flux and soil air concentration profiles in West Germany. Soil ^{222}Rn as tracer for gas transport in the unsaturated soil zone. *Tellus* 42B: 20–28.

Francey, R, ed 1984 *Baseline Atmospheric Program, Australia 1981–1982*. Canberra, Australia, Canberra Publishing and Printing Co: 83 p.

Heimann, M, Monfray, P and Polian, G 1990 Modeling the long-range transport of ^{222}Rn to subantarctic and antarctic areas. *Tellus* 42B: 83–99.

Koelzer, W 1988 *Lexikon zur Kernenergie*. Kernforschungszentrum Karlsruhe GmbH, Karlsruhe, FRG.

Kromer, B, Pfleiderer, C, Schlosser, P, Levin, I, Münnich, KO, Bonani, G, Suter, M and Wölfli, W 1987 AMS ^{14}C measurement of small volume oceanic water samples: Experimental procedure and comparison with low-level counting technique. *In* Gove, HE, Litherland, AE and Elmore, D, eds, Proceedings of the 4th International Symposium on Accelerator Mass Spectrometry. *Nuclear Instruments and Methods* B29: 302–305.

Kunz, C 1985 Carbon-14 discharge at three light water reactors. *Health Physics* 49(1): 25–35.

Levin, I 1987 Atmospheric CO_2 in continental Europe – an alternative approach to clean air CO_2 data. *Tellus* 39B(1–2): 21–28.

Levin, I, Kromer, B, Schoch-Fischer, H, Bruns, M, Münnich, M, Berdau, D, Vogel, JC and Münnich, KO 1985 25 years of tropospheric ^{14}C observations in Central Europe. *Radiocarbon* 27(1): 1–19.

Levin, I, Kromer, B, Wagenbach, D and Münnich, KO 1987 Carbon isotope measurements of atmospheric CO_2 at a coastal station in Antarctica. *Tellus* 39B: 89–95.

Levin, I, Münnich, KO and Weiss, W 1980 The effect of anthropogenic CO_2 and ^{14}C sources on the distribution of ^{14}C in the atmosphere. *In* Stuiver, M and Kra, RS, eds, Proceedings of the 10th International ^{14}C Conference. *Radiocarbon* 22(2): 379–391.

Levin, I, Schuchard, J, Kromer, B and Münnich, KO 1989 The continental European Suess-effect. *In* Long, A and Kra, RS, eds, Proceedings of the 13th International ^{14}C Conference. *Radiocarbon* 31(3): 431–440.

Manning, MR, Lowe, DC, Melhuish, WH, Sparks, RJ, Wallace, G, Brenninkmeijer, CAM and McGill, RC 1990 The use of radiocarbon measurements in atmospheric studies. *Radiocarbon* 32(1): 37–58.

Nydal, R and Lövseth, K 1983 Tracing bomb ^{14}C in the atmosphere 1962–1980. *Journal of Geophysical Research* 88: 3621–3642.

Rotty, RM 1987 Estimates of seasonal variation in fossil fuel CO_2 emission. *Tellus* 39B: 184–202.

Schmitt, R, Schreiber, B and Levin, I 1988 Effects of long-range transport on atmospheric trace constituents at the baseline station Tenerife (Canary Islands). *Atmospheric Chemistry* 7: 335–351.

Schoch, H, Bruns, M, Münnich, KO and Münnich, M 1980 A multi-counter system for high precision carbon-14 measurements. *In* Stuiver, M and Kra, RS, eds,

Proceedings of the 10th International ^{14}C Conference. *Radiocarbon* 22(2): 442–447.

Stuiver, M and Polach, HA 1977 Discussion: Reporting of ^{14}C data. *Radiocarbon* 19(3): 355–363.

Trivett, NBA 1989 A comparison of seasonal cycles and trends in atmospheric CO_2 concentration as determined from robust and classical regression techniques. *In* Proceedings of the 3rd International Conference on Analysis and Evaluation of Atmospheric CO_2 Data, Present and Past. *WMO Report* 59.

Trivett, NBA and Worthy, DEJ 1989 Analysis and interpretation of trace gas measurements at Alert, NWT, with emphasis on CO_2 and CH_4. *In* Proceedings of the 3rd International Conference on Analysis and Evaluation of Atmospheric CO_2 Data, Present and Past. *WMO Report* 59.

Wagenbach, D, Görlach, U, Moser, K and Münnich, KO 1988 Coastal Antarctic aerosol: The seasonal pattern of its chemical composition and radionuclide content. *Tellus* 40B: 426–436.

Wahlen, M, Tanaka, N, Henry, R, Deck, B, Zeglen, J, Vogel, JS, Southon, H, Shemesh, A, Fairbanks, R and Broecker, WS 1989 Carbon-14 in methane sources and in atmospheric methane: The contribution from fossil carbon. *Science* 245: 286–290.

32

ENVIRONMENTAL IMPACT OF ATMOSPHERIC CARBON-14 EMISSIONS RESULTING FROM THE NUCLEAR ENERGY CYCLE

R L OTLET, M J FULKER and A J WALKER

INTRODUCTION

The Nuclear Energy Cycle comprises all the stages involved in the generation of power from fission (and possibly fusion in the future), and presently includes mining and milling, fuel fabrication, its utilization during reactor running, post-reactor handling (including reprocessing) and the subsequent disposal or storage of waste. During reactor running, ^{14}C is produced in the fuel, moderator, coolant and structural materials. A small amount (estimated at only a few percent of the total) may be produced during the fission process itself by ternary fission, but the major fraction is produced by the neutron activation reactions given in Table 32.1. The target nuclei exist in the basic components of the reactor, such as the coolant and moderator, or in impurities in the fuel and reactor materials generally. The most important contribution to the total ^{14}C produced comes from nitrogen impurities in these materials *via* the familiar $^{14}N(n,p)^{14}C$ reaction. Consequently, although the activation cross-sections are well known for the neutron spectra of the different reactor types, estimation of the quantities of ^{14}C actually produced is often complicated by difficulty in assessing precise levels of nitrogen impurity present in the various target materials (Bush, Smith & White 1984).

Some of the total ^{14}C produced may be released through coolant leakage while the reactor is running, or on shut-downs during gas-purging operations. Releases will be in the form of carbon dioxide, carbon monoxide or hydrocarbons (principally methane). Graphite-moderated reactors and boiling water reactors (BWRs) release ^{14}C mainly as CO_2; pressurized water reactors (PWRs) release ^{14}C predominantly as hydrocarbons (Kunz, Mahoney & Miller 1974; Levin, Munnich & Weiss 1980; Bush, Smith & White 1984; Hertelendi, Uchrin & Ormai 1989). ^{14}C not released during the reactor operations is retained within the matrix of the materials until reprocessing, in the case of the fuel, or decom-

TABLE 32.1. Arisings

Activation reaction		Main sources
$^{13}C(n,\gamma)^{14}C$		Graphite moderated reactors and CO_2 coolant
$^{17}O(n,\alpha)^{14}C$		Water moderated reactors and oxide fuels
$^{16}O(n,^{3}He)^{14}C$		Not significant except in Fast reactor neutron spectra
$^{15}N(n,d)^{14}C$		Not significant except in Fast reactor neutron spectra
Ternary fission		Only a few percent of the total
$^{14}N(n,p)^{14}C$	*****	**Principal contributor from nitrogen impurities mainly in the fuel, but also in moderator, coolant and structural materials**

missioning, in the case of the more immobile structural materials. On reprocessing, the releases are almost exclusively CO_2.

Initially, researchers thought that the greatest contribution to the global inventory of ^{14}C released to the atmosphere, as a result of the nuclear energy cycle, would occur during fuel reprocessing; Bush, Smith and White (1984), for example, predicted that nearly 70% of the total ^{14}C arisings from the world nuclear program in the year 2000 would be from fuel reprocessing. More recent data from UNSCEAR (1988), weighted according to reactor and reprocessing plant type and noting that only about 5% of the annual energy equivalent of fuel is reprocessed, suggest that the releases from the reactors are considerably greater, possibly by a factor of 40.

Whatever the source or the original form of the ^{14}C, however, the principal route to humans is as CO_2 taken into the animal/human food chain *via* photosynthesis during plant uptake. This nuclear-produced ^{14}C adds to the natural, cosmogenic ^{14}C also present in atmospheric CO_2, and as such, makes its mark both locally and eventually, globally.

GLOBAL IMPACT

When assessing the radiological impact of this additional ^{14}C contribution, it is not usual to make comparisons with the magnitude of the natural background component as a criterion of acceptability, but it is, nevertheless, of interest to put the two components into scale. Bush, White and Smith (1983) estimated the natural cosmogenic production rate as approximately 9.2×10^5 GBq yr^{-1}, a rate that supports a total global inventory of 8.3 × 10^9 GBq (Schmidt 1979), a figure remarkably close to the 1.33 × 10^{10} GBq proposed by W F Libby in 1952 (converting his figure to present-day SI units) or, in more familiar units, 13.56 dpm g^{-1} (226 Bq kg^{-1}) the natural specific activity of carbon, which this production supports (Karlen *et al* 1964). The estimated current and future predictions of the total ^{14}C releases have taxed numerous researchers over the years and, in general, have tended to produce rather lower estimates. In 1983, Bush, White and Smith predicted a total annual arising by the year AD 2000 of 2.55 × 10^6 GBq, of which 9.25 × 10^5 GBq would be released to the atmosphere. This level means artificial production would approximately equal and, hence, multiply the natural production rate, mentioned above, by a factor of 2. More recently, Briggs and Hart (1988), using a more up-to-date judgement of the expected fruition of the originally projected nuclear power programs, reduced the predicted arisings, gaseous and other, during the years 1985 to 2000, to a total of 5.87 × 10^6 GBq. They converted this to a total increase to the global inventory, by the year 2000, of the order of 0.04% to 0.07%. They did not, however, predict the specific activity rise this increase would produce, but quoted an interesting statistic regarding the effect of fossil-fuel combustion (Suess effect) in countering the predicted specific activity increase. The authors calculated that the CO_2 produced by a 2000 MW coal burning station is equivalent to negative ^{14}C discharge of 9 × 10^3 GBq yr^{-1}. Using data from UN Economic Commission for Europe (1983), we show in Table 32.2 that the net ^{14}C enhancement between the years 1980 and 2000 is consistently negative. This is also the conclusion reached by McCartney, Baxter and Scott (1988a) who, considering multiple scenarios in multi-box models, conclude that the global ^{14}C specific activity will most probably continue to reduce at least until after 2015. Clearly, by any estimate, the expected enhancements due to the nuclear energy cycle will be significantly lower than the increase in atmospheric ^{14}C attributable to the nuclear bomb tests which, in 1963, enhanced the natural background level of the northern hemisphere by around × 2.

TABLE 32.2. Equivalent Arisings From Power Generation

	1980		1985		1990		2000	
	GWa	GBq ^{14}C	GWa	GBq ^{14}C	GWa	GBq ^{14}C	GWa	GBq ^{14}C
Gross	5.0 10^3		5.5 10^3		6.3 10^3		7.8 10^3	
Nuclear	1.6 10^2	1.5 10^5	3.8 10^2	3.6 10^5	6.2 10^2	5.8 10^5	1.0 10^3	9.7 10^5
Fossil	4.5 10^3	-2.0 10^7	4.9 10^3	-2.2 10^7	4.7 10^3	-2.1 10^7	6.3 10^3	-2.8 10^7
Net ^{14}C		-2.0 10^7		-2.2 10^7		-2.1 10^7		-2.7 10^7

The radiation dose received by humans (*per capita*) from the naturally occurring ^{14}C has been estimated at 15 microsieverts (μSv) yr^{-1} (UNSCEAR 1988). The increase due to the releases from the nuclear energy cycle is clearly small compared with this but, by convention, there is no place, when considering the radiological impact, for comparisons with background or concepts of *de minimis* doses, meaning individual annual doses so small, as they might well be from ^{14}C, that they can be neglected. Both the rules of the ICRP system (ICRP 26, 1977), which has been adopted for calculating collective dose estimates, and the application of the ALARA principle therein, which aims to reduce levels of exposure to 'as low as reasonably achievable', require the dose from ^{14}C to be considered when making assessments of dose to the population. In fact, under this system, ^{14}C, because of its long half-life, comes out rather prominently in the collective effective dose equivalent commitment, which is assessed from the summation of the estimated individual doses to all the population (10^{10}+ in the whole world) and integrated over all time (see ICRP 42, 1984 for definitions).

Webb *et al* (1986), for example, estimated collective doses to the European population (700 million) due to discharges from Sellafield, the United Kingdom reprocessing plant, in 1975, and predicted the increase to 1989. These are reproduced in Figure 32.1, where the fraction of ^{14}C in the total collective dose resulting from the 1989 discharges was predicted to be around 83%. However, based on more recent data (UNSCEAR 1988), Table 32.3 shows that the main contribution to collective dose from ^{14}C, integrated to 10,000 years, is pres-

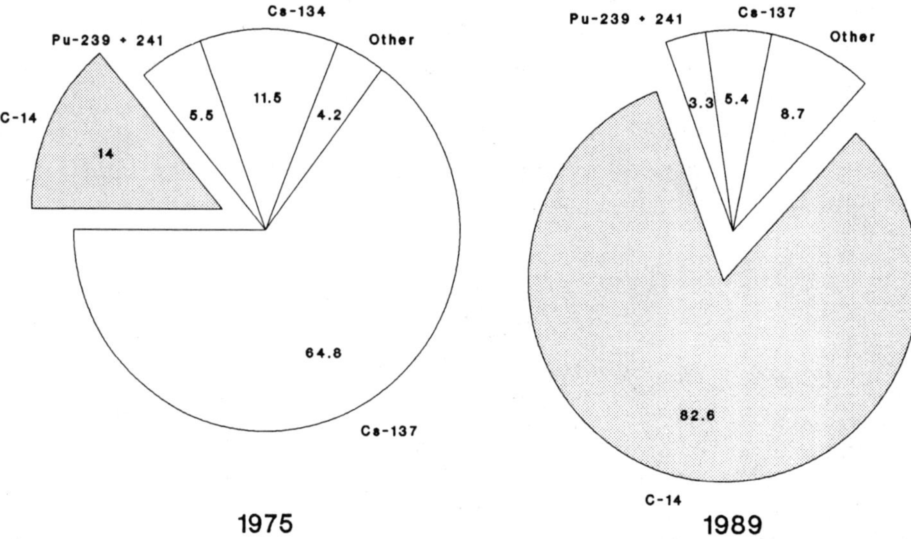

Fig 32.1. Percent contributions by nuclide to collective dose from 1975 and 1989 discharges (after Webb *et al* 1986).

TABLE 32.3. Collective Doses for the Nuclear Fuel Cycle

	Man Sv/GWa	
	^{14}C	All isotopes (inc ^{14}C)
Local and regional collective dose from the nuclear fuel cycle		4
Reactors	1.6	
Reprocessing	0.04	
Mine and mill tailings and fuel fabrication		150
Waste disposal		0.5
Reactors (intermediate level waste		0.05
Reprocessing (solid waste)		
Globally dispersed ^{14}C	63	63
Totals	65	218

(Data after UNSCEAR 1988)

ently from the globally dispersed component (63 man Sv/GWa). This amount comprises about 30% of the collective dose from the complete fuel cycle. (It may be noted, however, that most of the remainder (150 man Sv/GWa) is from arisings of other radioisotopes from mining, milling and fuel fabrication at the front end of the fuel cycle.)

Although it is of interest to note the position of ^{14}C in the collective dose inventory, its estimation is theoretical, and it may be some years hence before the actual increment becomes discernible through the standard measurement processes for ^{14}C. The position locally, *ie*, in the vicinity of reactor power stations or reprocessing plants, has provided more scope for the experimentalists in recent years, and is discussed in detail below.

LOCAL EFFECTS

Studies have been made of the effects on the local environment of ^{14}C discharges to atmosphere over a number of years.

Radioactivity Monitoring

In this category lies the monitoring of enhancements in ^{14}C caused by the general operation of nuclear installations (eg, Cimbak *et al* 1986; Hertelendi, Uchrin & Ormai 1989). This work is carried out at nuclear establishments around the world as part of the routine monitoring program required by licensing authorities. General background levels have also been widely studied since the occurrence of the atmospheric weapons trials (eg, Munnich & Vogel 1963; Baxter, Ergin & Walton 1969; Vogel & Marais 1971; Tans, de Jong & Mook 1979; Tans 1981; Nydal & Lovseth 1983; Barrette *et al* 1980; Levin *et al* 1985, 1992; Povinec, Chudý & Šivo 1986). These studies provide the data base against which the enhanced levels, including the Chernobyl event, can be compared (eg, Kuc 1987; Salonen 1987; Florkowski, Kuc & Rozanski 1988).

Chronological Studies

This work borrows directly from the established techniques widely used in studies for calibrating the ^{14}C time scale (eg, de Vries 1958; Suess 1970; Damon, Long & Wallick 1973; Ralph, Michael & Han 1973; Stuiver & Pearson 1986; and many more) as well as bomb test effects and other non-nuclear pollution studies (Cain 1978). We believe that Levin, Munnich and Weiss (1980) were the first to use tree rings to establish a chronology of emissions from a nuclear site. They measured the ^{14}C level in individual rings from 1968 to 1979 from samples close to the Obrigheim power reactor in Germany, and demonstrated an average enhancement at that time of 2.7%. Later, the research was extended to other locations in Germany (Segl *et al* 1983) and, more recently, similar studies have been reported in Czechoslovakia (Povinec, Chudý & Šivo 1986) and Yugoslavia (Obelić *et al* 1986).

We began our work on tree rings in 1979 with the objective of reconstructing a chronology of enhancements at the Sellafield reactor/reprocessing plant in northern England (Otlet, Walker & Longley 1983). In that year, a sycamore tree, over 100 years old, was felled to allow for road widening at a point 1.6 km northeast and inland of the nuclear site. We carried out ^{14}C measurements on the last 30 of its annual rings. The 30 years spanned from before the establishment of the nuclear site through the period when only reactors were running, and then to the start-up and subsequent operation of the reprocessing work. The rings, which were 1 mm to 10 mm thick, were readily pared off as single-year samples from the trunk section, approximately 0.7 m in diameter, and individually measured for ^{14}C by liquid scintillation counting.

Figure 32.2 illustrates the resultant ^{14}C values compared with the well-known atmospheric $^{14}CO_2$ line for the northern hemisphere. For 1952 to 1963, the levels are slightly above, but generally follow the atmospheric line. The small peak in 1957/8 may be, in some way, connected with the Windscale pile fire which occurred in November 1957, but this can be only speculation. After 1963, as the fuel reprocessing plant came into operation, the line shows major departures from the atmospheric level, with clear peaks in 1973 and 1976 and minima in 1968 to 1969, 1974 and 1978.

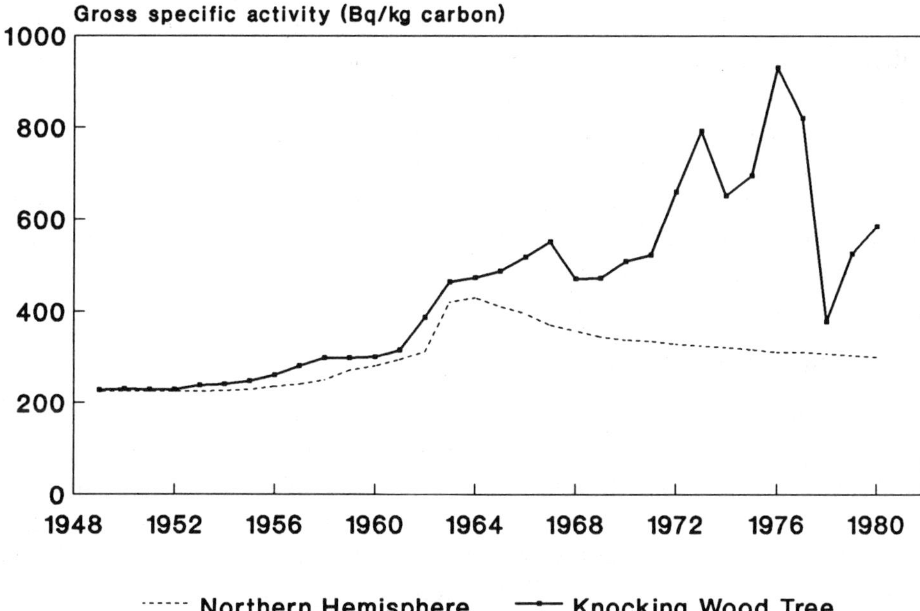

Fig 32.2. ^{14}C in tree rings from Knocking Wood sycamore, 1950–1980, compared with northern hemisphere levels

It was tempting to try to correlate the tree-ring profile with estimated plant throughput data. Indeed, a measure of agreement was obtained with our results (Otlet, Longley & Walker 1989) as it was with those of McCartney *et al* (1986) and McCartney, Baxter and Scott (1988b), who later carried out a similar exercise on a tree from the same area. (The precise location of their tree is not specified but the profile is very similar to the Harwell tree.) Perhaps precise agreement should not be expected, however, as a number of limiting factors are inherent in the correlation process. As mentioned, any estimates of releases from plant operations depend, quite crucially, on the initial level of nitrogen impurities in the fuel and fuel cladding materials, data which were not always available. Also, as regards the tree record, we must remember that photosynthesis takes place only in the daylight hours of the spring/summer growing season, whereas the reprocessing plant may run 24 hours a day for most of the year (Sellafield usually briefly shuts down for maintenance each midwinter, however). Finally, a single tree can record only the prevailing ^{14}C level at its location, which may or may not be subject to wind fraction effects (eg, summer sea breezes during daylight hours), a phenomenon highlighted by further work described below.

The problem of location became more obvious when we attempted to correlate tree-ring profiles from a number of trees lying along a coastal transect (Otlet,

Walker & Fulker 1990). Three groups of profiles emerged from the results which showed approximate agreement within themselves but little agreement overall. The original intention in carrying out this multi-tree exercise was not, however, to look for such correlations; we hoped it might show the dispersion characteristics along that line with the view of trying to validate the current theoretical dispersion model. In the event, the process of cutting individual tree rings, even from core samples and using small sample techniques, was recognized as too slow for a comprehensive survey of the area and alternative methods were sought as discussed below.

Dispersion Measurements

This area of research examines the dispersion of CO_2 from nuclear establishments through measurement of ^{14}C directly in air or as taken up by plants. Levin, Munnich and Weiss (1980) reported the use of tree leaves (in addition to tree rings, mentioned above, and direct air collectors) as integrating samplers covering the approximate period, beginning of April to end of May, in combined studies of the admixture of fossil-fuel-derived CO_2 and CO_2 released to the atmosphere from a number of German power reactors. The enhancements observed in the leaves and the tree rings were used to calculate the ^{14}C source strengths using an average dispersion factor. In this case, the calculated ^{14}C emissions agreed well with the direct emission measurements made in the power plant. In further, more detailed work (Levin *et al* 1988), the meteorological dispersion parameters were actually measured and the ^{14}C source strength calculated using the Gaussian plume model. For the Philipsburg location, which lay in the flat Rhine valley, good agreement, within a factor of 2, was obtained between the calculated (theoretical) source strength and that measured. At the second site, Isar/Ohu, which lies in hilly terrain, the model calculations agreed well with those observed up to a distance of 1 km, but failed at greater distances.

This conclusion concurs with that from our dispersion studies (Otlet, Walker & Longley 1983; Walker, Otlet & Longley 1986) around Sellafield, for which we chose to use hawthorn berries (*Crataegus* sp), rather than leaves, as the fractions of carbon in a leaf due to contemporaneous photosynthesis or earlier plant storage of carbon are not precisely known. The carbon of hawthorn berries, a fruit akin to the apple family, originates from photosynthesis in the berry's neighboring leaves and, thus, is contemporary with the atmospheric CO_2 during the growing period of the berry. Also, as an interesting spin-off from historical times, hawthorn has long been used as a hedging material to enclose field boundaries. Thus, it is possible to find hawthorn almost anywhere in the region, a handy resource for a collector! Another advantage is that the growing season is longer than for leaves (3–4 months) and can be reasonably well defined. We carried out two growth experiments, one in the north and one in the south of England, by taking random grab samples throughout the growing season, and found that the take-up of carbon is approximately exponential with a halving time of approximately three weeks. We also found about one week's starting time difference between the north and south of the country.

In a preliminary study, in 1981, dispersion characteristics were obtained from hawthorn berry measurements along five transect lines, two coastal and three inland, radiating from the epicenter (discharge stacks) of the reprocessing plant (Otlet, Walker & Longley 1983). Attempts to model these results were complicated by the nature of the surrounding terrain, which was relatively flat along the northwest-to-southeast coastal regions, but with hills and valleys on the inland transects. In a wider hawthorn survey two years later, it was possible, using a contour-fitting computer package, to construct the specific activity isopleths over the area. This enabled some of the effects of the terrain to be traced (eg, valley lines going up through the hills), but it also showed us that terrain was not the only factor, and we realized that wind fractions must also be studied. Wind fractions in any direction during the daylight hours of the summer season are very variable and can be quite different from year to year.

Since these initial dispersion experiments, we have studied variations each year from annually collected samples along a 4 km arc drawn around the center of the site. Results of the tests have been reported in detail elsewhere (Fulker, Otlet & Walker 1987), and only a brief résumé is given here. The measured specific activity profiles of the hawthorn berries were compared with predicted profiles developed using the R-191 (Clarke 1979) dispersion model assuming an average stack height for the releases (120 m) and a constant release rate. The relevant meteorological data (Pasquill stability, wind direction, etc), for the area was fed into the model and integrated over the daylight hours of the growing season, enabling the theoretical estimate to be built up at a rate consistent with the experimental carbon uptake curve for the region.

Over the six years of the study, the results showed only occasional agreement between the predicted and experimental results with the suggestion of dependence for good agreement on the frequency, during the period, of the stable dispersion categories. In 1984, there was a preponderance of high-stability periods with the dominant wind fraction caused by daytime sea breezes blowing into the 120° sector. Agreement was good that year between the measured and predicted profiles (Fig 32.3A). Alternatively, in 1983 and 1981, which were characterized by extended periods of instability, the predicted profile still shows the sea breezes peak towards the 120° sector, but also a much broader, less well-defined peak, some 50° to the left of the predicted peak in the

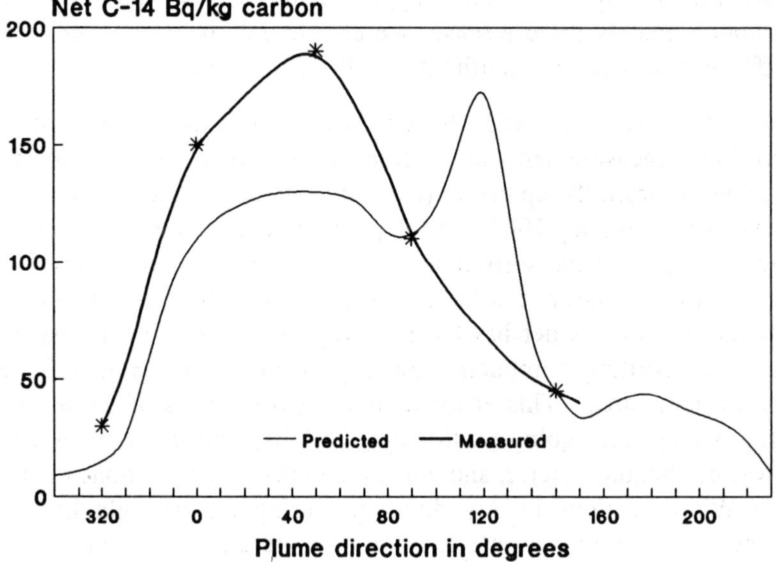

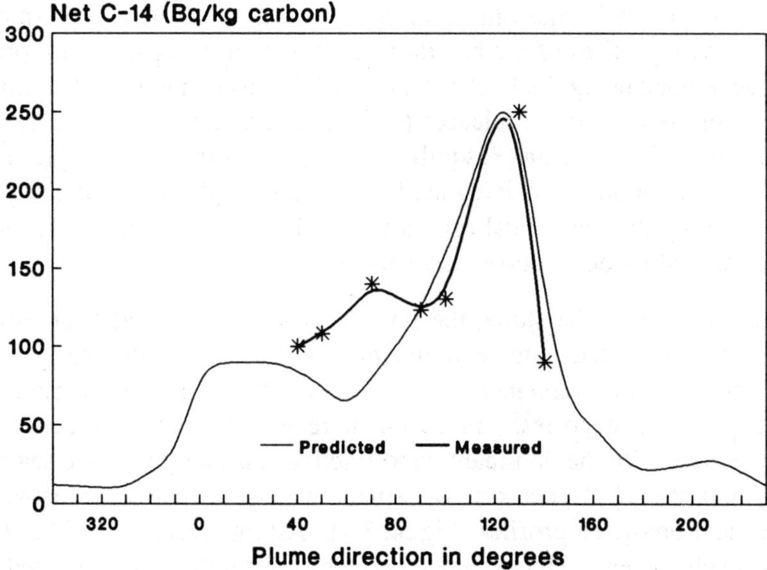

Fig 32.3. Net ^{14}C in hawthorn berries at 4 km, compared with predicted angular distribution (Fulker, Otlet & Walker 1987). Reprinted by permission of Elsevier Applied Science Publishers Ltd.

measured profile (Figure 32.3B). It has been suggested that, under these conditions, plume curvature, resulting from terrain influences, has modified the experimental results. We are currently studying this hypothesis in more detail in a separate windfield experiment.

Correlating predicted and measured results is important for validating models, which are used to calculate doses to the critical groups, *ie*, people for whom local produce forms a significant part of their diet. Killough and Rohwer (1978) considered the effects of applying the actual dispersion categories relevant to the daylight hours growing season for an inland site at Knoxville, Tennessee. They found that, because the frequency of unstable dispersion categories in the daylight hours was significantly greater than the average 24 hours value, the uptake doses to food consumers were increased by a factor of three, compared with the 24-hour average dispersion conditions. The introduction of the topic of dose estimation leads into the final section of the regional ^{14}C effects, the estimation of doses received by the critical groups consuming the local produce.

Validation of the Specific Activity Model for Collective and Critical Dose Assessments

The specific-activity model for ^{14}C, generally used in the calculations of dose received by the population, makes the primary assumption that the specific activity of the carbon in the organs, skeletal material, etc, of the body is identical to the specific activity of the carbon in the atmospheric carbon dioxide of the air. This is because the main pathway is *via* plants and animal-derived products. Of course, the very principle of ^{14}C dating relies on this assumption. The applicability of the model has been discussed by various authors (eg, Wirth 1982; Fischer & Muller 1982). Apart from the well-known modifications introduced by isotopic effects in plant material, which heads the food chain (there is an intrinsic difference of a few percent between specific activities in the C_3- and C_4-type plants and also in marine-derived foodstuffs), it is usually concluded that the specific-activity model is an adequate approximation. In a garden plot experiment in the vicinity of two Magnox and two AGR-type power reactors at Hinkley Point in southwest England, Kluczewski, Nair and Bell (1986) inferred, from the measured ^{14}C in the produce, that the specific-activity approach was probably adequate, in fact slightly pessimistic, for calculating the concentration of ^{14}C, especially when variation in growing rate was taken into account.

To follow this up, we set up a plot in the vicinity of Sellafield, and grew a variety of crops in the 1988 season in the natural soil and in neighboring sub-plots, which contained soil imported from a region with normal, *ie*, non-enhanced air ^{14}C concentrations. Again, only a résumé of the results, which

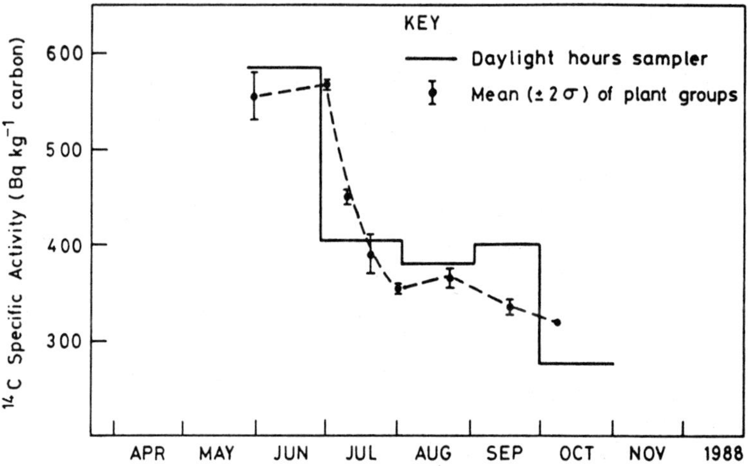

Fig 32.4. Comparison of air and crop concentrations (Otlet, Walker & Fulker 1989). Reprinted by permission of IOP Publishing, Ltd and the authors.

are given in more detail elsewhere (Otlet, Walker & Fulker 1989), can be given here. Basically, no significant difference was found between the root or leaf species in either plot. 1988 was an interesting year for this study, however, as the $^{14}CO_2$ fell sharply in air concentration during part of the growing season. The effect is seen in Figure 32.4, where the monthly average ^{14}C values for air samplers run only during daylight hours are compared with mean ^{14}C values for crops harvested at the same time and observed to have grown over the same time period. Quite good agreement is observed between the air values and those in the crops, particularly for the early part of the year. Values for the later groups fall significantly below the air values, a phenomenon which is being examined further. Thus, although the results can be said broadly to support the specific-activity model, a further refinement is needed to allow more precise modeling of dose assessment.

Conclusion

^{14}C derived from the nuclear energy cycle may be a relative newcomer among the many and varied applications of ^{14}C dating, but it is assuming an increasingly important role. Although ^{14}C has been, for some time, a 'poor relation' amongst the radionuclides included in the collective dose inventory and critical group assessments, it is now moving to a position of greater significance. From the point of view of global impact, however, the position of ^{14}C could change if a concept of *de minimis* doses were ever adopted in the ICRP method of assessing collective equivalent dose equivalent. This is because ^{14}C is included more by virtue of its long half-life adding to the dose commitments of the population in future generations than to an increase in the predicted carbon specific activity, which, despite nuclear energy cycle releases, may continue to fall in the short term due to fossil-fuel combustion.

The position may be rather different, however, in critical group assessments. Taking Sellafield as an example, improved technology has reduced the discharges of particulate radioactivity to the atmosphere, and of actinides and radiocesium to the sea. As a result, ^{14}C now features as a significant contributor to the dose to the critical group of people living close to the site and consuming locally produced food. In the year 1988, out of an estimated total committed effective dose equivalent of 110 μSv to this group, ^{14}C, mainly through the milk pathway, accounted for 29 μSv (Fig 32.5). This level of significance makes it important that the parameters controlling ^{14}C uptake during the growing periods of the year, eg, meteorological effects, such as sea breezes and plume curvature on the local terrain, should be well understood and justify continued research effort.

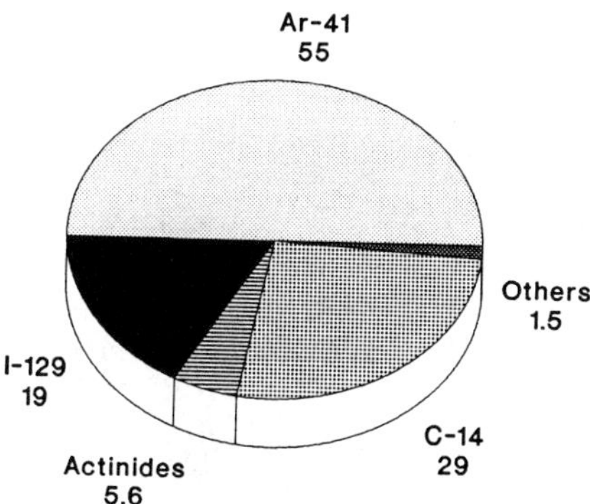

Fig 32.5. Critical group dose estimate in microsieverts, Sellafield 1988 (Jones *et al* 1991). Reprinted by permission of Nuclear Technology Publishing.

REFERENCES

Barrette, L, Lasalle, P, Martel, Y and Samson, C 1980 Variations of ^{14}C in oats grown from 1957 to 1978 in Quebec. *In* Stuiver, M and Kra, RS, eds, Proceedings of the 10th International ^{14}C Conference. *Radiocarbon* 22(2): 410–414.

Baxter, MS, Ergin, M and Walton, A 1969 Glasgow University radiocarbon measurements I. *Radiocarbon* 11(1): 43–52.

Briggs, A and Hart, D 1988 The Management of Carbon-14 and Iodine-129 Wastes, a Site-Specific Survey of Current and Future Arisings, Possible Management Options and Potential Impact with Respect to the United Kingdom. *DOE Report* DOE/RW/89.031, UKAEA, Harwell Laboratory.

Bush, RP, Smith, GM and White, IF 1984 *Carbon-14 Waste Management*. EUR 8749 EN, Commission of the Economic Community, Luxembourg.

Bush, RP, White, IF and Smith, GM 1983 *Carbon-14 Waste Management*. AERE – R10543, AERE Harwell, Oxfordshire, UK.

Cain, WF 1978 Carbon-14, tree-rings and urban pollution. *Environment International* 1: 167–171.

Cimbak, S, Cechova, A, Grgula, M, Povinec, P and Sivo, A 1986 Anthropogenic radionuclides ^{3}H, ^{14}C, ^{85}Kr and ^{133}Xe in the atmosphere around nuclear power reactors. *Nuclear Instruments and Methods* B17: 560–563.

Clarke, RH 1979 *A Model for Short and Medium Range Dispersion of Radionuclides Released to the Atmosphere*. NRPB-R91, National Radiological Protection Board, Oxfordshire.

Damon, PE, Long, A and Wallick, EI 1973 Dendrochronologic calibration of the carbon-14 time scale. *In* Rafter, TA and Grant-Taylor, T, eds, *Proceedings of the 8th International ^{14}C Conference*. Wellington, Royal Society of New Zealand: 28–71.

Fischer, E and Muller, H 1982 Kohlenstoff 14 in der terrestrischen nahrungskette. *Atomkernenergie, Kerntechnik* 41: 124–126.

Florkowski, T, Kuc, T and Rozanski, K 1988 Influence of the Chernobyl accident on the natural levels of tritium and radiocarbon. *Applied Radiation Isotopes* 39(1A): 77–79.

Fulker, MJ, Otlet, RL and Walker, AJ 1987 A comparison of model predictions of gaseous dispersion with environmental measurements of ^{14}C around Sellafield, UK. *In* Desmet, G, ed, *CEC Conference on Reliability of Radioactive Transfer Models*. London, Elsevier: 30–37.

Hertelendi, E, Uchrin, G and Ormai, P 1989 ^{14}C release in various chemical forms with gaseous effluents from the Paks nuclear power plant. *In* Long, A and Kra, RS, eds, Proceedings of the 13th International ^{14}C Conference. *Radiocarbon* 31(3): 754–761.

ICRP Pub 26, 1977 Recommendations of the ICRP. International Commission on Radiological Protection. *Annals of the ICRP* 1(3). Oxford, Pergamon Press.

ICRP Pub 42, 1984 *A Compilation of the Major Concepts and Quantities in use by ICRP. A Report of a Task Group of the ICRP Committee* 4. Oxford, Pergamon Press.

Jones, SR, Fulker, MJ, McKeever, J and Stewart, TH 1991 Aspects of population exposure consequent on discharges of radionuclides to the environment from the nuclear reprocessing plant at Sellafield in Cumbria. In *Radiation Protection Dosimetry*. London, Nulcear Technology Publishing.

Karlen, I, Olsson, IU, Kallenberg, P and Kilicci, S 1964 Absolute determination of the activity of two ^{14}C dating standards. *Arkiv foer Geofysik* 4: 269.

Killough, GG and Rohwer, PS 1978 A new look at the dosimetry of ^{14}C released to the atmosphere as carbon dioxide. *Health Physics* 34: 141–159.

Kluczewski, SM, Nair, S and Bell, JNB 1986 *A Field Study of the Uptake of ^{35}S and ^{14}C into Crops Characteristic of the UK Diet.*

TPRD/B/0735/R86, Central Electricity Generating Board, Berkeley Nuclear Laboratories, UK.

Kuc, T 1987 ^{14}C traced in Krakow after the Chernobyl accident. *Radiocarbon* 29(3): 319–322.

Kunz, CO, Mahoney, WE and Miller, TW 1974 ^{14}C gaseous effluents from pressurized water reactors. *In* Population Exposures, Proceedings of the 8th Midyear Topical Symposium of the Health Physics Society. *US Atomic Energy Commission Report* CONF-74018. Springfield, Virginia: 229–234.

Levin, I, Bössinger, R, Bonani, G, Francey, RJ, Kromer, B, Münnich, KO, Suter, M, Trivett, NBA and Wölfli, W 1992 Radiocarbon in atmospheric carbon dioxide and methane: Global distribution and trends. *In* Taylor, RE, Long, A and Kra, RS, eds, *Radiocarbon After Four Decades: An Interdisciplinary Perspective.* New York, Springer-Verlag, this volume.

Levin, I, Kromer, B, Barabas, M and Munnich, KO 1988 Environmental distribution and long-term dispersion of reactor CO_2 around two German nuclear power plants. *Health Physics* 54(2): 149–156.

Levin, I, Kromer, B, Schoch-Fischer, H, Bruns, M, Munnich, M, Berdau, D, Vogel, JC and Munnich, KO 1985 25 years of tropospheric ^{14}C observations in central Europe. *Radiocarbon* 27(1): 1–19.

Levin, I, Munnich, KO and Weiss, W 1980 The effect of anthropogenic CO_2 and ^{14}C sources on the distribution of ^{14}C in the atmosphere. *In* Stuiver, M and Kra, RS, eds, Proceedings of the 10th International ^{14}C Conference. *Radiocarbon* 22(2): 379–391.

Libby, WF 1952 *Radiocarbon Dating.* Chicago, Chicago University Press: 124 p.

McCartney, M, Baxter, MS, McKay, K and Scott, EM 1986 Global and local effects of ^{14}C discharges from the nuclear fuel cycle. *In* Stuiver, M and Kra, RS, eds, Proceedings of the 12th International ^{14}C Conference. *Radiocarbon* 28(2A): 634–643.

McCartney, M, Baxter, MS and Scott, EM 1988a Carbon-14 discharges from the nuclear fuel cycle: 1, Global effects. *Journal of Environmental Radioactivity* 8: 143–155.

_____1988b Carbon-14 discharges from the nuclear fuel cycle: 2, Local effects. *Journal of Environmental Radioactivity* 8: 157–171.

Munnich, KO and Vogel, JC 1963 Investigations of meridional transport in the troposphere by means of carbon-14 measurements. In *Proceedings of the Symposium on Radioactive Dating.* Vienna, IAEA.

Nydal, R and Lovseth, K 1983 Tracing bomb ^{14}C in the atmosphere 1962–1980. *Journal of Geophysical Research* 88(C6): 3621–3642.

Obelić, B, Krajcar-Bronić, I, Srdoč, D and Horvatinčić, N 1986 Environmental ^{14}C levels around the 632 MWe nuclear power plant Krsko in Yugoslavia. *In* Stuiver, M and Kra, RS, eds, Proceedings of the 12th International ^{14}C Conference. *Radiocarbon* 28(2A) 644–648.

Otlet, RL, Longley, H and Walker, AJ 1989 Studies of environmental radioactivity in Cumbria Part 13: Measurements of carbon-14 in tree-rings from trees growing in the Sellafield area. *AERE R* 12362, London, HMSO: 24 p.

Otlet, RL, Walker, AJ and Fulker, MJ 1989 The transfer of ^{14}C to foodstuffs from experiments on a controlled plot in Cumbria. *In* Goldfinch, EP, ed, *Radiation Protection Theory and Practice.* Society for Radiological Protection, Proceedings of the 25th Anniversary Symposium. Bristol and New York, Institute of Physics.

_____1990 Survey of the dispersion of ^{14}C in the vicinity of the UK reprocessing site at Sellafield. *Radiocarbon* 32(1): 23–30.

Otlet, RL, Walker, AJ and Longley, H 1983 The use of ^{14}C in natural materials to establish the average gaseous dispersion patterns of releases from nuclear installations. *In* Stuiver, M and Kra, RS, eds, Proceedings of the 11th International ^{14}C

Conference. *Radiocarbon* 25(2) 593–602.

Povinec, P, Chudý, M and Šivo, A 1986 Anthropogenic radiocarbon: Past, present and future. *In* Stuiver, M and Kra, RS, eds, Proceedings of the 12th International ^{14}C Conference. *Radiocarbon* 28(2A): 668–672.

Ralph, EK, Michael, NH and Han, MC 1973 Radiocarbon dates and reality. *MASCA Newsletter* 9: 1–20.

Salonen, L 1987 Carbon-14 and tritium in air in Finland after the Chernobyl accident. *Radiochimica Acta* 41: 145–148.

Schmidt, PC (ms) 1979 Alternative ways of reducing ^{14}C emission in high temperature reactor fuel processing. Dissertation, Technische Hochschule Aachen, Julich (Kernforschungsanlage, Nr Jul–1567) 108 p.

Segl, M, Levin, I, Schoch-Fischer, H, Munnich, M, Kromer, B, Tschiersch, J and Munnich, KO 1983 Anthropogenic ^{14}C variations. *In* Stuiver, M and Kra, RS, eds, Proceedings of the 11th International ^{14}C Conference. *Radiocarbon* 25(2): 583–592.

Stuiver M and Pearson, GW 1986 High-precision calibration of the radiocarbon time scale, AD 1950–500 BC. *In* Stuiver, M and Kra, RS, eds, Proceedings of the 12th International ^{14}C Conference. *Radiocarbon* 28(2B): 805–838.

Suess, HE 1970 Bristlecone pine calibration of the radiocarbon time-scale 5200 BC to the present. *In* Olsson IU, ed, *Radiocarbon Variations and Absolute Chronology*. Proceedings of the 12th Nobel Symposium. Stockholm, Almqvist and Wiksell: 595–605.

Tans, PP 1981 A compilation of bomb ^{14}C data for use in global carbon model calculations. *In* Bolin, B, ed, *Scope 16: Carbon Cycle Modelling*. New York, John Wiley & Sons.

Tans, PP, de Jong, AFM and Mook, WG 1979 Natural atmospheric ^{14}C variation and the Suess effect. *Nature* 280: 826–828.

UNSCEAR 1988 *Sources and Effects of Ionizing Radiation*. Report to the General Assembly, United Nations, New York.

UN Economic Comm for Europe 1983 *An Efficient Energy Future: Prospects for Europe and North America*. London, Butterworths.

Vogel, JC and Marais, M 1971 Pretoria radiocarbon dates I. *Radiocarbon* 13(2): 378–399.

Vries, H, de 1958 Variation in concentration of radiocarbon with time and location on earth. *Koninklijke Nederlandse Akademie van Wetenschappen Proceedings Series* B61: 1–9.

Walker, AJ, Otlet, RL and Longley, H 1986 Applications of the use of Hawthorn berries in monitoring ^{14}C emissions from a UK nuclear establishment over an extended period. *In* Stuiver, M and Kra, RS, eds, Proceedings of the 12th International ^{14}C Conference. *Radiocarbon* 28 (2A): 681– 690.

Webb, GAM, Cooper, JR, Dionian, J, Haywood, SM and Jones, JA 1986 The radiological impact on the public of nuclear in the UK, 1970–1990. In *Nuclear Energy of Today and Tomorrow*. Proceedings of the Foratom IX (ENC 4/ENC 86) International Conference. Geneva, European Nuclear Society.

Wirth, E 1982 The applicability of the ^{14}C specific activity model. *Health Physics* 43: 919–922.

33

MANKIND'S PERTURBATIONS OF PARTICULATE CARBON

LLOYD A CURRIE

INTRODUCTION

A central challenge for contemporary environmental research is the assessment of human impacts on the natural chemical cycles. The balance between natural and anthropogenic sources of carbonaceous particles represents a critical part of perhaps the most influential of these cycles, the carbon cycle. In this chapter, I examine the contributions of radiocarbon to our understanding of this system, especially in light of major advances in isotopic measurement (notably accelerator mass spectrometry (AMS)), environmental sampling, trace and microanalysis, and computation and modeling. Following a brief review of the history of atmospheric particulate carbon and ^{14}C measurements, I consider: the particulate carbon life cycle (formation, transport and reaction, sinks and natural archives); chemical information content of the particles (isotopic, elemental, molecular and structural); and case studies of local and global particle source apportionment based on univariate and multivariate compositional patterns.

Significance of Particulate Carbon

The particle phase, sometimes described as "the fourth state of matter," comprises one of the most interesting, important and complex forms of carbon in the environment. Interesting, because carbonaceous particles carry potential information on the formation, age and transport history of the constituent molecules. Important, because the chemical and physical properties of the particles imply effects on health, visibility and climate. Complex, because the carbonaceous material exhibits a wide range of structural and morphological characteristics, plus an enormous range in chemical composition. It is this very complexity, in fact, that contributes to the possibility of applying chemical and isotopic pattern recognition techniques to extract information on sources and environmental history – *ie*, to employ the particles, themselves, as tracers.

The deleterious effects of carbonaceous particles, and more generally, combustion particles, serve as the driving force for research on the formation and atmospheric life cycle of these particles. Primary health effects of carbonaceous particles, and/or their gaseous precursors, relate to the formation of tropospheric ozone (Chameides *et al* 1988) and genotoxicity (Tuominen *et al* 1988). Ozone formation results from a complex series of photochemical reactions involving reactive hydrocarbons, nitrogen oxides and the hydroxyl radical. Mutagenic and carcinogenic effects are associated with the biological activity of the organic combustion products. Visibility tends to be a local or regional consequence of fine particle formation and transport. In the eastern US and other regions with at least moderate humidity and sulfur-containing air masses (originating from nearby or distant fossil fuel power plants), the primary fine-particle aerosol is sulfate; in the western US and other arid regions assaulted by combustion particles, the main component is carbonaceous soot (White & Macias 1989). It is fitting that the "Four Decades of Radiocarbon Studies" conference took place in the mountains atop metropolitan Los Angeles, for we could directly see the brown, carbonaceous haze from the city. In the Great Smoky Mountains (Shaw 1987), by contrast, the haze is white due to its major sulfate content. (A century ago, before industrialization, the Smokies were said to have had a "blue haze" associated with gas-to-particle conversion of natural terpenes.)

Potential climate effects are well appreciated. Sulfate particles are important for cloud nucleation and important increases in the earth's albedo, and therefore, tropospheric cooling (Hansen *et al* 1980), whereas black carbon (soot) has important effects on the radiation balance due to both absorption and scattering of solar radiation (Shaw 1982). This is of special consequence for the Arctic, where long-range transport of anthropogenic combustion particles forms the brown "Arctic Haze" each spring (Rahn & McCaffrey 1980). Natural experiments (volcanic emissions) and smaller scale anthropogenic experiments and theory (major fires, nuclear winter scenarios) provide semi-quantitative data on temperature effects (Fields *et al* 1989).[1]

[1]Note added in proof: Major, global-scale anthropogenic and natural injections of particles occurred during 1991, even as this chapter was going to press. The massive man-made oil conflagration, involving 600 wells in Kuwait, is predicted to continue burning for years, and already record-low temperatures and record-high toxic particle concentrations have been reported for the region (Flam 1991). This was completely overshadowed by the natural event, however. On 15, 16 June 1991, Mount Pinatubo in the Philippines sent 20 megatons of gas and particulate debris into the stratosphere, probably constituting the largest volcanic eruption of the century. The volcanic plume circled the globe in three weeks, and it is expected to persist for a few years, decreasing global temperatures and stratospheric ozone. The main culprit in this case is sulfate rather than carbonaceous particles (Kerr 1991).

The influence of human activities on the production and life cycle of atmospheric particles has become, therefore, the focus of international research programs (Galbally 1989). Although the global production of natural aerosol exceeds that of anthropogenic aerosol by a factor of 5–10, regional and urban aerosol can be overwhelmingly anthropogenic (Prospero 1984). Measurements of Los Angeles aerosol, combined with emissions inventories and receptor modeling, have shown that particulate carbon accounts for 40% of the average fine particle mass concentration (Larson, Cass & Gray 1989), and that 40%–70% of this arises from motor vehicles (Pratsinis *et al* 1984). Direct measurement of the fossil fraction of particulate carbon, by means of radiocarbon, has shown typical values of 60%–70% in a number of urban areas; particles from locations subject primarily to natural emissions or to woodburning show typically 80%–90% biospheric carbon (Currie, Klouda & Voorhees 1984).

FOUR DECADES OF (R)EVOLUTION

Early History

The earliest measurements of ^{14}C in atmospheric particles actually took place nearly four decades ago, some eight years after the discovery of ^{14}C dating. Documentation of mankind's perturbations of particulate carbon came much earlier, however, during the Middle Ages. A fascinating account of documented English air pollution history, beginning in 1257, has been given by Peter Brimblecombe (1981). Table 33.1, which has been adapted from this work, shows that the societal impacts of pollutant carbon (and sulfur) from fuel combustion transcend time. Both today and in 18th century England, shortages of cleaner burning fuels have resulted in increased particulate carbon. Severe consequences, such as major adverse health effects, are commonly required to stimulate appropriate legislation or other broad, corrective measures. Two extreme cases of health effects occurred in early and modern England: the high incidence of cancer in chimney sweeps two centuries ago (Goldberg 1985: viii), and the disastrous London fog of 1952 (Brimblecombe 1981).

The source of "soot" pollution in ancient England was eminently clear; it came from the "seacoal" brought into London to replace the diminishing supplies of firewood. The more complex origins of carbonaceous pollution in 20th century cities led to the initial interest in applying ^{14}C measurement as a means for discriminating fossil from biogenic sources. Two "heroic" experiments were performed in the late 1950s by two of the pioneers of radiocarbon dating, Jim Arnold and Hans Suess, both of whom participated in this conference at Lake Arrowhead. It is noteworthy that Arnold's paper of 35 years ago begins with the statement that, "The Los Angeles smog is probably the most publicized of any air pollution problem"; and notes that, "There exists an unequivocal method

TABLE 33.1. Early Observations of the Effects of Air Pollutants*

Pollution problem	Earliest reference
Health/annoyance	1257 Annals of Dunstable
Economic loss	1377 Assize of Nuisance
Damage to furnishings	1512 Earl of Northumberland
Damage to gardens	1603 Platt
Damage to buildings	1620 James I
Increased death rate	1658 Digby; 1662 Graunt
Damage to silverware	1658 Digby
Increased corrosion	1661 Evelyn
"Great stinking fog"	1691 Gadbury

•Little Ice Age: industry (rural) → London
•(Black) umbrellas: response "sootfall"
•Disastrous fog of 1952 → 4000 deaths

*Adapted from Brimblecombe (1981)

for distinguishing between carbonaceous material arising from biological sources and that from fossil fuels ..." (Clayton, Arnold & Patty 1955). This historic paper suggests also the possibility of employing ^{7}Be to estimate the rate of subsidence of the Los Angeles atmosphere, and the application of isotopic tracers to label suspected (polluting) fuel supplies, topics that I shall return to, later.

The experiments were heroic from the perspective of the sampling and ^{14}C counting capabilities of the time. For example, taking 5 g carbon as a reasonable sample for counting, and 25 μg m^{-3} carbon as a typical pollution concentration, one would need to sample 200,000 m^3 air. In fact, Clayton, Arnold and Patty (1955) did sample about 175,000 m^3 of Los Angeles air over a period of 4.3 days, using a huge filtering apparatus capable of operating at 1000 ft^3 min^{-1}. The resulting sample, containing 6 g carbon (equivalent to 34 μg m^{-3}), was found by liquid scintillation counting to contain 25.7 ± 1.6 % contemporary carbon (now called percent modern carbon (pMC)). The work performed by Lodge, Bien and Suess (1960) was performed on a similarly grand scale (3.8 g atmospheric carbon collected), but it went beyond the earlier study in that selected sub-chemical fractions were counted (using acetylene gas proportional counting). Similar results were obtained, showing that the urban carbonaceous particles were mostly derived from fossil-fuel combustion.

Recent Advances

Advances in our understanding of global and local sources, and the atmospheric life cycle of carbonaceous particles have been fueled by (r)evolution in at least four areas: 1) societal concern with energy and the environment; 2) advances in atmospheric and geochemical sampling; 3) enhanced measurement sensitivity for stable and long-lived radionuclide isotope mass spectrometry (μg levels), micro-organic analysis (pg levels), and individual particle microanalysis (fg levels); and 4) computational capabilities, especially as applied to atmospheric and chemical modeling, and statistical graphics. The driving forces have been both societal and scientific. Beyond the sociopolitical needs (and funds), which have helped to establish major technological advances, the past two decades have seen important societal awareness of global fuel depletion, local and regional insults to health and the environment, and potential long-term climatic consequences. Carbonaceous particles (and gases) and the need to discriminate anthropogenic from natural fluxes have been central to these societal concerns. Work in our own laboratory, for example, began in response to the suggestion that eastern US (mid-Atlantic) urban (carbonaceous) pollution was due primarily to natural forest emissions (Currie & Murphy 1977). Fortunately, pioneering work by Oeschger on ^{14}C in ice cores had shown the way to make measurements on carbon samples two orders of magnitude smaller than those of Arnold and Suess (Oeschger *et al* 1972). Two further developments followed in short order: the dramatic switch to woodburning, partly as a result of the oil embargo and perceived petroleum shortages; and the discovery of AMS (Muller 1977; Purser *et al* 1977). ^{14}C has become *the* tracer for woodburning (Ramdahl *et al* 1984), as well as the means for developing and calibrating more routine chemical tracers for monitoring (Currie *et al* 1989). At the same time, AMS has extended the sensitivity by an *additional* 2–3 orders of magnitude, so that we may now assay ^{14}C in individual compounds or classes of compounds in just 10–20 μg of atmospheric carbon (Verkouteren *et al* 1987).

Advances in computation have had a major impact on studies of carbonaceous particles in that they: 1) permit the linking of source, reaction and transport modeling with ambient measurement; and 2) make possible the *direct incorporation* of isotopic data with chemical and structural data into appropriate multivariate theoretical and empirical receptor models. In this way, we can transcend the fossil/biogenic carbon discriminating power of ^{14}C, and determine the identities and fluxes of the manifold sources of particles. An illustration of the breadth of environmental problems that can be attacked through the combined uses of elemental and organic measurements at the ng to μg levels or individual particle measurements at the sub-pg level, combined with ^{14}C-AMS at the μg level, is given by Currie *et al* (1989). Table 33.2, adapted from this work, summarizes the recent progress in both decay and atom (AMS) counting of ^{14}C. A crucial point, conveyed in this table, is that *practical application to environmental carbon research is limited by the chemical blank*, not simply counting statistics or accelerator background.

TABLE 33.2. Micro-^{14}C Environmental Measurement: State of the Art*

Capabilities**					
Method	Conventional sample	Signal limit (100 c)	Background equiv mass	Blank† mass	Blank† recovery
Decay	5–10 mg	2.4 mg	5 mg	40 μg	95%
AMS	*0.5–2 mg*	*1.3 μg*	*0.2 μg*	15 μg	30% (1984)
				1.4 μg	*95%* (1987)

*Adapted from Currie *et al* (1989)

**Columns 2 and 5 indicate the mass of carbon commonly taken for measurement, and introduced as a chemical blank, respectively; columns 3 and 4 give signal (100 count) and background measurement limits, expressed as the equivalent mass of modern carbon

†*Total* blank (1989) = 1–2 μg carbon, *including* major organic chemical separations.

The Past Decade. A partial summary of ^{14}C atmospheric particle results from our laboratory is given in Table 33.3 (Currie, Klouda & Voorhees 1984). Four classes of environmental issues are represented: "normal" urban pollution; agricultural and/or woodburning pollution; rural concentrations (isotopic and chemical); and remote manifestations. The summary results, given as medians, tell only part of the story – *viz*, that: 1) the carbon in (US) urban particles is typically predominantly fossil in origin (eg, Houston); 2) areas of significant usage of woodstoves and fireplaces exhibit high contemporary carbon (eg, Elverum, Norway – winter); 3) rural and forested regions have mostly contemporary carbon particles; and 4) special studies of remote areas can show large fossil carbon contributions (Barrow, Alaska). Not shown are the concentrations (μg m^{-3}) of the carbon or the variations in percent contemporary carbon. In fact,

TABLE 33.3. Ambient Particle Samples (^{14}C)

Sample	Location	Contemporary C (%) (median)
(Urban)	Denver	29 ("brown cloud")
	Salt Lake City	28
	Houston	34
	Los Angeles	38
	Portland	77 (vegetative burning)
	Elverum	69 (wood burning)
(Rural)	Desert – Utah	88
	Forest – US (Shenandoah)	92
	Forest – USSR (Abastumani)	80
(Remote)	Point Barrow	27 ("Arctic haze")

the urban concentrations far exceed the rural and remote concentrations, and f_C (fraction of contemporary carbon) variability is also higher for urban regions, except for special cases of long-range transport (Barrow). Figure 33.1 shows the rather different patterns of variability for the Houston and Shenandoah series. The ability of ^{14}C to reliably apportion fossil and biogenic carbon was extremely important in these two studies: it had been presumed that the carbon in Houston was essentially all fossil, due to refinery operations; the emissions inventory on which dispersion modeling was based showed no component for a strong, varying biogenic source; the ^{14}C "surprise" (of up to 60+ pMC) led to a re-examination and recognition of long-range transport of soot from agricultural burning. The Shenandoah ^{14}C surprise was of the opposite sort: large concentrations of coal-burning sulfate in the rural, forested region led to the presumption that the C-particles would, therefore, be primarily fossil. The ^{14}C data proved that to be false; the carbonaceous particles, which had relatively low concentrations, were of biospheric origin.

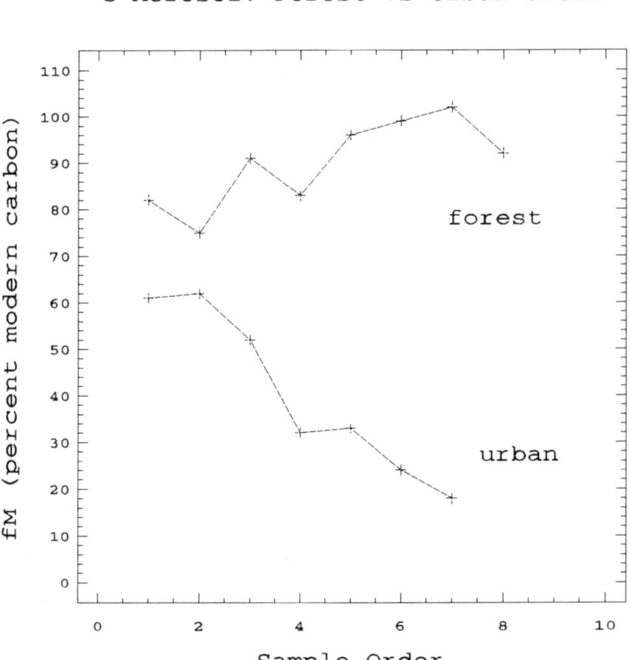

Fig 33.1. Aerosol radiocarbon – patterns of variability. The 7 sequential urban (Houston) samples spanned a period of one week; the 8 sequential forest (Shenandoah) samples spanned a period of two weeks. SURPRISES: Shenandoah – high sulfate concentrations (fossil energy), yet mostly modern (biogenic) carbon. Houston – no biogenic carbon in the emissions inventory, yet up to 60 pMC in aerosol.

Measurements of ^{13}C form a very interesting complement to those of ^{14}C, for they can distinguish particles arising from marine sources or tropical vegetation, as opposed to the isotopically lighter carbon (relatively less ^{13}C) particles from petroleum or temperate-zone wood. An illustration, showing the utility of ^{13}C for discriminating distant sources of carbonaceous aerosol in the marine atmosphere, is given in Figure 33.2 (Cachier 1989). Particles from southern-hemisphere (mostly tropical) sources are well separated, in isotopic composition and atmospheric concentration, from those arising from northern-hemisphere (mostly temperate and industrial) sources. Still more interesting, for environmental source discrimination, are dual isotope measurements. Important investigations of this type have been published by Court *et al* (1981), Kaplan and Gordon (1989), and Tanner and Miguel (1989). Polach, following the initial work with Court *et al* (1981), has employed the dual-isotope technique for the discrimination of particles affecting the Sydney and Canberra atmospheres from C_3 and C_4 plants, oil combustion and marine sources. Tanner and Miguel (1989) performed an analogous study in Rio de Janiero, with the interesting outcome that biogenic alcohol motor-vehicle fuel constituted an important source com-

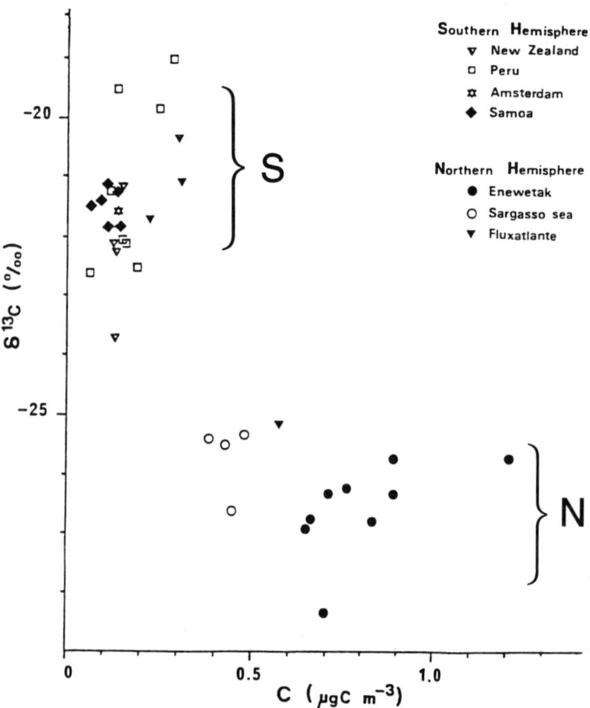

Fig 33.2. Bulk carbonaceous aerosol samples in the marine atmosphere. ^{13}C serves as an effective discriminator for particles arising from the southern hemisphere (S: mostly tropical sources) *vs* those from the northern hemisphere (N: mostly temperate and industrial sources). (Adapted from Cachier 1989 by permission of the author.)

ponent. This work followed earlier, dual-isotope research by Gaffney, Tanner and Phillips (1984). Work by Kaplan and Gordon (1989) led to a number of interesting conclusions from dual isotopic measurements in Los Angeles, such as: 1) natural gas combustion was *not* a primary source of carbon particles; and 2) significant contemporary carbon contributions came from particle-forming photochemical reactions of organic vapors (emitted from vegetation and in cooking). Earlier studies of ^{14}C in Los Angeles aerosol have been reported by Berger, McJunkin and Johnson (1986) and Currie *et al* (1983). Further information on ^{14}C in atmospheric particles can be found in Cooper, Currie and Klouda (1981) and Ramdahl *et al* (1984). Finally, an important series of conferences, "Carbonaceous Particles in the Atmosphere," constitutes the prime colloquy for progress in their measurement, transformations, composition, effects and apportionment (Carbonaceous Particles in the Atmosphere 1978, 1983, 1987, 1991).

Isotopic standards are essential for the control of measurement accuracy in environmental carbon research. Figure 33.3 shows the ^{13}C-^{14}C plane for several

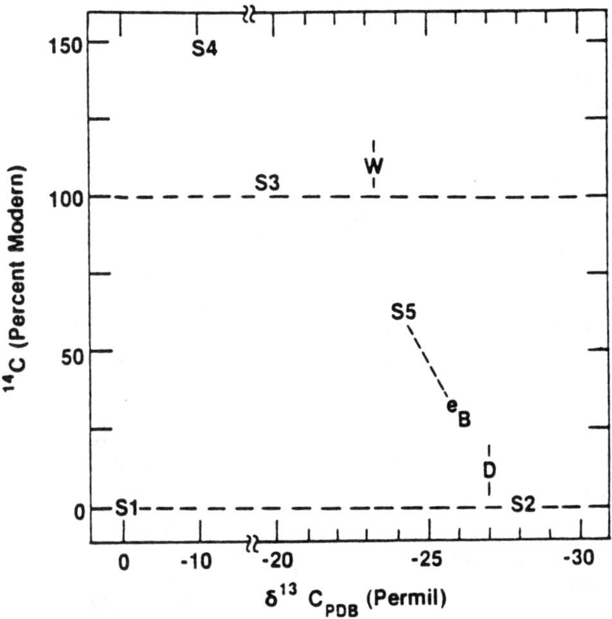

Fig 33.3. ^{13}C-^{14}C plane – standards (Currie 1984). The NIST (formerly NBS) urban particulate SRM #1648 illustrates isotopic heterogeneity: S5 (total carbon), e (elemental/graphitic) carbon. The proximate sample B represents aerosol collected in Barrow, Alaska (Arctic Haze). Other standards shown: for ^{13}C, S1 (PBD), S2 (RM21 graphite); for ^{14}C, S4 (ANU sucrose), S3 (SRM #4990B, oxalic acid dating standard). Other samples: W (woodsmoke aerosol); D (diesel aerosol).

of the more important standards, together with dual-isotope data for carbonaceous particles from woodburning (W), diesel exhaust (D) and the forementioned Barrow sample (B) (Currie 1984). Points beginning with *S* denote standards or reference materials used in measurement calibration for natural levels of ^{13}C (S1, S2) and ^{14}C (S3, S4), or for validation of atmospheric particulate carbon measurements (S5). The calibration materials generally indicate the end points for the respective isotopic concentrations in natural samples. The intermediate isotopic values for the urban particle reference material *S*5, for example, are consistent with this material deriving from a mixture of sources. This material exhibits a vital characteristic that must be recognized in environmental studies, namely "isotopic heterogeneity." The point is that the isotopic composition of complex environmental materials, such as atmospheric particles, may differ in different chemical fractions. For the standard reference material *S5* the elemental carbon fraction (e) differs substantially in both ^{13}C *and* ^{14}C from the average isotopic composition. Interestingly, the isotopic composition of the elemental carbon fraction, e, is remarkably similar to that of the Barrow sample (B), which represents long-range transport of industrial/urban soot.

PARTICULATE CARBON LIFE CYCLE

A broad review of the carbonaceous particle life cycle was presented at a conference convened a decade ago (Wolff & Klimisch 1982). More recently, NASA and the National Academy of Science have presented reviews of gas and particle cycles in the atmosphere, with a major goal of distinguishing and quantifying anthropogenic and natural sources (National Research Council 1984). A thorough examination of the abundance and history of black carbon in the environment has been given by Goldberg (1985), and a review of organic particulate sources, with emphasis on anthropogenic and natural source characterization and source apportionment has been prepared by Daisey, Lioy and Cheney (1986). Emphasis on the atmospheric cycle for natural aerosol was given in an article by Jaenicke (1980). Most recently, interest in global warming and biomass burning resulted in an international topical (Chapman) conference in March 1990 (Levine 1990).

Particle lifetimes in the atmosphere range from days to years, depending on particle size and composition, altitude of injection, physical and chemical interactions, and the nature of sinks. As a result, aerosol history and forecasting are strongly influenced by the distribution of source strengths. This means that we may have the opportunity to influence the future history (the anthropogenic component), and we may read the past history from the aerosol record in glaciers and sediments. Table 33.4 gives an overview of global sources (Prospero 1984). Here we see that the global natural aerosol exceeds the anthropogenic component by about a factor of five, but, as shown in the original publication, each component is uncertain by at least a factor of two. Total source strengths can be a little misleading, however, in view of spatial and compositional heterogeneity. Spatially, anthropogenic sources are concentrated in inhabited regions, especially the cities or other densely populated areas; compositionally, the impacts are strongly dependent on chemical toxicity, particle albedo, cloud interactions, etc. With these factors in mind, we see from Table 33.4 that anthropogenic emissions are dominated by sulfates (secondary aerosol), whereas natural emissions are dominated by sulfates, sea salt and dust. Carbonaceous aerosol derives primarily from hydrocarbon emissions and combustion, both of which have significant natural and anthropogenic sources. Hydrocarbon emissions have a major influence on visibility and regional ozone formation; combustion particles ("soot") have important albedo and toxicity effects. Carbon isotopes are most important for quantifying sources in each category. Complementing the global aerosol source inventory presented in Table 33.4 is an urban carbonaceous aerosol source inventory for Los Angeles, prepared by Cass, Boone and Macias (1982). The contrast is striking. This work suggests that anthropogenic emissions are responsible for at least 94% of the urban (Los Angeles) carbonaceous particle production, with mobile sources, alone, contributing 55%. About one-third of these urban emissions are ascribed to non-volatile (elemental) carbon, with 60% of that deriving from diesel vehicles. Total production is given as 46 metric tons of carbon per day. Next, we shall take a brief look at the principal stages in the life cycle of carbonaceous aerosol.

TABLE 33.4. Estimates of Global Particle Production (10^6 tons year^{-1})*

	Man-made	Natural
Direct production	30	780
	(industrial processes – 12)	(sea salt – 500)
	(stationary power – 10)	(windblown dust – 250)
	(transportation – 2)	(volcanoes – 25+)
	(miscellaneous - 6)	(forest fires – 5+)
Secondary (from gases)	250	470
	(sulfates – 200)	(sulfates – 335)
	(nitrates – 35)	(nitrates – 60)
	(hydrocarbons – 15)	(hydrocarbons – 75)
Total	280	1250

*Adapted from Prospero (1984). See the original publication for more detailed source information and uncertainty ranges.

Formation

Carbonaceous particles derive from: 1) direct ("primary") combustion processes, plus abrasion/suspension of mineral dust and vegetative material; and 2) "secondary" photochemical reaction processes on volatile precursors. The abrasion/suspension component (including vegetative particulate emissions, such as plant waxes and pollen) tends to be larger (≥ μm in diameter), and hence, has a more rapid settling time than the fine (0.1–1 μm) photochemical and combustion aerosol. Thus, these latter aerosol components are more consequential with respect to residence time, light scattering and inhalation (toxic) effects. Photochemical (and free radical) formation of particles is important with unsaturated organic species, such as alkenes and especially the naturally emitted terpenes and isoprene. Considerable debate continues over the influence of these naturally emitted vapors on urban and rural ozone formation. In the final section of this chapter, I shall refer to an experiment, currently underway, to help solve this question through the use of ^{14}C.

Primary combustion particles carry the memory of the carbonaceous source material (fuel) and/or that of the combustion process. Gentle processes, such as pyrolytic dehydrogenation, provide a good chance of source identification by examining the aerosol structure, which then reflects the source composition. A classic example is the dehydrogenation of abietanes, which constitute an important structural component of softwood rosin. An end product, retene, thus serves as a characteristic tracer for softwood combustion (Fig 33.4A) (LaFlamme & Hites 1978; Ramdahl 1983). High-temperature combustion produces acetylenic-free radicals that polymerize to form polycyclic aromatic hydro-carbons (PAH), characterized by larger and larger numbers of aromatic rings, terminating in graphitic microcrystals (Fig 33.4B) (Badger 1962; Charlson & Ogren 1982). Such structures occur also as a result of diagenesis (Blumer 1976; Mauney, Adams & Sine 1984). The distribution of ring sizes, and the extent and position of alkyl side chains, hold great promise for source apportionment and characterization (eg, temperature) of the formation process.

Transport, Reaction and Removal

An excellent introduction to aerosol transport appears in Jaenicke (1980). The essentials are given in Figure 33.5, where we see that the removal processes and the (inert) particle lifetimes are governed by the particle size. From the figure, and from Table 1 in Jaenicke (1980), we see that the decade of greatest consequence covers the size range of 0.1 to 1.0 μm. Such particles may be transported over distances of 8000 km and to altitudes of 20 km during their 2–3 week residence times. Smaller particles are removed by coagulation, whereas larger particles undergo rapid sedimentation. (Fine particles injected into the

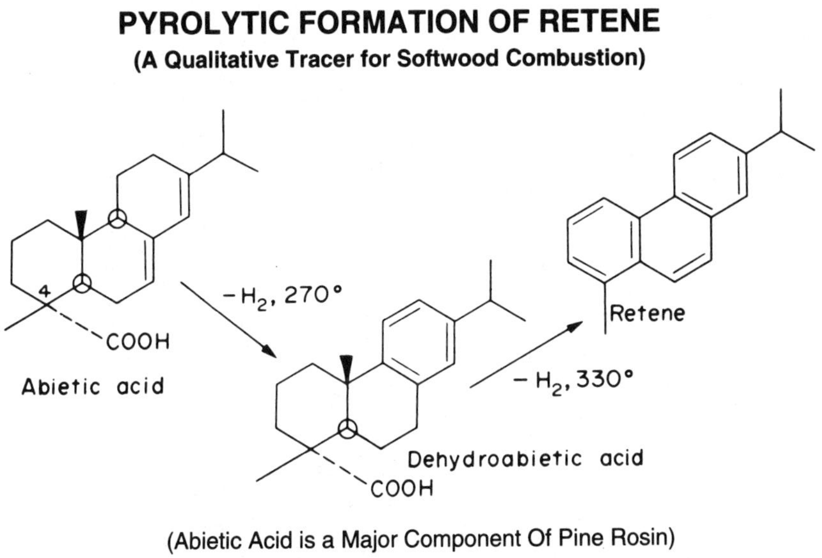

PYROLYTIC FORMATION OF RETENE
(A Qualitative Tracer for Softwood Combustion)
4
COOH
Abietic acid
−H2, 270°
Dehydroabietic acid
COOH
−H2, 330°
Retene
(Abietic Acid is a Major Component Of Pine Rosin)

Fig 33.4A. Pyrolytic formation of retene (LaFlamme & Hites 1978). (Reprinted by permission of the authors and Pergamon Press plc © 1978.)

PYROSYNTHESIS OF PAH AND SOOT
[Badger, 1962]
Polybenzenoid Radicals
[Kamens, 1982]
• OH
Cl
• R
Soot
[Graphitic Macromolecules (~99%-C)]

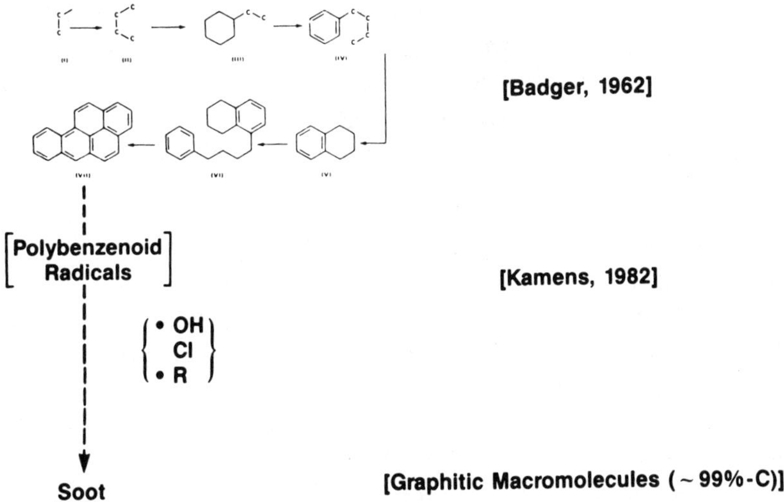

Fig 33.4B. Pyrosynthetic formation of PAH (Badger 1962) and soot (Kamens 1982)

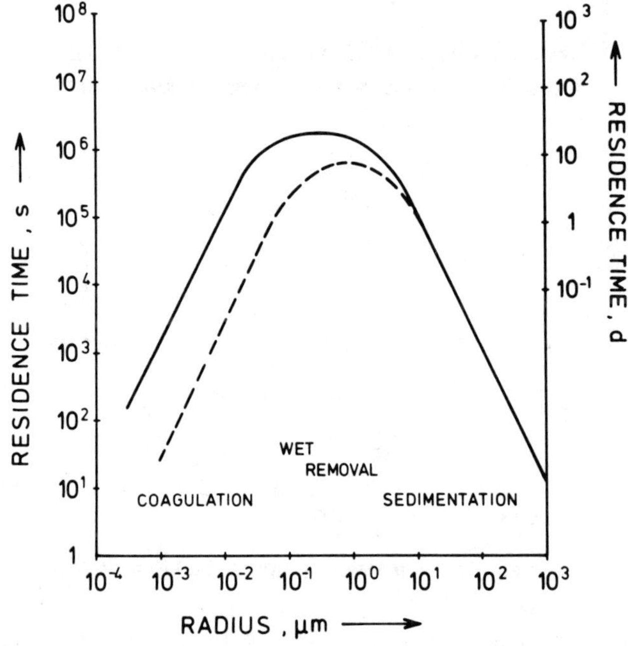

Fig 33.5. The residence time of aerosol particles. Two different cases have been calculated: background aerosol (——), number density (N) = 300 cm^{-3}, residence time τ_{wet} = 21 d; continental aerosol (– – –), N = 15,000 cm^{-3}; τ_{wet} = 8 d. In both cases, the height of the homogeneous aerosol layer was assumed to be 1.5 km. This only has effects on the sedimentation branch of the curve (Jaenicke 1980). (Reprinted by permission of the author and the Annals of the New York Academy of Sciences.)

stratosphere, eg, from nuclear weapon tests or volcanic action, may linger for several years.) Thus, aerosol from combustion and photochemical reaction is clearly a matter of global (or at least hemispheric) concern. An illustration of such long-range transport of combustion aerosol is the "Arctic Haze," which has been ascribed to industrial sources deep in the Soviet Union and central Europe.

Removal through cloud processing and chemical reaction is extremely complex, and obviously highly dependent on particle composition. This is one reason that conservative quantities, such as isotope ratios, are so important for reliable source apportionment. Cloud cycling will not be discussed here (for background information and references, see National Research Council (1984) and Rowland and Isaksen (1988)). Transformation through photochemical and oxidative reaction constitutes a major field of study, performed both in the laboratory and in polluted atmospheres. Accounting for such transformations, for example, through trajectory analysis and hybrid (non-linear) modeling permits us to: 1) identify reactive precursors, resulting in noxious products, such as the highly mutagenic nitro-aromatics; 2) pinpoint less reactive (and less volatile) species, such as certain of the PAH, which may serve as reliable source tracers; and 3) build chemical transformations into the source apportionment (receptor) modeling. An illustration of the last, given in Friedlander (1981), involves the use of first-order reaction constants for atmospheric PAHs.

Archival Records

Aerosol history is preserved in natural archives, such as ocean, lake and bog sediments, as well as glaciers and the polar ice caps. Because the resident concentrations are generally quite low, single particle or very sensitive methods of organic analysis are generally required. Illustrations of the records of combustion particles are found in the work of Hites (1981) and Griffin and Goldberg (1981). In the former work, trace analysis of PAH was performed on sediment cores taken from the Pettaquamscutt River and the Grosser Plöner Sea. In Figure 33.6, we see a dramatic rise in anthropogenic PAH in both of the locations, coinciding with the onset of the Industrial Revolution. The latter research utilizes single carbonaceous particle morphology, as opposed to the former's bulk particle chemical (PAH) analyses. Figure 33.7, from Griffin and Goldberg (1981), shows the same anthropological phenomenon (Industrial Revo-

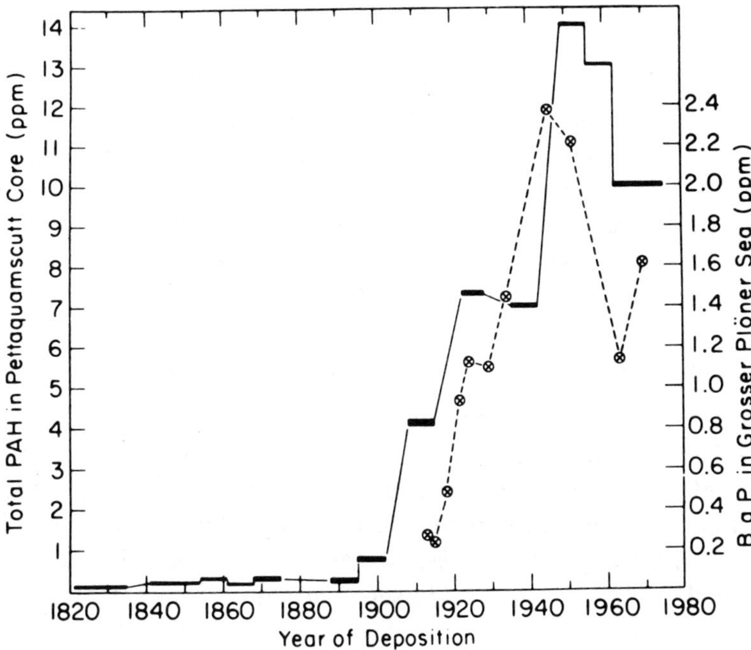

Fig 33.6. Combustion carbon (PAH, soot) from human activities: growth of PAH in Pettaquamscutt River sediment (—— left scale); growth of benzo(a)pyrene (BaP) in the Grosser Plöner Sea (– – – right scale). (Reprinted by permission of Hites (1981). © 1981 American Chemical Society.)

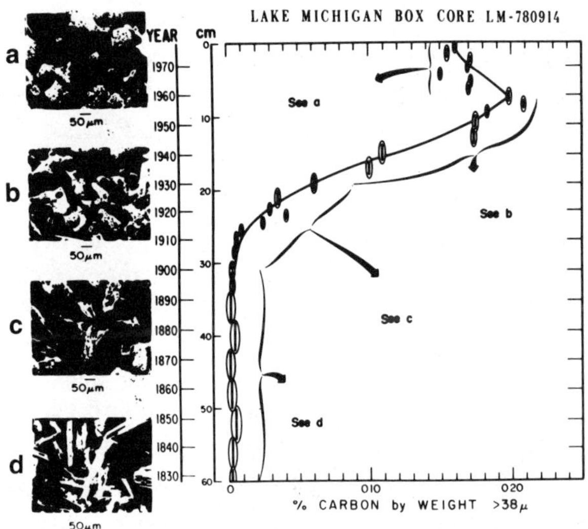

Fig 33.7. Charcoal particles in Core LM 780914 collected Sept 14, 1978 from SE Lake Michigan at 43°00′N and 86°22′W. The scanning electron micrographs show the particle morphologies common to four periods: a. 0–8 cm interval represents the particles in the sediments of the post-1960 period; b. 12–14 cm, 1930–1960; c. 27–28 cm, 1900–1930; d. 30–32 cm, pre-1900 (Griffin & Goldberg 1981). (Reprinted by permission of the authors and Pergamon Press plc © 1981.)

lution), this time in a box core from Lake Michigan. Of special interest in this case is the fact that the particle morphology preserves the structure or combustion fingerprint of the original fuel. Thus, we see not only the record of accelerated industrialization, we are able also to discern transitions in fuel type. Future studies of this sort, especially in polar ice cores, have enormous promise, as a result of advanced techniques of single-particle *compositional* analysis, such as laser microprobe mass spectrometry (Denoyer *et al* 1982).

INFORMATION CONTENT OF CARBONACEOUS PARTICLES

It was impossible to review the particulate carbon life cycle without making a number of references to characteristic carbonaceous particle composition and structure reflecting the respective sources, reactivity and modes of formation. This section, therefore, will complement the preceding one by including some systematic information on isotopes, elements, molecules and structural features that may aid in assigning origins to such particles.

Isotopes

Nothing is quite so pertinent to the origin and apportionment of environmental carbonaceous material as the carbon isotopic composition. For the biogenic/fossil source dichotomy, ^{14}C is the discriminator *par excellence*. Since atmospheric particles derive almost entirely from "living" or "dead" fuel (in terms of the 8300-year mean life of ^{14}C), explicit and quantitative apportionment of the carbonaceous material extracted follows directly. (Because of nuclear testing, the age distribution of "living" fuel, such as firewood, must be taken into account for second-order corrections (Currie, Klouda & Gerlach 1981). Limitations in the application of ^{14}C are twofold: ability to perform sufficiently sensitive measurements, with respect to sample size; and ability to discriminate within the respective (fossil, biogenic) source classes. Miniature counters, and more recently AMS, have largely overcome the first limitation – though the ability to determine ^{14}C in ng of individual organic compounds or fg of individual carbonaceous particles would be very interesting. The second limitation is overcome, in part, by ^{13}C. Marine carbon, inorganic carbon and carbon deriving from different photosynthetic cycles can be distinguished with this stable isotope. The pair of isotopes, as shown in the first part of the chapter, serves as a very useful two-dimensional discriminating tool. Direct source attribution with carbon isotopes depends on extracting the organic material of primary concern, such as benzene, graphitic soot, PAH, etc. Measurement of ^{14}C then gives direct information on the material extracted (Currie *et al* 1989). Indirect methods, where ^{14}C is employed to "calibrate" chemical patterns, or as an intrinsic variable in multivariate analysis, will be discussed in the next section.

Non-carbonaceous isotopes do not trace carbon, except by association. There have been some notable illustrations, however, where this associative link has been effectively used. To the extent that emissions from individual sources "travel together," it is meaningful to utilize alternative isotopic or molecular tracers to better understand the overall system. If all of the tracers involved are conservative, *ie*, the system can be described by a linear model, then the associative assumption is valid. Caution is in order, however, in attempting to combine data across chemical species, or between gas and particle phases. Excellent reviews of the application of a number of stable and radioactive isotopes to geochemical and atmospheric problems can be found in Fritts and Fontes (1980). The underlying principle, of course, is that different source materials exhibit different isotopic compositions, notable examples being oceanic *vs* combustion sulfate sulfur, and synthetic fertilizer *vs* atmospheric nitrogen. Lead isotopes are quite interesting also, in that ore deposits having different geologic history exhibit different Pb-206/207/208 ratios.

A fascinating illustration of the application $^{206}Pb/^{207}Pb$ to the apportionment of lead aerosol between the US and Canada has been given by Sturges and Barrie (1987). The aerosol lead sources, attributed to motor vehicles in both nations, showed significant differences in this isotopic signature, examples being 1.213 ± 0.008 (SD) for the eastern US, *vs* 1.148 ± 0.007 for Ontario sites. The differences were ascribed to different Pb ore bodies. A remarkably similar study took place in Turin, for the purpose of identifying aerosol lead sources in the local atmosphere. This study differed from the one above, in that the stable isotope tracer was *purposely added* to the motor vehicle fuel (using again a different source for the lead ore), in order to test for consequent changes in aerosol lead and serum lead of inhabitants (Facchetti & Geiss 1984). The change in the $^{206}Pb/^{207}Pb$ ratio in adult blood, following the tracer injection, showed that about 30% of the serum lead was of local airborne origin.

An analogous study, just now nearing completion, involved the labeling of fuel oil with an unnatural mixture of neodymium isotopes, as part of a multisource combustion particle source apportionment study in Roanoke, Virginia (W R Kelly, personal communication 1989). This urban study was especially interesting since it *combined information* from ^{14}C data with Nd isotopic data for quantitative apportionment of aerosol carbon among the three principal (wintertime) sources: motor vehicles, fuel oil combustion and woodburning.

Other cosmogenic isotopes bring additional information. For example, ^{7}Be, already suggested by Arnold in 1955, has been applied to the study of aerosol transport and deposition by Olsen *et al* (1985). By combining data from this cosmogenic isotope with that from ^{210}Pb, which also attaches to atmospheric aerosols following its radiogenic formation, these authors drew interesting conclusions about wet *vs* dry fallout.

Elemental and Molecular Tracers

Information concerning aerosol sources is most frequently drawn from the elemental composition of the ambient particles, with organic composition becoming more and more used. The entire field of aerosol source apportionment (or "receptor modeling") evolved through the use of linear (least squares or factor analysis) modeling of multiple elemental variables, using "source signatures" as identifying features (Hopke 1985; Gordon 1988). The reason this method succeeds is that geochemical and biological processes enrich certain elements in the respective fuels and mineral and plant material (vegetation). Further enrichment may then take place in the emission or combustion process, depending on chemical properties and volatility. An illustration is given in Figure 33.8 (McElroy *et al* 1982). An example of elemental enrichment is given by potassium. Its concentration in vegetation is enhanced compared to certain fossil fuels, and aerosol potassium from (wood) combustion is contained primarily in much smaller particles than those from resuspended mineral dust. Thus, potassium, combined with particle size discrimination, can serve as an effective tracer for woodburning in the presence of motor vehicle emissions and resuspended mineral dust (Lewis & Einfeld 1985). A problem is that the potas-

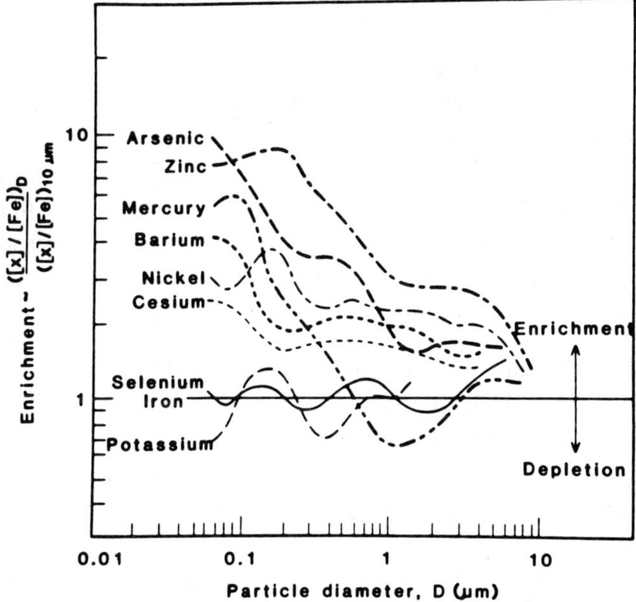

Fig 33.8. Enrichments of trace metals in emissions from coal-burning boiler outlets. Enrichment is defined as the concentration ratio of the element to iron, at a specific particle diameter, divided by the ratio at 10 μm (McElroy *et al* 1982). (Reprinted by permission of the authors and *Science*. © 1982 by the AAAS.)

sium content of wood is not fixed. It varies with geographic location and the nature of the vegetation burned. Concurrent measurements of ^{14}C, an absolute biomass burning tracer, have been central to the calibration and validation of potassium as a regional secondary woodburning tracer (Currie, Beebe & Klouda 1988). Molecular patterns ("molecular markers"), especially those associated with congeners and homologous series of organic molecules, are showing increasing promise for carbonaceous aerosol source identification. The class of PAH compounds has received the most study, perhaps because they are produced in all (incomplete) combustion; they form a very large class having a number of subtle structural differences; and some of their members are extremely important from the perspective of mutagenic and carcinogenic effects. A discussion of a "calibration" field experiment, combining ^{14}C and PAH measurements, was described recently by Currie *et al* (1989). Besides the PAH, and the special member retene, there appears to be great promise in woodburning tracing with methoxylated phenols (Hawthorne *et al* 1988), and odd-even alkane carbon abundance patterns for tracing naturally-emitted plant material (Simoneit 1984).

Structural Information

Structural insights on carbonaceous particles can be derived from several types of analytical data such as chemical shifts seen with NMR (Cooper & Malek 1981), microRaman spectroscopy (Schrader 1986), and laser microprobe mass spectrometry (LAMMS) (Denoyer *et al* 1982). All of these techniques provide information on large-scale, polymeric molecular structure, and hence, they are useful in giving structural clues for organic material formed in pyrodegration and pyrosynthesis. Some additional information on these techniques, and their combination with ^{14}C, will be given in the next section.

Information derived from large assemblages of individual atmospheric particles, using microanalytical electron and ion microprobe techniques, has also yielded important insight into particle sources in both urban and remote settings. A factor-analytic method known as "particle class balance" has successfully combined morphological data (shape, density) on individual particles with microanalytical elemental data to apportion particles in urban atmospheres (Johnson & McIntyre 1983). Pure morphological individual particle source identification has been possible in some situations, one of the most notable being the recognition of unique morphologies from coal, oil and wood combustion particles found in lake sediment (Griffin & Goldberg 1979) (see Fig 33.7).

EXTRACTING THE INFORMATION – CASE STUDIES

From the preceding discussion, it becomes clear that multifaceted information on the origins and history of carbonaceous particles is captured in their isotopic, elemental, organic and structural composition. The ability to sort out particle sources, and especially to distinguish the anthropogenic and natural components, can be greatly enhanced by astutely bringing together these several types of information. To briefly illustrate such multivariate means for augmenting the source apportionment information deriving from ^{14}C is the main objective of the present section. In much of the prior work with ^{14}C as a source identifier, it has been employed to apportion the carbon *in* a given material or sample by *direct* measurement. The multivariate techniques differ in that isotopic data are used in *parallel* with other kinds of compositional data for calibration (eg, using multiple linear regression) or for multivariate correlation studies (eg, using factor analysis). The parallel approach is especially interesting, in that it 1) uses all of the compositional information, and 2) indirectly extends the influence of ^{14}C to the level of nanograms and even individual carbonaceous particles. An introduction to the direct (or serial) *vs* indirect (or parallel) application of ^{14}C was presented at the Dubrovnik Radiocarbon Conference (Currie *et al* 1989).

The Arctic Haze

In the spring of each year, the remote, Arctic atmosphere loses some of its pristine character, and shows large increases in particle loading, the (carbonaceous) particles being remarkably like those found in urban, industrial regions. In Figure 33.3, we saw already that the ^{14}C content of such particles from Barrow, Alaska, showed the particulate carbon to be largely fossil in origin, at a level quite comparable to that of typical urban aerosol (Table 33.3). Even more remarkable is the near coincidence of the dual (^{13}C, ^{14}C) isotopic composition of this material with the elemental or "graphitic" component of the urban particle standard reference material (SRM) (collected in St Louis), also shown in Figure 33.3. At this point, it is interesting to take a closer look at the graphitic structural information. Figure 33.9 shows the twin Raman spectral bands that characterize the solid-state, graphitic structure (Rosen *et al* 1982). It is quite clear that polycrystalline graphite, an end-product of high-temperature pyrolytic combustion, is present in both the Arctic and urban (St Louis) soot.

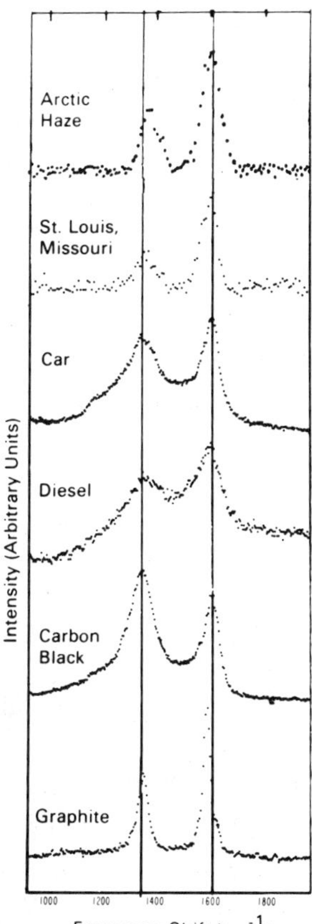

Fig 33.9. Raman spectrum of an Arctic sample compared to those of urban particulates, various source emissions, carbon black and polycrystalline graphite (Rosen *et al* 1982). (Reprinted (with Fig 33.10A) by permission of the authors and Plenum Publishing Corporation.)

MicroRaman spectroscopy is one of several individual particle analytical techniques that can perform synergistically with ^{14}C measurements on the bulk particulate material. Another, also suited to individual carbonaceous particle analysis, is LAMMS. In this technique, neutral (photon) excitation and ionization produces positive and negative mass spectra of even the most refractory material. As shown in Currie *et al* (1989), carbon ion cluster patterns are indicative of combustion particle sources, and show some promise for ^{14}C calibration of selected individual particle LAMMS spectra.

Returning to the Arctic Haze, it is useful also to consider time series, compositional data. Figures 33.10A and 33.10B give two such records (Rosen *et al* 1982; Rahn & McCaffrey 1980). Three indicators of industrial particulate pollution: graphitic carbon, excess sulfate and excess vanadium, show similar seasonal trends – all indicative of an influx of urban pollutants during late winter and early spring. ("Excess" sulfate and vanadium denote non-marine and non-crustal components, respectively.) Not shown is the total aerosol concentration, but this, too, exhibits a major seasonal trend, decreasing by a factor of ten or more during the summer. The clear, industrial pollution of the Arctic is an example of long-range transport. It is made possible by two factors: the long residence time of the fine combustion aerosol (Fig 33.5), and meteorological patterns that transport emissions from central Europe or central Asia to the remote Arctic. This link has been established using back trajectory analysis.

Designed, Multitracer Urban Studies

In contrast to the evolving evidence of long-range transport of aerosol pollution to the Arctic, are carefully planned studies of urban regions, where the goal is to achieve quantitative apportionment of carbonaceous particles, and the associated aerosol mutagenicity (more correctly: genotoxicity). Such is the objective of the *Integrated Air Cancer Project* of the USEPA (Stevens *et al* 1990). The nature of the design, and the selection of multiple tracers (including ^{14}C), set this investigation apart from many others. Cities and seasons (*viz* winter) were selected, such that primarily two (Albuquerque, Boise) or three (Roanoke) combustion sources predominated, with particulate pollution levels frequently exceeding prescribed limits. In the dual source studies, ^{14}C gave unique, quantitative discrimination between motor vehicle carbon (aerosol) and that produced by woodburning. The later, triple source experiment utilized enriched Nd isotopes, added to local fuel oil in Roanoke, to provide further quantitative discrimination of that source from the other two (Kelly & Ondov 1990). Thus, stable isotope spectrometry was linked with AMS to obtain unambiguous apportionment of the three primary carbonaceous aerosol sources.

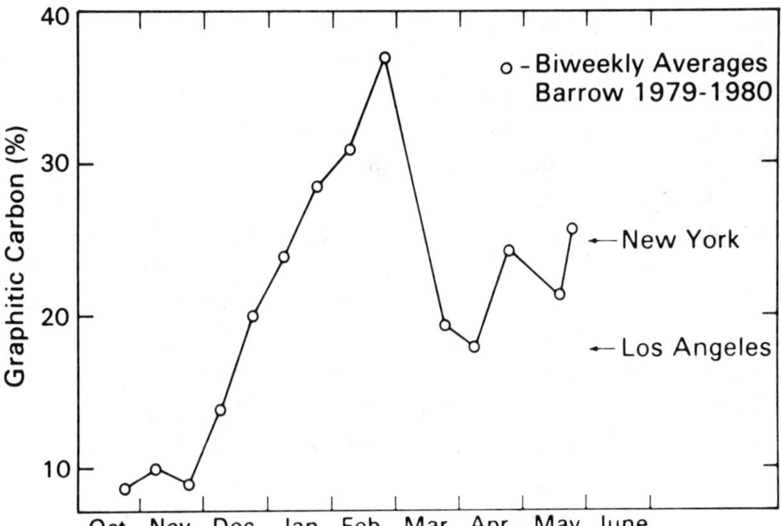

Fig 33.10A. Seasonal variation of graphitic carbon as a percentage of the total carbon content of the aerosol at Barrow from October 1979 to May 1980. These values are compared to the average values obtained in New York City and Los Angeles (Rosen *et al* 1982).

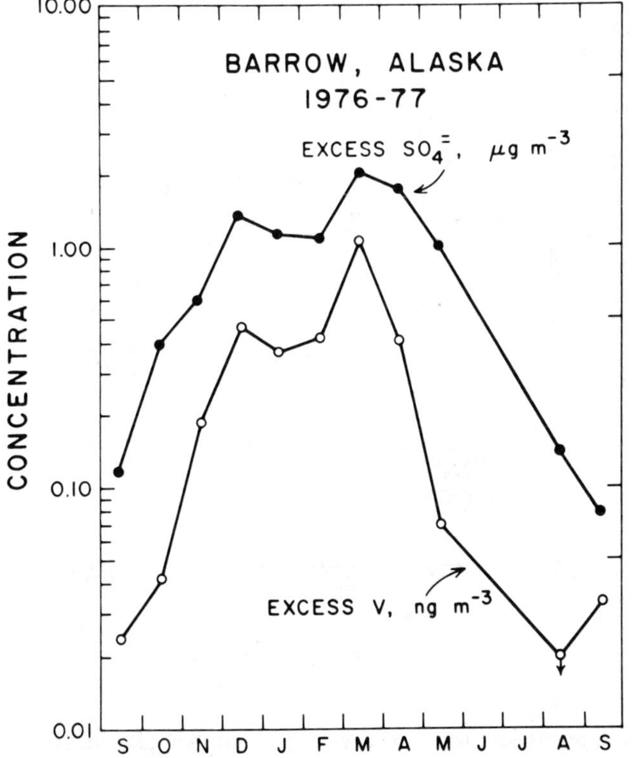

Fig 33.10B. Monthly mean concentrations of excess sulfate and vanadium at Barrow, Alaska (1976–1977) (Rahn & McCaffrey 1980). (Reprinted by permission of the authors and the Annals of the New York Academy of Sciences.)

In addition to the isotopic data were meteorological data, considerable information on elemental and organic (especially PAH) composition, plus mutagenicity; all of this produced not only quantitative apportionment of combustion sources and their mutagenic impact, but also new information on source "fingerprints" and calibration/validation data on inexpensive elemental tracers (soil-corrected K, Pb) by ^{14}C (Lewis & Einfeld 1985). Total particulate ^{14}C showed large differences in source impacts: at the Albuquerque traffic intersection site (daytime), there was only 35% contemporary carbon in the aerosol; at the residential site (nighttime), the corresponding figure was 95%. Illustrations of the "calibration" (regression) function of ^{14}C – for 1) mutagenicity, 2) an elemental source tracer, and 3) an organic source tracer – are given in Figure 33.11. Figure 33.11A shows the relation between mutagenicity and fraction of wood carbon (^{14}C derived); this type of analysis was critical in deriving the con-

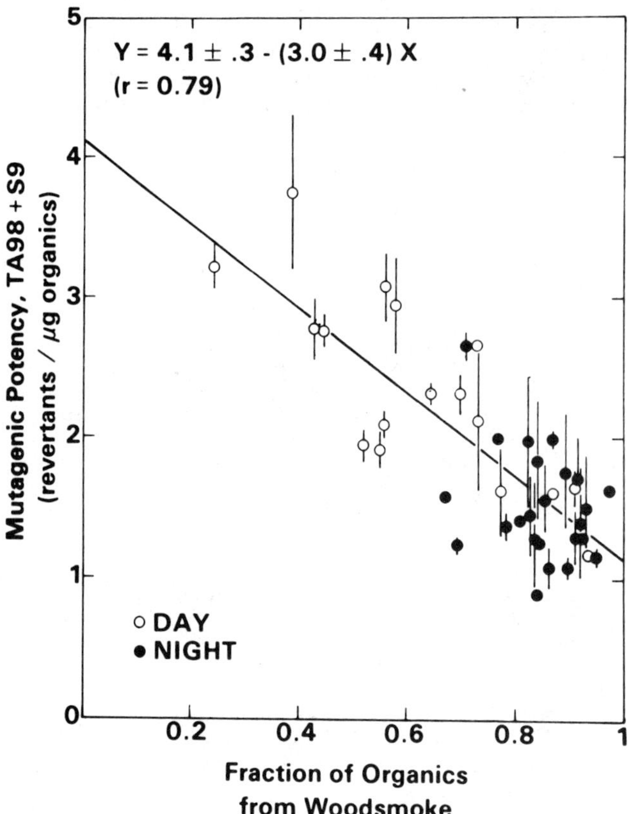

Fig 33.11A. Linear regression, aerosol mutagenicity *vs* fraction of extractable (aerosol) organic matter from woodsmoke. (Reprinted by permission of Lewis *et al* (1988). © 1988 American Chemical Society.)

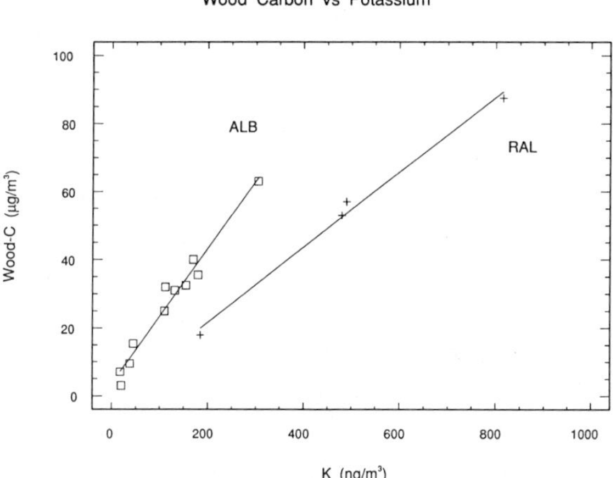

Fig 33.11B. Linear regression, aerosol carbon from wood *vs* mineral-corrected potassium; Albuquerque and Raleigh

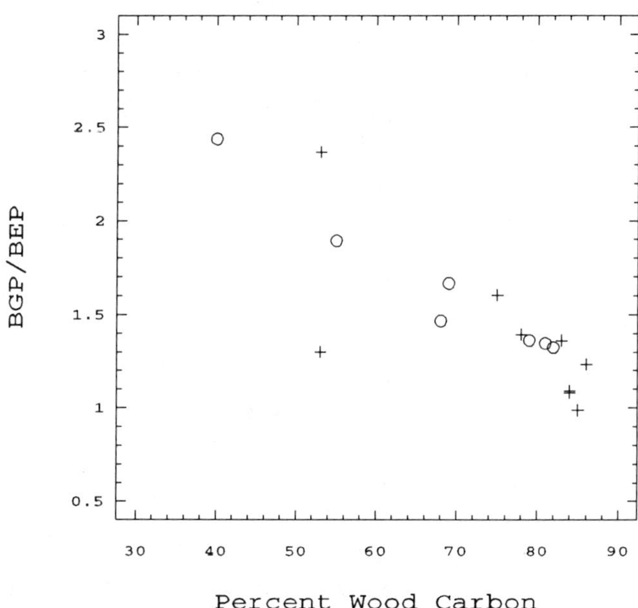

Fig 33.11C. Normalized PAH, Benzo(ghi)Perylene/Benzo(e)Pyrene [BGP/BEP] *vs* percent wood carbon; Albuquerque aerosol (adapted from Currie *et al* 1989) (o = traffic intersection sites; + = residential)

clusion that the mutagenic potency of carbonaceous aerosol (extracted organic matter) from motor vehicles was about three times that from woodburning (Lewis *et al* 1988). Figure 33.11B shows similar calibration curves for woodburning potassium from two cities. These curves support the use of potassium as a woodburning tracer, but they also indicate that it must be calibrated (eg, with ^{14}C), because of varying wood-potassium contents in different regions. Figure 33.11C, adapted from Currie *et al* (1989), shows a similar linear relation between ^{14}C source apportionment and a normalized 6-ring PAH, benzo(ghi)perylene [B(ghi)P]. Benzo(e)pyrene, which is found in the soot from both motor vehicle exhaust and woodburning, was used for normalization. Analogous behavior of this ratio was found in a study (not involving ^{14}C) comparing Bavarian cities having quite different motor vehicle population densities (Steinmetzer, Baumeister & Vierle 1984). In each case, based on quite different models, B(ghi)P appears to be a tracer for motor vehicle aerosol.

Multivariate Source Patterns

Multivariate methods of aerosol source apportionment are the norm, especially in their graphical (exploratory) mode, and in urban and regional studies where severe particulate pollution levels must be diagnosed and controlled. Such methods are very important because they are suited to the multispecies synergism mentioned above; and they have the potential to give estimates for the number of sources, their chemical profiles and their intensities – all from a single data set. In nearly all cases, source apportionment is achieved by means of multi-element chemical analysis followed by receptor modeling techniques based on regression or factor analysis (Hopke, 1985). Unique source tracers, like ^{14}C, are rare; though Pb, V and mineral-corrected K are often used as indicators of motor vehicles, fuel oil and woodburning, respectively. Also rare, is the direct incorporation of ^{14}C data into the multivariate data analysis process – though it has been shown to have a potentially large impact on the accuracy of the results, because of multicollinearity reduction (Lowenthal *et al* 1987).

To illustrate, very briefly and graphically, the combination of ^{14}C with multi-element data, I refer to a study discussed from a univariate perspective at the Trondheim Radiocarbon Conference (Currie *et al* 1986). A more complete, quantitative analysis will be presented elsewhere. The pollution problem giving rise to the study involved significant use of wood as a fuel in the town of Elverum, Norway. Even though the town (and its motor vehicle) population was but a few percent of that of Oslo, the wintertime PAH pollution levels were comparable (Ramdahl *et al* 1984). The question for ^{14}C, addressed in Currie *et al* (1986) was, how much of the carbonaceous pollution was due to woodburning.

TABLE 33.5. Multivariate Particulate Data: Elverum, Norway*

Sample	f_w	C	K	Pb	Fe	SO_4	Mn	V
6	63	15	121	263	20	730	4	1
7	78	50	424	607	33	1540	6	1
8	64	30	328	497	62	1620	10	2
9	64	9	197	213	27	2710	4.5	8
10	32	5.1	218	54	33	2800	6	6
11	62	6.9	140	160	72	2320	6.5	14
12	54	12	179	262	57	910	4.5	2
13	65	26	246	547	179	1300	8	2
14	83	22	231	325	56	1490	2.5	2
15	67	25	245	419	110	1410	4	2
16	26	11	339	94	150	5340	15.5	17
17	61	10	99	70	59	310	2	1

*Units: f_w (percent woodcarbon); C ($\mu g/m^3$); all others (ng/m^3)

The multivariate perspective on the Elverum study begins with the data (Table 33.5) for ^{14}C, carbon and six other chemical species. One can imagine numerous ways to seek relations among the several variables, such as correlation coefficients, and 2- and 3-dimensional scatterplots; but such low dimensional comparisons can be misleading as well as tedious. For example, to examine all 3-variable relations in the data, one must construct 56 three-dimensional plots; and even these would not convey relations among four or more variables. A remarkably efficient way to visualize high-dimensional space is found in principal component analysis (PCA), and its offspring, factor analysis (Malinowski & Howery 1980). A complementary technique, which examines samples for multivariable similarity, is cluster analysis. Both methods have been applied in an exploratory sense to the data in Table 33.5, *after normalizing* all species to carbon. Element ratio normalization, which is analogous to isotope ratio normalization, reduces the effective number of variables from 8 to 7. The result is that seven of the samples (#6–8 and #12–15) appear to form an "urban cluster," the remaining samples apparently being associated with a background site and with long-range transport of pollutants. Applying PCA, to understand the internal structure of this cluster of samples, we find a pattern in two dimensions, as shown in Figure 33.12. (Vectors represent the normalized variable directions; points represent the individual samples.)

What Figure 33.12 represents is the "best" planar view of 7-variable space. In fact, if there were only three primary sources leading to this cluster of particles, the 2-dimensional projection shown here would convey essentially all of the systematic (source apportionment) information for these samples. The clustering

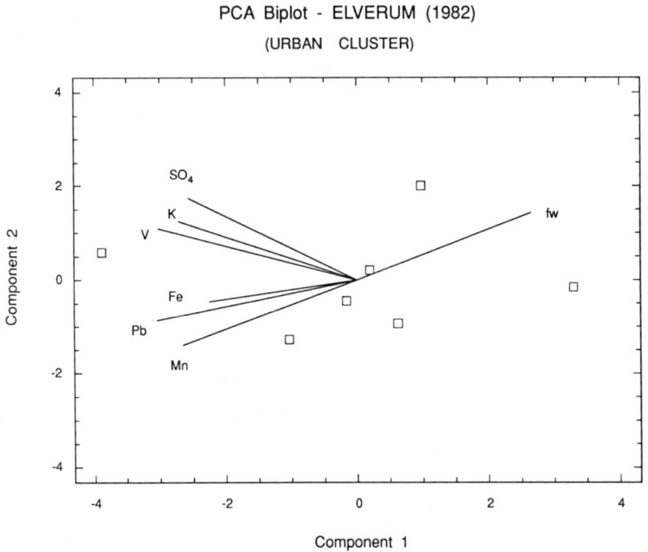

Fig 33.12. Principal component display (2 dimensions) of the Elverum aerosol "urban cluster," based on 7-dimensional (elemental + ^{14}C) data. The three sets of correlated variables imply motor vehicle (Mn, Pb, Fe), fuel oil (V, K, SO_4) and wood (f_w) aerosol sources, respectively.

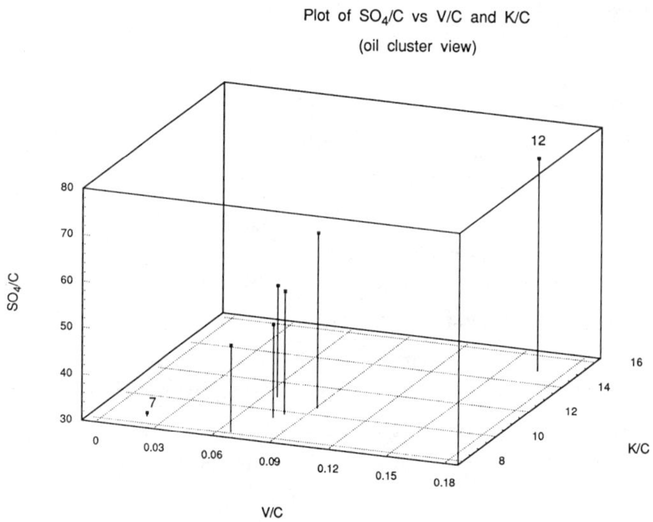

Fig 33.13. Three-dimensional "fuel oil" subspace (of the full 7-dimensional space) for the Elverum aerosol (urban cluster). A 1-dimensional linear (mixture) model is implied. (A 4th dimension/variable, f_w, is indicated: samples 7 and 12 have large and small values, respectively, for this variable.)

of variables in Figure 33.12 gives some visual clues as to the source signatures. The known tracers for wood (^{14}C), motor vehicles (Pb) and fuel oil (V) give some credibility for identification of these sources – consistent with external information about the locale (Ramdahl *et al* 1984). The importance of ^{14}C in combination with the other variables is underlined by its location at an extreme of the first principal component (x-axis). The rays in Figure 33.12 also suggest less mysterious (real variable) projections. We may examine the 7-variable data, for example, using the 3 normalized variables in the fuel oil direction, V, K and SO_4. This is shown in Figure 33.13, which supports the idea of a linear relationship for these 3 variables and the relative contribution of the fuel oil source. (Note that the probability of selecting the "right set" of 3 variables from 7, by chance, is only about 3%.) These exploratory, graphical analyses show the importance of treating the multivariable data as a whole; and they point the way toward quantitative, algebraic multicomponent source apportionment.

OUTLOOK - BEYOND PARTICULATE CARBON

This review has dealt primarily with application of radiocarbon, often in concert with other isotopes or other chemical or physical data, to quantify sources of carbonaceous particles. Discrimination between anthropogenic sources, which we may be able to control, and natural sources, which we cannot, is one of the main objectives. The actual system with which we must deal, however, is not neatly compartmentalized into particulate carbon sources and gaseous carbon sources. Nor is it characterized simply by conservative chemical species. The particulate carbon cycle is, in fact, linked to the gaseous carbon cycle through gas-to-particle conversion involving volatilization, condensation and photochemical reaction. These real-world complications are especially prominent in the non-methane hydrocarbon (NMHC)/ NO_x/ OH/ tropospheric O_3 system. Major concerns exist regarding the influence of natural NMHCs *vs* anthropogenic NMHCs on urban and rural photochemical ozone production. A major field experiment is underway (summer 1990), in Atlanta (Stevens & Meyers 1989). Radiocarbon has been assigned a critical role in apportioning NMHC source carbon in this very complicated, non-linear system.

The tropospheric ozone problem, and its relation to natural and anthropogenic carbonaceous emissions, was mentioned at the beginning of this chapter. At this point, it is appropriate to note that this is one of the most important regional problems that carbon isotope measurements can help resolve. The basic issue relates to the relative importance of biogenic emissions – terpenes and isoprene, from forested areas – in regions also having large quantities of anthropogenic (fossil) NMHC emissions. The complexity of the system and the diversity of views regarding ozone control are well captured by Chameides *et al* (1988), who suggest that a prescribed 30% anthropogenic NMHC reduction in Atlanta, using emission controls, might fall short by a factor of two or more, because of the non-linearity of the system plus uncertainties in the natural NMHC component.

The tropospheric ozone study is perhaps an indicator for future atmospheric carbon isotope research. Radiocarbon, with its unique fossil-biospheric carbon discriminating power, will surely be applied to the whole range of volatile and particulate carbon species. Large-scale measurements of this sort will provide increasingly reliable information on humankind's influence on the natural carbon environment.

ACKNOWLEDGMENT

This article is a contribution from the National Institute of Standards and Technology, and is not subject to US copyright.

REFERENCES

Badger, GM 1962 Mode of formation of carcinogens in human environment. *National Cancer Institute* 9: 1–6.

Berger, R, McJunkin, D and Johnson, R 1986 Radiocarbon concentrations of California aerosols. *In* Stuiver, M and Kra RS, eds, Proceedings of the 12th International ^{14}C Conference. *Radiocarbon* 28(2A): 661–667.

Blumer, M 1976 Polycyclic aromatic compounds in nature. *Scientific American* 234: 34–45.

Brimblecombe, P 1981 Environmental impact of fuel changes in early London. *In* Cooper, JA and Malek, D, eds, *Residential Solid Fuels*. Beaverton, Oregon Graduate Center: 1–11.

Cachier, H 1989 Isotopic characterization of carbonaceous aerosols. *Aerosol Science and Technology* 10: 379–385.

Carbonaceous Particles in the Atmosphere 1978, 1983, 1987, 1991 Conference Proceedings (1978) Novakov, T, ed, Berkeley/ Los Angeles, University of California Press; (1984) Malissa, H, Puxbaum, H and Novakov, T, eds, *Science of the Total Environment* 36; (1989) Novakov, T, Malissa, H and Hansen, ADA, eds, *Aerosol Science and Technology* 10(2); the next in the series is scheduled for April 1991.

Cass, GR, Boone, PM and Macias, ES 1982 Emissions and air quality relationships for atmospheric carbon particles in Los Angeles. *In* Wolff, GT and Klimisch, RL, eds, *Particulate Carbon. Atmospheric Life Cycle*. New York, Plenum Press: 207–244.

Chameides, WL, Lindsay, RW, Richardson, J and Kiang, CS 1988 The role of biogenic hydrocarbons in urban photochemical smog. Atlanta as a case study. *Science* 241: 1473– 1475.

Charlson, RJ and Ogren, JA 1982 The atmospheric cycle of elemental carbon. *In* Wolff, GT and Klimisch, RL, eds, *Particulate Carbon. Atmospheric Life Cycle*. New York, Plenum Press: 3–18.

Clayton, GD, Arnold, JR and Patty, FA 1955 Determination of sources of particulate atmospheric carbon. *Science* 122: 751–753.

Cooper, JA, Currie, LA and Klouda, GA 1981 Assessment of contemporary carbon combustion source contributions to urban air particulate levels using C-14 measurement. *Environmental Science and Technology* 15: 1045–1050.

Cooper, JA and Malek, D, eds 1981 *Residential Solid Fuels*. Beaverton, Oregon Graduate Center: 1271 p.

Court, JD, Goldsack, RJ, Ferrari, LM and

Polach, HA 1981 The use of carbon isotopes in identifying urban air particulate sources. *Clean Air*, February: 6–11. (This was the first of a series of Sydney and Canberra air pollution studies by Polach and coworkers.)

Currie, LA 1984 ^{14}C as a tracer for carbonaceous aerosols: Measurement techniques, standards, and applications. *In* Liu, BYH, Pui, DYH and Fissan, HJ, eds, *Aerosols*. New York, Elsevier: 375–378.

Currie, LA, Beebe, KR and Klouda, GA 1988 What should we measure? Aerosol data: Past and future. *In* Proceedings of the 1988 EPA/APCA International Symposium on Measurement of Toxic and Related Air Pollutants. *Air Pollution Control Association*: 853–863.

Currie, LA, Klouda, GA, Continetti, RE, Kaplan, IR, Wong, WW, Dzubay, TG and Stevens, RK 1983 On the origin of carbonaceous particles in American cities: Results of radiocarbon "dating" and chemical characterization. *In* Stuiver, M and Kra, RS, eds, Proceedings of the 11th International ^{14}C Conference. *Radiocarbon* 25(2): 603–614.

Currie, LA, Klouda, GA and Gerlach, RW 1981 Radiocarbon: Nature's tracer for carbonaceous pollutants. *In* Cooper, JA and Malek, D, eds, *Residential Solid Fuels*. Beaverton, Oregon Graduate Center: 365–385.

Currie, LA, Klouda, GA, Schjoldager, J and Ramdahl, T 1986 The power of ^{14}C measurements combined with chemical characterization for tracing urban aerosol in Norway. *In* Stuiver, M and Kra, RS, eds, Proceedings of the 12th International ^{14}C Conference. *Radiocarbon* 28(2A): 673–680.

Currie, LA, Klouda, GA and Voorhees, KJ 1984 Atmospheric carbon: The importance of accelerator mass spectrometry. *Nuclear Instruments and Methods* 233: 371–379.

Currie, LA and Murphy, RB 1977 Application of isotope ratios to the determination of the origin and residence times of atmospheric pollutants *In* Proceedings of the Eighth Materials Research Symposium, National Bureau of Standards. *NBS Special Publication* 464: 439–447.

Currie, LA, Stafford, TW, Sheffield, AE, Klouda, GA, Wise, SA, Fletcher, RA, Donahue, DJ, Jull, AJT and Linick, TW 1989 Microchemical and molecular dating. *In* Long, A and Kra, RS, eds, Proceedings of the 13th International ^{14}C Conference. *Radiocarbon* 31(3): 448–463.

Daisey, JM, Lioy, PJ and Cheney, JL 1986 Profiles of organic particulate emissions from air pollution sources: Status and needs for receptor source apportionment modeling. *Journal of the Air Pollution Control Association* 36: 17–33.

Denoyer, E, Van Grieken, R, Adams, F and Natusch, DFS 1982 Laser microprobe mass spectrometry 1: Basic principles and performance characteristics. *Analytical Chemistry* 54: 26A–41A.

Facchetti, S and Geiss F 1984 Isotopic lead experiment. EEC, Brussels, *Report EUR 8352 EN*. CEC Joint Research Centre, Ispra.

Fields, DE, Cole, LL, Summers, S, Yalcintas, MG and Vaughn, GL 1989 Generation of aerosols by an urban fire storm. *Aerosol Science and Technology* 10: 28–36.

Flam, F, ed 1991 Briefings – A brighter forecast from Kuwait. *Science* 253: 28–29.

Friedlander, SK 1981 New developments in receptor model theory, Chapter 1. *In* Macias, ES and Hopke, PK, eds, Atmospheric Aerosol: Source/Air Quality Relationships. Washington, DC, American Chemical Society, *Symposium Series* 167.

Fritts, P and Fontes, J Ch 1980 *Handbook of Environmental Isotopic Geochemistry* 1. Amsterdam, Elsevier: 545 p.

Gaffney, JS, Tanner, RL and Phillips, M 1984 Separating carbonaceous aerosol source terms using thermal evolution, carbon isotopic measurements, and C/N/S determinations. *Science of the Total Environment* 36: 53–60.

Galbally, IE, ed 1989 *The International Global Atmospheric Chemistry (IGAC) Programme*. Commission on Atmospheric

Chemistry and Global Pollution of the International Association of Meteorology and Atmospheric Physics: 55 p.

Goldberg, ED 1985 *Black Carbon in the Environment.* New York, Wiley-Interscience.

Gordon, GE 1988 Receptor models. *Environmental Science and Technology* 22: 1132–1142.

Griffin, JJ and Goldberg, ED 1979 Morphologies and origin of elemental carbon in the environment. *Science* 206: 563–565.

_____1981 Sphericity as a characteristic of solids from fossil fuel burning in Lake Michigan sediment. *Geochimica et Cosmochimica Acta* 45: 763–769.

Hansen, JE, Lacis, AA, Lee, P and Wang, W-C 1980 Climatic effects of atmospheric aerosols. *In* Kneip, TJ and Lioy, PJ, eds, *Aerosols – Anthropogenic and Natural, Sources and Transport.* Annals of the New York Academy of Sciences 338: 575–587.

Hawthorne, SB, Miller, DJ, Barkley, RM and Krieger, MS 1988 Identification of methoxylated phenols as candidate tracers for atmospheric wood smoke pollution. *Environmental Science and Technology* 22: 1191–1196.

Hites, RA 1981 Sources and fates of atmospheric polycyclic aromatic hydrocarbons, Chapter 10. *In* Macias, ES and Hopke, PK, eds, Atmospheric Aerosol: Source/Air Quality Relationships. Washington, DC, American Chemical Society, *Symposium Series* 167: 187–196.

Hopke, P 1985 *Receptor Modeling in Environmental Chemistry.* New York, John Wiley & Sons.

Jaenicke, R 1980 Natural aerosols. *In* Kneip, TJ and Lioy, PJ, eds, *Aerosols – Anthropogenic and Natural, Sources and Transport.* Annals of the New York Academy of Sciences 338: 317–329.

Johnson, DL and McIntyre, BL 1983 A particle class balance receptor model for aerosol apportionment in Syracuse, NY: Receptor models applied to contemporary pollution problems. *In* Dattner, SL and Hopke, PK, eds, *Receptor Models Applied to Contemporary Pollution Problems.* Pittsburgh, Air Pollution Control Association: 238–248.

Kamens, RM 1982 An outdoor exposure chamber to study wood combustion emissions under natural conditions. *In* Frederick, ER, ed, *Proceedings of the Residential Wood and Coal Combustion.* The Air Pollution Control Association. Mars, Pennsylvania, Choice Book Manufacturing Company.

Kaplan, IR and Gordon, RJ 1989 *Contemporary Carbon in Atmospheric Fine Particles Collected in Los Angeles During the 1987 SCAQS.* Anaheim, California, Air and Waste Management Association.

Kelly, WR and Ondov, JM 1990 A theoretical comparison between intentional elemental and isotopic atmospheric tracers. *Atmospheric Environment* 24(A): 467–474.

Kerr, RA 1991 Huge eruption may cool the globe. *Science* 252: 1780.

LaFlamme, RE and Hites, RA 1978 The global distribution of polycyclic aromatic hydrocarbons in recent sediments. *Geochimica et Cosmochimica Acta* 42: 289–303.

Larson, SM, Cass, GR and Gray, HA 1989 Atmospheric carbon particles and the Los Angeles visibility problem. *Aerosol and Science and Technology* 10: 118–130.

Levine, J, ed 1991 *Chapman Conference on Global Biomass Burning.* Cambridge, Massachusetts, MIT Press, in press.

Lewis, CW, Baumgardner, RE, Stevens, RK, Claxton, LD and Lewtas, J 1988 Contribution of woodsmoke and motor vehicle emissions to ambient aerosol mutagenicity. *Environmental Science and Technology* 22 (8): 968–971.

Lewis, CW and Einfeld, W 1985 Origins of carbonaceous aerosol in Denver and Albuquerque during winter. *Environment International* 11: 243.

Lodge, JP, Bien, GS and Suess, HE 1960 The carbon-14 content of urban airborne particulate matter. *International Journal of Air Pollution* 2: 309–312.

Lowenthal, DH, Hanumara, RC, Rahn, KA

and Currie, LA 1987 Effects of systematic error, estimates and uncertainties in chemical mass balance apportionments. Quail Roost II Revisited. *Atmospheric Environment* 21: 501–510.

Malinowski, ER and Howery, DG 1980 *Factor Analysis in Chemistry*. New York, John Wiley & Sons.

Mauney, T, Adams, F and Sine, MH 1984 Laser microprobe mass spectrometry of environmental soot particles. *Science of the Total Environment* 36: 215–234.

McElroy, MW, Carr, RC, Enson, DS and Markowski, GR 1982 Size distribution of fine particles from coal combustion. *Science* 215: 13–19.

Muller, RA 1977 Radioisotope dating with a cyclotron. *Science* 196: 489–494.

National Research Council 1984 *Global Tropospheric Chemistry*. Washington, DC, National Academy of Sciences Press.

Oeschger, H, Stauffer, B, Bucher, P, Frommer, H, Moll, M, Langway, CC, Hansen, BL and Clausen, H 1972 ^{14}C and other isotope studies on natural ice. *In* Rafter, TA and Grant-Taylor, T, eds, *Proceedings of the 8th International Conference on Radiocarbon Dating*. Wellington, Royal Society of New Zealand: D70–D90.

Olsen, CR, Larsen, IL, Lowery, PD, Cutshall, HN, Todd, JF, Wong, GTF and Casey, WH 1985 Atmospheric fluxes and marsh-soil inventories of ^{7}Be and ^{210}Pb. *Journal of Geophysical Research* 90: 10487–10495.

Pratsinis, S, Novakov, T, Ellis, EC and Friedlander, SK 1984 The carbon containing component of the Los Angeles aerosol: Source apportionment and contributions to the visibility budget. *Journal of the Air Pollution Control Association* 34: 643–650.

Prospero, JM 1984 Aerosol particles. *In Global Tropospheric Chemistry*. Washington, DC, National Academy of Sciences Press: 136–140.

Purser, KH, Liebert, RB, Litherland, AE, Buekens, RP, Gove, HE, Bennett, CL, Clover, MR and Sondheim, WE 1977 An attempt to detect stable atomic nitrogen (–) ions from a sputter ion source and some implications of the results for the design of tandems for ultra-sensitive carbon analysis. *Revue de Physique Appliquée* 12(10): 1487–1492.

Rahn, KA and McCaffrey, RJ 1980 On the origin and transport of the winter Arctic aerosol. *In* Kneip, TJ and Lioy, PJ, eds, *Aerosols – Anthropogenic and Natural, Sources and Transport*. Annals of the New York Academy of Sciences 338: 486–503.

Ramdahl, T 1983 Retene – a molecular marker of wood combustion in ambient air. *Nature* 306: 580–582.

Ramdahl, T, Schjoldager, J, Currie, LA, Hanssen, JE, Müller, M, Klouda, GA and Alfheim, I 1984 Ambient impact of residential wood combustion in Elverum, Norway. *Science of the Total Environment* 36: 81–90.

Rosen, H, Hansen, ADA, Dod, RL, Gundel, LA and Novakov, T 1982 Graphitic carbon in urban environments and the Arctic. *In* Wolff, GT and Klimisch, RL, eds, *Particulate Carbon. Atmospheric Life Cycle*. New York, Plenum Press: 273–294.

Rowland, FS and Isaksen, ISA 1988 *The Changing Atmosphere*. Proceedings of the Dahlem Workshop. Chichester, England, Wiley-Interscience.

Schrader, B 1986 Micro raman, fluorescence, and scattering spectroscopy of single particles, Chapter 19. *In* Spurny, KS, ed, *Physical and Chemical Characterization of Individual Airborne Particles*. New York, John Wiley & Sons: 358–379.

Shaw, GE 1982 Perturbation to the atmospheric radiation field from carbonaceous aerosols. *In* Wolff, GT and Klimisch, RL, eds, *Particulate Carbon. Atmospheric Life Cycle*. New York, Plenum Press: 53–73.

Shaw, RW 1987 Air pollution by particles. *Scientific American* 257: 96–103.

Simoneit, BHT 1984 Application of molecular marker analysis to reconcile sources of carbonaceous particulates in tropospheric aerosols. *Science of the Total Environment* 36: 61–72.

Steinmetzer, H-C, Baumeister, W and Vierle, O 1984 Analytical investigation on the contents of polycyclic aromatic hydrocarbons in airborne particulate matter from two Bavarian cities. *Science of the Total Environmental* 36: 91–96.

Stevens, RK, Lewis, CW, Dzubay, TG, Baumgardner, RE, Zweidinger, RB, Highsmith, VR, Cupitt, LT, Lewtas, J, Claxton, LD, Currie, LA, Klouda, GA and Zak, B 1990 Mutagenic atmospheric aerosol sources apportioned by receptor modeling. *In* Zielinski, WL, Jr and Dorko, WD, eds, *Monitoring Methods for Toxics in the Atmosphere*. Philadelphia, American Society for Testing and Materials: 187–196.

Stevens, RK and Meyers, E, eds 1989 *Proceedings of the Quail Roost III Volatile Organic Receptor Modeling Workshop*. Raleigh, North Carolina, USEPA.

Sturges, WT and Barrie, LA 1987 Lead 206/207 ratios in the atmosphere of North America as tracers of US and Canadian emissions. *Nature* 329: 144–146.

Tanner, RL and Miguel, AH 1989 Carbonaceous aerosol sources in Rio de Janeiro. *Aerosol Science and Technology* 10: 213–223.

Tuominen, J, Salomaa, S, Pyysalo, H, Skyttä, E, Tikkanen, L, Nurmela, T, Sorsa, M, Pohjola, V, Saurl, M and Himberg, K 1988 Polynuclear aromatic compounds and genotoxicity in particulate and vapor phases of ambient air: Effect of traffic, season, and meteorological conditions. *Environmental Science and Technology* 22: 1228–1234.

Verkouteren, RM, Currie, LA, Klouda, GA, Donahue, DJ, Jull, AJT and Linick, TW 1987 Preparation of microgram samples on iron wool for radiocarbon analysis via accelerator mass spectrometry: A closed-system approach. *Nuclear Instruments and Methods* 29: 41–44.

White, WH and Macias, ES 1989 Carbonaceous particles and regional haze in the western United States. *Aerosol Science and Technology* 10: 111–117.

Wolff, GT and Klimisch, RL, eds 1982 *Particulate Carbon. Atmospheric Life Cycle*. New York, Plenum Press: 411 p.

BIOMEDICAL APPLICATIONS

PREFACE

AUSTIN LONG

The past, present and future of radiocarbon up to this point in this book have been about time. In the context of the development and evolution of the ^{14}C dating technique and its growing applications, the time frame is historical, with distinct beginnings. In terms of the extent of time that radiocarbon as a dating tool usefully spans, the scale is in tens of millennia. In this last section, we see radiocarbon playing a less direct time role, and we learn that ^{14}C is not the only isotope called "radiocarbon." In biomedical applications, rather than determine ages of dead things, radiocarbon helps extend ages of living humans. As a radioactive tracer in human organisms and in medical experiments, ^{14}C and its sister radioisotopes help diagnose illnesses and understand biochemical processes. Radiocarbon is the ideal biological tracer, because nearly all biochemicals are primarily composed of carbon. Radiocarbon dating technology is well poised for the current trend, driven by health and safety interests and by radioactive waste disposal concerns, toward using lower levels of tracers.

Martyn and Margaret Jope summarize the development of radiocarbon as a biological tracer and reveal the important, perhaps even critical, part radiocarbon portrayed in understanding the fundamental reactions in photosynthesis, those comprising the Calvin Cycle. The interesting connection here is that if photosynthesis did not do its part, cosmogenic ^{14}C may not be spread so nicely around the Earth, and radiocarbon dating might not work so well.

Jay Davis well illustrates that while AMS has already opened new avenues for research in radiocarbon dating methods and applications, it is just beginning to do the same in biomedical fields. AMS applies to biology and medicine the same advantages it has in dating: small sample size and rapid analysis. Speed is even greater in these applications where requirements for precision are less stringent. Small size has clear advantages in the case of biopsy samples and in disposal. AMS adds a third advantage: extended range of detectable ^{14}C concentration with resilient recovery from samples with high levels of ^{14}C.

Radiocarbon is not limited to historical, cultural and geological endeavors. Here is where radiocarbon affects our future.

34

RADIOCARBON IN THE BIOLOGICAL SCIENCES

MARTYN JOPE and MARGARET JOPE

We tend nowadays to equate radiocarbon research with Libby and dating, but it had, in fact, a considerable pre-Libby prehistory in the decade 1935–1945, in the hands of chemists and physicists turning increasingly towards biological problems.

Stable isotopes (^{15}N, ^{2}H) had been effectively used from 1934 at Columbia University College of Physicians and Surgeons by Schoenheimer (from Freiburg) and Rittenberg, joining him from Urey's group (Schoenheimer 1942), and ^{18}O by Ruben's group a few years later. Radioactive isotopes (^{11}C, ^{31}S, ^{35}P, ^{125}I) were used very soon after (Schoenheimer & Rittenberg 1940; Ruben, Hassid & Kamen 1939).

The radiocarbon tracer work was developed at Berkeley, where in 1935, Ruben and Kamen were setting out to investigate photosynthesis, the complex sequence of processes whereby all plants (and hence, animals) derive their food from atmospheric CO_2 through solar energy (Ruben, Hassid & Kamen 1939). Ruben's team needed a radioactive carbon as a tracer to chart the metabolic pathway of the CO_2 ingested from the atmosphere. The Berkeley cyclotron, then recently built, was able to provide artificial radioactive ^{11}C by bombarding ^{10}B with deuterons (^{2}H), and this ^{11}C was used for some years in pioneer work by the Ruben and Kamen team, in spite of its restrictively short half-life of 21.5 minutes (Cockcroft, Gilbert & Wilson 1935), giving only a few hours of experimentation time (Ruben, Hassid & Kamen 1939; Ruben, Kamen & Hassid 1940). Seaborg (1940) says five hours. ^{11}C is, in fact, now being used again (McCallum *et al* 1981).

The search for a longer-lived radiocarbon isotope had, however, been on since the reality of ^{14}C had been put forward in 1936, by Goldhaber working with Chadwick in Cambridge (Bonner & Brubaker 1936). The Berkeley cyclotron proved able to produce artificial ^{14}C (by deuterium bombardment of graphite: Ruben, Hassid & Kamen 1939), and its much longer half-life (then estimated at 10^3–10^5 years: Ruben, Kamen & Hassid 1940; Ruben & Kamen 1941) made radioactive-C tracer manipulation much more practicable. The key role of Lawrence's Berkeley 'Rad-Lab', and of his brother John, who was a physician working intensely on the medical application of this work on particle beams, is discussed by Heilbron and Steidel (1989).

After 1947, when Libby showed the reality of ^{14}C actually present in nature (formed in the atmosphere when cosmic-ray-generated neutrons collide with ^{14}N atoms), the usual source of ^{14}C for tracer work became the enrichment of this natural ^{14}C. It is worth noting that Libby was at that time (from 1935) working in the chemistry department at Berkeley, mainly on isotopes of much heavier elements, but clearly involved in the early biological isotope work.

The results of this pioneer tracer work on photosynthesis by Ruben and Kamen and their co-workers, using first the cyclotron ^{11}C and then ^{14}C, and by Schoenheimer and Rittenberg on stable isotopes, penetrated deeply into biochemical thinking – and teaching and textbooks – in the early 1940s (eg, Harrow 1944). Isotope tracer work might indeed be seen as playing a key role in the rise of biochemistry as a truly independent scientific discipline – the study of the dynamic flux of biomolecular processes, beyond the merely structural study of biomolecules.

The complexities of photosynthesis began, however, to appear rather intractable, and a review article of 1953 asserted that "one can arrive at wrong conclusions even more easily with tracers than without them" (Brown & Frenkel 1953). This slur was handsomely countered by some years of research by Calvin and coworkers, again at Berkeley, who with the constant aid of ^{14}C tracers, were able to formulate the Calvin cycle (Calvin 1962), which earned the Nobel Prize for chemistry in 1961, and which has held its position to this day as the basic sequence of 14 biosynthetic processes underlying photosynthesis. The conclusions were further supported by work on chloroplasts (Jensen & Bassham 1966).

From this Calvin cycle work, still making much use of ^{14}C tracers, but based also on ^{13}C studies, grew the biological concept of two basic photosystems for green plant growth: C_3 (or I), and C_4 (or II); I operates through the Calvin cycle unaided, but II requires additional enzymes or other biomolecules to be transported to the site of photosynthesis. Bacteria have a much simpler photosynthetic system, which does not evolve O_2, though related to green plant photosystem II (Feher *et al* 1989). These concepts of photosynthesis can now be seen in their biological context, as involving chloroplast membrane proteins, and they can ultimately be set out more fundamentally as energy (= electron) transfers on a quantized basis (Feher *et al* 1989: 115), a significant trend in biochemical thinking to which ^{14}C has made its contribution.

We may note one particular remarkable multi-function chloroplast enzyme in the Calvin cycle – Ribulose Bisphosphate Carboxylase-Oxygenase (RUBISCO), the most abundant protein on the face of the earth, on which most living organisms depend directly or indirectly for their energy (Ellis & Gray 1986). RUBISCO both catalyzes the key process in biosynthesis and preserves the balance by photorespiration (expulsion of CO_2). RUBISCO activity is measured through [^{14}C]$NaHCO_3$ (Goloubinoff *et al* 1989). Further, and perhaps most important for future generations in work on this enzyme, we may see some prospects for improving the efficiency of photosynthesis by genetic engineering, thus enhancing agricultural activity (Ellis & Gray 1986; Anderson *et al* 1989).

In a biogeochemical context, the radioactive ^{14}C was instrumental in drawing attention to the fractionation of carbon isotopes during photosynthesis, the green plant absorbing selectively the lighter CO_2. By analogy between the $^{14}C/^{12}C$ and $^{13}C/^{12}C$ ratios, it became an accepted principle that can be applied to fossil material to demonstrate the biological origin of many mineral hydrocarbons (Nier & Gulbranson 1939; Abelson & Hoering 1961; Park & Epstein 1966; Calvin 1969: 88-91).

Another aspect of C-isotope research in relation to biology should be mentioned here, although it involves the stable isotope ^{13}C – Nuclear Magnetic Resonance (NMR). ^{13}C NMR has revealed the dynamic 'breathing' state of the collagen supra-molecular aggregate (Torchia & Vanderhart 1976), and can be very informative on the stacking patterns of purine and pyrimidine bases, and so help to raise DNA study from a largely two-dimensional base-pair-sequence view to a fully spatial conception (Richards 1989; Schwartz & Koval 1989).

Radioactive isotopic markers are indispensable to current biological research. ^{14}C is used in several distinct ways: as metabolic pathway and mechanism tracers (Cornforth & Ryback 1967); in autoradiography to identify and locate specific tissues, cells or biomolecules (as on chromatograms); in substrates for measuring enzyme activity; in probes for locating sensitive domains in biomolecules; or quite simply to provide a scale for biomolecular mass (eg, Olsen, Jonsson & Normark 1989).

Radiocarbon must be seen here in its place among other isotopic markers; carbon however has a special place as the main universal backbone constituent of biomolecules, with which they can be radiomarked without changing their properties and behavior. ^{3}H can sometimes be made to serve, but we have no convenient radioisotopes of N or O, and S and P are of only specialized occurrence in biomolecules.

The isotope work on photosynthesis is one of the major contributions of radiocarbon to modern biology. Early in the story, however, other topics vital to medicine as well as to biochemistry were being investigated with the aid of ^{14}C (and even ^{11}C). Aspects of metabolism of carbohydrates, cholesterol (Wright 1961; Popjak & Cornforth 1960; Cornforth & Ryback 1967), and fat (which had proved especially difficult (Rommel & Goebell 1976)) were thus clarified. Cornforth's work in this field earned the 1975 Nobel Prize for Chemistry. There are many applications to clinical medicine, such as work on gall stones (Druffel & Mok 1983), ^{14}C breath tests (on respiratory CO_2), and many others discussed by Davis (1992). The scope of ^{14}C contributions to the biological sciences is summarized in Figure 34.1.

After more than 40 years, ^{14}C still continues to play its vital part (alongside ^{3}H, ^{32}P, ^{35}S and ^{125}I) in the most forward-looking biological research. The value of ^{14}C in current research is made clear by the commercial availability of so many ^{14}C-marked biochemicals (eg, Sigma Catalog 1990: 1661–1683). ^{14}C is used, for instance, in charting the protein gradients that can regulate morphogen transcription in *Drosophila* embryos, the body segmentation pattern being laid down

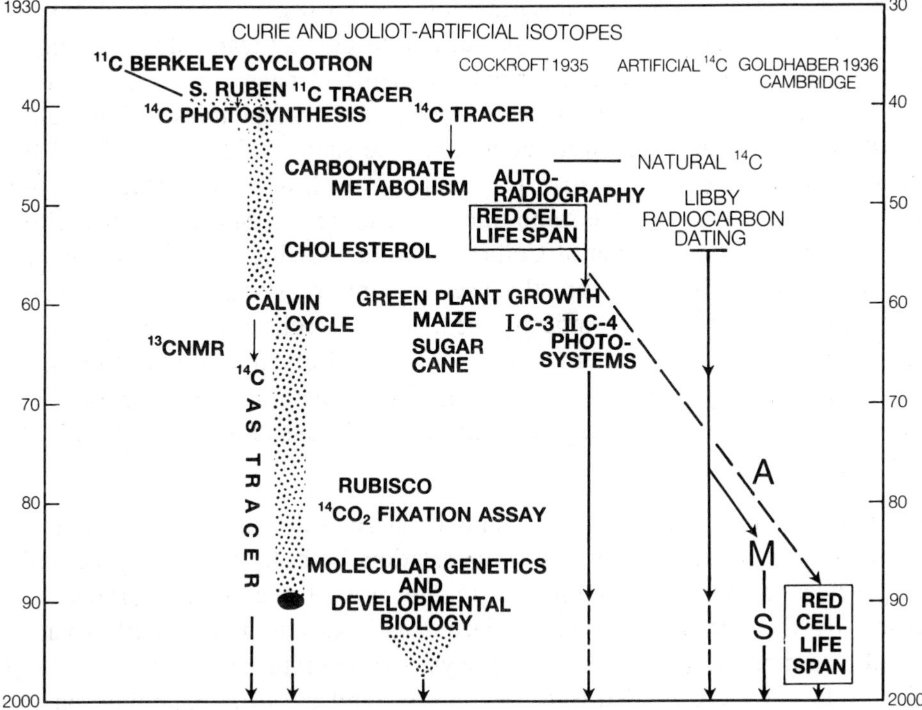

Fig 34.1. Summary of the contributions of ^{14}C to research in the biological sciences. Items of biological significance are shown bold. For top line, see Joliot and Curie (1934).

in response to opposing morphogen gradients emanating from localized sources at each pole of the embryo (Struhl 1989; Driever & Nüsslein-Volhard 1989; Irish, Lehmann & Akam 1989). There is currently a race between the fruit fly *Drosophila melanogaster* and the nematode worm *Caenorhabditis elegans* to provide the clearest model of how the information stored in the genome can effect morphogenic control over the patterns of growth of a developing organism, one of the most pressing topics in modern biology. ^{14}C has been used also for identifying molecular species in molecular recognition of transfer-RNA (Perret *et al* 1990); biological molecular recognition is again a most pressing topic.

We conclude by discussing a long-term research program in which we took part at an early stage, as a case study which illustrates how ^{14}C has been used to clarify problems as they arose – the precisely programmed life-span of the red blood corpuscle. The need to define the hemoglobin (Hb) standard in wartime Britain (Macfarlane *et al* 1944) led us to study severely cyanosed patients on sulphonamide drugs and workers in contact with TNT in shell-filling factories. In most of these people, the cause of cyanosis was methemoglobin (MHb), which the body systems can quickly reduce back to functional Hb. But a few were far more intractable, due to sulfhemoglobin (SHb), which the body systems do not revert to functional Hb. This seemed to give a means of estimating the life-span of the red blood cells in which SHb had been formed (Jope 1945; 1946). In Oxford, Witts and his team were at work on anemias and the fate of transfused blood, and they were addressing the same problem of red cell life-span by differential agglutination (Callender, Powell & Witts 1945; Callender, Loutit & Jope 1946). At the same time, Rittenberg and Shemin, in the laboratory set up by Schoenheimer at Columbia, were showing with ^{15}N that glycine (including its carbon) is directly used in the biosynthesis of hemin for hemoglobin (Shemin & Rittenberg 1945), and they followed the disappearance of ^{15}N in red blood cells after the ingestion of ^{15}N glycine in man, to estimate the human red blood cell life-span (Fig 34.2; Shemin & Rittenberg 1946). These three varied methods agree well: 116 ± 5 days, 120 ± 5 days, 127 ± 3 days, respectively.

But this was for man; what about animals, larger and smaller? The subject is of some evolutionary importance, in relation both to the emergence of a corpuscular circulatory system and to the rise of the human protein (and especially the immune) systems. Earlier estimates of red cell life-span for smaller animals (dog 114, and rat 90 days) seemed rather long. So we fed ^{14}C glycine to one rat, and tried to follow the rise and fall of radioactivity on very small successive blood samples from this one rat (Shemin & Rittenberg (1946) had used one rat per point). Autoradiograph photometry of the rat red cells gave

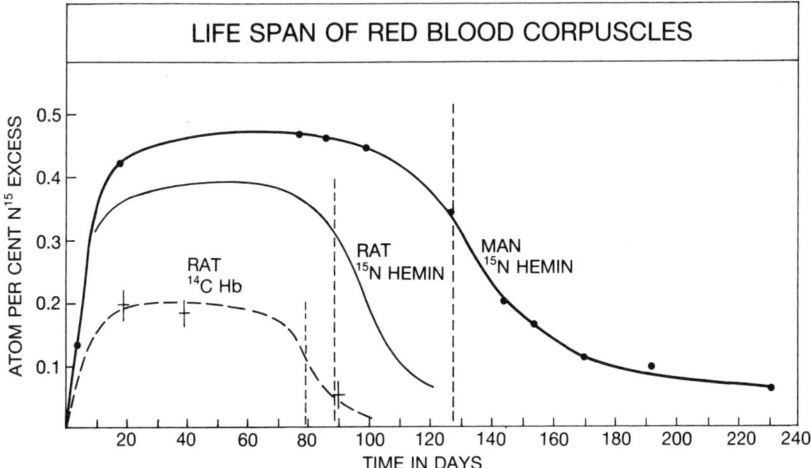

Fig 34.2. Estimation of the sharply programmed life-span of red blood corpuscles in humans and rats, by feeding ^{15}N-glycine (measuring the life-span of the hemin – 127 and 90 days) and ^{14}C-glycine (rat only, measuring the life-span of the globin and other cell proteins as well as hemin in the corpuscle – 80 days). The ^{14}C as obtained (Jope & Jope 1950, unpublished notes) from successive very small blood samples from an individual rat (no value was available for the 60–80-day range). For the ^{15}N data (Shemin & Rittenberg 1946), one rat was desanguinated for each point. The ^{14}C vertical scale is arbitrary. The small sample of ^{14}C-glycine came to us through the good offices of Prof D Rittenberg; colleagues at AERE Harwell cooperated in the estimation of the ^{14}C radioactivity.

too weak a signal, but a Geiger counter registered a moderate signal after 20 days, weaker after 40 days, and low after 90 days (Fig 34.2: the 65-day point was missed). This tentative result suggested that the rat red cell life-span is somewhat under 90 days. Clearly, red cell life-span does not directly reflect the animal's life expectancy, nor its body weight; much more comparative study (at the genetic level) is needed.

There, in the early 1950s, this feasibility study had to rest, but now, after 40 years, this whole subject of red blood corpuscle life-span through the whole animal kingdom might be profitably reopened, for AMS has the sensitivity needed for small creatures, and the fact that AMS, in measuring a ratio ^{14}C/^{12}C (and its return to normality), makes it independent of Hb sample size (difficult to measure meaningfully on the tiny blood samples), which is a great advantage.

As to the evolutionary significance of red blood corpuscle life-span, the lamprey, for instance, is one of the most primitive creatures having a corpuscular circulatory system (Thomson 1929: 6, 579–585, 803; Young 1962). Lamprey red corpuscles in circulation are biconcave discs (as mammalian; *cf* frog (Fig 34.3; Jope 1949: 214–215, figs 11–14)). Have they got a specific life-span and pre-

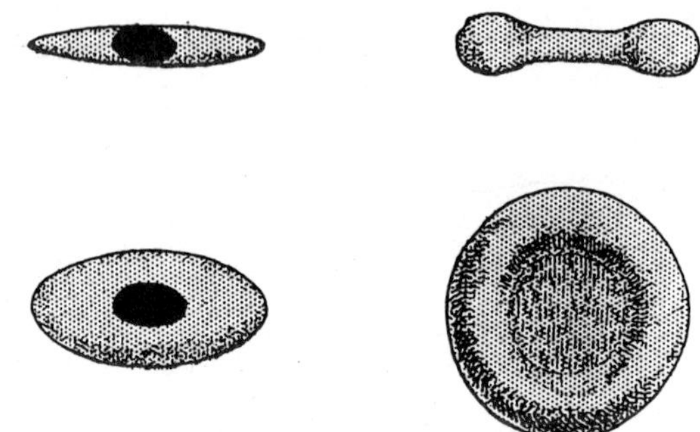

Fig 34.3. Red blood corpuscle shapes in circulation: right, human (and most other mammals), circular, bi-concave in profile, without nucleus; left: frog, salamander, etc, flattened, ovoid, nucleated in circulation. Scale: × 1000 (after Thomson 1929).

cisely programmed death? Birds and fish seem to be again somewhat different. All this might well be profitably investigated, with the aid of ^{14}C and ^{15}N, as a zoological program by AMS. The whole idea of a precisely prescribed life-span, continually overrunning within an overall life-span, is not quite the same as renewal or repair (Young 1971: 146–148), and seems a novel concept in biological thinking.

Finally, how is this precisely programmed death of red blood corpuscles actually operated? What are the biomolecular mechanisms? Red corpuscles are nucleated during erythropoiesis, but the nucleus disintegrates as the cell begins its circulatory functional life, and this must mark the inception of cell death. Nuclear fragments do, however, survive in the cytoplasm, and it could be from these that proteolytic control is exercised over proteins of ankyrin type, a newly recognized group which connect between structural elements within the corpuscle, and which help to maintain its distinctive biconcave shape (Fig 34.3; Lux, John & Bennett 1990). Once this biconcave shape goes, the corpuscle is unstable and well on the way to destruction.

But regulation of red corpuscle death is yet more complex; as iron has to be conserved, destruction can proceed only in certain parts of the circulatory system (eg, the spleen), and signals must pass both ways across the corpuscle wall. Such cellular disruption has been studied in an analogous case of neurotransmitter release from synapse vesicles, making most elegant use of ^{14}C-marked probes to reveal a sensitive domain along a protein chain, and hence, how the protein can position itself in penetrating the vesicle wall, through which it can then transmit a signal (Perin *et al* 1990; *cf* Parker *et al* 1989). The death of a red blood corpuscle would be initiated when the nucleus is fragmented, but final disruption must be implemented by a signal from outside, transmitted presumably by a protein inserted through the corpuscle wall, activating the expression of genes producing proteases that disrupt ankyrins. Much work lies ahead (cytological and immunological as well as the biomolecular considered here) before we have a definitive view of red corpuscle death mechanisms (*cf* MacDonald & Lees 1990, for T-lymphocytes), and there are a number of ways in which ^{14}C can assist (eg, Robbins, Horowitz and Mulligan (1990) show how gene expression may be quantified with a ^{14}C-marked substrate).

In this brief survey we have shown how, during nearly six decades, the radioactive carbons, first the short-lived ^{11}C, then ^{14}C, have been used in various ways to advance a wide range of biochemical and biological research, from charting biochemical pathways to clarifying molecular-genetic and evolutionary problems, or medical diagnosis and treatment. ^{14}C continues to be of ever-increasing importance in the biological sciences and has been involved in three Nobel Prize awards. But the whip in the tail comes when we see that radiocarbon dating is not, after all, an independent physical measure of absolute time, but has to be calibrated against the annual increment scale of tree-ring growth, and is thus, ultimately dependent for its accuracy upon the biological sciences. We have stressed how isotopic tracer work has done much to create biochemistry as an autonomous scientific discipline – the study of biomolecular dynamic systems, beyond merely of biomolecular structures. We have noted the pioneer work with the short-lived ^{11}C and the influence of the Berkeley cyclotron, and Libby's role in bringing ^{14}C to the fore. Now AMS, which was developed as a facility very much in response to the needs of geochronologists and archaeologists to use ^{14}C for dating purposes, has its own advantages to offer also in charting ^{14}C for the biological sciences.

REFERENCES

Abelson, PH and Hoering, TC 1961 Carbon isotope fractionation in formation of amino acids by photosynthetic organisms. *Proceedings of the [US] National Academy of Sciences* 47: 623.

Anderson, I, Knight, S, Schneider, G, Lindqvist, Y, Lindqvist, T and Lorimer, GH 1989 Crystal structure of the active site of ribulose-Bisphosphate carboxylase. *Nature* 337: 229–234.

Bonner, TW and Brubaker, WM 1936 The disintegration of nitrogen by slow neutrons. *Physical Review* 49: 778.

Brown, AH and Frenkel, AW 1953 Photosynthesis. *Annual Review Biochemistry* 22: 423–458.

Callender, ST, Loutit, JF and Jope, EM 1946 Discussion on the life and death of the red blood corpuscle. *Royal Society of Medicine Proceedings* 39: 755–762.

Callender, ST, Powell, EO and Witts, LJ 1945 The life-span of the red cell in man. *Journal of Pathological Bacteriology* 57: 129–139.

Calvin, M 1962 The path of carbon in photosynthesis. *Science* 135: 897.

_____1969 *Chemical Evolution*. Oxford, Clarendon Press.

Cockcroft, J, Gilbert, E and Wilson, TCR 1935 Disintegration of a short-lived carbon isotope. *Proceedings of the Royal Society of London* 148A: 225.

Cornforth, JW and Ryback, G 1967 The stereospecificity of enzyme reactions. *Reports on Progress in Chemistry* 42: 428.

Davis, JC 1992 New biomedical applications of radiocarbon. *In* Taylor, RE, Long, A and Kra, RS, eds, *Radiocarbon After Four Decades: An Interdisciplinary Perspective.* New York, Springer-Verlag, this volume.

Driever, W and Nüsslein-Volhard, C 1989 The bicoid protein is a positive regulator of *hunchback* transcription in the early *Drosophila* embryo. *Nature* 337: 138–143.

Druffel, EM and Mok, HYI 1983 Time history of human gallstones: Application of the post-bomb radiocarbon signal. *In* Stuiver, M and Kra, RS, eds, Proceedings of the 11th International ^{14}C Conference. *Radiocarbon* 25(2): 629–636.

Ellis, RJ and Gray, JC 1986 Ribulose bisphosphate carboxylase-oxygenase. *Philosophical Transactions of the Royal Society of London* B 313: 1–165.

Feher, G, Allen, JP, Okamura, MY and Rees, DC 1989 Structure and function of bacterial photosynthetic reaction centres. *Nature* 339: 111–116.

Goloubinoff, P, Christeller, JT, Gatenby, AA and Lorimer, GH 1989 Reconstitution of active dimeric ribulose bisphosphate carboxylase from an unfolded state depends on two chaperonin proteins and Mg-ATP. *Nature* 342: 884–889.

Harrow, B 1944 *Textbook of Biochemistry,* 3rd edition. Philadelphia, WB Saunders Co.

Heilbron, JR and Steidel, RW 1989 *Lawrence and His Laboratory.* Berkeley, University of California Press.

Irish, V, Lehmann, R and Akam, M 1989 The *Drosophila* posterior group gene *nanos* functions by repressing *hunchback* activity. *Nature* 338: 646–648.

Jensen, RG and Bassham, JA 1966 Photosynthesis by isolated chloroplasts. *Proceedings of the [US] National Academy of Sciences* 56: 1095–1118.

Joliot, F and Curie, M 1934 Artificial production of a new kind of radio element. *Nature* 133: 201.

Jope, EM 1945 Some new evidence on the dynamics of red blood cell destruction. *Biochemical Journal* 39: liii–liv.

____1946 Disappearance of sulphemoglobin from the blood of TNT workers in relation to the dynamics of red cell destruction. *British Journal of Industrial Medicine* 3: 136–142.

____1949 The U-V absorption of hemoglobins inside and outside the red blood cell. *In* Roughton, FJC and Kendrew, JC, eds, *Haemoglobin.* London, Butterworth.

Lux, SE, John, KM and Bennett, V 1990 Analysis of cDNA human erythrocyte ankyrin indicates a repeated structure with homology to tissue differentiation and cell cycle control proteins. *Nature* 344: 36–42.

MacDonald, HR and Lees, RK 1990 Programmed death of autoreactive thymocytes. *Nature* 343: 342–344.

Macfarlane, RG, O'Brien, JRP, Douglas, GC, Jope, EM, Jope, HM and Mole, RH 1944 The Haldane haemoglobinometer. *British Medical Journal* 1: 248–255.

Nier, AO and Gulbranson, EA 1939 Fractionation of carbon isotopes in the atmosphere. *Journal of the American Chemical Society* 61: 697.

Olsen, A, Jonsson, A and Normark, S 1989 Fibronectin binding mediated by a novel class of surface organelles on *Escherichia coli. Nature* 338: 652–655.

Park, R and Epstein, S 1966 Carbon isotope fractionation during photosynthesis. *Geochimica et Cosmochimica Acta* 21: 110.

Parker, MW, Pattus, F, Tucker, AD and Tsernoglou, D 1989 Structure of membrane pore-forming fragment of colicin A. *Nature* 337: 93–96.

Perin, MS, Fried, VA, Mignery, GA, Jahn, R and Südhof, TC 1990 Phospholipid binding by a synaptic vesicle protein homologous to the regulatory region of protein kinase C. *Nature* 345: 260–263.

Perret, V, Garcia, A, Grosjean, H, Ebel, J-P, Florentz, C and Geige, R 1990 Relaxation of a transfer RNA specificity by removal of modified neucleotides. *Nature* 344: 787–789.

Popjak, G and Cornforth, JW 1960 The biosynthesis of cholesterol. *Advances in Enzymology* 22: 281–326.

Richards, T 1989 DNA molecules observed. *Nature* 338: 331–332.

Robbins, PD, Horowitz, JM and Mulligan, RC 1990 Negative regulation of c-*fos* expression by the retinoblastoma gene product. *Nature* 346: 668–671.

Rommel, K and Goebell, H 1976 *Lipid Absorption: Bio-Chemical and Clinical Aspects.* London, Clowes.

Ruben, S, Hassid, WZ and Kamen, MD 1939 Radioactive carbon in the study of photosynthesis. *Journal of the American Chemical Society* 61: 661–663.

Ruben, S, Kamen, MD and Hassid, WZ 1940 Photosynthesis with radioactive carbon. *Journal of the American Chemical Society* 62: 3443–3448.

Ruben, S and Kamen, MD 1941 Half-life of long-lived radioactive carbon. *Physical Review* 59: 349–354.

Schoenheimer, R 1942 *The Dynamic State of Body Constituents.* Cambridge, Massachusetts, Harvard University Press.

Schoenheimer, R and Rittenberg, D 1940 The study of metabolism with the aid of isotopes. *Physiological Review* 20: 218–245.

Schwartz, DC and Koval, M 1989 Conformational dynamics of individual DNA molecules during gel electrophoresis. *Nature* 338: 520–522.

Seaborg, GT 1940 Artificial radioactivity. *Chemical Review* 27: 196–285.

Shemin, D and Rittenberg, D 1945 Utilization of glycine for synthesis of a porphyrin. *Journal of Biological Chemistry* 159: 567–568.

_____1946 The life-span of the red blood cell. *Journal of Biological Chemistry* 166: 621–636.

Sigma Catalog 1990 *Biochemicals, Organic Compounds for Research and Diagnostic Reagents.* St Louis, Missouri, Sigma Chemical Company Ltd.

Struhl, G 1989 Differing strategies for organizing anterior and posterior body-pattern in *Drosophila* embryos. *Nature* 338: 741–744.

Thomson, JA 1929 *Outlines of Zoology.* Oxford.

Torchia, DA and Vanderhart, DL 1976 ^{13}C magnetic resonance evidence for anisotropic molecule motion in collagen fibrils. *Journal of Molecular Biology* 43: 315–321.

Wright, GP 1961 The metabolism of gmelin. *Royal Society of Medicine Proceedings* 54: 26–30.

Young, JZ 1962 *The Life of Vertebrates.* Oxford, Clarendon Press.

_____1971 *An Introduction to the Study of Man.* Oxford, Clarendon Press.

NEW BIOMEDICAL APPLICATIONS OF RADIOCARBON

J C DAVIS

INTRODUCTION

The development of accelerator mass spectrometry (AMS) and its rapid application to radiocarbon detection produced a revolution in archaeology, earth science and oceanography for two primary reasons: the sample size required to achieve a reliable date was reduced by >1000, permitting multiple dates and greater precision. Further, the detection ("machine") background was reduced to less than the contaminations caused by sample preparation, thus greatly increasing the sensitivity for routine analysis. These increases in sensitivity and precision have had qualitative, as well as quantitative effects on these disciplines. For similar reasons, considerable impact upon biomedical science was predicted, but not tested until recently. Radiocarbon is the premier tracer for following many biological reaction pathways. Radiocarbon is non-labile in many organic molecules, unlike tritium, and may be placed in specific sites with known binding properties. Unlike many other substitutional tracers, ^{14}C does not distort the structure of the host molecule. At the concentrations one can easily predict for AMS analyses, the danger of radiolytic perturbation of the biochemistry under study is essentially zero. Similarly, the radiation dose to the host organism from decay of ^{14}C is negligible in design of an experiment.

The potential of AMS and radiocarbon in biomedical work was suggested specifically at the first AMS conference (Keilson & Waterhouse 1978). Subsequently, a speculative description of possible biomedical uses of ^{14}C and other isotopes was provided by Elmore (1987). Despite the potential value of this use of radiocarbon detection, no application combining AMS analysis and radiocarbon in biomedicine was reported until this year (Turteltaub *et al* 1990a). The slowness to exploit this possibility has two explanations, one sociological and one technical. The time required for a technique to achieve maturity in one field and then jump disciplinary boundaries is frequently more than a decade. Given that AMS truly became a routine tool in archaeology and the earth sciences only in the mid-1980s, such a lag is perfectly normal. Additionally, there is a genuine technical obstacle. All the laboratories performing both AMS and conventional radiocarbon analyses for archaeology, oceanography and the earth sciences depend upon their ability to measure radiocarbon at or *below* natural abundance. If a single mistake is made with compounds or samples intentionally labeled 1000 × to 10^6 × natural abundance, the consequences of such ongoing programs are trivially easy to predict.

The motivation to overcome these obstacles exists within the biomedical research effort at Lawrence Livermore National Laboratory (LLNL). A specific assignment of the biomedical and environmental activities of the Department of Energy is the assessment of the risk and consequences of the generation and use of energy. A major component of such assessment is dosimetry for exposure to radiation and to carcinogenic and/or mutagenic materials. The dilemma of such assessment is illustrated in Figure 35.1. One wishes to know the response of an organism to some chemical insult at the environmentally relevant risk level. However, owing to limitations of technique, methodology, etc, one often must measure the dose-response curve at higher dose, then extrapolate to the much lower environmental dose. The extrapolation is uncertain as to whether the response is linear, whether damage repair mechanisms actually result in no response (and hence no risk) occurring at low dose, or whether there is some finite threshold risk at any dose. As legal rule-making and significant economic consequences follow from the assigned risk, however derived, the motivation to provide better diagnostic techniques at low exposures is powerful. We have begun a program at LLNL to exploit the sensitivity of AMS for radiocarbon to provide such information.

EXPERIMENTAL CAPABILITIES

The spectrometer used for measurement in the initial research applications of AMS to biomedical problems is the University of California/LLNL AMS beamline at the LLNL Multiuser Tandem Laboratory (Davis *et al* 1990; Southon *et*

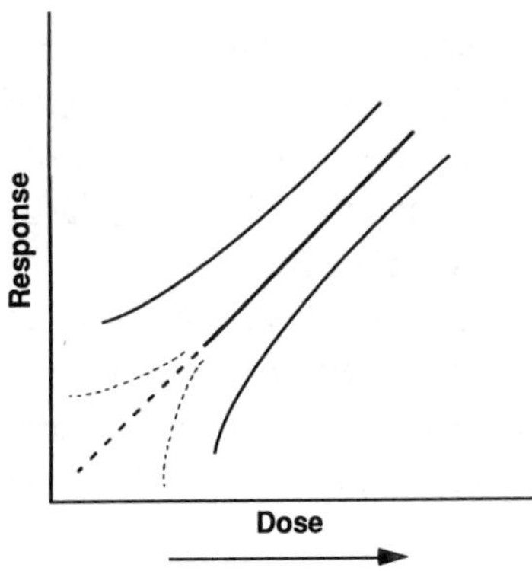

Fig 35.1. Hypothetical behavior of dose-response curves when extrapolated to low dose, beyond usual measurement thresholds

al 1990). This spectrometer was co-funded by LLNL for geochemistry research and by the University of California Regents for archaeological and anthropological research. The decision to attempt biomedical experiments was made in a proposal for internal discretionary research funds at LLNL intended for use in creating new program opportunities across disciplinary lines. The research proposal was a collaborative effort of nuclear physics, nuclear chemistry, biomedical sciences, and environmental sciences staff.

Unique aspects of this spectrometer include 5 cm magnet apertures, 10 cm optics elsewhere, and a 60 sample Cs sputtering ion source with spherical ionizer assembly (Proctor *et al* 1990). As samples in this ion source are stored in a cassette several cm behind the working point of the ion source, are individually exposed to the Cs and then replaced in the cassette, it was felt that this design would have adequate protection against cross contamination of samples with widely variant ^{14}C concentrations. An accidental contamination of the sample graphitizer and ion source of the Simon Fraser AMS group (Vogel, Southon & Nelson 1990) during preparatory tests for these experiments proved to be recoverable, so we were confident that both biomedical and archaeological programs could coexist on the same spectrometer.

Sample preparation for experiments utilizes the graphitization technique of Vogel (Vogel, Nelson & Southon 1987). The graphitization facility for biomedical samples is ~0.5 km downwind from the spectrometer and the preparation lab used for natural samples. The preparation protocol for all samples is to perform dilutions of extracted organic material before graphitization, so that neither the graphitizer nor the spectrometer ever sees material greater than 100 times Modern. Additionally, personnel performing dissections or handling high-level materials never work in the laboratory where low-level samples are handled. Sample handling and graphitization are performed from lowest to highest isotope concentration with calibration standards (typically ANU sugar) and backgrounds prepared between unknowns as a check against contamination. As described below, these protocols are adequate to monitor contamination, but instances of sample contamination have occurred, confirming the initial fears of other laboratories. These problems have arisen because the extreme sensitivity of AMS requires careful protocols in laboratories accustomed to using large amounts of ^{14}C. However, irrecoverable disasters have not occurred, and new protocols and experience should make AMS more readily available to biomedicine. We have successfully prepared and measured approximately 1000 samples each from high-level and low-level materials without any cross-contamination between them or degradation of background level in either low-level graphitizer or the spectrometer itself.

INITIAL EXPERIMENTS

Our initial experiment was a measurement of the dose-response curve for DNA damage caused in mouse liver cells by a specific carcinogen: 2-amino-3,8-dimethylimidazo[4,5-f]quinoxaline, or MeIQx. This compound is present in the human diet, from meat fried at high temperature. As most cancer is mediated or initiated by genetic fault, the covalent binding of a MeIQx metabolite to DNA, (DNA adduct), is taken to be indicative of genetic cell damage. We have measured (Turteltaub *et al* 1990a) the formation of adducts in mouse liver DNA 24 hours after administering a dose of MeIQx labeled at a single location with ^{14}C. The liver DNA was isolated, extracted and dialyzed to remove residual non-covalently bound metabolites, then diluted 10- to 1000-fold with sodium lauryl sulfate (SDS) solution previously determined to be 0.4 times Modern, graphitized, and the ^{14}C content determined by AMS. Figure 35.2 shows the dose- response curve expressed in terms of adducts per 10^{12} nucleotides on DNA for a dose range from 500 pg/kg to 5 mg/kg of body weight. Three separate

DNA adduct dose-response measured for a carcinogen exposure has been determined to one adduct per 10^{11} nucleotides by AMS

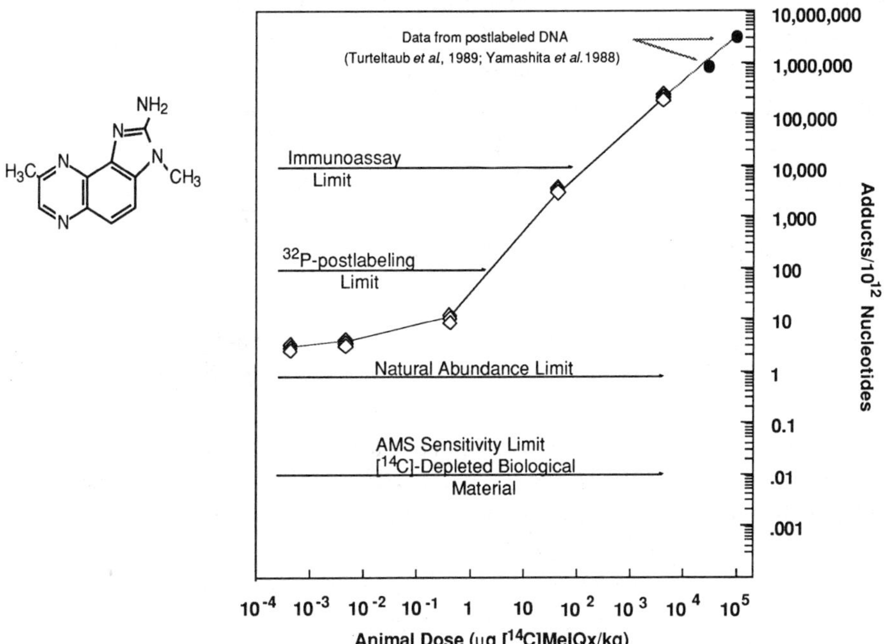

Fig 35.2. Dose-response curve for production of DNA adducts by MeIQx (Turteltaub *et al* 1990a)

animals were measured at each dose. The low doses approximate the chemical doses as low as those expected from the human consumption of several hamburgers.

As can be seen in Figure 35.2, the dose-response curve is linear from a dose of 500 ng/kg to 5 mg/kg of body weight. These results extrapolate linearly to other adduct concentrations reported previously from our laboratory and others (Yamashita *et al* 1988; Turteltaub *et al* 1990b) for animals given much higher doses of MeIQx. The measurements at higher dose use a ^{32}P-postlabeling technique that presents both radiological hazards for the experimenters and requires quantification through procedures which may have uncertain efficiencies. The present measurements are able to detect one ^{14}C-labeled adduct per 10^{11} nucleotides, an improvement in sensitivity of one order of magnitude over the best adduct measurements to date. These measurements are also 3 to 5 orders of magnitude better than traditional quantitative adduct measurement. In addition, the doses of radiation given are orders of magnitude less than those used in traditional and accepted tracer measurements in biomedical research. Reproducibility in these measurements is high (~10%), and may be limited by animal-to-animal variation. Instrument precision for this work was 2%, which should allow studies of the animal-to-animal variations of these responses.

There are two obvious limitations at the low-dose end of the experiment. First, we cannot expect to detect adducts at levels below one per 10^{12} nucleotide pairs because of the ^{14}C naturally present in tissue. The dose-response curve should go flat at that level because of the natural isotopic abundance of ^{14}C, not because of biology. In fact, we did not reach the natural abundance limit because of contamination of samples with ^{14}C from conventional scintillation counting experiments previously performed in the biomedical research areas. For example, a swipe from a hood in one sample extraction area used in the initial work was measured as 18,000 Modern.

Also shown in Figure 35.2 is the easily achieved background level for biomedical AMS, 0.5–1% Modern. If biological hosts could be grown depleted in ^{14}C, another two orders of magnitude in the sensitivity of detection of adducts will be made possible. For many materials, this sensitivity would allow determination of DNA damage at doses below those of environmental exposure, clearly defining one component of risk assessment without extrapolation. In initial experiments, methanotropic bacteria have been grown depleted to 1% Modern in ^{14}C. The first generation of mice grown on a diet some of whose components were synthesized from fossil material have been depleted to 25% of Modern. The effort to grow these animals is in itself regarded as a fundamentally interesting experiment in nutritional research.

As a check on the initial experiment and the more stringent protocols developed for sample handling and preparation, another set of mice were dosed with the toxic and potent carcinogen 2,3,7,8-tetrachlorodibenzo-p-dioxin, TCDD or dioxin. TCDD is rapidly absorbed, but the molecule does not readily react directly with DNA and should not form adducts. The TCDD used in this study was fully substituted with ^{14}C, giving us 12 times greater sensitivity than in the MeIQx adduct study. In Figure 35.3, the ^{14}C content of liver DNA from animals dosed with TCDD is plotted in units of Modern. For comparison, the results of the MeIQx experiment are shown as well. No evidence of formation of adducts is seen, except possibly at the highest dose, corresponding to the LD_{50} for this species. At that toxicity, off-normal metabolic effects might explain some binding to DNA. In comparison to our original measurements with MeIQx, improved protocols have minimized contamination of the DNA samples, and we are unaffected by the presence of unbound TCDD throughout the bulk tissue, except possibly at the highest dose: at the LD_{50} dose, the DNA isolate was 20

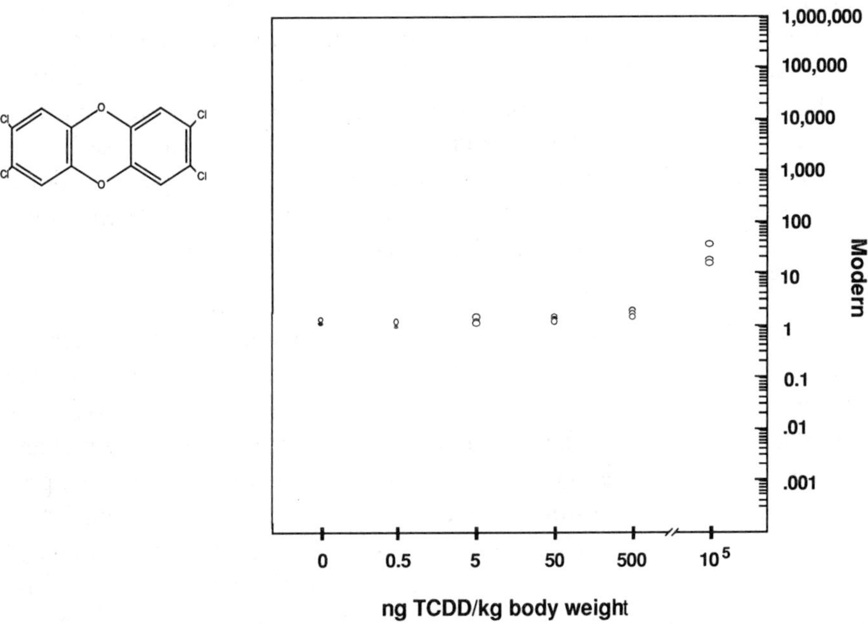

Fig 35.3. Radiocarbon content of mouse liver DNA exposed to MeIQx and TCDD (Felton *et al* 1990). Reprinted by permission of the authors and Elsevier Science Publishers.

times Modern whereas the whole livers themselves were approximately 10^6 times Modern.

WORK IN PROGRESS

Demonstration of the sensitivity and utility of AMS for detecting radiocarbon tags in biomedical applications has led to numerous follow-on experiments. Several topics being pursued at present include the following:

Carbon-14 Depleted Hosts – As pointed out above, hosts depleted in ^{14}C offer two orders of magnitude gain in sensitivity for these measurements. We have succeeded in growing methanotropic bacteria on methane generated from petroleum. The ^{14}C content of the bacteria has been measured at 1% Modern. We are currently growing mice on a special diet. The first generation of animals grown on this diet average 70% Modern in bulk tissue. Although we are pursuing development of such animals for toxicology studies, their utility for nutritional work cannot be overemphasized. If mice depleted to 1% in ^{14}C can be successfully grown, then any natural foodstock is inherently tagged, allowing detailed tracing of metabolic pathways. Experiments with these hosts could be safely performed by any AMS laboratory, greatly expanding the measurement base.

Dose and Time Dependence Studies – Our ability to measure extremely small amounts of extracted DNA (samples of ~1 μg before dilution) makes possible following the rates of damage and repair in organisms as a function of dose. Not just the shape of the dose-response curve but its time history are important in the assessment of risk. We have begun experiments with collaborators at the National Institutes of Health/National Cancer Institute to measure the adduct kinetics of carcinogen adducts in lymphocytes of primates, making maximal use of precious primate research subjects by extracting only small blood samples instead of taking larger biopsies.

Measurement of Chemotherapy Dose – Accurate adjustment of the treatment level in chemotherapy is important but is difficult, as one wishes to work near the cytotoxic limit in patients with weak and varying metabolisms. We have begun a collaboration to study the metabolism of anticancer drugs in normal and resistant tumors using mouse models and human cancer patients. It may be possible to design therapy regimes that can be monitored in real time to adjust dose optimally for individual patients. In such applications, the chemical toxicity of the drugs being used would be more significant than the small radiation dose incurred. For example, a normal 500 ng/kg dose level would produce a radiation dose to a human subject of 0.003 milliSieverts based on a 24-hour half-life for the compound, approximately 1% of the normal annual exposure.

Optimal Sample Preparation and Spectrometers – The graphitization technique most widely used for AMS studies (Vogel, Nelson & Southon 1987) was designed to produce the highest possible accuracy and repeatability for the measurement of natural samples. In the present area, biological noise and host-to-host variation may well limit absolute accuracy to 5% or more. At the same time, experiments can easily be envisioned in which hundreds of samples need to be measured. Development is in progress on one- or two-step batch techniques in which numerous organic samples are formed into solids compatible with satisfactory performance in our sputtering ion source. Initial tests have produced materials with current yields 25–50% of those prepared conventionally with measurement accuracy of 1–2%. Reduction of this technique to routine practice will greatly increase measurement throughput.

Similarly, spectrometers designed for archaeological work are overcomplex and too expensive for clinical fielding. As one may regard the initial ^{14}C level as somewhat of a free parameter, simpler devices may be adequate for research in this field. A notional design for a simple machine capable of both tritium and radiocarbon measurements is shown in Figure 35.4. By placing the ion source

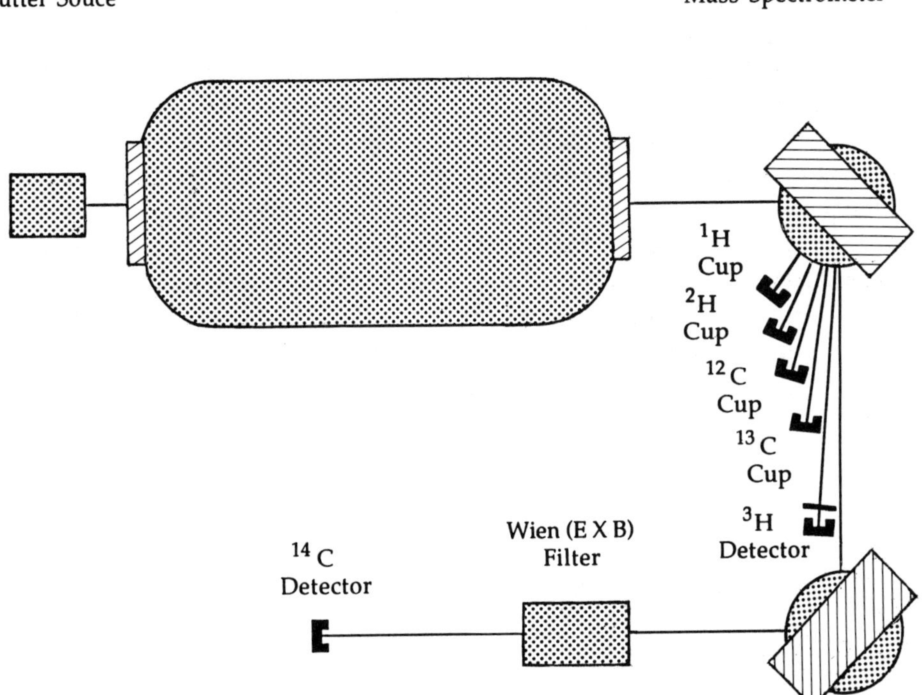

Fig 35.4. Schematic of small accelerator optimized as a spectrometer for simultaneous AMS measurement of tritium and radiocarbon

directly on the tandem accelerator, one achieves simultaneous injection of all isotopes in the simplest possible manner. Initial tests (Roberts *et al* 1991) of this configuration using our own spectrometer suggest that successful operation at the required accuracy is possible. If the scientific results of our work create the measurement demand expected, we will seek funding to build a prototype of such a device.

SUMMARY

The initial experiments applying AMS to the use of radiocarbon in biomedical research have demonstrated the expected sensitivity and utility. The ability to perform research with labeled compounds without interfering with work done on natural abundance materials has also been demonstrated. Sufficient follow-on activities have been identified in a wide variety of applications to ensure growth of this new subfield. While other isotopes, particularly tritium, may have considerable utility in specific research tasks, the chemical properties of carbon and the fully developed technology for measuring its isotopes assure that radiocarbon will remain at the center of all future work.

ACKNOWLEDGMENTS

Development of the biomedical research program to utilize AMS has involved open and enthusiastic collaboration and cooperation across disciplinary and organizational lines. John Vogel and Ken Turteltaub performed the initial experiments and developed the sample handling protocols; Ivan Proctor and John Southon brought the spectrometer into operation; Marc Caffee and Bob Finkel built the detector and data acquisition system; and Mark Roberts has developed specialized control algorithms for spectrometer operation. Erle Nelson, Bart Gledhill and Jim Felton have provided ideas, encouragement, support and critical review since the inception of this activity. Without the efforts of all of the above, little would have been possible.

References

Davis, JC, Proctor, ID, Southon, JR, Caffee, MW, Heikkinen, DW, Roberts, ML, Moore, TL, Turteltaub, KW, Nelson, DE, Loyd, DH and Vogel, JS 1990 LLNL/UC AMS facility and research program. *In* Yiou, F and Raisbeck, G, eds, Proceedings of the 5th International Conference on Accelerator Mass Spectrometry. *Nuclear Instruments and Methods* B52(3,4): 269–272.

Elmore, D 1987 Ultrasensitive radioisotope, stable isotope and trace-element analysis in the biological sciences using tandem accelerators. *Biological Trace Element Research* 12: 231.

Felton, JS, Turteltaub, KW, Vogel, JS, Balhorn, R, Gledhill, BL, Southon, JR, Caffee, MW, Finkel, RC, Nelson, DE, Proctor, ID and Davis, JC 1990 Accelerator mass spectrometry in the biomedical sciences: Applications in low-exposure biomedical and environmental dosimetry. *In* Yiou, F and Raisbeck, G, eds, Proceedings of the 5th International Conference on Accelerator Mass Spectrometry. *Nuclear Instruments and Methods* B52(3,4): 517–523.

Keilson, J and Waterhouse, C 1978 Possible impact of the new spectrometric techniques on ^{14}C tracer kinetic studies in medicine. *In* Gove, HE, ed, *Proceedings of the First Conference on Radiocarbon Dating with Accelerators*. University of Rochester: 391-397.

Proctor, ID, Southon, JR, Roberts, ML, Davis, JC, Heikkinen, DW, Moore, TL, Garibaldi, JL and Zimmerman, TA 1990 The LLNL ion source – past, present and future. *In* Yiou, F and Raisbeck, G, eds, Proceedings of the 5th International Conference on Accelerator Mass Spectrometry. *Nuclear Instruments and Methods* B52 (3,4): 334–337.

Roberts, ML, Southon, JR, Davis, JC, Proctor, ID and Nelson, DE 19901 A dedicated AMS facility for ^{3}H and ^{14}C. Proceedings of the Eleventh International Symposium on the Application of Accelerators in Research and Industry. *Nuclear Instruments and Methods* B56/57: 882–885.

Southon, JR, Caffee, MW, Davis, JC, Moore, TL, Proctor, ID, Schumacher, B and Vogel, JS 1990 The New LLNL AMS Spectrometer. *In* Yiou, F and Raisbeck, G, eds, Proceedings of the 5th International Conference on Accelerator Mass Spectrometry. *Nuclear Instruments and Methods* B52(3,4): 301–305.

Turteltaub, KW, Felton, JS, Gledhill, GL, Vogel, JS, Southon, JR, Caffee, MW, Finkel, RC, Nelson, DE, Proctor, ID and Davis, JC 1990a Accelerator mass spectrometry in biomedical dosimetry: Relationship between low-level exposure and covalent binding of heterocyclic amine carcinogens to DNA. Proceedings of the National Academy of Science. *USA* 87: 5288–5292.

Turteltaub, KW, Watkins, BE, Vanderlaan, M and Felton, JS 1990b Role of metabolism on the DNA binding of MeIQx in mice and bacteria. *Carcinogenesis* 11: 43.

Vogel, JS, Nelson, DE and Southon, JR 1987 ^{14}C background levels in an accelerator mass spectrometry system. *Radiocarbon* 29(3): 323–333.

Vogel, JS, Southon, JR and Nelson, DE 1990 Memory effects in AMS system: Catastrophe and recovery. *Radiocarbon* 32(1): 81–83.

Yamashita, K, Umemoto, A, Grivas, S, Kato, S and Sugimura, T 1988 In vitro reaction of hydroxyamino derivatives of the MeIQx Glu-P-1 and Trp-P-1 with DNA: ^{32}P-post labeling analysis of DNA adducts formed in vivo by the parent amine and in vitro by their hydroxyamino derivatives. *Mutagenesis* 3: 515.

INDEX